AEROTECH VI

Innovation in Aerospace Engineering and Technology

IOP Conference Series: Materials Science and Engineering
Volume 152

Kuala Lumpur, Malaysia
8 - 9 November 2016

ISBN: 978-1-5108-3175-9
ISSN: 1757-8981

Printed from e-media with permission by:

Curran Associates, Inc.
57 Morehouse Lane
Red Hook, NY 12571

Some format issues inherent in the e-media version may also appear in this print version.

Printed by Curran Associates, Inc. (2016)

For permission requests, please contact the Institute of Physics
at the address below.

Institute of Physics
Dirac House, Temple Back
Bristol BS1 6BE UK

Phone: 44 1 17 929 7481
Fax: 44 1 17 920 0979

techtracking@iop.org

Additional copies of this publication are available from:

Curran Associates, Inc.
57 Morehouse Lane
Red Hook, NY 12571 USA
Phone: 845-758-0400
Fax: 845-758-2633
Email: curran@proceedings.com
Web: www.proceedings.com

Table of contents

Volume 152

AEROTECH VI - Innovation in Aerospace Engineering and Technology
8–9 November 2016, Kuala Lumpur, Malaysia

Accepted papers received: 11 October 2016
Published online: 31 October 2016

Preface

Aerodynamics

Aerospace Structures and Propulsion

Aerospace Design

Space Systems

Aerospace Reviews

Supporting Technologies for Aerospace

IOP Conf. Series: Materials Science and Engineering **152** (2016) 011001 doi:10.1088/1757-899X/152/1/011001

Preface

The AEROTECH conference was founded in 2005 and first held in Putrajaya, Wilayah Persekutuan. Since then, subsequent AEROTECH conferences were held biennially and this year, it is the sixth edition in the series. The AEROTECH conference is organized by the Department of Aerospace Engineering, Universiti Putra Malaysia and was initiated to provide a platform for local and international aerospace communities to convene and share the recent development and progress in aerospace technology. For this edition, AEROTECH VI conference is co-organized with Politeknik Banting Selangor. Overall, AEROTECH VI organizer has received over 90 paper submissions from both local and international communities. However, only 69 were accepted for this conference. All accepted papers have undergone an extensive review process by at least two experts from the relevant field.

As one of the engineering fields with the highest technological density, aerospace engineering is always the spearhead of technological breakthrough. Its continual progression nourishes the other engineering fields and societies. Hence, AEROTECH VI adopts the theme "Innovation in Aerospace Engineering and Technology" with the aim of promoting creativity and advancement through the exchange of knowledge and status of cutting-edge research in aerospace engineering and technology. AEROTECH VI is also promoting industry-led research and technology activities among the academia and industry players. AEROTECH VI has attracted researchers and scientists not only from Malaysia but also from countries such as Iran, Iraq, Australia, Taiwan, Singapore, Thailand, Korea, India, Germany, USA and Spain. The conference covers research in the areas of aerodynamics, propulsion, aerospace structure and material, aerospace design, flight and system performance, space systems, aerospace management and operation, and aerospace reviews.

We, the editors, thank the authors for their contributions and the reviewers for their expertise and valuable comments. We also like to thank the conference organizing committee and Universiti Putra Malaysia for their commitment and tremendous support to the success of AEROTECH VI. Our deep gratitude is extended to Politeknik Banting Selangor for their cooperation in co-organizing this year's conference.

The Editors

IOP Conf. Series: Materials Science and Engineering **152** (2016) 011002 doi:10.1088/1757-899X/152/1/011002

Peer review statement

All papers published in this volume of *IOP Conference Series: Materials Science and Engineering* have been peer reviewed through processes administered by the proceedings Editors. Reviews were conducted by expert referees to the professional and scientific standards expected of a proceedings journal published by IOP Publishing.

Effect of surface roughness on helicopter main rotor blade

W N W Rohizan[1,2*], A S M Rafie[2], M Y Harmin[2] and C C Ciang[2]

[1]KD PANGLIMA HITAM – Unit 1 PASKAL, Royal Malaysian Navy, Malaysia
[2]Department of Aerospace Engineering, Faculty of Engineering, Universiti Putra Malaysia, Malaysia

*wan.norhafizan@yahoo.com.my

Abstract. This paper describes the effect of surface roughness when applied on helicopter main rotor blade. The aim is to prove that surface roughness can be used as a simple and inexpensive method to achieve better flight performance such as enhanced thrust and/or reduced power requirement. The research was done experimentally, using scaled model of Eurocopter AS350 Ecureuil. Smooth profile of the main rotor blade was modified by applying surface roughness on the upper and lower camber in transition and turbulent boundary layer region, starting from 25% of chord length and extending up to the trailing edge (TE). This study was conducted with the conditions of a vertical flight (particularly in hover condition) since this is a high power-consuming flight regime for helicopter. The experiment resulted in lower power requirement but at the expense of reduced thrust at the middle collective pitch level. At upper range of the collective pitch level, surface roughness was seen to delay the stall angle as well as increase the lift in the stall region. Meanwhile, at lower pitch level, there was an increase in thrust-to-power ratio.

1. Introduction

Given the complexity of the helicopter main rotor mechanism, it is practically impossible to have any active high-lift devices installed on the main rotor blade. Therefore, the blades remain basically solid, a design which is unchanged since the advent of helicopter despite the fact that many researches have been done to achieve improvement in flight performance. The desired improvement includes enhanced thrust (for better flight performance) and lower power requirement (for extended flight duration).

2. Literature review

Looking at previous research works, it is clear that surface modifications can enhance aerodynamic performance considerably as discovered by Reuss [1]. Surface modifications energise the flow by creating vortices, which makes the flow remains attached to the surface and hence delaying boundary layer separation. Meanwhile, the research by Srivastav [2] shows that surface roughness can reduce the drag drastically and improve lift. This can be efficient at high angles of attack and also can increase the stall angle to some extent. Based on the work done by Dhiliban *et al.*, surface modifications on lower surface gives better aerodynamic efficiency than on the upper surface [3]. A research by Noonan *et al.* using slotted airfoil tips found that at high coefficient of lift, the slotted airfoil profile gave lower coefficient of torque and high figure of merit [4]. In 2012, Jensen [5] conducted a study using airfoil flap model with vortex generator (VG) and cylinder. He found that, while VG showed notable increase in effects at low angles of attack, the pattern then declined near the stall angle. Furthermore, the use of cylinder showed improvement across the range and performed better than VG.

While all previous works were based on computational, 2-dimensional and in fixed wing condition, no notable work has been done for helicopter main rotor blades. Therefore, this research will explore the effect of surface roughness when applied to the main rotor blades. The values to be observed are the coefficient of thrust (C_T) and coefficient of power (C_P). It is decided to place the roughness starting from 25% of chord length away from leading edge (LE) and extending up to the trailing edge. This is done since this region is conceived to be where the flow separation will occur, forming transition and turbulent boundary layer region. In addition, Matheis found that adding roughness at LE resulted in drastic loss of aerodynamic performance due to premature flow transition that was caused by the roughness [6]. A study on NACA 0012 profile also showed a lower lift coefficient and a higher drag coefficient when roughness was placed at LE [7]. Moreover, another research found that the presence of roughness at LE of the wind turbine blades caused an early TE turbulent separation, hence reducing its aerodynamic efficiency [8].

According to Johnson [9] and Leishman [10], C_T and C_P are defined as in Equation 1 and Equation 2, respectively, in which V_T is the tip speed of rotor blade.

$$C_T = \frac{T}{\frac{1}{2}\rho V_T^2 A} \qquad (1) \qquad\qquad C_P = \frac{P}{\frac{1}{2}\rho V_T^3 A} \qquad (2)$$

As cited by Seddon and Newman[11], the rotor solidity is defined as ratio of blade area to disc area, known as *s* or sometimes σ, given by Equation 3. This ratio is unique to the rotary wing aerodynamic calculations and will be used later to determine the actual C_T and C_P.

$$s = \frac{NcR}{A} = \frac{NcR}{\pi R^2} = \frac{Nc}{\pi R} \qquad (3)$$

where N = number of blades, c = chord length and R = rotor swept radius.

3. Methodology

This research was done experimentally using an open circuit wind tunnel and a scaled down helicopter model. Although the test was performed using the wind tunnel, no frontal air flow was involved. Instead, only the test section is used, where the helicopter model was mounted on the six-component balance to see the effects in vertical flight.

3.1. Experimental setup

For the experimental model, the Eurocopter AS350 helicopter model (see Figure 1) was chosen since it was available off-the-shelf and can be readily used. The wind tunnel used for the experiment was the OWLT-1000 of Aerodynamic and Fluid Mechanics Lab, Faculty of Engineering, UPM with the test section of 1.0 m × 1.0 m x 1.0 m in dimension. The model was mounted in the wind tunnel and on the electronic six-component balance to measure the thrust and power. The mean rotational speed was measured using multi-range tachometer (see Figure 2) at all collective pitch levels. The measurements were taken in vertical direction at the root of rotor blade. The tachometer displayed different rotational speed at each pitch level. The blade pitch angle was determined visually. The operating parameter of helicopter model is tabulated in Table 1.

The surface roughness profile was prepared using sand paper strip and applied to the main rotor blade surface using double sided adhesive tape. Experimental analysis for smooth surface (baseline) was performed first before applying the surface roughness on the rotor blade. After that, the surface roughness was applied from the tip to near the root, starting from 25% of chord length and extended up until the trailing edge on the upper surface and lower surface (see Figure 3, Figure 4, Figure 5, Figure 6 and Figure 7).

IOP Conf. Series: Materials Science and Engineering **152** (2016) 012001 doi:10.1088/1757-899X/152/1/012001

Table 1. Helicopter model operating parameter

Blade Diameter (m): 0.72		**Rotor Disc Area (m^2)**: 0.43		
Chord Length (m): 0.024		**Rotor Area (m^2)**: 0.0266		
Rotor Solidity (%): 6.19				
Collective Pitch	**Blade Pitch Orientation (+ve/-ve)**	**Rotational Speed**		**Linear Speed (m/s)**
		rpm	**rad/s**	
-2	-ve	729.60	76.40	28.27
-1	-ve	926.60	97.03	35.90
0	0	890.00	93.20	34.48
1	+ve	839.40	87.90	32.52
2	+ve	754.46	79.01	29.23
3	+ve	651.90	68.27	25.26

Figure 1. Helicopter model mounted on six-component balance

Figure 2. Multi-range tachometer

Figure 3. Roughness at upper surface covering 25% of airfoil and located 25% from the leading edge

Figure 4. Roughness at lower surface covering 25% of airfoil and located 25% from the leading edge

Figure 5. Roughness at upper surface covering 50% of airfoil and located 25% from the leading edge

Figure 6. Roughness at upper surface covering 75% of airfoil and located 25% from the leading edge

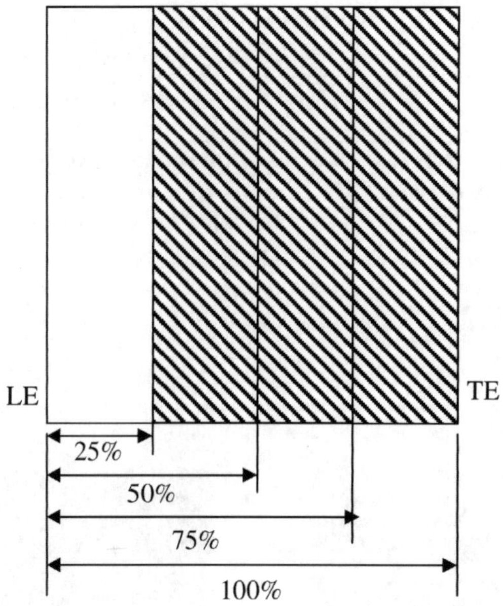

Figure 7. Region covered by roughness on the upper and lower surface

In order to get continuous power supply and subsequently consistent result, a direct power supply of maximum around 45W was used. The readings from six-component balance and power converter were recorded and computed into a simple Excel spreadsheet to determine the values of C_T and C_P. Because the rotor solidity ratio, *s* is instrumental in influencing the experimental results, it was also included in the calculation.

4. Results and discussion

The results for the research are presented in three plots: C_T/s vs. collective pitch, C_P/s vs. collective pitch and C_T/C_P ratio vs. collective pitch. Consistency analysis was also conducted, with repeatability within 1% to 5% for rotor power requirement. However, significant variations recorded for C_T/s value, with deviation up to 0.11 N (which translated to 86% value deviation from average) due to presence of resonance during experimental run.

4.1. Thrust force

Figure 8 shows the C_T/s values at different collective pitch levels for four different roughness models (baseline, 25%, 50% and 75%). Since the thrust was positive (upward) at zero pitch level, it was concluded that the main rotor blade profile was asymmetrical. At lower pitch (-2 and -1), the effect of surface roughness was insignificant, with deviation of only 0.01 in C_T/s value between all roughness condition. At middle range (0 until 2), baseline surface performed better than those with surface roughness. This is also the range where the difference of baseline model against roughness models started to show, where the deviation in C_T/s value increased from 0.02 – 0.03 between all models. Only at the maximum pitch level (3), the baseline model showed a decline in C_T/s value because it had exceeded the stall angle. At this pitch level, all cases of surface roughness outperformed the baseline condition by showing a higher C_T/s value with increment between 23% to 53% than the baseline and without any indication of stalling. One notable pattern could be seen where for all cases with surface roughness, the graphs show an increasing pattern even though at maximum pitch level.

From the graph, it is deduced that stall angle for all cases of applied surface roughness is higher than that of the baseline model. However, those higher values of C_T/s at high collective pitch range come at the expense of lower C_T/s values at middle pitch range. At negative pitch range, the difference was smaller and deemed insignificant even though all roughness models showed slightly higher C_T/s value, which suggests that higher positive thrust is achieved. Due to limitation of the model's control mechanism, the stall angle for other conditions cannot be determined. Nevertheless, from the result, it is shown that for both fixed wing and rotary wing, surface roughness is beneficial at higher angles of attack and can push stall angle to be higher to some extent.

4.2. Power requirement

Figure 9 shows the plot of C_P/s versus collective pitch level. The minimum C_P/s values were observed to be in the range between -1 and 0 pitch levels for all roughness models. The baseline recorded the lowest reading. Meanwhile, for other cases, the minimum C_T/s increased as roughness size increased, with increment between 1.05% to 8.06% compared to the baseline model.

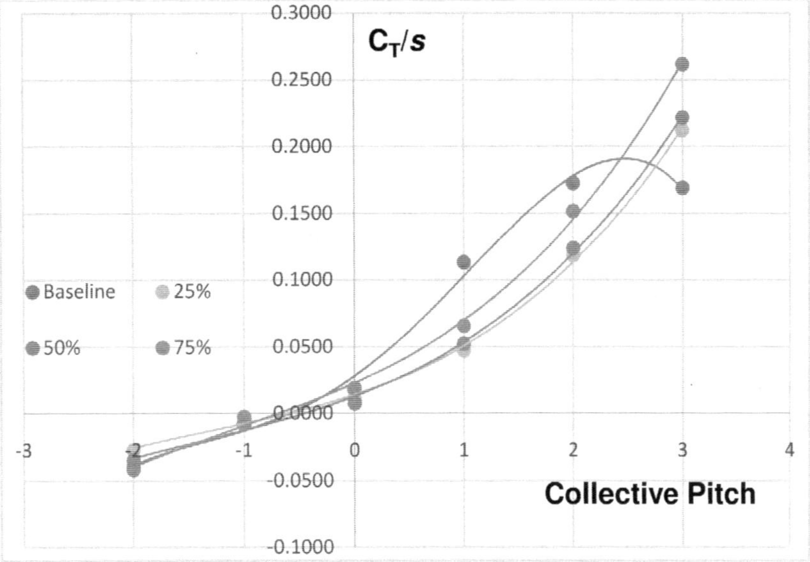

Figure 8. C_T/s vs. collective pitch level

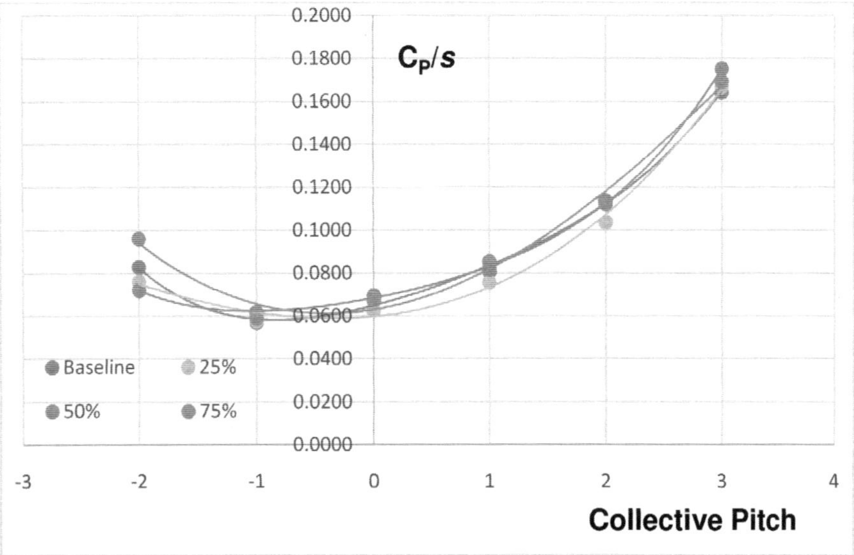

Figure 9. C_p/s vs. collective pitch level

The graphs show identical pattern for all roughness condition: as the blade pitch increases, the rotor power requirement also increases until it reaches the maximum blade pitch angle, regardless of the blade pitch orientation. From the graph as well, generally the power requirement for all roughness condition showed a higher reading than the baseline condition. This is due to higher power required to overcome resistance from the rougher surface conditions. However, an exception is for 25% roughness condition, which consistently showed lower power requirement than the baseline model in mid-range (at pitch level 0 until 2). Therefore, in case of lower power requirement becoming the priority, 25% roughness condition should be considered.

Given small values of C_P/s, a slight change will give a large difference in percentage as mentioned earlier. Nevertheless, throughout all roughness conditions, only small deviations recorded for C_P/s with the difference was between 2% to 5% at all pitch level (with exception of -2 level due to cut-in power required to rotate the blades from standstill). This was parallel with the power consumed during the test where it showed very consistent pattern. 40 W to 45 W of power was expended at any blade pitch level, with the average figure of 42.12W. The only exception was at -2 level, where the difference in C_T/s value across all roughness ranges were more significant (i.e. 12.48%) because of cut-in power mentioned above. Therefore, it is determined that surface roughness effect is insignificant and has little impact on the rotor power requirement.

4.3. Thrust to power ratio

C_T/C_P ratio curve provides a view on the thrust generated in proportion to the power required at a particular pitch level, while at the same time gives the optimum operating angle of an airfoil. It also shows the aerodynamic efficiency of the airfoil. As shown in Figure 10, the roughness models were deemed effective at negative pitch level (-2 and -1) and high pitch (level 3). However, it comes at the expense of lower C_T/C_P ratio at the middle range (0 to 2) where the baseline model gave a better performance. When the operating angle was pushed at higher pitch level and at maximum pitch (3), the improvement was 50% higher than the baseline model (whereas for baseline model, C_T/C_P ratio started to drop beyond pitch level 2). Between the three roughness models, the 75% roughness showed the highest aerodynamic efficiency where it outdid other models beginning from level -1 up to 3. It also outperformed other roughness models in average of between 14% - 21% across the pitch ranges.

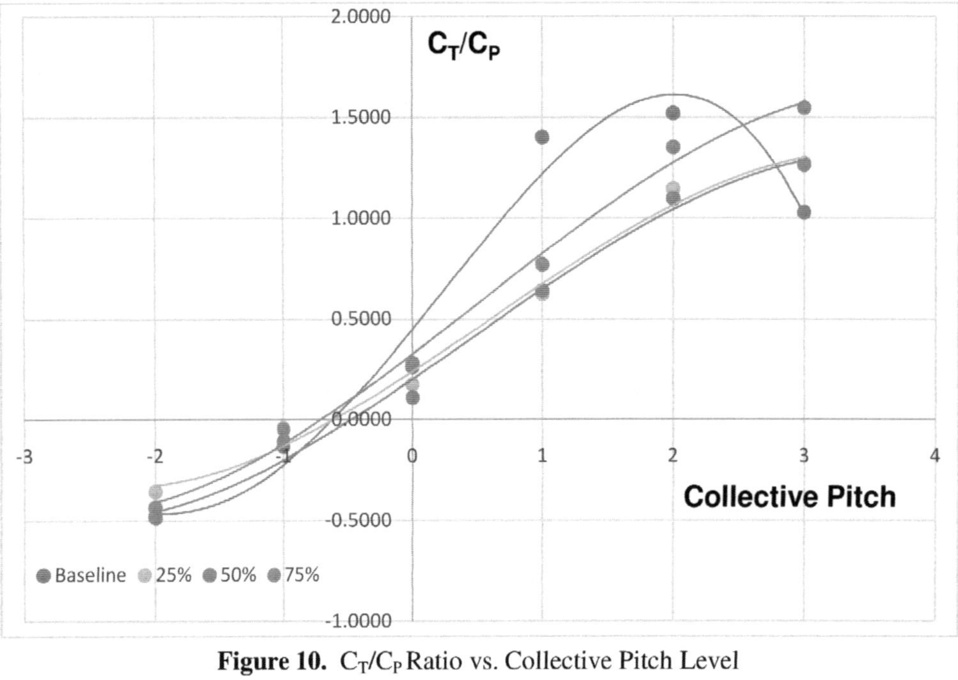

Figure 10. C_T/C_P Ratio vs. Collective Pitch Level

Interpreting the graph, it shows that with surface roughness applied, the operating envelope of the helicopter becomes wider (with more useful blade pitch range). With aerodynamic efficiency of the rotor blade is spread more evenly and improving at certain pitch level, it enables pilot to operate the helicopter to maximum pitch level without any risk of stalling the aircraft. In addition, given the graphs for all roughness models are less steep than that of the baseline, surface roughness gives more predictable flight profile that can contribute towards safer helicopter operation. It is also evident that the C_T/C_P ratio graph largely influenced by C_T/s vs collective pitch graph (refer to Figure 8).

5. Conclusion

Experimental work was conducted to understand the effect of surface roughness on helicopter main rotor blade. Surface roughness has a significant impact on thrust and power coefficients. Baseline model with smooth surface shows the lowest minimum drag coefficient. Meanwhile, in general, the increase of surface roughness will lead to increasing minimum drag coefficient value. For thrust, it is seen that surface roughness improves the coefficient of lift at lower and higher pitch levels, but it performs at middle range compared to the baseline. Subsequently, for thrust/power ratio, the same pattern is also observed. From here, it is clear that surface roughness is beneficial for certain range of operations. As a type of passive flow control method, it suffers from inability to adapt accordingly to flow condition. Nevertheless, this experiment proves that surface roughness gives higher stall angle and consequently widens helicopter operating envelope to the maximum without the risk of stall. Apart from giving a better flight performance in high angles of attack and full power, the surface roughness can give more predictable flight profile that contributes to safer helicopter operations. This surface roughness can be made as a controllable one, which will eliminate the use of active-based high lifting devices. Simply by using the surface roughness, the thrust and power values can be manipulated according to the required preference. This will enhance the vertical flight performance and have great impact on sophisticated military-purpose helicopter where manoeuvrability and endurance are of paramount importance.

Acknowledgement

The authors like to thank the management of Faculty of Engineering, UPM and the lab technicians of the Aerodynamic and Fluid Mechanics Lab, UPM for their assistance during the test. The gratitude is also extended to Royal Malaysian Navy, especially the Commanding Officer of KD PANGLIMA HITAM and ABEM Wan Nachat s/o See Ian from 501 Squadron, for the encouragement and support given to ensure that the research can be conducted successfully.

References

[1] Reuss R L, Hoffmann M J and Gregorek G M 1995 *Effects of Surface Roughness and Vortex Generators on the NACA 4415 Airfoil* National Renewable Energy Laboratory

[2] Srivastav D 2012 *International Conference on Fluid Dynamics and Thermodynamics Technologies*

[3] Dhiliban A, Meena P, Narasimhan P S, Vivek M, Pillai S N and Parammasivam K M 2013 *The Eighth Asia-Pacific Conference on Wind Engineering*

[4] Noonan K W, Yeager W T, Singleton J D, Wilbur M L and Mirick P H 2001 *Wind Tunnel Evaluation of a Model Helicopter Main-Rotor Blade with Slotted Airfoils at the Tip* National Aeronautics and Space Administration

[5] Jansen D P 2012 *Passive Flow Separation Control on an Airfoil-Flap Model (The Effect of Cylinders and Vortex Generators* Master Thesis Delft University of Technology

[6] Matheis B D, Huebsh W W and Rothmayer A P 2004 *RTO AVT Specialists' Meeting on Enhancement of NATO Military Flight Vehicle Performance by Management of Interacting Boundary Layer Transition and Separation*

IOP Conf. Series: Materials Science and Engineering **152** (2016) 012001 doi:10.1088/1757-899X/152/1/012001

[7] Chakroun W, Al-Mesri I and Al-Fahad S 2004 *Wind Engineering* **28** 547–64
[8] Marzabadi F R and Soltani M R 2013 Scientia Iranica **20** 508-15
[9] Johnson W 2013 *Rotorcraft Aeromechanics* (Cambridge: Cambridge University Press)
[10] Leishman G 2000 *Principles of Helicopter Aerodynamics* (Cambridge: Cambridge University Press)
[11] Seddon J and Newman S 2011 *Basic Helicopter Aerodynamics* (Hoboken: John Wiley & Sons)

IOP Conf. Series: Materials Science and Engineering **152** (2016) 012002 doi:10.1088/1757-899X/152/1/012002

Effect of cavity fore wall modifications in supersonic flow

S Jeyakumar[1*], K Arun Yuvaraj[1], K Jayaraman[2], F Cardona[3] and M T H Sultan[3]

[1]Department of Mechanical Engineering, Kalasalingam University, India
[2]Department of Aerospace Engineering, Veltech University, Chennai, India
[3]Aerospace Manufacturing Research Centre (AMRC), Universiti Putra Malaysia, Malaysia

*sjeyakumar1974@gmail.com

Abstract. Understanding the complex behaviour of the cavity flow is essential for the design of supersonic combustor. The characteristics of the axisymmetric aft ramp cavity with fore wall modification have been experimentally studied in a blow-down type supersonic flow facility. The facility consists of a conventional CD nozzle that issues a flow Mach number of 1.88 to a supersonic combustor of circular cross section, which is placed immediately downstream of the nozzle. The axisymmetric cavities are incorporated within the combustor. The cavities are of open type and their length is kept constant while their depth varies. The aft wall of the cavities is inclined with three ramp angles and the fore wall is provided with a constant fillet radius of 3 mm. The performance of the cavity is analysed based on wall static pressures, momentum flux distribution at the exit of the combustor and stagnation pressure loss across the combustor. The study reveals that fore wall fillet cavities provide less cavity drag and stagnation pressure loss for various aft ramp angles under identical operating conditions.

1. Introduction

Cavity is considered as one of the potential devices for mixing enhancement and also effective flame holding in scramjet engine combustor [1–6] because of the generation of less drag than other active devices like struts [7, 8] and pylons [9]. The cavity flow has been studied in the past for applications in wheel wells, aircraft weapon bays, scramjet combustors, etc. The renewed interest in the cavity flow is due to the complexity of flow field within it. Researchers have characterized the cavities based on the shear layer reattachment as either open or closed. In case of open cavity, the separated shear layer from the leading edge of the cavity reattaches at the aft wall. On the other hand, for closed cavity, the shear layer reattaches at the bottom wall of the cavity and this creates higher drag than open cavity. Even though the flow over the cavity is supersonic, the shear layer regime within the cavity is subsonic. The shear layer is unstable in subsonic regime and a small pressure disturbance in the shear layer causes pressure coupled feedback mechanism [10], which triggers a periodic vortex shedding that leads to the acoustic oscillations of the fluid stream. Open cavities are desirable for scramjet applications as they impose smaller drag penalty on the flow.

An experimental study by Yu *et al.* [6] on cavities in a Mach 2 reacting flow field has reported that flame holding is achieved by smaller aspect ratios than the longer one. In addition, the multiple cavity configurations provide an effective flame holding than the single cavity. Chung [11] investigated the characteristics of a rectangular cavity in supersonic flow field based on cavity geometry and Mach number of the flow. Kang *et al.* [12] studied the effect of zigzag cavity in Mach 2.5 flow as compared to a plain cavity. They observed that zigzag cavity generated transverse oscillations, which enhanced combustion within shortened combustor than plain cavity. Experimental and computational studies on different aft wall angles and off-set ratios of the cavities in a supersonic flow indicated that stable flow was achieved by decreasing the aft wall angle below 90 degrees [3]. Similar observations were found for varying cavity aft wall angles below 90 degree in a supersonic flow field [13, 14].

The above mentioned investigations are based on rectangular cavities in a 2D supersonic flow with and without cavity aft wall angle. Moreover, literatures on axisymmetric cavity flow are scant. In an on-going research, the features of axisymmetric cavity configurations for scramjet combustor have been reported [15]. The objective of this study is to investigate the performance of axisymmetric fore wall fillet cavities in a Mach 1.88 flow. The investigation on wall static pressures, momentum flux distribution at the exit of the combustor and stagnation pressure loss associated with the flow across the combustor are estimated for improving the scramjet combustor performance.

2. Experimental setup

A blow-down type supersonic flow facility is used in this study for the non-reacting flow experiments. The facility was consisted of a CD nozzle, followed by a circular cross sectional duct that acted as a supersonic combustor. The nozzle issued a flow Mach number of 1.8. The operating condition of the experiment is shown in Table 1. The supersonic combustor was of 26 mm in diameter and 95 mm in length. Cavities were incorporated at a distance of 30mm from the inlet of the combustor. Sectional view along the flow direction of the combustor is shown in Figure 1. The cavities were of open type and axisymmetric, with constant length, L but varying depths, D. The cavity details are illustrated in Figure 2. The aft wall of the cavity was inclined to three different angles: 15, 20 and 30 degrees with the horizontal direction. A fillet radius of 3mm was provided at the fore wall of the cavity for "with fillet cavity" configurations. Table 2 shows the geometric dimensions of the cavity used for this study.

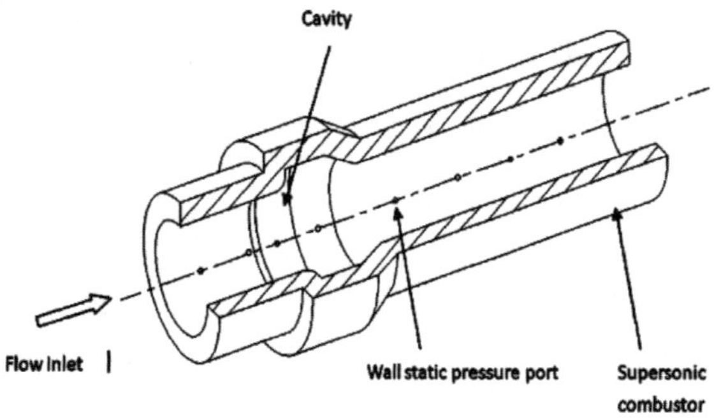

Figure 1. Schematic diagram of supersonic combustor

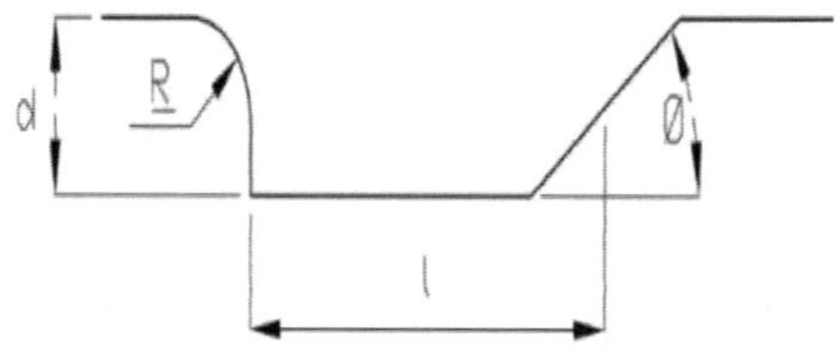

Figure 2. Schematic diagram of cavity layout

Table 1. Operating conditions of supersonic combustor

Parameters	Values
Stagnation pressure	800Kpa
Static pressure	110Kpa
Stagnation temperature	300K
Mach number	1.8
Mass flow rate of air	0.2 kg/s

Table 2. Geometrical information of cavity configuration

Cavity configuration	Length, L (mm)	Depth, D (mm)	L/D	Ramp angle, °
WF3θ15	15	3	5	15
WF4θ15	15	4	3.75	15
WF5θ15	15	5	3	15
WF3θ20	15	3	5	20
WF4θ20	15	4	3.75	20
WF5θ20	15	5	3	20
WF3θ30	15	3	5	30
WF4θ30	15	4	3.75	30
WF5θ30	15	5	3	30

The static pressure distribution along the axial length of the combustor was measured using 1-mm diameter ports that were placed along the combustor wall in the flow direction. Nine pressure taps were equipped along the combustor walls to acquire pressure data along the flow direction. In addition, a long cone static and pitot pressure probes were used to measure static and stagnation pressures at the exit of the combustor to examine the flow characteristics influenced by the cavities. Each experiment was executed three times for repeatability. The uncertainties were estimated to be less than 3% for the pressure measurements. The probes were moved in the radial direction of the flow field by a traversing mechanism.

3. Results and Discussion

3.1. Wall static pressure distribution

The surface static pressures along the axial length of the flow for various aft ramp angles of with and without fillet cavities are plotted in Figure 3 and Figure 4. In the plots, wall static pressure (P_w) to the inlet static pressure (P_i) of the combustor is normalized to the non-dimensional axial length (x/L) of the combustor. The 'x' denotes axial distance of the probe measured from the inlet of the combustor and L denotes the total length of the combustor. WF denotes cavity with fillet, third numeral denotes the depth of the cavity whereas the cavity ramp angles are mentioned as θ15, θ20 and θ30. The fillet radius, R of 3 mm was used for all the tests. The cavity without fillet is mentioned as WOF. In Figure 3a, a uniform wall static pressure profile is observed for no cavity over the entire length of combustor. For cavity aft ramp angle θ15 (Figure 3a), a rise in static pressure is seen at the leading edge of the cavity than the free stream pressure due to the shear layer separation, which led to a compression zone. The separated shear layer reattached at aft ramp of the cavity, resulting in increased static pressure.

The static pressure over the cavity region decreased with increase in aft ramp angle from 15 to 30 degrees of the cavity. A study on 2D flow over rectangular cavity by Gruber *et al.* [1] revealed that a decrease in aft ramp angle will reduce the static pressure at the bottom wall but will increase it at the reattachment region, i.e. at the aft ramp of the cavity. Moreover, the schlieren pictures revealed that the flow became stable in the cavity region by reducing the cavity aft ramp angle. A stable flow field with continuous source of radicals to stabilize the flame is essential for a good flame holder. In the present study, the flow visualization techniques cannot be implemented due to the selected cross section of the combustor. Higher static pressure was observed for θ15 and θ20 degrees due to the shock reflections that emerged from the cavity leading edge. A uniform static pressure profile was observed for θ30 in the cavity region. Similar trend was observed for the fillet cavities (see Figure 4a), with a marginal decrease in the static pressure profile at the leading edge of the cavity, indicating that cavity drag was slightly reduced by the fore wall fillet cavity.

In case of cavity with 4 mm depth (see Figure 3b), the static pressure at the leading edge of the cavity was less than 3 mm cavity depth (see Figure 3a) and this implies that the increase in depth of the cavity leads to an expansion zone. At x/L= 0.5, the static pressure has more value and this indicates that a stronger reattachment of shear layers at the cavity aft ramp. For fillet cavities (see Figure 4b), the flow at the cavity leading edge is expansive in nature and showing less static pressure. It has a peak value at x/L= 0.39, indicating that reattachment of shear layers occurred at the base of the aft wall that increased the entrainment rate of the flow within the cavity. Further increase in depth of the cavity led to decrease in static pressure profile for all cavity with (see Figure 4c) and without fillet (see Figure 3c) configurations. A reduction in wall static pressure values is observed in the cavity region for fore wall fillet cavities, which indicates less cavity drag than without fillet cavities. In case of without fillet cavities, the shear layer reattached at the aft wall of the cavity. In contrast, the reattachment occurred towards the bottom wall corner of the cavity for fillet cavities, which increased the entrainment rate of flow in the cavity and enhanced the recirculation within the cavity regime, making it more competent for flame holding in scramjet combustors.

IOP Conf. Series: Materials Science and Engineering **152** (2016) 012002 doi:10.1088/1757-899X/152/1/012002

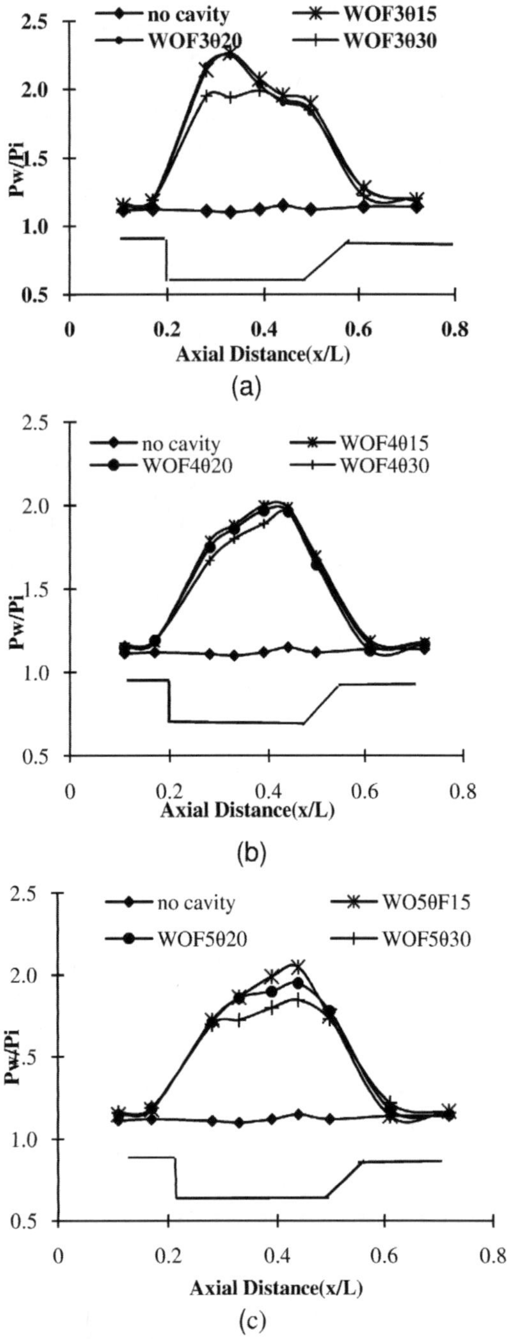

Figure 3. Wall static pressure profile of without fillet cavity for various aft ramp angles

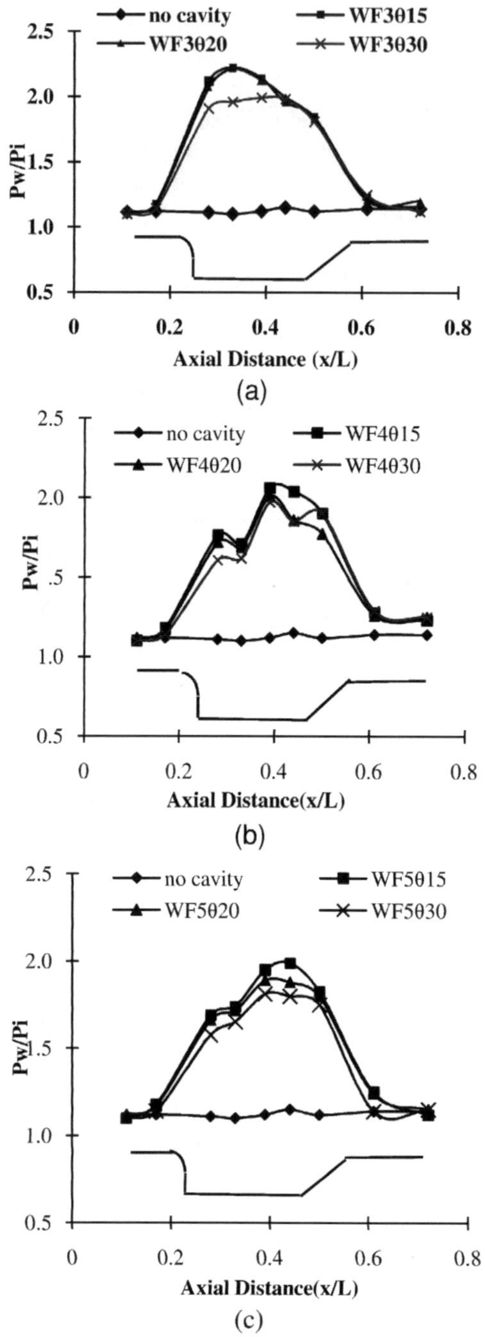

Figure 4. Wall static pressure profile of with fillet cavity for various aft ramp angles

3.2. Momentum flux distribution

The momentum flux distribution at the exit of the supersonic combustor in the radial direction is a measure of the extent of bulk mixing. The momentum flux is calculated as in Equation 1.

$$\mu = p\,(1 + \gamma M^2) \tag{1}$$

where p is the measured value of static pressure and Mach number, M is calculated from the measured values of static and stagnation pressures using Rayleigh-Pitot formula.

The momentum flux values for various cavity configurations at different radial distances at the exit of the combustor are depicted in Figure 5. In the plot, r/R denotes radial distance from the axis, r normalized by the radius, R of the supersonic combustor. Experimental results are presented for all cavity configurations operating under identical conditions and are compared with no cavity case. The momentum flux distribution was almost uniform for all cavity configuration compared to no cavity. In the case of no cavity, the momentum mixing was not uniform and it was showing poor mixing along the radial direction of the flow. For cavity configuration, the nature of the curve tended to be uniform from the centre towards the wall of the combustor. It was also observed that increasing the cavity ramp angle from 15 to 30 degrees would increase the momentum flux values at the combustor exit along the radial direction. A uniform momentum flux distribution from the center of the combustor till r/R of 0.8 was observed for fore wall fillet cavities, showing uniform mixing were provided by the fore wall fillet cavities.

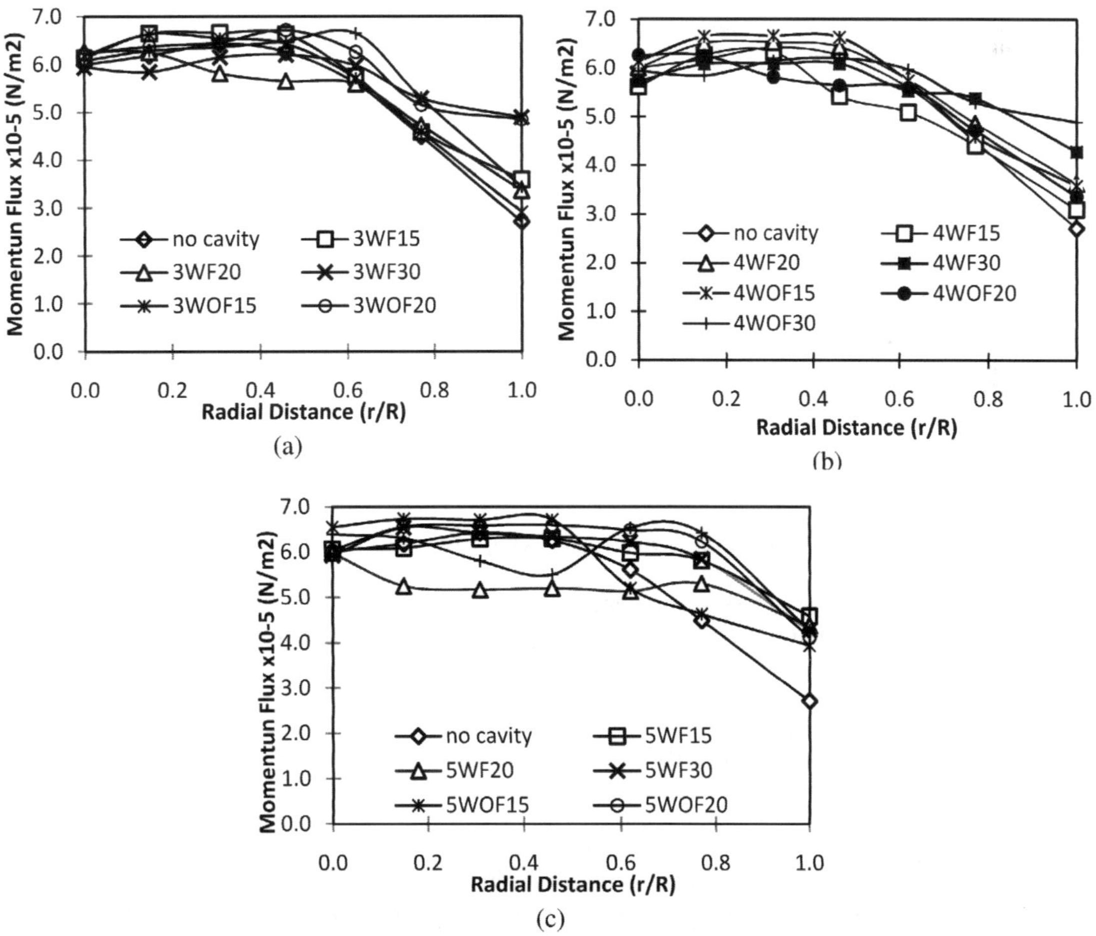

Figure 5. Radial distribution of momentum flux for various cavity configurations

IOP Conf. Series: Materials Science and Engineering **152** (2016) 012002 doi:10.1088/1757-899X/152/1/012002

3.3. Stagnation pressure loss

Stagnation pressure loss across the combustor is defined as the difference in stagnation pressures at the considered inlet and axial distance, normalized by the inlet stagnation pressure measured at the exit sections of the combustor along the radial direction. The change in stagnation pressure loss for various cavity configurations is shown in Figure 6. The stagnation pressure loss for no cavity was calculated to be 11%. From the plot, it is observed that with fillet cavities provided less stagnation pressure loss than without fillet cavity for all aft ramp angles. Moreover, increase in aft ramp angle increased the stagnation pressure loss due to stronger recompression of shear layers, which created a compressive zone downstream of the cavity. The strength of the shock waves emanating from the leading edge of the without fillet cavities was stronger than that of fillet cavities, which obstructed the main stream and increased the stagnation pressure loss. Fore wall fillet cavity of 4 mm depth and 15 degree ramp angle provided less stagnation pressure loss than other cavity aft ramp angles. From the observation, the fore wall fillet cavities of selected geometric dimensions will enhance fluid entrainment inside the cavity from the main stream. This leads to better mixing enhancement with marginal increase in stagnation pressure loss compared to cavity without fore wall fillet for varying cavity ramp angles.

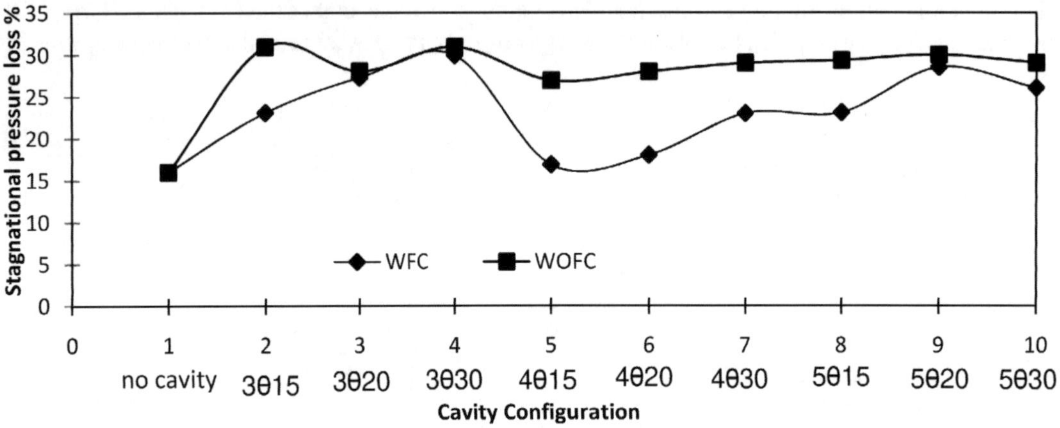

Figure 6. Stagnation pressure loss for with and without fillet cavities

4. Conclusion

The experiments were carried out in a blow-down type non-reacting supersonic flow facility, which can maintain a flow Mach number of 1.8 with an operating pressure of 0.7 MPa and at atmospheric temperature. Aft ramp cavities with constant fillet dimension at the fore wall were used for this study. In case of fore wall fillet cavities, the shear layer reattached at the aft wall bottom corner of the cavity, which increased the entrainment of the flow into the cavity and also enhanced the mixing of the flow. Decreasing aft ramp angle increased the static pressure profile due to stronger reattachment of the shear layer at the aft wall of the cavity. Increase in depth of the cavity decreased the static pressure profile for varying aft ramp angles. Momentum flux profile proved that fillet cavities provided uniform momentum mixing than without fillet cavities. Fore wall fillet cavities provided marginal variation in stagnation pressure loss than without fillet cavities. From the results, it is evident that fore wall fillet cavities with lower aft ramp angles can provide good mixing along with less stagnation pressure loss than without fillet cavities. Hence they are suitable as a potential flame holding device in scramjet combustors.

Acknowledgments
This work is supported by the Department of Science and Technology, Government of India, under the Grant no: SR/FTP/ETA-55. This work is also supported by UPM under GP-IPB 9441503.

References

[1] Mathur T, Gruber M, Jackson K, Donbar J, Donaldson W, Jackson T and Billig F 2001 *J Propuls Power* **17** 1305–12
[2] Ben-Yakar A and Hanson R K 2001 *J Propuls Power* **17** 869–77
[3] Gruber M R, Baurle R, Mathur T and Hsu K-Y 2001 *J Propuls Power* **17** 146–53
[4] Yu K H and Schadow K C 1994 *Combust Flame* **99** 295–301
[5] Yu K, Wilson K J, Smith R and Schadow K C 1998 *36th AIAA Aerospace Sciences Meeting and Exhibit*
[6] Yu K H, Wilson K J and Schadow K C 2001 *J Propuls Power* **17** 1287–95
[7] Tomioka S, Murakami A, Kudo K and Mitani T 2001 *J Propuls Power* **17** 293–300
[8] Tam C, Hsu K-Y, Gruber M and Raffoul C 2007 *43rd AIAA/ASME/SAE/ASEE Joint Propulsion Conference & Exhibit*
[9] Martis R R, Misra A and Singh A 2014 *AIAA J.* **52** 591–603
[10] Li W, Taku N, Oyama A and Fujii K 2013 *AIAA J.* **51** 253–7
[11] Chung K 2003 *J Aircr.* **40** 137–42
[12] Kang S H, Lee Y J, Yang S S and Choi B 2012 *J Propuls Power* **28** 739–46
[13] Ali M M and Kurian J 2008 *J Propuls Power* **24** 635–7
[14] Vikramaditya N S and Kurian J 2014 *Exp Thermal Fluid Sci.* **54** 102–9
[15] Jeyakumar S, Assis S M and Jayaraman K 2016 *Int J Turbo Jet Eng*

Computational analysis of aerodynamic characteristics for wing in ground effect craft in lateral stability

M A U Amir[1,2], A Maimun[2*], S Mat[3] and M R Saad[1]

[1]Faculty of Science and Defense Technology, Universiti Pertahanan Nasional Malaysia, Malaysia
[2]Marine Technology Centre, Faculty of Mechanical Engineering, Universiti Teknologi Malaysia, Malaysia
[3]Aeronautical Laboratories, Faculty of Mechanical Engineering, Universiti Teknologi Malaysia, Malaysia

*adi@fkm.utm.my

Abstract. Wing in-ground effect (WIG) crafts are becoming promising transportation over the last decade. However, stability and control problems faced by the WIG in earlier development are still unresolved. This paper objectively investigates the lateral stability of wing in ground effect craft. The wing encompasses a winglet at the end of the wingtip. Lift, drag and pressure were measured with the respect of the heeling angle of 10^0, 15^0 and 20^0, respectively, with the h/c of 0.3. Initial results from the computational studies show that the ground effect pressure distributions provide a natural righting moment when the WIG craft heels near ground. This initial result provides an insight to understand the current state of knowledge of stability for WIG, particularly on transverse or lateral stability of WIG where it plays important roles in the safety aspect. It is crucial to understand the stability and its component in order to avoid any unforeseen accident. This paper discusses the results obtained from the numerical studies.

1. Introduction

The dynamic behaviour of the wing in ground (WIG) effect is formed significantly by its stability and control properties. Today, WIG tends to be more concerned with wider issues of flying and handling qualities rather than the conventional, which are more on stability and control. WIG craft design faces many challenges in the aspect of aerodynamics. Many studies deal specifically with the ground effect aerodynamics, stability and control, and hydrodynamics. Until this moment, very few researchers have done studies on the lateral stability of WIG [1, 2, 3]. To date, surprisingly this topic has not yet been comprehensively studied and there is very little (or none) available scientific publication in any readily referenced form.

In recent years, there has been increasing interest in WIG, with the pace of development quickened and the progress evolves dramatically [4]. Nevertheless, the stability and control problems that faced by WIG in earlier development still continue until now. Many researchers study the dynamic stability of WIG during take-off, heave, pitch and surge motions (longitudinal stability) but very few researches conducted on roll, yaw and sway especially during cross wing (transverse and lateral stability). This is very important since WIG is commonly operated over the surface of seawater and in the air. These axes (i.e. roll, yaw and sway) are more common to WIG during the cruise compared to heave, pitch, and surge motion because the wind on the sea surface is more dominant compared to higher altitude.

The effect of lateral stability is still unresolved to date. Lateral-directional stability must be put into consideration in WIG development to ensure that roll of the WIG remains at wings level and the yaw tends to weathercock into the wind when the ailerons and rudder are at their zero or datum position. This way the WIG will naturally seek lateral-directional equilibrium without interference by the pilot and this is crucial because the distance between the WIG and the surface is very narrow, which will result with catastrophic incident if the wing or the body get caught with the surface due to roll.

In lateral condition when a WIG craft is heeled or rolled, the pressure on its lifting surfaces changes and redistributed. At one point, the lifting elements approach the ground while on the other, deviate from it. The coefficient of lift increases for the wing that approaches the ground and decreases for the portion that leaves the ground. This situation causes a generation of recovering transverse moment. Therefore, a WIG craft in lateral aspects has a natural aerodynamic stabilization of roll angle in flight close to the shield, which is an extremely important aspect in flight safety [5]. However, this theory never been studied for WIG in a proper scientific way.

2. Background of the study

A research team has been established in Universiti Teknologi Malaysia (UTM) on the WIG study. The aim of this research is to further investigate WIG at higher speed. Recent studies are focusing on the aerodynamic characteristics and stability during take-off, cruise and landing. This includes various variables such as angle of attack, with or without endplate, with or without wing tip, single hull and multi hull. Lift and drag data has been discussed in detail in previous studies [6, 7, 8, 9]. The previous studies are focusing only on the longitudinal stability. Hence, there is a need to extend the study on the transverse or lateral stability because its consequences is important and the fact that it is understudied is indeed a cause for concern.

3. Model

A numerical calculation on longitudinal analysis of a single hull WIG, equipped with the endplate, is performed on available UTM model. Several modifications have been done to the model. The profile for the wing root is changed to NACA 4415. It has been decided also to set the profile of the wing tip with NACA 6409 and, at the wing tip, the aerofoil's anhedral angle is set to be 13 degrees as shown in Figure 1. The wings tip is using NACA 6409. The overall design is adopted from previous model [6, 7, 9]. The detailed dimensions of the model are tabulated in Table 1.

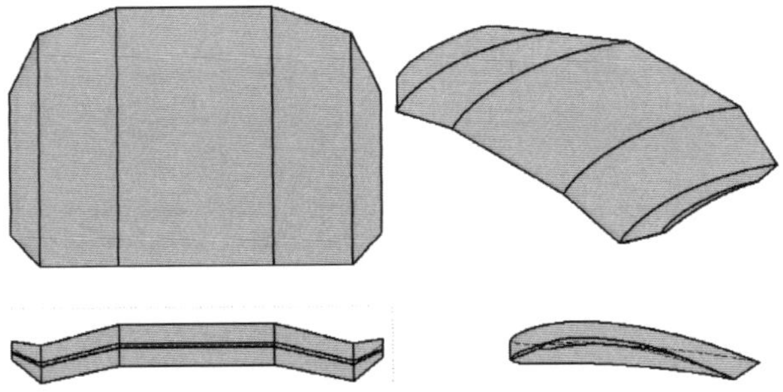

Figure 1. Initial wing design

Table 1. Principle dimensions of the WIG design

Parameter	Units
Wing span	166.8 mm
Wing root chord	112.35 mm
Middle wing span	83.4 mm
Aspect ratio	1.48
Anhedral angle	13 degree

4. Validation

There are two main parts in this study. The first is a CFD simulation and it is followed by wind tunnel experiment. For the simulation, the data was verified with the experimental data obtained previously. The results showed a consistent trend between the simulation and the experiment. The slope of lift and drag coefficients are presented in Figure 2 and Figure 3, respectively. The comparison showed that both lift and drag coefficients obtained from the simulation are higher than those from the wind tunnel experiments.

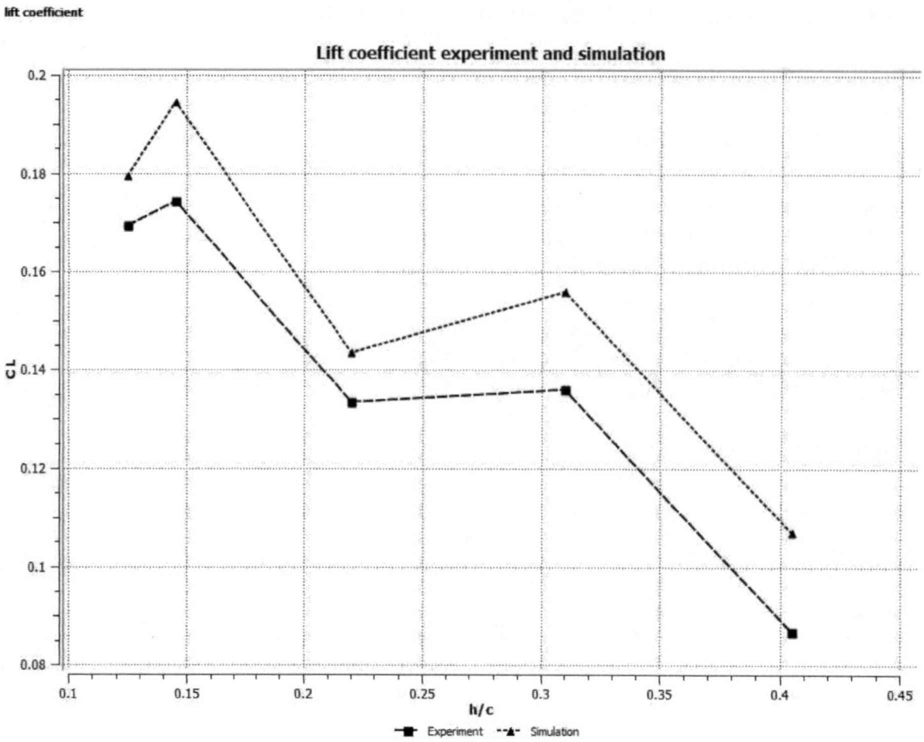

Figure 2. Lift coefficient vs h/s

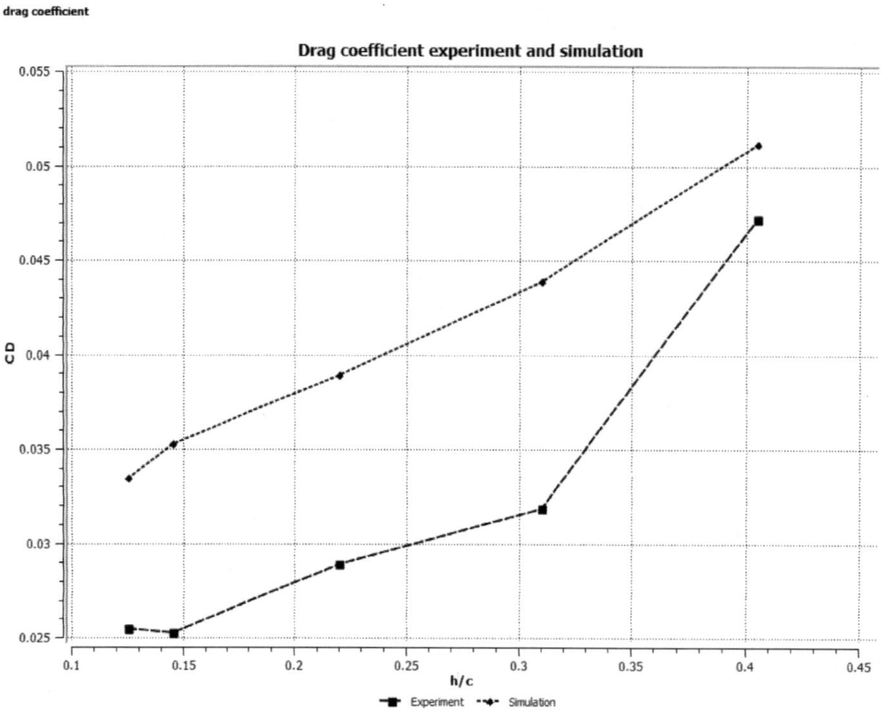

Figure 3. Drag coefficient vs h/s

5. Numerical

The analysis is carried out by solving the incompressible Reynolds average Navier Stokes equations and realizable k-ε turbulence model using FLUENT. The Reynolds number is 1.2 x 10^6 (based on the chord length, c) with the given velocity is 25 m/s. For incompressible flow, the governing equations (continuity and momentum) are solved by coupling method of two variables, which are pressure and velocity. The Semi-Implicit Method for Pressure Linked Equations (SIMPLE) will be used to achieve the objective. The governing equations are written as:

$$\rho\left(\frac{\partial \mathrm{u}_{\mathrm{i}}}{\partial \mathrm{x}_{\mathrm{i}}}+u_j\frac{\partial u_i}{\partial x_i}\right)=-\frac{\partial p}{\partial x_i}+\mu\frac{\partial}{\partial x_i}\left(\frac{\partial u_i}{\partial x_i}+\frac{\partial u_j}{\partial x_i}\right)+S_M \tag{1}$$

Equation 1 is known as the Navier-Stokes equation for incompressible fluid flow, where S_M is a source term that is used to include the smaller contributions to the viscous stress term by the second viscosity coefficient as well as the body force. The derivation of energy equation involves the balance calculations such as those performed for the momentum equation. The internal energy equation of incompressible fluid flow is given as:

$$\rho\left(\frac{\partial l}{\partial t}+(V.\nabla l)\right)=-p\nabla.V+k(\nabla.\nabla T)+\Phi+S_l \tag{2}$$

The left hand side of Equation 2 is the rate of change of energy embodied by the local and convective derivatives of internal energy, $l=h-\frac{p}{\rho}$. At the right hand side, it consists of the rate of work done on the fluid element by pressure forces, net rate of heat transfer to the fluid element, viscous dissipation term and source term, respectively.

6. Boundary condition

In this case, it is assumed zero relative velocity between the surface and the fluid. Thus the non-slip condition is applied on the model surface. The velocity perpendicular to the wall is zero because it is finite and tangent to the wall. In any case, if there is no friction between the surfaces, then the slip condition will apply. The boundary conditions that are set are:

- *Velocity inlet.* The inlet velocity is specified directly. However, the program will extrapolate the pressure from the adjacent cells using reconstruction gradients.
- *Pressure outlet.* The boundary pressure is specified but the boundary face velocity is extrapolated from the interior cells using reconstruction gradient.
- *Wall.* The surface of the wing profile is considered no-slip wall. The pressure is extrapolated from the adjacent cell using reconstruction gradients. The roughness of the wall is set as smooth.
- At inlet, the velocity is prescribed to be a uniform velocity. The ground is considered to be rigid and has same velocity as inlet.

7. Grid sensitivity validation and mesh quality

Table 2 shows the specifications of different grids used in grid sensitivity. The qualities of meshes are in good condition as they have a lower number of skewness and high number of orthogonal quality. Figure 4 shows the pressure coefficient distribution on the lower surfaces of the airfoil as computed by the three grids. Grid B and C have the same result and good agreement but the latter one required less time. Therefore, Grid C has been chosen to conduct the analysis presented hereafter.

Table 2. Specification of grids

Grid	No. of cells	Growth rate	Aspect ratio	Skewness	Orthogonal quality
A	2038301	1.1	1.8225	0.2188	0.8645
B	2746814	1.1	1.8232	0.2187	0.8644
C	3403968	1.1	1.8235	0.2187	0.8643

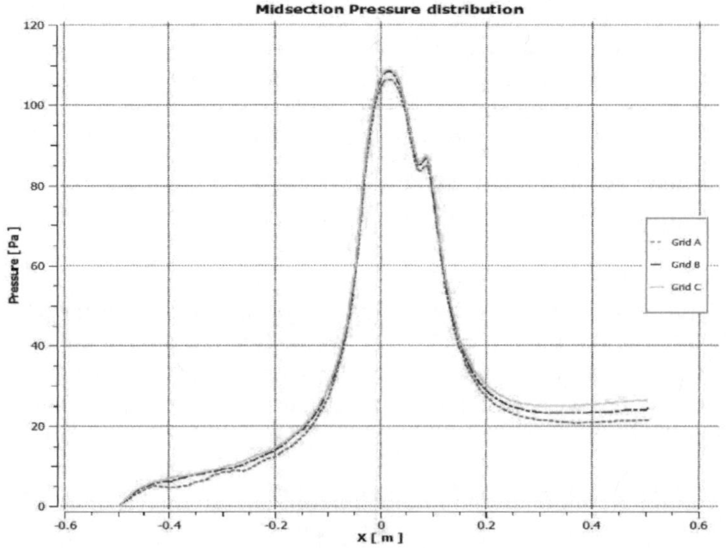

Figure 4. Pressure distribution on the wing during heeling angle

8. Results and discussions

Pressure coefficient distributions obtained from simulation studies on the upper and lower surfaces are shown in Figure 5 and Figure 6, respectively. Essentially, the results showed that the wing produced significantly higher lift and lower drag. The ground effect was evenly distributed along the wing span and wing chord. The lift was higher at the middle of the wing and decreased at the end of span.

During the unstable or heel condition of the WIG craft, the pressure distribution changed across the wing section. As the wing was located closer to the ground, the effect was higher on the wing (in this case, it was lift coefficient). The effect of in ground was more prominent on the descending section of the wing that was approaching the ground. This pressure generated a natural righting moment on the WIG craft as depicted in Figure 5 and Figure 6. Nonlinear pressure distribution on the wing between the rise and descending sides basically cancelled and balanced each other until it was in equilibrium condition since the righting moment react with the increases of lift due the heeling angle.

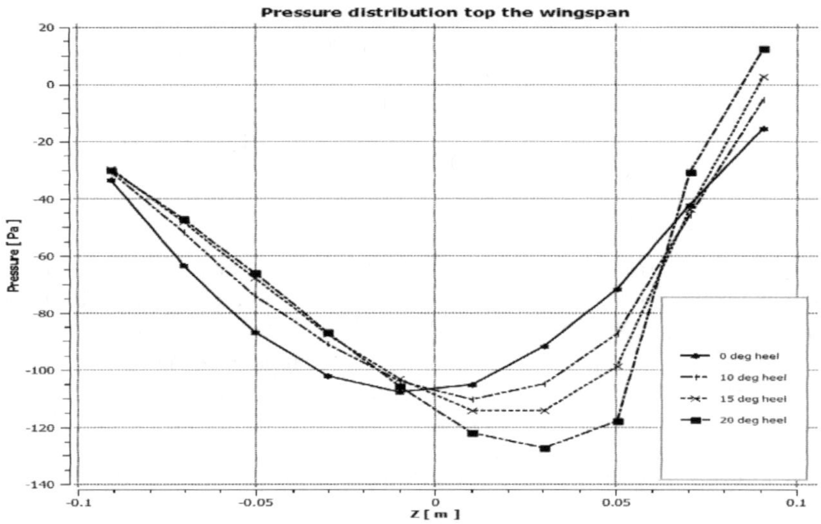

Figure 5. Pressure distribution top wingspan, the negative Z referring to the descending side

IOP Conf. Series: Materials Science and Engineering **152** (2016) 012003 doi:10.1088/1757-899X/152/1/012003

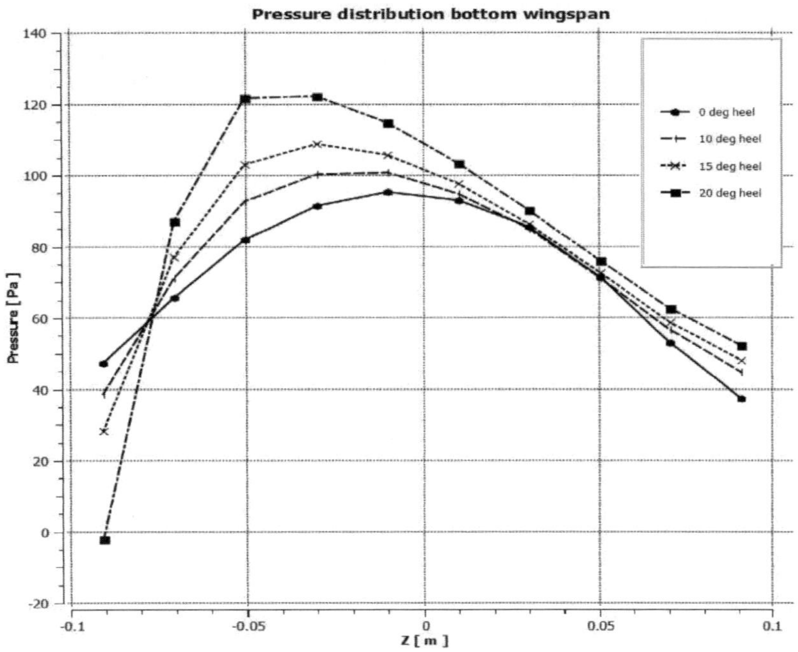

Figure 6. Pressure distribution bottom wingspan, the negative Z referring to the descending side

The anhedral increases lift pressure and forbids flow from escaping under the wing section easily to keep the pressure high. The effect of anhedral is quite similar to endplate as discussed scientifically [3, 10, 11]. The comparison on anhedral lift distribution during descending side at different heeling angle is shown in Figure 7. The vertical lift is increased with increment of heeling angle. It can be observed also that the anhedral angle and winglet further improved the lift increment. Furthermore, the righting moment was induced due to the lift pressure during descending side of the wing. In midsection area, not much different on the pressure distribution as shown in Figure 8, except at heeling angle 20^0. At this test condition, the pressure shows a slight increase in lift pressure. Several important observations can be seen in Figure 9. Simulation results show that the winglet has reduced vortex formation on the upper surface and increase the lift coefficient.

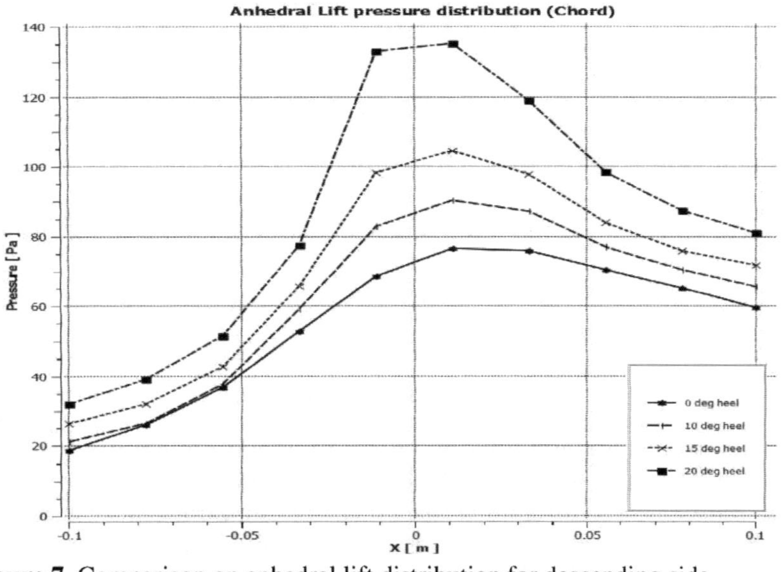

Figure 7. Comparison on anhedral lift distribution for descending side

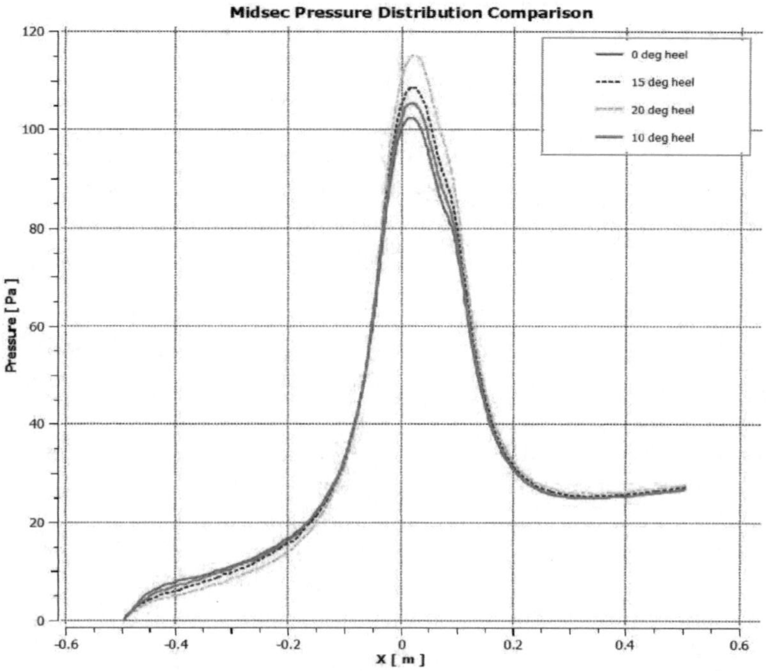

Figure 8. Midsection pressure distribution comparison for different heel angle

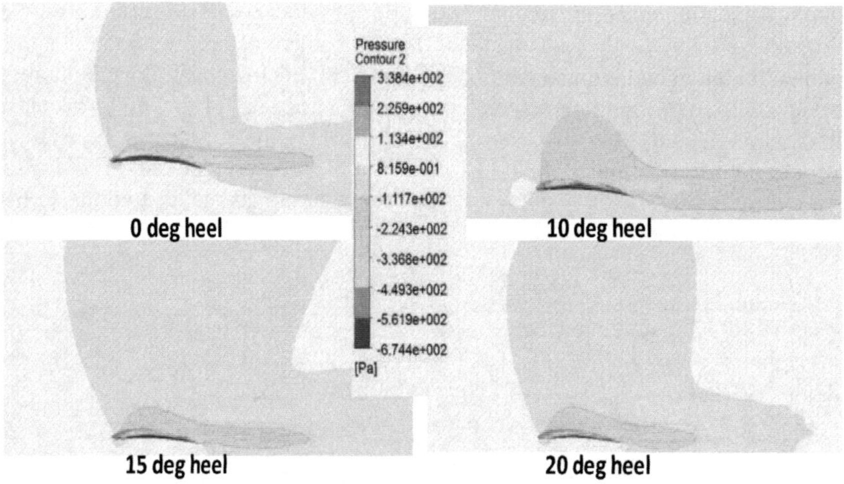

Figure 9. Winglet pressure distribution comparison for different heel angle

Pressure distributions of the upper and lower surfaces are shown in Figure 10. The results show that the descending side induced more pressure, particularly at 20^0 heel (close to the ground). WIG experienced an increase in vertical lift when heeling in ground effect. The situation is different from aircraft stability where heeling will decrease the vertical lift of the plane in air. Therefore, this means it requires less control to overcome the righting motion because of the existing natural righting moment. Another observation from the results shows the side force may be induced on the WIG craft. The pressure distribution on the descending side was higher compared to that on the rising side. It also had a tendency from a leak of high pressure downwash to the lower pressure on the outer surface pressure that will further increase a side force. More experimental data from the wing tunnel testing is required to verify this hypothesis.

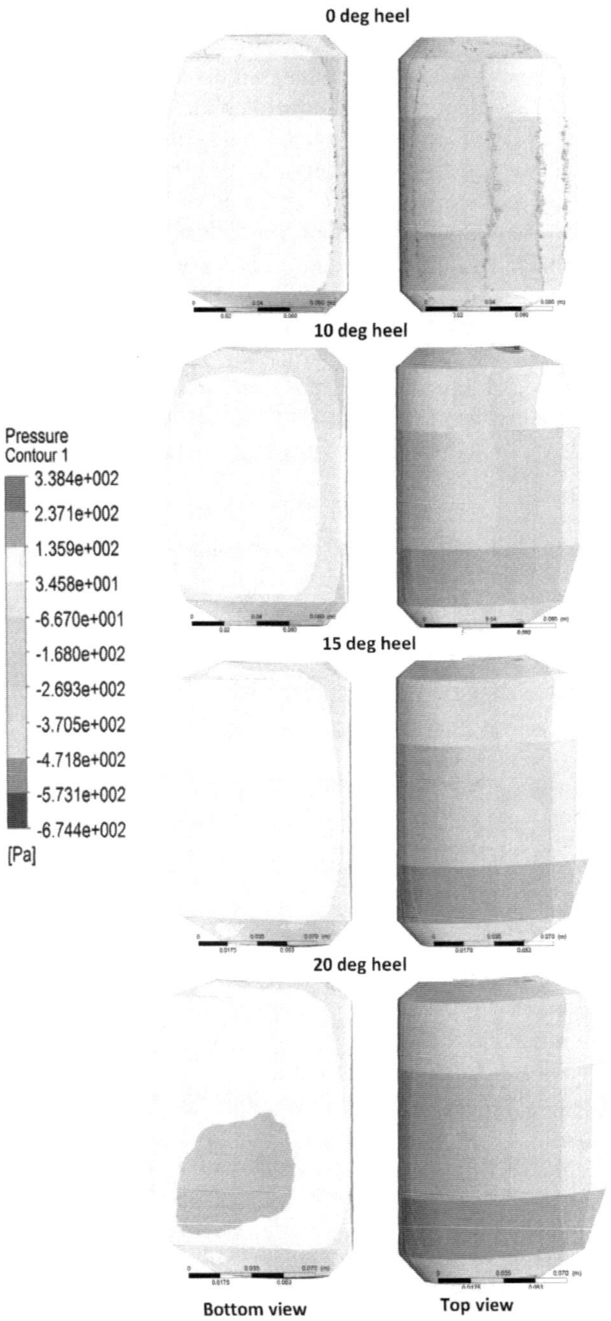

Figure 10. Pressure distribution on the wing during heeling angle

The operation of the WIG craft that is close to the ground at high speed requires a good stability in both longitudinal and lateral directions. In some cases, the stability of the WIG is improved by using a complex automatic height control system [12]. However, the design of this system is complicated as it requires information on environmental conditions that varies and uncertain. On the other hand, if such system is not available onboard the craft, the pilots must maintain the stability of the craft manually, and that will put extra cautions that will lead to their fatigue for a long flight duration. Thus, with the improvement of the design, the safety of the WIG can be assured with the natural righting moment can be an advantage of the in ground effect.

9. Conclusion

The research is driven by the importance of understanding the lateral stability of WIG. The simulation has been carried out to study the effect of heeling angle on lateral stability. The results show that the ground effect increases vertical lift when heeling on the craft's descending side. Thus a natural righting moment is introduced when the WIG approximately close to the ground, where pressure distribution on its lifting surfaces changes and redistributed. Coefficient of lift increases at wing that approaches the ground and decreases for the portion that leaves the ground. This situation causes a generation of recovering transverse moment. Therefore, WIG craft in lateral has a natural aerodynamic stabilization of roll angle in flight close to the shield. Because WIG naturally seek lateral-directional equilibrium, the stability can be sustained without interference by the pilot. On the other hand, side force occurs from the effect of different pressure distribution during heeling angle on the surface. To proceed with this study, a model of wind tunnel experiments will be designed and fabricated in the near future. The model will be tested in low-speed wind tunnel at UTM for several testing conditions. Data from this experiment will be compared with the simulation work that has been carried out.

Acknowledgement

The authors would like to thank Universiti Teknologi Malaysia (UTM) for supporting this research under the Research Industrial Grant Vot No: 3C007. All the expenses for the conference are using the allocation from this grant.

References

[1] Divitiis N D 2005 *Journal of Aircraft* **42** 148-57
[2] Kornev N and Matveev K 2003 *41st Aerospace Sciences Meeting and Exhibit, Aerospace Sciences Meetings* (Reno: USA)
[3] Qing Jia W Y and Yang Z 2016 *International Journal of Naval Architecture and Ocean Engineering* **8** 209-17
[4] Halloran M and O'Meara S 1999 *Wing in Ground Effect Craft Review* DSTO Aeronautical and Maritime Research Laboratory
[5] Cook M V 2007 *Flight Dynamic Principles: A linear systems approach to aircraft stability and control* (Elsevier)
[6] Tofa M M, Maimun A, Ahmed Y, Jamei S, Priyanto A and Rahimuddin 2014 *The Scientific World Journal* **2014**
[7] Rahimuddin, Maimun A, Tofa M M, Jamei S and Tarmizi 2014 *14th International Ship Stability Workshop (ISSW)* (Kuala Lumpur: Malaysia)
[8] Maimun A, Jamei S, Priyanto A and Azwandi N 2010 *Proceedings of the 9th WSEAS International Conference on Applications of Computer Engineering*
[9] Maimun A, Jamei S, Priyanto A and Azwandi N 2010 *WSeas Transactions on Fluid Mechanics* **5**
[10] Afshar H and Alishahi M M 2009 *J. Appl. Sci.* **20** 3752-57
[11] Yang W and Yang Z 2012 *Aircr. Eng. Aerosp. Technol.* **84** 4-12
[12] Daniel D 2008 *Proceedings of the 17th IFAC World Congress* (Seoul: Korea)

IOP Conf. Series: Materials Science and Engineering **152** (2016) 012004 doi:10.1088/1757-899X/152/1/012004

A computational study on effect of pitch difference in pure plunging tandem wings

M R Rasani, M Z Nuawi, W M F W Mahmood and Z Harun

Jabatan Kejuruteraan Mekanik dan Bahan, Fakulti Kejuruteraan dan Alam Bina, Universiti Kebangsaan Malaysia, 43600 Bangi, Malaysia

E-mail: rasidi@ukm.edu.my

Abstract. Flapping wing in tandem configuration may offer enhanced aerodynamic performance at low Reynolds number, in which micro air vehicles operate. The present study aims to investigate the effect of fore-hind wing pitch difference on the aerodynamic performance of tandem wings. To that end, two-dimensional, laminar flow around two thin flat airfoils that are sinusoidally plunging in phase with each other, were computationally simulated at a Reynolds number of 10000, using a flow solver in an Arbitrary Lagrangian-Eulerian framework. The fore wing pitch angle was fixed to $10°$, while the hind wing pitch angle was varied between $-10°, 0°, 10°$ and $20°$. Numerical results shows that aerodynamic performance of the fore wing may be affected by the hind wing pitch angle and that tandem wings may offer improved lift to drag efficiency at some optimal fore-hind wing pitch difference compared to twice the results of a similar single wing case. In addition, the complex fore-hind wing vortex interaction is also affected by the hind wing pitch angle.

1. Introduction

Research of flapping wings or airfoils have generated much interest in recent years, especially in development of bio-inspired micro air vehicles (MAV). Micro air vehicles and nano air vehicles are flying vehicles no larger than 15 cm in any of its dimensions and operate at low Reynolds numbers, over which flow is laminar and viscous [1, 2]. At these low Reynolds numbers, insects and small birds have been shown to generate enhanced aerodynamic performances by exploiting leading and trailing edge vortices in their flapping motion [3–5], which is advantageous for micro or nano air vehicles.

Aerodynamics of wings flapping in tandem have also attracted much attention [3, 6–8] and may offer improved aerodynamic efficiencies [4, 9] under certain configurations. Increased combined thrust were found when the fore and hind wings were flapping in phase with each other [2] or within $0 \pm 50°$ phases of each other [3], which is suitable for rapid forward flights. Previous investigations also indicate that improved efficiency may be obtained at some optimal fore-hind wing phase difference [4], which may be efficient during cruising or hovering flights. In terms of tandem wing spacing, Broering and Lian showed that maximum thrust was achieved when spacing is 0.5 of chord length and that propulsive efficiency may be increased by 40% compared to a single wing configuration [10]. More recently, Kok et al. [11] investigated their insect-inspired micro air vehicle using blade element analysis and showed increased aerodynamic efficiency when asymmetrical flapping occurs at a frequency below the natural frequency of the system.

From the literature review, the effect of flapping phase and spacing between fore-hind wing on aerodynamic performance of tandem wing have been investigated. Moreover, their flapping kinematics have typically involved plunging in combination with pitching of both wings. However, the effect of pitching has not been fully investigated and therefore, the present study aims to analyze the effect of fore-hind wing pitch difference on aerodynamic performance of tandem wings. Towards that end, we simulate unsteady flow around two tandem thin airfoils under pure in-phase fore-hind wing plunging motion with constant wing spacing, to isolate the effects of pitching angle. Both unsteady aerodynamic forces and vortex interactions at various pitch angles were analyzed using computational fluid dynamics, at Reynolds number $Re = 10000$ and flapping frequency of 10 Hz.

2. Methodology

2.1. Flow equations

In order to simulate this problem, the two-dimensional, incompressible and unsteady Navier-Stokes and continuity equations were solved. Although at $Re = 10000$ the flow may not be entirely laminar, a laminar flow is assumed, considering previous studies indicating negligible force differences at Re below 50000 [7, 12]. Furthermore, the present study begins by considering a two-dimensional model, as primary vortex interaction in tandem wings have been shown to be two-dimensional [9]. In order to account for fluid mesh deformation that follows tandem wing motions, the Navier-Stokes and continuity equations are expressed in an Arbitrary Lagrangian-Eulerian (ALE) description [13], as shown in equations (1) and (2):

$$\frac{\partial u_i}{\partial t} + (u_j - \widetilde{u_j})\frac{\partial u_i}{\partial x_j} = -\frac{1}{\rho}\frac{\partial p}{\partial x_i} + \frac{\partial}{\partial x_j}\left[\frac{\mu}{\rho}(\frac{\partial u_i}{\partial x_j} + \frac{\partial u_j}{\partial x_i})\right] \quad (1)$$

$$\frac{\partial (u_i - \widetilde{u_i})}{\partial x_i} = 0 \quad (2)$$

where u_i, $\widetilde{u_j}$ and p are the fluid velocities, fluid grid velocities and pressure respectively ($i, j = 1, 2$ represent the 2-D cartesian directions), $\rho = 1.185$ kg/m^3 is the constant air density used and $\mu = 1.831 \times 10^{-5}$ Pa s represents the air dynamic viscosity used in all simulations. The fluid grid velocity $\widetilde{u_j}$ is computed using a Laplacian Diffusion model, which smooths the grid velocity from a value matching the wing motion near the fluid-wing interface, to zero at the fluid domain boundaries. Time advancement was accomplished using a second order backward euler scheme, while a high-resolution scheme was employed to solve the advective terms.

2.2. Computational model

The computational domain spans 4 chord lengths upstream of the fore wing and 20 chord lengths downstream of the fore wing, with another 4 chord lengths separating respectively, top and bottom walls from both wings. Figure 1 shows the computational domain and boundary conditions imposed on the present flow problem, where a uniform velocity U was specified at the inlet and atmospheric (zero gauge) pressure condition was defined at the outlet boundary. No-slip wall conditions are imposed on both fore and hind wings. Both wings are idealized as thin, flat airfoils with a chord length of 50 mm and a thickness of 2 mm. Inlet velocity U was set to 3 m/s, giving a chord-based Reynolds number $Re = 10000$.

The flow domain was discretized using predominantly 8-noded hexahedral elements, with finer mesh surrounding both wings and finer 8-noded prismatic elements in regions close to wing surfaces, to better resolve flow and vortex patterns in the vicinity of both wings as shown in figure 2(a), yielding a total of 75168 elements. In order to maintain grid quality in regions close to both wings (as they move according to prescribed wing motion), separate domain for these regions (as shown by shaded area in figure 2(a)) were considered, so that grids in both regions

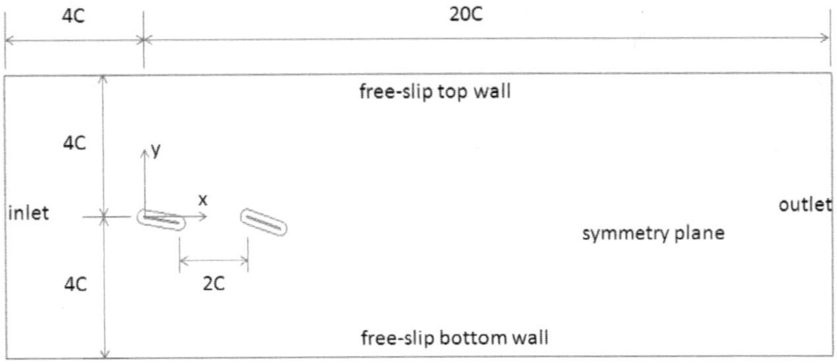

Figure 1. Computational domain and conditions at respective boundaries used in this study. C represents chord length of fore or hind wing. For a single wing configuration, similar computational model and domain is employed but without hind wing.

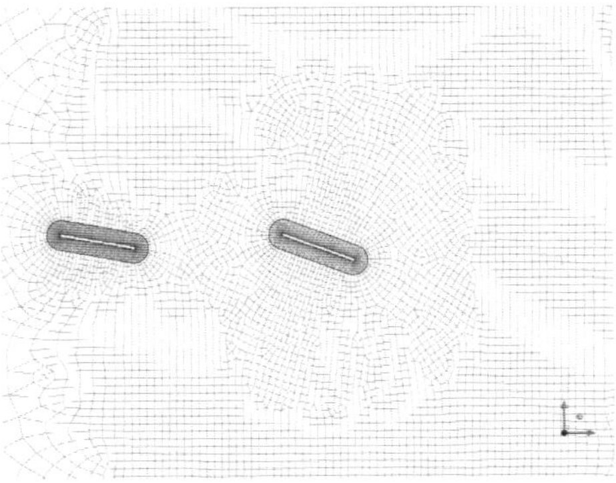
(a)

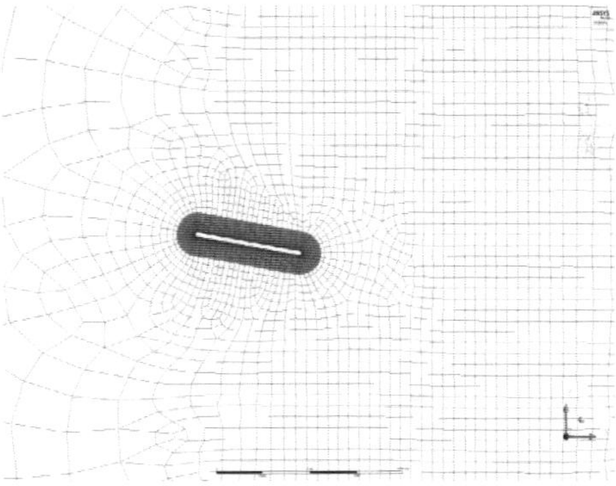
(b)

Figure 2. Discretization of flow computational domain: (a) tandem wing, with shaded regions representing separate domain containing fore and hind wings (b) single wing configuration, where shaded region represents separate domain containing fore wing only.

may move uniformly together. Flow variables were conserved across the domain interfaces using an implicit and conservative grid interface algorithm [14].

2.3. Case studies

In the present study, a fore-hind wing spacing equal to 2 chord lengths and a pure sinusoidal plunging motion of the form $y(t) = y_0 \sin(2\pi ft)$, where plunging amplitude y_0 was set to 0.5 chord length and plunging frequency f was set to 10 Hz were considered, giving a Strouhal number of approximately 0.17. In addition, we focus our attention to in-phase plunging between the fore and hind wing. Four cases were simulated, where the fore wing pitch angle was fixed to $\alpha_1 = 10^\circ$ in all cases, while the hind wing pitch angles were varied between $\alpha_2 = -10^\circ, 0^\circ, 10^\circ$ and 20°. An additional single wing, plunging with a pitch angle of 10°, was also simulated for comparison purposes. Considering the flapping frequency, a timestep size of 0.001 s was chosen in all cases and simulations were run to a total time of 1 s, giving 10 complete flapping cycles. Results reported herein were extracted from the final 10th cycle.

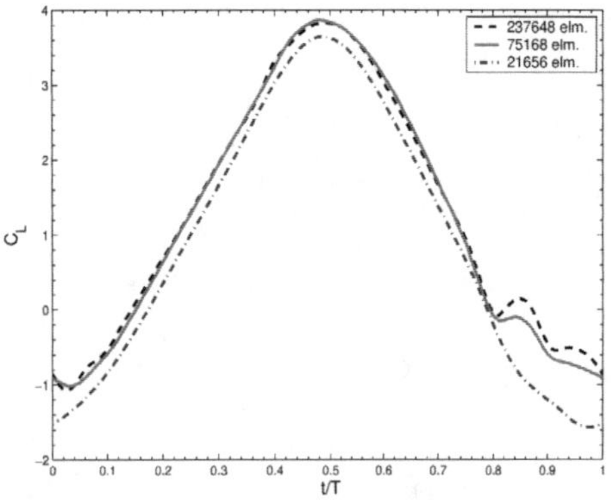

Figure 3. Lift coefficient history over time of fore wing at different grid resolution.

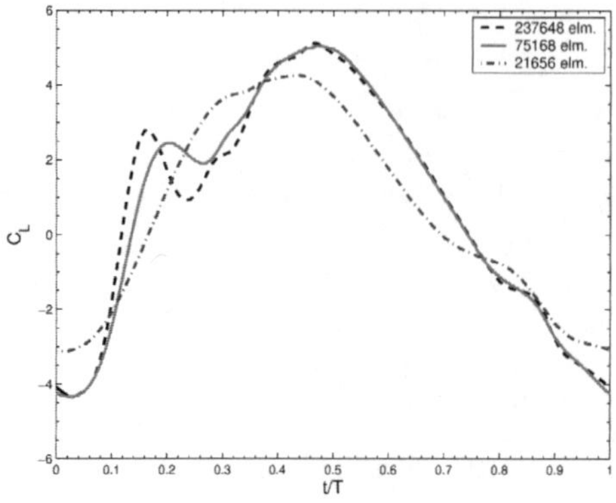

Figure 4. Lift coefficient history over time of hind wing at different grid resolution.

3. Results and discussion

3.1. Preliminary analysis

We first present aerodynamic force results between three different grid resolutions i.e. the baseline grid that was used in this study (total 75168 elements), a finer grid resolution where the grid is refined everywhere, especially in regions surrounding and between both wings (resulting in total 237648 elements) and a coarser grid where the grid size was approximately doubled the baseline grid (resulting in total 21656 elements). Comparison of the lift coefficient history at fore and hind wing, respectively shown in figures 3 and 4, suggest small difference between the baseline grid and the finer grid. Therefore, the baseline grid was used in the present simulation.

Table 1. Comparison of time-averaged lift and drag coefficients of tandem wings between present model with experimental results adapted from [15, 16].

$\overline{C_L}$		$\overline{C_D}$		
exp. [15]	present model	exp. [15]	exp.^{a}	present model
0.39	0.34	0.10	0.07	0.06

^{a} based on single wing data in [16]

In addition, lift and drag coefficients from a fixed tandem wing configuration experiment by Jones et al. [15] at Reynolds number $Re = 100000$ and $10°$ angle of attack , is presented in table 1 for comparison. Present model with similar configuration shows reasonable agreement with experimental results. Discrepancy in drag coefficient is likely due to the squared leading and trailing edge geometry employed in [15] compared to a circular edge in the present model. Further comparison considering single wing experimental data taken from more rounded geometry edges in Pelletier and Mueller [16], suggest reasonable agreement in drag coefficient and discrepancy is likely due to difference in airfoil thickness used (4% of chord in present model vs. 2% of chord in [16]).

3.2. Aerodynamic forces

Table 2 summarizes time-averaged lift and drag coefficients ($\overline{C_L}$ and $\overline{C_D}$) over a single flapping period for all tandem cases, together with a single wing configuration. Although the fore wing was maintained at pitch angle $\alpha_1 = 10°$ (similar to the single wing case), both its $\overline{C_L}$ and $\overline{C_D}$

Table 2. Summary of time-averaged lift and drag coefficient for single and all tandem cases. Total single wing values were based on twice its lift and drag coefficients.

	fore wing		hind wing		single wing		total tandem			total single		
α_2	$\overline{C_L}$	$\overline{C_D}$	$\overline{C_L}$	$\overline{C_D}$	$\overline{C_L}$	$\overline{C_D}$	$\overline{C_L}$	$\overline{C_D}$	L/D	$\overline{C_L}$	$\overline{C_D}$	L/D
−10	1.03	0.14	−0.57	0.02	-	-	0.46	0.16	2.9	-	-	-
0	1.14	0.16	0.15	−0.08	-	-	1.30	0.08	16.0	-	-	-
10	1.23	0.18	0.77	0.04	1.21	0.18	2.00	0.21	9.3	2.41	0.35	6.9
20	1.33	0.20	1.40	0.40	-	-	2.74	0.60	4.6	-	-	-

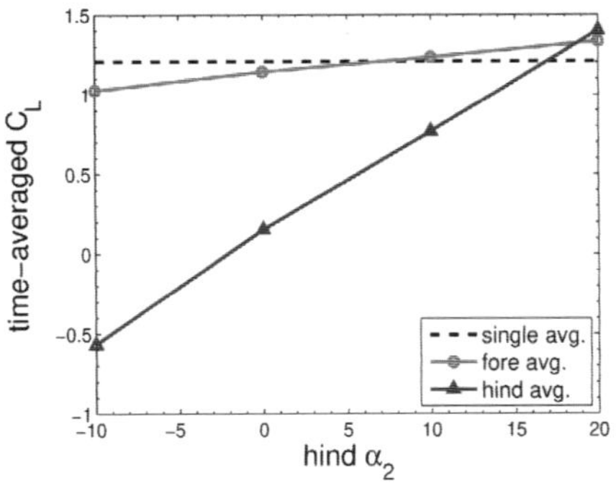

Figure 5. Variation of time-averaged lift coefficient against hind wing pitch angle α_2 (fore wing pitch angle fixed at $\alpha_1 = 10°$).

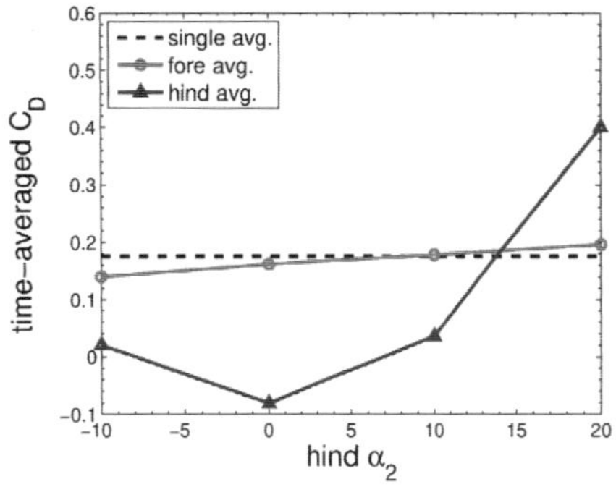

Figure 6. Variation of time-averaged drag coefficient against hind wing pitch angle α_2 (fore wing pitch angle fixed at $\alpha_1 = 10°$).

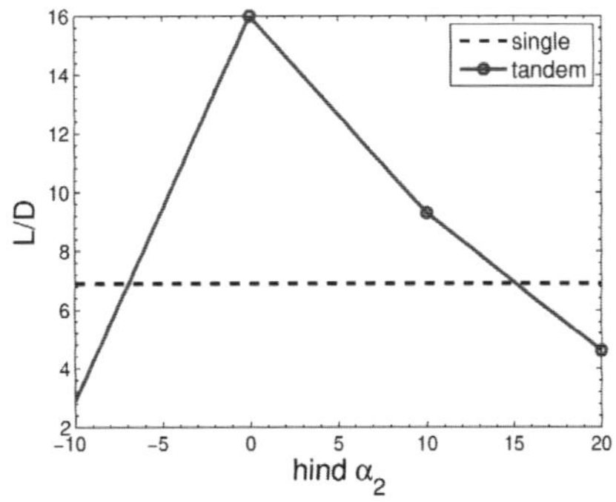

Figure 7. Variation of aerodynamic efficiency (L/D) against hind wing pitch angle α_2 (fore wing pitch angle fixed at $\alpha_1 = 10°$).

shows increasing trend with increasing hind wing pitch angle α_2, which is better illustrated in figures 5 and 6. At $\alpha_2 = 20°$, fore wing $\overline{C_L}$ is higher than the single wing, but its $\overline{C_D}$ is also increased compared to similar single wing configuration, indicating that hind wing α_2 may influence aerodynamic performance of the fore wing. In contrast, at $\alpha_2 = 10°$, the hind wing does not generate as much lift and drag to the similar single wing case, which may be explained by its interaction with the fore wing.

Table 2 also presents total or combined force coefficients from both fore and hind wings, as these represent the total lift and drag experienced by a tandem wing flyer. In order to fairly compare the performance of combined fore-hind wing to those of a single wing, the equivalent total force for a single wing flyer were based on twice its lift and drag coefficients, as shown in table 2. Summing $\overline{C_L}$ and $\overline{C_D}$ from both fore and hind wings, the combined lift and drag for tandem cases increases as the hind wing α_2 is increased. More importantly, the combined lift to drag ratio in table 2 and illustrated in figure 7, shows a non-linear trend with increasing α_2, suggesting an optimal phase difference may be found for the best lift to drag efficiency. This is likely due to non-linearity in the hind wing drag coefficient, which is minimal at low angles of attack where drag is mainly frictional, and increases at higher pitch angles as greater induced drag profile is generated. In addition, although the tandem case where both fore and hind wing are pitched at 10° shows lower combined lift than twice the lift of similar single wing configuration, their combined lift to drag ratio suggest better performance in comparison to ratio of twice the lift and drag, of similar single wing configuration.

3.3. Vortex interactions

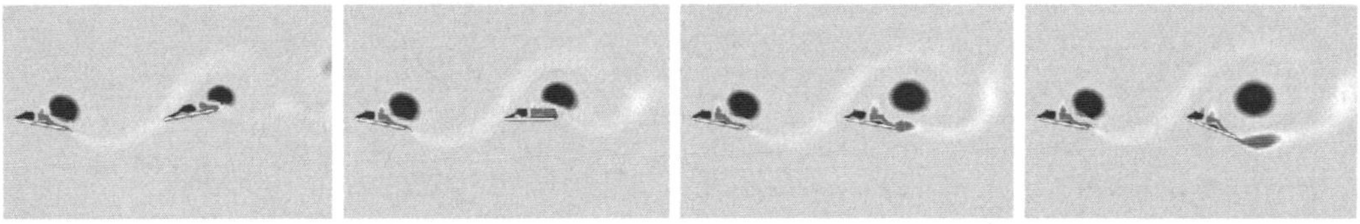

(a) (b) (c) (d)

Figure 8. Contour of vorticity slightly after start of upstroke (80% of flapping cycle) for: (a) hind wing pitch $\alpha_2 = -10°$ (b) hind wing pitch $\alpha_2 = 0°$ (c) hind wing pitch $\alpha_2 = 10°$ and (d) hind wing pitch $\alpha_2 = 20°$.

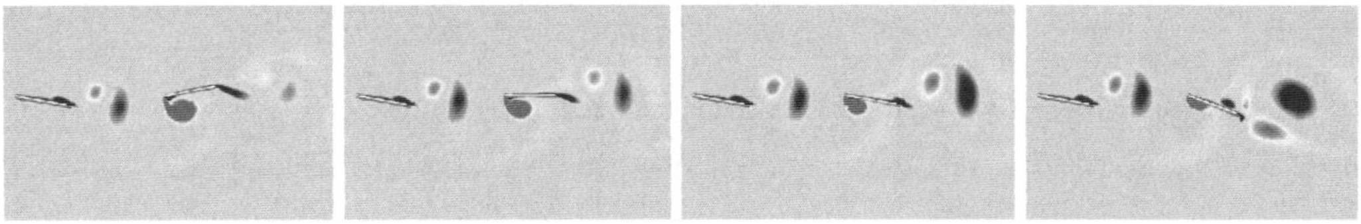

(a) (b) (c) (d)

Figure 9. Contour of vorticity at mid of upstroke (0% of flapping cycle) for: (a) hind wing pitch $\alpha_2 = -10°$ (b) hind wing pitch $\alpha_2 = 0°$ (c) hind wing pitch $\alpha_2 = 10°$ and (d) hind wing pitch $\alpha_2 = 20°$.

Figure 8 to figure 11 presents contours of vorticity (where red contours represent positive (counterclockwise) vorticity and blue contours represent negative (clockwise) vorticity) for all tandem cases, from start of upstroke to the middle of downstroke. In the beginning of the upstroke, a trail of counterclockwise (CCW) vortex from the trailing edge of the fore wing is about to bisect the hind wing, as shown in figure 8. At larger hind wing pitch angle, the strength of the trailing vortex that bisects the leading edge of the hind wing does not appear as pronounced as at lower pitch angles (comparing figure 8(d) to figure 8(a)). This may explain the decreasing size of counterclockwise leading edge vortex (LEV) forming underneath the hind wing with increasing α_2, as indicated in figure 9, which may then generate lower negative lift or downforce during the upstroke.

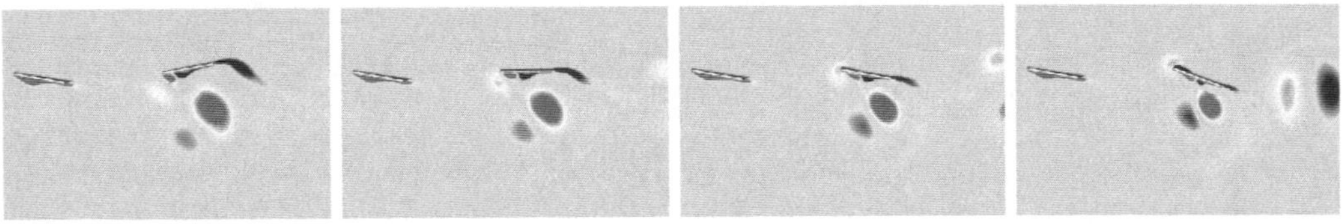

(a) (b) (c) (d)

Figure 10. Contour of vorticity nearing end of upstroke (20% of flapping cycle) for: (a) hind wing pitch $\alpha_2 = -10°$ (b) hind wing pitch $\alpha_2 = 0°$ (c) hind wing pitch $\alpha_2 = 10°$ and (d) hind wing pitch $\alpha_2 = 20°$.

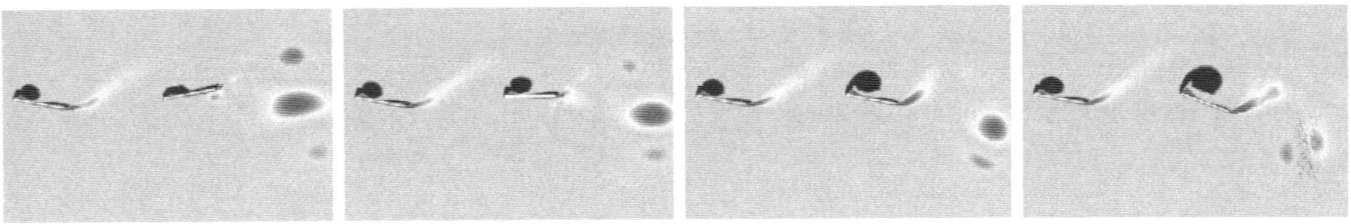

(a) (b) (c) (d)

Figure 11. Contour of vorticity in the mid of downstroke (50% of flapping cycle) for: (a) hind wing pitch $\alpha_2 = -10°$ (b) hind wing pitch $\alpha_2 = 0°$ (c) hind wing pitch $\alpha_2 = 10°$ and (d) hind wing pitch $\alpha_2 = 20°$.

Consequently, with smaller CCW LEV at larger α_2, it is hypothesized that the incoming clockwise LEV shed from the fore wing is less deflected away from the hind wing, allowing stronger interaction with the leading edge of the hind wing at larger α_2 during downstroke. This may generate larger LEV at the hind wing for larger α_2, as indicated in figure 11, which may generate larger lift during the downstroke. Unlike a single wing or the fore wing, the hind wing generates LEV during both upstroke and downstroke that leads to a lower time-averaged lift coefficient in a single flapping cycle.

4. Conclusion

A computational study on the effect of fore-hind wing pitch difference to aerodynamic performance of tandem wings at *Re* = 10000 was presented. Although, in the present tandem configuration, their combined lift may not better twice the lift of similar single wing case, tandem wing may offer improved lift to drag efficiency at some optimal fore-hind wing pitch difference. In addition, vortex interaction between fore and hind wings may also explain the variation in lift at different hind wing pitch angles. As the present study had limited its attention to a fixed set of configuration, further investigation is recommended to examine effect of pitch difference but at smaller wing spacing, larger Strouhal numbers and different fore-hind wing phasing.

Acknowledgments

Acknowledgments The authors would like to gratefully acknowledge the financial support provided by Universiti Kebangsaan Malaysia under project grant GGPM 2013-092.

References

[1] Platzer M F, Jones K D, Young J and S Lai J C 2008 *AIAA Journal* **46** 2136–2149
[2] Broering T, Lian Y and Henshaw W 2010 *Aerospace Sciences Meetings* (American Institute of Aeronautics and Astronautics) AIAA 2010-865
[3] Warkentin J and DeLaurier J 2007 *Journal of Aircraft* **44** 1653–1661
[4] Usherwood J R and Lehmann F O 2008 *Journal of The Royal Society Interface* **5** 1303–1307
[5] Sirohi J 2013 *Proc. SPIE: Bioinspiration, Biomimetics, and Bioreplication* vol 8686 86860U
[6] Kumar A and Hu H 2011 *Fluid Dynamics and Co-located Conferences* (American Institute of Aeronautics and Astronautics) AIAA 2011-3120
[7] Broering T M 2013 *The Characterization of Wing-Wing Vortex Interactions of A Tandem Flapping Wing Configuration and Its Relationship to the Phase Angle and Wing Spacing.* Electronic theses and dissertations paper 157. University of Louisville
[8] Gravish N, Peters J M, Combes S A and Wood R J 2015 *Phys. Rev. Lett.* **115**(18) 188101
[9] Rival D, Schönweitz D and Tropea C 2011 *Bioinspiration & Biomimetics* **6** 016008
[10] Broering T and Lian Y 2010 *Fluid Dynamics and Co-located Conferences* (American Institute of Aeronautics and Astronautics) AIAA 2010-4385
[11] Kok J M, Lau G K and Chahl J S 2016 *Journal of Aircraft* **53** 800–810
[12] Young J, S Lai J C and Germain C 2008 *AIAA Journal* **46** 918–924
[13] Donea J, Huerta A, Ponthot J P and Rodriguez-Ferran A 2004 Arbitrary Lagrangian-Eulerian Methods *Encyclopedia of Computational Mechanics* vol 1 ed Stein E, Borst R D and Hughes T J R (United Kingdom: John Wiley & Sons) chap 14
[14] ANSYS Inc 2014 *ANSYS CFX Solver Theory Guide - Release 14.5*
[15] Jones R, Cleaver D J and Gursul I 2015 *Experiments in Fluids* **56** 1–25
[16] Pelletier A and Mueller T J 2000 *Journal of Aircraft* **37** 825–832

IOP Conf. Series: Materials Science and Engineering **152** (2016) 012005 doi:10.1088/1757-899X/152/1/012005

Ordered roughness effects on NACA 0026 airfoil

Z Harun*, A A Abbas, R Mohammed Dheyaa and M I Ghazali

Department of Mechanical and Materials, Universiti Kebangsaan Malaysia, Malaysia

*zambri@ukm.edu.my

Abstract. The effects of highly-ordered rough surface – riblets, applied onto the surface of a NACA 0026 airfoil, are investigated experimentally using wind tunnel. The riblets are arranged in directionally converging – diverging pattern with dimensions of height, h = 1 mm, pitch or spacing, s = 1 mm, yaw angle $\alpha = 0^{\circ}$ and 10°. The airfoil with external geometry of 500 mm span, 600 mm chord and 156 mm thickness has been built using mostly woods and aluminium. Turbulence quantities are collected using hotwire anemometry. Hotwire measurements show that flows past converging and diverging pattern inherit similar patterns in the near-wall region for both mean velocity and turbulence intensities profiles. The mean velocity profiles in logarithmic regions for both flows past converging and diverging riblet pattern are lower than that with yaw angle $\alpha = 0^{\circ}$. Converging riblets cause the boundary layer to thicken and the flow with yaw angle $\alpha = 0^{\circ}$ produces the thinnest boundary layer. Both the converging and diverging riblets cause pronounced outer peaks in the turbulence intensities profiles. Most importantly, flows past converging and diverging pattern experience 30% skin friction reductions. Higher order statistics show that riblet surfaces produce similar effects due to adverse pressure gradient. It is concluded that a small strip of different ordered roughness features applied at a leading edge of an airfoil can change the turbulence characteristics dramatically.

1. Introduction

Studies of turbulent boundary layer have been an important focus in the past few decades. Reducing fuel usage would be monumental if these efforts could be commercially realized, where high ordered surface roughness called riblets are applied in aerospace, automotive and energy [1, 2, 3]. There have been a lot of technological advancements in getting energy efficient vehicles or systems, such as the lighter composite and more fuel-efficient jet engines for aviation industry, energy efficient vehicles by employing hybrid technology, using electric motors and hydrogen fuel cells for automotive industry and many more. These efforts have been helpful but there are more rooms for improvement to reduce drag by controlling surface turbulence.

Skin-friction drag comes from a thin region of slow moving fluid immediately adjacent to a solid surface. Energy consumption to overcome skin-friction drag is relatively costly, for example: around 50% of the total drag in modern aircraft is caused by skin-friction drag [4, 5, 6], in large ships such as container ship and Very Large Crude Carrier (VLCC), it is up to 80%. One area to control skin drag is by controlling the turbulence levels on the surface and one of the methods is by applying surface roughness. A few decades ago, scientists have realized that turbulent flows have repeating patterns or features known as coherent structures [7] and small-scale near-wall structures dominate turbulent production [8], which are responsible for drag. Most engineering applications deal with high geometry, high velocities, and therefore high Reynolds numbers. As Reynolds number increases, small-scale

IOP Conf. Series: Materials Science and Engineering **152** (2016) 012005 doi:10.1088/1757-899X/152/1/012005

structures are replaced by large-scale features that reside further away from the wall [9, 10]. It is believed that large features may influence the near-wall structures, which are directly correlated to the skin friction drag. One of the surface roughness types is the directional riblets.

Originally, bio-inspired riblets placoid scales-like topology have been observed from nature [11], especially over the whole body of shark so-called "shark skin" [12]. Concepts of featured surfaces leading to skin friction reduction were studied to control turbulent boundary layers, for instance the herringbone riblets of birds is a passive method used to reduce drag up to 16% in turbulent flow [13]. Generally, the friction in the surface is proportional to total kinetic resistance at more than 60% for a cargo ship [12]. Riblets can contribute up to 10% drag reduction compared to flows over a flat surface [14]. More importantly, for a better understanding of boundary layer control, riblets arranged in a 'herringbone' pattern have been studied in zero pressure gradient (ZPG) with different parameters. Riblets could alter flow characteristics in the near-wall region [3, 8] similar to pressure gradient effects [16]. Nano-coating with microstructural riblets has already been applied in aerospace and maritime applications, resulting in drag reduction of 5.2% for torpedo-shaped and 6.2% for wing profile [17]. Furthermore, different types of riblets have been applied in very wide applications such as nature gas pipeline, racing car, high-speed train, windmill, swimming suit and airplane [11]. Last but not least, riblets have been tested on airfoils, namely NACA 0012 [18, 19]. In this research, high order riblet is applied by conducting the experiment with NACA 0026 airfoil [20].

2. Methodology

2.1. Flow facility

The experiment has been conducted at low-speed turbulent boundary layer wind tunnel in Faculty of Engineering and Built Environment, Universiti Kebangsaan Malaysia (UKM), Malaysia. This wind tunnel is referred to as Pangkor-LST, which is an open-circuit blown type [21]. Main geometries of the wind tunnel are shown in Figure 1.

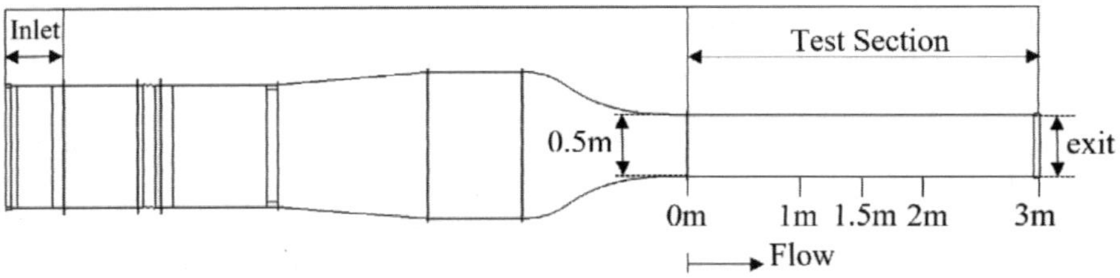

Figure 1. Illustration of wind-tunnel geometry (side view)

The wind tunnel consists of a 1.2 m wide, 0.5 m high and 3 m long test section. The traverse is newly built, with a ball screw pitch of 1.6 mm coupled with Vecta stepper motors model PK266, ensuring a traversing resolution well below 0.1 mm in the spanwise and vertical directions. The traverse is located on the test section and it can be moved manually along the test section (in the streamwise direction). Velmex model VXM-3 controllers used were sequenced automatically with the acquisition system. Further details of design and construction low-speed wind tunnel are given in [21].

2.2. Experimental procedure

The hotwire anemometry system used is the National Instrument's Multi-Channel CTA system model 54N80. The hotwire data is collected using module NI 9215 while data of all other sensors such as temperature, static pressure (pitot-tube), atmospheric pressure, room humidity and dew point were collected using Comet, model H7331 [21]. This system allows measurement performed at very high frequency. It is important that high frequency is employed so that turbulence characteristics could be gathered [16]. The hotwire calibration was performed in-situ. Some information on data acquisition are described in Table 1.

Table 1. Experiment equipment

Equipment	Specification
Acquisition system	National Instrument's model NI 9215
Hotwire anemometry	Multi-Channel CTA 54N80, 8-channel
Sensor	Handmade, Wollaston Pt core probes, Dantec 55P05, 5 μm, 1.5 mm, exposed section = 1.06 mm Overheat ratio 1 – 1.5
Temperature, atmospheric pressure and room humidity sensor	Comet, model H7331
Pressure differential	LSI-Lastem model ESP024

2.3. *Wollaston wire probes*
Experimenting with hotwires require hotwire fabricating bench and other related equipment; otherwise waiting for a repaired sensor can be frustrating. For this, a hotwire fabricating bench was fabricated and it consists of the small traverse to control the sensor position, etching set up, cutter, welding equipment and a microscope. Our newly built three-dimensional (3D) traverse system comes with mounting for etching feature. The 3D traverse jigs are used to traverse and hold both hotwire support with probe manually in three directions (x, y, and z). It provides ability and control for the movement under the microscope and this is especially important during etching and soldering processes. A fine grit sandpaper is required before soldering processes. Soldering flux is applied at the portion between the Wollaston wire and the prong tip. The primary purpose of flux is to prevent oxidation [22]. The in-house etching facility allowed the experiment to be conducted and have full control of the exposed portion of the sensor. This was especially important to address the attenuations of signal due sensor length [26, 27]. Distilled water was used to clean the newly-built sensor. A multimeter was used to measure the resistance of the new sensor and finally, the sensor was balanced to obtain an acceptable response during operation. For the experiment, Wollaston wire produced by Sigmund Cohn Corp. with platinum core diameter of 5 micron (μm) was used.

2.4. *NACA 0026 airfoil*
A purpose-built NACA 0026 airfoil has been fabricated using mostly wooden structure, plywood surface and stainless steel support [20]. The chord, c is 600 mm and the span, s_p is 500 mm. The thickness, t is 156 mm. The chord line falls on the mean camber line, therefore this is a symmetrical airfoil. Wind tunnel test section bottom and top surface work as plates to minimize two-dimensionality issues. A 16 mm diameter stand made of stainless steel is attached to one side of the airfoil to ensure a rigid support. The riblet surface was applied on one side of the airfoil using glue and this is shown by the black strip in Figure 2. The length of the riblet surface is 50 mm. The riblets' surfaces were applied here because of the needs of a sufficient length where the after-effect could develop in the boundary layer towards the rest of the airfoil (until the trailing edge).

An airfoil is chosen to run the effect of riblets because study on riblets has been documented mostly only on the ZPG turbulent boundary layer or flat surfaces. Vortex generators (VG) in the form of triangular shapes are most often used to delay flow separation. To accomplish this, VGs are often placed on the external surfaces of vehicles and wind turbine blades. They are usually installed quite close to the leading edge of the airfoil in order to maintain steady airflow over the control surfaces at the trailing edge. To replicate similar effects, riblet surfaces are applied near to the leading edge of the airfoil. Furthermore, a symmetry airfoil such as NACA 0026 is chosen because two dimensionalities issues could be kept minimum. The length or chord, c = 60 cm was chosen so that sufficient locations of measurement could be performed. In the earlier experiment [3], the thickness of boundary layer is approximately $\delta \approx 5$ cm and therefore, it is possible to have approximately ten places of measurement points along the streamwise direction (chord, c / boundary layer thickness, δ :60 cm / 5 cm = 12, i.e. order of ten measurement points per streamwise direction).

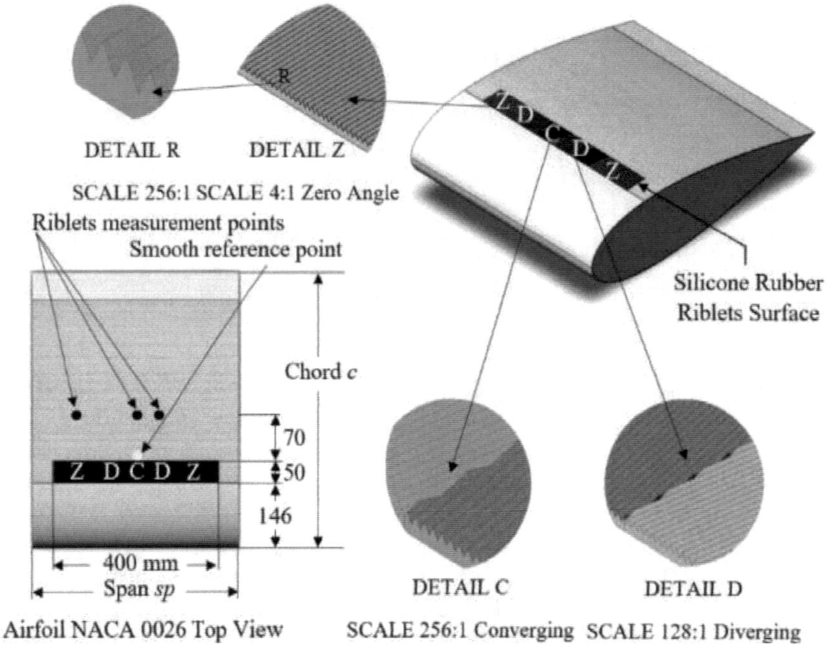

Figure 2. NACA 0026 airfoil dimension and application of riblet

2.5. Converging – diverging riblets fabrication

The riblets were made from silicone rubbers. A mould fabricated from aluminium has been machined using a high-precision CNC machine [20]. Silicone rubber material and hardening materials were mixed to reduce hardening process. The mixing ratio was 450 g of silicon rubber to 5.5 ml hardener. The properly-mixed materials were transferred into a pressure chamber to remove air bubble [3] for one minute. For this experiment, a purpose-built pressure chamber connected to a vacuum pump was also constructed. The materials now, free of bubbles, were poured evenly onto the aluminum mould. This pouring process may produce some small bubbles on the surface. Applying a gentle brush on the surface was helpful to ensure the materials fill in triangular-shaped mould surface. It took four days for the material to be completely hardened and ready to be cut: these were called riblets. The riblets were cut by a cutter such that converging and diverging pattern with yaw angle 10° could be formed. The thickness has been controlled so as to match with the rectangular cavity provided on the airfoil, hence riblet surface was flushed with the adjacent airfoil surface.

Prior to using silicone rubber, different riblet materials such as resin-epoxy mixtures have been tested. However, their surface quality was not as good as the silicone rubber. The resin-epoxy mixtures were more flexible but small particles could be found on the riblets' surface. In summary, a riblet strip onto an airfoil was produced and applied, where the angle of flow is zero and 10° (α = 0° and 10°), relative to the air flow. The height of the riblet is h = 1 mm and the spacing, s = 1 mm. Figure 3(a) shows s and h, while Figure3(b) shows the riblets' arrangement. Note that, in this figure, s appears longer than h due to the bending of the riblet surface to obtain a clear photograph. The general arrangement of the experiments contains the NACA0026 airfoil, a strip of riblets with diverging, converging and zero alignment with the flow, a hotwire sensor and a Pitot tube attached to a fine-resolution traverse. This arrangement is shown in Figure 4.

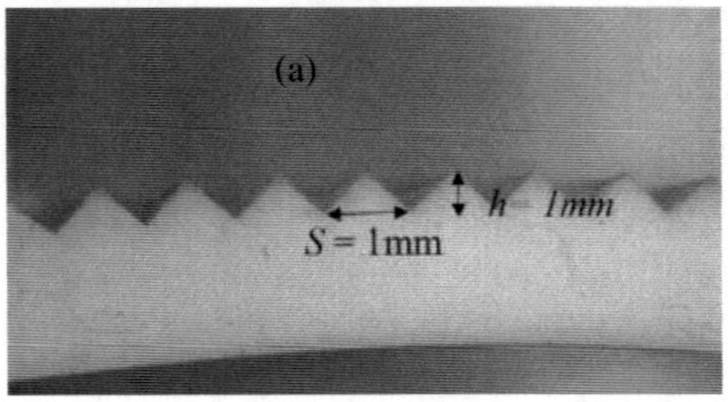

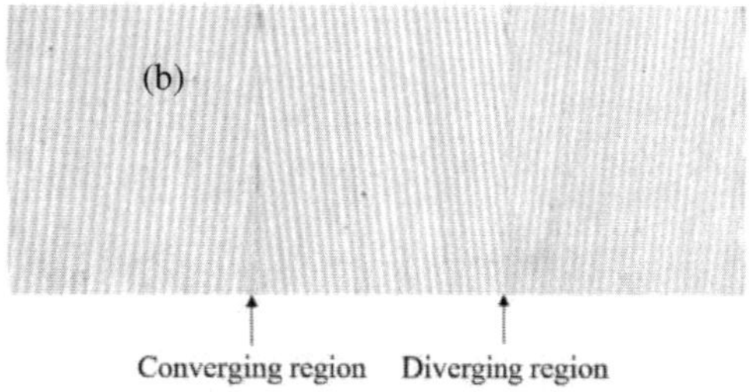

Figure 3. (a) Silicone rubber riblet dimension,(b) Converging–diverging riblet pattern arrangement

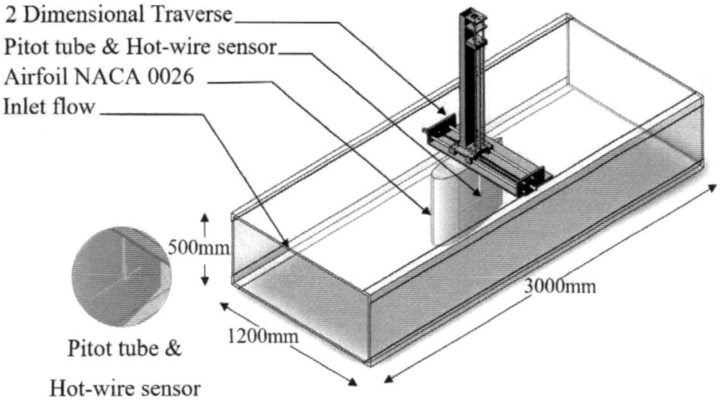

Figure 4. General Arrangement of NACA 0026 airfoilin UKM-LST

3. Result and analysis

3.1. Hotwire experiment and experimental parameters

The locations of hotwire sensor, Pitot tube, two dimensional traverse system were properly arranged so that the flows were not obstructed. The distance of the rod of the traverse to the airfoil surface is at minimum 300 mm away. This way, sufficient areas were provided for boundary layer to develop without obstructions. Prior to performing the measurement, the hotwire was calibrated in-situ using the Pitot tube. The calibration was carried out by positioning the hotwire in the center of test section of the wind tunnel (free-stream). The temperature, along with the atmospheric pressure and room humidity, was recorded during the hotwire calibration. This is called the pre-calibration. Another calibration process at the end of the experiment was also performed, which is called post-calibration. The calibration curves were compared and, if necessary, temperature compensations might have to be implemented.

The parameters and key flow properties for the smooth and rough-wall case with a converging-diverging angle $\alpha = 10^{\circ}$ and zero angle. Table 2 contains skin friction velocity (U_{τ}), friction Kármán number (Re_{τ}) and the boundary layer thickness δ. The experimental codes indicate converging *C*, diverging *D* and zero angle *Z* for the flow exposed with rough/riblet surfaces, *S* for smooth surface.

Table 2. Experimental parameters for smooth and rough surfaces, $l^+ = lU_\tau/\nu \approx 35 - 40$ for all cases

Exp Code		Location (mm)	U_∞ (m/s)	h (m)	s (m)	δ (m)	U_τ (m/s)	Re_τ	$C_f = 2(U_\tau/U_\infty)^2$
C	converging	270	9.38	0.001	0.001	0.0120	0.36	280	0.00147
D	diverging	270	8.93	0.001	0.001	0.0120	0.36	280	0.00162
Z	zero degree	270	9.21	0.001	0.001	0.0089	0.41	240	0.00197
S	smooth	200	10.22			0.0073	0.52	250	0.00258

3.2. Mean velocity

The traverse was set with logarithmic spatial movement. Figure 5 illustrates the application of the Clauser method to obtain skin-friction velocity over the rough and smooth surfaces. The vertical axis represents velocity, the overbar indicates mean value. The velocity is scaled with the friction velocity U_τ obtained from the Clauser method while the surface-normal position is scaled with inner variables U_τ/ν to produce $z^+ = zU\tau/\nu$. In this case, it was found that $\kappa = 0.25$ and $B = -1.5$. The value of κ seems to deviate from the normally accepted values $0.38 < \kappa < 0.41$. The product of Kármán constant, κ and intercept, B here is κB=0.25x-1.5 = - 0.375, appears in the lower end of the proposed empirical relationship of κB vs. B [24]. This is however acceptable because all flows considered here are equivalent the adverse pressure gradient flow i.e. as the fluids move pass the 146 mm from the leading edge of the airfoil (refer Figure 2), the fluids experience adverse pressure conditions as the results of thickening flow area. The exposed sensor part for all measurements is l = 1.5 mm, results in non-dimensionalised sensor length $l^+ = lU_\tau/\nu \approx 35 - 40$ for all flows.

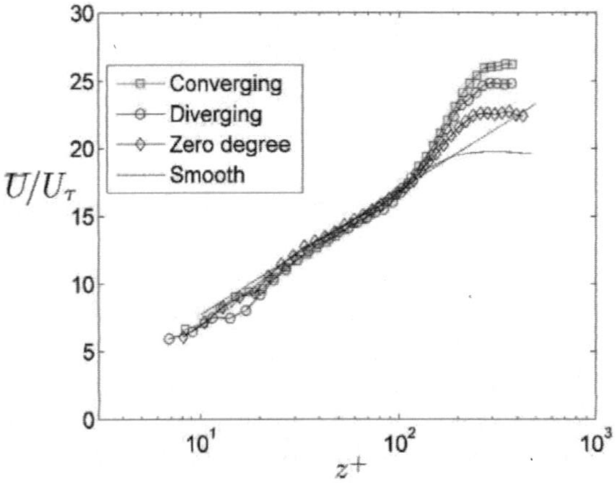

Figure 5. Mean velocity profile, the dashed-line indicates $1/\kappa \ln(z^+) + B$, here $\kappa = 0.25, B = -1.5$

The skin friction is calculated by a generalized form $C_f = 2(U_\tau/U_\infty)^2$. The skin friction produced by C and D flows are approximately the same i.e. $C_f \approx 0.00150$. Skin frictions are however 30% and 70% higher for zero degree, Z and smooth surface, S flows, respectively. Higher drag (results of higher skin friction) attributed to converging surface and vice versa as reported earlier on flat plate wind tunnel [3] was not observed here. The shifting of mean velocity profiles in the logarithmic region for C and D was also not observed as reported before [3].These results however are in agreement with other surface roughness studies, i.e. a riblet surface lowers the drag by 6.2% as compared with smooth surface [6] and simulation with shark skin surface produced drag reduction up to 27% as compared with smooth surface [23]. The latter hypothesized that "the existence of the riblets makes the viscous sub-layer thicker than that of the flat surface, and the shear stress and local friction coefficient on riblet surface are reduced. Besides, drag reduction of riblet surface mainly concentrates on the bottom of riblets" [23].

3.3. Turbulence intensities

The Reynolds number is approximately the same for all flows. Therefore, the turbulence intensities shown in Figure 6 do not contain Reynolds number effects. In Figure 6(a), the turbulence intensities, $\overline{u^2}$ are normalised with U_τ and the wall normal distance, z^+. In the near wall region, both flows exposed to converging *C* and diverging *D* riblets exhibit similar features i.e. maximum turbulence intensities in the near wall i.e. $\overline{u^2}/U_\tau^2|_{max}\approx 12$, this occurs at $z^+ = 20$. It is possible that the position at which the inner peak occurs higher than the normally accepted values of $z^+ = 12 - 15$ due influences of the large scale features in the outer region for thin boundary layer called imprint [27].

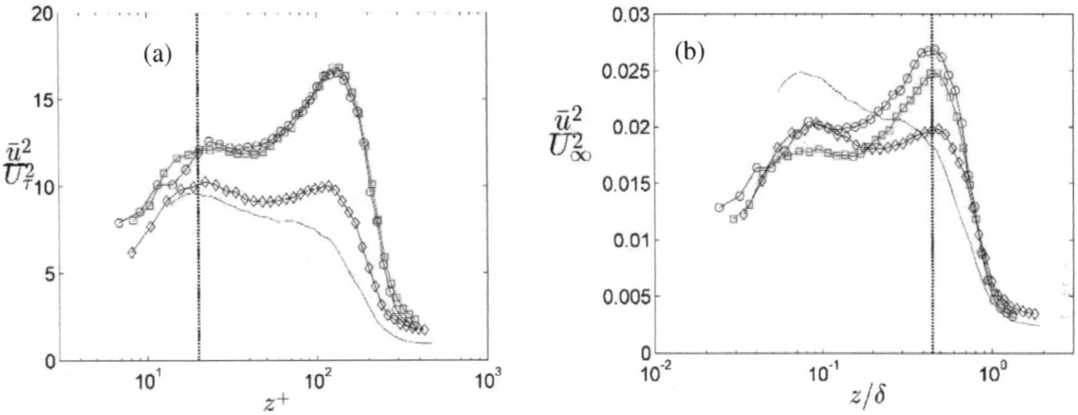

Figure 6. Turbulence intensities (a) scaled with inner variables, vertical dotted line indicate $z^+ = 20$, (b) scaled with outer variables, the vertical dotted line indicates $z/\delta = 0.45$

Figure 6 (b) shows turbulence intensities scaled with outer variables free-stream velocity, U_∞ and boundary layer thickness, δ. Nearer to the wall, in contrast to inner variable-scaled data, this scaling suggests that turbulence intensities decrease with pressure gradient [16, 25, 28] as the smooth wall profile was acquired at $x\approx 200$ mm from the leading edge and the *C*, *D* and *Z* flows were located at $x\approx$ 270 mm from the leading edge (pressure gradient increases from $x\approx 200$ mm to $x\approx 270$, as the flow area increases).

3.4. Higher-order statistics

In this section, more turbulence quantities are analysed through observations in the third and fourth order moments. The third order moment shows the skewness (asymmetry) while the fourth order moment describes the kurtosis (flatness) of the velocity signal. Figure 7 shows the skewness of streamwise velocity fluctuations for the entire fluid flow layer for the smooth-wall, converging, diverging regions and riblet with zero angle. Skewness is defined as $S_k = \overline{u^3}\left(\overline{u^2}\right)^{-\frac{3}{2}}$. A positive value of S_k implies that large positive values of u are more frequent than large negative values.

To start with, sensor length and Reynolds number effects [29] are ruled out because the experiment has been conducted within the range of controlled l^+ and Re_τ. The wall normal distance is normalised with their respective inner variable z^+. For all four cases, they exhibit a non-Gaussian distribution, particularly in the near wall and wake region. All four flow cases exhibit a similar pattern at the edge of the boundary layer, the skewness plot shows that they have sharp negative skew and then change suddenly to positive. The *C* and *D* flows share similar trend from near the wall until the outer region. The *Z* flow however shows lower S_k value, especially in the logarithmic region. In general, the plot reveals that in the logarithmic region, the fluctuating velocities are skewed positively indicating that the velocity distribution is concentrated on the right of PDF distribution, meaning that majority of the velocity fluctuations are high. High S_k values indicate higher degree of interactions between the small and large-scales [30]. It is proposed here that positive S_k in the logarithmic region is due to increased interactions of the small-scales in the inner region with the large-scales in the outer region, especially for the C and D cases, which is similar to pressure gradient effects [17].

Figure 8 shows kurtosis factors of the streamwise velocity component for all four cases. Flatness is defined as $F = \overline{u^4}\left(\overline{u^2}\right)^{-2}$. A rise in flatness is often attributed to a rise in intermittency. Similar with skewness, the plot shows a departure from the Gaussian characteristics in the near-wall and wake region. The kurtosis figure shows that they have very sharp peak at the edge of the boundary layer. The sharp change at this particular location is associated with the intermittency of the turbulent and non-turbulent interface.

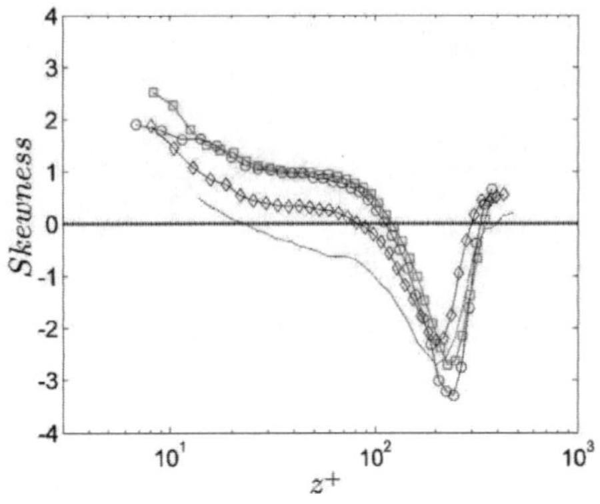

Figure 7. Skewness, S

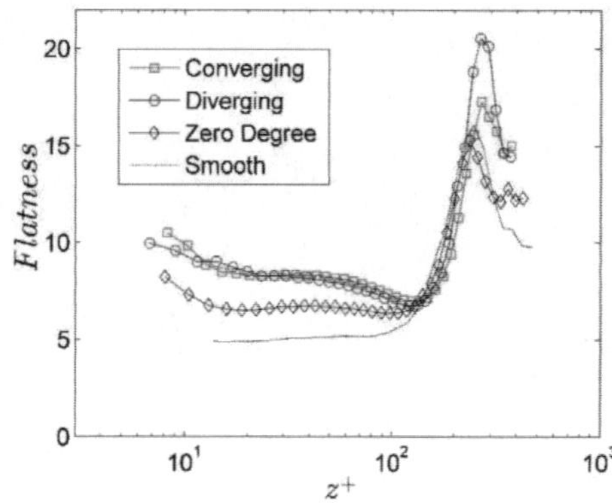

Figure 8. Flatness, F

4. Conclusion

The effect of converging–diverging riblet-type surface roughness (riblets arranged in a 'herringbone' pattern) is investigated experimentally on NACA 0026 airfoil. For this initial parametric investigation, three different parameters of the surface are analysed in detail: the converging–diverging riblet with yaw angle $\alpha = 10°$, riblet $\alpha = 0°$ and smooth surface. Hotwire anemometry, with single wire, is used for experiment. The mean velocity profile is fitted to the logarithmic law with a constant $\kappa = 0.25$ and an intercept of $B = -1.5$. In the inner region, the mean velocities appear jagged due to the effects of the rough surface. In the wake region, the mean velocity scaled with friction velocity, U_τ for flow exposed to converging riblet pattern is higher than that exposed to the diverging pattern. In the near wall region, the inner wall maximum value is 12 and it occurs at $z^+ = 20$. For turbulence intensities scaled with U_∞ and δ, the effect of surface roughness is visibly more pronounced in the outer region. The turbulence intensities produced by the flow past the converging and diverging riblet show clear outer hump at $z/\delta = 0.45$. It is clear that the large scale features dominate in the outer region for the converging and diverging cases in comparison with the zero degree case. It is also shown that the skin friction for flows past the converging and diverging pattern have 25 – 30% less than that of the flow past riblets aligned with the flow ($\alpha = 0°$). The results are consistent with vortex generator studies, which have reported an increase in turbulent intensities in rough surface of the flow field and a decrease in turbulent intensities within the smooth region. The higher order turbulence statistics also show marked different between these flows. These results are very encouraging in the sense that a small strip of vortex generator applied at a leading edge of an engineering application can have up to 20 – 25% reduction in skin friction, hence reduction in total drag, and more importantly reduction in energy or fuel.

Acknowledgments
We would like to express our gratitude for the financial supports provided by UKM research grant FRGS/1/2013/TK01/UKM/03/1.

References

[1] Koeltzsch K, Dinkelacker A and Grundmann R 2002 *Exp. in Fluids* **33** 346-50
[2] Nugroho B, Kulandaivelu V, Harun Z, Hutchins N and Monty J P 2010 *17th Proc. Australasian Fluid Mechanics Conference* (Auckland)
[3] Nugroho B, Hutchins N and Monty J P 2013 *Int. J. of Heat and Fluid Flows* **41** 90–102
[4] Saric W S, Carpenter A L and Reed H L 2011 *Phil. Trans. R. Soc. A* **369** 1352–64
[5] Lee J H, Sung H J and Krogstad P-A 2011 *J. Fluid Mech* **669** 397–431
[6] Stenzel V, Wilke Y and Hage W 2011 *Prog. Organic Coatings* **70** 224–9
[7] Robinson S K 1991 *Annu. Rev. Fluid Mech* **23** 601–39
[8] Kline S J, Reynolds W C, Schraub F A and Rundstadler P W 1967 *J. Fluid Mech.* **30** 741–73
[9] Marusic I, Mathis R and Hutchins N 2010 *Intl. J. of Heat and Fluid Flow* **31** 418–28
[10] Smits A J and Marusic I 2013 *Physics Today* **66** 25–30
[11] Luo Y, Yuan L, Li J and Wang, J 2015 *Micron* **79** 59-73
[12] Dean B and Bhushan B 2010 *Phil. Trans. R. Soc. A* **368** 4775-806
[13] Chen H, Rao F, Shang X, Zhang D and Hagiwara I 2013 *J. of Bionic Engineering* **10** 341-9
[14] Martin S and Bhushan B 2016 *Journal of Colloid and Interface Sc.* **474** 206-15
[15] García-Mayoral R and Jiménez J 2011 *Phil. Trans. R. Soc. A* **369** 1412-27
[16] Harun Z, Monty J P, Mathis R and Marusic I 2013 *J. of Fluid Mech.* **715** 477-98
[17] Stenzel V, Wilke Y and Hage W 2011 *Progress in Organic Coatings* **70** 224-9
[18] Sareen A, Deters R W, Henry S P and Selig M S 2014 *J. of Solar Energy Eng.* **136** 021007
[19] Viswanath P R 2002 *Progress in Aerospace Sci.* **38** 571-600
[20] Harun Z, Abbas A A, Etminan A, Nugroho B, Kulandaivelu V and Khashehchi M. 2014 *25th Intl. Symp. on Transport Phenomena*
[21] Harun Z *et al.* 2016 *J. Teknologi* **78** 63-70
[22] Lomas C G 2011 *Fundamentals of hot wire anemometry* (New York: Cambridge University Press)
[23] Hu H-B, Song B-W, Liu G-S, Guo X-J and Pan G 2009 *J. of China Ordnance* **5** 168–74
[24] Nagib H M and Chauhan K A 2008 *Phys. of Fluids* **20** 101518
[25] Monty J P, Harun Z and Marusic I 2011 *Intl. J. of Heat and Fluid Flow* **32** 575-85
[26] Hutchins N, Nickels T B, Marusic I and Chong M S 2009 *J. of Fluid Mech.* **635** 103-36
[27] Hutchins N and Marusic I 2007 *Phil. Trans. R. Soc. A* **365** 647–64
[28] Nagano Y, Tsuji T and Houra T 1998 *Int. J. Heat Fluid Flow* **19** 563–72
[29] Johansson A V and Alfredsson P H 1982 *J. Fluid Mech.* **122** 295–314
[30] Metzger M M and Klewicki J C 2001 *Phys. Fluids* **13** 692–701

IOP Conf. Series: Materials Science and Engineering **152** (2016) 012006 doi:10.1088/1757-899X/152/1/012006

Aerodynamic performances of cruise missile flying above local terrain

A Ahmad, M R Saad*, A Che Idris, M R A Rahman and S Sujipto

Department of Mechanical Engineering, Faculty of Engineering, Universiti Pertahanan Nasional Malaysia, Malaysia

*rashdan@upnm.edu.my

Abstract. Cruise missile can be classified as a smart bomb and also Unmanned Aerial Vehicle (UAV) due to its ability to move and manoeuvre by itself without a pilot. Cruise missile flies in constant velocity in cruising stage. Malaysia is one of the consumers of cruise missiles that are imported from other nations, which can have distinct geographic factors including their local terrains compared to Malaysia. Some of the aerodynamic performances of missile such as drag and lift coefficients can be affected by the local geographic conditions in Malaysia, which is different from the origin nation. Therefore, a detailed study must be done to get aerodynamic performance of cruise missiles that operate in Malaysia. The effect of aerodynamic angles such as angle of attack and side slip can be used to investigate the aerodynamic performances of cruise missile. Hence, subsonic wind tunnel testings were conducted to obtain the aerodynamic performances of the missile at various angle of attack and sideslip angles. Smoke visualization was also performed to visualize the behaviour of flow separation. The optimum angle of attack found was at $\alpha=21°$ and side slip, $\beta=10°$ for optimum pitching and yawing motion of cruise missile.

1. Introduction

Missile invention started during World War II when the first V-1 flying bomb was launched and hit London in 13th June 1944 by German forces. Then, the German army engineering team developed a new era of ballistic missiles through the successor of V-1 called V-2 (Vengeance Weapon 2) that can reach supersonic speed with over Mach 3, which means it can travel 270 km in only five minutes [1]. During the cold war between USA and Russia, weapon technology has developed rapidly between the two superpower nations that represent the capitalists and the communists block, producing a lot of modernized weapon during that time [2]. The race of technology was also seen happening until now, including in the field of aeronautics and missile systems [1, 2, 3].

Cruise missile motion is different from artillery shell like mortar. This is because the artillery shell behaves as a projectile, where its initial velocity and projection angle must be predetermined in order to deliver the shell to the target based on the range required. On the other hand, cruise missiles manoeuvre in trajectory motion that has its own flight path based on the range and speed capability. Cruise missiles operate in four stages: climb, cruise, dive and run-in [4]. Climbing is the phase when the missile flies in vertical axis to achieve the required altitude before cruising. Cruising is when the missile travels in constant velocity and close to zero acceleration. Diving stage is when the missile manoeuvres in yaw motion towards lower altitude to avoid detection by radar and being intercepted by the air defence system. This movement requires high-g manoeuvres and decreases the total mission range capability due to high consumption of fuel. The final stage is run-in phase before hitting the target. Missiles are also capable to travel on very low altitude to avoid interception with the air defence system.

Every flying object that has symmetrical shape and fuselage needs angle of attack to generate lift [5]. Similarly, missile requires an angle of attack to launch it to the air. Basically, most of the cruise missile have symmetrical shape. Thus the presence of the angle of attacks, α is vital in producing lift for the initial movement at the launch platform to ensure that the missile will travel at the required altitude [6]. The incoming air velocity will push the surface area of an object that has an angle of attack and produce lift that is beneficial for the object to increase its altitude. High velocity of air will generate high value of lift force. This will result in high rate of climb of the flying object [6]. The values of angle of attack also affect the lift force generated. The value of angle of attack is directly proportional to the lift coefficient until at a certain angle where aerodynamic stall causes the decrease in lift coefficient [5]. Angle of slip, β is an angle between the center of fuselage and the direction of air flow. This angle occurs when the missile performs yaw motion in manoeuvre [8]. When avoiding the obstacles in the trajectory path, missile usually uses yaw and pitch motions involving angle of slip and attack [9]. Angular motion that occurs on y-axis is called pitching motion, which involves angle of attack. On the other hand, angular motion on z-axis is called yawing motion, which involves the sideslip angle [9].

The main objective of this study is to investigate the effects of angle of attack and side slip on the aerodynamic performances of cruise missile. In addition to that, the optimum angle of attack and side slip for cruise missile are also investigated, which will then lead to the determination of the suitability of the cruise missile when flying above local terrain.

2. Methodology

2.1. CAD of Tomahawk cruise missile model

Solidworks software was used to generate CAD drawing for Tomahawk Cruise Missile. The drawing was based on the real dimensions for Tomahawk cruise missile. Scale 1:20 was used in order to suit the model for print scale in the 3D printer. The wingspan for this missile was 178 mm and the overall length was 275mm. The diameter of the fuselage of the missile was 31 mm and the fin length was 90 mm. The isometric view for this missile is shown in Figure 1.

2.2. Model fabrication

3D printing, also known as Additive Manufacturing (AM), relates to many processes used to make and fabricate a three-dimensional object. In 3D printing, successive layers of material are formed under computer control to create an object. These objects can be of almost any shape or geometry, and are produced from a 3D model but with limited size of model. Poseidon-X 3D Printer was used to print the missile model with the same scale as the drawing in Figure 1. CURA software was used to import the CAD file from Solidworks. The printing process took 6 hours to complete. PLA type filaments were used as the base material in producing this model. Figure 2 shows the completed missile model with support and bracket.

2.3. Wind tunnel testing

Longwin LW-9300R subsonic wind tunnel shown in Figure 3 was used for the experiments. The speed of air that can be generated by this wind tunnel is from 1 m/s to 105 m/s. The speed of the flow was set based on the desired Reynolds number of 173,749. Dimensions for the test section of the wind tunnel is 30 cm in wide, 30 cm in height and 100 cm in length. This wind tunnel is an open-loop suction type wind tunnel. Upstream flow is sucked from outside by the rotor of wind tunnel located downstream to generate flow towards the test section. This wind tunnel is capable to measure static pressure, total pressure and speed by Pitot static pressure probe, and differential pressure meter with accuracy of

<0.25%. Lift and drag forces are measured by a three-component force balance system located at the bottom of the test section. 2D flow visualization can be conducted by the laser and smoke generator configuration as shown in Figure 4. The mounting of the model inside the test section is as shown in Figure 5.

Figure 1. CAD model of Tomahawk cruise missile model

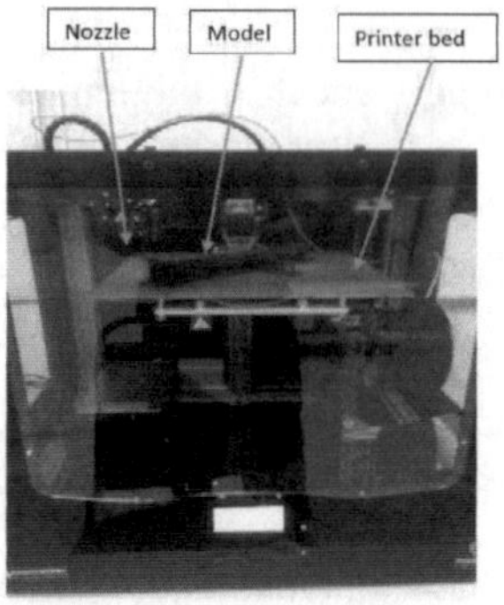

Figure 2. Fabrication of model using 3D printer

Figure 3. LW-9300R subsonic wind tunnel in UPNM

Figure 4. Recording process of flow visualization

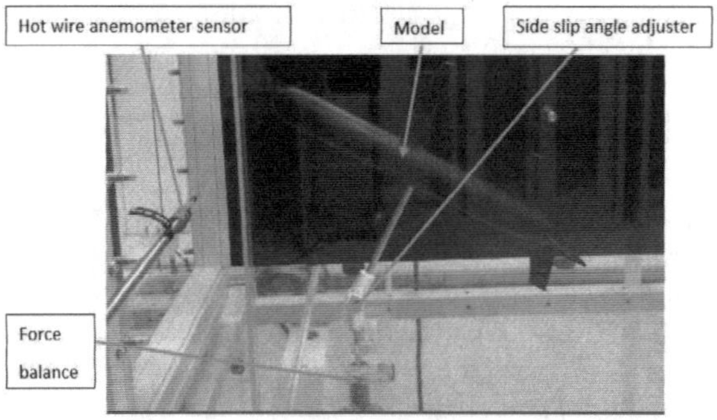

Figure 5. Installation of missile model at side slip angle, β=20° in 3 component force balance

3. Results and discussions

3.1. Validation analysis

The validation process is conducted in order to ensure that the experimental setup is able to reproduce a similar result from the established data. Figure 6 shows the graph of F_L/F_D against angle of attack, α of the experiments being compared with theoretical results obtained from Fleeman [10]. The results from Fleeman were selected based on the similarities of the fuselage geometry used in this study. From the comparison, it can be observed that the results matched the theoretical data with slight errors as calculated in Table 1.

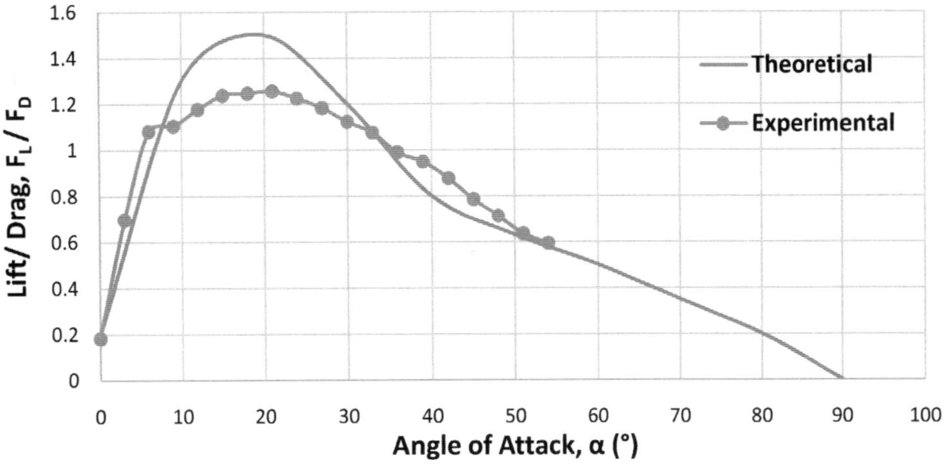

Figure 6. Comparison of F_L/F_D against angle of attack, α between experiment and theoretical

Table 1. Percentage of error between theoretical and experimental

Angle of Attack, α (°)	F_L/F_D Theoretical	F_L/F_D Experimental	Error, ϵ (%)
0	0.20	0.18	9.5
10	1.30	1.18	9.2
20	1.50	1.25	16.7
30	1.20	1.12	6.7
40	0.80	0.86	7.5
50	0.63	0.64	0.8

A few reasons that might have caused the error. Firstly, it might be due to the angle of attack of the missile model inside the wind tunnel. There was no definite method in positioning the missile model at 0° angle of attack, so an estimation was done when placing the missile model on the force balance. In addition to that, during the calibration of the force balance, there was a small error margin. This could also had an effect on the drag force results. Furthermore, the surface finishing of the missile model after fabrication was not completely smooth. Due to this, the surface had to be sand-papered. However even after using sand paper, the surface was still not completely smooth. Because of this, the results' accuracy might have reduced slightly. From another point of view, some of the minor dimensions of the model from Fleeman [10] were unknown, hence careful estimations were made for the actual wind tunnel model. This could also contribute to the errors. The highest percentage error recorded was at the angle of attack, α=20° that corresponded to 16.7% of error between the theoretical and experimental values. This might be due to the slight vibration on the model setup because of the drag induced at that specific angle of attack. On other hand, the lowest percentage of error was found at angle of attack, α=10°, accounting for 0.8% of error between the theoretical and experimental values. Overall, the errors recorded were considered to be within acceptable range.

3.2. Effect of angle of attack on lift and drag

Table 2 shows the tabulation of the results obtained for angles of attack (from 0° to 54°) of the model inside the wind tunnel for Reynolds number 173,749. C_L and C_D values were calculated from F_L and F_D that were obtained from the force balance, respectively. Please note that the uncertainty of the readings of the F_D and F_L is ± 0.01 N.

Table 2. Tabulation of aerodynamic coefficient in the variation of angle of attack

Angle of Attack, α (°)	Drag Force, F_D (N)	Lift Force, F_L (N)	Drag Coefficient, C_D	Lift Coefficient, C_L	F_L/F_D
0	0.21	0.04	0.31	0.06	0.19
3	0.23	0.16	0.32	0.23	0.70
6	0.25	0.26	0.35	0.38	1.08
9	0.28	0.31	0.41	0.45	1.10
12	0.33	0.39	0.48	0.56	1.18
15	0.37	0.46	0.54	0.66	1.24
18	0.44	0.54	0.63	0.78	1.25
21	0.50	0.63	0.72	0.0	1.26
24	0.56	0.70	0.82	1.00	1.22
27	0.65	0.76	0.93	1.10	1.18
30	0.73	0.81	1.04	1.17	1.12
33	0.78	0.84	1.13	1.21	1.08
36	0.87	0.86	1.25	1.24	0.99
39	0.95	0.90	1.37	1.30	0.95
42	1.03	0.90	1.48	1.29	0.88
45	1.14	0.89	1.64	1.28	0.78
48	1.23	0.87	1.76	1.25	0.71
51	1.31	0.83	1.89	1.20	0.63
54	1.32	0.78	1.90	1.13	0.59

Based on Figure 7, the trend of lift-to-drag ratio that increased proportionally to the angle of attack from 0° to 9° can be observed. The values then slightly increased between the angles of attack from 9° to 21°. After that, it started to decrease constantly from angle of attack of 21°. From the figure, the optimum angle of attack can be determined based on the highest value of lift to drag ratio, which is at 21°. This angle corresponds to the highest value of F_L/F_D and it can be taken as the best angle for pitching movement of missile when flying. The range of angles of attack from 15° to 24° is found to be the most effective for pitching movement of cruise missile. From Figure 8, value of lift coefficient increased in direct proportional manner with angles of attack from 0° to 39°. It started to decrease constantly after angle of attack of 39° and this phase is known as the stalling phase. Stalling phase is the loss of lift due to the factors of beyond the limitation of angle of attack in pitching.

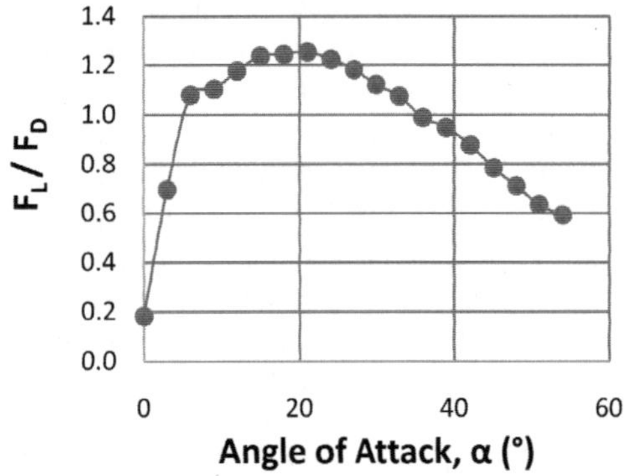

Figure 7. Graph of lift to drag ratio against angle of attack

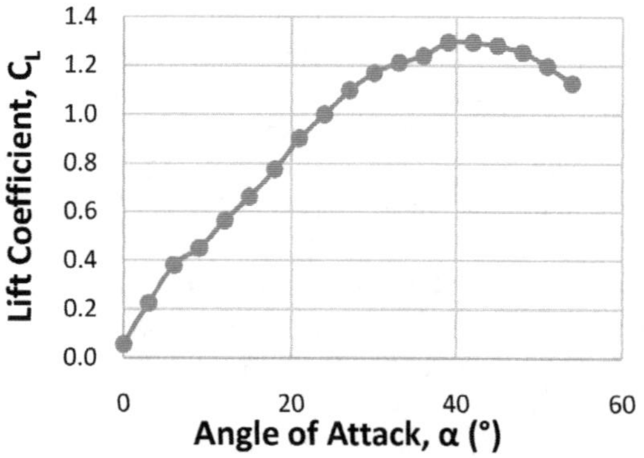

Figure 8. Graph of relationship between lift coefficient and angle of attack

3.3. Effect of side slip angle on lift and drag

Lift and drag coefficients are heavily influenced not only by the angle of attack but also the side slip angle. This is proven from the data collected and tabulated in Table 3, which shows the tabulation of aerodynamic coefficients on the variation of side slip angle experiment where the angle of attack was set constant at 0°. The side slip angle of the model was adjusted inside the wind tunnel using the side slip adjustor, connecting the model to the mounting rod as previously shown in the setup in Figure 8.

Table 3. Tabulation of aerodynamic coefficient in the variation of side slip angle

Angle of Side Slip, β (°)	Drag Force, F_D (N)	Lift Force, F_L (N)	Drag Coefficient, C_D	Lift Coefficient, C_L	F_L/F_D
0	0.09	0.02	0.13	0.03	0.22
5	0.10	0.02	0.14	0.03	0.20
10	0.11	0.05	0.16	0.07	0.45
15	0.13	0.01	0.18	0.01	0.08
20	0.28	0.00	0.41	0.00	0.00
25	0.30	-0.01	0.44	-0.01	-0.03
30	0.32	-0.02	0.47	-0.03	-0.06

Figure 9 shows the relationship between lift coefficient and side slip angle from angle of attack of 0° to 30°. Based on the figure, initially the lift coefficient slightly decreased from the side slip angle of 0° to 5°. Then, it increased rapidly starting from 5° to 10°. After that, the value went down as the side slip angle increased. The peak value recorded for lift coefficient was at the side slip angle, β= 10°. Beyond the value of β = 20°, the missile lost its lift and entered the stall region. The graph of drag coefficient with relation to side slip angle is shown in Figure 10. Based on the figure, drag coefficient value in overall increased with respect to the side slip angle. The drag coefficient increased marginally from side slip angle of 0° to 15°. Then, it suddenly underwent rapid increase from side slip angle of 15° to 20°. Finally, it slowly increased from side slip angle of 20° to 30°.

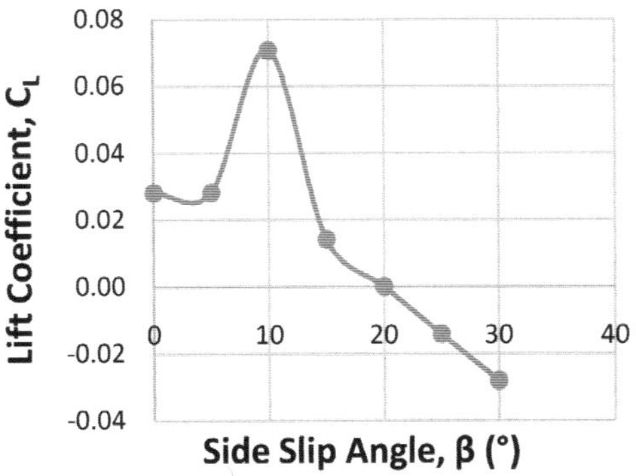

Figure 9. Graph of relationship between lift coefficient and side slip angle

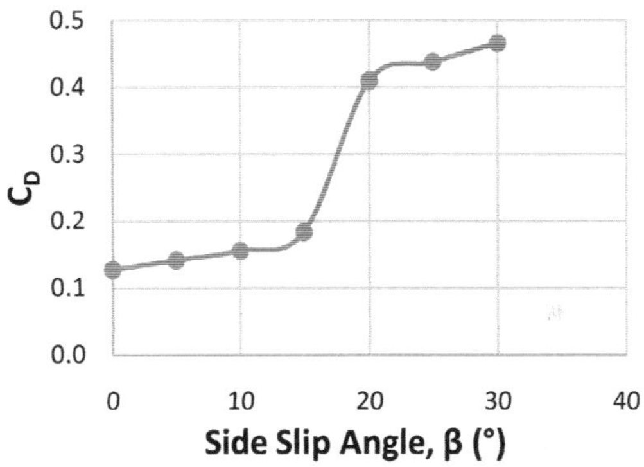

Figure 10. Graph of relationship between drag coefficient and side slip angle

Figure 11 shows the lift-to-drag ratio plot against the side slip angle. Based on the figure, there are three types of trend that can be observed from the plotted values. The first one is the slightly decrease of lift-to-drag ratio value from side slip angles of 0° to 5°. Then, the lift-to-drag ratio value increasing rapidly starting from side slip angles of 5° to 10°. After that, the value went down as the side slip angle increases. Hence, the optimum side slip angle for yawing determined from Figure 11 is at 10° due to the highest value of lift to drag ratio.

As a summary for this analysis, the optimum value for side slip angle, β was able to be determined, which was at 10°. This was useful in configuring the yawing motion of the missile. Besides that, the limitation angle for the missile to yaw before reach a stalling phase was also observed from the data, where side slip angle above 20° can be considered as not safe for missile manoeuvres.

IOP Conf. Series: Materials Science and Engineering **152** (2016) 012006 doi:10.1088/1757-899X/152/1/012006

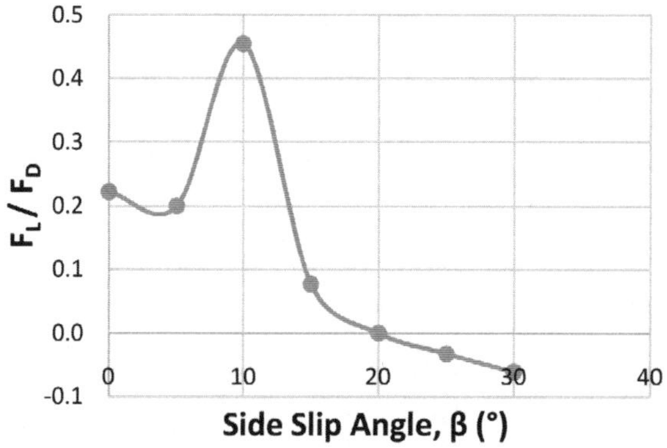

Figure 11. Graph of lift to drag ratio against side slip angle

3.4. Flow visualization results

The flow visualization experiment was conducted at velocity of 1.6 m/s to visualize the behaviour of the boundary layer on the missile body at various angles of attack ranging from 10° to 30°. Figure 12 indicates the location of the separation points on the model at different angles of attack.

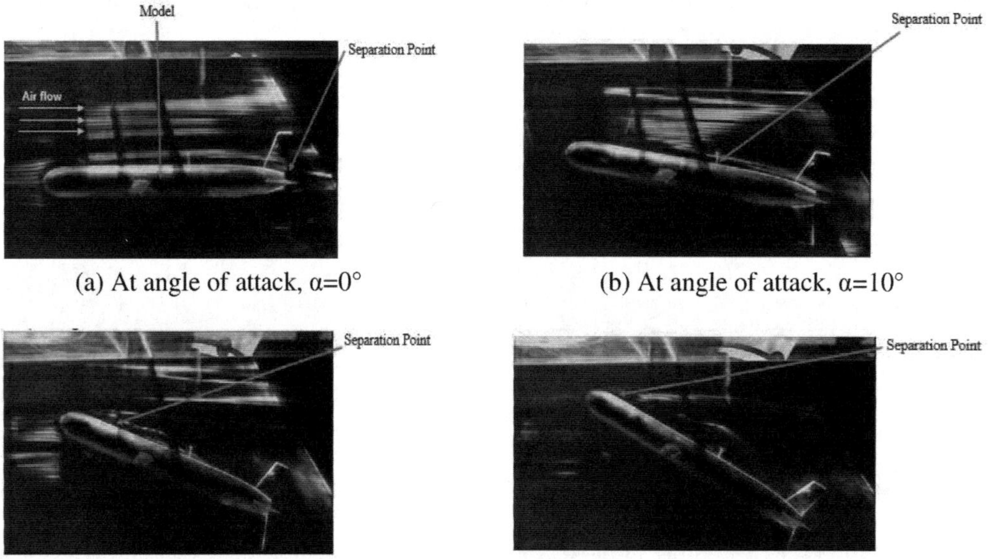

(a) At angle of attack, α=0° (b) At angle of attack, α=10°

(c) At angle of attack, α=20° (d) At angle of attack, α=30°

Figure 12. Photos of flow visualization experiments on missile model at various angles of attack

From the images in Figure 12, it can be seen the difference of flow trajectory between each of angle of attack increment. As an example, the separation point for 0° of angle of attack was at the rear region near the fin but by increasing angle of attack, the separation point was found to shift to the upstream of the fuselage. This in turn will cause large separation region that eventually contributed to the increase of the drag force. The separation point marks the starting point of turbulent eddies that occurred on the fuselage of missile model. This is the main reason why the increased of angle of attack also increases the value of drag coefficient as shown by the data in Figure 10.

3.5. Pitching motion analysis

Malaysian terrain has no permanent pasture or prairie land. Hills and mountains mostly dominate the Malaysian terrain that have an average elevation of 1,829 meters. This will cause a lot of manoeuvring for the missile to climb and avoid itself from hitting any obstacles such as hills or mountains. Thus, by determining the optimum angle of attack based on Table 4, the missile is able to plan its movements by avoiding obstacles and manoeuvring at the most efficient aerodynamic performances. Figure 13 presents a simple pitching analysis done on the combination of angle of attack and the climb that the missile needs to make in order to avoid an obstacle of 1,829 m high. From the figure, the distance, D can be calculated that translates into the point before the obstacle where the missile needs to pitch up.

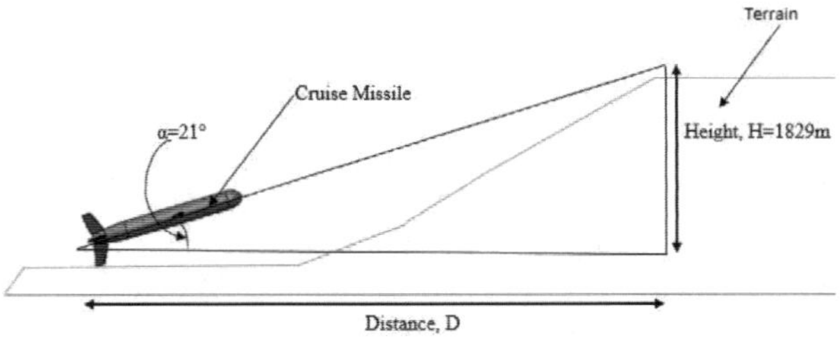

Figure 13. Pitching analysis on optimum angle of attack

Table 4. The optimum aerodynamic angles

Type of Angle	Optimum Angle (°)
Angle of Attack	21
Side Slip Angle	10

4. Conclusion

The experiments conducted in the wind tunnel provides the data for aerodynamic performances of the cruise missile model. The optimum values found for angle of attack was at $\alpha=21°$ and side slip angle at $\beta=10°$ for pitching and yawing motion of the missile from maximum ratio of lift to drag data for this two aerodynamic angles. The flow visualization shows the existence of flow separation that represents the starting point of vortices and turbulent eddies that causes induced drag due to increment of angle of attack. Finally, a simple dynamic analysis was performed by calculating the required distance as a guideline for the missile to start climbing for the optimum angle of attack to avoid from hitting high obstacles like hills and mountains when flying above Malaysian terrain.

Acknowledgements

The authors would like to thank Ministry of Education Malaysia for financial support under the RAGS (RAGS/1/2014/TK09/UPNM/1) and RACE (RACE/F3/PK6/UPNM/13) research grant schemes. Also, an appreciation to technical staffs of Faculty of Engineering, Universiti Pertahanan Nasional Malaysia for providing the assistances to the equipment.

References

[1] De Vorkin D H and Neufeld M J 2011 *Endeavour* **35** 187-95
[2] Eckersley C F 2009 *47th AIAA Aerospace Sciences Meeting*
[3] Adams D, Criss T B and Shankar U J 2008 *IEEE Aerospace Conference*
[4] Krieger R 1982 *9th Atmospheric Flight Mechanics Conference*
[5] Zastawny M, Mallouppas G, Zhao F and Wachem B 2012 *International Journal of Multiphase Flow* **39** 227-39
[6] Ma Y, Guo J and Tang S 2015 *Journal of Aerospace Science and Technology* **45** 324-34
[7] Riedel F 1941 *Guidance and Control Conference*
[8] Jouannet C and Krus P 2007 *25th AIAA Applied Aerodynamic Conference*
[9] Anderson J D 2016 *Introduction to Flight* (New York: McGraw-Hill)
[10] Fleeman E L 2006 *Tactical Missile Design* (Reston: AIAA)

IOP Conf. Series: Materials Science and Engineering **152** (2016) 012007 doi:10.1088/1757-899X/152/1/012007

Experimental study of UTM-LST generic half model transport aircraft

M I Ujang, S Mat*, K Perumal and M N Mohd. Nasir

Faculty of Mechanical Engineering, Universiti Teknologi Malaysia, Malaysia.

*shabudin@utm.my

Abstract. This paper presents the experimental results from the investigation carried out at the UTM Low Speed wind tunnel facility (UTM-LST) on a half model generic transport aircraft at several configurations of primary control surfaces (flap, aileron and elevator). The objective is to measure the aerodynamic forces and moments due to the configuration changes. The study is carried out at two different speeds of 26.1 m/s and 43.1 m/s at corresponding Reynolds number of 1 x 10^6 and 2 x 10^6, respectively. Angle of attack of the model is varied between -2° to 20°. For the flaps, the deflection applied is 0°, 5° and 10°. Meanwhile, for aileron and elevator, the deflection applied is between -10° and 10°. The results show the differences in aerodynamic characteristics of the aircraft at different control surfaces configurations. The results obtained indicate that a laminar separation bubble developed on the surface of the wing at lower angles of attack and show that the separation process is delayed when the Reynolds number is increased.

1. Introduction

Semi-span aircraft model consists of only half of the complete wing or tail of an aircraft. As an aircraft is a symmetric vehicle, investigation using a semi span half model is acceptable since the effect will be similar on the other side of the wing during straight and level longitudinal flight conditions. In general, most wind tunnel testing research requires that the correct flight conditions is simulated and the most accurate data is obtained. These issues are important in order to develop the accurate performance characteristics, especially at low-speed flight conditions such as takeoff, approach and landing that are usually encountered by subsonic transport aircraft. Typically, the Reynolds numbers achievable at the speed appropriate for takeoff and approach conditions in the available testing facilities are below the desired full-scale Reynolds number [1-3]. The need to extend Reynolds number testing capabilities up to full-scale conditions can be satisfied with the development of a semi-span testing capability.

A semi-span model testing offers few advantages over full span testing. By fabricating a semi-span model, a larger and more detailed design can be used. Due to the larger model size provided by semi-span testing, not only is the desired increased Reynolds number testing capability is produced, but the larger model size also improves data quality due to improved model strength, stiffness and overall fidelity. Constructing only half the model yields further benefits in terms of reduced model's cost. The complex high-lift systems and controls will only need to be produced for one wing. Another advantage of semi-span testing is that it only requires simple mounting on the balance and turntable. One of the disadvantages of semi-span testing is the wind tunnel's wall interference effects due to the model size. In this study, the model is fabricated and mounted on a splitter plate, which is offset from the tunnel wall, such that it is outside the wall's boundary layer. The height of the model is also increased to avoid the boundary layer effect [1-3]. One of the main advantages of the half model is because it has a large wing area. Therefore, the flow visualization experiments can be carried out more effectively. A generic half model transport aircraft has been built in Universiti Teknologi Malaysia (UTM). The main objective of this project is to investigate the effect of Reynolds number on the performance of the aircraft when the control surface is changed.

2. Test facility and model description

The investigation is conducted in Universiti Teknologi Malaysia Low Speed Wind tunnel (UTM-LST) Aeronautical Laboratory facility. This is a closed-circuit, single-return, atmospheric wind tunnel that is capable of producing a maximum speed of 80 m/s. Cross sectional dimension of the test section is 2 m (breadth) x 1.5 m (height) x 5.8 m (length). A generic half-span model of transport aircraft has been chosen for this experiment. The model is designed so that its cross-sectional profile is NACA 2213. This semi-span model is consisted of a port wing from the full span model and a semi-fuselage, which is fabricated from a mold of the full-span fuselage [5,6]. The wing employs control surfaces of flap and aileron system, and tail with horizontal stabilizer and moveable elevator. The model is fabricated without engine nor vertical stabilizer. The model's detail dimensions is shown in Table 1 and Figure 1.

Table 1. Specification of the model

Parameter	Dimension
Fuselage Length, L	2.362 m
Wing Area, S	0.252 m^2
Mean Aerodynamic Chord	0.339 m
Half-span, b/2	0.983 m
Wing Volume, V_{wing}	0.00072 m^3
Fuselage Volume, $V_{fuselage}$	0.58 m^3

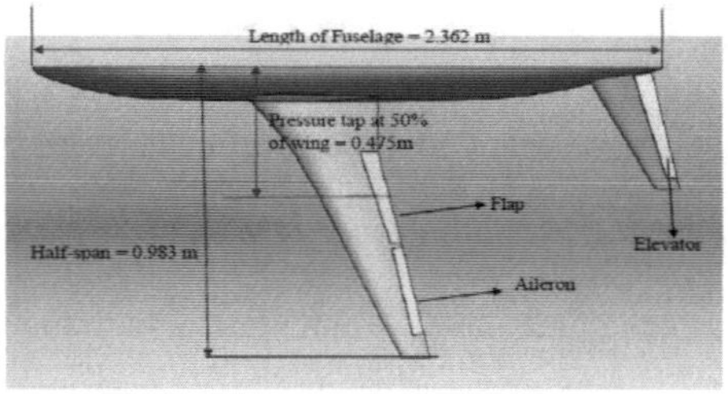

Figure 1. Dimension and location of the control surfaces of UTM-LST Half Model

There are two main external balances used in Aerolab to measure the aerodynamic characteristics of the half-span model. The first one is called *semi span balance*, which has a high capacity of up to 2000 N. Several previous projects [5, 6] applied this balance to measure aerodynamic characteristics but Aerolab found the balance to be unsuitable for relatively light aircraft model. Therefore, a new balance has been purchased called *JR3*. The capacity of this balance is about 1000 N in each channel. This is the first project to measure aerodynamic characteristics of UTM half-span model using JR3. The model is installed on JR3 balance to measure the forces and moments due control surfaces at different deflection angles. The full model was mounted on a 1.6-meter diameter turntable on the floor of the tunnel, approximately 1.9 meter aft of the tunnel floor's boundary layer (upstream) removal system (BLRS). The modified installation is depicted in Figure 2. Additionally, in order to further understand the flow physics involved in semi-span testing, surface pressure measurements and simple flow visualization using tuft are conducted. The model has 18 pressure ports (9 on each side to and bottom) located 50% of the wingspan for the pressure measurement.

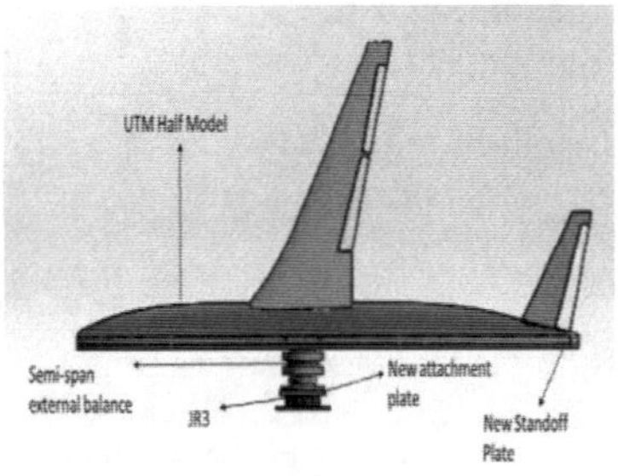

Figure 2. Modified installation with standoff

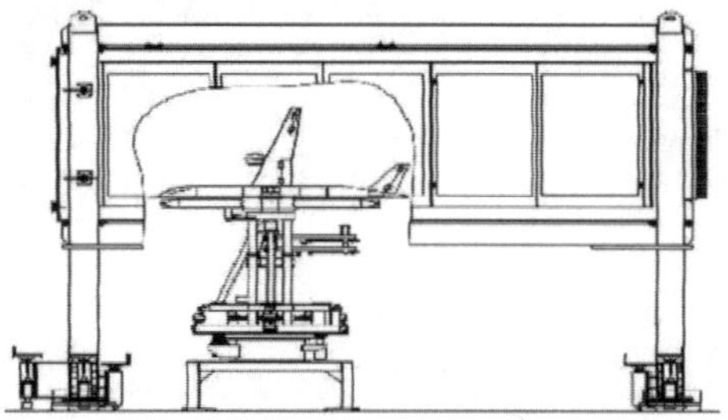

Figure 3. Original installation of UTM-LST half-span model with semi-span balance in UTM-LST wind tunnel

A standard wall correction as discussed in [7] is applied in this project. The wall correction consists of solid blockage and wake blockage. Solid blockage occurs due to the presence of the model inside the section, which creates velocity increase across the model. Calculation of the solid blockage is done using the simple equation for solid-blockage correction for two dimensional case as in Equation 1 [7].

$$\varepsilon_{sb} = \frac{K_1(model\ volume)}{C^{\frac{3}{2}}} \tag{1}$$

where K_1 is a spanning configuration factor which is value of 0.52 for a model spanning across the tunnel height or in vertical direction. The model volume is 0.00073 m^3 and c is the tunnel parameters. Meanwhile, the wake blockage correction is also done in two dimensional case, using the simplified equations as in Equation 2, Equation 3 and Equation 4.

$$\varepsilon_{wb} = \frac{{}^{c}/_{h}}{2} C_{du} \tag{2}$$

$$\varepsilon_{wb} = \frac{{}^{0.5}/_{2}}{2} C_{du} \tag{3}$$

$$\varepsilon_{wb} = 0.125 C_{du} \tag{4}$$

where C_{du} is the uncorrected drag coefficient from the balance reading while h is the wind tunnel height, which is 1.5 meter. The total blockage factor for this model is about 12% at the angle of attack of $\alpha = 23^{o}$.

3. Wind tunnel testing

The installation of the UTM-LST half-span model in the test section is shown in Figure 4. The wind tunnel experiment is conducted in the UTM Low Speed Tunnel at Universiti Teknologi Malaysia's Aeronautics Laboratory with cross sectional 2.0m width x 1.5m height. The model was tested at 26.1 m/s and 43.1 m/s that corresponds to 0.5 x 10^6 and 1.0 x 10^6 Reynolds number based on the mean aerodynamic chord, respectively. The model has an overall length of 2.362 m and the wing area of 0.252 m^2. During the experiment, angle of attacks were varied from -2° to 20°. The experiments are conducted in two phases: first, the model with no control surfaces deflection is tested and then the model with several control surfaces deflection is tested. Three measurements techniques are employed on the model, surface pressure measurement and steady balance data. In the final stage, threaded tuft flow visualization method is employed on the model at angle of attack varies from -2° to 20°.

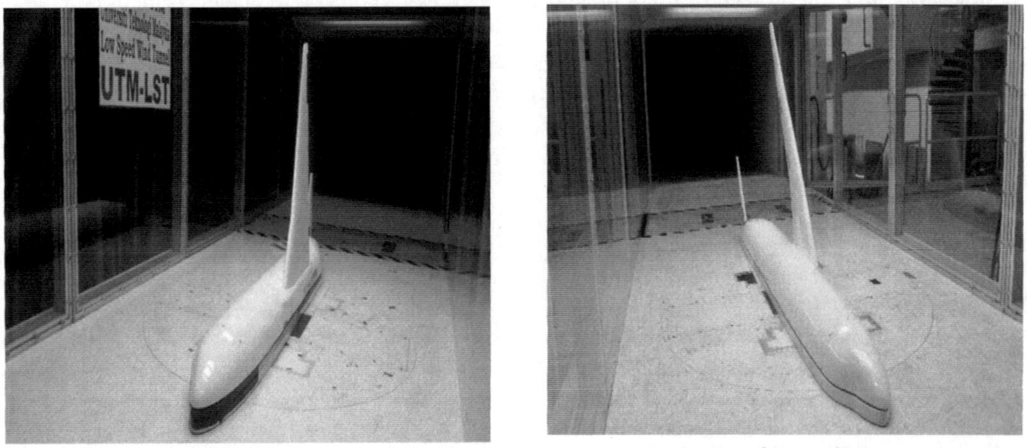

a) $\alpha = 8^{o}$ b) $\alpha = 0^{o}$

Figure 4. The installation of half-span model in UTM-LST

4. Results

4.1. Steady balance data

Figure 5, Figure 6 and Figure 7 show the results obtained from JR3 balance experiments. Blockage correction factor of 13% has been taken into consideration in the calculation. Lift coefficient in Figure 5 shows that the change of the Reynolds number has no impact on the results of the flap, aileron and stabilizer. The lift coefficient often increases once the angle of flap, aileron and stabilizer is increased. The results obtained show that stalling angle for this aircraft is at about $\alpha = 8°$. However, increasing the stabilizer angle can give an impact on lift (see Figure 5(iii)). The characteristics of drag and pitching moment are shown in Figure 6 and Figure 7, respectively. Similar trend is observed for the drag and pitching moment coefficients. In all cases, δ_F is the angle of flap deflection, δ_{Ae} is the aileron angle while δ_{St} is the stabilizer angle.

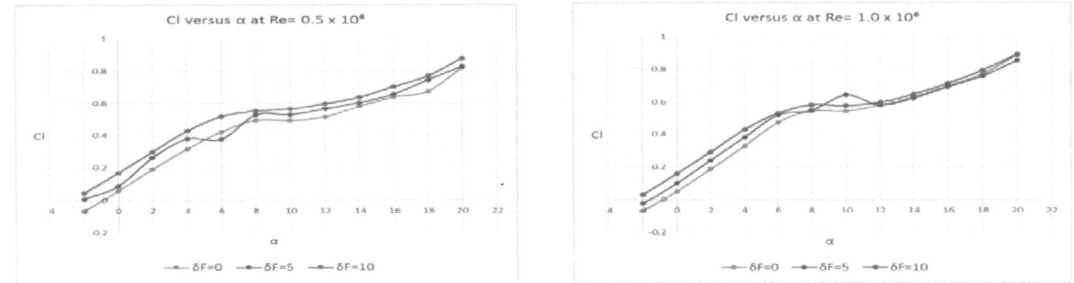

a) $R_{Mac} = 0.5 \times 10^6$ b) $R_{Mac} = 1.0 \times 10^6$

(i) Effect of flap deflections on C_L

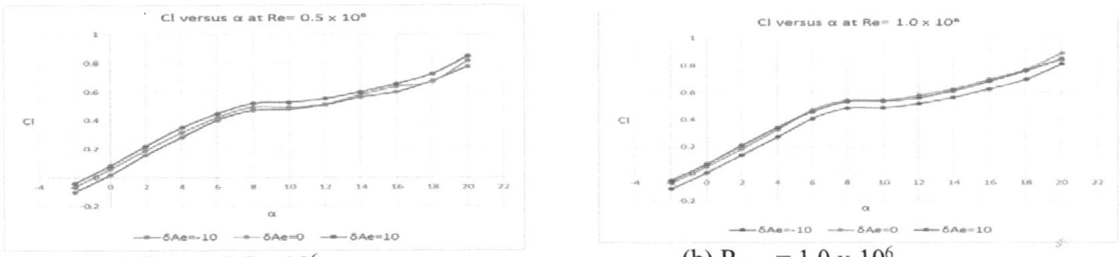

a) $R_{Mac} = 0.5 \times 10^6$ (b) $R_{Mac} = 1.0 \times 10^6$

(ii) Effect of aileron deflections on C_L

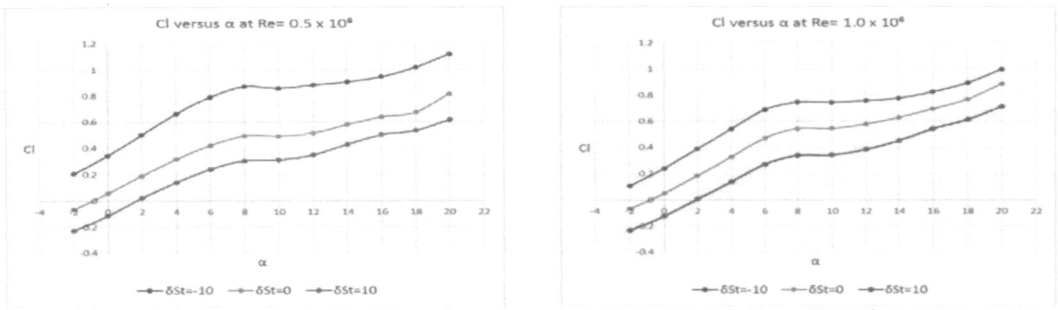

(a) $R_{Mac} = 0.5 \times 10^6$ (b) $R_{Mac} = 1.0 \times 10^6$

(iii) Effect of stabilizer deflections on C_L

Figure 5. The effects of control surfaces on C_L

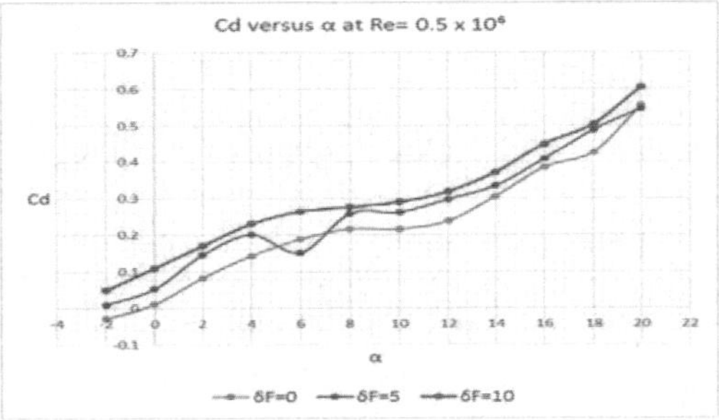

(a) $R_{Mac} = 0.5 \times 10^6$

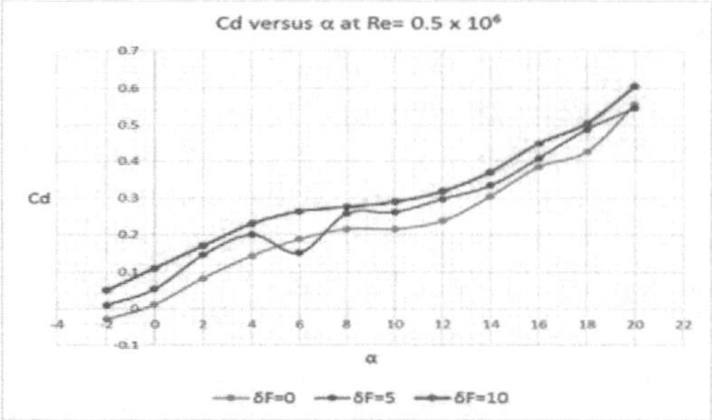

(b) $R_{Mac} = 1.0 \times 10^6$

(i) Effect of flap deflections on C_D

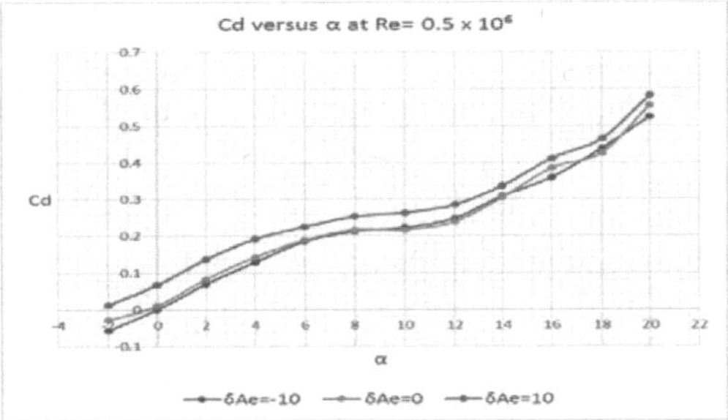

(a) $R_{Mac} = 0.5 \times 10^6$

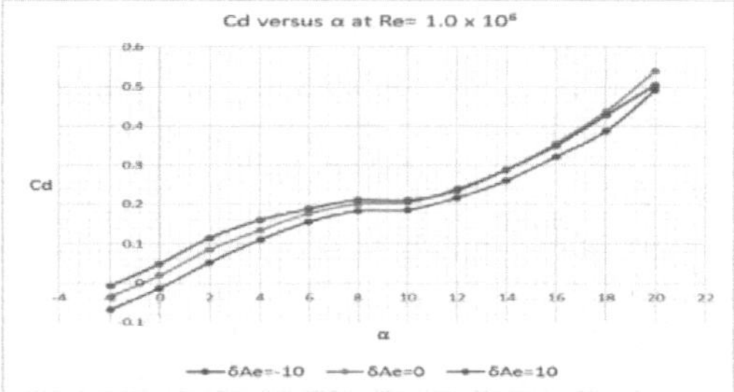

(b) $R_{Mac} = 1.0 \times 10^6$

(ii) Effect of flap Aileron on C_D

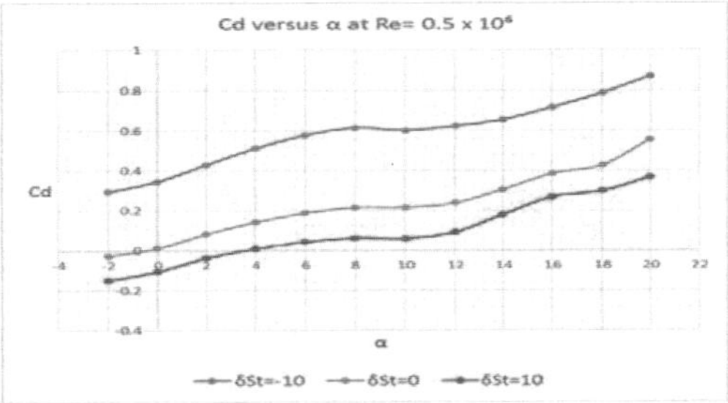

(a) $R_{Mac} = 0.5 \times 10^6$

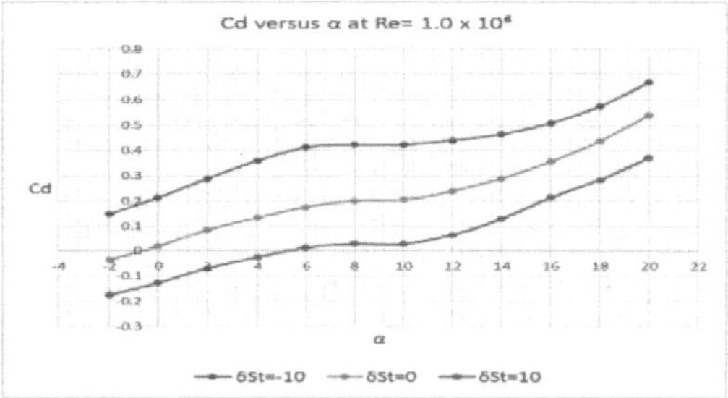

(b) $R_{Mac} = 1.0 \times 10^6$

(iii) Effect of flap Stabilizer on C_D

Figure 6. The effects of control surfaces on C_D

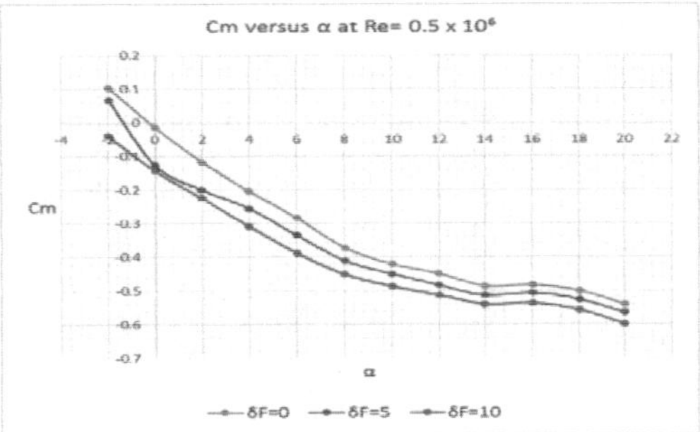

(a) $R_{Mac} = 0.5 \times 10^6$

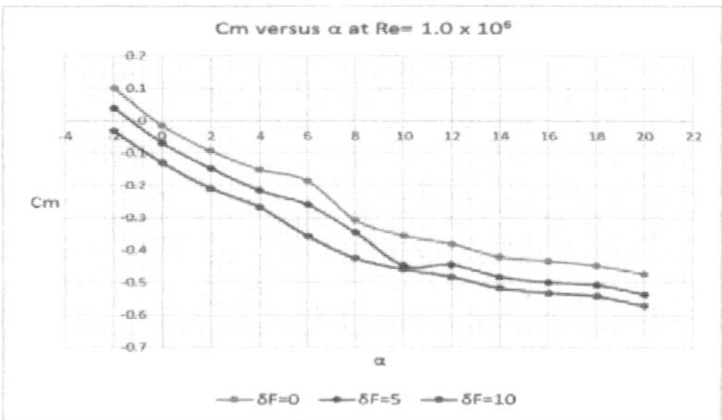

(b) $R_{Mac} = 1.0 \times 10^6$

Figure 7. C_M at Different Flap deflection

4.2. Surface pressure measurements data

Sample data of surface pressure measurement at an angle of attack of 0°, 4^0 and 8^0 are shown in Figure 8. The data obtained shows that only the flap is affecting the flow separation compared to aileron and stabilizer. Therefore, only data involving flap is discussed here. Figure 8(i) shows the distribution of pressure at α =0°. The diagram shows that the flow attaches to the wing on the entire wing. For both Reynolds number, it can be observed that the lift increases if the flap angle is increased. At Reynolds number 1 x 10^6, it can be noticed that the laminar separation bubble might develop at 15% to 25% of the wing if the flap is set at 10°. Laminar separation bubble can be predicted to occur on the wing if gradient of Cp against x/c is zero. From Figure 8(i) at the Reynolds number of 1 x 10^6, it may develops from x/c = 15 – 25 % of the wing.

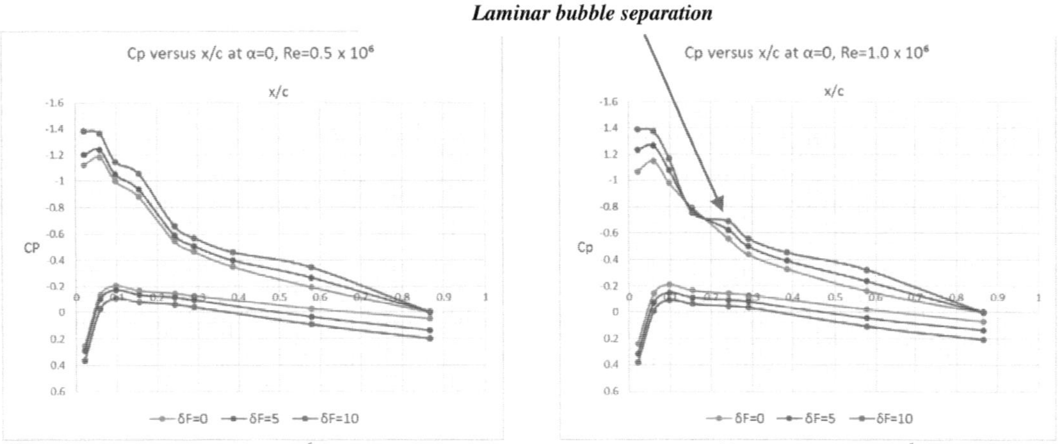

(a) $R_{Mac} = 0.5 \times 10^6$ (b) $R_{Mac} = 1.0 \times 10^6$

(i) Pressure coefficient at $\alpha = 0^0$

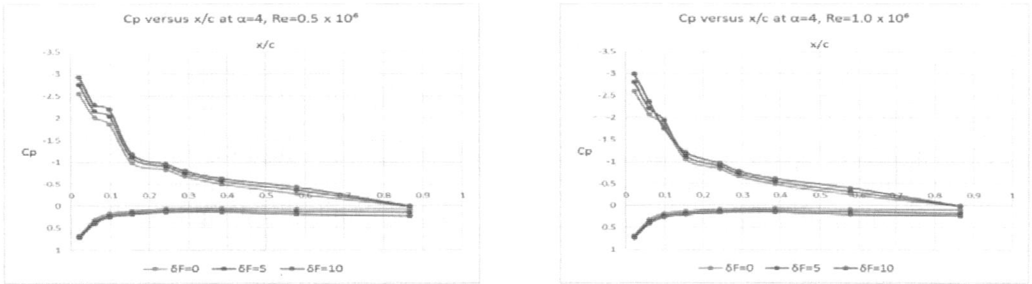

(a) $R_{Mac} = 0.5 \times 10^6$ (b) $R_{Mac} = 1.0 \times 10^6$

(ii) Pressure coefficient at $\alpha = 4^0$

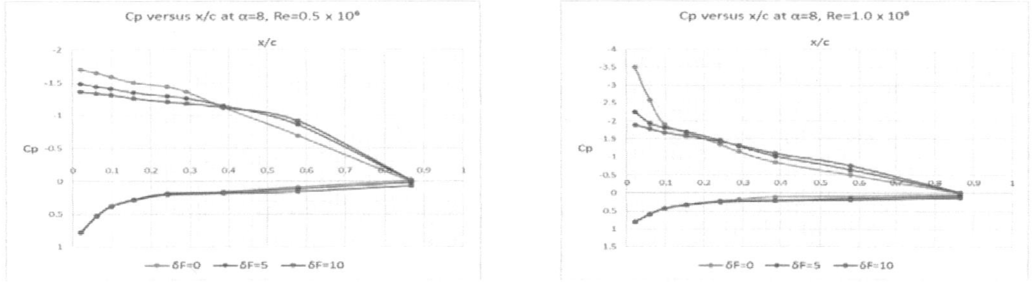

(a) $R_{Mac} = 0.5 \times 10^6$ (b) $R_{Mac} = 1.0 \times 10^6$

(iii) Pressure coefficient at $\alpha = 8^0$

Figure 8. Surface pressure measurement data at $\alpha = 0^0$, 8^0 and 10^0

When the angle of attack is increased to $\alpha = 4^0$, not much of a difference can be observed compared to the result obtained for $\alpha = 0^0$. The airflow is still attached to the wing to the wing surface. Laminar separation bubble might develop at position 8% of the wing at Reynolds number of 0.5 x 10^6. There is a significant difference when the angle of attack is increased to $\alpha = 8^0$. Figure 8(iii) indicates that the flow separation has developed on the wings. At Reynolds number of 0.5 x 10^6, it can be observed that the increase in flap angle has promoted the flow separation. At higher Reynolds number of 1 x 10^6 (see Figure 8(iii)), it can be observed that the flow separation has been delayed. The attached flow still exist in front of the wing and this indicates that the increase of Reynolds number has delayed the process of separation. When Reynolds number is increased, this means that the viscosity of the airflow has also increased. The airflow becomes more turbulent in this situation. Stronger ability of turbulent boundary layer to sustain the adverse pressure gradient has delayed the separation process [8].

4.3. Thread flow visualization results

This section discussed the thread flow visualization that has been carried out on the model. During the experiment, a camera was placed on the roof of the UTM test section. Images of the flow separation were recorded for every five seconds. This process is done for all specified angles of attack. Sample images of surface flow visualization using thread performed on the clean wing configuration is shown in Figure 10. This experiment was performed at angles of attack between -2° to 20° and the Reynolds number of 0.5 x 10^6. In Figure 9(a), it can be seen that the thread is fully attached to the surface of the wing. It shows in this position that the airflow is attached to the wing surface and separation is not observed. When the angle of attack is increased to $\alpha = 5°$, the airflow in the leading edge is attached to the wing surface to a certain chordwise position. In the trailing edge region, it can be observed that the flow separation begins to develop. Increasing the angle of attack further causes the separation point to move forward as shown in Figure 9(c). Finally, at $\alpha = 5°$, the entire wing is covered by the flow separation (see Figure 9(d)).

(a) The flow separation at α=0° (b) The flow separation at α=5°

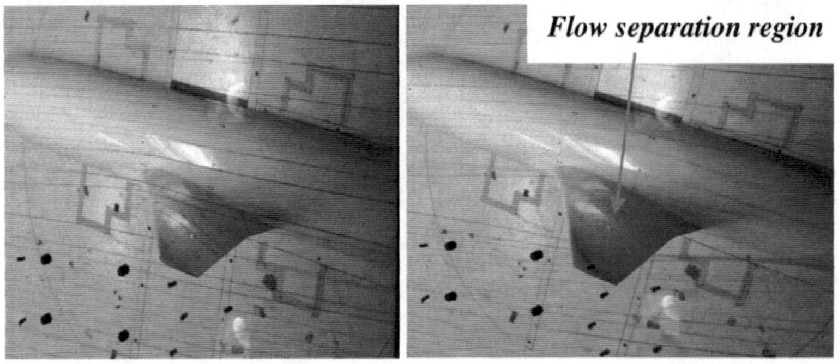

(c) The flow separation at α=8° (d) The flow separation at α=15°

Figure 9. Thread tuft flow study at Reynolds number of 0.5 x 10^6

5. Conclusion

This experiment is conducted with the aim of providing a detailed explanation on the aerodynamic characteristics of half-span model of transport aircraft at certain control surface deflections. In this experiment, three measurement techniques have been carried out on the aircraft model, i.e. balance measurement, surface pressure measurement and flow measurement visualization. It is deduced that when flaps are deflected at larger angle, the flow is still attached and the lift increases with increasing angle of attack. This paper discusses in detail the effect of control surfaces on lift, drag and moment. The results obtained also show that the airflow is sensitive to Reynolds number. It is found that the separation process is delayed if the Reynolds number is increased.

Acknowledgements

This research is funded by research grant from Ministry of Higher Education and Universiti Teknologi Malaysia (UTM Research University Grant Q.J130000.2524.12H06). It should be noted that the data presented, the statement made and views expressed are solely the responsibility of the author.

References

[1] Yokokawa Y, Murayama M, Uchida H and Tanaka K 2010 *AIAA Aerospace Science Meeting*
[2] Gatlin G M and McGheet R J 1997 *Journal of Aircraft* **34** 500-5
[3] Milholen W E, Chokani N and McGhee R J 1996 *Journal of Aircraft* **33** 115-22
[4] Mansor S 2008 *Low Speed Wind Tunnel* Universiti Teknologi Malaysia
[5] Ishak I S, Mat S, Mat Lazim T and Awang M Z *2003 International Conference on Mechanical Engineering*
[6] Sunthrasakaran R 2015 *Journal of Aeronautical Science and Engineering* **1**
[7] Pope A and Rae W 1984 *Low-speed wind tunnel testing* (New York: Wiley)
[8] Mat S 2010 *The Analysis of Flow on Round-Edged Delta Wings* PhD Thesis University of Glasgow

Finite element analysis of metal matrix composite blade

R Isai Thamizh[1*], R Velmurugan[2] and R Jayagandhan[3]

[1]Gas Turbine Research Establishment, DRDO, Bangalore, India
[2]Aerospace Engineering Dept, IIT Madras, Chennai, India
[3] Engineering Design, IIT Madras, Chennai, India

[*]isaithamizh.isai@gmail.com

Abstract. In this work, compressor rotor blade of a gas turbine engine has been analyzed for stress, maximum displacement and natural frequency using ANSYS software for determining its failure strength by simulating the actual service conditions. Static stress analysis and modal analysis have been carried out using Ti-6Al-4V alloy, which is currently used in compressor blade. The results are compared with those obtained using Ti matrix composites reinforced with SiC. The advantages of using metal matrix composites in the gas turbine compressor blades are investigated. From the analyses carried out, it seems that composite rotor blades have lesser mass, lesser tip displacement and lower maximum stress values.

1. Introduction

Composite materials can be defined as a material with two or more distinct macroscopical phases [1]. The constituents of composite material are easily distinguishable as individual components. Composites are used in various industries and in this paper, the application of composites in gas turbine blades has been studied. There are various structural scales and types of analysis in composite materials. The first level of analysis is known as micromechanics, which deals with the fiber and matrix, their arrangement and lamina properties. The next level of analysis is macromechanics, which deals with number of layers, thickness and other laminate properties. This is followed by the structural analysis that deals with the behavior of composite laminate structures under loading. The behavior of composite compressor rotor blades under loading has been studied using the finite element structural analysis approach. Finite element (FE) analysis has been performed on both composite and metallic compressor blades, and the advantages of one over the other have been established.

Composite materials are used in several industries because of its versatile application. The composites could serve as potential replacement of Ti alloy as they are lighter, stronger, tougher and have a higher corrosion resistant [2]. The deformation characteristics of the composite can be tailored to improve its performance to be better than that of Ti alloy. For example, stiffness and yield strength of the composites can be varied by altering the orientation of the fibers. There are different types of composites based on the matrix material: polymer matrix composite, ceramic matrix composite and metal matrix composite. Metal matrix composites (MMC) offers improved elevated temperature strength and modulus. It has an edge over alloys as the properties are tailorable for specific requirements [3]. Therefore, MMC is considered in the present study.

2. Literature survey

Jabbar *et al.* [4] have carried out a comparative study on structural analysis of first stage rotor blade made of titanium, steel and aluminum. Furthermore, Kim and Lee [5] discussed the modal characteristics and endurance strength of compressor rotor blade while Singerman and Jackson [6] described the state-of-the-art technology of Ti MMC aerospace fabrications, the potential payoffs and recent advances in processing that now lead to high quality, affordable Ti MMC components. The literatures on analysis of compressor blade made of Ti matrix composites reinforced with SiC are scarce. Therefore, the present work has been focused to investigate the performance of composites as compressor blade through static stress analysis using FE.

3. Material selection

This paper deals with the FE analysis of compressor rotor blades of gas turbine engine made of metal and metal matrix composites. The metal considered for this study is Ti 64, which constitutes 6% Al and 4% V. Meanwhile, the metal matrix composite considered for study is Ti 15-3 matrix with SiC reinforcements. Ti 15-3 constitutes 15% V, 3 % Cr, 3% Sn and 3 % Al and it has been chosen as the matrix material because of its two major advantages: it has better strength at elevated temperature and it is amenable to cold rolling for making foils that is not possible with Ti 64.The fibre volume fraction considered for analysis is 35%. Material properties considered for analysis are presented in Table 1 [8, 9].

Table 1. Material properties of Ti 64 and TMC

Material Properties	Metal	TMC				
	Ti 64	Particulate	$[0]_8$	$[0/90]_{2S}$	$[0/\pm45/90]_S$	$[0_2/\pm45]_S$
Density (g/cc)	4.54	3.86	3.86	3.86	3.86	3.86
Poisons ratio	0.3	0.28	0.33	0.25	0.33	0.38
Modulus of elasticity (GPa)	117	225	194	148	136	161
Thermal coefficient of expansion (x 10^{-6}/°C)	8.6	6.75	6.75	6.75	6.75	6.75
UTS (MPa)	930	1689	1500	1000	1150	1250

Static stress analysis and modal analysis have been carried out to obtain the stresses, deformation and natural frequency of the blade. These analyses were done on compressor rotor blade made of Ti alloy and TMC. In TMC, two cases were considered for analysis: particulate composites and laminate composites. Different types of laminates were considered for the analysis and they included uni-directional, cross-ply and quassi- isotropic laminates. Brief outline of various analyses carried out are given in Figure 1.

4. Finite element analysis

The finite element method is a numerical technique for analyzing structures with large number of degrees of freedom. It is extremely advantageous for analyzing structures with complex geometry and loading, which cannot be solved exactly using analytical methods. In this present work, ANSYS software was used for the analysis. Analysis of composite blades is different from the analysis of metallic blades because in case of composites, the concepts of layering, orientation of fibres, stacking sequence and also directional properties have to be taken into consideration. Static stress analysis and modal analysis had been carried out for metallic and composite blade.

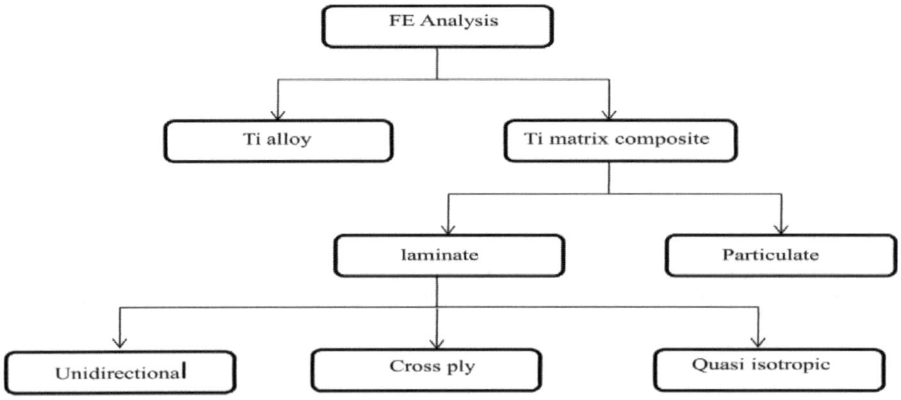

Figure 1. Outline of analysis for composite rotor blade

4.1. Mesh generation

The finite element model was developed using HYPERMESH software. The 3D model of the blade was divided into several zones and the 2D mesh was generated. 2D mesh was then converted into 3D mesh using appropriate commands. Solid 185 elements were used in the analysis of the metallic blades. The same elements cannot be used for the analysis of composites because non-availability of layering feature. Therefore, layered Solid 185 elements were used in the analysis of composite blades. Quality checks such as Jacobian, warpage and aspect ratio were ensured to be within permissible limits.

In FE analysis, a finer mesh results in a more accurate solution. However, as the mesh becomes finer, the computation time also increases. From convergence study, an optimized mesh that can give accurate results at considerably lesser time was obtained. H-Method was chosen to perform the convergence in ANSYS. This method involves running the analysis using a coarse mesh and the results obtained are compared with that obtained by analysing the same model but with finer mesh. In this method, the number of elements is increased and mesh refinement is done locally. This process is repeated until the difference between two successive solutions is less than 2 %. The final configuration of mesh that arrived from H-Method of convergence had 100526 nodes and 90784 elements.

The surfaces and 2D elements of the model were deleted. The 3D elements of the model were exported to ANSYS. User profile was selected in the preference and the element type was then set as Solid 185 in the utility menu. The components were deleted and the elements were renumbered. Finite element model of the blade was then exported to ANSYS. The 3D elements were now imported into ANSYS and the preference was set to be structural. The FE model of the blade is shown in Figure 2.

4.2. Loads and boundary conditions

The loads acting on the compressor rotor blades were centrifugal, gas and thermal loads. The major load acting on the blade is the centrifugal load as it is subjected to RPM of 16000. It acts along the radial direction of the blade. Gas loads act in the axial and tangential direction of the blade due to the difference in velocity and pressure between the leading edge and trailing edge of the blade. They act along X-axis and Y-axis, respectively. In this study, the blade was subjected to uniform temperature of 300°C. The boundary conditions that were provided to simulate the attachment of root on to the disc are shown in Figure 3. In the analysis of laminate composites, the sequencing of layups was carried out as shown in Figure 4 and the thickness of each layer was 0.5 mm.

IOP Conf. Series: Materials Science and Engineering **152** (2016) 012008 doi:10.1088/1757-899X/152/1/012008

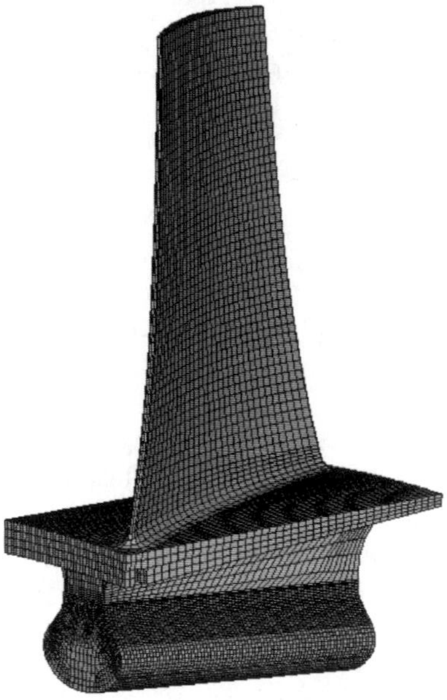

Figure 2. Finite element model of the compressor rotor blade

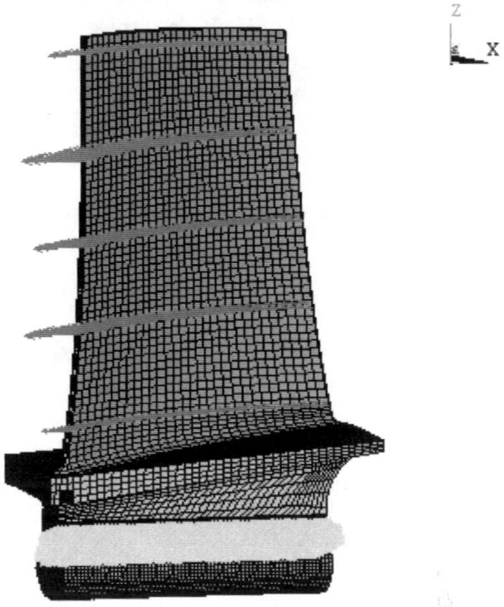

Figure 3. Loads and boundary conditions of compressor rotor blade

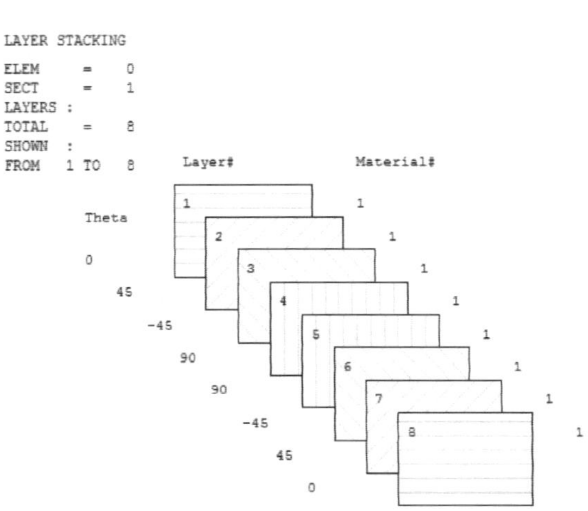

Figure 4. Sequence of lay up of composite laminate

4.3. Static stress analysis

Static stress analysis was carried out to determine the displacements and stresses on the compressor rotor blade. The stress plots of metallic and composite blade are given in Figure 5 and the analysis results are provided in Table 2. From Table 2, it is observed that the mass of the TMC blade was 27.6 g while the mass of the metallic blade was 32.5 g. From the analysis, it is evident that the radial tip displacement of Ti alloy was greater than that of the composites. This is because modulus of elasticity of Ti alloy is smaller when compared to that of TMC and the coefficient of thermal expansion of Ti alloy is greater than that of the TMC. The displacement is the least for unidirectional laminates and was 0.28 mm. The reason being, in unidirectional laminates, the fibres are oriented along the direction of loading. In the case of compressor rotor blade, the major load is the centrifugal load and it acts in the radial direction. As the fibres of the unidirectional laminates are along the radial direction, they have the highest stiffness in the same direction, resulting in minimal displacement. On the other hand, the particulate composites' displacement was 0.29 mm and was slightly higher than unidirectional laminates. In case of particulate composites, the stiffness was the same in all directions. The displacements of cross ply and quasi isotropic laminates were 0.37 mm and 0.38 mm, respectively. They were greater than unidirectional and particulate composites, but smaller than that of Ti alloy. This is because, in cross ply and quasi isotropic composites, fibres in few lamina are oriented in the direction of loading while few are not.

Considering Von mises stress, it is seen from Figure 5 that the stresses in composites are lesser than that in metal alloy. The maximum stress in case of Ti alloy is on the pressure side hub section whereas in the case of particulate composites, it is seen on the suction side hub section. Both Ti alloy and composites show gradual gradation in stress pattern. The red patch refers to the location of stresses greater than 180 MPa and it is comparatively smaller in case of particulate composites. The stress patterns of $[0]_8$, $[0/90]_{2S}$, $[0/\pm45]_S$, and $[0/\pm45/90]_S$ are more or less similar with a small variation in the maximum stress value. Considering the maximum principal stress, the maximum stress location is at the pressure side hub section for Ti alloy and the composites. The maximum principal stress of Ti alloy was greater than that of TMC. It is also seen that the stress patterns of all the composites were similar and the difference in maximum stress was not more than 2 MPa.

Table 2. Results of stress analysis of the compressor rotor blade

Properties	**Metal**	**TMC**				
	Ti 64	**Particulate**	$[0]_8$	$[0/90]_{2S}$	$[0/\pm45/90]_S$	$[0_2/\pm45]_S$
Mass of blade (g)	32.5	27.6	27.6	27.6	27.6	27.6
Von Mises stress (MPa)	225	207	219	202	201	212
Maximum principal stress (MPa)	233	200	201	197	197	199
Maximum radial displacement (mm)	0.70	0.29	0.28	0.37	0.38	0.32

4.4. Analytical estimation of stress

To validate the results of FE analysis, analytical estimation of the stresses was carried out. It is difficult to theoretically obtain the exact stresses acting on the rotor blade due to the complexity in blade structure and loading. However, the results can be obtained by considering the blade as a cantilever beam and the loads were uniformly distributed. This does not depict the exact structure and loading, but it has been used in an average sense to obtain the results. Centrifugal stresses acting on the blade depends on the size of the rotor blade and also on the rotational speed of the rotor. This is given by Equation 1.

$$\sigma_c = \frac{F_c}{A_h} = \int_{r_h}^{r_t} \rho\omega^2 \frac{A_b}{A_h} r\,dr \qquad (1)$$

where σ_c = centrifugal stress, A_b = area of interest at the required radius from the centre, ρ = density of the material, A_h = area at the hub, N = number of revolutions per minute, ω = angular speed, r_h = radius at hub and r_t = radius at tip.

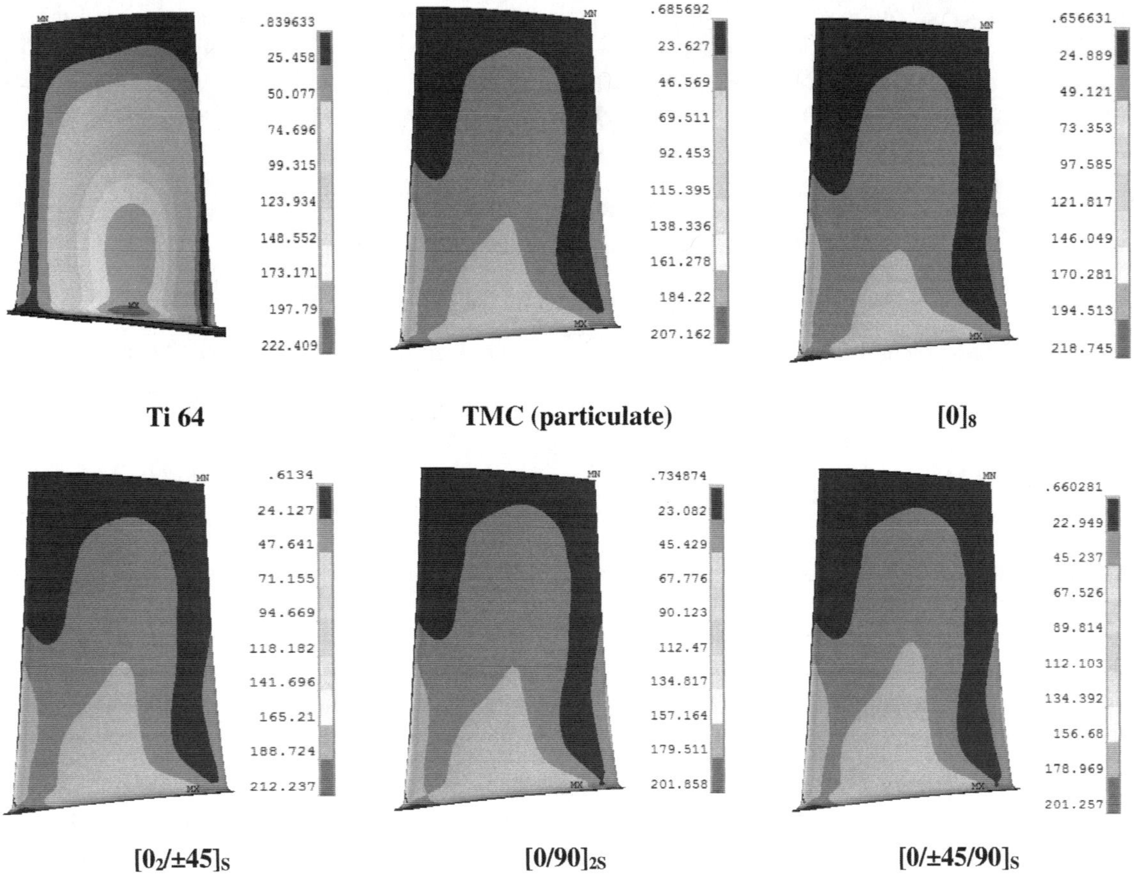

Figure 5. von Mises stress plot (MPa) of compressor rotor blade

Equation 2 shows a more simplified form of Equation 1, in which most parameters are pre-calculated. This shows that centrifugal stress is directly proportional to the $\rho A\omega^2$.

$$\sigma_c = \frac{\rho\omega^2 A}{4\pi}\left[2 - \frac{2}{3}\left(1 - \frac{A_t}{A_h}\right)\left(1 + \frac{1}{1+\frac{r_h}{r_t}}\right)\right] \tag{2}$$

$$\omega = \frac{2\pi N}{60} \tag{3}$$

$$N = 16028 \quad \text{and} \quad \omega = \frac{2\times\pi\times16028}{60} = 1678.4$$

Centrifugal stress is the major stress acting on the blade. The centrifugal stress was estimated using Equation 2 for the compressor blade made of Ti alloy. The blade is assumed to be a cantilever beam with uniformly distributed load acting on it. The bending moment was estimated and it was used in calculating the bending stresses, which is given by Equation 4. The stresses were estimated at five different sections from hub to tip along the radius of the blade. The stresses acting at different sections of the blade are given in Table 3 and they clearly indicate that the magnitude of stresses decreases from hub to tip. This implies that the stress pattern matches with the results obtained from numerical analysis. The area of cross section and the radius were obtained from the solid model.

$$\sigma_b = \frac{My}{I} \tag{4}$$

where σ_b = bending stress, M = bending moment and I / y = section modulus.

Table 3. Variation of stress along the radius of the blade

Blade Section	Radius (mm)	Area (mm sq)	Stress (MPa)
Hub	215	260	406
Below Mean	225	74	103
Mean	235	56	71
Above Mean	245	47	58
Tip	255	33	38

The maximum stress obtained from FEA was 233 MPa and that obtained from analytical estimation was 406 MPa. The order of magnitude of stress matches with that obtained from FE analysis. The results did not exactly match because of the many assumptions that had been used in the analytical estimation of the stresses. The variation in the thickness and profile of the blade had not been taken into consideration. Thus these analytical results gave a gross estimation of stress and not as accurate as the results obtained from FE analysis.

4.5. Modal analysis

In gas turbines, when the operating speed of engine is equal to the natural frequency of the blade system, each of the individual blade begins to resonate. At resonance, a vibratory deflection of large amplitude is induced by relatively small amplitude stimuli. The result is severe mechanical stresses leading to failure. Modal analysis is used to determine the vibration characteristics such as natural frequencies and mode shapes of a structure, which are important parameters in the design of a structure for dynamic loading conditions. Modal analysis has been carried out with the pre-stressed compressor rotor blade and the first five natural frequencies have been obtained and are provided in Table 4. Natural frequency is directly proportional to stiffness and inversely proportional to the density of the body. As far as the metals are concerned, as the stiffness increases, the density also increases. The greatest advantage of composites here is that the increase in stiffness comes with decrease in density, resulting in increased natural frequency of the system.

Modal analysis has been carried out with the pre-stressed compressor rotor blade to capture the effects of change in stiffness due to centrifugal stresses. While estimating the natural frequency of the blade, the analysis is carried out using the compressor rotor blade without any loads acting on it. However, during actual operation of the engine, the blades will be subjected to centrifugal stress that results in a change of the blade's stiffness. Therefore, to capture this pre-stress effect, the blade was first subjected to the static analysis and then followed by modal analysis with pre-stress option on. The first five natural frequencies were obtained for Ti alloy, particulate and laminate composites. The results are provided in Table 4 and the mode shapes are provided in the Figure 6.

Table 4. Result of modal analysis of the compressor rotor blade

Frequency	Metal	TMC				
(Hz)	Ti 64	Particulate	$[0]_8$	$[0/90]_{2S}$	$[0/\pm45/90]_S$	$[0_2/\pm45]_S$
Mode 1	1393	1658	1677	1503	1463	1570
Mode 2	2628	3414	3431	3057	2886	3110
Mode 3	4903	5215	5286	4612	4442	4810
Mode 4	6420	9580	7731	6814	6489	7047
Mode 5	7016	10459	8776	7751	7337	7953

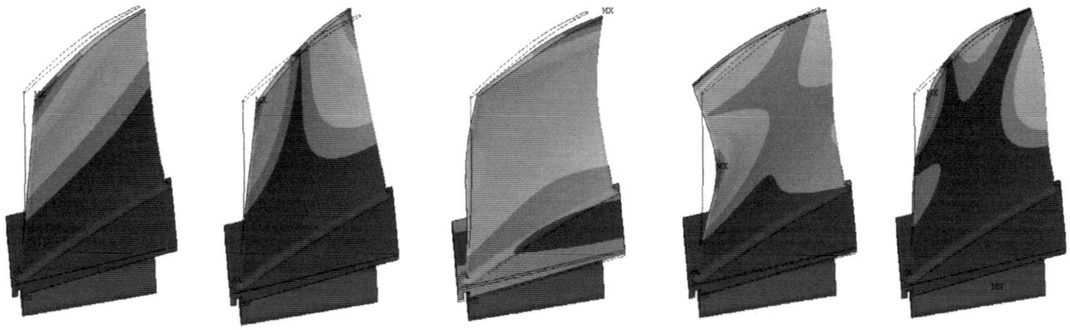

Mode 1 **Mode 2** **Mode 3** **Mode 4** **Mode 5**

Figure 6. Mode shapes of Ti particulate composite rotor blade

It can be seen in Figure 6 that the mode shapes obtained for Ti alloy and Titanium matrix composites are similar. The first mode is a flexural mode, also known as 1F, in which there is only one nodal point. Animation in ANSYS shows the blade oscillating back and forth about its hub. The second mode is a torsional mode, also known as 1T, where the blade twists about its central line. The third mode is again a flexural mode, it is 2F because it has two nodal points wherein the animation shows a sine wave motion. The fourth and fifth modes could not be clearly distinguished as flexural or torsional mode because the animation showed a complex motion that was a combination of both flexure and torsion.

However, there is a considerable difference in natural frequencies of compressor rotor blade made of Ti alloy, particulate composites and laminates with various orientations in Table 4. Blades made of Titanium matrix composites have higher natural frequency compared to those made of Ti alloy. There is a difference of 300 Hz when comparing the first natural frequency of Ti alloy against that of TMC. For higher modes, the difference in frequencies increases. Unidirectional laminates have the highest natural frequency. The difference in natural frequencies of composites with various orientations is not more than 100 Hz.

5. Failure criteria

5.1. Maximum stress theory

Although there are several theories derived to identify the failure of a component, there is no single theory that is suitable for all kinds of material and loading. The selection of failure criteria is mainly based on the type of material and nature of loading. Hence it is difficult to conclude on only one of such theory. The maximum stress criterion had been employed in the case of TMC, which considers that the composite fails when the stress in the component exceeds the respective allowable limit. This is a simple and direct way to predict failure of composites as interaction between the stresses acting on the lamina is not considered [12]. The maximum allowable stresses of various laminates considered for analysis are in given in Table

5. Due to limitations in availability of material properties, the maximum allowable stress in compression was assumed to be the same as in tension. The failure criteria option was checked in ANSYS and these allowable stresses were provided as input after which the analyses were carried out. The stresses in each lamina were checked against the allowable limits. It was found the stress experienced by the particulate, uniaxial, cross ply and quasi isotropic laminates were all within the allowable limits. Thus the safety of the compressor rotor blades was ensured under static loading condition. Maximum stress criteria is given by the following equations:

$$-\sigma_{11}^{C} < \sigma_{11} < \sigma_{11}^{T} \quad (5)$$

$$-\sigma_{22}^{C} < \sigma_{22} < \sigma_{22}^{T} \quad (6)$$

$$-\sigma_{33}^{C} < \sigma_{33} < \sigma_{33}^{T} \quad (7)$$

$$|\tau_{12}| < \tau_{12}{}^{y} \quad (8)$$

$$|\tau_{13}| < \tau_{13}{}^{y} \quad (9)$$

$$|\tau_{23}| < \tau_{23}{}^{y} \quad (10)$$

Table 5. Maximum allowable stress of composite laminates [13]

Maximum Allowable Stress (MPa)	0°	45°	90°
σ_L	1518	1275	380
σ_T	380	623	1518
τ_{LT}	326	569	326

5.2. Tsai – Wu criterion

The maximum stress failure criterion does not account for the interaction of stresses, thus Tsai Wu failure criterion that accounts for stress interaction was also employed. It is given by Equation 11. It was found that the resulting value of failure criteria check for composite laminates was less than 1. Thus the safety of the compressor rotor blades is ensured under static loading condition.

$$X_1 \sigma_{11} + X_2 \sigma_{22} + X_3 \sigma_{33} + X_{11}\sigma_{11}^2 + X_{22}\sigma_{22}^2 + X_{33}\sigma_{33}^2 + X_{44}\tau_{23}^2 + X_{55}\tau_{13}^2 + X_{66}\tau_{12}^2$$
$$+ 2X_{12}\sigma_{11}\sigma_{22} + 2X_{13}\sigma_{11}\sigma_{33} + 2X_{23}\sigma_{22}\sigma_{33} < 1 \quad (11)$$

$$X_1 = \frac{1}{\sigma_1^T} - \frac{1}{\sigma_1^C} \quad (12)$$

$$X_{11} = \frac{1}{\sigma_1^T . \sigma_1^C} \quad (13)$$

$$X_{12} = -\frac{1}{2}\sqrt{X_{11} . X_{22}} \quad (14)$$

6. Results and discussion

It is mandatory to provide a sufficient tip clearance such that the tip of the blade does not rub against the casing during operation. This tip clearance is provided based on the radial growth of the blade. When the tip displacement of the blade is more, it implies that the tip clearance required is also more. As the tip clearance increases, the tip loses also increases and this results in decreased efficiency of the gas turbine.

Therefore, it is advantageous to have a lesser radial tip displacement that will require lesser tip clearance and eventually better efficiency. It is evident from the results of static stress analysis of the TMC (as in Table 2) that the maximum radial tip displacement has decreased by 0.4 mm compared to Ti alloy. This implies that for composite blades, the tip clearance can be smaller and hence during off design conditions, the losses would be minimized leading to increased efficiency.

The mass of the composite blades is less by 4 grams compared to the metallic blade, which is another major advantage. This is because of the lesser density of TMC against the Ti alloy. This reduction in mass is vital because it will result in weight reduction of nearly 15% per stage. If all stages of compressor are replaced with composites, the reduction in mass will be tremendous. The reduced mass of the blades could allow further weight saving in discs as well. This weight reduction will result in an overall engine weight reduction and reduced fuel consumption, thus providing operational cost savings.

As far as stresses are concerned, reduction in stress of around 30 MPa is seen in TMC compared to that of metallic blade. It is observed from the stress plots (as in Figure 5) that the major part of the composite airfoil reveals stresses less than 70 MPa against the metallic blade. In case of laminate composite blades, the stresses of individual laminas were also obtained. Tsai Wu criterion and maximum stress criterion have been employed as failure criteria for Ti alloy and TMC, respectively, and the stresses were found to be within the allowable limit. The increase in stiffness was made possible through fibre reinforcement and it provides blade's designers with an opportunity to tune the performance of the blade under load. The increasing material stiffness also changes the resonant frequency of the blade, thereby allowing removal of any damaging vibration modes from the engine running range without any excessive section thickening or added weight.

7. Conclusion

The static stress and modal analysis of Ti alloy and TMC for the compressor blade application have been carried out using FE analysis. The stresses of composite blades obtained in static analysis are within the allowable stress limit. First five natural frequency of the rotor blades made of Ti alloy and TMC were obtained. It is concluded that TMC blades have lower weight, lesser maximum radial tip displacement, better strength, better stiffness and higher natural frequency. The strength of composite has been compared with Ti alloy and its advantages are substantiated quantitatively in the present work. Failure criteria checks show that the metallic and composite rotor blades are both safe under static loading condition. However, a detailed study has to be carried out considering HCF and LCF. Further testing is essential to conclude on the replacement of metallic blades with composites.

References

[1] Barbero E J 2010 *Introduction to Composite Materials Design* (New York: Wiley-Interscience)
[2] Hooker J A and Doorbar P J 2000 *Materials Science and Technology* **16** 725-31
[3] Zweben C 2014 *Advanced Materials by Design*
[4] Jabbar A, Rai A K, Ravinder R P and Dakhil M H 2014 *International Journal of Mechanical and Production Engineering Research and Development* **4** 73-94
[5] Kim K and Shin Lee Y S 2014 *Journal of Mechanical Science and Technology* **28** 1421-29
[6] Singerman S A and Jackson J J 1996 *Proceedings of 8th International Symposium on Superalloys*
[7] Campbell F 2006 *Manufacturing Technology of Aerospace Structures* (Elsevier)
[8] Johnson W S, Lubowinski S J and Alton L Highsmith 1990 *Thermal and Mechanical Behavior of Composites* (Philadephia: ASTM)
[9] Balasubramanian M 2013 *Composite Materials and Processing* (Boca Raton: CRC Press)
[10] Dharmadhikari S R, Mahakalkar S G, Giri J P and Khutafale N D 2013 *International Journal of Modern Engineering Research* **3**

[11] ANSYS Mechanical APDL introductory tutorials
[12] Camanho P P 2002 *Failure Criteria for Fibre-Reinforced Polymer Composites*
[13] Bull J W 1996 *Numerical analysis and modeling of composite materials*

Aeroelastic effects of a simple rectangular wing-box model with varying rib orientations

M S Othman[1], O T Chun[1], M Y Harmin[1,2*] and F I Romli[1,2]

[1]Department of Aerospace Engineering, Faculty of Engineering, Universiti Putra Malaysia, Malaysia
[2]Aerospace Manufacturing Research Centre (AMRC), Faculty of Engineering, Universiti Putra Malaysia, Malaysia

*myazdi@upm.edu.my

Abstract. A simple rectangular wing-box model with varying individual rib orientations has been analysed with the interest on studying their impact on flutter and divergence instabilities. The varying rib orientations offer an alteration in bending-torsional coupling characteristics, resulting in significant effect on aeroelastic behaviour of the wing. Substantial improvement is found in comparison with the baseline and parallel ribs orientation configurations, which leads to the possibility of significant weight reduction.

1. Introduction

Air transportation has emerged as a vital element in ensuring continuous growth of both economic and social benefits. However, due to rising jet fuel cost and increasing demand for air transportation, the cost efficiency has now become a major issue in the airline industry to sustain their quality standard of customer service and continue their market survival. Hence, there have been a great drive and demand in aerospace industry to produce more efficient and cost effective aircraft designs. The primary goal is to ensure that the improved aircraft will deliver a better efficiency in operating cost per miles without having to compromise its overall performance. In recent years, the approach of aeroelastic tailoring on the aircraft design has become an auspicious solution in improving its overall efficiency. This can be idealised through active or passive control of aeroelastic wing system, or both.

Significant research efforts have been expanded into active aeroelastic wing concepts with the main goal is to provide an optimal wing aerodynamic shape that can be adapted at any flight conditions [1, 2]. Even with the current advances in smart material and adaptive structure capabilities, the practical solution using this concept is still uncertain. This is mainly due to the conflicting structural condition where the aeroelastic wing has to be stiff but also flexible at the same time. On the other hand, current research and development in the use of aeroelastic tailoring in passive approach have been much more devoted in the use carbon-fibre composite structures.

Over the past decade, composite technology has progressively become an important area in the construction of aerospace structural components. Substantial advancements have already been made in improving their properties and also manufacturing processes. Following this trend, the commercial aerospace industry nowadays has utilised composite materials to greater extent in the new generation commercial aircraft designs like Boeing 787 and Airbus A350. The primary advantage from the use of the composite materials is their superior strength-to-weight ratio, which enables a significant weight reduction without having to compromise strength and durability. In addition to this, their anisotropic characteristic also allows the exploitation of bending-torsional properties through altering the stacking sequence of the plies [3, 4, 5]. Nevertheless, aircraft configurations are still more or less the same such that conventional configuration of engines under-slung on pylons attached to swept back wings, spars in spanwise direction, ribs in perpendicular direction to spars and passengers carried in a cylindrical-like fuselage.

There have been several attempts on aeroelastic tailoring of interior wing configuration such that it behaves similar to anisotropic manner [6, 7, 8]. Most of these works are considering the ribs to vary at the same orientation, resulting in a parallel ribs arrangement at any desired orientation and leading to changes in the bending-torsion coupling behaviour. This notably improves the aeroelastic performance with no additional weight penalty. Inspired by their findings, further investigation is carried out in this study by assessing all possible orientations for each individual rib separately. The impact of variation in ribs orientation on the aeroelastic instabilities is the main focus in this work.

2. Methodology

A wing model loosely based upon a Goland rectangular wing-box [6, 9] is considered in this study and shown in Figure 1. The wing-box consists of front and rear spars, and six ribs. It is measured 1000mm in span, 200mm in width and 50mm in height with the spar's thickness of 10mm. The ribs are equally divided along the span with 200mm margin between each of them. By default, the ribs are of 10mm thick and they are positioned in perpendicular direction to the spar. Both spars and ribs are modelled as a simple flat plate using MSC NASTRAN's four-node quadrilateral shell elements (CQUAD4) with material properties entry of aluminium alloy 2024 [10] and a fully clamped boundary condition at the root of the wing. A mesh convergence study is performed in order to establish the finite element (FE) model with a mesh density that is sufficiently fine to permit a good approximation of the solution.

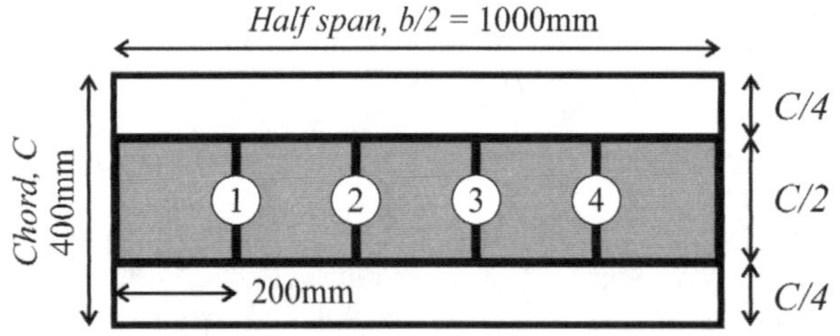

Figure 1. Planform view of the wing with the baseline wing-box represented by the shaded area

Figure 2 illustrates the FE representation of the studied aeroelastic wing model. It should be noted that the structural FE model only considers the wing-box section, with the total planform of the wing is completed by the doublet lattice aerodynamic panels. This allows the problem to be simplified while retaining its essential characteristics. The varying rib orientations only consider the ribs that are located between the tip and root of the wing, which are numbered in sequence from 1 to 4 in Figure 2. Figure 3 and also Table 1 describe a number of considered rib orientations with the centre of rotation is defined at the mid chord. Each of these ribs is allowed to vary with a constraint that they will not overlap each other. Clockwise rotation is considered as positive orientation whereas counter clockwise rotation is considered as negative orientation. In addition, thickness of the ribs is also varied as they are oriented in order to ensure that the total weight of the wing remains constant in all cases. All the procedures involved have been coded in MATLAB and the system has been fully integrated with the MSC NASTRAN solutions to simulate the parametric studies.

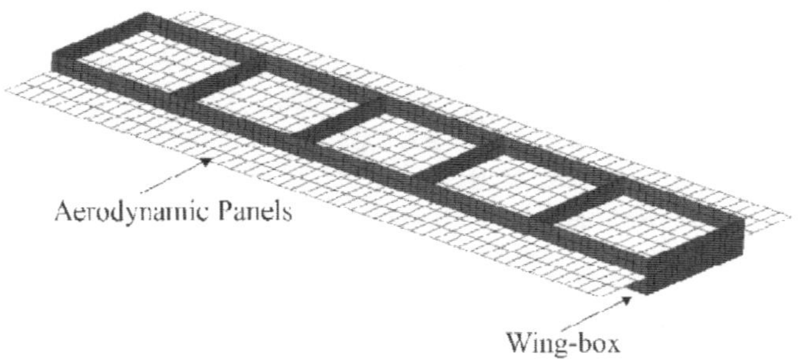

Figure 2. Aeroelastic wing model

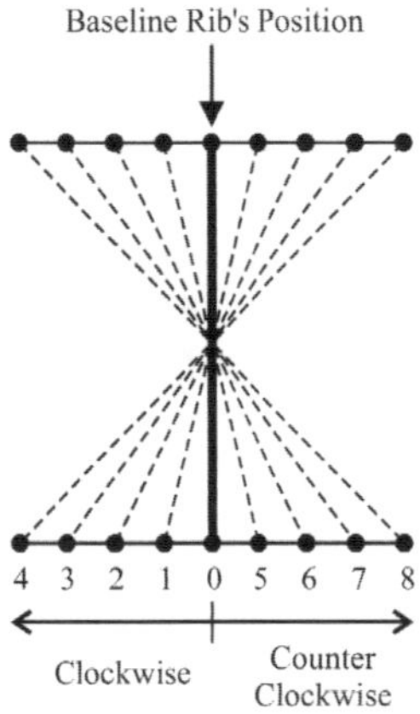

Figure 3. Possible rib's orientation

Table 1. Considered rib's angle of orientation

Positive Orientation (clockwise)		Negative Orientation (counter clockwise)	
Position	Angle, θ (°)	Position	Angle, θ (°)
1	14.04	5	-14.04
2	26.57	6	-26.57
3	36.87	7	-36.87
4	45.00	8	-45.00

3. Results and discussion

Normal modes analysis are conducted for all possible single oriented rib configurations leading a total of 32 configurations. Since aeroelastic instabilities are mainly related to the bending-torsional mode coupling, only bending and torsional modes are taken into account and highlighted in this work. The bending and torsional frequencies variation for each of the considered rib orientations are shown in Figure 4. These are mainly due to the change in structural mass and stiffness distribution as well as the position of shear centre along the wing.

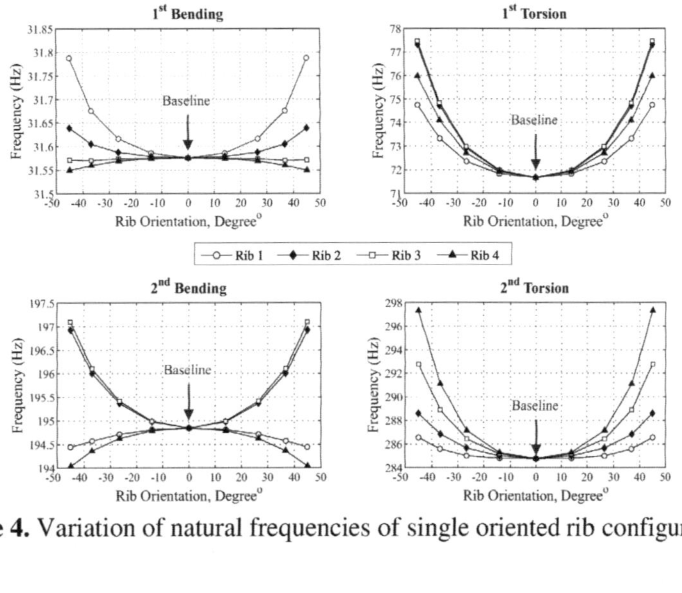

Figure 4. Variation of natural frequencies of single oriented rib configurations

In terms of bending modes, the variation in natural frequencies is very small and can be assumed negligible. In case of oriented ribs positioned towards the root of the wing, both bending frequencies are increased slightly as the rib is oriented away from its baseline position. This is possibly due to the dominant influence in structural stiffness over the mass distribution that acting in the direction of the bending related axes. Meanwhile, in the case of oriented rib positioned towards the tip of the wing, the bending frequencies decrease slightly but with a much reduced magnitude and this imposed opposite characteristic than the previous condition. The resulting torsional modes are in greater scale than the bending modes. The results clearly shows that both torsional frequencies are increased with increasing orientation of the ribs, which shows significant influence in structural stiffness over mass distribution that acting in the direction of torsional related axes. Interestingly, the impact is greater in the case of rib that is oriented towards the middle of the wing. Furthermore, the symmetrical trend in natural frequencies are due to the symmetrical impact for a similar orientation magnitude of both positive and negative directions. These findings suggest possibilities to increase the frequency spacing between the coupled flutter modes, which may result in significant improvement of the aeroelastic performance. Hypothetically, the bigger the frequency gap of the coupled modes, the weaker the interaction among them, resulting in increase in instability speeds.

A series of aeroelastic analyses are performed using well-known p-k method [11] that is available in MSC NASTRAN to assess the impact of the variation in ribs orientation on flutter and divergence instabilities. Figure 5(a) shows V-ζ and V-ω plots for the first two modes of the baseline aeroelastic model, depicting both flutter and divergence instabilities. These instabilities can be identified at the point where the damping changes in sign as the air speed increases, with the flutter represent the oscillatory instability while the divergence represent the non-oscillatory instability. Note that the flutter is a coupled effect between the first bending and torsional modes, whereas the divergence is obtained from a non-oscillatory solution of the first bending mode. Figure 5(b) compares the baseline system with the oriented configuration system of second rib oriented at -45°. It can be clearly seen that there is an increase in frequency spacing, causing both flutter and divergence instabilities to be further delayed. Meanwhile, Figure 6 shows that almost all the considered oriented individual rib cases are able to improve the aeroelastic instabilities with both flutter and divergence speeds can be increased by nearly 11 percent compared to the baseline system. Another important outcome that emerges from the result is that the individual rib orientation shows approximately a two-fold increase in aeroelastic stabilities than the parallel ribs orientation.

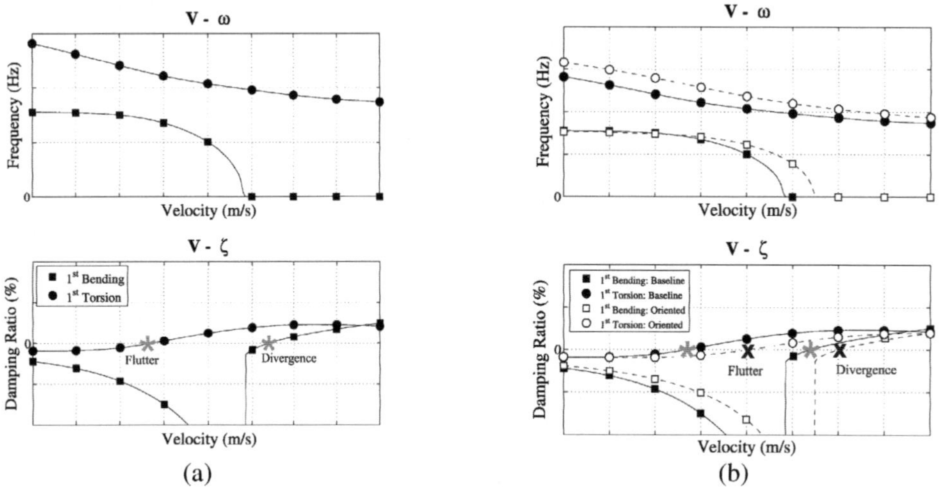

Figure 5. V-ζ and V-ω plots of baseline and selected oriented rib configurations: (a) baseline system, (b) baseline and oriented rib systems

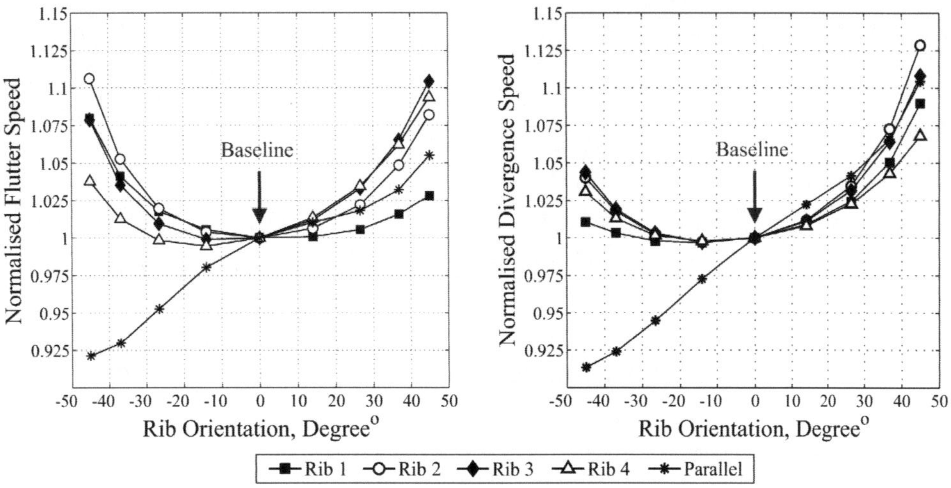

Figure 6. Normalised flutter and divergence speeds for varying individual rib and parallel ribs orientations

4. Conclusion

This study is set out to determine the potential of wing rib modification via aeroelastic tailoring. It is found that variation of single rib's orientation resulted in an increase of frequency spacing between the bending-torsional modes, which led to a remarkable improvement in aeroelastic performance when compared to the baseline and parallel ribs orientation systems. This encouraging outcome provides a promising potential in reduction of the aircraft weight without having to compromise its aeroelastic performance. Further research should be undertaken to explore all possible combinations of the rib's orientation that can be idealised through the use of optimisation techniques.

Acknowledgements

The authors acknowledge the support from Universiti Putra Malaysia (UPM) through research grant GP-IPM-9401100.

References

[1] Pendleton E W, Bessette D, Field P B, Miller G D and Griffin K E 2000 *Journal of Aircraft* **37** 554–61
[2] Weisshaar T A 2013 *Journal of Aircraft* **50** 337-53
[3] Shirk M H, Hertz T J and Weisshaar T A 1986 *Journal of Aircraft* **23** 6-18.
[4] Manan A, Vio G A, Harmin M Y and Cooper J E 2010 *Engineering Optimization* **42** 171–84
[5] Jun T K, Harmin M Y and Romli F I 2014 *Applied Mechanics and Materials* **629** 182-8
[6] Harmin M Y, Ahmed A T, Cooper J E and Bron F 2011 *52nd AIAA/ASME/ASCE/AHS/ASC Structures, Structural Dynamics and Materials Conference*
[7] Francois G, Weaver P M and Cooper J E 2015 *56nd AIAA/ASME/ASCE/AHS/ASC Structures, Structural Dynamics and Materials Conference*
[8] Francois G, Weaver P M and Cooper J E 2016 *57nd AIAA/ASME/ASCE/AHS/ASC Structures, Structural Dynamics and Materials Conference*
[9] Goland M 1945 *Journal of Applied Mechanics* **12** A-197 - A-208.
[10] Bauccio M 1993 *ASM Metal Reference Book* (Novelty: ASM International)
[11] Rodden W and Johnson E 1994 *MSC/NASTRAN Aeroelastic Analysis User's Guide* (Santa Ana: The Macneal-Schwendler Corporation)

IOP Conf. Series: Materials Science and Engineering **152** (2016) 012010 doi:10.1088/1757-899X/152/1/012010

Shape memory alloy actuation effect on subsonic static aeroelastic deformation of composite cantilever plate

A M H Hussein, D L Abdul Majid* and E J Abdullah

Department of Aerospace Engineering, Faculty of Engineering, Universiti Putra Malaysia, Malaysia

*dlaila@upm.edu.my

Abstract. Shape memory alloy (SMA) is one of the smart materials that have unique properties and used recently in several aerospace applications. SMAs are metallic alloys that can recover permanent strains when they are heated above a certain temperature. In this study, the effects of SMA actuation on the composite plate under subsonic aeroelastic conditions are examined. The wind tunnel test is carried out for two configurations of a cantilever shape memory alloy composite plate with a single SMA wire fixed eccentrically. Strain gage data for both bending and torsional strain are recorded and demonstrated during the aeroelastic test for active and non-active SMA wire in two locations. The cyclic actuation of the SMA wire embedded inside the composite plate is also investigated during the aeroelastic test. The results show reduction in both bending and torsional strain of the composite plate after activation of the SMA wire during the wind tunnel test.

1. Introduction

In the last few decades, smart structures become an interesting new area of research in aerospace engineering fields. Many researchers' work on adaptive structures embedded with smart materials and investigate their ability to reduce the aeroelastic difficulties and improve the wing performance. Static aeroelasticity is one of these difficulties where the aerodynamics load induced by steady flow interacts with elastic force of the wing structure. This phenomenon has a great impact on the performance, structure load distribution and the stability of the aircraft [1, 2]. The reason behind using an adaptive structure is because this type of structure has the ability to modify the material characteristics or its shape by using actuators [3], such as piezoelectric or shape memory alloys (SMAs) that are available in the industry.

Shape memory alloy is one of the commercialized smart materials and it has unique characteristics as actuator where it can change its physical geometry and mechanical properties, and has self-healing capacity [4, 5]. This type of material could provide new control methods with less complexity and small weight actuators. SMAs mainly undergo two different phases: martensite phase (M) that occurs in low temperature and the stiffness is low, and austenite phase (A) that occurs in high temperature and the stiffness is higher than that in martensite. In terms of exhibition, SMA can demonstrate two unique properties: pseudoelasticity and shape memory effect [5]. SMAs in the form of wires or thin films can be embedded in a host composite material to form a smart composite structure. The potential benefits of embedded actuators are the ability of controlling external shape, changing stiffness and providing vibration control of the composite.

There are two methods widely used to fabricate composite plate embedded with shape memory alloy. The first one is direct embedded where the SMA wires are separately clamped in jig during the fabrication of composite plate and then cured [6, 7]. On the other hand, the second method is by using sleeves embedded with dummy wire and replace with SMA wired after the composite has cured [8, 9]. It is difficult to fabricate a structure with direct embedded SMA with high reliability as the electrical resistive heating required for actuator activation may cause cracks in the structure or de-bonding of SMA actuator from the structure. One of the reasons that the second method by sleeve tubes is used in this test is because it is easy to replace the SMA wire if there is a need, which is it not applicable in the first method.

In control point of view, SMA wire can be used passively or actively to control static or dynamic properties of the composite structure. Passive method is mainly used to increase the strength of the composite. However, in active method, SMA is activated to change the stiffness, shape or vibration frequency of the structure. Active property tuning (APT) and active strain energy tuning (ASET) are two ways to actively control the static or dynamic properties of the structure. For APT, the SMA wires are not pre-strained and don't generated recovery forces. Meanwhile, the second technique is ASET where SMA are pre-strained and the recovery stress is generated during activation [10].

In terms of commercial exploitation, nickel-titanium alloys and copper-based alloys are the two most successful and each displays different properties. Nickel-titanium alloys are known to have a greater shape memory strain and thermally stable, with excellent corrosion resistance and higher ductility. On the other hand, copper-base alloys are much less expensive, and they can be melted and extruded in air easily, plus have a wide range of potential transformation temperatures. Characteristics of both alloy systems must be considered for any particular application. Thus far, nickel-titanium has proven to be the most flexible and beneficial in engineering applications [26]. Hence, nickel-titanium alloy is exploited in this current work.

Static aeroelastic is one branch of aeroelasticity where the forces involved are only elastic and aerodynamic forces. These forces can be lead to difficulties that should be taken into consideration in design process of wing such as on lift distribution, divergence and control system reversal phenomena. The analysis of the spanwise aeroelastic lift distribution is important mainly in two ways: to fulfill the requirements of performance analysis that requires the knowledge of the total moment and force on the aircraft as a function of altitude and flight condition, and to ensure the integrity of the aircraft structure for specific flight condition such that the stress analysis can be conducted [1]. The bending and torsion moment on the wing can be found and the maximum stress generally will be located on the wing root. Several research works have been carried out on static aeroelasticity. For example, Bea Jae-Sung *et al.* [11] studied the aeroelastic deformation and conducted static aeroelastic analysis of a variable-span morphing wing. The results show increased bending moment at the root of the wing due to an increase wingspan, hence larger bending stiffness is required in the wing structure. For low aspect ratio wing, a study conducted by Standford *et al.* [12] investigated the static aeroelasticity of micro air vehicle wing, which is presented with an experimental and numerical analysis. Also studies on both static and limit cycle oscillation aeroelastic analyses for high aspect ratio wing model have been done by Tang and Dowell [13] where theoretical modal was developed and compared with experimental test. On the other hand, there are earlier studies on using smart material for static aeroelastic. Weisshaar and Ehlers [14] investigated the effect of piezoelectric layer attached on laminated wing structure and observed an increase in divergence speed.

Adaptive structures are defined as structures that can modify their material properties and shape. They can be also formed either by passive, active or intelligent smart structures [3], and become a new technique to design the wing structure. Nam *et al.* [16] studied the effects of SMA spar design in composite wing on the aeroelastic performance improvement. On dynamic aeroelasticity, literature studies like Barzegari *et al.* [15] studied numerically the limit cycle oscillation amplitude of a wing embedded with pre-strained SMA wire and illustrated the significant improvement in the in-plane load effect of the SMA wire on aeroelastic behaviors. Additionally, SMA is used in applications where it is required to control the structure's vibration and natural frequencies. Some of these studies include that of Ostachowicz *et al.* [17], in which they presented a detailed finite element analysis of SMA fibers embedded inside a composite plate and the results showed the great influence of the SMA on natural frequency. They concluded that there is a great potential effectiveness to control the vibration of the composite structures by embedding SMA actuator under applied heat. Lau *et al.* [18] evaluated the natural frequencies using analytical model of composite beam embedded with SMA wires clamped at both ends while Zhang *et al.* [19] investigated both theoretically and experimentally the vibration characteristics of composite plate embedded with SMA in different arrangement. They reported that in the high temperate phase transformation (austenite phase) of the SMA wire, the natural frequency is higher than that in the lower phase. There are also other studies on dynamic properties of composite plate such as John and Harir [20] where they mounted the SMA wire over the plate.

Other applications of the shape memory alloy in aerospace include using it to control the shape of wing or airfoil. Baz *et al.* [21] introduced the mathematical model on shape control of NITINOL strips embedded inside composite beam and the obtained results showed possibility of controlling the shape of the composite beam without compromising their structural stiffness or frequencies. Icardi and Ferrero [22] studied the feasibility of using SMA torsion actuator in adaptive wing with flexible ribs and skin to change the airfoil shape and they presented in finite element method. There are several studies on using the SMA actuator in improving the aerodynamic performance on of the wing [23, 24]. Some studies also remarked on control of the shape memory alloy by constructing control system. One of these studies is by Abdullah *et al.*[25] whereby a Proportional–Integral–Derivative (PID) control system for SMA actuator wire that was attached on the top of composite plate was developed and the results showed an effective response.

In this present work, the focus is on using active embedded shape memory alloy to reduce the stress that is generated on the top plate during aeroelastic effect by measuring the strain. When aerodynamic loads start to twist the plate tip, the angle of attack is consequently increased which generates more lift until it reaches the equilibrium condition. That additional lift adds more moments and torsion forces on the plate root, which leads to more stress on the structure and eventually decrease the life span of the wing. Thus experimental tests on two different configurations of SMA cantilever plates are conducted. SMA wires inside the composite plates are activated by DC power supply during the test and the effect of the eccentric force that is generated during the activation of SMA wire is observed and recorded through the strain gage reading for both bending and torsional strains.

2. Material and methodology

2.1. Material

The actuators used in this experiment are Flexinol® 70C° (Dynalloy, Inc), SMA wire ready-to-use pre-strained 5-8 % and the pull force of single wire is 1250 g. The wire properties are listed in Table 1. For the composite plate in the experiment, four layers of E-glass fiber (with dimension 35 x 25 cm) are used as reinforcement since the fiber glass has poor electric conductivity compared to carbon fiber or Kevlar. For the matrix, EpoxAmite®100 with 103 Hardener is used. For the sleeve, silicone tubes Nexsilz (Tomo Heater & Rubber Sdn. Bhd.) with inner and outer diameters of 1.2 mm and 2.2 mm, respectively, are used.

2.2. Fabrication

To accommodate the alignment of the sleeves embedded within the composite layers, a rectangular jig is used with one side is fixed for holding one end of the sleeve and the other side is adjustable for tightening the other end of the sleeve. Vacuum infusion technique is used to fabricate the composite plate to improve the consolidation of the structure. Firstly, a smooth waxed surface was prepared and laid two layer of E-glass. The silicone tubes were then embedded with dummy wires in each tube and aligned in straight lines, and the two remaining layers of the fiber glass were then placed on top. The resin feed lines, peel ply and vacuum bag were installed. The resin was prepared where the epoxy and the hardener were added with a ratio of 1:3 and infused inside the bag with pressure at 100 bars and then cured at room temperature.

Table 1. Flexinol wire properties

Parameter	Value
Diameter (mm)	0.30
Density (g/cm^3)	6.45
Specific Heat (cal/g/°C)	0.20
Melting Point (°C)	1300
Latent Heat of Transformation (cal/g)	5.78
Thermal Conductivity (W/cm/°C)	0.18
Thermal Expansion Coefficient (°C^{-1})	
Martensite	6.6 x10^{-6}
Austenite	11.0 x10^{-6}
Poisson Ratio	0.33
Linear Resistance (Ω/m)	12.2
Electrical Resistivity (approx.) (μΩ.cm)	
Martensite	80
Austenite	100

Two specimens with a dimension of 285 mm x 40 mm are cut by vertical band saw after de-mold. The composite plate with one embedded silicone tube for each specimen is as shown in Figure 1. The dummy wire was removed and replaced carefully with single Flexinol® SMA wire that was free to move inside the sleeve during activation. The numbers in the specimen "SE05" and "SE15" indicate the distance of the SMA wire from the leading edge (L.E.), which was 5 and 15 mm, respectively. Two metal c-shape barriers with 1 mm hole in the middle were clipped at both ends of each specimen and then crimped the end of the SMA wire. Since the structure was a flat plate, it is assumed that the location of the aerodynamic center at which the aerodynamic lift acted on it was located in the quarter of the chord measured for the leading edge and the elastic axes were at the center mid of the chord. Therefore, the SMA was located in two locations in between the leading edge and elastic axes to study the effect of the force that countered the aerodynamic force and moments on the plate.

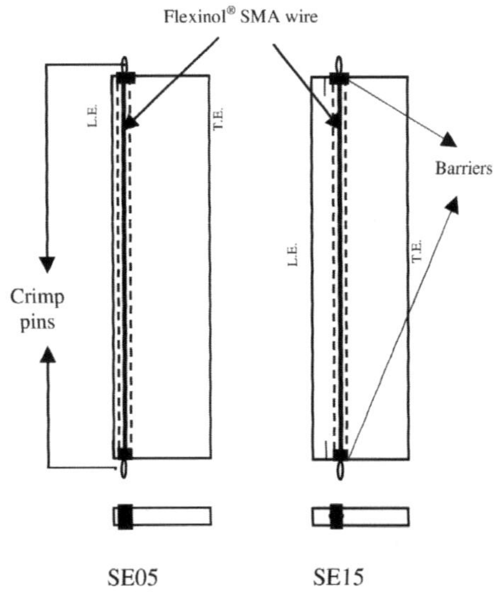

Figure 1. Schematic diagram of two composite plates embedded with SMA

2.3. Method
The research was carried out with two main types of test. First test, the SMA wire inside the cantilever composite plate was activated with absence of airspeed, which means there was no aerodynamic load acting on the plate and both bending and torsional strain gages were observed. This test was done to analyze the effect of the contraction force that was generated when the SMA was activated on the cantilever plates. In the second test, static aeroelastic test was conducted in which the airspeed inside the wind tunnel was increased up to 30 m/s and the strain gages responses were recorded, initially with non-active SMA and later with the SMA wire embedded to the plate was activated and de-activated, and these were repeated three times during the operated wind speed to examine the cyclic effect. In the beginning, correlation between the SMA wire temperature and the voltage that was generated from DC power supply was made to set the amount of electricity to activate the shape memory alloy wire.

2. Experiment setup

2.1. Temperature measurement
Temperature measurements of SMA wire with 285 mm were carried out to correlate the temperature with the voltage. SMA wire was hanged in test stand and a dead weight of 5 N was connected at the end of it as depicted in Figure 2. Thermocouple (K type) was attached in the middle of the SMA wire to measure the temperature during heating by electric power. In this test, DC power supply (GW Instek GPS- 3030D) was used and the data acquisition was executed via NI DAQ -9211 module associated with LabVIEW. Isolated tape was used in contact surface during the test.

2.2. Wind tunnel test
In this present study, the experiment was conducted in open circuit subsonic wind tunnel at Aerospace Lab, Aerospace Engineering Dept., Universiti Putra Malaysia. The cross section of the wind tunnel is (1 x 1 m) and its maximum operating airspeed is 40 m/s with variable driving system. Side-wall test rig was mounted on the side wall of the wind tunnel [26], which consists of three main parts: rigid frame that supports all the parts, rotating desk that allows the plate to rotate with different angles of attack and a clamped system to fix the plate in the frame.

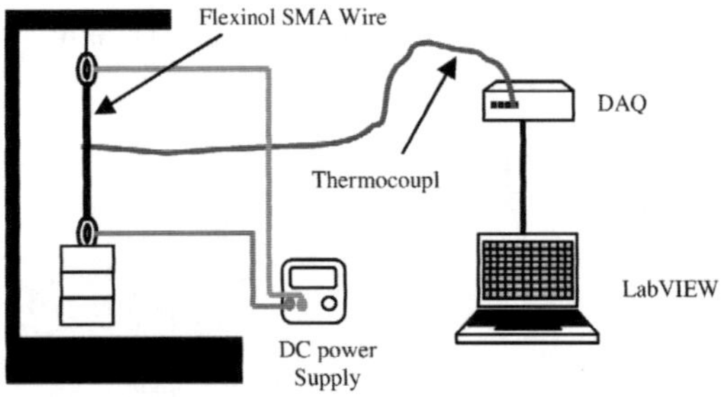

Figure 2. Temperature measurement of Flexional SMA wire setup

Cantilever plates with effective length of 250 mm (aspect ratio= 6.25) were fixed first to thick steel arm as shown in Figure 3, which was used for holding the plates and placed it to the middle stream of the wind tunnel to avoid the boundary layer of the wall . The arm and plates were mounted straight (un-swept) in the side-wall test rig with angle of attack ($\alpha = 1^{o}$). Strain gauges (KYOWA Tri-axial KFRP-5-120-D22) were located and attached on the top surface of the plate at 100 mm from fixed end. The data for both bending and torsional strain were monitored by using NI DAQ associated with LabVIEW. Two vinyl-coated electric wires connected to DC power supply (GW Instek GPS- 3030D) were attached to both end of shape memory wire. The experiment set-up is shown schematically in Figure 4. The wind tunnel speed was monitored by digital manometer instrument (Testo 510) and Figure 5 shows the composite plate embedded with SMA wire inside the wind tunnel.

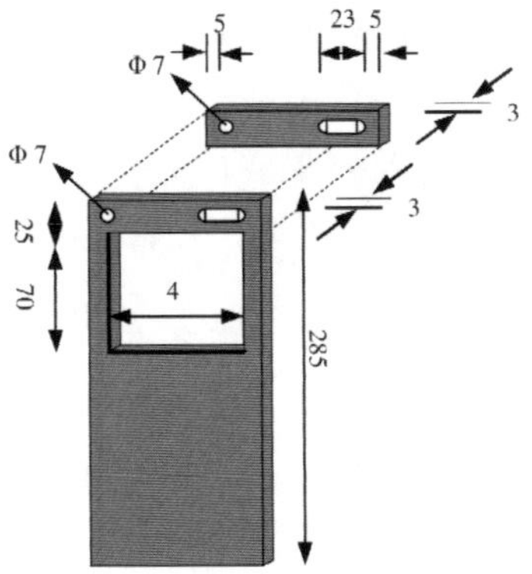

Figure 3. Steel arm all dimension in [mm]

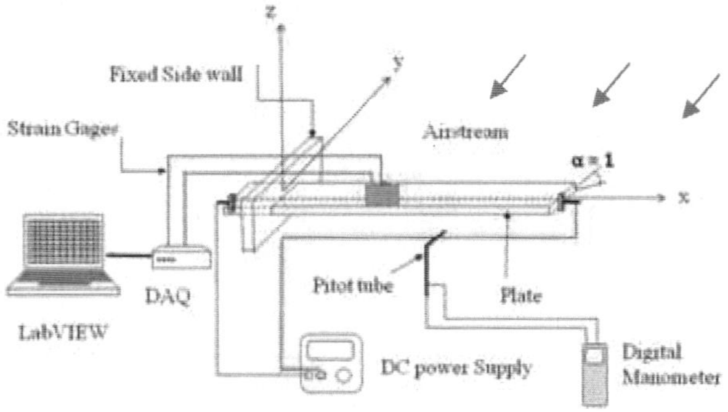

Figure 4. Schematic illustration of wind tunnel experiment for composite plates embedded with SMA

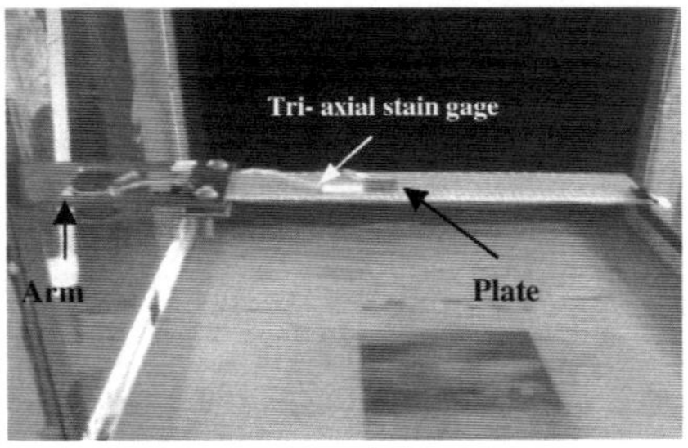

Figure 5. Inside wind tunnel section

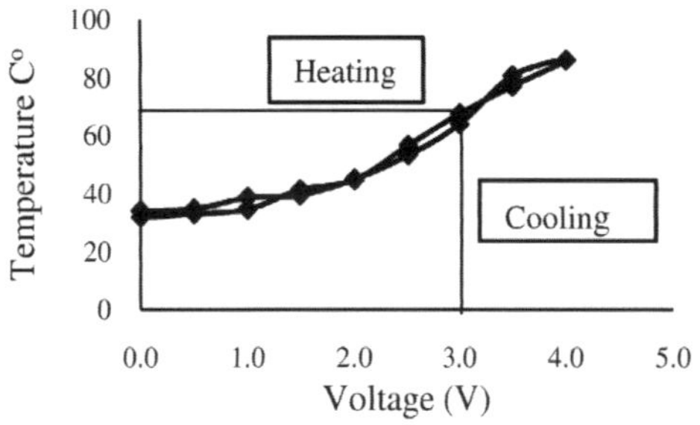

Figure 6. Relationship between the voltage and temperature of SMA wire

3. Results and discussion

3.1. Correlation of temperature with voltage

For this test, the relation between the temperature and the voltage of the SMA wire is shown and the results are plotted in Figure 6. Starting with room temperature where the SMA undergoes in martensite phase, the temperature data are recoded with a voltage interval of 0.5V up to 4V. It can be observed from the data that the temperature increases gradually with voltage. The recommended temperature from the Flexinol manufacture to operate the SMA wire is at 70 °C. As a result, electrical DC power up to 3V (~ 69 °C) is maintained.

3.2. Activating plate with absence of airspeed

In this test, the SMA wire inside the cantilever plates (SE05 &SE15) was activated with absence of airspeed or dynamic pressure. The outcomes of the test are plotted in Figure 7. In the beginning, all the strain gauges were calibrated and started from zero strain. SMA activation wire generated force on the plates increases the bending strain gradually up to 1.5 V and subsequently increases rapidly after that until it reached 3V (this voltage correlated with ~ 69 °C). The bending strain of plate "SE05" recorded 10.8 micron-strain at 3V but it was higher for plate "SE15" at the same voltage, which reached up to 16.6 micron-strain. On the other hand, torsion strain data shows negative strain during increasing the voltage and plate "SE15" recorded the highest torsional strain of -37.0 micro-strain. Other observation was during the cooling down of the SMA wire by de-activation, the strain gauges did not reach to their initial status that was recorded before activating the SMA wire at zero volt. The positive bending and torsion signs are shown in Figure 8.

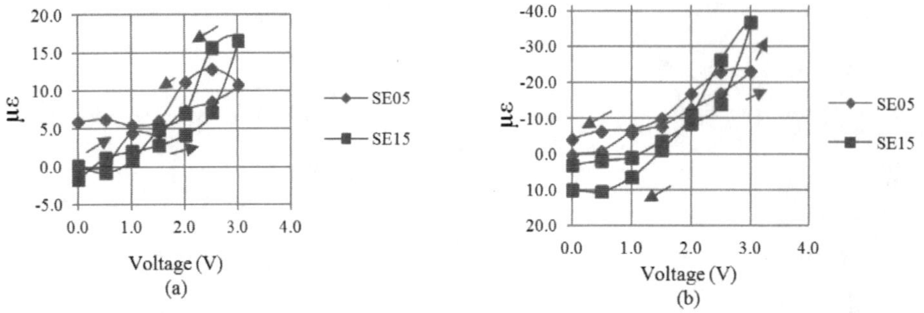

Figure 7. SMA actuation effect on strain gauges (a) Bending, (b) Torsion

3.3. Static aeroelastic deformation test

In this aeroelastic test, the airspeed inside the wind tunnel was operated at 30 m/s and the plate started to deform under the aerodynamic forces (Figure 8). The variation of strain gauge readings before and after activating the SMA wire inside the cantilever composite plates is illustrated in Figure 9 and Figure 10. The first observation during examining the data shows that each configuration of the plates "SE05" and "SE15" has higher strain in bending in terms of magnitude than torsion strain. Moreover, the initial bending strain gauges of non-active SMA at 30 m/s of wind speed recorded negative values for both composite plates embedded with SMA. The bending strain reading for plate "SE05" and "SE15 was -162.10 and - 106.14 micro-strain, respectively. On the other hand, torsion strain gauge readings for the both showed positive values, the first plate "SE05" scored 102.05 micro-strain while the other remaining plates "SE15" scored 35.31 micro-strain. The initial strain gauge of both plates in this test is not the same and this may be due to the vortex generated on the tip of the plates because the clip that held the SMA wire varied from one plate to another.

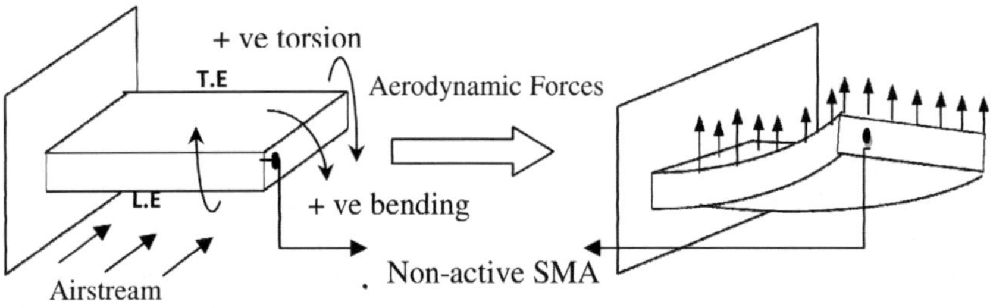

Figure 8. Plate bending and twist sketch due to aerodynamic forces with non-active

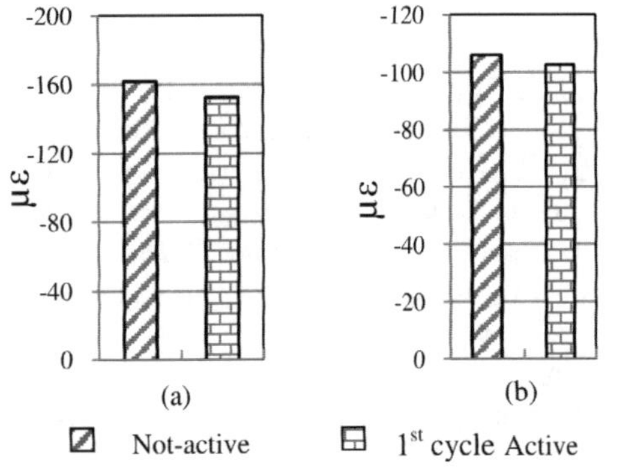

Figure 9. Bending strain gauges at airspeed 30 m/s for plates: (a) SE05, (b) SE15

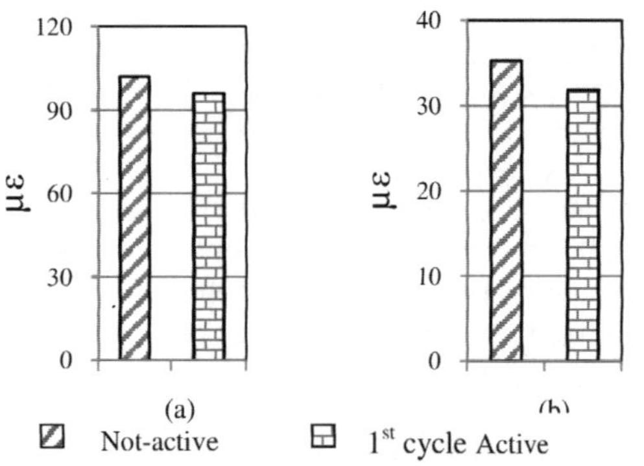

Figure 10. Torsional strain gauges at airspeed 30 m/s for plates: (a) SE05, (b) SE15

After that, the SMA wire inside the composite plates was activated up to 3V while the wind speed was operating at same speed. The eccentric force started to generate at the activation temperature of the SMA. For plate SE05, the results showed that the bending strain was reduced from its initial status before activation by 5.82% where the strain reading reach up to -152.7 micro-strain. Meanwhile, for plate SE15, there was also reduction of around 3.23%. For torsional effect, the activation of the SMA wire generated a considerable force and started to reduce the strain readings for both plates. Plate "SE15" scored the highest strain reduction by around 10%.

The experimental values of the bending and torsional strain during SMA cyclic activation wire at airspeed of 30 m/s are tabulated in Table 2. The cyclic data are normalized by dividing the strain gauge readings of each active plate (SE05 and SE15) with their initial non-active SMA at the same condition of wind speed, which is plotted in Figure 11. The cycle number (0) in x-axis in the figure indicates the SMA wire is non-active. ε_i indicates the strain gauge at each cycle number i = 1, 2 and 3 are active whereas ε_0 represents initial reading of strain gauge during non-active SMA and k are for plate SE05 and SE15.

In the cyclic effect test, the first plate SE05 bending strain was reduced at the first cycle by 5.8% as mentioned earlier . However, in the second and third cycle, the reduction reached around 5.46 % and 4.35 %, respectively. On the other hand, in the torsional strain, the reduction was fluctuating with 5.91% at the first cycle and then declined to 4.66 % at the second cycle. The reduction then increased at the third cycle and reached 11.2%. For the second plate SE15, both bending and torsional strain readings scored the highest reduction at the second cycle with 10.92 % and 25.1%, respectively, and then the reduction decreased around to 7% in the bending strain and 18 % in the torsional strain.

Table 2. Bending and torsional strain reading at airspeed of 30 m/s

Sample	Strain	Bending				Torsional			
		Not Active	Active cycle			Not Active	Active cycle		
			(1)	(2)	(3)		(1)	(2)	(3)
SE05	Micro	-162.10	-152.66	-153.25	-155.05	102.05	96.01	97.29	90.62
	*Normalize**	*1.0*	*0.94*	*0.95*	*0.96*	*1.0*	*0.94*	*0.95*	*0.89*
SE15	Micro	-106.14	-102.71	-94.54	-98.93	35.31	31.92	26.48	28.95
	*Normalize**	*1.0*	*0.97*	*0.89*	*0.93*	*1.0*	*0.90*	*0.75*	*0.82*

**Normalize = [Active SMA cycle $_k$ (i) / Not Active SMA $_{k}$] ; where i is cycle 1, 2 and 3 & k is plate SE05 and SE15*

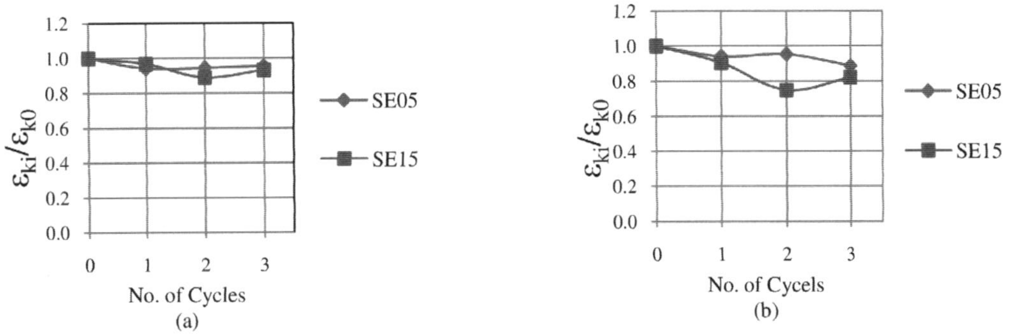

Figure 11. Normalized SMA activation cyclic effect on: (a) bending (b) torsion strain gauges at airspeed 30 m/s

4. Conclusion

Two configurations of composite plate embedded with SMA wire: SE05 and SE15 were fabricated. Experimental evaluation on strain gauges of active cantilever plate was conducted in two cases: first without aerodynamic load where the airspeed was zero and second with airspeed at 30 m/s. During the second case, the cyclic effect of the SMA actuator on the structure was also studied. In the first case, it is found that by activating the pre-strained SMA wire inside the composite plate, both bending and torsional strain readings are changed due to actuation of the shape memory alloy wire that produced large recovery force that was generated during the heating of the SMA wire by electricity. From the results, positive bending strain values are observed in both plates, which indicates that tension stress was generated on the top surface of the composite plate. On the other hand, torsional strain readings were negative and that shows that twist occurred during the activation of the SMA wire and change started to take place when the voltage was around 1.5- 2V.

In the static aeroelastic deformation test, when the airspeed inside the wind tunnel reached up to 30 m/s, the two non-active cantilever composite plates started to deform due to aerodynamic forces that acted on the wing. With relatively large aspect ratio, each plate started to deform due to the aeroelastic effect that could be observed from the strain gauge readings. In addition, the plates had bending strain greater than torsional and this is expected since the aspect ratio is 6.25. However, during the activation of the embedded SMA wire, the strain gauge readings (bending and torsional) for both plates were reduced. The active SMA, besides increase in the overall stiffness (from austenite phase to martensite phase), it also produced recovery force that operated as eccentric force that started to counter balance the aerodynamics moment and forces generated during the wind tunnel test.

On the cyclic effect, it can be observed that the effect of SMA force was reducing in subsequent cycles, except in the plate SE05 where the torsional strain in the third cycle was higher than the second cycle. However, for the remaining cases, the effect was reducing. This result highlights the limitation in the number of times that SMA wire can be used in one operation set before it was trained or reset it again. Furthermore, there was a small accumulated residual strain to the plate, which it can be seen on the first test in absence of airspeed where the strain gauge did not reach to zero strain when the SMA was de-activated. This condition may lead to the discrepancy in trend between the cycles.

In the end, reduction on the strain gage during static aeroelastic deformation test and considerable effects of the embedded active SMA have been shown for both configuration plates. The results show that both configurations have near effect change in bending and torsional strain gauge for each plate separately in the first and the last activation. Further studies are required to develop full understanding of the SMA effect on composite plates with low angles of attack by aligning the SMA wire in different location or configuration that could lead to a significant role in adaptive wing design in the future.

References

[1] Hodges D H and Pierce G A 2002 *Introduction to Structural Dynamics and Aeroelasticity* (USA: Cambridge University Press)
[2] Wright J R and Cooper J E 2008 *Introduction to Aircraft Aeroelasticity and Loads* (England: John Wiley & Sons)
[3] Vepa R 2010 *Dynamics of Smart Structures* (England: John Wiley & Sons)
[4] Jani J M, Leary M, Subic A and Gibson M A 2014 *Mater. Design* **56** 1078-113
[5] Lagoudas D C 2008 *Shape Memory Alloys: Modeling and Engineering Applications* (New York: Springer)
[6] Turner T L, Lach C L and Cano R J 2001 *Proc. of SPIE Smart Struct.. and Mater*
[7] Schrooten J, Michaud V, Parthenios J, Psarras G C, Galiotis C, Gotthardt R, Månson J A and Humbeeck J V 2002 *Mater. Trans.* **43** 961-73
[8] Baz A, Poh S, Ro J and Gilheany J 1995 *J. Sound Vib.* **185** 171-85
[9] Chopra I 2002 *AIAA J.* **40** 2145-87
[10] Wei Z G, Sandstrom R and Miyazaki S 1998 *J. Mater. Sci.* **33** 3763-83
[11] Bae J S, Seigler T M and Inman D J 2005 *AIAA Journal.* **42** 528-34
[12] Stanford B, Sytsma M, Albertani R, Viieru D, Shyy W and Ifju P 2007 *AIAA J.* **45** 2828-37
[13] Tang D, Dowell E H 2001 *AIAA J.* **39** 1430-41
[14] Weisshaar T A and Ehlers S M 1990 *AIAA Journal* **28** 1611-23
[15] Barzegari M M, Dardel M, Fathi A and Ghadimi M 2012 *Acta Astronaut.* **79** 189-202

[16] Nam C, Chattopadhyay A and Kim Y 2002 *Proc. of SPIE Smart Struct. and Mater.: Smart Struct. and Integrated Sys.*
[17] Ostachowicz W, Krawczuk M amd Żak A 1999 *Finite Elem. Anal. Des.* **32** 71-83
[18] Lau K T, Zhou L M and Tao X M 2002 *Compos Struct.* **58** 39-47
[19] Zhang R X, Ni Q Q, Masuda A, Yamamura T and Iwamoto M 2006 *Composite Structures* **74** 389-98
[20] John S and Hariri M 2008 *Comps. Part A Appl. Sci. Manuf.* **39** 769-776
[21] Baz A, Chen T and Ro J 2000 *Composites Part B* **31** 631-42
[22] Icardi U and Ferrero L 2009 *Mater Design* **30** 4200-10
[23] Dileep E, Nebish M and Loganathan V 2013 *Res J Recent Sci.* **2** 17-22
[24] Balaguru I and Sendhilkumar S 2013 *IJRET.* **5**
[25] Abdullah E J, Majid D L, Romli F I, Gaikwad P S, Yuan L G and Harun N F 2015 *Int. J. Mech. Mater. Des.* **11** 25-39
[26] Mavroidis C 2002 *J. Res. Nondestruct. Eval.* **14** 1-32
[27] Mohd Rafie A S, Attaran A, Majid D L and Basri S 2006 *Proc. of 1st Regional Conf. on Vehicle Eng. & Tech*

An analytical study on the performance of the organic Rankine cycle for turbofan engine exhaust heat recovery

S Saadon[1*] and A R Abu Talib[1,2]

[1]Department of Aerospace Engineering, Faculty of Engineering, Universiti Putra Malaysia, Malaysia
[2]Aerospace Manufacturing Research Centre, Faculty of Engineering, Universiti Putra Malaysia, Malaysia

[*]mimisaadon@upm.edu.my

Abstract. Due to energy shortage and global warming, issues of energy saving have become more important. To increase the energy efficiency and reduce the fuel consumption, waste heat recovery is a significant method for energy saving. The organic Rankine cycle (ORC) has great potential to recover the waste heat from the core jet exhaust of a turbofan engine and use it to produce power. Preliminary study of the design concept and thermodynamic performance of this ORC system would assist researchers to predict the benefits of using the ORC system to extract the exhaust heat engine. In addition, a mathematical model of the heat transfer of this ORC system is studied and developed. The results show that with the increment of exhaust heat temperature, the mass flow rate of the working fluid, net power output and the system thermal efficiency will also increase. Consequently, total consumption of jet fuel could be significantly saved as well.

1. Introduction

Due to the world's growing population, the requirement for energy is increasing very fast, resulting in a higher primary fossil fuels consumption and enormous emission of waste. According to the research done by Saleh *et al.* [1], the consumption of the worldwide energy had gone up over 40% these past decades. This results in a significant environmental and also economical complications. The misused energy of the fuel from exhaust engine is over 30-40% and only a small fraction of the energy, which is roughly 12-25%, are converted to useful work [2]. Therefore, in recent years, studies linked to waste heat recovery have pinched a greater awareness. Specifically, there is motivating force in the aviation industry to achieve a better propulsion system performance, principally due to the fuel cost issues and stricter future laws. Besides causing a drop in fossil fuels market, waste heat recovery will also cause a drop in greenhouse gases and thus making it easier to supply the future with a better environment.

There is a lot of debates proving that this waste heat recovery energy as a practical resource of a growing attention due to its large quantity [3]. Heat Recovery Steam Generators (HRSGs) is one type of steam-based recuperative system that have been used on various applications but are not generally fit to be applied in smaller gas turbines engine due to problems of weight as well as supply of water [4]. However, organic Rankine cycle (ORC) has become one of the favourable waste heat recovery energy technologies due to its small size and its undoubtedly great benefit for integration in the power production systems in the next few years. A Rankine cycle is a heat engine thermodynamic cycle that can convert the heat into mechanical work. The heat is supplied externally to a closed loop. An organic Rankine cycle is named for its use of organic fluid with liquid-vapor phase change occurring at a lower temperature than the phase change in water-steam, thus allowing Rankine cycle heat recovery from lower temperature. The low temperature heat is converted into useful work that can itself be converted into electricity. The researchers also provide the reasons behind the use of an ORC with an aircraft gas turbine engine sized and indicates an improvement of performance in the neighbourhood of 16% [4]. In a Rankine cycle, the system makes use of the heat to lift up the temperature and pressure of organic fluid. ORC technology has already been widely exploited in various low-grade waste thermal energy applications such as biomass energy, solar energy, geothermal energy and also waste heat energy [5]. In power plant applications and marine diesel engine, there have been several researches concerning the thermodynamic analysis and optimization method of an ORC using waste heat source [6, 7, 8] and the study has also been proven experimentally recently [9]. Nonetheless, several factors must be taken into consideration when assessing the idea on an aircraft engine. Firstly, the objective of the system for a power plant is to draw out as much as possible energy from the exhaust engine to produce electricity. However, in a jet engine, the main point is not to supply power but to produce thrust and to reduce as much as possible the fuel consumption.

As part of the normal operation of a jet engine, some chambers inside it contain air at very high pressure. Most of this air goes through the combustion chamber and exits at the nozzle to provide thrust. A small amount is bled off by an inlet or duct, and channelled by pipes and used to pressurize the cabin by supplying air to the environmental control system. Additionally, bleed air is also been used to keep critical parts of the aircraft ice-free. Bleed air is essentially loss of engine thrust as some air is taken out of it, thus this air cannot be used to provide thrust. This has led to current technologies shift in the development of the bleed-less aircraft (or low bleed use) such as engine driven electric compressor. However, there is a trade off as electric functions need higher electricity generation from engine.

An ORC waste heat recovery system basically recuperates waste heat out of the core turbofan's engine exhaust system and exploits it to generate electrical power to operate the aircraft environmental control system or other systems. It is a recovery system that searches to expand the performance of the overall engine. The waste heat in this application is used to produce electricity in order to operate the external air compressors to supply flow to the aircraft environmental control system. Therefore, less electricity generation is needed from the aircraft engine, hence less fuel consumption. This reduces the engine bleed air and consequently provides higher thrust for the aircraft engine.

Unfortunately, there was not a lot of studies concerning thermodynamic analysis of an ORC waste heat recovery system in an aircraft and their potential benefits, except for one by C A Perullo *et al.* [10]. They examined the possibility as well as the advantages of an ORC recuperative system in order to be applied to in-flight aircraft power generation. They also assumed a good ORC system design and some explanations are given to sum up the complete ORC system. In this study, the researchers used the Environmental Design Space (EDS) as the investigation tool to examine the possible advantages of an ORC system integrated to an aircraft engine [11, 12]. The technique applies an improved boiler located inside the nozzle walls of an existing aircraft engine to withdraw heat out of engine exhaust. An organic fluid named R245fa was used as the working fluid in order to examine the system because it appeared to have the highest cycle energy performance throughout various functioning pressures compared to the other three potential working fluids. However, they decided to make several modeling assumptions due to the interest of time and resources. First, in order to estimate off-design conditions, the researchers decided to work with fixed heat transfer coefficients. Secondly, it was assumed that there is no pressure losses other than across the turbine. Though the core engine fluid flow evolves in respect to the engine running status and throttle situation, they supposed a fixed core exhaust flow with a percentage of 30% that connects with the evaporator. Furthermore, before the expansion process in the nozzle, the heat is withdrawn out of the engine exhaust circulation. As a result, fuel burn savings of about 2.3% is possible depending on the real ORC-jet engine system weight. The purpose of their study is not to evaluate the design procedure in terms of the detailed heat transfer involved and the thermodynamic performance of an ORC system, hence a very short explanation of the model was given. The aim of this study is to carry out the detailed design process of the ORC system through analytical study of their thermodynamic behavior and to investigate the total fuel consumption and the system performance.

2. Analysis of thermodynamic and heat transfer

Figure 1 shows the schematic diagram of a basic ORC system integrated with the engine core of an aircraft. Composition of an ORC system normally consists of an evaporator, a turbine, a condenser, a working fluid pump and a regenerator to preheat the working fluid before entering the evaporator in order to increase thermal efficiency of the system. Basically the system is integrated to the jet engine in the middle of the low pressure turbine exit duct and the core nozzle. Figure 1 below illustrates the connection between the Brayton cycle and the ORC cycle, which is located at the evaporator. The system begins to operate at the outlet of the cold side of the regenerator (station 1) with a specific heat transfer from the exhaust jet engine and a specific evaporator outlet temperature. An inlet temperature of the working fluid to evaporator, $T_{EV,WF\_In}(i)$ is guessed and the system is analysed once through.

The ORC system diagram is shown in Figure 2, of which the thermal process can be described as follows. The evaporator used in this study is a shell-tube heat exchanger. It is the most common type of heat exchanger in many industrial applications and is suited for higher-pressure applications. The heat exchanger of this type consists of a shell with several tubes inside it. One fluid flows inside the tubes while the other flows outside the tubes within the shell to transfer heat between the two fluids. Heat is then transferred from one fluid to the other through the tube walls.

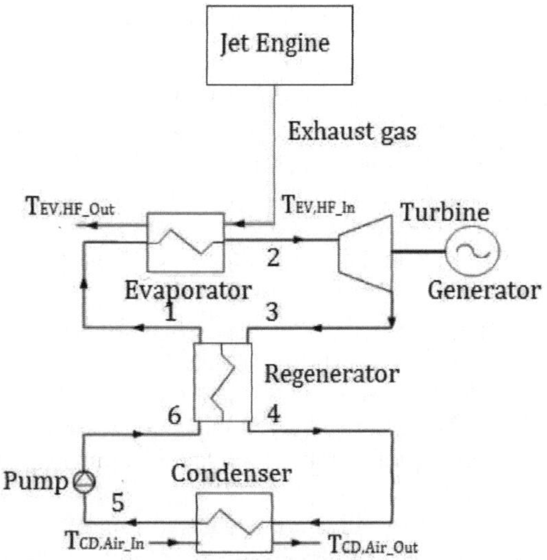

Figure 1. Schematic diagram of the ORC system with regenerator integrated to a jet engine

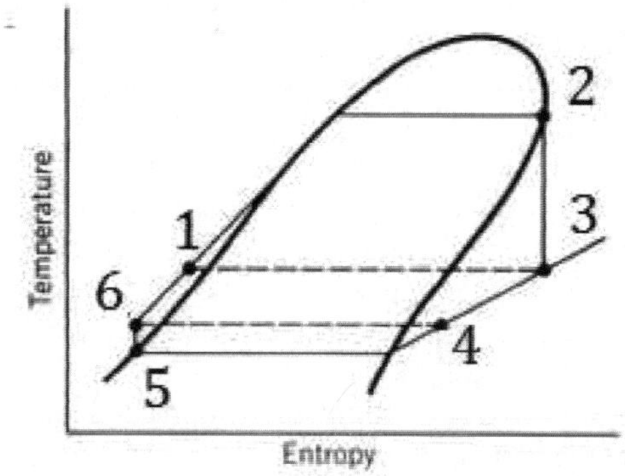

Figure 2. T-s diagram of the ORC system with regenerator

According to J Sun and W Li [8], there are three types of modelling approaches that can be applied to envisage the performance of the evaporator, which include distributed modelling, zone modelling and single node lump modelling methods. The single node lump modelling concept is normally not suitable when the phase change occurs since the method assumes that the temperature difference inside the "lump" is negligible, thus implies that constant specific heat inside is unchanged too. Meanwhile, for the second method which is zone modelling, the evaporator is divided into three zones: super-heat, two-phase and sub-cool. Subsequently, in each zone, the single node lump modelling can be applied.

However, a robust iterative algorithm is demanded in order to achieve the value of the heat transfer surface area of each zone [8]. The third choice of modelling method is distributed modelling, which considers the evaporator as small segments from the inlet of the flow to the outlet. Accordingly, the single node lump modelling, for example NTU-ε (Number of Transfer Units) and LMTD (Log-Mean Temperature Difference) methods, can be applied in each segment in order to predict the heat and mass transfer. This distributed modelling method has a better precision in contrast to the first two modelling with a robust approach. In this analytical study, the distributed modelling is applied in order to simulate the two heat exchangers, which are the condenser and the evaporator.

Figure 3 is showing three successive segments, noted i-1, i and i+1. As the precision is sufficient to suppose that heat capacity of the exhaust heat from jet engine and working fluid are roughly invariable in each discrete segment, to form the conservation equations, the single node modelling which is here, the NTU-ε method is applied. For segment i, the enthalpy increases at the working fluid side due to the heat absorbed from the exhaust heat of the jet engine.

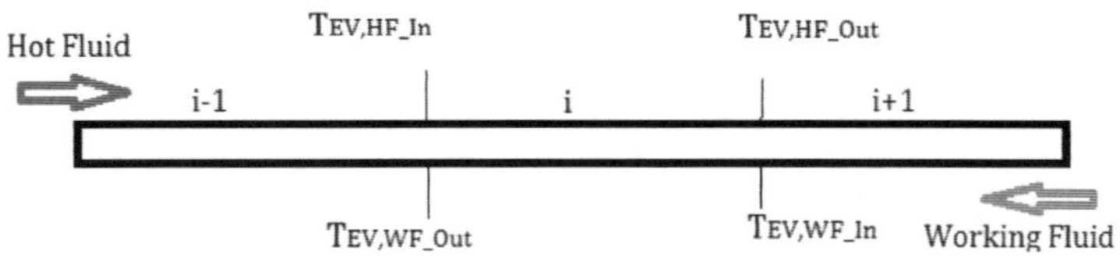

Figure 3. Discrete segments of evaporator

$$q_{EV,WF}(i) = \dot{m}_{EV,WF}\left[h_{EV,WF\_Out}(i) - h_{EV,WF\_In}(i)\right] \tag{1}$$

According to Equation 1, $q_{EV,WF}(i)$ is the heat transfer rate of the evaporator, $\dot{m}_{EV,WF}$ is the mass flow rate of the working fluid through the evaporator and $h_{EV,WF\_Out}(i)$ as well as $h_{EV,WF\_In}(i)$ are the entrance and exit enthalpies of the working fluid for the segment i. In the same manner, heat transfer rate at the hot side, $q_{EV,HF}(i)$ (exhaust heat of the jet engine) is noted as:

$$q_{EV,HF}(i) = \dot{m}_{EV,HF}\, C_{p,EV,HF}(i)\left[T_{EV,HF\_In}(i) - T_{EV,HF\_Out}(i)\right] \tag{2}$$

Here, $C_{p,EV,HF}(i)$ is the specific heat of the exhaust heat, $\dot{m}_{EV,HF}$ is the mass flow rate of the exhaust heat through the evaporator and $T_{EV,HF\_In}(i)$ as well as $T_{EV,HF\_Out}(i)$ are the entrance and exit temperatures of the exhaust heat for the segment i.

The maximum energy transferred, $q_{EV,MAX}(i)$ along the evaporator can be defined as:

$$q_{EV,MAX}(i) = C_{EV,MIN}(i)\left[T_{EV,HF\_In}(i) - T_{EV,WF\_In}(i)\right] \tag{3}$$

$$\text{with } C_{EV,MIN}(i) = MIN\left\{\dot{m}_{EV,WF}\, C_{p,EV,WF}(i), \dot{m}_{EV,HF}\, C_{p,EV,HF}(i)\right\} \tag{4}$$

Meanwhile, the effectiveness $\varepsilon_{EV}(i)$ can be written as:

$$\varepsilon_{EV}(i)=\left\{2\times\left\{\left[1+C_{r,EV}(i)+\left(1+C^2{}_{r,EV}(i)\right)^{0.5}\right]\times\frac{1+\exp\left[-NTU_{EV}(i)\times\left(1+C^2{}_{r,EV}(i)\right)^{0.5}\right]}{1-\exp\left[-NTU_{EV}(i)\times\left(1+C^2{}_{r,EV}(i)\right)^{0.5}\right]}\right\}\right\}^{-1} \quad (5)$$

$$\text{where} \quad C_{r,EV}(i)=\frac{C_{EV,MIN}(i)}{C_{EV,MAX}(i)} \quad (6)$$

$$C_{EV,MAX}(i)=MAX\left\{\dot{m}_{EV,WF}\,C_{p,EV,WF}(i),\dot{m}_{EV,HF}\,C_{p,EV,HF}(i)\right\} \quad (7)$$

The rate of heat transfer in the evaporator $NTU_{EV}(i)$ is calculated and defined as:

$$NTU_{EV}(i)=\frac{U(i)A_{EV}(i)}{C_{EV,MIN}(i)} \quad (8)$$

The total heat transfer rates of the segments Q_{EV},

$$Q_{EV}=\sum_{i=1}^{N_{EV}}q_{EV}(i) \quad (9)$$

where N_{EV} is the total of evaporator segments.

The pump power comsumption in process 5 to 6 is defined by:

$$W_{pump}=\frac{\dot{m}_{WF}\,Cp_{WF}\left(\Pi^{\frac{\gamma-1}{\gamma}}-1\right)}{\eta_{pump}} \quad (10)$$

where $\Pi=\frac{P_2}{P_1}$, η_{pump} is the efficiency of the pump and γ is the specific heat ratio of the working fluid.

When the working fluid exits the pump as a high pressure saturated or superheat fluid and flows through the turbine, the power is produced and exits from the turbine as low pressure superheat fluid. Then the turbine power output is obtained from:

$$W_{\exp}=\dot{m}_{WF}\,Cp_{WF}\,\eta_{\exp}T_{\exp\_in}\left(1-\Pi^{\frac{1-\gamma}{\gamma}}\right) \quad (11)$$

where is $T_{\exp\_in}$ the turbine inlet temperature and $\eta_{\exp}$ is the efficiency of the turbine.

The net power output of the ORC system is:

$$W_{net} = W_{\exp} - W_{pump} \tag{12}$$

The thermal efficiency of the ORC system is defined as:

$$\eta_{net} = \frac{W_{\exp} - W_{pump}}{Q_{EV}} \tag{13}$$

Consequently, the total of conventional fuel (kerosene) that could be saved is:

$$\dot{m}_{fuel} = \frac{Q_{EV}}{HV_{fuel}} \tag{14}$$

where HV_{fuel} is the heating value of kerosene, HV_{fuel} = 43 000 kJ/kg.

3. Results and discussions

This section walks through the potential benefits of ORC system, which is integrated to a CFM56-7B turbofan engine on an aircraft size of 737-800. The working fluid chosen is R245fa. Design parameters are listed in Table 1 below based on the study done by other researchers [10]. The specific heat of the working fluid R245fa is assumed to be 1.36 kJ/kg.K in the range of the design temperature value whereas the one for kerosene is 2.01 kJ/kg.K. A calculation is run according to the equations defined in Section 2.

Table 1. Main parameter of the ORC system for turbofan engine [10]

Property	Unit	Value
Mass flow rate of R245fa	kg/s	1 – 5.4
Mass flow rate of hot gas	kg/s	0.1 – 4.4
Exhaust heat temperature	K	700 - 1140
Inlet temperature of R245fa	K	100 - 540
Inlet temperature of turbine	K	100 - 500
Turbine inlet pressure	MPa	1.2
Turbine outlet pressure	MPa	0.053
Pump inlet pressure	MPa	0.045
Pump outlet pressure	MPa	1.36
Turbine efficiency	-	0.776
Pump efficiency	-	0.577

The exhaust heat temperature from the jet engine will vary with the aircraft engine's performance. The variations of the working fluid mass flow rate with the exhaust heat temperature are illustrated in Figure 4. It shows that a higher exhaust engine temperature of the jet engine, the working fluid mass flow rate will also increase due to the increment in the total heat load of the ORC system. This can be justified with the increasing demands of working fluid that results in more electricity being produced due to a higher available heat resulting from a higher exhaust heat temperature.

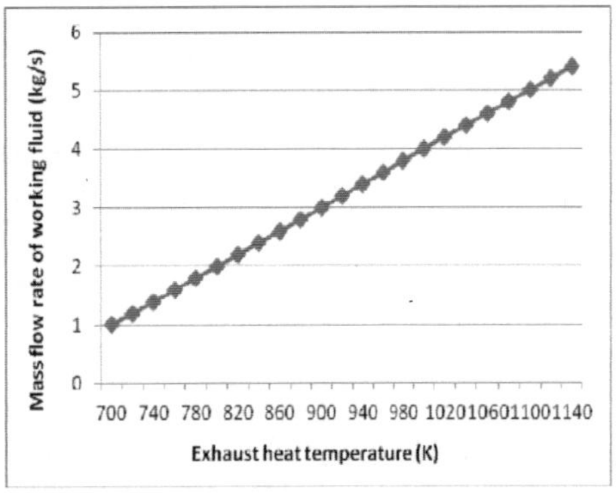

Figure 4. Variations of working fluid mass flow rate with the exhaust heat temperature

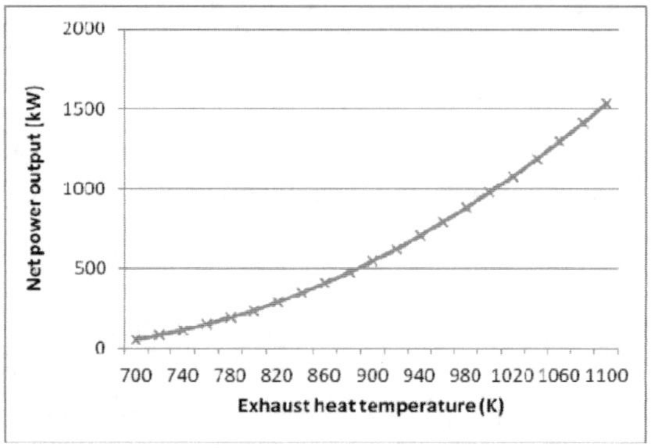

Figure 5. Variations of net power output with the exhaust heat temperature

In addition, the results show that both net power output (Figure 5) and system efficiency (Figure 6) are improved with the increment of exhaust heat temperature, which indicates that the working fluid mass flow rate and the heat source temperature play a major role in optimizing the power generation and the system's thermal performance. This is because an increase in evaporation temperature and working fluid mass flow rate would eventually simulate a better energy load. This implies an increase in the turbine power output and subsequently the overall power output. Thus the ORC heat recovery efficiency is increased too due to the relation between the heat load and the exhaust heat temperature to the engine load capacity. However, a much less mass flow rate of the working fluid is demanded at the beginning, thus a slightly decrease of system thermal efficiency at temperature around 720-740K. The reason is that, in spite of the fact that more heat are used to generate more power with growing exhaust waste heat temperature, there could be a situation where the rising electric power is counter-balanced by the working fluid pump power consumption as a means to make use of the heat. As a result, the system performance could decrease a little at a certain point.

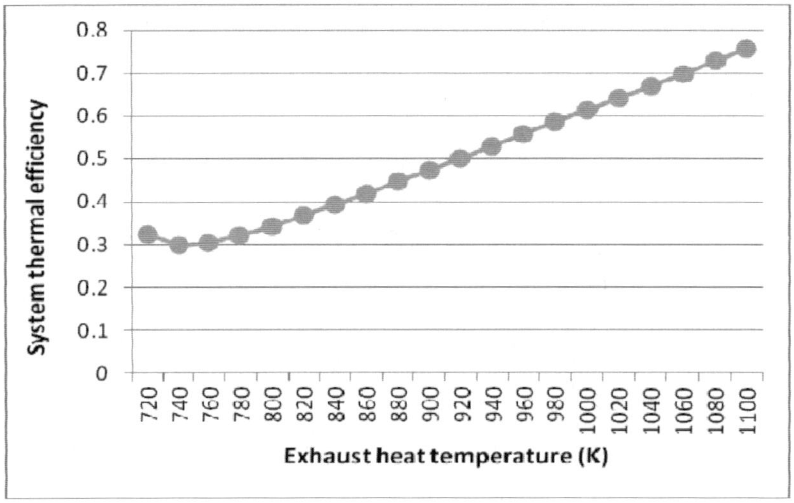

Figure 6. System thermal efficiency vs exhaust heat temperature

In order to show how the ORC impacts the Brayton cycle, according to this analytical method, the overall waste heat extracted out of the exhaust heat of the jet engine in evaporator is 1244.25 kW at

IOP Conf. Series: Materials Science and Engineering **152** (2016) 012011 doi:10.1088/1757-899X/152/1/012011

exhaust heat temperature of 843 K and mass flow rate of working fluid of 2.4 kg/s of the ORC system. Suppose that the cycle is run for 300 days a year, the total extracted waste heat energy annually will be 3.225 x 10^{10} kJ/year. Eventually, it could conserve the conventional fuel (kerosene) usage of 7.50 x 10^5 kg/year.

4. Conclusion and the way forward

Starting from a comprehensive review of the design of ORC system integrated to an aircraft's engine and the mathematical model as well as the heat transfer process involved, this paper has presented a rigorous derivation of the net power output and the energy efficiency gained by ORC heat recovery technology from basic principles. This study also indicates the potential of reducing fuel consumption by integrating the ORC system to turbofan engine. Such results will be very useful in future to develop in details the numerical simulation of ORC system, and consequently to study the system performance with different variations of parameters and to determine the irreversibility losses through calculation of energy of each ORC component to achieve an optimized system performance and eventually higher thrust power.

References

[1] Saleh B, Koglbauer G, Wendland M and Fischer J 2007 *Energy* **32** 1210-21
[2] Hasanuzzaman M, Rahim N A, Saidur R and Kazi S N 2011 *Energy* **36** 233-40
[3] Ziviani D, Beyene A and Venturini M 2014 *Appl. Energy* **121** 79-95
[4] Bronicki L Y and Schochet D N 2005 *Proc. ASME Turbo Expo* 79-86
[5] Tchanche B F, Lambrinos G, Frangoudakis A and Papadakis G 2011 *Renew, Sustain. Energy Rev.* **15** 3963-79
[6] Song J, Song Y and Gu C W 2015 *Energy* **82** 976-85
[7] Yang M H and Yeh R H 2015 *Appl. Energy* **149** 1-12
[8] Sun J and Li W 2011 *Appl. Thermal Eng.* **31** 2032-41
[9] Pu W, Yue C, Han D, He W, Liu X, Zhang Q and Chen Y 2016 *Applied Thermal Engineering* **94** 221-7
[10] Perullo C A, Mavris D N and Fonseca E 2013 *Proc. of ASME Turbo Expo: Turbine Technical Conf. and Exposition*
[11] Kirby M and Mavris D 2008 *26th Int. Cong. of the Aeronautical Sciences*
[12] Schutte J, Tai J and Mavris D 2012 *48th AIAA/ASME/SAE/ASEE Joint Propulsion Conf. & Exhibit*

IOP Conf. Series: Materials Science and Engineering **152** (2016) 012012 doi:10.1088/1757-899X/152/1/012012

Effect of multiple jet impingement plate configurations on Reynolds Number in a pipe

M S Ramli[1], A R Abu Talib[1, 2*], M Y Harmin[1,2] and H R Mohd Saiah[1]

[1] Department of Aerospace Engineering, Universiti Putra Malaysia, Malaysia.
[2] Aerospace Manufacturing Research Centre, Universiti Putra Malaysia, Malaysia.

* abdrahim@upm.edu.my

Abstract. Experimental investigations were carried out to study the effect of varying multiple jet impingement plate configurations on Reynolds Number (*Re*) in a closed conduit. Air was considered as the working fluid. There were six multiple impingement plates used for this experiment where each plate has a different hole configurations that include the hole diameter, hole orientation, pitch in x-direction and pitch in y-direction. Four sets of orifice plate with diameter of 0.02, 0.03, 0.04, and 0.05 m were used to get the mass flow rate in the pipe. Air was sucked through the impingement plate for five different settings of suction fan with an interval of 10Hz from 10 to 50Hz. By taking the data for constant suction fan setting at 50Hz, it was found that the impingement hole orientation for both in-line and staggered does not give any effect on the *Re* obtained since the differences was considerably small and fell within the accepted errors. Meanwhile, impingement hole diameter was found to be directly proportional with the *Re* obtained. It was also found that the different pitch in multiple hole impingement plate resulted in changes of *Re*. The results show that the *Re* was decreasing with higher pitch. The uncertainty analyses for the *Re* were also presented.

1. Introduction

In the field of pipes flow measurement, Reynolds number (*Re*) is one of the important and critical parameters to be determined. The effect of *Re* as one of the important parameters in jet impingement heat transfer field has been widely discussed and can be found in numerous literatures and books. The effect of *Re* on the development of flow structures in round jets has been explored by Dimotakis [1] who came out with the proposed critical value of *Re* = 10000, beyond which flow properties such as mixing become a weak dependence of *Re*.

Kwon and Seo [2] investigated the effect of *Re* on the behaviour of round jets using particle image velocimetry (PIV). Applying a range of low *Re* = 177–5142, they found out that the length of the zone of flow development decreased with increasing *Re* centerline velocity decayed more rapidly. Lee *et al.* [3] investigated the effects of jet-to-target plate distance and *Re* on the heat transfer from an array of jets impinging on a flat plate. They found out that the local and spatially averaged Nusselt numbers (*Nu*) showed strong dependence on the impingement jet *Re* for all situations examined. Goldstein *et al.* [4] investigated an injection of air perpendicular to the target surface. They dealt with high range of *Re* (60000 – 100000). They concluded that even with high jet *Re*, the dependency of *Nu* over jet-to-target distance did not change. They also gave the correlation of heat transfer as $Nu/Re^{0.76}$ for all jet-to-target distance values. Meanwhile, Goldstein and Seol [5] concluded that the dependence of *Nu* on the jet *Re* could be approximated by $Nu/Re^{0.7}$. Furthermore, Mohd Saiah [6] investigated single jet impingement heat transfer with *Re* ranging from 20000 to 30000. It was found that the optimal jet impingement *Re* for 5 mm hole diameter was at 20567 and any increase in *Re* would result in overdeveloped potential core, which lowered the heat transfer process [6]. The optimal jet impingement *Re* for 10 mm hole diameter was found to be at 28567 and lower *Re* would result in an underdeveloped potential core.

All the aforementioned literatures has proven that the effect of *Re* is very important to fluid flow experiments. Inspired by the work of Mohd Saiah [6], which involved a single hole jet impingement heat transfer, this study will investigate the effect of multiple hole impingement configurations on the pipe *Re*. There are three main configurations for impingement plate, which are the hole diameter, pitch and orientation. The effect of these three configurations will be investigated and examined.

2. Methodology

Re for internal flow in a circular pipe can be expressed as:

$$\mathrm{Re} = \frac{VD}{\nu} \tag{1}$$

where V = mean velocity in m/s, D = characteristic length of the geometry (diameter in this case) in m and $\nu = \mu/\rho$ = kinematic viscosity of the fluid in m^2/s.

There are three types of flow for Re, which are laminar, transition and turbulent flow. According to Yunus and Robert [7], the flow in a circular pipe is laminar for Re < 2300, turbulent for Re > 4000, and transitional in between.

$\mathrm{Re} < 2300$	laminar flow
$2300 \le \mathrm{Re} \le 4000$	transitional flow
$\mathrm{Re} > 4000$	turbulent flow

The design of the experimental rig was based on Son *et al.* [8] and Mohd Saiah [6]. Although the objective of this study is not similar with the aforementioned researchers, design of their experimental set-up can still be used since it can determine the flow characteristics such as fluid mass flow rate, fluid velocity and *Re*.

The British standard BS1042 [9] was chosen as a reference document since it provides the detailed geometry parameters for fabrication, instalment requirements and working conditions for primary devices. The conduit used in this study is a straight cylindrical passage of constant cross-sectional area and has a length that ensures the upstream fluid flow approaching the orifice plate is fully developed profile and free of swirl. This length is also known as hydrodynamic entry length, L_h that is commonly taken to be the distance from the pipe inlet to the point at which the velocity boundary layer merges at the centreline as shown in Figure 1 [7].

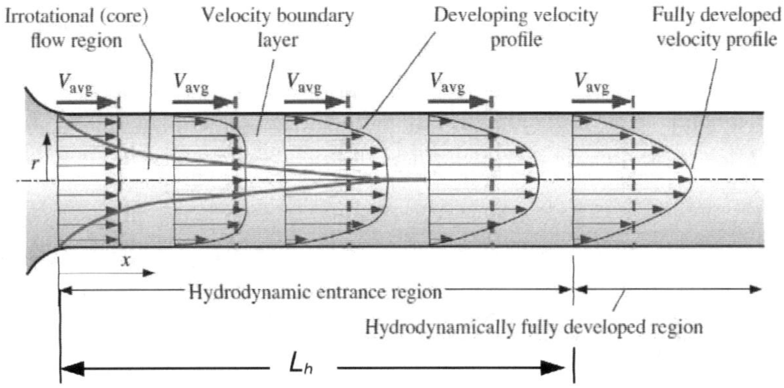

Figure 1. The development of the velocity boundary layer in a pipe [7]

The hydrodynamic entry length is different for both laminar and turbulent flow. According to Kays and Crawford [10], Shah and Bhatti [11] and also Zhi-qing [12], the approximate length of L_h for laminar and turbulent flow can be approximated as $115D_l$ and $10D_l$, respectively. However, the pipes used in practice are commonly several times the length of the entrance region, so that the flow through the pipes is fully developed. In this study, the recommended length for upstream and downstream sections based on BS1042 were $20D_I$ and $6D_I$, respectively [9], where the D_I for the pipe was equal to 0.1 m. Figure 2 shows a schematic diagram of the experimental rig set-up.

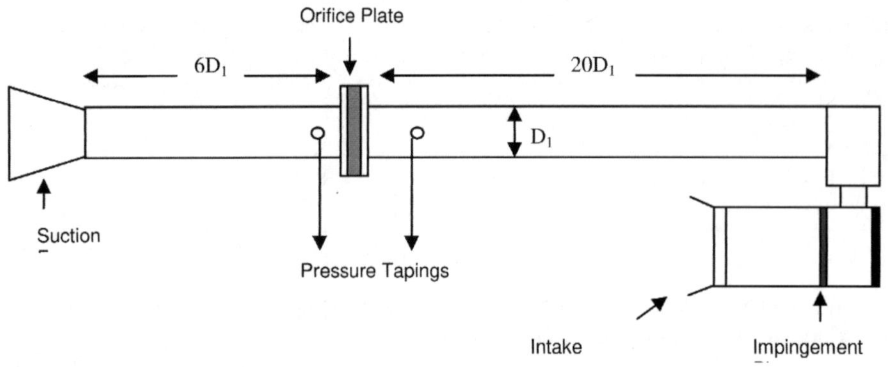

Figure 2. Schematic diagram of the experimental rig set-up

2.1. Intake nozzle
In this study, the chosen type for intake nozzle was ISA1932. The intake nozzle, as the name implies, directs the fluid into the test rig in an appropriate manner. The intake nozzle was situated at the most front section of the test rig. The earlier section of the test rig was a rectangular shape conduit, hence the cross-sectional diameter of the conduit at this section will be replaced by the hydraulic diameter, D_H that Holman [13] defined as:

$$D_H = \frac{4A}{P} \tag{2}$$

$$D_H = \frac{4(ab)}{2a+2b} \tag{3}$$

where in Equation (2), A is the cross-sectional area of the conduit and P is the wetted perimeter of the conduit. Equation (2) can also be rearranged to Equation (3), in which a and b are both the sides of a rectangular. The inner cross-sectional area for the intake nozzle was 100 mm x 100 mm.

2.2. Impingement plates
There are six multiple holes impingement plates that were used in this study. Each impingement plate had a different configuration in terms of hole diameter, hole orientation as well as pitch in *x*- and *y*-direction as listed in Table 1. Figure 3a and Figure 3b illustrate the overview of the impingement plate. Note that the pitch in x- and y-direction is the center-to-center spacing of holes.

2.3. Orifice plates
Orifice plate can be defined in various ways [9, 14]. The main important characteristic of this primary device is that it provides an abrupt change in the cross-sectional area of the conduit and thus results in a change of energy. Orifice plate had become the most preferable flow meter due to its versatility, cost effectiveness and simple construction.

Table 1. Six multiple holes impingement plates with different configurations

Impingement Plates	Hole Diameter (mm)	Hole orientation	Pitch *x*-direction (mm)	Pitch *y*-direction (mm)	Number of holes
A	7.0	in-line	21.0	26.6	15
B	7.0	Staggered	21.0	26.6	17
C	5.0	in-line	15.0	20.0	25
D	5.0	staggered	15.0	20.0	23
E	5.0	in-line	20.0	25.0	15
F	5.0	staggered	20.0	25.0	17

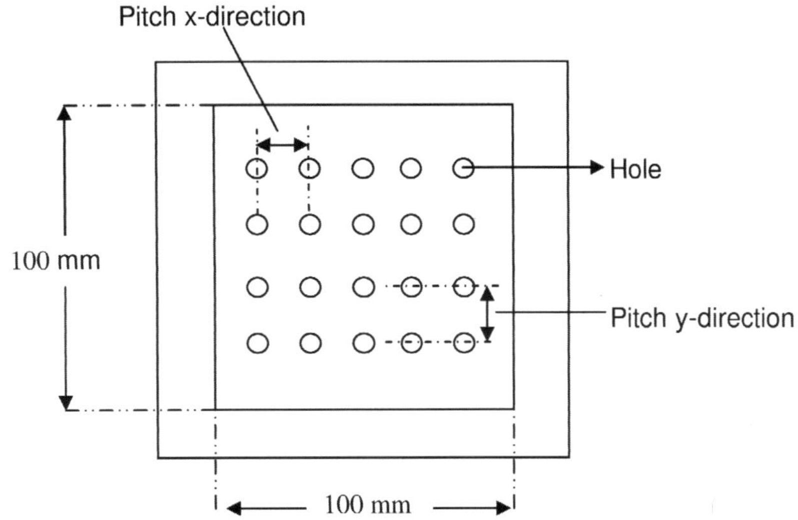

(a)

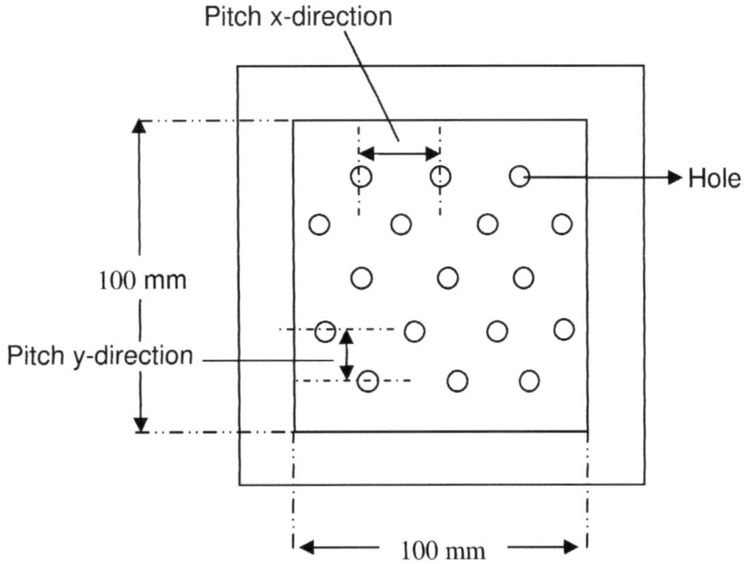

(b)

Figure 3. (a) Multiple holes impingement plate with in-line orientation, (b) Multiple holes impingement plate with staggered orientation

IOP Conf. Series: Materials Science and Engineering **152** (2016) 012012 doi:10.1088/1757-899X/152/1/012012

There is a total of four orifice plates used in this study, which each of them has a diameter of 0.02, 0.03, 0.04 and 0.05 m, respectively. Pressure measurements on orifice plate can be divided into three different approaches [9]. For this study, one approach was selected due to its simplicity, which was the $D_1/2$ and D_1 pressure taps, where D_1 is the pipe's inner diameter and D_2 is the orifice diameter as shown in Figure 4.

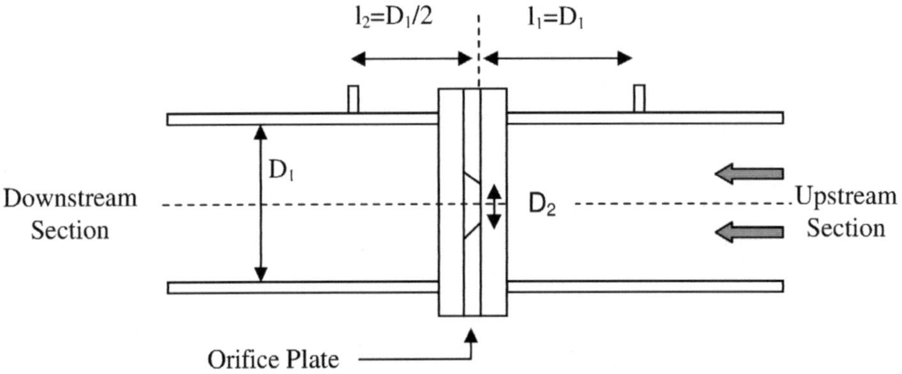

Figure 4. Orifice plate pressure tapings

The principle of an orifice plate is based on the measurement of static pressure difference between the upstream and downstream sides. Based on ISO5167 [15] and ASME MFC-3M [16], the mass flow rate, q_m in pipe by using an orifice plate can be determined with the following equation:

$$q_m = CA_2\left(\frac{2\Delta p.\rho_1}{1-\beta^4}\right)^{1/2} \tag{4}$$

where C is the discharge coefficient, A_2 is the orifice area, ρ_1 is the density at upstream side and β is the diameter ratio between pipe and orifice which is D_2/D_1. Here the differential pressure, Δp can be measured by means of a digital manometer HHP 2021. Note that C is dependant of the diameter ratio, β and the pipe's Re. Equation (5) shows the governing relationship for discharge coefficient, C [9].

$$C = 0.5959 + 0.0312\beta^{2.1} - 0.1840\beta^8 + 0.0029\beta^{2.5}\left[\frac{10^6}{\text{Re}}\right]^{0.75} + 0.09L_1\beta^4\left(1-\beta^4\right)^{-1} - 0.0337L'_2\beta^3 \tag{5}$$

in which L_1 is a ratio of l_1/D_1, L_2 is a ratio of l'_2/D_1 with l'_2 is the distance of the downstream pressure tap from the downstream face of the orifice plate. Whenever $L_1 \geq \frac{0.0390}{0.0900}$, 0.0390 will be used for the term $\beta^4\left(1-\beta^4\right)^{-1}$. In this study, L_1 equals to 1, therefore the orifice discharge coefficient yields:

$$C = 0.5959 + 0.0312\beta^{2.1} - 0.1840\beta^8 + 0.0029\beta^{2.5}\left[\frac{10^6}{\text{Re}}\right]^{0.75} + 0.00351L_1 - 0.0337L'_2\beta^3 \tag{6}$$

The calculation of C is an iterative process since Re is needed to calculate C, but the velocity in the pipe needed for Re cannot be determined until C is known. However, this iterative process converges quite rapidly in just one or two iterations.

2.4. Uncertainty analysis

Predictions of errors are vital for researchers, especially for those who are dealing with experimental works. Experimentation is not just about taking the data. Each instrument has its own accuracy, hence the prediction of uncertainties is dominated by the accuracy errors that are also known as the random errors. The uncertainty analysis will provide the information for experimenters in knowing the scatter and how far the experimental data may deviate from the true value.

For this study, uncertainty analyses for *Re* obtained have been done by using Taylor Series Method (TSM) for propagation of uncertainties. Equation (7) is given by Hugh and Glenn [17] that describes the propagation of overall uncertainties in measured variables into the overall uncertainty of the result.

$$U_r = \left[\sum_{i=1}^{J} \left(\frac{\partial r}{\partial X_i} \right)^2 U_i^2 \right]^{1/2} \tag{7}$$

where r is the function of two or more variables and U_i is the uncertainty for variable X_i.

Referring to Equation (8) for *Re*, there are only two variables that have been considered as sources of error, which are air velocity, V and pipe diameter, D while air density, ρ and air viscosity, μ are assumed to be constant.

$$\mathrm{Re} = \frac{\rho . V . D_1}{\mu} \tag{8}$$

The air velocity, V can be defined from the mass flow rate, q_m as shown in Equation (9).

$$V = \frac{q_m}{\rho . A_1} \tag{9}$$

To get the uncertainty for V, the uncertainty for q_m needs to be determined first. From the Equation (4), by using the general uncertainty analysis in Equation (7), the uncertainty for q_m is determined as:

$$U_{q_m} = \left(\frac{\partial q_m}{\partial A_2} \right)^2 U_{A_2}^{\ 2} + \left(\frac{\partial q_m}{\partial \Delta p} \right)^2 U_{\Delta p}^{\ 2} + \left(\frac{\partial q_m}{\partial \beta} \right)^2 U_{\beta}^{\ 2} \tag{10}$$

By taking the derivatives of q_m with respect to the variables, then substituting it into Equation (10) and dividing by q_m^2 yields:

$$\left(\frac{U_{q_m}}{q_m} \right)^2 = (1)^2 \left(\frac{U_{A_2}}{A_2} \right)^2 + \left(\frac{1}{2} \right)^2 \left(\frac{U_{\Delta p}}{\Delta p} \right)^2 + (2)^2 \left(\frac{U_{\beta}}{\beta} \right)^2 \tag{11}$$

Then, by applying the Equation (7) again to the Equation (9), the uncertainty for V was determined as:

$$\left(\frac{U_V}{V} \right)^2 = (1)^2 \left(\frac{U_{q_m}}{q_m} \right)^2 + (1)^2 \left(\frac{U_{A_1}}{A_1} \right)^2 \tag{12}$$

Thus, the uncertainty for *Re* was determined as:

$$\frac{U_{\mathrm{Re}}}{\mathrm{Re}} = \sqrt{ \left(\frac{U_V}{V} \right)^2 + \left(\frac{U_{D_1}}{D_1} \right)^2 } \tag{13}$$

Table 2 shows the accuracy of the measuring devices that have been used in this study.

Table 2. Measuring devices used with their accuracy

Measuring Devices	Accuracy
Digital Manometer HHP 2021	0.1% reading +0.1 % full scale +1 digit (full scale of 13 kPa)
Digital Caliper 500-752-10	± 0.001" / 0.01 mm
Ruler	± 0.05 cm

3. Results and discussions

Table 3 shows the results of *Re* with their uncertainties for each setting of orifice diameter and suction fan without impingement plate being attached. Meanwhile, Table 4 to Table 9 show the results of *Re* with their uncertainties for each orifice diameter and suction fan setting for every impingement plate. Since pressure gives the most contribution to uncertainties values, the uncertainties percentage values obtained for *Re* were considerably high at smaller pressure difference, which was up to ±115%. This was due to the fact that the range of the digital manometer used was quite high for this experiment and it only has better accuracy for larger pressure difference measurements. For that reason, only *Re* with uncertainties below 30% were selected and *Re* with uncertainties more than 30% were neglected.

Table 3. Reynolds number with uncertainties for each orifice diameter and suction fan setting without impingement plate

	Orifice Diameter (m)			
	0.02	0.03	0.04	0.05
Suction Fan Setting (Hz)	Reynolds Number, *Re*			
10	918 ± 57.57 %	1760 ± 115.06 %	null	null
20	1710 ± 16.55 %	3290 ± 23.1 %	4620 ± 38.41 %	4350 ± 115.06 %
30	2570 ± 7.41 %	5070 ± 9.76 %	7010 ± 16.55 %	8540 ± 28.84 %
40	3400 ± 4.44 %	6840 ± 5.5 %	9870 ± 8.41 %	11200 ± 16.55 %
50	4210 ± 3.15 %	8500 ± 3.76 %	12100 ± 5.74 %	14000 ± 10.62 %

Table 4. Reynolds number with uncertainties for each orifice diameter and suction fan setting for impingement plate A

	Orifice Diameter (m)			
	0.02	0.03	0.04	0.05
Suction Fan Setting (Hz)	Reynolds Number, *Re*			
10	652 ± 115.06 %	null	null	null
20	1450 ± 23.1 %	2550 ± 38.42 %	2700 ± 115.06 %	null
30	2320 ± 9.03 %	3880 ± 16.55 %	4620 ± 38.42%	4350 ± 115.06 %
40	3080 ± 5.29 %	5270 ± 9.03 %	6500 ± 19.28%	7420 ± 38.42 %
50	3800 ± 3.69 %	6530 ± 6.01 %	7940 ± 12.92 %	8540 ± 28.84%

Table 5. Reynolds number with uncertainties for each orifice diameter and suction fan setting for impingement plate B

	Orifice Diameter (m)			
	0.02	0.03	0.04	0.05
Suction Fan Setting (Hz)	Reynolds Number, *Re*			
10	652 ± 115.06 %	1760 ± 115.06 %	null	null
20	1580 ± 19.28 %	2550 ± 38.41 %	2700 ± 115.06 %	null
30	2410 ± 8.41 %	4150 ± 14.51 %	5320 ± 28.84 %	4350 ± 115.06 %
40	3210 ± 4.91 %	5470 ± 8.41 %	7010 ± 16.55%	7420 ± 38.41 %
50	3960 ± 3.46 %	6840 ± 5.5 %	8760 ± 10.62 %	9530 ± 23.10 %

Table 6. Reynolds number with uncertainties for each orifice diameter and suction fan setting for impingement plate C

	Orifice Diameter (m)			
	0.02	0.03	0.04	0.05
Suction Fan Setting (Hz)	Reynolds Number, *Re*			
10	652 ± 115.06 %	1760 ± 115.06%	null	null
20	1450 ± 23.10 %	2550 ± 38.41 %	2700 ± 115.06 %	null
30	2320 ± 9.03 %	3600 ± 19.28 %	4620 ± 38.41 %	4350 ± 115.06 %
40	3020 ± 5.51 %	4850 ± 10.62 %	5940 ± 23.10%	6090 ± 57.57%
50	3750 ± 3.78 %	6200 ± 6.62 %	7490 ± 14.51 %	7420 ± 38.41 %

Table 7. Reynolds number with uncertainties for each orifice diameter and suction fan setting for impingement plate D

	Orifice Diameter (m)			
	0.02	0.03	0.04	0.05
Suction Fan Setting (Hz)	Reynolds Number, *Re*			
10	652 ± 115.06 %	null	null	null
20	1450 ± 23.1 %	2090 ± 57.57 %	2700 ± 115.06 %	null
30	2230 ± 9.76 %	3600 ± 19.28 %	3790 ± 57.57 %	4350 ± 1.53 %
40	3020 ± 5.51 %	4850 ± 10.62 %	5320 ± 28.84 %	6090 ± 57.57 %
50	3750 ± 3.78 %	6020 ± 6.99 %	7010 ± 16.55 %	7420 ± 38.41 %

Table 8. Reynolds number with uncertainties for each orifice diameter and suction fan setting for impingement plate E

	Orifice Diameter (m)			
	0.02	0.03	0.04	0.05
Suction Fan Setting (Hz)	Reynolds Number, *Re*			
10	652 ± 115.06 %	null	null	null
20	1290 ± 28.84 %	1490 ± 115.06 %	null	null
30	2040 ± 11.6 %	2940 ± 28.84 %	2700 ± 115.06 %	null
40	2730 ± 6.63 %	3880 ± 16.55 %	3790 ± 57.57 %	4350 ± 115.06 %
50	3400 ± 4.44 %	4630 ± 11.66 %	5320 ± 28.84 %	4350 ± 115.06 %

Table 9. Reynolds number with uncertainties for each orifice diameter and suction fan setting for impingement plate F

	Orifice Diameter (m)			
	0.02	0.03	0.04	0.05
Suction Fan Setting (Hz)	Reynolds Number, *Re*			
10	652 ± 115.06 %	null	null	null
20	1290 ± 28.84 %	2090 ± 57.57 %	2700 ± 115.06 %	null
30	2140 ± 10.62 %	2940 ± 28.84 %	3790 ± 57.57 %	4350 ± 115.06 %
40	2800 ± 6.3 %	4150 ± 14.51 %	4620 ± 38.41 %	4350 ± 115.06 %
50	3520 ± 4.19 %	5070 ± 9.76 %	5940 ± 23.10 %	6090 ± 57.57 %

Figure 5 shows the *Re* at 50 Hz without impingement plate being attached. These values of *Re* can be used as a benchmark to other *Re* values obtained with impingement plates.

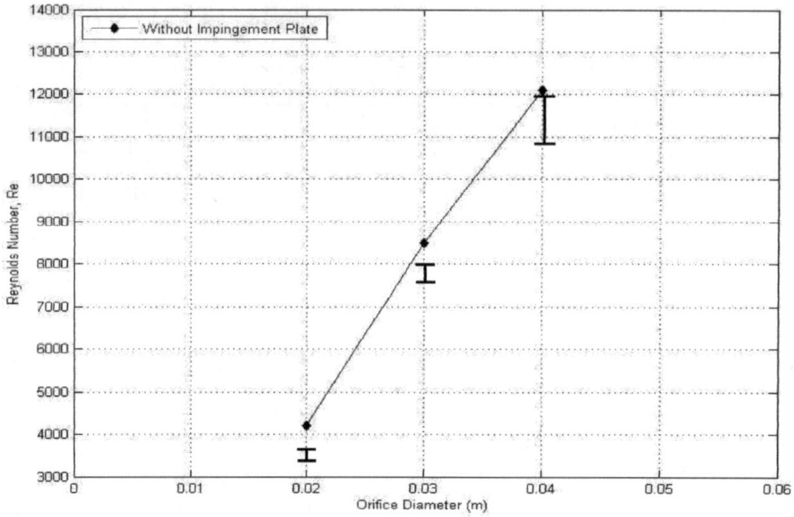

Figure 5. *Re* without impingement plate being attached

3.1. Effect of impingement plate hole orientations

Figure 6 (a), (b) and (c) show the comparison of *Re* values for impingement plates that have a similar hole diameter and pitch but different holes orientation. All three graphs obviously give a quite similar trend. At suction fan of 50 Hz, the *Re* for impingement plate A and B, C and D, and E and F increased similarly with an increase of orifice diameter regardless of the hole orientation. Though there were some slight differences between *Re* values at orifice diameter 0.04 m, they were considerably small (not more than 10%) and fell within the uncertainties. The small differences were due to higher holes number in the impingement plate B, which resulted in higher open area and fluid mass flow rate.

3.2. Effect of impingement plate hole pitch

Figure 7 (a) and (b) show the comparison of *Re* values for impingement plates that have similar hole diameter and orientation but different pitch. All two graphs obviously give a quite similar trend. At suction fan of 50 Hz, *Re* for impingement plate C and E, and D and F increased in different increments with an increase of orifice diameter. This shows that different pitch on the impingement plates resulted in different *Re*. This was due to the fact that by having higher hole pitch, it will reduce the number of hole in the fixed area of impingement plate. This will result in a lower open area in the impingement plate and thus the fluid mass flow rate through the impingement plate will be reduced.

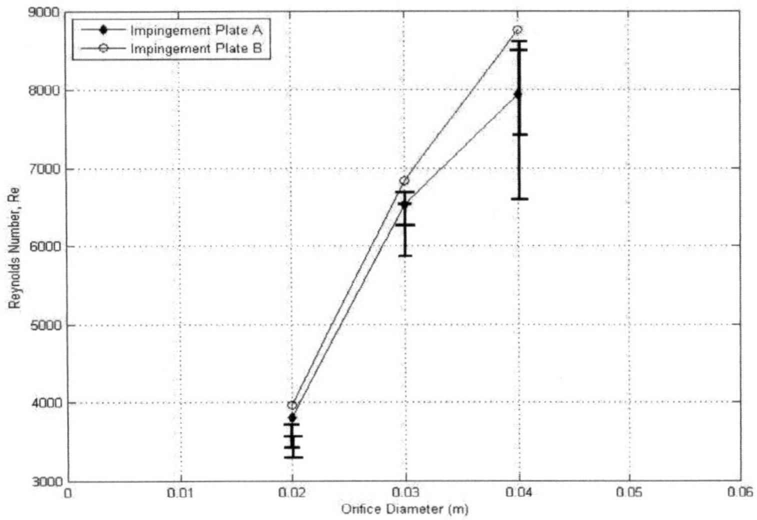

Figure 6. (a) Comparison of *Re* between in-line and staggered impingement plates A and B for 7mm hole diameter

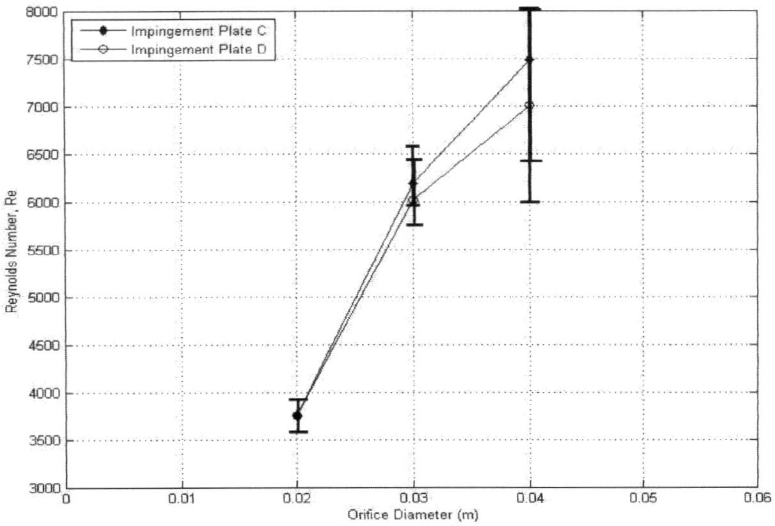

Figure 6. (b) Comparison of *Re* between in-line and staggered impingement plates C and D for 5mm hole diameter with 15mm pitch-x and 20mm pitch-y

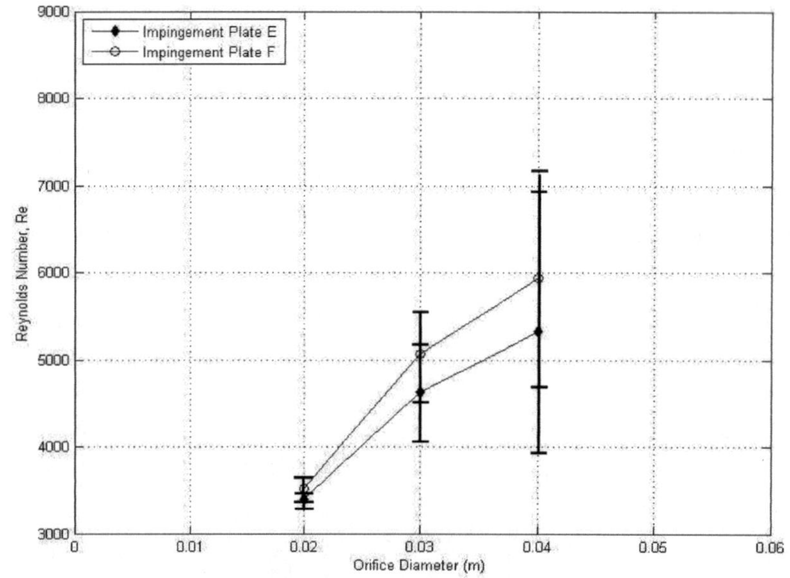

Figure 6. (c) Comparison of *Re* between in-line and staggered impingement plates E and F for 5mm hole diameter with 20mm pitch-x and 25mm pitch-y

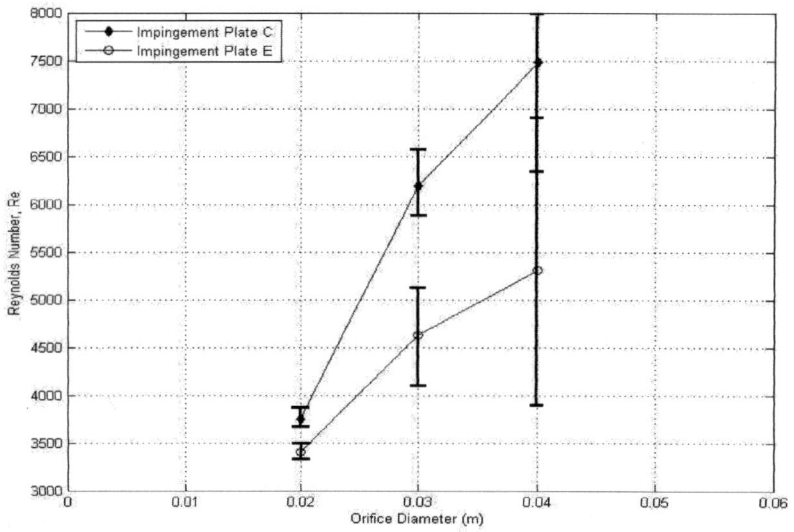

Figure 7. (a) Comparison of *Re* between 15mm pitch-x, 20mm pitch-y and 20mm pitch-x, 25mm pitch-y impingement plates C and E for in-line orientation

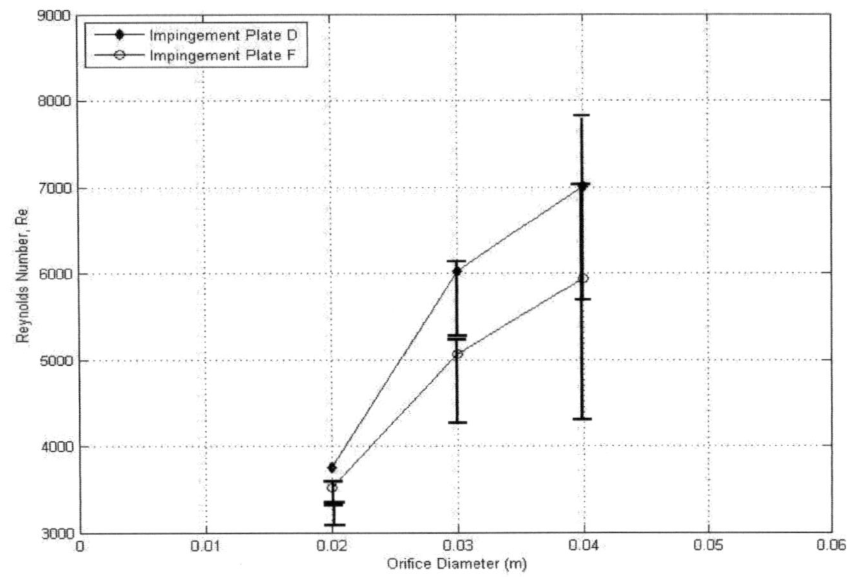

Figure 7. (b) Comparison of *Re* between 15mm pitch-x, 20mm pitch-y and 20mm pitch-x, 25mm pitch-y impingement plates D and F for staggered orientation

3.3. Effect of impingement plate hole diameter
Figure 8 (a) and (b) show the comparison of *Re* values for impingement plates that have similar hole pitch and orientation but different diameter. Though the hole pitch between impingement plate A and E, and B and F were not exactly the same, their differences were small and would not give significant effect to the overall *Re* since the numbers of hole were the same. Thus the effect of hole diameter can be investigated by using these impingement plates. All two graphs obviously show a rather similar trend. At suction fan of 50Hz, *Re* for impingement plate A and E, and B and F increased in different increments with increase of orifice diameter. This shows different hole diameter on the impingement plates resulted in different *Re*. This was due to the fact that by having larger hole diameter will result in a higher open area of the impingement plate and thus higher fluid mass flow rate through the hole. Consequently, higher *Re* will be obtained.

4. Conclusion
The overall goal of this experimental study was to examine the effect of having different configuration of multiple impingement plate on *Re* in a pipe. The configurations include the holes diameter, hole pitch as well as hole orientation that were in-line and staggered. Based on obtained data for constant suction fan setting at 50Hz and for orifice diameters of 0.02m, 0.03m and 0.04m, it was found that for the fixed area of impingement plates, orientation of the impingement plate does not give different *Re* as long as the number of impingement hole remains the same. Meanwhile, *Re* was found to be lower when the holes pitch is higher. It shows that for fixed area of impingement plate, the holes pitch was inversely proportional with the *Re*. Also found in this study was that the impingement hole diameter gives a significant effect on *Re* in a pipe. There was a directly proportional relationship between holes diameter with *Re* obtained. The larger the holes diameter, the higher the *Re* would be.

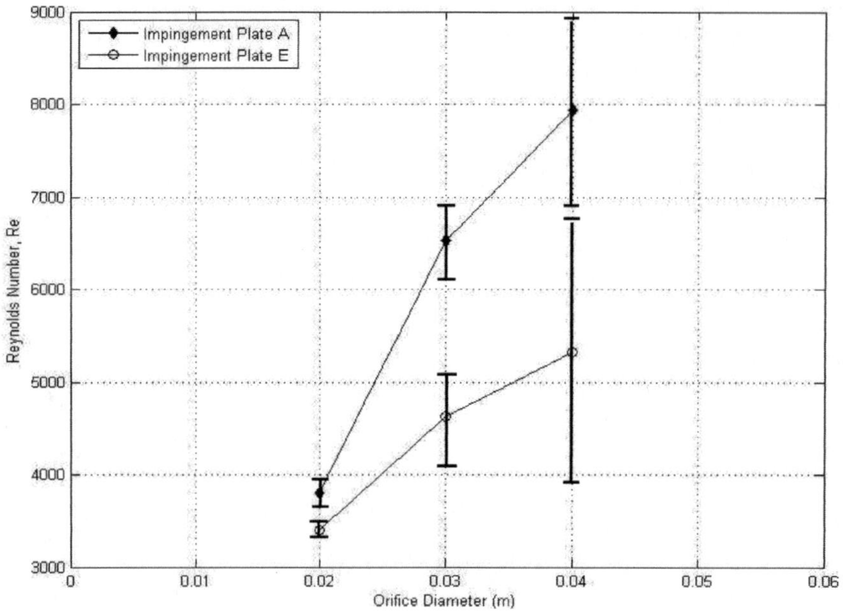

Figure 8. (a) Comparison of *Re* between hole diameter of 7mm and 5mm impingement plates A and E for in-line orientation

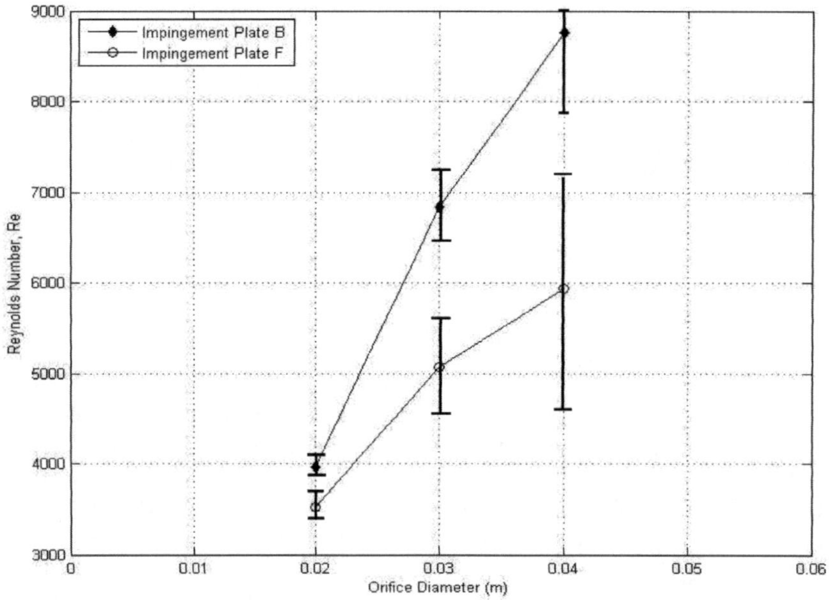

Figure 8. (b) Comparison of *Re* between hole diameter of 7mm and 5mm impingement plates B and F for staggered orientation

Acknowledgement
The first author likes to thank the Ministry of Higher Education Malaysia for provision of MyBrain15 scholarship.

References

[1] Dimotakis P E 2000 *J. Fluid Mech* **409** pp. 69–98
[2] Kwon S J, Seo I W 2005 *Exp. Fluids* **38** pp. 801–812
[3] Lee J, Ren Z, Ligrani P, Fox M D and Moon H 2015 *International Journal of Thermal Sciences* **88** pp. 7-18
[4] Goldstein R J, Behbahani A I and Heppelmann K K 1986 *International Journal of Heat Transfer* **29** pp. 1227-1235
[5] Goldstein R J and Seol W S 1991 *International journal of Heat and Mass Transfer* **34** pp. 2133-2147
[6] Mohd Saiah H R 2011 *Effects of Non-Isothermal Single Circular Impinging Jet on a Quasi-Adiabatic Flat Plate* Master of Science Thesis Universiti Putra Malaysia, Malaysia
[7] Yunus A C and Turner R H 2005 *Fundamentals of Thermal-Fluid Sciences* (New York: McGraw-Hill)
[8] Son C, Gillespie D, Ireland P and Dailey G M 2001 *Journal of Turbomachinery* **123** pp. 154-160
[9] BS1042 British Standard 1981 *Measurement of fluid flow in closed conduit*
[10] Kays W M and Crawford M E 1993 *Convective Heat and Mass Transfer* (New York: Mc-Graw Hill)
[11] Bhatti M S and Shah R K 1987 *Handbook of Single-Phase Convective Heat Transfer* (New York: Wiley Interscience)
[12] Zhiqing W 1982 *Applied Mathematical Mechanic* **3** p. 433
[13] Holman J P 1997 *Heat Transfer* (New York: McGraw-Hill)
[14] Daniel 1997 *Measurement and control, fundamentals of orifice meter measurement* http://www2.emersonprocess.com/siteadmincenter/PM Daniel Documents/Fundamentals-of-Orifice-Measurement-techWpaper.pdf
[15] ISO5167 2003 *Measurement of fluid flow by means of pressure differential devices inserted in circular cross-section conduits running full – part 2: Orifice plate*
[16] ASME MFC-3M 2004 *Measurement of fluid flow in pipes using orifice, nozzle, and venturi*
[17] Coleman H W and Steele W G 2009 *Experimentation, Validation, and Uncertainty Analysis for Engineers* (New Jersey: John Wiley & Sons)

IOP Conf. Series: Materials Science and Engineering **152** (2016) 012013 doi:10.1088/1757-899X/152/1/012013

Mould design and manufacturing considerations of honeycomb biocomposites with transverse fibre direction for aerospace application

N H Manan, D L Majid* and F I Romli

Department of Aerospace Engineering, Faculty of Engineering, Universiti Putra Malaysia, Malaysia

*dlaila@upm.edu.my

Abstract. Sandwich structures with honeycomb core are known to significantly improve stiffness at lower weight and possess high flexural rigidity. They have found wide applications in aerospace as part of the primary structures, as well as the interior paneling and floors. High performance aluminum and aramid are the typical materials used for the purpose of honeycomb core whereas in other industries, materials such as fibre glass, carbon fibre, Nomex and also Kevlar reinforced with polymer are used. Recently, growing interest in developing composite structures with natural fibre reinforcement has also spurred research in natural fibre honeycomb material. The majority of the researches done, however, have generally emphasized on the usage of random chopped fibre and only a few are reported on development of honeycomb structure using unidirectional fibre as the reinforcement. This is mainly due to its processing difficulties, which often involve several stages to account for the arrangement of fibres and curing. Since the use of unidirectional fibre supports greater strength compared to random chopped fibre, a single-stage process in conjunction with vacuum infusion is suggested with a mould design that supports fibre arrangement in the direction of honeycomb loading.

1. Introduction

Over the years, a large amount of researches has been conducted on the green and eco-friendly materials. However, fundamental researches in regards to natural fibre have deepened only in the recent years due to increasing demands for better environmental protection. The natural fibre reinforced polymer composites become high valuable materials due to their low cost and satisfactory mechanical properties, which make them attractive due to availability and renewability of raw materials. Normally, natural fibres are used as reinforcing material or else known as fillers for polymer based matrices [1]. They are renewable, cheap, completely or partially recyclable and biodegradable. Natural fibres have been proven to be the alternative to synthetic fibre in transportation including aerospace industry [2]. They have advantages of low density, high strength-to-weight ratio and most importantly, they are abundantly available and low cost compared to synthetic fibres. However, the primary disadvantage of natural fibre composites is their relatively high moisture absorption but this condition can be overcome by conducting chemical treatment to modify the fibre surface properties [3].

One of many applications in using natural fibres is in the form of honeycomb structures. Honeycomb structures are man-made structures that resemble beehives, which consist of hexagonal cells. Applications of the honeycomb structures in aircraft parts include doors, wing flap, floors, wing-body fairings, rudders, overhead stowage bin, ceiling or sidewall panels, engine cowls, spoilers, nacelles and radomes [4]. The main function of using this structure is to reduce the sum of materials used, and to reach minimum weight and materials' cost in the manufacturing process. Honeycomb core with reinforced polymer composites offer greater specific stiffness and strength properties in flexure compared to their monolithic counterparts [5]. The most common materials used in manufacturing the honeycomb core include aluminum, fibreglass, carbon fibre, Nomex and also Kevlar reinforced with polymer. As the research goes on, the materials of honeycomb structures had been varied from time to time and there is an increasing demand to use natural fibre as the reinforcement to produce honeycomb biocomposites.

Natural fibre composite that is transformed into honeycomb cores improves weight saving, mechanical properties and also functional capabilities as heat dissipation, vibration control and energy dissipation characteristics. This has extended their application from predominantly non-structural components to load bearing structures. Recent researches explored the potential application of honeycomb structures from natural fibres in industries and several techniques were proposed to develop the honeycomb from natural fibres. These techniques include fabrication of corrugated sandwich structures and single core structure. In both techniques, the type of resin dictates the design of the mould. For example, closed mould is needed for thermosetting resins whereas thermoplastics resin needs a mould that is made up of material that can withstand high temperature.

2. Existing fabrication process of honeycomb biocomposites

There are two types of process that can be done to fabricate honeycomb core depending on which type of resin used. They are corrugation process and single core process. Based on the literature review, normally single core process used thermosetting resin in a closed mould. On the other hand, corrugation process used thermoplastics resin in a compression mould. This is due to the characteristic of resin itself, either it is in solid or liquid form. However, in corrugation process, several processing steps are necessary such as compression, bonding, heating and cutting before it can be formed into a honeycomb core. The following are reviews on the development of natural fibre honeycomb core at the present time.

2.1. Corrugation process

Figure 1 shows the setup used for corrugation process by G. Petrone *et al.* (2013). Two types of fibres, which are unidirectional flax fibre mat and also random fibre mat, are used to make the core. They are fabricated using film stacking process where dry flax max and Low Density PolyEthylene (LLDPE) films are interleaved and laid between half hexagonal matched-dies that have the ability of heating and cooling process. The matched-die is heated up to 165°C after closing the mould at 500mm/s. A constant forming pressure of 0.5 MPa is maintained throughout the heating cycle for consolidation and the formed part is cooled within the die to avoid spring-back. A few other processes involved in producing the core are the laser cutting process of the corrugations to desired core thickness, and their assembly and bonding process using ultrasonic methods [6]. In addition, Rao et al. fabricated honeycomb core of sheet sisal fibre and polypropylene composites using the thermoforming process and assembling process, including ultrasonic bonding [7]. The sheets are thermoformed into half hexagonal and sinusoidal profiles using the matched-die thermoforming technique with forming rate of 500mm/min and 165°C die temperature. The formed corrugations are then extracted after they are left to cool to room temperature in the die. The corrugated sheets are cut to desired height and bonded using ultrasonic welder.

From both works, it is concluded that using corrugation process has a main disadvantage of involving too many processes, which increases manufacturing costs.

Figure 1. Matched dies for forming and ultrasonic bonding of flax-PE honeycomb core [5]

2.2. Single core process

From Ariel S. *et al.* (2014), a single core of jute fabric reinforced with vinylester composites is fabricated using two methods [5]. The first method is using a mould with fixed inserts as shown in Figure 2. The jute fabrics were placed between the inserts in zigzag pattern in ribbon direction and the vinylester were then poured into the mould at 20°C. The ensemble was molded under 50MPa pressure in hydraulic press for an hour at 80°C. The core was later post cured for 2 hour at 140°C in an oven. This method however results in different average fibre volume contents for the longitudinal and diagonal walls.

Meanwhile, the second method is by lateral compression method that allows for the lateral compression of the ensemble as shown in Figure 3. The resin wetted jute fabrics were intercalated between combs in ribbon direction, and the ensemble was then laterally pressed and clamped to remove excess resin and to drag entrapped air bubbles. Similar to the method for mould with fixed inserts, there is no longitudinal force applied to the fibre while disposing it into the mould. Both methods were completed with thermal treatments to cure and post-cure the resin. Using the mould with lateral compression requires the user to manually insert the fiber in between the hexagonal insert, which usually results in the lack of consistency of fiber volume throughout the honeycomb core structure.

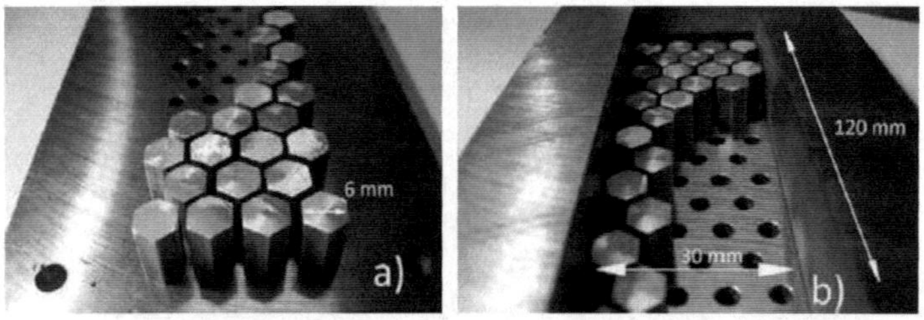

Figure 2. Mould with fixed inserts

Figure 3 : Mould with lateral compression

2.3. Comparison of corrugation process and single core process

The corrugation process contains several procedures such as cutting and ultrasonic bonding as mentioned before. Corrugation process is mainly about producing honeycomb core layer by layer, and then assembles them into a sandwich structure. It involves the use of many machines and labor works, thus increases the manufacturing cost. Many precaution steps need to be taken while fabricating honeycomb core structure with corrugation process such as the distribution of fiber volume in every layer and consistent length of the corrugation core (or otherwise, another cutting process has to be done to maintain the core length). Corrugated sandwich may have defects where the layer of sandwich does not perfectly bond to each other, or the cutting size of the layer is not the same, which results in inconsistent length of honeycomb core.

In contrast, single core process does not involve any other processes and it is easier to be used. Single process will readily produce a honeycomb core without the process of cutting, bonding and assemble. It is obvious that single core process is more reliable compared to corrugation process. This method can save a lot of time because it involves only one process for each core.

3. Methodology

Figure 4 shows a flowchart of this work. After the research requirement is identified, the next step taken in order to achieve the objective of this research work is by conducting a literature review. Before proceeding with CAD design using Solidwork software, the literature review is done to gather the information through research of associated fields that comprise of the manufacturing of honeycomb structure, type of mould used, suitable materials for the mould and the dimensions involved in designing the mould in accordance with ASTM standard, including all the detail size of honeycomb core structure such as wall thickness, core thickness and cell size. In this project, epoxy will be used as the resin. From preliminary work, it is found that epoxy and aluminum does not bond well. This is the reason aluminum is chosen as the material of the mould. The kenaf/epoxy composites can be easily taken out as it will not stick to the aluminum mould.

The process parameters highlighted in this work are pressure of vacuum infusion, weight ratio of kenaf to epoxy composite, temperature and number of inlet and outlet of the mould. The flow rate of resin into the mould depends on the pressure used. For example, in this work, the initial pressure should be around 60 psi. Initially, the vacuum process is run without the usage of resin in order to remove trapped air inside the mould. Once the reading of pressure is stable, the vacuum pressure will be increased to 80-100 psi and resin is infused into the mould. The epoxy has three types of curing rate which are slow, medium and fast curing. The one used in this experiment is a type of slow curing epoxy. Therefore, the pressure set must be suitable for the curing rate. If a higher pressure is used, the flow rate of resin will increase but air bubbles are most likely to form in between the fibres. However, if the pressure is too low, the resin will cure before it completely spread out into the mould. The weight ratio of kenaf fiber to epoxy is 10:90.

The next parameter is the position of inlet and outlet at the mould. Resin will be infused through the inlet while the outlet is to remove excess air and resin during infusion process. The position of inlet is important to make sure that the resin is spread evenly throughout the mould. For this research, the mould is designed to have five inlets and one outlet as shown in Figure 6. Vacuum infusion and curing process will take place under room temperature. The curing process usually takes 24 hours before epoxy hardened.

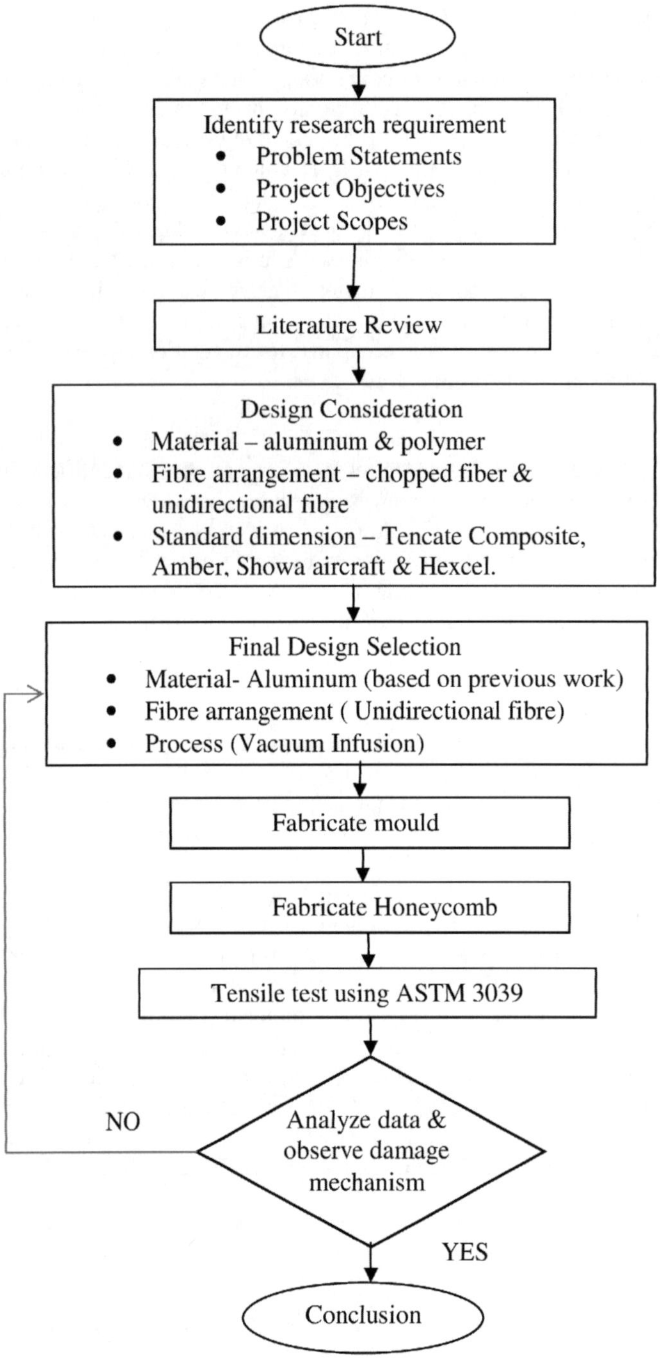

Figure 4. Project flowchart

4. Initial design concept

A preliminary work has been done on a single core process using open mould with manual fixed insert. The work was focused on fabrication of mould to produce honeycomb core of rice husk reinforced with polypropylene composite. The mould was fabricated using aluminum and the hexagonal inserts are made from Allen key of harden mild steel. The mould is shown in Figure 5. The mould is fabricated using the processes of CNC Milling, EDM wire cut and EDM Die Sinker. The functionality of the mould has been demonstrated with composites powder blend that was manually compressed between gaps of inserts in the mould and cured in the furnace with temperature up to 250°C. However, it was found that this method is not suitable for thermoplastic resin as it comes in pallet form and difficult to compress them into the gap of hexagonal inserts. Furthermore, heat transfer to the mould was uneven at the surface and at the centre of the mould, causing the thermoplastic's failure to melt completely as shown in Figure 5b. The difference between this initial design concept and previous work by Ariel S. *et al.* (2014) is the type of resin used. The resin used previously was thermoset while in the initial design was thermoplastic.

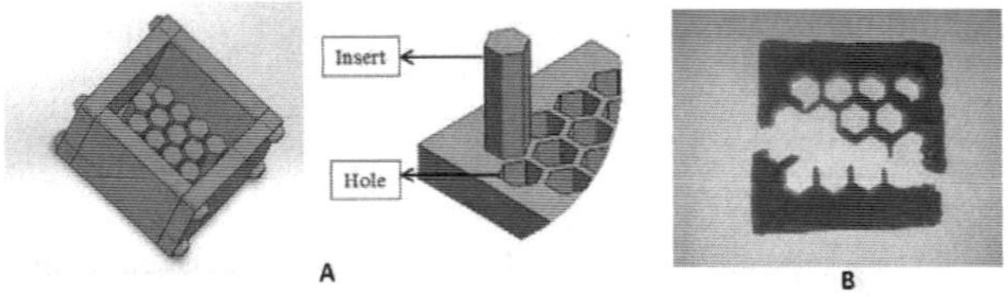

Figure 5. (a) Drawing of open mould of inserts, (b) composites failed to melt completely

5. Proposed single core process with vacuum infusion

In this work, a honeycomb core of kenaf fibre/epoxy with transverse fibre direction is introduced by using the vacuum infusion using single core process. Single core is proposed because it uses less manufacturing process and using vacuum infusion method will increase the consistency of resin flow into the mould, thus ensure the resin spread evenly throughout the mould. Unidirectional fibre arrangement provides a greater load bearing as compared to random chopped fibre. In this work, transverse fibre orientation is suggested since honeycomb core mainly supports transverse compression. The preparation is started by fabricating the mould using aluminum to prevent the composites from stick to the mould. The mould is designed to support the use of continuous unidirectional kenaf fibre in honeycomb structure. All the dimensions of the mould are referred to the standard size of honeycomb structure used in aircraft. A single core mould with fixed inserts is designed based on the aerospace specifications for honeycomb core structures from four companies: Tencate Advance Composites, Amber Composites, Showa Aircraft Industry and Hexcel [8–13]. It is found that the cell size ranges from 3.2 mm to 19 mm and the core wall thickness is in the range of 2 mm to 100 mm with a tolerance of ±0.125 mm. The material for honeycomb core is kenaf fibre reinforced with epoxy. The epoxy is chosen due to the fact that it is more suitable to be used in vacuum infusion process compared to the thermoplastic polymer. Furthermore, as compared to other resins, epoxy composites exhibit higher strength values, thus contribute a higher ultimate tensile strength in the structure [14].

The honeycomb core will be fabricated using vacuum infusion process, thus a closed mould as shown in Figure 6 will be used. The black arrows indicate the inlet where resin will be infused and the red arrow indicates the outlet to the vacuum pump. A gasket rubber is mounted at the surrounding of top cover and base to prevent air from entering the mould. The height of the hexagonal inserts is depending on the core thickness. In this research, a honeycomb core thickness of 10 mm and wall thickness is 2 mm is suggested. Kenaf fibre is arranged in transverse direction in the mould before infusing the resin. Vacuum infusion has the advantages of consistent flow of resin and not depending on the workers' concentration and skills [15].

The vacuum pressure is maintained at 80 psi while infusing the resin into the mould to maintain medium flow rate of the resin in order to prevent air bubble trapped between the hexagonal inserts. Furthermore, the mould is designed with five injector pins at the bottom of the base as shown in Figure 7 (as indicated by the arrows) to ease the dismantle process of honeycomb structure from the mould. In addition, more injector pins can be added in the event that the honeycomb core is still difficult to be dismantled from the mould.

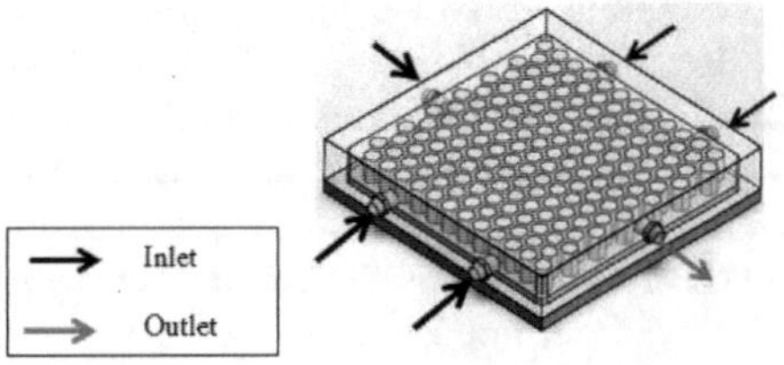

Figure 6. The sketching on vacuum infusion mould

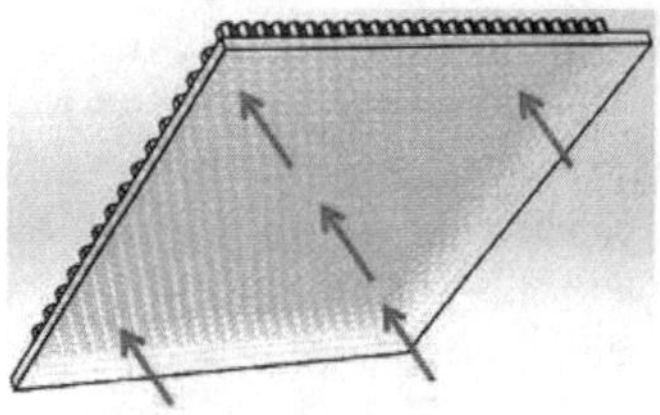

Figure 7. Schematic drawing of injector pins at the bottom of mould

6. Conclusion
A single mould has been designed properly such that it can be used to produce honeycomb core structure with transverse fibre direction using vacuum infusion method. This method can reduce manufacturing cost besides having a major advantage of consistent flow of resin into the mould. The presence of injector pin will ease the dismantling process compared to the first design concept. Furthermore, the mould is easy for maintenance as the inserts are not mounted permanently to the base.

Acknowledgement
The authors would like to acknowledge Universiti Putra Malaysia for the fund under Geran Putra IPS GP-IPS/2016/9487900. In addition, we wish to acknowledge staff from Laboratory of Composites Material, Faculty of Engineering, Universiti Putra Malaysia for all technical guidance and facilities provided.

References
[1] Al-oqla F M and Sapuan S M 2014 *J. Clean. Prod.* **66** 347–54
[2] Riedel U and Nickel J 2005 *Biopolym. Online* **272** 34–40
[3] Bongarde U S and Shinde V 2014 *Int. J. Eng. Sci. Innov. Technol.* **3** 431–6
[4] Liu L, Meng P, Wang H and Guan Z 2015 *Compos. Part B Eng.* **76** 122–32
[5] Stocchi A, Colabella L, Cisilino A and Álvarez V 2014 *J. Mater.* **55** 394–403
[6] Petrone G, Rao S, De Rosa S, Mace B R, Franco F and Bhattacharyya D 2013 *Compos. Part B* **55** 400–6
[7] Rao S, Yadama V and Bhattacharyya D *18th Int. Conf. Compos. Mater.* p1–6
[8] Tencate Advance Composites *TenCate Advanced Composites*
[9] Antibody C D 2011 *Product Datasheet* 2011–2014
[10] Amber Composites 2005 *Aac amber aluminum commercial honeycomb*
[11] Showa Aircraft Industry Co. Ltd. 2012 *PAA-SHH*
[12] Cell H and Specifications S 2012 *Aramid Fiber/Phenolic Resin Honeycomb* 1–2
[13] Hexcel Composites 1999 *Honeycomb Data Sheets* 1–40
[14] Mahjoub R, Mohamad J, Rahman A, Sam M and Raftari M 2014 *J. Mater.* **64** 640–9
[15] Gaetzi R 2008 *Why vacuum infusion benefits your quality, budget and environment*

IOP Conf. Series: Materials Science and Engineering **152** (2016) 012014 doi:10.1088/1757-899X/152/1/012014

Design and mechanical analysis of a 3D-printed biodegradable biomimetic micro air vehicle wing

E Salami[1, 3*], P B Ganesan[1], T A Ward[2], R Viyapuri[1] and F I Romli[3]

[1] Department of Mechanical Engineering, Faculty of Engineering, University of Malaya, Kuala Lumpur, Malaysia
[2] School of Engineering and Physical Science, Heriot-Watt University, Putrajaya, Malaysia
[3] Department of Aerospace Engineering, Faculty of Engineering, Universiti Putra Malaysia, Malaysia

*erfansalami@hotmail.com

Abstract. The biomimetic micro air vehicles (BMAV) are unmanned, micro-scaled aircraft that are bio-inspired from flying organisms to achieve the lift and thrust by flapping their wings. There are still many technological challenges involved with designing the BMAV. One of these is designing the ultra-lightweight materials and structures for the wings that have enough mechanical strength to withstand continuous flapping at high frequencies. Insects achieve this by having chitin-based, wing frame structures that encompass a thin, film membrane. The main objectives of this study are to design a biodegradable BMAV wing (inspired from the dragonfly) and analyze its mechanical properties. The dragonfly-like wing frame structure was bio-mimicked and fabricated using a 3D printer. A chitosan nanocomposite film membrane was applied to the BMAV wing frames through casting method. Its mechanical performance was analyzed using universal testing machine (UTM). This analysis indicates that the tensile strength and Young's modulus of the wing with a membrane is nearly double that of the wing without a membrane, which allow higher wing beat frequencies and deflections that in turn enable a greater lifting performance.

1. Introduction

Bionic or biomimetic relates to the creation of artificial structures inspired by those found in nature [1]. This multi-dimensional observational information includes a wide assortment of fields, such as biology, physics, chemistry, mathematics, material science and also biomechanics. Biomimetic micro air vehicle (BMAV) is a micro-scaled aircraft that mimics the flapping wing movement of insects [2]. Dragonflies are highly agile flyers [3, 4] and they are capable of migrating across oceans, moving in any direction and suddenly changing course. They also can hover and accelerate quickly from a dead stop as well as from a hovering attitude. Their non-planar dynamic wings are long and narrow, and significantly larger than their body. The wings possess high power/weight ratio and exhibit excellent functional characteristics through the coupling and synergy of material, configuration and structure. In addition, the grooves in the wing hold an airfoil of air around it, which brings down the friction. The wings rotate around several axes, responding both to muscle actions and inertia effects. Dragonflies can have different flight maneuvers such as gliding, taking off, hovering and flapping. The wings enable the dragonflies to perform a very smooth and stable flight with a considerable agility. Indeed, they are one of the quickest and most agile flying insects in the living world. Dragonflies have one of the most interesting wing structures and remarkable features of nature in the science of engineering design. Both the micro and macro structures of dragonfly wings play a significant role in their performance, and this is the reason numerous researchers have attempted to concentrate on the impact of these parameters on wing controllability and efficiency.

The structural characteristics possessed by dragonfly wings provide a biological inspiration for the investigation and development of flapping-wing micro air vehicles. A study conducted by Chen *et al.*[5] reported that a dragonfly can accelerate at 4G (G-force) linearly and 9G in sharp turns while pursuing prey. Sunada *et al.*[6] and Dickinson *et al.* [7] revealed that the motion of dragonfly wing is controlled by the wing root muscle only, which is easier to mimic in a BMAV than the wing motions used by other insects. These qualities make dragonfly a preferred candidate to be bio-mimicked in a BMAV. BMAVs are typically imagined for use on military and civil missions, which have a limited duration, including police or military surveillance, indoor video mapping and remote sensing of hazard sites. The artificial wings of BMAV must be strong enough and flexible to tolerate the different forces created by flapping motion closely similar to a real insect. The main types of deformation that the wings experience during flight are twisting and bending, and these wings will have to be able to cope with the forces that operate on its surface. Several fabrication procedures for small insect artificial wings have been suggested. The microelectromechanical systems (MEMS) photolithography and etching method has been presented by Pornsinsirirak *et al.*, in which the membrane and vein of the wing are fabricated using parylene-C and titanium alloy (Grade 5), respectively [8]. Akin *et al.* presented non-natural wings that mimic the main venation structure of a beetle using thin Kapton film for the membrane and pre-impregnated carbon/epoxy fiber for the framing structure [9]. Meanwhile, Bao *et al.* described the design and micro-machining of the three-dimensional BMAV wing from stainless steel (SU-8) material using microelectromechanical system (MEMS) lithography [10]. Very limited articles have been written about the BMAV wing frame structures that encrust the membrane.

Earlier studies have demonstrated that insect's wings are one of the best samples of design integrity [2, 11-13] because they are super lightweight composite structures with complex shapes at both the micro and macro stages. The utilization of lightweight structural materials is essential in numerous applications such as biomedical applications, ground vehicles, aircraft and packaging to enhance its efficiency and cost [14]. Recent studies on biomimetic air vehicles (BMAV), which are taken as a kind of micro-scaled flapping winged aircraft that imitate the flapping wing movement of birds, bats or insects [15], have interested scientists to search for high efficiency lightweight materials. Nowadays, the researchers are studying about technologies in determining flying patterns of a group of BMAV [16]. Accordingly, the BMAV must be mass-produced using reasonable materials for executed in a cost effective manner. In addition, due to the range and power limitations, missions may be performed one-way, with no intention to return the BMAV to its original launch point [17]. These requirements lead to the need for BMAV to be fabricated from bio-degradable materials (similar to insects). The wing structure and membrane is generally the largest part of the BMAV design. Therefore, there is much interest in fabricating lightweight biodegradable wings.

2. Methodology

2.1. Material

PLA(Polylactic acid) is one of the two most commonly used 3D printing materials (with the other being ABS). This biodegradable material has the virtue of being odorless and does not warp significantly like the other 3D printing materials. Due to its low melting point, using PLA as a 3D printing material does not require a heated bed. 3D printer can be used to fabricate complex structures. MakerBot Replicator 2X experimental 3D printer that was used in this study was purchased from Makerbot (Brooklyn, USA). Grey PLA printing filament was purchased from MyDuino and used as received. CAD designs were drawn and visualized in Solidworks Software and transferred to 3D printable format using the Makerbot Desktop Software supplied with the Makerbot Replicator 2X.

2.2. CAD design and manufacturing method

Dragonfly wings have a very complex and fine structure, which is hard to replicate. Sivasankaran *et al.* used the Canny edge detection method on digital images of dragonfly fore and hind wings, and the spatial network analysis method to create simplified model from the detailed model [18]. They demonstrated that the spatial network analysis is a suitable approach to simplify a complicated insect wing structure such as dragonfly. This simplified dragonfly wing (for use on a BMAV) was modelled using Solidworks computer aided design (CAD) software and transferred into a 3D printer compatible file. A MakerBot Replicator 2X experimental 3D printer was configured to "fine printing method" for this research since the sample size was approximately 8 cm with a thickness of 0.5 mm. Both fore and hind wings were fabricated using bio-based Polylactic Acid (PLA). Figure 1 shows the 3D model of a simplified dragonfly wing structure and Figure 2 presents the fabricated dragonfly wing model.

Lin *et al.* studied the potential of utilizing 3D printing technologies to fabricate bio-inspired shapes and they had successfully fabricated 3D printed polymer based bio-inspired shapes [19]. It was mentioned that geometrically structured shapes and designs occurred naturally for a reason and they possessed enhanced mechanical properties and provided inspiration for materials design. Kao *et al.* fabricated PLA-based scaffolds for bone tissue engineering using 3D printing technique [20]. They fabricated high-resolution scaffolds that were comparable to readily available scaffolds by utilizing the advantages of PLA. In addition, Stansbury and Idacavage reviewed the possibilities of 3D printing method to be expanded into a conventional production method for polymers [21]. They mentioned that 3D printing technologies had already altered many industrial and academic operations with their robustness and simplicity in fabrication of any design without requiring molds or dies. Therefore, it will be interesting to study the possibilities of producing a 3D printed the PLA-based wing structure and evaluate its performance.

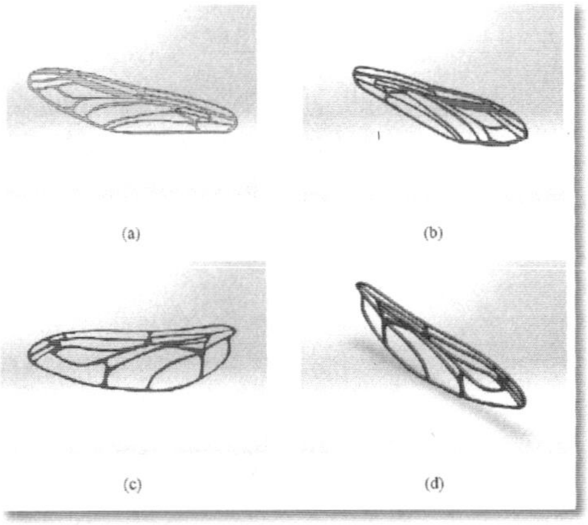

Figure 1. 3D CAD design of dragon wing: (a) Top view of fore wing, (b) Iso view of fore wing, (c) Top view of hind wing, (d) Iso view of hind wing

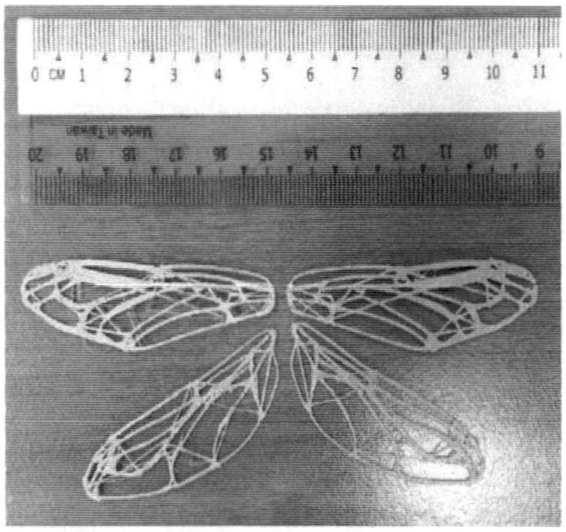

Figure 2. Fabricated dragonfly wing

2.3. Membrane

The dragonfly's wing membrane is primarily composed of chitin. Although chitin is naturally available in large quantities, it is underutilized in commercial applications due to its insolubility in common organic solvents and its intractable bulk structure. Chitosan was examined to solve these problems. Chitosan is a biodegradable and biocompatible natural polysaccharide composed of linear polysaccharide 1,4 -linked 2-amino-deoxy-b-D-glucan [22]. Once the wing structure was printed out, the exposed top surface layer required sanding to remove jagged material remains on the structure. The structures were then submerged in one of the selected chitosan nanocomposite film solutions in a Petri dish. A drying time of 48 hours was required. Specification of the wings is listed in Table 1 and Figure 3 depicts the complete design process.

2.4. UTM machine parameter setting

Mechanical properties were analysed for all composite material samples by using the Shimadzu AGS-X series tensile tester machine according to the ASTM D638-14 method (i.e. standard test method for tensile properties of plastic). A low capacity load cell was used to increase the sensitivity of the break due to the wing structure thickness. The wing structures were pre-conditioned in a dry cabinet 48 hours before the testing was conducted. A crosshead speed of 5 mm/min was used for each test. Figure 4 illustrates the experiment setup.

Table 1. Specification of wing structure

Specification	Fore wing	Hind wing
Base width (mm)	7.00	8.50
Centre width (mm)	12.50	14.40
Tip width (mm)	7.23	8.50
Length (mm)	56.00	48.00
Mass without membrane (g)	0.12	0.08
Mass with membrane (g)	0.18	0.15

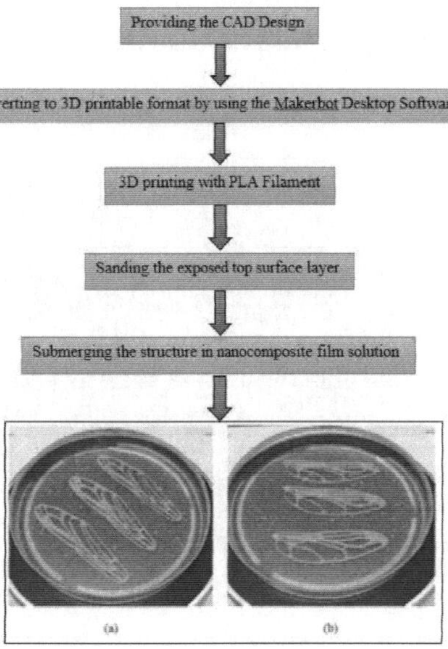

Figure 3. Design process: (a) Forewing, (b) Hindwing

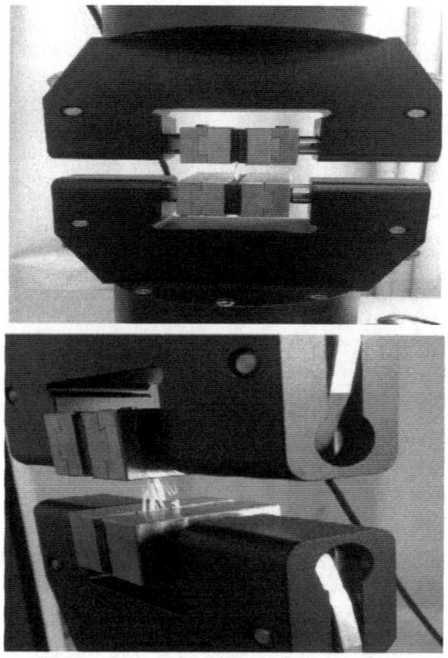

Figure 4. Experimental setup

According to the principle of tensile testing, the material strain, ε can be taken as:

$$\varepsilon = \frac{\Delta L}{L_0} \tag{1}$$

where ΔL is the increase of length in the tensile force direction and L_0 is the original sample gauge length. Correspondingly, the material stress, σ can be calculated as follows:

$$\sigma = \frac{F}{A_0} \tag{2}$$

where F is the tensile force applied on the sample and A_0 is the cross-sectional area of the sample. By employing Hooke's law, the Young's modulus, E of the material is:

$$E = \frac{\sigma}{\varepsilon} \tag{3}$$

Due to the nonlinear mechanical characteristics of biomaterials, the elastic force and the deformation do not exhibit a clearly defined elastic limit. Therefore, the offset method was used to compensate errors. The value of 0.02% was set off on the extension axis and a line was drawn parallel to the straight-line portion of the loading curve. By this principle, tensile strength, Young's modulus and elongation at the break were calculated. At least three samples were tested for each wing structure.

3. Result and discussion

All samples of the PLA fore wing and hind wing, with and without membrane, were tested and analyzed. The forewing PLA sample without membrane exhibited tensile strength of 3.25 MPa as shown by Figure 4a. Addition of the nanocomposite film as membrane increased the tensile strength to 6.280 MPa as illustrated by Figure 5a. A similar trend can be seen for the hindwing PLA sample without membrane, which obtained the lowest tensile strength of 2.96 MPa, and the value was increased to 4.89 MPa after adding the membrane to the wing structure as shown by Figure 5b. On the other hand, for the Young's Modulus as depicted by Figure 5c, PLA forewing without membrane recorded the lowest stiffness of 134.90 MPa and it was significantly increased up to 334.51 MPa with the addition of the membrane. As the Figure 5d displayed, the PLA hindwing stiffness was equal to 158.95 MPa and, as expected, a considerable raise up to 223.64 MPa was achieved by adding the membrane. All in all, addition of nanocomposite membrane increases the tensile strength and Young's modulus of the samples. The increment of tensile strength could be linked to the formation of more stable network due to the attractive interactions between the wing structure and the nanocomposite film. The compaction of this network also leads to increase in stiffness.

4. Conclusion

Biomimetic structures and designs are always exciting as they are imitation of the living surroundings. In this work, a biodegradable dragonfly inspired wings (forewing and hindwing) for BMAV application were designed using Solidworks software. 3D geometric PLA wing samples were printed using the Makerbot 3D printer. Constructing the wings using 3-D printer method had expedited the design fabrication period as it took less than 15 minutes. Therefore, several wings could be printed per day for testing purposes and ultimately for developing a BMAV. The printed wings were then casted with nanocomposite membrane. The mechanical test performed yielded a greater tensile strength and stiffness for the reinforced wings samples with the membrane, which would permit higher wing beat frequencies and deflections that in turn allowed greater lifting performance.

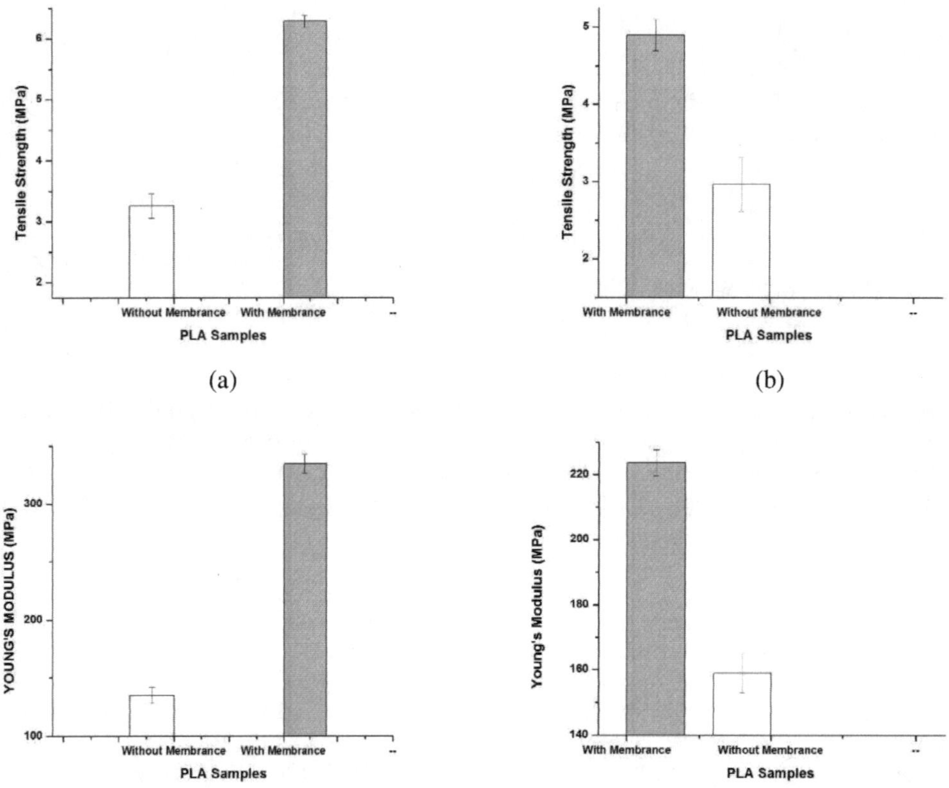

(a) (b)

(c) (d)

Figure 5. Average mechanical properties of PLA wing structure: (a) tensile strength of forewing, (b) tensile strength of hindwing, (c) Young's Modulus of forewing, (d) Young's Modulus of hindwing

Acknowledgments
The authors would like to thank to the University of Malaya for financial support under the High Impact Research MoE Grant: UM.C/625/l/HIR/MoE/ENG/53 (D000053-16001) from the Ministry of Education, Malaysia and the University of Malaya Research Grant: RG155-12AET.

References

[1] Ward T A, Rezadad M, Fearday C J and Viyapuri R 2015 *International Journal of Micro Air Vehicles* **7** 375-94
[2] Nair P, Ward T A, Viyapuri R and Johan M R 2015 *4th International Conference on Material Science and Engineering Technology*
[3] Nagai H, Nishi A, Fujimoto T and Isogai K 2009 *Proceedings of the 4th International Symposium of Aero Aqua Bio-Mechanisms*
[4] Thomas A L, Taylor G K, Srygley R B, Nudds R L and Bomphrey R J 2004 *Journal of Experimental Biology* **207** 4299-323
[5] Chen Y, Skote M, Zhao Y and Huang W 2013 *Journal of Fluids and Structures* **40** 115-26

IOP Conf. Series: Materials Science and Engineering **152** (2016) 012014 doi:10.1088/1757-899X/152/1/012014

[6] Sunada S, Zeng L and Kawachi K 1998 *Journal of theoretical Biology* **193** 39-45
[7] Dickinson M H, Lehmann F-O and Gotz K 1993 *Journal of Experimental Biology* **182** 173-89
[8] Pornsin-Sirirak T N, Tai Y, Nassef H and Ho C 2001 *Sensors and Actuators A: Physical* **89** 95-103
[9] Akın D, Kasgoz A and Durmus A 2014 *Composites Part A: Applied Science and Manufacturing* **60** 44-51
[10] Bao X, Dargent T, Grondel S, Paquet J-B and Cattan E 2011 *Microelectronic Engineering* **88** 2218-24
[11] Rajabi H and Darvizeh A 2013 *Chinese Physics B* **22** 088702
[12] Rajabi H, Moghadami M and Darvizeh A 2011 *Journal of Bionic Engineering* **8** 165-73
[13] Darvizeh, Darvizeh, Rajabi and Rezaei 2009 *The International Journal of Multiphysics* **3** 101-10
[14] Goede M, Stehlin M, Rafflenbeul L, Kopp G and Beeh E 2009 *European Transport Research Review* **1** 5-10
[15] Orlowski C T and Girard A R 2012 *Progress in Aerospace Sciences* **51** 18-30
[16] Kushleyev A, Mellinger D, Powers C and Kumar V 2013 *Autonomous Robots* **35** 287-300
[17] Rubentheren V, Ward T A, Chee C Y and Nair P 2015 *Cellulose* **22** 2529-41
[18] Sivasankaran P N and Ward T A 2016 *Aerospace Science and Technology* **49** 259-68
[19] Lin E, Li Y, Ortiz C and Boyce M C 2014 *Journal of the Mechanics and Physics of Solids* **73** 166-82
[20] Kao C-T, Lin C-C, Chen Y-W, Yeh C-H, Fang H-Y and Shie M-Y 2015 *Materials Science and Engineering: C* **56** 165-73
[21] Stansbury J W and Idacavage M J 2016 *Dental Materials* **32** 54-64
[22] Badawy M E and Rabea E I 2014 *Cellulose* **21** 3121-37

Design and implementation of a vision-based hovering and feature tracking algorithm for a quadrotor

Y H Lee[1] and J S Chahl[1,2]

[1]Division of Information Technology and Engineering, The School of Engineering, University of South Australia, Australia
[2]Joint & Operations Analysis Division, Defence Science and Technology Group, Australia

*javaan.chahl@unisa.edu.au

Abstract. This paper demonstrates an approach to the vision-based control of the unmanned quadrotors for hover and object tracking. The algorithms used the Speed Up Robust Features (SURF) algorithm to detect objects. The pose of the object in the image was then calculated in order to pass the pose information to the flight controller. Finally, the flight controller steered the quadrotor to approach the object based on the calculated pose data. The above processes was run using standard onboard resources found in the 3DR Solo quadrotor in an embedded computing environment. The obtained results showed that the algorithm behaved well during its missions, tracking and hovering, although there were significant latencies due to low CPU performance of the onboard image processing system.

1. Introduction

Significant number of research programs on the vision-based control for Unmanned Aerial Vehicles (UAVs) are ongoing actively as an alternative sensor to GPS [1]. In fact, in case of ground vehicles, many vision-based control algorithms have already been proposed so far. However, when it comes to UAVs, specific additional challenges should be considered such as rigid body objects moving in 3D, dynamic effects and external perturbations, and relatively poor quality of video sequences filmed from the vibrating movement common to UAVs [2]. Simultaneous localization and mapping (SLAM) is one of the more popular technologies for position control for UAVs [1]. One of the popular approaches to the visual simultaneous localization and mapping (VSLAM) algorithm was introduced by Klein and Murray [3]. Their approach splits the task into two threads: one for tracking and the other for mapping. The advantage of this separation is that by running both threads at different rates (the tracking thread runs faster than the mapping thread), the tracking thread can update the position more quickly while the mapping thread can employ a slower but more powerful technique. In addition, since the mapping algorithm does not process all frames, the image processing work can have more resource allocated, especially when the UAV moves slowly or hovers [4]. When VSLAM uses a single camera, it cannot provide the scale factor that is a crucial parameter for calculating absolute velocity and position in 3D of the vehicle. In order to acquire this scale factor, several methods were proposed such as employing air pressure sensor or online multi rate extended kalman filter (EKF) based on inertial sensors [5].

Another vision-based control area that is attracting many research teams is the object tracking field. A research team in University of Toronto has carried out vision-based hovering and object tracking, and tested several related algorithm such as optical flow, colour-based image processing, kalman filter tracking [6]. Their work achieved quite desirable results despite some limitations such as the camera was fixed on the ground staring upward to film the quad-rotor and the image processing tasks were done on ground station. Other interesting research in object tracking is a quad-rotor UAV platform that tracks a moving target on the ground based on visual information from a camera [7]. An interesting point is that, in case of loss of track of the moving target, the quad-rotor increases its altitude so as to get wider field of view. However, a limitation was that their algorithm was only verified for a moving object with purely translational motion. Similar UAV vision-based on-board object tracking research has also been conducted in [8]. The research employed a low-cost on-board monocular vision system and designed a closed-loop object tracking controller. With colour detect vision algorithm and closed-loop controller, the object tracking system showed a stable result though it has some limitations such as stationary object, light condition and restricted background around the object.

The aim of this work is to enable hovering and object tracking for UAVs using vision data from a monocular camera. For this goal, the vision system should possess following capabilities: detecting the target object, identifying the current pose $(x,y)^T$ and also orientation θ relative to the target object, and approaching the object and holding the position around the object. For the first and second capabilities, image processing algorithms provided by OpenCV libraries were employed such as SURF [9] and moments [10] based image processing. For the third capability to steer the UAV, the 'dronekit' python library for 3DR Solo was employed. Robust detect algorithms, filtering and compensating corrupted position data due to external environment such as wind, precise control algorithm to steer the UAVs were key challenges for this project.

2. Methodology

2.1. Quadrotor platform

3DRobotics has introduced the Solo quadrotor, which is a Linux based drone and shown in Figure 1. Even though its main design purpose is a high-tech toy for entertainment, it can be a solid and well-integrated platform for drone-related research due to its standard development environment – Linux, OpenCV, dronekit. The hardware specification of the Solo quadrotor is listed in Table 1.

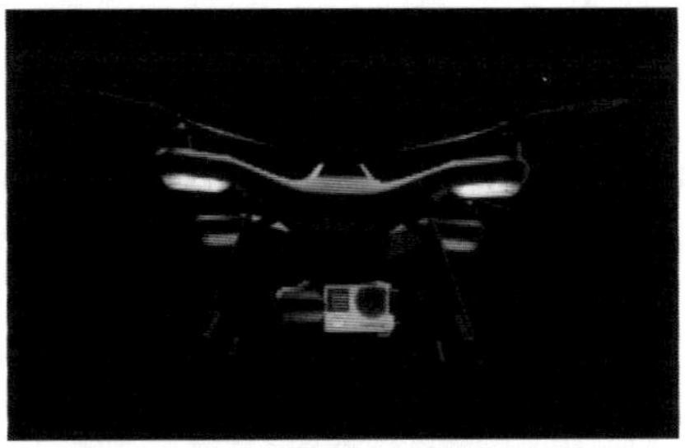

Figure 1. Front picture of the solo with GoPro camera

Table 1. The solo hardware specification

Component	Specification
Propeller	24cm diameter, 144cm pitch
Weight	1.5kg
Dimension	25cm tall, 46cm motor to motor
Flight battery	Lithium poly 5200mAh
Max Payload	420g
Autopilot	Pixhawk 2
Software	APM copter

Solo's control is carried out by two embedded boards: Pixhawk 2 and IMX6, as shown in Figure 2. The Pixhawk 2 is responsible for low-level control like motor, sensors and flight control. Meanwhile, the IMX6 performs the role of high-level controller such as video streaming and communication with the remote controller. All control messages from the IMX6 are passed to the Pixhawk 2 in the form of mavlink protocol commands. Specification for both Pixhawk 2 and IMX6 is provided in Table 2.

2.2. Image processing

Several image processing algorithms were employed in this study for object detection, pose estimation and tracking. Speed Up Robust Features (SURF) algorithm was used for target detection, then after applying image processing, the pose information of the target relative to the quadrotor was estimated. OpenCV libraries provided a standard framework and architecture for the above image processing algorithms that the software used to implement the algorithms.

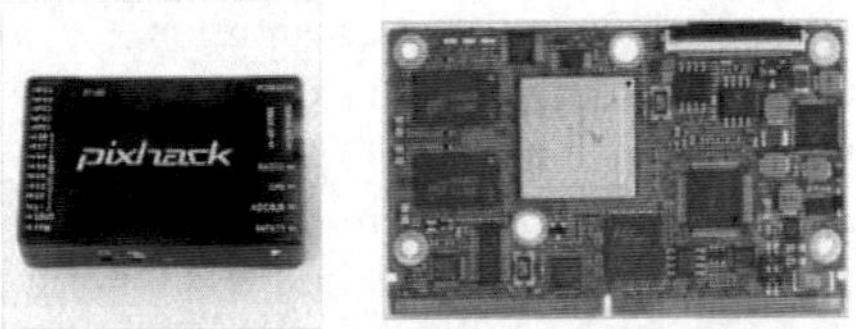

Figure 2. Image of Pixhawk 2 and IMX6

Table 2. Specification for embedded boards in the solo

	CPU	Memory	OS
Pixhawk2	32 bit ARM Cortex M4 168MHz	256KB	NuttX RTOS
IMX6	32bit ARM Cortex A9 1GHz	512MB	Yocto embedded Linux

SURF algorithm extracts some keypoints and descriptions from an image, and then using them to detect the same scene or object from other images [9]. It uses an intermediate image representation known as the integral image. Given a input image *I* and point (*x* , *y*), integral image *IΣ* is calculated by the sum of the values between the point and the origin as in Equation 1.

$$I\Sigma(x,y) = \sum_{i=0}^{i\le x}\sum_{j=0}^{j\le y} I(x,y) \tag{1}$$

The SURF detector is based on the determinant of the Hessian matrix [9]. The Hessian matrix H is the matrix of partial derivatives of function, f as given in Equation 2.

$$H\big(f(x,y)\big) = \begin{bmatrix} \frac{\partial^2 f}{\partial x^2} & \frac{\partial^2 f}{\partial x\,\partial y} \\ \frac{\partial^2 f}{\partial x\,\partial y} & \frac{\partial^2 f}{\partial y^2} \end{bmatrix} \sigma \tag{2}$$

The determinant of this matrix is calculated by Equation 3.

$$\det(H) = \frac{\partial^2 f}{\partial x^2}\frac{\partial^2 f}{\partial y^2} - \left(\frac{\partial^2 f}{\partial x\,\partial y}\right)\left(\frac{\partial^2 f}{\partial x\,\partial y}\right) \tag{3}$$

This determinant value is used to classify the maxima and minima of the function. Then, calculate the Hessian matrix H, as function of both space X = (x, y) and scale σ as shown in Equation 4.

$$H(X,\sigma) = \begin{bmatrix} L_{xx}(X,\sigma) & L_{xy}(X,\sigma) \\ L_{xy}(X,\sigma) & L_{yy}(X,\sigma) \end{bmatrix} \tag{4}$$

where $L_{xx}(X,\sigma)$ refers to the convolution of the second order Gaussian derivatives $\frac{\partial^2 g(\sigma)}{\partial x^2}$ with the image at point X = (x, y) and similarly for L_{xy} and L_{yy} [11].

In order to achieve invariance to image orientation, SURF uses wavelet responses in horizontal and vertical directions for neighbourhood of size 6s, and applies Gaussian weights. The feature descriptors, the wavelet responses in the horizontal and vertical directions are used. The algorithm constructs a square window around the interest point. Then the window is divided into 4x4 pixel sub-regions. The horizontal and vertical wavelet sample points are collected in the form of Equation 5.

$$v_{subregion} = [\Sigma dx, \Sigma dy, \Sigma|dx|, \Sigma|dy|] \tag{5}$$

Using the binary image resulting from the above image processing, the pose of the target object can be calculated. First, the spatial moments of intensity distribution for binary images are calculated using Equation 6.

$$M_{i,j} = \sum_{x,y}(I(x,y)x^i y^j) \tag{6}$$

Then, central moments are used to calculate the total area of the intensity distribution:

$$\mu_{i,j} = \sum_{x,y}(I(x,y)(x - x_c)^i(y - y_c)^j) \tag{7}$$

The coordinate (x_c,y_c), which is the centre of the target object, can be calculated as in Equation 8 [12].

$$x_c = \frac{M_{1,0}}{M_{0,0}}, y_c = \frac{M_{0,1}}{M_{0,0}} \tag{8}$$

It is highly possible that taking pictures from constantly moving quadrotor will also generate noisy information. Thus some filtering algorithms were needed to compensate for this noise and determine the true value. In OpenCV, the Kalman filter algorithm can be used for this purpose. In addition, it can be used to predict future position in case of temporary loss. The Kalman filter estimates the state of a process by using a form of feedback controller. Firstly, the filter estimates the states and then obtains feedback in the form of measurements. Equations for the Kalman filter can be categorized into two groups [13]: the time update equations that are responsible for obtaining the *a priori* estimates for the next time step -> *predictor* and also the measurement update equations that are responsible for the feedback -> *corrector.*

Suppose X is a discrete-time process, Z is a measurement, and W and V represent the process and measurement noise, respectively. The state of X and measurement Z equations can be expressed as follow [14]:

$$X_{k+1} = AX_k + W_k \tag{9}$$

$$Z_k = HX_k + V_k \tag{10}$$

Then, the Kalman filter estimates the state X of time k and corrects the prediction using measurement Z using the following equations [13] :

Time update (prediction):

$$\hat{X}_k^- = A\hat{X}_{k-1} + Bu_{k-1} \tag{11}$$

$$P_k^- = AP_{k-1}A^T + Q \tag{12}$$

Measurement update (correction):

$$K_k = P_k^- H^T(HP_k^- H^T + R)\text{-}1 \tag{13}$$

$$\hat{X}_k = \hat{X}_k^- + K_k(Z_k - H\hat{X}_k^-) \tag{14}$$

$$P_k = (I - K_K H)P_k^- \tag{15}$$

3. Implementation

For this project, two processes were developed that both processes run on the IMX6 board embedded in the Solo – the flight controller and the vision controller. In the case of the flight controller, it is a python script using dronekit python library. Therefore, there is no need to be concerned about platform dependency. However, when it comes to vision controller, it was developed using C++ language and using OpenCV C++ libraries. Cross development environment should be set up as following: build OpenCV libraries for the arm platform and set-up cross development environment to build the vision controller to make it run on the arm platform. Firstly, the complier should be the same one that has been used to build the root-file-system on the IMX6 board. To find this, the following command was run with a binary file in current root-file-system in the solo.

```
$ readelf –l 'binary file name' | grep interpreter
```

As a result of above command, it was found that 'arm-linux-gnueabihf-gcc-4.8' was employed to build the root-file-system in the Solo.

To install OpenCV libraries into the IMX6 board, first OpenCV libraries were built with the above cross complier. Then, OpenCV static and shared libraries were generated. These libraries were then copied into an arbitrary directory in the IMX6. After that, the following was set up to ensure that the application can refer to the shared libraries correctly during run time.

```
$ vi /etc/ld.so.conf // open shared library configuration file
    Add a directory where you copied the shared libraries
$ ldconfig // reconfigure with newly added directory
$ ldconfig –p // check the opencv libraries correctly registered
```

Dronekit-python, which is already installed in the IMX6, allows developers to create applications that can run on-board and communicate with the auto-pilot flight controller (Pixhawk 2) [15]. It provides application program interfaces in the forms of mavlink protocol that applications programs running on on-board companion computer are able to receive current state, parameters information of a vehicle, as well as controlling the vehicle's movement and operation. By controlling the UAV from an on-board application using the dronekit API, researchers are able to add their enhanced functionalities such as vision-based control and artificial intelligence mission decision.

3.1. Vision controller

The vision controller was developed using C++ language and used OpenCV libraries to perform image processing tasks. It ran as a process and continued to run until it received a termination message from the root user. This is shown in Figure 3.

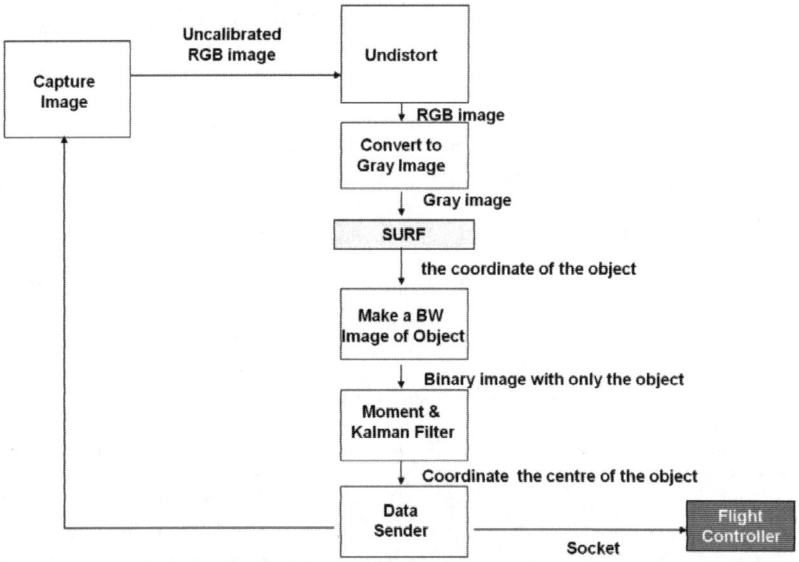

Figure 3. The vision controller flow chart

To capture images, GoPro camera was employed. It was connected to the IMX6 board via HDMI interface and used linux V4L2 (video 4 Linux 2) capture driver. The image format from the camera was YUYV. An issue was found that the capture driver in the IMX6 board allowed only 640 x 480 resolution. So, regardless of the resolution setting of the camera, all captured images are 640 x 480 resolution. For better detection, we needed to capture 720p (1280 x 720) resolution image. To resolve this issue, we did not use OpenCV capture library but ported a framework that directly controlled the capture driver using linux IOCTL kernel interface. As a target object, the symbol of the university as shown in Figure 4 is set. Every detection cycle required two images: one is a target object image and the other is a image that includes the target object. Then, the process tried to find the keypoints and descriptions for both images. Finally, it tried to find any matched keypoints and descriptions from the latter image. As a result, the target object in the latter image can be found as in Figure 5.

Figure 4. The target object

Figure 5. The detection of the target object

Technically, from previous step, what actually acquired was the coordinates of two points: upper left corner and lower right of the target object. Using this coordinate information, a binary image can be made as following Figure 6. Now, using moments functionality, the coordinates of the centre of the object can be extracted as shown in Figure 7. By calling the kalman filter API with the coordinate as an input, the filtered and predicted final coordinates was generated. Throughout the image processing task, the main purpose of vision controller to acquire the coordinate of the target object was achieved. Next, the coordinate information should be transferred to another process, which was flight controller. In order to achieve inter-process communication between these two linux processes, local UDP socket interface was selected due to its simplicity and low overhead. The structure for data communication is as in Table 3.

Figure 6. Binary image of the object

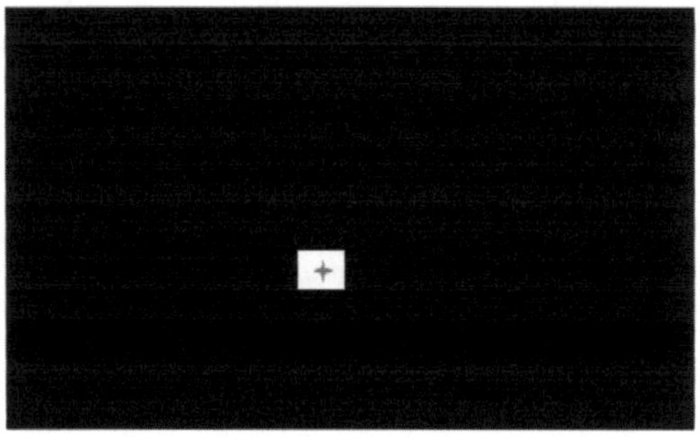

Figure 7. Finding of the coordinate of centre

Table 3. Communication data structure

Type	Name	Description
Unsigned int	Position_x	X coordinate
Unsigned int	Position_y	Y coordinate
Unsigned short	Heading	Angle between quad-rotor heading and the object position

3.2. Flight controller

The flight controller was developed using python script that primarily used dronekit python APIs to control the quadrotor. As explained previously, dronekit python generates mavlink messages to control the behaviours of the Pixhawk 2, the autopilot. Figure 8 shows an overview of the flight controller's behavior. After taking off, it waited for the pose information from the vision controller. As soon as it received the pose information, it made a decision for its next behavior based on its current position status relative to the target object. The first test point was whether the object was in the desired area, which was set as shown in Figure 9. If the object was in the area, the quadrotor hovered. If the object was not in the desired area, the next test point was the angle between the object and the heading of the quadrotor. If this angle was not in the margin, then it tried to adjust the heading by commanding the quadrotor to yaw. Next, if the angle was now in the margin, the only thing left was to command the quadrotor to translate so as to place the object in the desired part of the image. The algorithm for this translational and angular movement is as follows.

```
Input : X,Y,Heading
Output: Turn Angle, Velocity
Constant: Max Speed, Angle Target, PositionX Margin, PositionY Margin, TargetX
Position, TargetY Position, proportional P
Variables : PositionX Difference, PositionY Difference

Repeat
  PositionX Difference = ||TargetX Position|  - |X||
  PositionY Difference = ||TargetY Position|  - |Y||

  If Angle Target > |Heading|
     If PositionX Margin > PositionX Difference and PositionY Margin > PositionY
Difference
        Hovering
     else
       Velocity = Max speed * math.tanh(P * math.fabs(PositionY Diffence))
       send_mavlink_set_velocity_msg(Velocity)
  else
     Turn Angle = Heading * P
     send_mavlink_set_yaw_msg(Turn Angle, direction)
Until receive termination message;
```

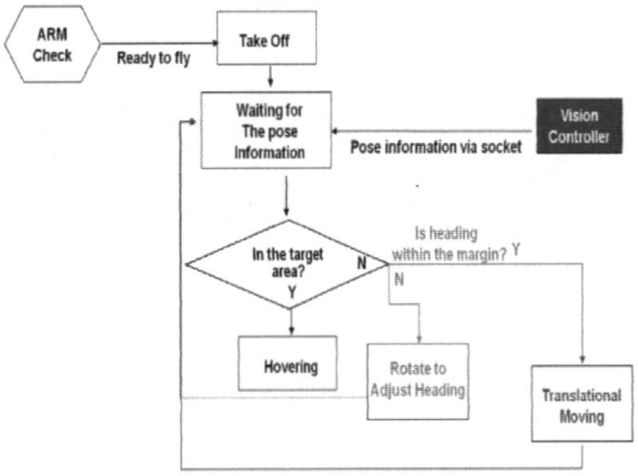

Figure 8. The flow chart of flight controller

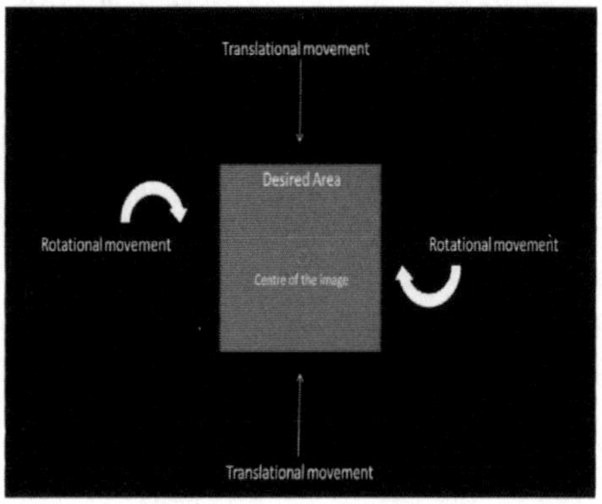

Figure 9. The desired area

4. Experiment

4.1. Hovering stabilization

In this evaluation, the test started with the condition that the target object was in the designed area and the quadrotor took off until it reached 2 m altitude. This is illustrated in Figure 10. When the quadrotor detected that the target object was in the desired area, it started hovering and was required to stay in hovering state since the target object was static. Figure 11, Figure 12 and Figure 13 show the centre pose $(x, y)^T$ of the target object during the hovering.

4.2. Tracking the target object

In this evaluation, the performance of tracking capability was verified. Due to performance limitations of the main embedded board (IMX6), the real time tracking was virtually infeasible. Therefore, in the test, the moving target object moved only a short distance (>1 m) and then waited until the quadrotor detected the new pose of the target object. Figure 14, Figure 15 and Figure 16 highlight the tracking examples. The initial pose of the object was to the left of the quadrotor. Hence the quadrotor made an angular movement first. After the heading of the quadrotor and the object was in the margin, it made translational motion backward.

Figure 10. Quadrotor experiment set up

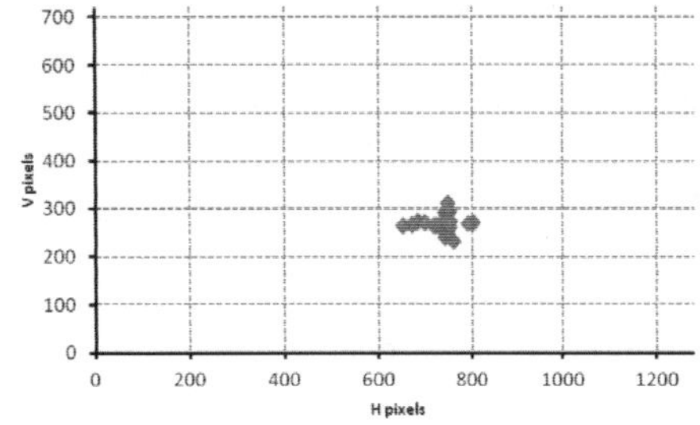

Figure 11. Trajectory of the centre of the target object

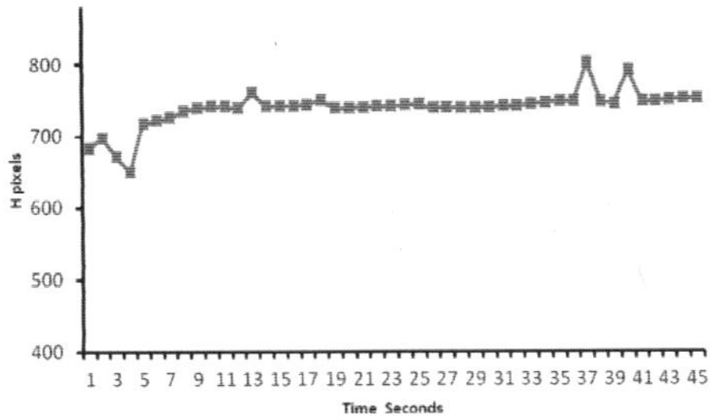

Figure 12. Trajectory of the centre (x axis) of the target object

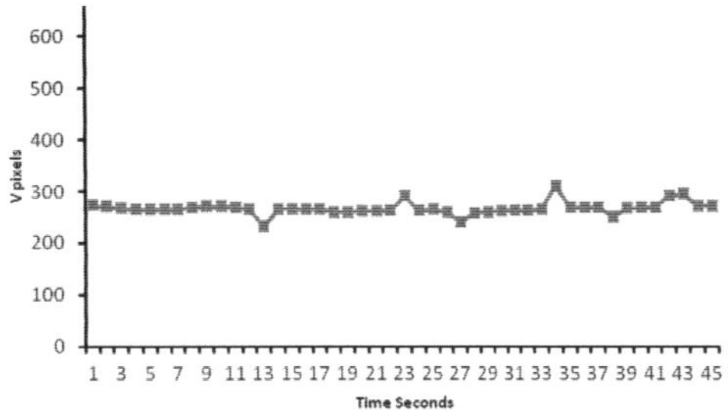

Figure 13. Trajectory of the centre (y axis) of the target object

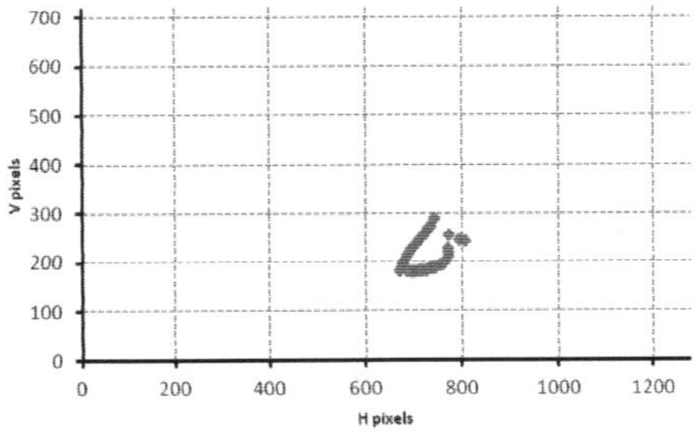

Figure 14. Trajectory of the centre of the target object – tracking

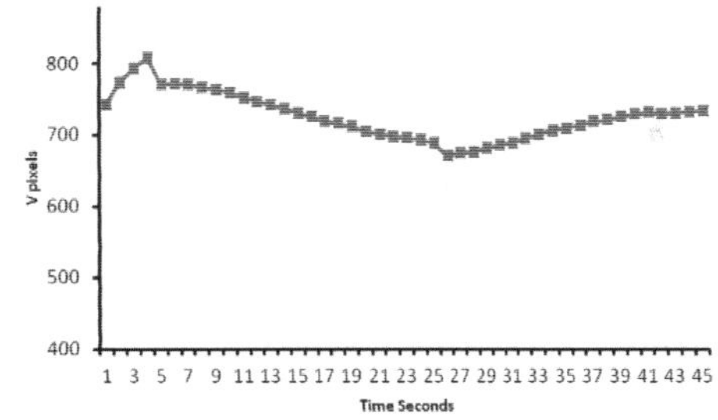

Figure 15. Trajectory of the centre (x axis) of the target object – tracking

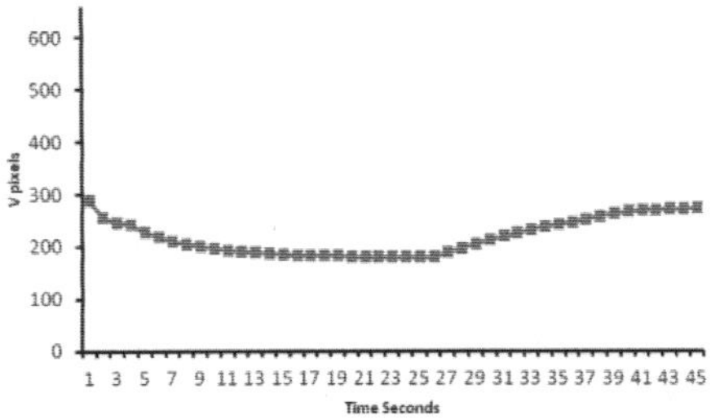

Figure 16. Trajectory of the centre (x axis) of the target object – tracking

5. Conclusion

In this paper, design and implementation of a vision based hovering and tracking algorithm for UAVs was investigated. The main goal was to implement the above capabilities on-board using the minimal reserve capacity in deeply embedded processor of the 3DR Solo quadrotor (IMX6). The algorithm was successfully implemented. The vision controller, which played the role of image capture, calibration, target detection and estimation of target pose, was developed and verified successfully. Using the pose information generated by the vision controller, the quadrotor did its task, hovered and tracked, and steered by the flight controller. Limitations existed mainly in the low performance of on-board system, especially for image processing with OpenCV where significant delays compromised several core functions such as real time tracking.

Acknowledgment

This work was supported by Tyche, the Defence Science and Technology Group's Trusted Autonomy initiative.

References

[1] Caballero F, Merino L, Ferruz J and Ollero A 2008 *Journal of Intelligent and Robotic Systems* **54** 137-61
[2] Courbon J, Mezouar Y, Guenard N and Martinet P 2010 *Control Engineering Practice* **18** 789-99
[3] Karlsson N, Bernado E D, Ostrowski J, Goncalves L, Pirjanian P and Munic M E 2005 *Proceedings of the 2005 IEEE International Conference on Robotics and Automation*
[4] Achtelik M W, Weiss S and Siegwart 2011 *IEEE International Conference on Robotics and Automation (ICRA)*
[5] Nutzi G, Weiss S, Scaramuzza D and Siegwart R 2011 *Journal of Intelligent and Robotic Systems* **61** 287-99
[6] Bohdanov D 2012 *Quadrotor UAV control for vision-based moving target tracking task* Master Thesis University of Toronto
[7] Gomez-Balderas J E, Flores G, Garcia Carrillo L R and Lozano R 2012 *Journal of Intelligent and Robotic Systems* **70** 65-78
[8] Kendall A G, Salvapantula N N and Stol K A 2014 *International Conference on Unmanned Aircraft Systems*
[9] Bay H, Ess A, Tuytelaars T and Gool L V 2008 *Computer Vision and Image Understanding* **110** 346-59
[10] Wikipedia *Image moment* [Accessed online: 21 March 2016]
[11] Bay H, Ess A, Tuytelaars T and Gool LV 2006 *9th European Conference on Computer Vision*
[12] Bohdanov D 2012 *Quadrotor UAV control for vision-based moving target tracking task* Master Thesis University of Toronto
[13] Welch G and Bishop G 2006 *An Introduction to the Kalman Filter* University of North Carolina at Chapel Hill
[14] I. Corporation 2001 *Open Source Computer Vision Library - Reference Manual*
[15] D. Robotics 2015 *About Dronekit*

Design of automatic rotor blades folding system using NiTi shape memory alloy actuator

M I F Ali and E J Abdullah*

Department of Aerospace Engineering, Universiti Putra Malaysia, Malaysia

*ermira@upm.edu.my

Abstract. This present paper will study the requirements for development of a new Automatic Rotor Blades Folding (ARBF) system that could possibly solve the availability, compatibility and complexity issue of upgrading a manual to a fully automatic rotor blades folding system of a helicopter. As a subject matter, the Royal Malaysian Navy Super Lynx Mk 100 was chosen as the baseline model. The aim of the study was to propose a design of SMART ARBF's Shape Memory Alloy (SMA) actuator and proof of operating concept using a developed scale down prototype model. The performance target for the full folding sequence is less than ten minutes. Further analysis on design requirements was carried out, which consisted of three main phases. Phase 1 was studying the SMA behavior on the Nickel Titanium (NiTi) SMA wire and spring (extension type). Technical values like activation requirement, contraction length, and stroke-power and stroke-temperature relationship were gathered. Phase 2 was the development of the prototype where the proposed design of stepped-retractable SMA actuator was introduced. A complete model of the SMART ARBF system that consisted of a base, a main rotor hub, four main rotor blades, four SMA actuators and also electrical wiring connections was fabricated and assembled. Phase 3 was test and analysis whereby a PINENG-PN968s-10000mAh Power Bank's 5 volts, which was reduced to 2.5 volts using LM2596 Step-Down Converter, powered and activated the NiTi spring inside each actuator. The bias spring (compression type), which functions to protract and push the blades to spread position, will compress together with the retraction of actuators and pull the blades to the folding position. Once the power was removed and SMA spring deactivated, the bias spring stiffness will extend the SMA spring and casing and push the blades back to spread position. The timing for the whole revolution was recorded. Based on the experimental analysis, the recorded timing for folding sequence is 2.5 minutes in average and therefore met the required criteria.

1. Introduction

Rotor Blades Folding (RBF) system is a part of the mechanism that has been designed for shipborne operating aircraft, especially helicopter, in order to significantly reduce the rotor disc diameter by transition process from spread to folding position. Suitable disc diameter will enable an aircraft to be stored inside the ship's hangar space. Instability of deck-floor, exposure to the salty environment and confined space working conditions will lead to an increase in safety concern on both personnel and the aircraft. Therefore, fast transition between folding to spread position before flying operation or from spread to folding position for storage is very important. However, most existing helicopters' main rotor blades folding or spread design mechanism in the market are using manual techniques, which fully utilizes human effort, excessive use of special tools and long lists of procedures. There are many researches and studies by the experts from different backgrounds such as aircraft manufacturers and academicians who aimed to improve the design and techniques. Consequently, an Automatic Rotor Blades Folding System (ARBF) is introduced on modern state-of-the-art helicopters such as Agusta NH90, SH60 Sea Hawk, Eurocopter EH101 and many more [1]. However, the development of such system is only done for newly developed helicopters and not for existing fleet, which are in operation worldwide. Due to high cost of research and development, some manufacturers like Agusta Westland still retain the conservative manual technique approach for their newly produced helicopter model AW159 Wild Cat. The Royal Malaysian Navy (RMN) Super Lynx Mk 100 was the older version of AW159. A conventional hydraulic operated ARBF systems requires 3000 psi utility hydraulic pressure and 28 VDC electrical powers, which adaptation to this system means an introduction of a bulky and complex new hydraulic system to the aircraft [2]. Various published and patented documents have also shown that the development of ARBF was evolving from time to time [2-5].

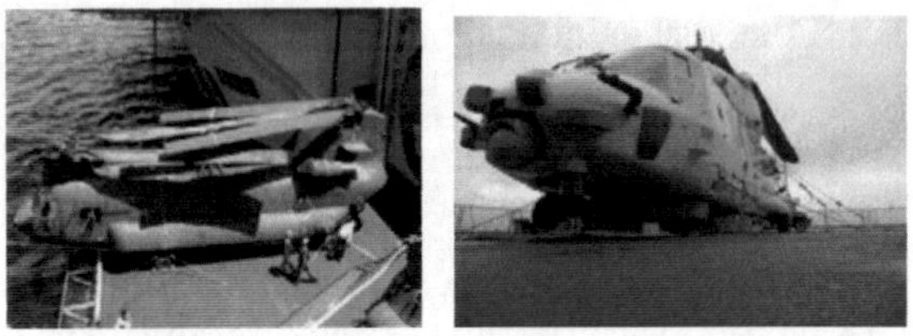

Figure 1. Typical helicopter with folded main rotor blades capability [5]

2. Problem statement

It was observed that the OEM operating design mechanism and techniques for the manual fold main rotor blade of Super Lynx have a few drawbacks, which would affect the overall performance of the aircraft and operations in term of aircraft mission readiness, crew fatigue, safety working condition and aircraft exposure to salty environment. In average, a full folding sequence by utilizing five trained technical crews could took approximately 30 to 45 minutes. As a comparison, a hydraulic operated automatic rotor blade folding system onboard NH90 would take less than two minutes or 93.3% more efficient for a complete folding sequence. For the five technical crew onboard, their duty is not only to fold or spread the helicopter blade before or after flight operations but also to position the 5330 kilograms machine to the spot or storage hangar. If an average five daily flying sortie was planned with certain interval, repetitious fold-spread sequence has to be carried out to fulfill the requirements and hence will cause the crew fatigue and increase the risk of incident or accident.

Obviously, fully automatic ARBF is not a new technology. Most modern naval aircraft like EH101, Sea Hawk, etc. have utilized the motored driven ARBF. Nonetheless, the present systems' principle is using hydraulic concept and custom made for specific aircraft in term of size and also functionality. Therefore, high weight penalty or it may be totally impossible to impose the technology to present in operation aircraft for upgrading programs because it may involve replacement of major components like main rotor blades and hub, and redesigning of new hydraulic system architecture. Thus cost for retro-fit or upgrade is very high. Another aspect to consider is the compatibility issue in term of size and complexity. Therefore, re-designing a new system with simplified concept seems to be a better option.

3. Methodology

Conceptual system design proposed in this study was derived mainly from accumulated information gathered from need analysis study via e-survey to experience end-user as target group, feasibility study of on-the-shelf systems and various researches that have been published till date. Design parameters selection, which encompasses pulling mechanism, controls, power sources and locking mechanism, was done using Morphological and Pugh Matrix. However, the main focus of this study is design of pulling mechanism or actuators, which have to consider several vital factors like weight, contraction length, shape, complexity and reliability. Since the weight and contraction is part of important factors, application of SMA materials is found to be the best alternative as compared to conservative hydraulic or pneumatic system mechanism. Although the venture of SMA materials application for large product design is still outnumbered, potential success for this design is found to be bright. Four combinations of actuator's preliminary design concept were proposed and each of the concept was credited based on the previously stated criteria.

- Concept 1 – Combination SMA wire and SMA spring (series)
- Concept 2 – Combination SMA spring and bias spring (series)
- Concept 3 – Combination SMA wire and SMA spring (see diagram below)
- Concept 4 – Combination SMA wire and bias spring (retractable casing)

Concept 4, which is shown in Figure 2, was chosen as the best combination due to largest contraction and meeting the complexity criteria. Its operational concept mechanism is shown in Figure 3.

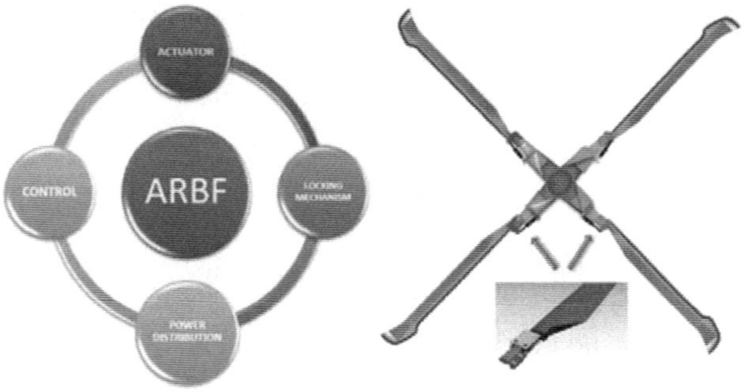

Figure 2. Basic architecture of the developed SMA ARBF system

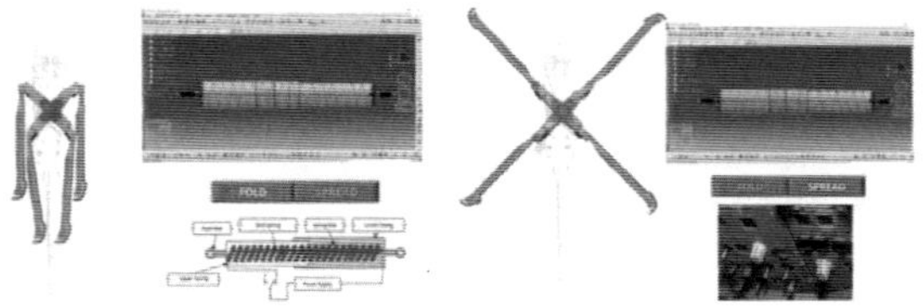

Figure 3. Operational concept mechanism of SMA ARBF system

4. Design calculation and verification

4.1. Power requirement

In this ARBF, SMA will be the main materials application for actuators and Super Lynx will be the subject of concern. The design requirements to be considered have to be calculated and verified. The average weight of one Super Lynx's main rotor blades is 45 kg. Thus, the load required to displace the blades to either approximately 20 degrees or 70 degrees to the back must encountered the down force of approximately 441.45 N. A very special aspect about SMA, with the diameter of 0.51mm, one string is capable to pull up to 3.5 kg. By rough calculation, only less than 20 thin strings are required to do the job, which is such an amazing breakthrough. The electrical power required in activating the SMA is very minimum and can be calculated based on the string diameter. In this actuator design, the SMA is best to be shaped as a spring that will perform dual functions once activated to pull or push the blades to the desired position and act as a spring bias to hold the actuators and blades at locking positions or back to spreading positions. For electrical activation of SMA spring type, the following equations can be used.

$$\text{Resistance: } R = \frac{\rho.L}{A} \quad (1)$$

where ρ = SMA resistivity, L = free length of element and A = cross sectional area.

Voltage (Ohm's Law): $V = I \cdot R$ (2)

Power Requirements: $W = I \cdot V = I^2 \cdot R$ (3)

4.2. Spring calculation

4.2.1. SMA spring (extension type). A SMA extension spring will behave as shown in Figure 4. At the low temperature the spring will be extended and when heated will contract providing a pulling force.

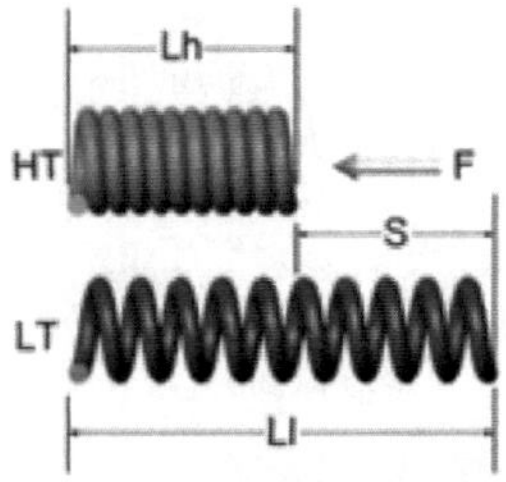

Figure 4. SMA extension spring actuation

With the exception of spring length, the following equations can be used to determine the spring diameters, number of turns, spring rates and reset force. Since the actuation for a spring working in extension is reversed from that in compression, the lengths must be re-evaluated. The spring body length describes a fully compressed spring as in Equation 4.

$$L_b = d\,.\,(n + 1) \tag{4}$$

where n is the number of turns of the spring. The free length of the spring or the length at which the shape setting takes place can now be evaluated as a function of the body length as in Equation 5.

$$L_f = L_b + 2\,.\,ID \tag{5}$$

When heated the spring will have a length as calculated by Equation 6.

$$L_h = L_f + \delta_h \tag{6}$$

Since the stroke must be equal to the difference in the high and low temperature lengths, then the low temperature length can be calculated as in Equation 7.

$$L_l = L_h + S \tag{7}$$

where S is the stroke.

4.2.2. Bias spring (compression type). For each actuator, the compression type of spring was used as a bias spring in order to re-position the blades at spreading configuration. The selection of the spring properties is calculated as follows.

For compression spring, the spring deflection: $\delta = \frac{8N_aD^3P}{Gd^4} = \frac{8N_aD^3P}{Gd^4}$ (8)

The stiffness constant: $k = \frac{P}{\delta} = \frac{Gd^4}{8N_aD^3}$ (9)

Torsional stress: $\tau_0 = \frac{8DP}{\pi d^3}$ (10)

$$\tau_0 = \frac{Gd\delta}{\pi N_a d^2} \tag{11}$$

IOP Conf. Series: Materials Science and Engineering **152** (2016) 012016 doi:10.1088/1757-899X/152/1/012016

Corrected torsional stress: $\tau = \chi\tau_0$ where stress correlation factor, $\chi = \frac{4c-1}{4c-4} + \frac{0.615}{c}$ and spring index, $c = \frac{D}{d}$.

Diameter of material: $d = \sqrt[3]{\frac{8DP}{\pi\tau_0}} = \sqrt[3]{\frac{8\chi DP}{\pi\tau}} = \text{mm}$ (12)

Number of active winding: $N_a = \frac{Gd^4\delta}{8D^3P} = \frac{Gd^4}{8D^3k}$ (13)

Alternatively, the number of active winding can be determined as: $N_a = N_t(X1+X2)$ (14)

where X1 and X2 are the number of turns at each end of the coil.

Spring-retained energy: $U = \frac{P\delta}{2} = \frac{k\delta^2}{2}$ (15)

For this study, the parameters involved in the spring properties are tabulated in Figure 5.

5. Design analysis

5.1. SMA behaviour

In order to assist in SMA design selection and application, series of experiments have been performed. The specifications of the SMA is listed in Table 1 [6]. A consistent load of 1.206 kg were used for the experiment. All load were hanged vertically to each SMA (wire and spring) with both ends connected to the power supply. The SMA parameters (initial length, length with respect to supplied voltage and SMA samples temperature) was recorded with increment of voltage and current. Temperature sensor of SMA sample were detected via thermocouple wire which connected to LabView Software. The experimental setup is illustrated in Figure 6.

5.2. Prototype development

A prototype of the selected concept of ARBF was fabricated for proof of operational concept, which encompassed the following major components:

- Main rotor hub: made from ply-wood in 1:45 scale. The hub was attached to a base for stability.
- Main rotor blades: made from wood in 1:45 scale. Attached with metal coupler with two holes each to functions as pivot point and lock.
- SMA actuator

The prototype is as shown in Figure 7 and Figure 8. The main components was assembled and tested. The evolution time for fording a spreading sequence was recorded for comparison.

Symbol	Value	Unit
N_a	17	
D	18	mm
G	78x10³	N/mm²
P	6.8	N
d	7.5	mm
δ	9.22	mm
k	0.74	N/mm
τ		N/mm²
τ_0	0.739	N/mm²
c	2.4	
χ	1.0875	
U	31.35	N/mm

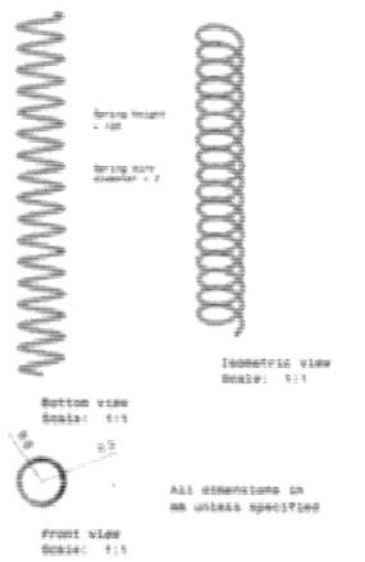

Figure 5. Symbols used in spring design formula

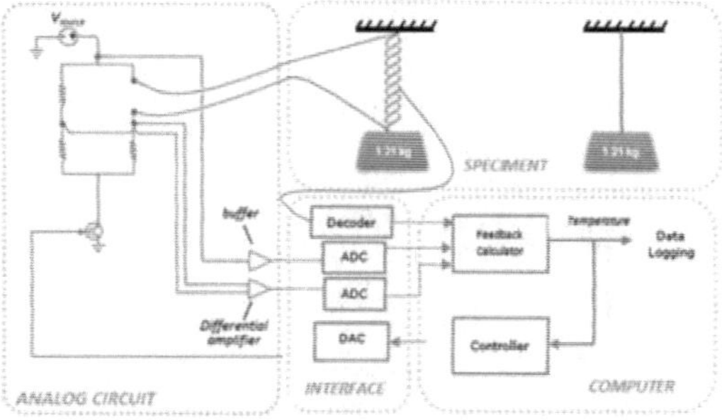

Figure 6. Experimental setup

Table 1. Specification of SMA samples

TYPE	WIRE	SPRING (Tension)
Material	Nickel Titanium (NiTi)	Nickel Titanium (NiTi)
Pull Force	1280 grams	Not Stated
Resistance	12.2 Ohm/meter	Not Stated
Length	35 cm	2 cm
Max Deformation	Not Stated	15 cm
Activation Temperature	70 – 90 °C	45 – 50 °C
Outer Diameter (Wire)	0.31 mm	0.75 mm
Outer Diameter (Winding)	N/A	6.5 mm
Number of Windings (Spring)	N/A	21

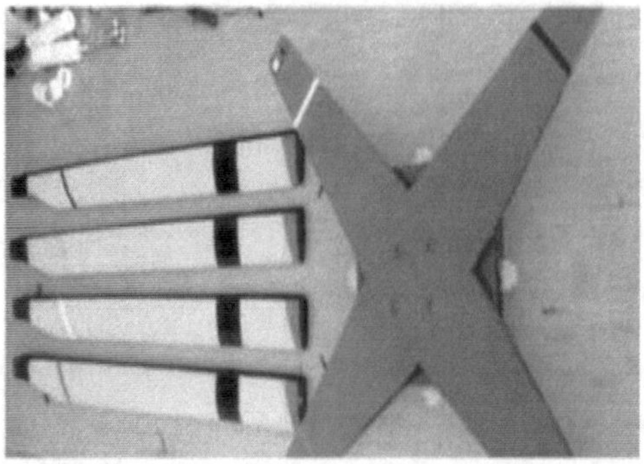

Figure 7. Prototype of main rotor hub and blades

(a) (b)

Figure 8. (a) SMA actuators and power supply and (b) ARBF assembly

5.3. Hardware in the loop (operation and control)

5.3.1. Experimental block diagram (operation). The analysis of concept of the operations for SMART ARBF was carried out with the layout in Figure 9. Main power supply was supplied by a PINENG PN-968s-10000mAH universal power bank unit with output of 5V DC. Since the power requirement for SMA activations obtained from experiment is 2.4V, a DC-DC Step-Down Converter Module was used to obtain desired output for prototype activations. The activation of folding/spread mechanism was controlled via an ON-OFF switch. All four actuators were connected to power supply module in parallel. The timing for all four blades to fold or spread were recorded and tabulated.

5.3.2. Experimental block diagram (control). For the purpose of proof of the operational concept, the activation was only done via an ON-OFF switch. However, for a better and more accurate control of SMA temperature, the used of PID controller is much better. Simulation of PID control block diagram as shown in Figure 10 is using LabView software to ensure that there is no oversupply of voltage and current or over-heat the SMA spring, which could degrade the overall performance for a long period of operations.

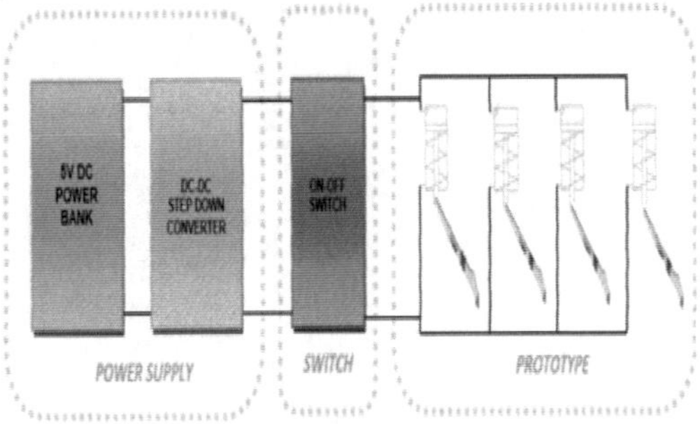

Figure 9. Block diagram (operation)

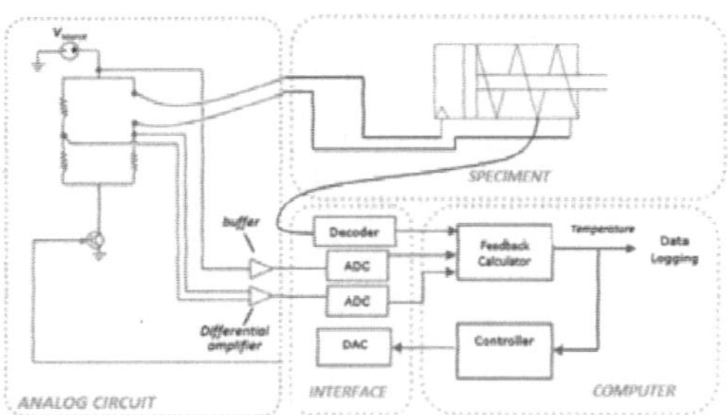

Figure 10. Block diagram (controls)

6. Results and discussion

A series of testing has been carried out using the fabricated and assembled prototype. The results of testing are presented in Table 2. It can be deduced that folding or spreading time of main rotor blades was not proportional to the voltage supplied. This is probably due to the inaccurate control of SMA temperature whereby the value of the temperature might also be influenced by the room temperature. However, the average time still met the design criteria, which is below five minutes. Furthermore, time for spreading sequence was longer than folding and this might be due to the internal resistance inside the actuators casing caused by friction of upper and lower casing, insufficient stiffness coefficient of spring bias in order to counter the stiffness of SMA spring and also slow decreasing of SMA spring temperature with regards to releasing of its pulling force.

Table 2. Average voltage for folding and spreading

TEST	VOLTAGE	TIME (MINUTE)	
		FOLD	SPREAD
RUN 1	2.40	2.3	3.2
RUN 2	2.38	3.1	3.1
RUN 3	2.41	2.2	3.6
RUN 4	2.45	2.2	3.2
RUN 5	2.50	2.1	3.2
RUN 6	2.39	3.3	3.1
RUN 7	2.40	2.6	2.9
RUN 8	2.38	3.3	3.6
RUN 9	2.44	2.5	3.5
RUN 10	2.50	2.4	3.2
AVERAGE	2.425	2.6	3.26

7. Conclusion

This paper presented the requirements for development of a new ARBF systems, which could possibly solve the availability, compatibility and complexity issue of upgrading a manual to fully automatic rotor blades folding system of a helicopter. The Royal Malaysian Navy Super Lynx Mk 100 was chosen as a baseline model. The design of the SMART ARBF's SMA actuator and proof of operating concept using a developed scale down prototype model have been presented. The performance target for full folding sequence is less than 10 minutes. Based on the experimental and numerical analysis, the recorded timing for folding sequence is 2.4 minutes in average and thus met the required criteria.

References

[1] Agusta Westland Official Website [Accessed on April 2016]
[2] Integrated Publishing *Automatic Rotor Blades Folding System* [Accessed on April 2016]
[3] *Main Rotor Blade Folding* Compound Integrated Electronic Technical Publication
[4] Chopra I 2002 *AIAA Journal* **40** 2145–87
[5] University of Maryland 2005 *Rotary-Wing Flight Control System* Annual AHS International Student Design Competition
[6] Dynalloy *Technical Characteristics of Actuator Wires: Flexinol–Dynaloy* [Accessed on April 2016]

Development of tilt-rotor unmanned aerial vehicle (UAV): material selection and structural analysis on wing design

M F Saharudin
Jabatan Penyenggaraan Pesawat, Politeknik Banting, Selangor, Malaysia

firdaus.saha@polibanting.edu.my

Abstract. This paper presents the design of a tilting rotor unmanned aerial vehicle (UAV), evaluation of flight loads based on the standard requirement, structural analysis to determine stress and sizing of the wing, and flight test of the UAV. The main objective is to perform structural analysis to size the UAV's wing section. The analysis shows that the structure design of the wing is safe to be used.

1. Introduction

Aeronautic industries have long begun and have been growing over decades with newer inventions, technologies as well as rapid research in this promising field [1]. There are a lot of researches have been made throughout the world including aerial mapping, unmanned aerial vehicle (UAV), etc. The search for an aircraft type with vertical take-off and landing (VTOL) capabilities has triggered the imagination of designers and inventors to produce numerous configurations using a wide variety of lifting and propulsion devices [3]. A summary of some of these configurations is available in the V/STOL (Vertical or Short Take-off and Landing) concepts illustration prepared by the McDonnell Aircraft Company in the 1960s [6]. For various aircraft types considered, one of the key distinguishing features is associated with the device used for providing the vertical lift. The development of VTOL is very important as it has many usage such as in the field of safe and rescue, military application and even some commercial applications [2]. The usage of VTOL cuts the runaway distance, making it easy to take-off and land on aircraft carrier as well as other constrained areas where the paved runaway or space is limited.

Throughout their long history, UAVs have played such an important role whether in warfare or surveillance. The idea of unmanned is to control a machine digitally and replace the needs of human intelligence on-board, saving the human capital for much better use. When compared to manned aerial vehicles, UAVs are believed to provide two vital benefits – they are cost effective and reduce the risk to a pilot's life [4]. However, accident rates in today's UAVs are over 100 times compared manned aircraft. Therefore, improved safety and reliability are still required. The requirement of this research is to design and fabricate a remote controlled aircraft, which is also known as Tilt-Rotor UAV, that can take-off and land vertically. Besides, the study also done to identify the material selection for the UAV and demonstrate the analysis on the wing design of the UAV. In order to achieve the whole benefits and purposes out of Tilt-Rotor UAV, there are a lot of obstacles need to be overcome. Since the rotors can be configured to be more efficient for propulsion (e.g. with root-tip twist) and it avoids a helicopter's issues of retreating blade stall, the tilt-rotor can achieve higher speeds than helicopters. A tilt-rotor aircraft differs from a tilt-wing in that only the rotor pivots rather than the entire wing. This method trades off efficiency in vertical flight for efficiency in STOL/STOVL operations.

Nevertheless, the scope of this research paper is limited to material selection and structural analysis on the wing design. The primary objectives of this project are to design VTOL of Tilt-Rotor UAV by surveying the material selection and to perform structural analysis to determine the sizing of wing.

2. Materials and methods

This study is focused on the development of multi tilt-rotor UAV by concentrating on the structure and design analysis. The information obtained will guide the UAV to move at tilting concept mechanism towards the respective object of interest. In accomplishing the outlined objectives, several guidelines have been observed: reviewing the previous related VTOL UAV projects and performing a standard structural analysis by several methods in aircraft structural analysis design concept.

2.1. Material selection and conceptual design

In this part, it is basically focused on selecting the suitable material to fabricate the tilt-rotor UAV and adhesive needed for this project. Some reference reading is made based on material selection for the main body such fuselage, wing and tail, type of propeller, type of adhesive and type of motor. Table 1 highlights some of the optional materials that can be selected for the main body of the UAV.

Table 1. Justification material selection for the main body [8]

Materials	Pros	Cons
Balsa wood	• Porous • Less glue required • Lightweight • Widely available • Stiff • Easy to sand	• Varying strength • Very expensive
Basswood	• Won't crush • Lightweight	• Hard to sand • Not widely available • More expensive
Depron foam	• Very lightweight • Easy to shape • Flexible • Very cheap	• Not very strong
Plastic	• Strong • Rigid	• Rigid • Hard to work with • Expensive • Relatively heavy
Metal	• Very strong • Rigid	• Very heavy • Expensive • Hard to work with • Not widely available
Fibre glass	• Very strong • Very lightweight	• Very expensive • No previous experience • Not widely available

Depron foam was chosen as the base frame material due its low cost and easy fabrication properties, despite Fiber Glass having a significant strength to weight advantage. The aluminum box are put in the wing for tilt rotor and the body is covered by fibre glass composite. Table 2 tabulates some of pros and cons of alternatives for the propellers.

IOP Conf. Series: Materials Science and Engineering **152** (2016) 012017 doi:10.1088/1757-899X/152/1/012017

Table 2. Justification for propeller selection [8]

Propellers	Pros	Cons
Dual Blade	• Easily available • Very efficient • Easy to use • Fairly cheap	• Larger diameter
Multi Blade	• Smaller diameter	• Less available • Less efficient
Wood Blade	• Very rigid • Efficient • Light	• Breaks easily
APC Blade (Metal)	• Don't break as easily • Efficient	• Heavy

For this tilt-rotor VTOL UAV, the location of the propeller is placed at the tip of the wing. The thrust must be at the *cg* location to make it more stable. When the aircraft experience gust load during hover, the tail rotor will produce thrust to make it stable. It was decided to use two-blade propellers as compared to three-bladed system used by the real V-22 Osprey [2]. This is due to increased efficiency of the two-bladed propellers over the three-bladed propellers. Moreover, justification for the adhesive selection can be seen in Table 3.

Table 3. Justification for adhesive selection [8]

Adhesives	Pros	Cons
Wood Glue (Urea)	• Easiest to use • Low cost • Light color	• Poor heat resistance • Poor moisture resistance • Bond not very strong
Hot Glue	• Quick cooling time • Relatively easy to use • Low cost	• Bond not strong • Leaves residue • Visible on plane
Gorilla Glue	• Very light • Expands while setting • Best for wood than other materials • Waterproof	• Hard to work with • Contains air bubbles • Somewhat expensive
Pro-bond Glue	• Expands when dry • Less glue required • Cheap • Water-resistant	• Heavy
Rubber Cement	• Strong flexible bond • Easy to peel off • Not brittle	• Flammable • Highly toxic • Expensive
Super Glue	• Very strong bond • Often used for model aircraft • Versatile • Water resistant	• Expensive • Can become brittle • Long cure times

To make the aircraft in stable and solid form, a strong adhesive connection must be used to ensure that the aircraft does not broken during flight or VTOL. On the other hand, some considerations for the motor selection is presented in Table 4.

Table 4. Justification for motor selection [8]

Motors	Pros	Cons
Electric	• Cheap • Easy to run • Clean • Doesn't require gasoline • Lightweight	• Low power / torque
Nitro	• Relatively cheap • Wide availability • High torque and power	• Special mixture of fuel • Heavy
Gas	• High torque and power	• Not as available • Heavy • Special mixture of fuel • Expensive
Jet	• Extreme power	• Extremely expensive • Not as available

Based on the above justification, the main body will be made from depron foam board due its flexibility, easy to shape, price tag and light in weight. For the propeller, the dual propeller is chosen due to its availability, efficiency and price tag. As there are a lot of parts that needs to be mounted, the glue will be the best solution due to the low operating cost and high availability. Last but not least, for the propulsion part, electric motor is chosen as it is very cheap to purchase, environmental friendly and easy to operate.

2.2. Conceptual design of tilt-rotor UAV

Aircraft design is basically a combination of science and art together in producing a beautiful machine. In the design phase, whether for normal configuration like the common civilian transport aircraft or for new configuration, there are few steps that need to be followed in order to design and build an aircraft. These design phases are in chronological order: conceptual design, preliminary design, detail design. In conceptual design phase, it basically starts by setting a few specifications for the new aircraft. Every designer has their own approach in developing their conceptual design. However, there are some basic guidelines that are vital and significant for conceptual design. They are design requirement, estimation of the weight of aircraft, identification of some critical performance parameters, type of configuration, performance analysis and optimization. For this study, the mission specifications of the tilt-rotor UAV are presented in Table 5. Some initial conceptual design estimations are then presented in Table 6.

Table 5. Preliminary mission specifications of the present UAV [7 – 8]

Mission Specifications	Mission Profile
V_{max} = 30 m/s V_{climb} = 49 ft/s Rate of climb = 10 ft/s V_{stall} = 39 ft/s V_{cruise} = 69 ft/s Range = 10000 ft Altitude = 1000 ft Endurance = 20 min Rho (sls) = 0.00238 slug/ft^3 Rho (cruise) = 0.002308 slug/ft^3	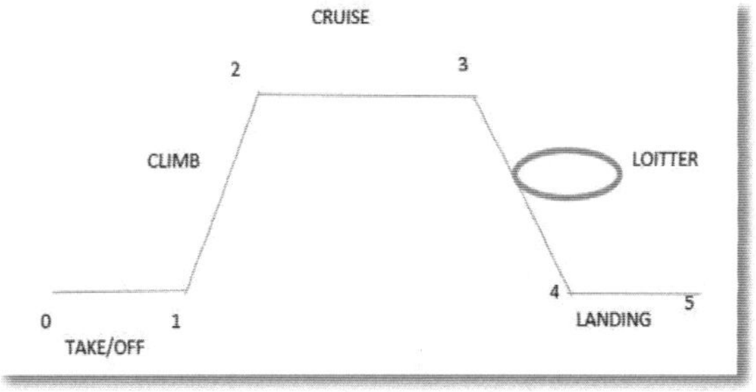

Table 6. Design calculations

	FORMULA/DESIGN	CALCULATION VALUE
MAXIMUM TAKE OFF WEIGHT	STALL: $\frac{W}{S} \ll \frac{1}{2}.\rho.Vstall^2.C_{lmax}$ TAKE-OFF: $\frac{W}{S} \ll TOP.\sigma.C_{lTO}.\frac{Hp}{w}$ CLIMB: $\frac{W}{S} = \frac{\left(\frac{T}{W}-G\right) \pm \sqrt{\left(\frac{T}{W}-G\right)^2 - \left(\frac{4Cdo}{\pi e}\right)}}{2/q\pi Ae}$ CRUISE: $\frac{W}{S} = q\sqrt{\pi AeCdo}$	STALL: $\frac{W}{S} \ll 1.79\ lb/ft2$ TAKE-OFF: $\frac{W}{S} \ll 1.411\ lb/ft2$ CLIMB: $\frac{W}{s} = 39.896\ lb/ft2$ CRUISE: $\frac{W}{S} = 1.582\ lb/ft2$
WING LOADING	The lowest wing loading is the one to be selected.	$\frac{W}{S} = 1.411\ lb/ft2$
FUSELAGE SIZING	$L = aWo^c$ 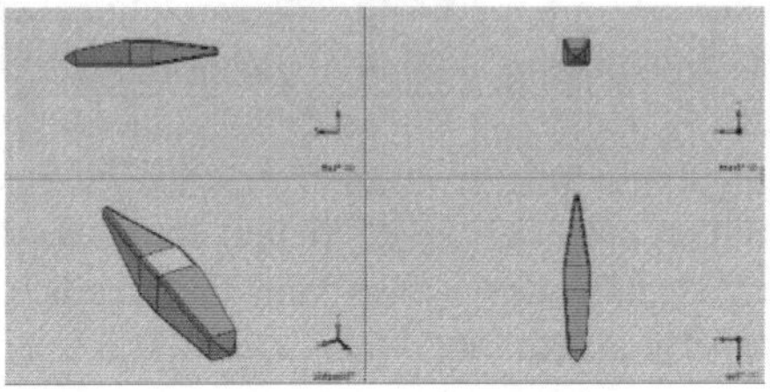 **Figure 1.** Fuselage sizing	$L = 2.31ft = 700\ mm$
WING SIZING	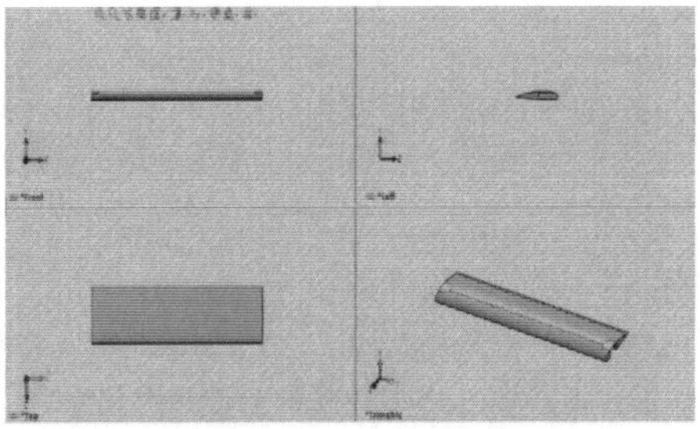 **Figure 2.** Wing Sizing	$\frac{Wo}{S} = 1.411,\ s = \frac{Wo}{1.411} = 1.79\ ft2$ $b = \sqrt{A \times S} = 3.3\ ft$ $Cr = \frac{2S}{b(1+\lambda)} = 0.54$ $\bar{c} = \frac{2}{3}.\mathrm{Cr}.\left(\frac{1+\lambda+\lambda^2}{1+\lambda}\right) = 0.54$ $Ct = \lambda Cr = 0.54\ ft,$ $\bar{y} = \frac{b}{6}\left(\frac{1+2\lambda}{1+\lambda}\right) = 0.82$
TAIL SIZING	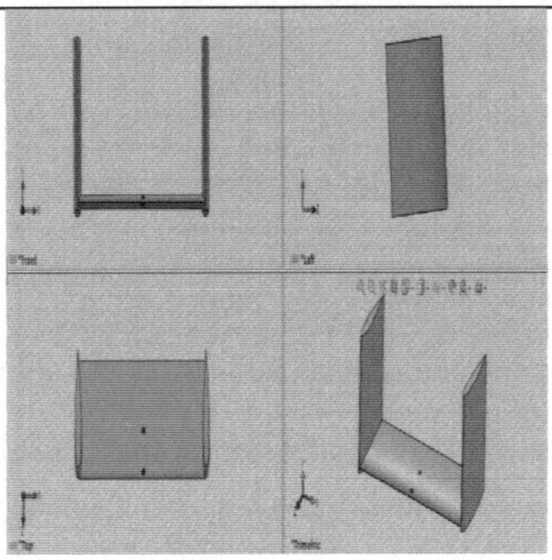 **Figure 3**. Tail Sizing	<u>VERTICAL TAIL</u> $Cvtnew = 0.95(0.04) = 0.038$ $Svt = \frac{Cvtnew.bw.sw}{Lvt} = 0.162ft$ $Lvt = \frac{60}{100}L = 1.386ft$ $b = \sqrt{A \times S} = 0.44ft$ $Cr = \frac{2S}{b(1+\lambda)} = 0.368ft$ $Ct = \lambda Cr = 0.368\ ft$ $\bar{c} = \frac{2}{3}.Cr.\left(\frac{1+\lambda+\lambda^2}{1+\lambda}\right) = 0.368ft$ $\bar{y} = \frac{b}{6}\left(\frac{1+2\lambda}{1+\lambda}\right) = 0.11\ ft$

Table 6. Design calculations (continued)

	FORMULA/DESIGN	**CALCULATION VALUE**
TAIL SIZING		HORIZONTAL TAIL $Chtnew = 0.95(0.5) = 0.475$ $Sht = \frac{Chtnew.Cw.sw}{Lvt} = 0.361ft$ $Lvt = \frac{30}{100}L = 0.693ft$ $b = \sqrt{A \times S} = 1.46ft$ $Cr = \frac{2S}{b(1+\lambda)} = 0.38ft$ $Ct = \lambda Cr = 0.114ft$ $\bar{c} = \frac{2}{3}.Cr.\left(\frac{1+\lambda+\lambda^2}{1+\lambda}\right) = 0.27ft$ $\bar{y} = \frac{b}{6}\left(\frac{1+2\lambda}{1+\lambda}\right) = 0.27\ ft$
CENTRE OF GRAVITY	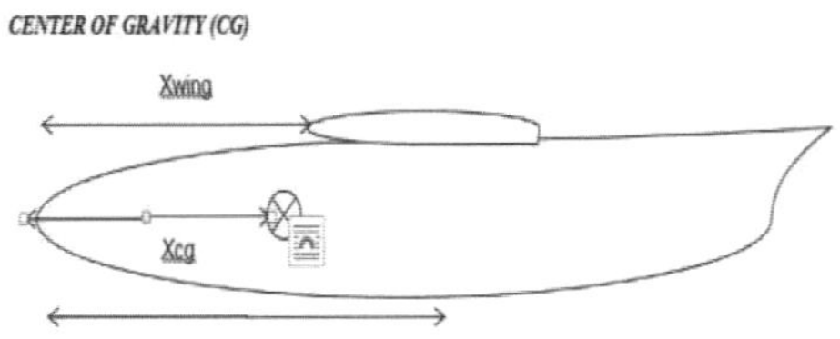 **Figure 4.** Centre of gravity	Weight distribution : Wo= 2.52 lb , Wwing= 0.25 kg , Wengine=0.20 kg, Wpl = 0.4 kg , Wtail= 0.25 kg $\sum Mcg = \sum Mle$ without wing : $x = \frac{0.20(0.9) + 0.4(0.3) + 0.25(2)}{0.2 + 0.4 + 0.25} = 0.941\ ft$ with wing : $Xcg = 0.97\ ft$ $Xcg = 0.97ft \simeq 296\ mm$ $static\ margin = \frac{(Xn - Xcg)}{c} = 0.1$ $0.1 = \frac{Xn - 0.97}{0.54} = Xn = 1.024$ $Xac = Xn - Xvt = 0.794\ ft$

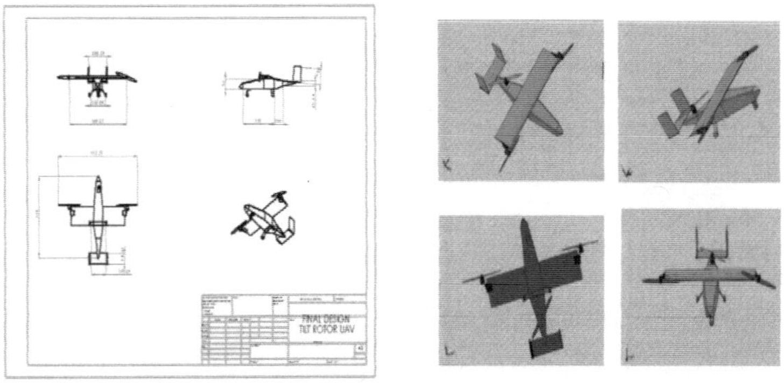

Figure 5. Layout and 3D view of the UAV design after achieved the design requirements

IOP Conf. Series: Materials Science and Engineering **152** (2016) 012017 doi:10.1088/1757-899X/152/1/012017

3. Results and discussion

Before the structure of the wing can be designed, there is a need to determine the loads that will be imposed on the aircraft's wing. This section deals with the general issue of aircraft loads and how they are predicted in the early stages of the design process. The load factor is the multiplier used on limit load to determine the design load. In this case, since there are specific requirements for the factor of safety, 4.5 will be used as the factor of safety as obtained from the STANAG USAR requirement [14]. The safety factor, N_z from the STANAG requirement is calculated as follows, taking the weight to be 1.3 kg or 2.866 lb:

$$Nz = 2.1 + (\frac{10900}{W+4536}) = 2.1 + (\frac{10900}{2.866+4536}) = 4.5$$

3.1. Structural analysis of wing design

In this section, the structural analysis is mainly about investigation of shear force and bending moment at the wing of the tilt rotor UAV. This theory will be applied at the calculation part for the tilt-rotor UAV project.

Table 7. Justification for structural analysis of wing design

ITEM ANALYSIS	RESULT	FIGURE
Load Factor	$Nz = 4.5$	
Shear force		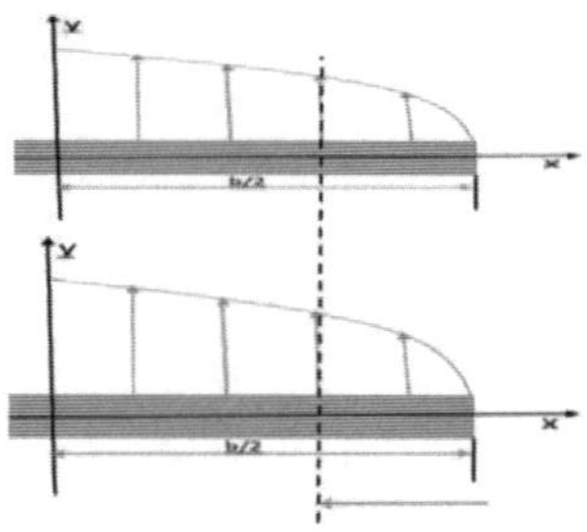
Va (aerodynamic shear force)	$Va = 0.2088\ N$	
Vp (propeller shear force)	$Va = 7.505N$	
$V(Total)$	$V = 7.7138N$	
Bending moment		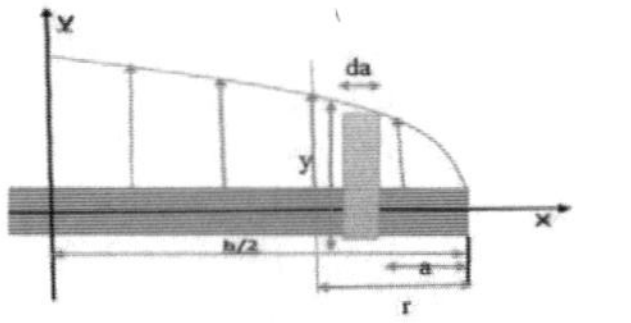
Ma (aerodynamic moment)	$Ma = 0.02724\ N.m$	
Mp (propeller moment)	$Mp = 1.3133\ N.m$	
$M(Total)$	$M = 1.34054$ N.m	
Moment of Inertia	$I = 2.62083 \times 10^{-9}$	
Shear Stress	$\tau = 344.91\ KPa$	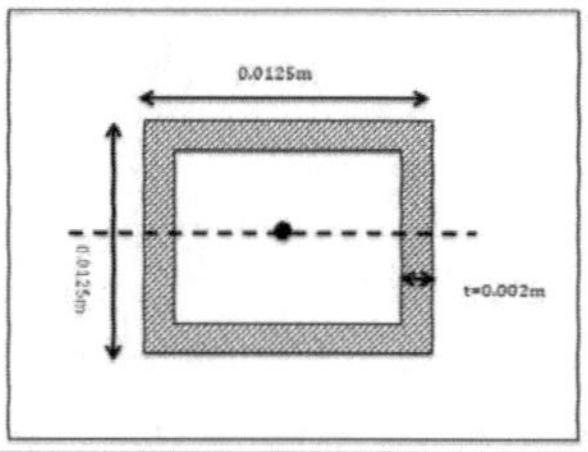
Bending stress	$\sigma = 4.260\ MPa$	$\sigma_x = -\frac{M_z y}{I_{zz}} + \frac{M_y z}{I_{yy}}$

Aluminium 2014-T6 was used as the reference, which $\sigma y = 414$ MPa. To compare it :

$$\sigma = 4.260\ MPa < \frac{414}{1.5} \quad \rightarrow \sigma = 4.260\ MPa < 276\ MPa$$

where σ is $4.260\ MPa$. This shows that the wing bending stress is less than bending stress yield, even though it is very small but may increase the weight later. For the shear stress:

$$\tau = \frac{VQ}{It} = 344.91\ KPa$$

where τ is 344.91 kPa, V is 7.7138 N , Q is $2.34375X10^{-7}$, I is 2.62083×10^{-9} and t is 0.002. In aircraft structure design, one of the most important factors is load factor. Each design of aircraft has its own *V-n* diagram. Here the *V-n* diagram is the same as RV-9 [8]. Since the airspeed is no more than 25 kts, a load factor of 1.5 is given, so the maximum stress is 276 MPa. For aluminum, the stress strain curve is shown in Figure 6. According to the calculation, the maximum stress is $4.118\ MPa$, which is far smaller than the upper yield point for aluminum, hence aluminum is safe to use.

3.2. Wing design simulation analysis

After the theoretical structural analysis has been performed by proving design requirement calculation and mechanical structure on wing parts, Solidworks and Catia V5 software have been used to design and simulation for this research. The load distribution is assume to be elliptical and the fixed support is at the edge of the wing, as shown in Figure 7. The wing is arranged as high wing location. The load distribution pressure is 7.784 N/m^2 as calculated by using area under the elliptical graph at the wing. The material of the wing is fixed foam, which has tensile strength of 7 lbs/sq.in. The properties of the material are 1.2 lb/ft^3 of the weight and the density of the foam is 1.2 lbs/cu.ft.

Mesh analysis on the wing is done to get the grid solid region on the wing and to justify structural wing design for doing further element structural analysis. This is shown in Figure 8. Figure 9 shows that the value of bending stress of the wing using this software simulation is 5.42 MPa. The diagram shows that the structure is safe because most region of the wing is blue, which not in critical condition. The wing is safe to use in the UAV after doing structural analysis. The UAV wing span is 1.96 ft and the chord length is 0.54 ft. After doing simulation by using Solidworks 2014 software, it shows that the bending stress is 5.42 MPa. From manual calculation, the theoretical bending stress is found to be 4.26 MPa. There is a small difference with approximately 10% between both values.

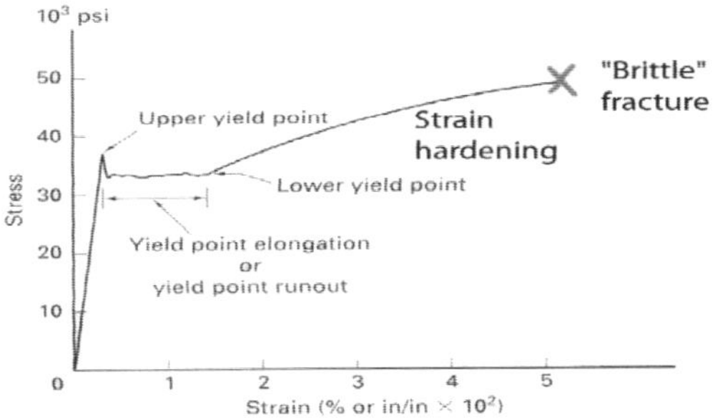

Figure 6. Stress-strain curve diagram [8]

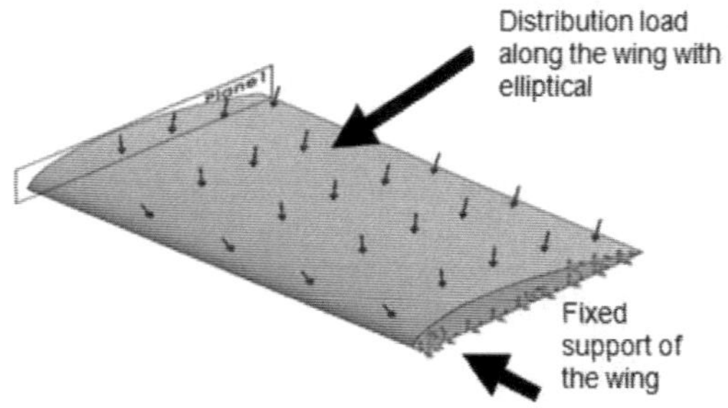

Figure 7. Load distribution applied along the wing

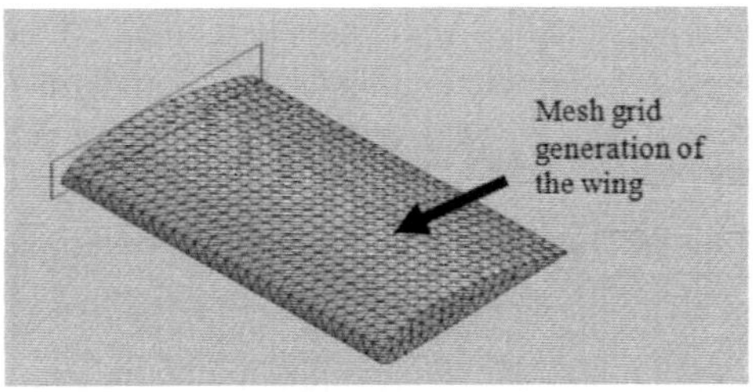

Figure 8. Mesh generation for wing

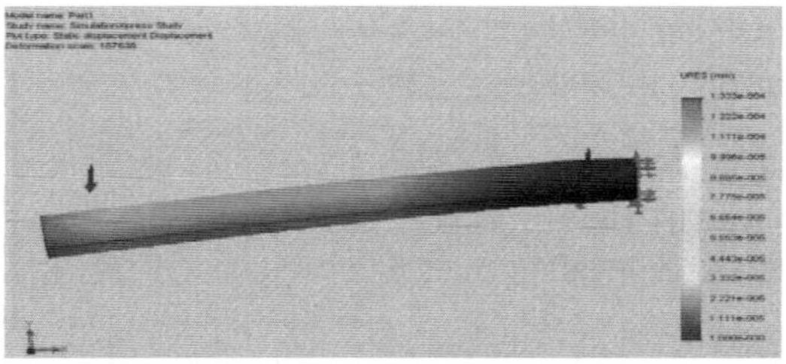

Figure 9. Von Mises stress analysis for wing

IOP Conf. Series: Materials Science and Engineering **152** (2016) 012017 doi:10.1088/1757-899X/152/1/012017

4. Conclusion

In this paper, the details of design, material selection and structural analysis on the wing of rotor UAV platform have been outlined. The designed Tiltrotor UAV is economical, moderately functional and an excellent platform for academic research purpose. By performing the structural analysis of the model, the results show that the design is safe for use. All in all, the objective of this project is achieved and it is ready to be fabricated for a flight test.

References

[1] Boeing Official Website [Accessed on April 2016]
[2] Wikimedia Commons *V-22 Osprey tiltrotor aircraft* [Accessed on April 2016]
[3] Panther UAS Specifications [Accessed on April 2016]
[4] Cetinsoy E, Dikyar S, Hancer C, Oner K T, Sirimoglu E, Unel M and Aksit M F 2012 *Mechatronics* **22** 723–45
[5] Holsten J, Ostermann T and Moormann D 2011 *CEAS Aeronautical Journal* **2** 69-79
[6] Liebeck H R and Page M A 1996 *NASA/McDonnell Douglas Blended-Wing-Body* McDonnell Douglas Aerospace
[7] McCormick B W 1995 *Aerodynamics, Aeronautics and Flight Mechanics* (New York: Wiley)
[8] Raymer D P 1999 *Aircraft Design: A Conceptual Approach* (Reston: AIAA Education Series)
[9] Landolfo G 2008 *Aerodynamic and Structural Design of A Small Nonplanar Wing UAV* Master Thesis University of Dayton
[10] Miletic M and Arnold A 2011 *Proceedings in Applied Mathematics and Mechanics* **11** 681–2
[11] O'Connell R F 1980 *Design, Development and Implementation of an Active Control System for Load Alleviation for a Commercial Transport Airplane* Advisory Group for Aerospace Research and Development Neuilly-Sur-Seine (France)
[12] Rauscher E 1984 *Structural and Materials Panel – Specialist Meeting I*
[13] Nonami K, Kendoul F, Suzuki S, Wang W and Nakazawa D 2010 *Autonomous Flying Robots* (Tokyo: Springer Japan)
[14] NATO Official Website [Accessed on April 2016]

IOP Conf. Series: Materials Science and Engineering **152** (2016) 012018 doi:10.1088/1757-899X/152/1/012018

Development of unmanned aerial vehicle (UAV) based high altitude balloon (HAB) platform for active aerosol sampling

S Lateran, M F Sedan, A S M Harithuddin and S Azrad*

Department of Aerospace Engineering, Faculty of Engineering, Universiti Putra Malaysia, Malaysia

* syaril@upm.edu.my

Abstract. The knowledge on the abundance and diversity of the minute particles or aerosols in the earth's stratosphere is still in its infancy as aerosol sampling at high-altitude still possess a lot of challenges. Thus far, high-altitude aerosol sampling has been conducted mostly using manned flights, which requires enormous financial and logistical resources. There had been researches for the utilisation of high altitude balloon (HAB) for active and passive aerosol samplings within the stratosphere. However, the gathered samples in the payload were either brought down by controlling the balloon air pressure or were just dropped with a parachute to slow the descend speed in order to reduce the impact upon landing. In most cases, the drop location of the sample are unfavorable such as in the middle of the sea, dense foliage, etc. Hence a system that can actively sample aerosols at high-altitude and improve the delivery method in terms of quality and reliability using unmanned aerial vehicle (UAV) is designed and tested in this study.

1. Introduction

Aerosol particles are found on the earth's surface and are transported both vertically and horizontally in the atmosphere [1]. Generally, aerosol particles appear in almost any shape or size, and can either be solid particles or liquid droplets. These particles can be defined as fine and coarse particles, which are smaller and larger than 1 micrometer, respectively [2]. Alternatively, they can also be categorized into small and big particles. Big particles' size usually ranges between 2.5 and 10 micrometers (from about 25 to 100 times thinner than a human hair) [3]. These particles are called PM10 (stands for Particulate Matter up to 10 micrometers in size). In contrast, the small particles' size is smaller than 2.5 micrometers (100 times thinner than a human hair). These particles are also called PM2.5 (Particulate Matter up to 2.5 micrometers in size) [3].

Aerosol particles sampling at high altitude has contributed much to meteorological researches such as the study of atmospheric pollution, health effects, global warming and ozone loss. Apart from that, multiple studies on microbial aerosol samplings have also been conducted to understand the dispersal of microorganisms from one part of the earth to another. Studies regarding aerosol particle samplings at high altitude are typically conducted by either manned (aeroplane) or unmanned vehicle (HAB). Table 1 below shows the comparison between these two methods. Based on Table 1, balloon is more cost effective and feasible method for aerosol samplings at high-altitudes compared to manned flight. The researches on aerosol sampling using balloons include tethered or non-tethered methods [1]. A tethered balloon based sampler provides the advantage of longer duration of the sampling process. The

IOP Conf. Series: Materials Science and Engineering **152** (2016) 012018 doi:10.1088/1757-899X/152/1/012018

drawback would be that it could only conduct sampling around a certain altitude due to the limitation posed by the tether length and weight. British Antarctic Survey from Natural Environment Research Council, Cambridge, United Kingdom used a tethered blimp and the sampling process was done at maximum altitude of 300 meters [4]. According to the requirement, a non-tethered balloon can be used so that the sampling process can be conducted at higher altitude. Without active control of the balloon altitude, the duration of the sampling would be much lesser as compared to the tethered-balloon based platform [5].

Table 1. Comparison between high altitude balloon and manned flight

Aspects	High Altitude Balloon (Unmanned flight)	Aeroplane (Manned flight)
Cost	Low	High
Effectiveness	Less Effective	Very effective
Range	Wide range (limited to vertical range only)	Wide range (unable to go much higher altitude)
Feasibility	Easy	Complicated

Aerosol sampling at higher altitudes using balloon have been conducted passively during ascend by Bryan *et al.* [1] and Harris *et al.* [6] from Louisiana State University (LSU). The LSU team developed a bioaerosol sampling payload, Life's Atmospheric Microbial Boundary (LAMB) to passively collect microbial aerosols from defined air mass volumes at different altitude ranges. LAMB was transported to high-altitude using latex sounding balloon, and was released and descended by parachute. Besides real-time tracking capability of the system provided by the embedded beacons on the payload, there is no way to ascertain a safe landing site for payload recovery. Meanwhile, Yang *et al.* [7] used a pump actively in the stratosphere above Pacific Ocean, adjacent to Honshu Island of Japan. In this research, which involved the Tokyo University of Pharmacy and Life Science (TUPLS) and Japan Aerospace Exploration Agency (JAXA) team, a payload was developed to actively sample bioaerosol at high-altitude from 12 km to 30 km in three phases. The payload was mounted on a gondola that hanged from a balloon. The balloon was eventually landed on the sea by exhausting the balloon gas.

Passive aerosol sampler rod

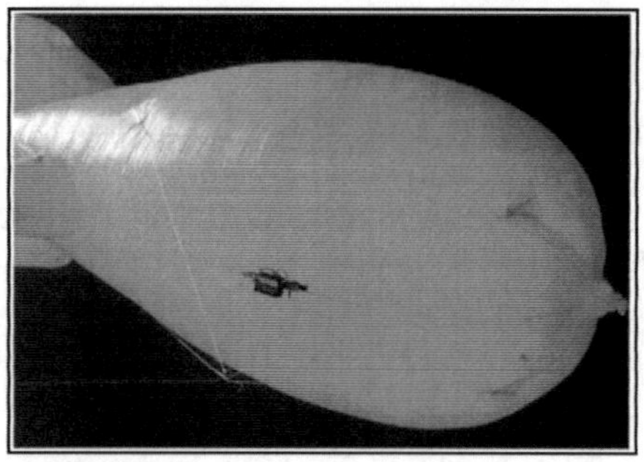

Figure 1. LAMB payload [1] (left) and the blimp [3] (right)

Manned flights for aerosol sampling may have higher reliability in terms of securing the sample from damages and lost. However, it would be too costly to perform in higher frequency. HAB-based sampling is of lower cost but there are risks sample contamination and lost during payload retrieval. Payload recovery mission such as the one conducted on the sea by Yang *et al.* [7] would require massive resources and complex logistical coordination. Given the current background of high-altitude aerosol sampling technology, where there is a gap between manned and unmanned methods in terms of cost and reliability, a UAV-based HAB platform for active aerosol sampling to reduce the risk and uncertainty of the landing site of the payload is developed and demonstrated. The development of such platform is essential to accelerate the high-altitude aerosol sampling based researches that would also

IOP Conf. Series: Materials Science and Engineering **152** (2016) 012018 doi:10.1088/1757-899X/152/1/012018

contribute in the fields such as high-altitude microbiology, weather prediction, pollution studies, etc., where current approaches require large-scale financial and human resources.

2. Method of aerosol sampling

Aerosol sampling method can be further divided into active and passive. Passive air sampling such as inertial impactors and gravity sampling rely on the physical process such as diffusion or permeation [8]. Inertial impactor depends on particle motions and the sampling surface motions while gravity sampling method allows air particle to fall onto the collecting surface due to gravitational force [8]. LAMB payload applied inertial impactor in their sampling process by allowing bioaerosol particles to impact onto the rod surface as the balloon ascended. This method effectiveness relies on constant wind speed and direction. On the other hand, the active sampling method involves vacuum pump or fan in drawing air onto the impaction surface such as filter paper and it is independent of wind speed.

In this research work, the active sampling method with filtration sampling was applied. Filtration sampling that involves inertial impaction and diffusion can be more effective at collecting particles from air stream than the passive sampling method [8]. The design of sampling device was taken from Andersen cascade sampler. Cascade sampler or cascade impactor is a suction sampler that has two or more stages of air collection with the nozzle as shown in Figure 2. This sampler can sample various size of air particle depending on the number of stages of sampling such as Andersen cascade sampler that has six-stage of sampling that differs by their hole sizes [9]. Low speed and a larger hole in the first stage makes it possible to capture larger air particles (greater inertia) and the smaller the hole size becomes, the smaller the particle size that is able to collect in subsequent stages [9]. Furuchi *et al.* used the same concept to develop a nanosampler, which is capable of collecting particles from PM10 to PM0.5. This sampler consists of four stages, each stage contains a different type of filter paper [10].

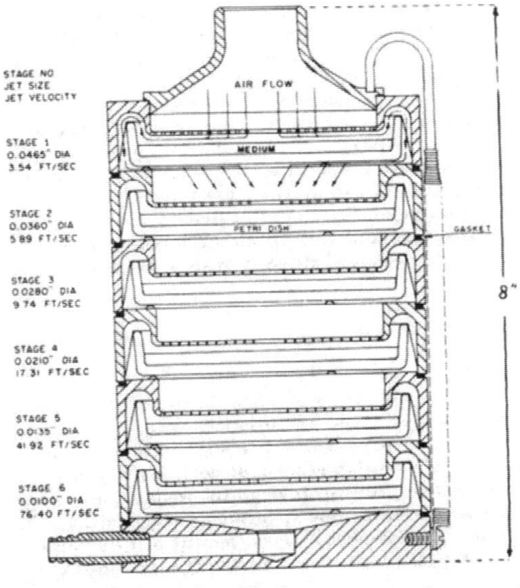

Figure 2. Andersen six-stage cascade sampler. Each stage contain a petri disc below the perforated plate that contain sticky surface to collect particles form air stream that flow through [9]

3. Materials and method

3.1. High altitude balloon vehicle

A 1600 g latex balloon filled with helium was used to bring the sampling device to the stratospheric altitudes, with a glider as the payload of the balloon. A parachute was used to slow down the free fall during descent. The total weight suspended below the balloon was about five kilograms. The balloon carried the payload up to its burst altitude. When the balloon burst, the parachute was deployed until the payload reached the parachute release altitude and the glider began to glide for recovery. The ratio of lift produced by the balloon was twice of the payload weight.

3.2. The fixed wing payload

The biggest concern of HAB mission is the landing site. To increase the chance to retrieve the payload and avoid water landing, a glider with a wing span of two meters was used as a payload to carry all experimental apparatus as shown in Figure 3. All electronics and sampling devices were placed inside the glider as illustrated in Figure 4. The internal compartment was covered with aluminium tape to prevent extreme internal temperature drop, which could cause failure of the electronics and power supply. This glider was pre-programmed to glide back to desired landing coordinate and would be then landed manually using RC controller.

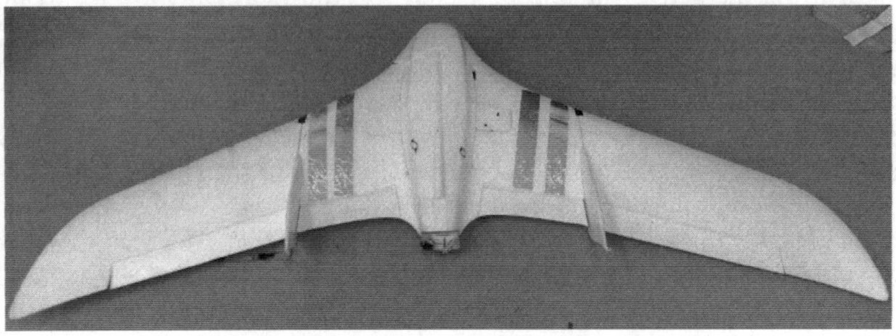

Figure 3. FX79 fixed wing with 2 m wingspan

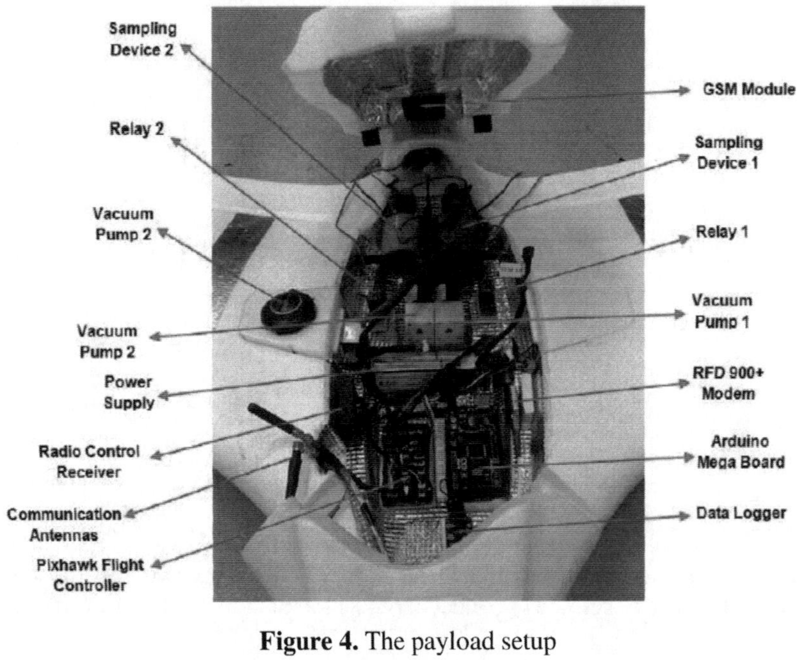

Figure 4. The payload setup

3.2.1. Sampling collector. The design of the sampling collector is shown in Figure 5. This sampling device was designed using CATIA software and printed using a 3D printer. The body was made of ABS (Acrylonitrile Butadiene Styrene). This sampling device used cellulose nitrate membrane filter with diameter of 47 mm and pore size of 0.45 μm. The collectors consisted of four stages: inlet stage, distribution stage, compression part, and outlet part. Nozzle shape of these collectors increased the speed of airflow. The small holes distribute the airflow evenly inside the nozzle. A 12 Volt pump was used to collect the aerosol particles.

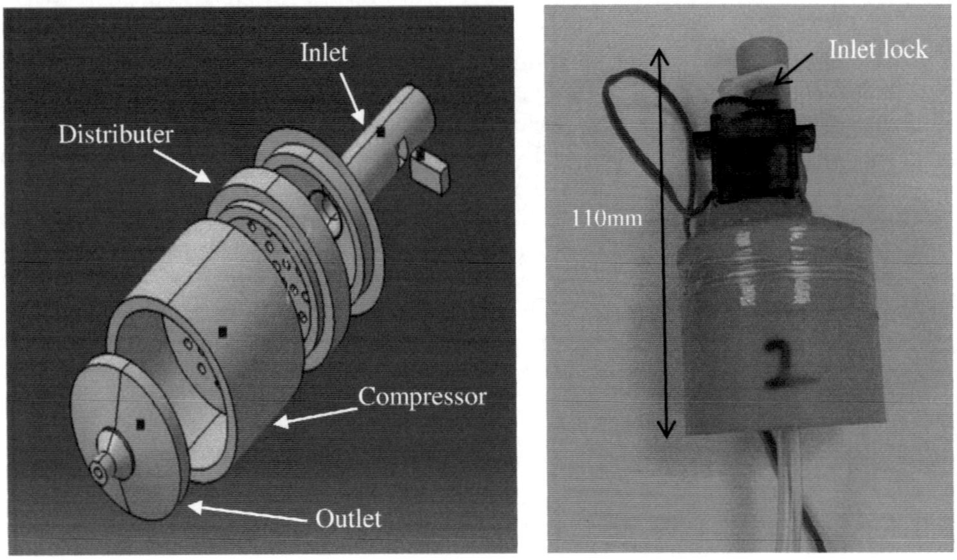

Figure 5. 3D view of sampling device (left) and sampling device after printed using 3D printer.

3.2.2. Electronic and power supply. The payload electronics consisted of power module and control system. The control system consisted of 3DR Pixhawk PX4 flight controller and Arduino Mega 2560 microcontroller. This system was responsible for monitoring the payload altitude and making decision of activation and deactivation of the pump. The flight controller was used to monitor the payload condition such as coordinate, altitude and attitude of the payload. During gliding mode, 3DR Pixhawk with GPS was used as a flight controller that control the glider and also responsible for parachute release mechanism. When the balloon burst at the maximum altitude, the glider began its free fall. Parachute was used to slow down the descent speed. Parachute release mechanism would be activated if the difference between current descent altitude and maximum altitude was 2 km, and the glider began gliding to desired landing coordinate autonomously. All important data from flight controller such as temperature inside the payload, speed, altitude and pressure was stored in data log inside the flight controller. The data was also transmitted to the ground station for backup when the onboard data log encountered errors.

For tracking and recovery, the payload used two tracking systems. The primary system used RFD 900+ modem radio transmitter to transmit GPS data to the ground station and secondary tracking system used GSM (Global System for Mobile Communication) network to send coordinate to ground station. Primary tracking system used 902-915 MHz frequency and broadcasted the data directly via Mavlink protocol to ground station. This system enabled the balloon to be continuously monitored from the ground station. The advantages of this tracking system include its independence of the other broadcasting method such as the Automatic Packet Reporting System (APRS) and cost savings for subscribing the broadcast system. The main reason of using two tracking system was to minimize the risk of payload lost. The primary system was used to report the real-time coordinate and altitude to the ground station while the secondary system functioned as backup tracking system when primary system failed or the glider encountered crash landing. The disadvantage of using secondary tracking system was its full dependence on the network coverage.

Located at the front of the glider was a Mobius wide-angle HD camera. This camera recorded the video of the surrounding (aerial view) throughout the flight. The payload was powered by three Lipo batteries. Two 11.4-V Lipo batteries were used to power the flight controller module and the pump respectively. Arduino Mega was powered by a 7.4-V Lipo battery.

3.3 Feedback control system design

Control system of this sampling device was done using Arduino Mega microcontroller by receiving the PWM (Pulse Width Modulation) signal from Pixhawk PX4 every 500 ms through serial port and triggered if the desired altitude obtained from GPS (Global Positioning System) was achieved. To preserve sample after sampling process, two 180 degree servos acted as lock and unlock mechanisms to open and close the inlet of the sampling device. The relay functioned as an on-and-off switch to control the vacuum pump. A pre-programmed Arduino Mega microcontroller was used to transmit digital signals to both relays and servo in order for activation. Pixhawk (with GPS) was used to give input (altitude reading) to the Arduino, which would then decide whether to send the PWM signal to activate the pump and servos for aerosol sampling, and the opening of the collector chambers. The schematic of the sampling device and the flow chart of the feedback control system are shown in Figure 6 and Figure 7, respectively. Figure 6 shows the schematic diagram of the platform system for aerosol sampling. The aerosol sampler contained two sampling chambers to collect aerosol particles with the help of vacuum pump.

The algorithm for sampling device is shown below. This algorithm was applied in Arduino Mega to make the decision to activate or terminate the sampling device using ch_6 (channel 6) and ch_7 (channel 7) from Pixhawk to Arduino digital pin.

```
//ch_6 and ch_7 connected to Pixhawk
If ch_6 ≥ 1800 or ch_7 ≥ 1800 (pump activation altitude reached)
    Then
        Turn on Relay
        Unlock inlet door
Else
        Turn off Relay
        Lock inlet door
```

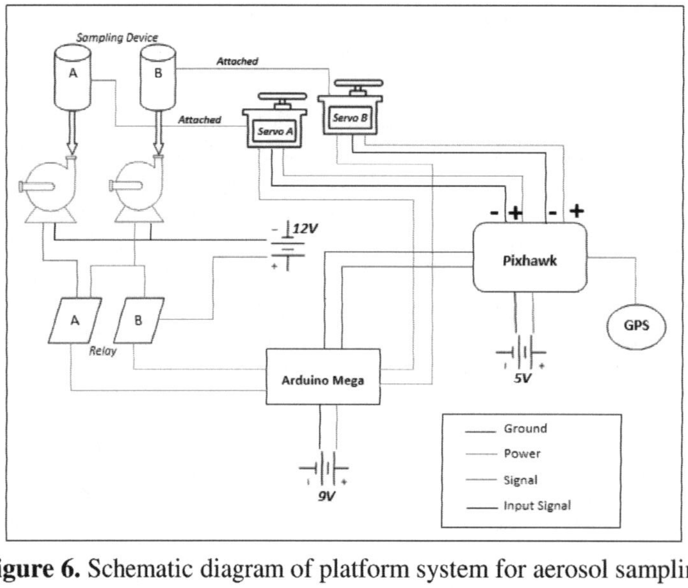

Figure 6. Schematic diagram of platform system for aerosol sampling

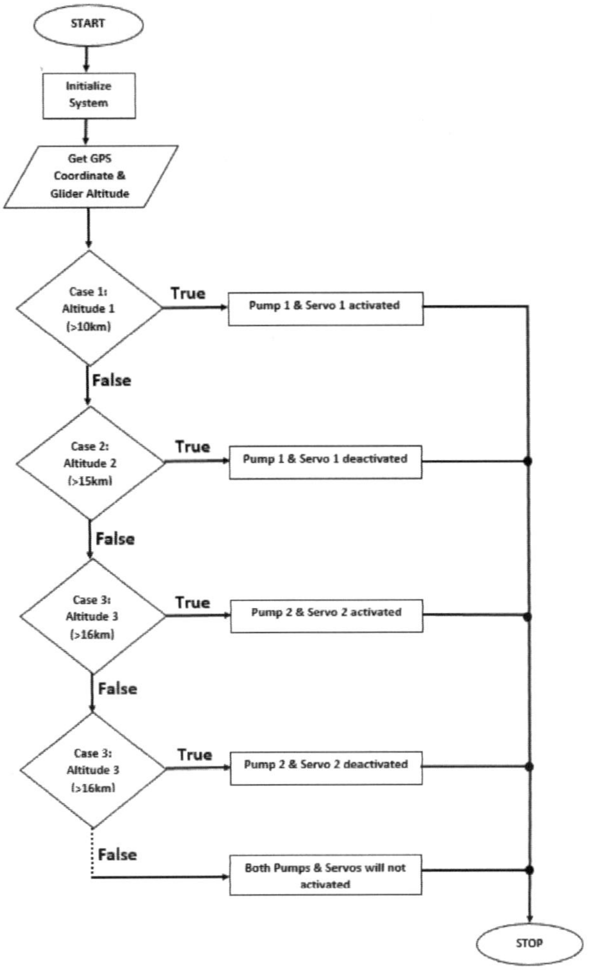

Figure 7. Switch case flowchart of platform system for aerosol sampling

Figure 7 shows the switch case flowchart of the platform system for aerosol sampling process. The flowchart of the system started with initialization of the on-board system, which included the Pixhawk, GPS and Arduino Mega. After the initialization process was completed, the balloon was released manually for ascend. During the ascending process through the atmosphere, the platform system was set with four conditions. These four conditions were used to activate and deactivate the vacuum pumps and servos for aerosol sampling process. For the sampling device number 1, as the balloon reached an altitude of 10 km, the system activated vacuum pump 1 and servo 1.This allowed the first sampling process to take place. This went the same to sampling device number 2. When the balloon reached an altitude of 16 km, the system activated vacuum pump 2 and servo 2, and both terminated at 20 km altitude. The classification of the samples is listed in Table 2.

In triggering process, if current ascent altitude more than activation altitude, Pixhawk would send 1800 PWM signal to Arduino mega. Arduino mega would switch on the 5-V relay by changing the relay state from low to high and unlock the sampling door chamber. The PWM value would stay constant at 1800 during sampling process until current altitude was more than deactivation altitude. Default PWM value that Pixhawk sent to Arduino Mega was 1100 and the inlet of sampling device was locked. For smooth and reliable data, Kalman filter was applied to filter all data including altitude data.

Table 2. Reference samples altitude

Sample	Altitude (m)
A	1000 – 15000
B	16000 – 20000
C	N/A (Reference)

3.4. Aerosol analysis

3.4.1. Analysis through weight of filter paper. Amount of aerosol particles collected was determined by using the initial weight and final weight of filter papers. In order to determine the weight of the filter paper, a highly sensitive weighing scale was used. In this experiment, the micro weighing scale was used to weigh the filter paper. The difference of weight between A and B filter papers with the reference filter paper (filter paper C) was used to determine the existence of other particles on the filter paper that made the weight of filter paper increased than the original weight.

3.4.2. SEM/DEX analysis method. Analysis of sample was done using scanning electron microscopy with energy dispersive X-ray spectroscopy (SEM/EDX) analysis, as shown in Figure 8. This method was used to determine the type of elements found on the filter paper. In this analysis, sample C was used as reference filter paper, which was compared to sample A and B. In SEM/DEX analysis, high-resolution images of surface topography with excellent depth of field were produced using a highly focused scanning electron beam. Besides, most elements in the periodic table (ranging from carbon to uranium) can be easily detected by using SEM/DEX analysis and a maximum sample size of 50 mm diameter (~20 mm thick) can be accommodated [11]. Additionally, SEM/EDX is also useful to capture images at 100 000 times magnification or greater, which is great advantage for the analysis of aerosol particles and some elemental information such as element weight percentage, element atom percentage and element compound percentage can also be obtained by using this analysis [11]. Lastly, SEM/EDX analysis is considered as a fast, cost-saving and non-destructive approach for surface analysis and it is often used to survey the surface analytical problems before proceeding to more specialized techniques. Since the filter paper used in this research work was 47 mm in diameter, it was compatible with the SEM/EDX.

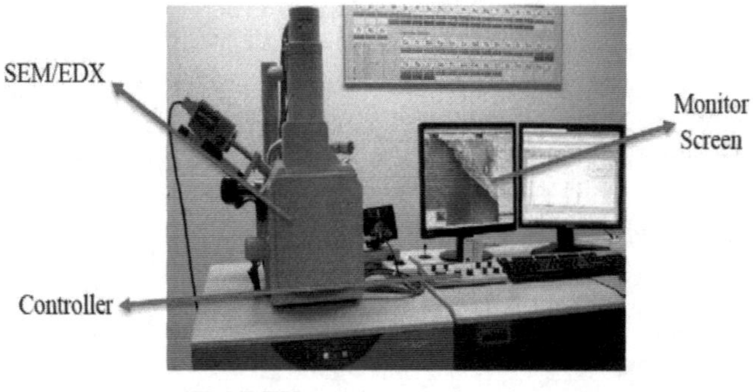

Figure 8. SEM/DEX machine used in sample analysis

4. Result and discussion

4.1. Flight test result

This subsection explains the results analysis of a flight test that was done, that includes analysis of mission profile, flight path, altitude profile, internal temperature, pressure and parachute release mechanism. On the 2nd of May 2016, the sampling device was flown with High Altitude Fixed Wing

Experimental Platform using 1600 g balloon and 2.5 kg payload. It was launched from the Paddy field of Kampung Chui Chak, Langkap Perak, Malaysia (4° 2'31.62"N 101°10'4.45"E) at 12:40 noon. The flight path of the flight is as shown in Figure 9. This flight was fully controlled by the flight controller system autonomously. After approximately four hours, the fixed wing (glider) was safely landed at a paddy field Felcra Seberang Perak, Perak, Malaysia (4° 3'31.51"N 100°51'54.00"E) that was about 50 m from the ground station. Mission profile and other important parameters were plotted against the Mission Elapsed Time in unit milliseconds (ms) as shown in Figure 10.

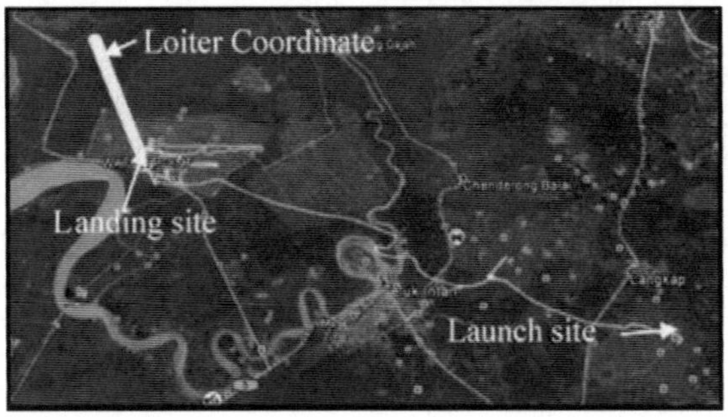

Figure 9. Flight path of May 2016 mission

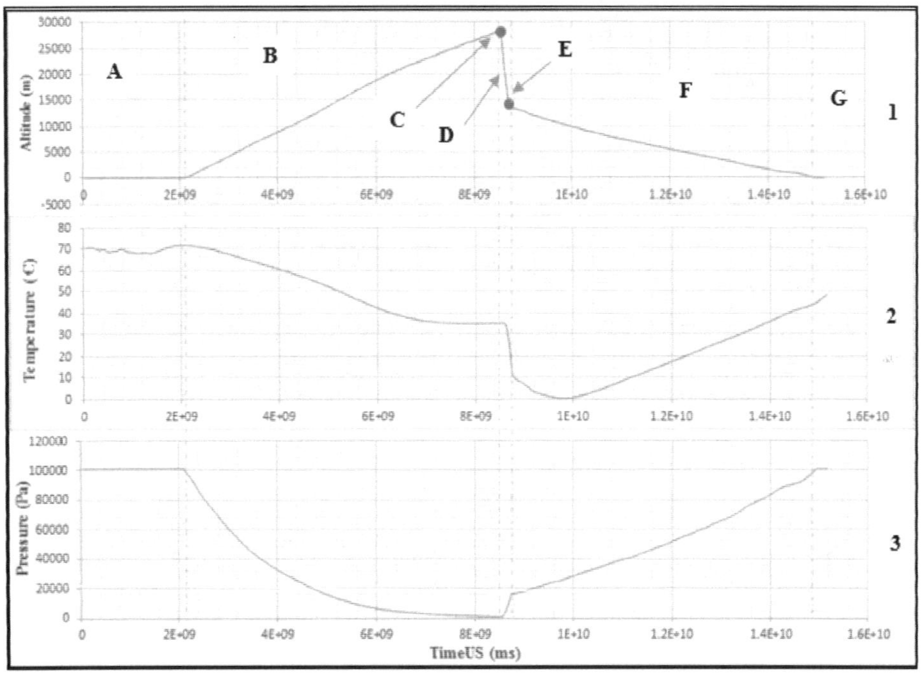

Figure 10. Mission Profile of 2nd May 2016 flight test. Region A indicated preparation time. Region B indicates launching and ascending period. Point C represents where the balloon burst (Maximum altitude). Region D indicates where the glider experience free fall. Point E indicates where the glider begins to glide. Region F represents loitering time and region G after the glider landed.

In Figure 10, graph 1 represents the whole mission profile that is presented as Altitude (m) against Time (ms), graph 2 represents Internal temperature (°C) against time (ms) while graph 3 represents Pressure (Pa) against time (ms) graph. Figure 10 is divided into 5 regions, which are preparation (A), ascending (B), free fall (D), loiter (F) and landing (G). The balloon took about 1 hour 40 minutes to reach the maximum altitude. As the balloon went up, helium gas expanded until the balloon reached its burst altitude (Point C) that was 28,000 m. At maximum altitude, the internal temperature was 32°C and zero pressure. After balloon burst, the glider experienced free fall and slowed down by the help of a parachute. After approximately 26,000 m, the parachute was released. At approximately 14,000 m, glider began to gain lift (Point E) and began its autonomous mission to desired landing coordinate (4° 3'31.51"N 100°51'54.00"E). After arriving at landing coordinate, the glider began loitering and at approximately 100 m altitude, ground control took over the control surface and landed the glider manually using radio controller. The glider loitering was about 1 hour 45 minutes as it descent slowly in a circular path. During free fall, internal temperature dropped gradually from 32°C to 10°C and during loitering, it dropped until 0°C, which was the lowest internal temperature recorded. The temperature drop was due to high velocity during free fall and loiter that increased the humidity inside the payload and led to condensation process. After 10,000 m, the temperature and pressure increased constantly. The important mission parameters are summarized in Table 3.

Table 3. Simplification of important mission parameter

Parameter	Value
Maximum altitude	28,000 m
Minimum internal temperature	0 °C
Minimum pressure	0 Pa
Average ascent speed	5 m/s
Average loiter speed	11 m/s
Flight time	≈ 4 hours 20 minutes

Figure 11 shows the PWM value that was sent by Pixhawk to the servo in the parachute release mechanism. To lock the parachute, 1800 PWM value supplied to the servo to output an equivalent rotation angle. This PWM value supplied constantly until release parachute altitude reached. After approximately 1 hour 45 minutes, the parachute release mechanism was activated by changing the PWM value to 1100, which is represented by point B. The PWM value represented by point A is due to system testing during the preparation process.

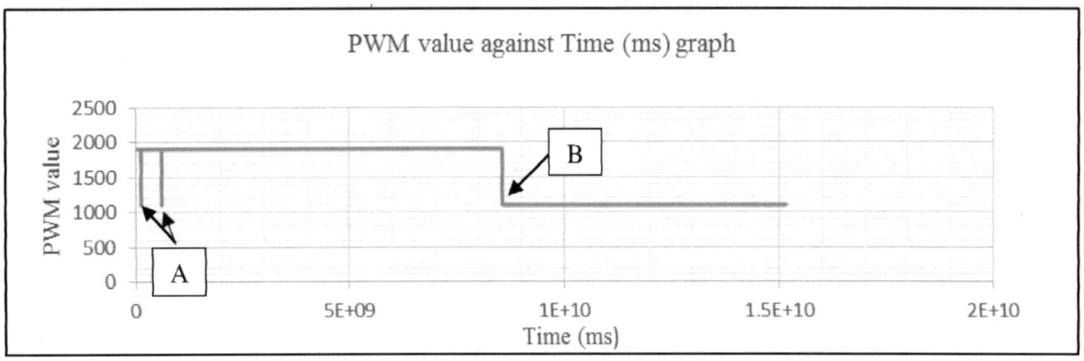

Figure 11. PWM value against Time (ms) graph of parachute release mechanism

4.2. Platform system

An analysis was done on the platform system by extracting the output data from the Pixhawk. One of the digital output data, PWM was extracted and analysed by plotting the data against pressure altitude. Figure 12 shows the graph of PWM against pressure altitude.

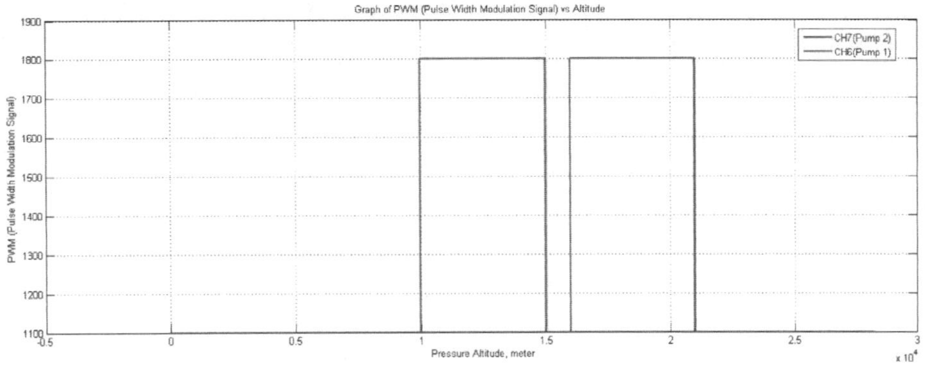

Figure 12. Graph of PWM versus Pressure

For the testing of platform system, a glider was used as a payload to carry the sampling device and all system onboard. Based on Figure 12, it shows the graph of pulse width modulation (PWM) versus pressure altitude. From this graph, it shows that both vacuum pumps were functioning properly according to the coding that has been designed. The red line of the graph represents vacuum pump number 1 and the green line of the graph represents vacuum pump number 2. From the graph, it shows that the vacuum pump number 1 was activated at an altitude of 10 km until 15 km and for vacuum pump number 2, it was activated at an altitude of 16 km until 21 km.

4.3. Weight analysis method

For aerosol sampling process, three filter papers were used where one of them was the reference filter paper and the other two were tested at different altitudes. Two sampling devices were used and both of them were run simultaneously. After sampling process was done, analysis and comparison between the two filter papers and the reference filter paper was conducted. For the analysis and comparison process, two methods were used. The first method was to determine the difference value between the initial and final weight of filter paper, and the second method was by SEM/EDX analysis.

In this experiment, cellulose nitrate membrane filter paper was used as the filter paper. This filter paper was 47 mm in diameter and had the pore size of 0.45-micron meter. For the sampling process, a vacuum pump with a flow rate of 45 L/min and negative pressure of 90kpa was used. A calculation for the duration of sampling process was determined by dividing the altitude range with ascending speed of the balloon (constant ascend speed of 5 m/s was used). Table 4 below shows comparison between the initial weight and final weight of the three filter papers with Petri dish.

Table 4. Comparison of initial weight and final weight of different filter paper

Sampling Altitude (km)	Sample	Readings	
		Initial weight (g)	Final weight (g)
10-15	A	18.45	18.50
16-21	B	18.46	18.47
N/A	C (Raw)	18.47	N/A

The duration for both sampling process A and B was calculated to be 1000 seconds. From the table above, the amount of aerosol particles collected was determined by using the initial and final weight of the filter papers. As mentioned earlier, a high sensitivity weighing scale was used to determine the weight of the filter paper. In this experiment, the micro weighing scale was used to weight the filter paper. Based on Table 4, the difference in weight between initial and final of sample A and sample B was 0.05 g and 0.01 g, respectively. This was the exact weight of the air particles that was collected from the sampling process. Based on sample A and sample B, the result shows that sample A has the

IOP Conf. Series: Materials Science and Engineering **152** (2016) 012018 doi:10.1088/1757-899X/152/1/012018

highest value in weight different between initial and final. From this observation, it was assumed that at an altitude of 16 km until 21 km, the size of the air particles is smaller than 0.45-micron meter while at an altitude of 10 km until 15 km, the size of the air particles is larger than 0.45-micron meter. Thus, sample A was able to collect more air particles than sample B. For sample C, it was set as a reference filter paper for further analysis.

4.4. SEM/DEX Analysis method

The collected particles on the filter paper were analysed using Scanning Electron Microscopy (SEM) with Energy Dispersive X-Ray (EDX) device, and the sample results are shown in Figure 13, Figure 14 and Figure 15 for first (Sample A), second (Sample B) and third (Sample C), respectively.

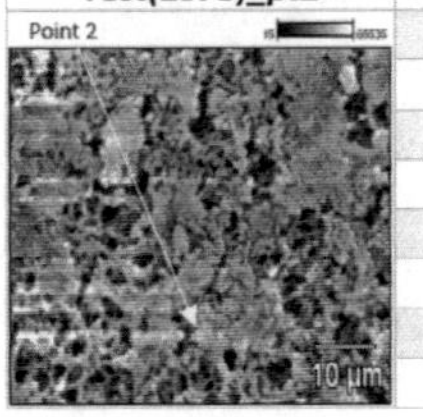

Test(1573)_pt2	C-K	N-K	O-K
		Weight %	
	23.10	26.59	50.30
		Atom %	
	27.61	27.25	45.13
		Formula	
	C	N	O
		Compound %	
	23.10	26.59	50.30

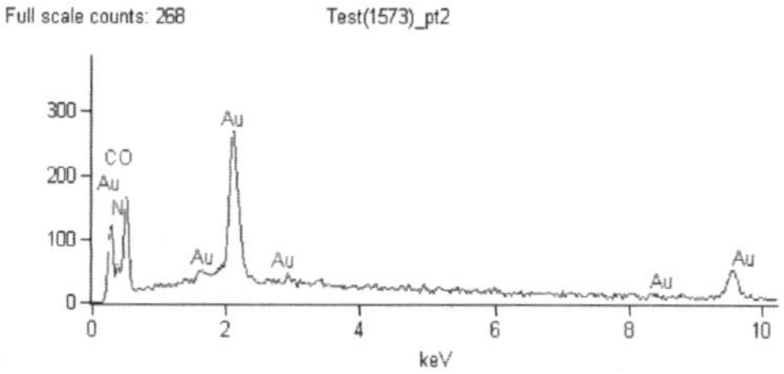

Figure 13. SEM/EDX result analysis of sample A

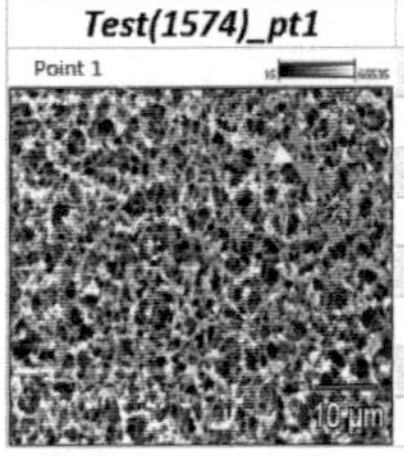

Test(1574)_pt1	C-K	O-K	K-K
		Weight %	
	97.64	0.40S	1.96
		Atom %	
	99.08	0.31	0.61
		Formula	
	C		K2O
		Compound %	
	97.64	0.00	2.36

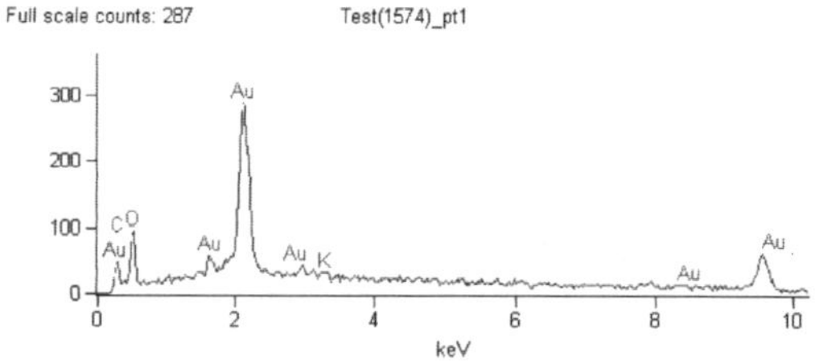

Figure 14. SEM/EDX result analysis of sample B

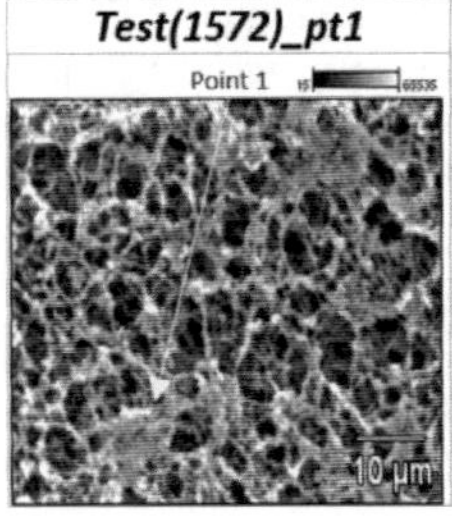

Test(1572)_pt1	C-K	O-K
	Weight %	
	38.17	61.83
	Atom %	
	45.12	54.88
	Formula	
	C	O
	Compound %	
	38.17	61.83

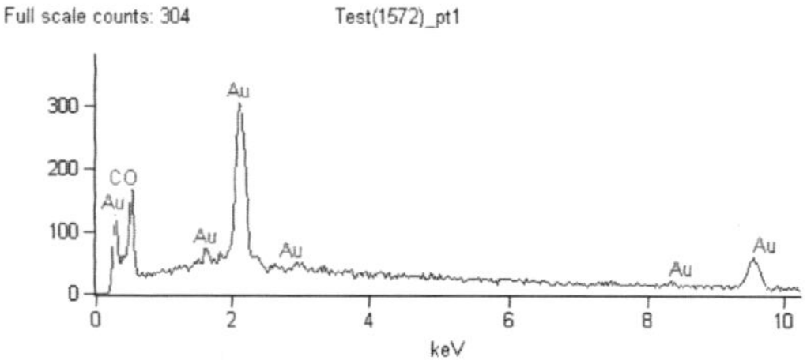

Figure 15. SEM/EDX result analysis of sample C

From the results of SEM/EDX analysis, it shows that all the samples have the same basic elements on the filter papers which were Carbon Monoxide (CO) and Gold (Au). Based on Figure 13, it shows the graph of elements (at Point 2) that was found on sample A. For sample A, which was sampled at an altitude of 10 km until 15 km, another element that is Nitrogen (N) was found. Based on Figure 14, it shows the graph of elements (at Point 1) that was found on sample B. For sample B, which was sampled at an altitude of 16 km until 21 km, another element that is Potassium (K) was found. In addition, from Figure 13, Figure 14 and Figure 15, they show all of the elements composition, which included the percentage of weight, percentage of the atom, percentage of compound and formula for elements that was found at both point 1 and point 2. Based on all the results, the platform system and the sampling device were verified to function properly. Even though the outcome was a success, it was only limited to particles with a size of 0.45-micron meter and above only. In order to collect aerosols particles that are smaller than 0.45-micron meter, an advanced sampling device or another method should be considered. Despite the limitation in the aerosol sampling process, the platform system successfully proved its reliability, feasibility and effectiveness in the sampling process. Hence, with a good consideration of sampling device and sampling method, a large range of aerosols particles size and various types of microbes can be obtained at high altitudes by using high altitude balloon (HAB) based platform system.

5. Conclusion and recommendations for future work

Unmanned Aerial Vehicle (UAV) based High Altitude Balloon for active aerosol sampling system was successfully developed and tested. It is also shown that automatic control works properly and the glider that carried the sample was retrieved safely at the desired landing coordinate autonomously. The system was able to collect aerosol particles (as in Figure 13 and 14 were Nitrogen (N) and Potassium (K) gas particles collected at different altitudes). It is hoped that the developed system can be utilised with different balloon settings, as well as with other filter types, to collect microbial and aerosols to characterise their distributions at the high-altitude atmosphere. Despite the positive results obtained through this research, several limitations to the system platform and sampling process existed. Thus further study is required before this platform system can be applied for aerosol sampling or other environmental/biological survey.

The following areas could benefit from further development of the system platform for aerosol sampling. Throughout this research, sampling process was only done for pore size of 0.45-micron meter. Any air particles that is larger than 0.45-micron meter was trapped on the filter papers but any air particle that is smaller than 0.45-micron meter was able to pass through. Furthermore, in this research, only one filtering chamber for one device, thus it limited the size of the particle that can be filtered. Hence for future work, for certain meteorological or bioaerosol research requirement, a suitable and precise sampler needs to be developed. More chambers are needed in the sampler to fit more filter papers of various sizes to increase efficiency. Other type of sampler that can be considered for application is the Rotorod Sampler. The Rotorod Sampler is commonly used to collect spores in the air by allergists [12,13]. The Rotorod Sampler consists of a U-shaped metal rod covered with sticky tapes, attached to a spindle to a battery-powered electric motor. Rotating the metal rod at high speed will trap spores and other air particles on the tapes. For analysis, the tapes are removed and examined microscopically to identify the spores and other particles such as pollen grains in the air. Modifications need to be made on the glider compartment to suit the sampler design and power requirements. Using such sampler with the current platform would accelerate studies on microbes or spores at high-altitude, which are abundant and diverse. A good communication and autopilot system must be developed and maintained throughout the mission profile of the HAB. The risk of payload lost could be lowered if lost of communications occurrence between the payload computer and ground station can be further reduced. During the test flight of the system, the signal between a ground station and HAB was sometimes lost due to some interference from other devices and dissipation of signal by clouds. Communications could also be cut off due to power issue, as batteries could not longer power devices in an unanticipated long flight or unexpected power consumption by other devices such as camera and sampler. Improvement in antenna design, introduction of satellite based communication system as well as installation of other redundant communication systems for backup such as low-powered beacons is needed in future platform to increase reliability during payload recovery. Studies on power management are required as merely increasing the number of batteries will lead to higher payload and may lead to undesirable glider flight performance.

References

[1] Bryan N C, Stewart M, Granger D, Guzik T G and Christner B C 2014 *Journal of Microbiological Methods* **107** 161-8

[2] Willeke K and Whitby K T 1975 *Journal of the Air Pollution Control Association* **25** 529-34

[3] Yang H-C, Chang S-H, Lu R and Liou D-M 2016 *Computer Methods and Programs in Biomedicine*

[4] Rankin A M and Wolff E W 2002 *Journal of Atmospheric and Oceanic Technology* **19** 1978-85

[5] Hara K, Osada K and Yamanouchi T 2013 *Atmospheric Chemistry and Physics* **13** 9119-39

[6] Harris M J, Wickramasinghe N C, Lloyd D, Narlikar J V, Rajaratnam P, Turner M P, Al-Mufti S, Wallis M K, Ramadurai S and Hoyle F 2002 *Proc. SPIE 4495, Instruments, Methods, and Missions for Astrobiology IV*

[7] Yang Y, Yokobori S, Kawaguchi J, Yamagami T, Iijima I, Izutsu N, Fuke H, Saitoh Y, Matsuzaka Y, Namiki M, Ohta S, Toriumi M, Yamada K, Seo M and Yamagishi A 2008 *JAXA Research and Development Report JAXA-RR-08-001*

[8] Harriet A B and William R S 1987 *Atmospheric Environment (1967)* **21** 451-6

[9] Paolo M, Giuliani C and Cristina S 2003 *Cultural Heritage and Aerobiology: Methods and Measurement Techniques for Biodeterioration Monitoring* (Berlin: Kluwer Academic Publishers)

[10] Furuuchi M, Eryu K, Nagura M, Hata M, Kato T, Tajima N and Otani Y 2010 *Aerosol Air Qual. Res.* **10** 185-92

[11] Vineyard M F, LaBrake S M, Ali S F, Nadareski B J, Safiq A D, Smith J W and Yoskowitz J T 2015 *Nuclear Instruments and Methods in Physics Research Section B: Beam Interactions with Materials and Atoms* **350** 77-80

[12] Frenz D A 1999 *Annals of Allergy, Asthma & Immunology* **83** 341-9

[13] Frenz D A and Guthrie B L 2001 *Annals of Allergy, Asthma & Immunology* **87** 390-3

IOP Conf. Series: Materials Science and Engineering **152** (2016) 012019 doi:10.1088/1757-899X/152/1/012019

Torsional thrust stand for characterization of microthrusters

K H Cheah[1*] and K S Low[2]

[1]School of Engineering, Taylor's University, Malaysia.
[2]Satellite Research Centre, School of EEE, Nanyang Technological University, Singapore.

*keanhow.cheah@taylors.edu.my

Abstract. This paper describes the setup of a precise thrust stand based on torsional pendulum design for characterizing the performance of microthrusters. Calibration has been carried out by using an improved version of electrostatic calibrator, which produces a wide range of accurate and repeatable calibration force. After the calibration, the thrust stand can resolve constant force from 40μN to 3.4mN and impulse bit from 7μNs to 340μNs. The usefulness of the thrust stand has been demonstrated by measuring the performance of two different microthrusters: a pulsed plasma thruster that produces impulse bit of 23.15μNs and a vaporizing liquid microthruster that produces steady state thrust of 633.5μN.

1. Introduction

In recent years, nanosatellites (<10 kg) have demonstrated their capability to carry out a wide range of missions at a rapid and cost effective way. The space community is very eager to send these satellites to perform more complex missions. Low thrust propulsion systems, which are often known as micro-propulsion systems, will benefit the complex missions in many ways. The propulsion systems can be used for routine activities such as attitude control, station keeping and drag compensation. They can also serve as the key enabling technology for advanced missions such as formation flying, proximity operations, rendezvous and docking with one another. In conjunction to this, thrust measuring tool that is able to resolve the forces in the order of micronewton is required for supporting the development of the various micropropulsion systems.

The most widely used instrument to directly measure the thrust from the micropropulsion systems is pendulum-type thrust stand. Three commonly used pendulum configurations are hanging pendulum, inverted pendulum and torsional pendulum. The hanging pendulum thrust stand [1, 2] is the easiest to set up. It is stable and least affected by the test conditions, e.g. change in flexural stiffness due to the thermal effect. However, it is also the least sensitive among the three configurations. To improve the sensitivity, a long pendulum arm is required. This poses difficulties to house the thrust stand in small facilities. In contrast, inverted pendulum thrust stand is highly sensitive and compact with the length of the pendulum arm as short as 20cm has been constructed [3]. Nevertheless, the compactness of the thrust stand makes it vulnerable to thermal expansion. In addition, the inverted design is very sensitive to the shift in centre of gravity of the thrust stand. Meanwhile, torsional pendulum thrust stand offers a good balance between the pros and cons of the other two configurations. It is relatively easy to set up and achieves high sensitivity with a reasonable pendulum arm. The arm rotational axis is parallel to the gravity vector, making it less sensitive to the environmental noise. It is the most widely used pendulum thrust stand for performance characterization of various micropropulsion systems [4, 5-10].

This paper presents the setup of a torsional thrust stand based on improved version of electrostatic (ES) calibration system that has been previously developed in this research centre. By clustering a few heat sinks, generation of calibration force up to several tenths of mN is possible. This will improve the range of calibration force to four orders of magnitude. Accurate and repeatable calibrations forces are produced by precisely controlling the voltages supplied to the calibrator. Practical applications of the thrust stand are demonstrated by using it to measure the performance of a pulsed plasma thruster and a vaporizing liquid microthruster.

2. Principle of thrust measurement

The motion of a typical torsional pendulum can be described by a second order differential equation as

$$J\ddot{\theta} + \lambda\dot{\theta} + k\theta = F_t r_t \tag{1}$$

where J is the moment of inertia about the rotational axis, θ is the angular displacement of the pendulum, λ is the damping coefficient, k is the torsional elastic constant and F_t is the externally applied force at a distance of r_t from the rotational axis.

When an impulsive force with magnitude of F_i is applied on the pendulum for a duration of t_i, the external force can be expressed as:

$$F_t = \begin{cases} 0 & t < 0 \\ F_i & 0 \le t \le t_i \\ 0 & t > t_i \end{cases} \tag{2}$$

The general solution for the motion of pendulum after applying impulsive force, $t>t_i$, is given as:

$$\theta(t) = \frac{r_t F_t t_i}{J\omega_o} e^{\frac{-\lambda t}{2J}} \sin(\omega_o t) \tag{3}$$

where ω_o is the natural frequency of the pendulum.

From the sinusoidal function in (3), the maximum angular displacement is given by:

$$\theta_{max} = \frac{r_t F_t t_i}{J\omega_o} \tag{4}$$

Comparing to angular displacement, the linear displacement of the pendulum, Δx, can be measured easily using an optical sensor. From small angle approximation, displacements can be correlated by:

$$\Delta x = r_{LDS}\theta \tag{5}$$

where r_{LDS} is the distance between the linear displacement sensor and the rotational axis

Combining (4) and (5), the impulse bit can be expressed by:

$$I_{bit} = F_t t_i = \frac{J\omega_o}{r_t r_{LDS}} \Delta x_{max} \tag{6}$$

When constant force, F_c, is applied on the pendulum, the pendulum motion is simply described as:

$$F_c r_t = k\Delta\theta \tag{7}$$

where $\Delta\theta$ is the steady state angular displacement of the pendulum caused by the force. Rearranging and considering (5), the constant force is expressed as:

$$F_c = \frac{k}{r_t r_{LDS}} \Delta x \tag{8}$$

where Δx is the steady state deflection of the pendulum caused by the force.

From (6) and (8), it is evident that a thrust stand based on torsional pendulum principle can be set up to evaluate the impulse bit and constant thrust from different thrusters with the prior knowledge of J, ω_o and k. These parameters are sensitive to the changes in configuration of the thrust stand. For example, distribution of weight in the thrust stand affects J and ω_o while the effective k is influenced by electrical wiring.

Instead of evaluating each of these parameters separately, a calibration process is carried out to establish the relationship between the thrust stand response and the applied impulse bit or constant force. Using the calibration approach, all of these parameters are accounted for under the parameter of sensitivity, S_{cal}. It simplifies the measurement of impulse bit from a thruster as:

$$I_{bit} = \frac{(L_{cal}/L_t)}{S_{cal,I_{bit}}} \Delta x_{range} \tag{9}$$

where L_{cal} and L_t are the distance between rotational axis to calibrator and thruster, respectively, $S_{cal,Ibit}$ is the sensitivity of thrust stand to impulse bit, Δx_{range} is the maximum peak-to-peak displacement of the thrust stand.

Similarly, the constant thrust from a thruster can be evaluated by:

$$F_c = \frac{(L_{cal}/L_t)}{S_{cal,F_c}} \Delta x_t \tag{10}$$

where $S_{cal,Fc}$ is the sensitivity of thrust stand to constant force and Δx_t is the steady state deflection of the thrust stand.

In a particular test set up, the L_{cal} and L_t are always fixed. To find out the sensitivity, a number of known calibration impulse bits or constant forces are applied to the thrust stand. The corresponding thrust stand responses (Δx_{range} or Δx_t) are recorded and plotted as a calibration curve. The linear relationship between the calibration force and the thrust stand response allows the sensitivity of the thrust stand to be computed from the gradient of the curve.

3. Setup of torsional thrust stand

A horizontal pendulum type thrust stand in symmetrical configuration is designed and set up. The balance consists of a rotating platform, a central pillar which houses the pivot and a heavy base plate for mounting of other accessories. The rotating platform is made of aluminium alloy beam of 60cm long. The thruster is mounted on one end while a counterbalance mass and an electrostatic calibrator are placed on the other end. The platform is attached to a pivot through the central pillar. This limits the motion of the platform to one degree of freedom only as shown schematically in Figure 1.

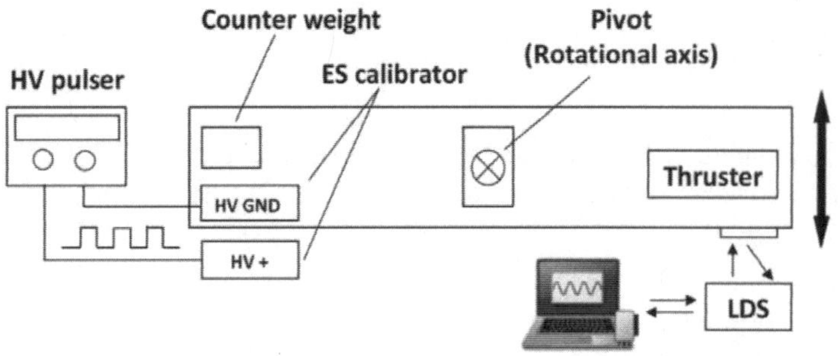

Figure 1. Schematic of the setup of torsional thrust stand

A stainless steel double-ended flexural pivot (6012-400, Riverhawk) was chosen. It demonstrates significant advantages over conventional bearings. The pivot is frictionless, requires no lubricant and almost maintenance-free. A strong permanent magnet with adjustable height was placed on one end of the platform to induce passive magnetic damping force of different magnitudes on the rotating platform.

An optical sensor (ILD 2300-2, MicroEpsilon) based on the principle of laser triangulation was used for linear displacement measurement. The non-intrusive and optical nature of the sensor is advantageous as it is not susceptible to electromagnetic interference [11]. This sensor has measuring range of 2mm and resolution as low as 30nm at measuring rate of 20kHz.

A non-contact calibration technique based on the principle was implemented. An improved version of the system based on 2D array of electrostatic fins was developed in our centre. The system is very compact and weighs only 23g. A detailed description of the improved system and its capability to produce a wide range of accurate and repeatable calibration force has been reported in our previous work [12]. In general, the ES calibrator system consists of a set of two 2D ES fin assemblies which engages for a particular distance as illustrated in Figure 2. By controlling the applied voltage precisely, it has been experimentally demonstrated that the calibration forces produced at 2mm difference of engagement distance are within 2% of deviation.

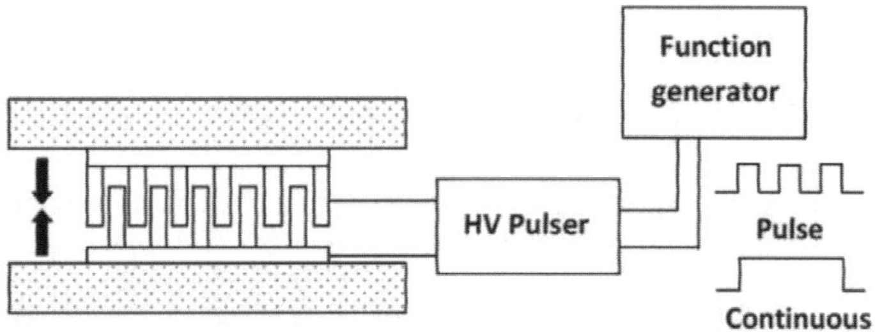

Figure 2. Schematic of electrostatic calibrator

A high voltage (HV) pulser (PVM-4210, DEI) supplies the required potential difference between the ES fin assemblies. The maximum voltage was limited to 900V to avoid arcing between the fin assemblies. Measuring with precision weighing balance, the ES calibrator system produces calibration force in the range of 40μN to 3.4mN consistently with deviation error less than 2%. Calibration impulse bit in the range of 7μNs to 340μNs was produced with error within 5%. For setting up the calibrator system on the torsional balance, one fin assembly was mounted on one end of the rotating platform while another fin assembly was fixed externally.

A vacuum facility has been set up not only to produce a vacuum environment required for the operation of electric thrusters but also to simulate the space environment for future testing of other thruster systems. The vacuum chamber is 70cm in diameter, 100cm in length and made of stainless steel. Vacuum solution for the chamber consists of a turbo pump (HiPace400, Pfeiffer Vacuum) and backed by a rotary vane pump (DUO65M, Pfeiffer Vacuum). An ultimate vacuum level in the order of 10^{-6}mbar is achievable. Throughout the testing, the chamber pressure is maintained at the level of 10^{-5} mbar.

4. Characterization of microthrusters

The versatility of the thrust stand to measure both impulse bit and constant thrust was demonstrated. It was used to characterize the performance of two different micropropulsion systems developed in our centre.

4.1. Pulsed plasma thruster

Pulsed plasma thruster (PPT) is an electric propulsion system that operates in pulse mode. When the PPT operates, it discharges the energy stored in the capacitor bank to ablate and ionize the solid teflon propellant to become plasma. Simultaneously, the flowing of high intensity electrical current during the discharge induces a strong magnetic field which accelerates the plasma to a velocity as high as a few km/s. From Newton's laws of motion, this produces an impulse bit in the opposite direction to the accelerated plasma.

Before the impulse bit measurement, the calibration process was carried out. The voltage input to the ES calibrator was fixed at 500V. The HV pulse width was varied from 16μs to 50μs for producing different impulse bits on the thrust balance. The resulting calibration curve as shown in Figure 3(a) was plotted after recording the thrust stand responses. Subsequently, a single PPT shot of unknown impulse bit was triggered. A typical thrust balance response to the PPT shot is shown in Figure 3(b). 10 measurements were taken and the average maximum displacement is 15.07±0.51μm. From the average maximum displacement and the sensitivity of the balance derived from the calibration curve, the impulse bit was evaluated as 23.15±0.79 μNs.

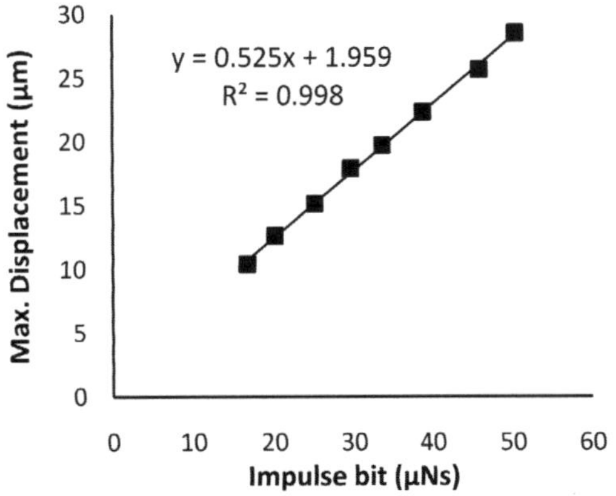

(a)

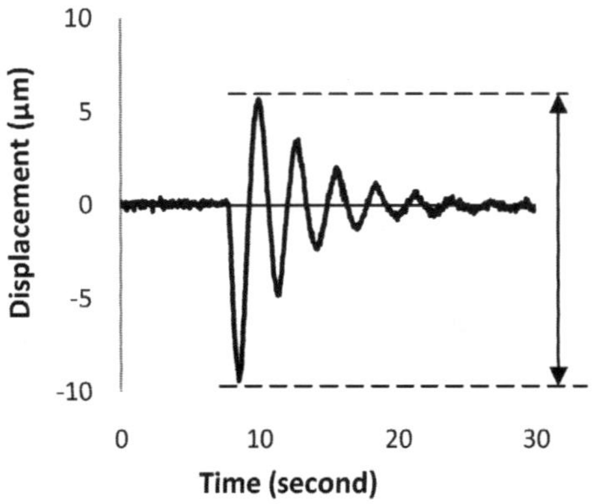

(b)

Figure 3. (a) Calibration curve for PPT impulse bit measurement, (b) thrust stand response to a single PPT shot. The arrow shows the maximum displacement

4.2. Vaporizing liquid microthruster

Vaporizing liquid microthruster (VLM) is a liquid propulsion system. Figure 4(a) shows a MEMS-scale VLM made of ceramic material [13]. Liquid propellant is injected into the microthruster using a syringe pump at a controlled flow rate. When sufficient electrical power is applied, the liquid propellant was vaporized. The hot vapor is ejected from the microthruster through the micronozzle as shown in Figure 4(b). A continuous force is produced when the liquid propellant is continuously fed into the microthruster while sufficient electrical energy is supplied.

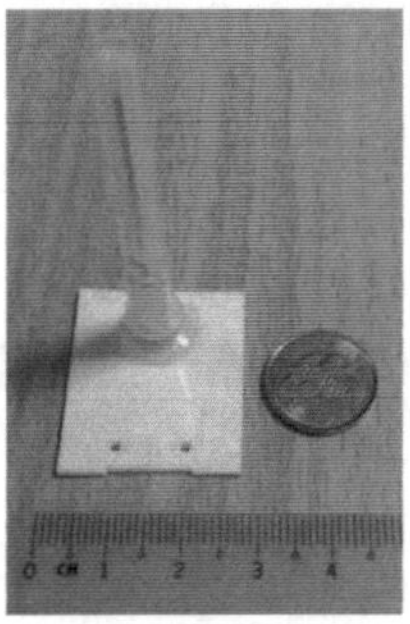

(a)

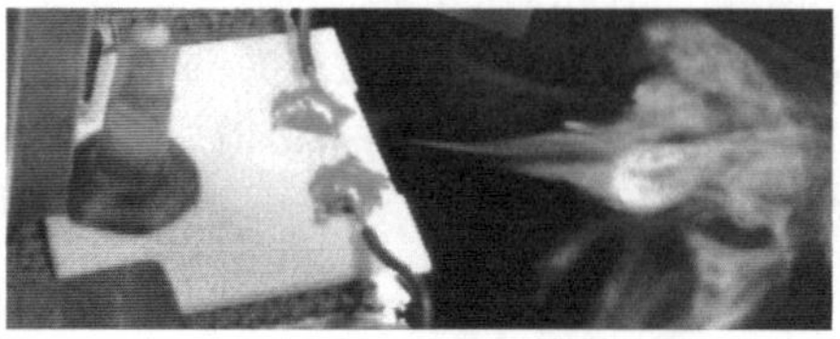

(b)

Figure 4. (a) Ceramic MEMS-scale VLM developed in SaRC, NTU. (b) Operation of VLM using water as propellant. Ejected water vapor is made visible using an appropriate lighting.

Similarly, calibration process was carried out prior to the thrust measurement. Different known calibration forces were applied on the thrust balance. The corresponding steady state deflection were recorded and plotted into the calibration curve in Figure 5(a). Then, the liquid propellant was fed in and the power supply was switched on. After approximately 10 seconds of operation, a steady state operation is achieved. The thrust balance was deflected for 128.56μm as shown in Figure 5(b). Considering the sensitivity from the calibration curve and using (9), maximum thrust of 633.5μN was produced at 1μl/s of flow rate.

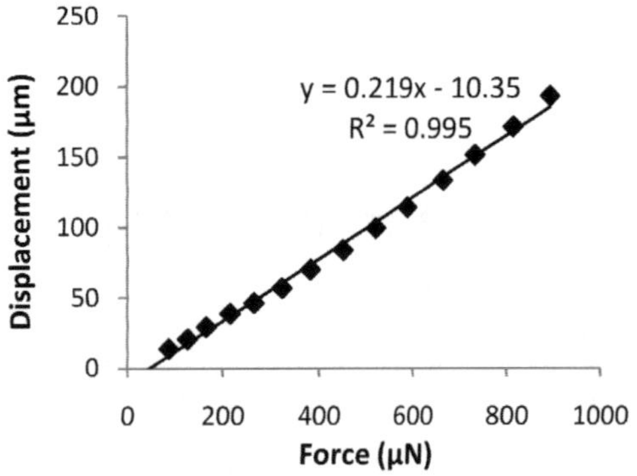

(a)

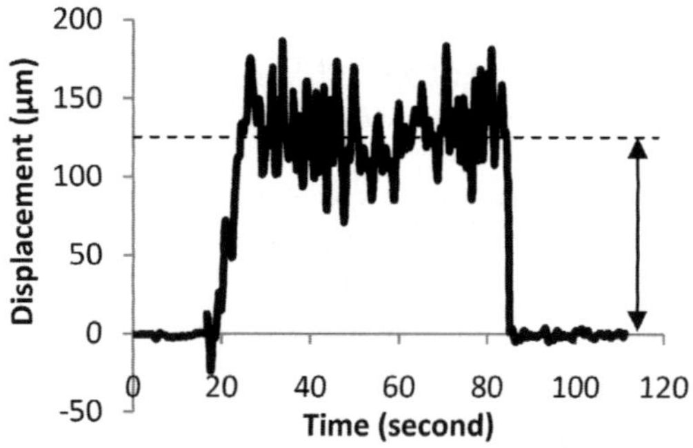

(b)

Figure 5. (a) Calibration curve for VLM constant thrust measurement, (b) Thrust stand response to operation of VLM

5. Conclusion

Development of a precise torsional pendulum thrust stand that can resolve forces in the order of micronewton has been presented. The thrust stand was set up as diagnostic tool for characterizing the performance of micropropulsion systems developed in SaRC, NTU. The thrust stand has been used to measure the performance of two microthrusters developed in our centre. The pulsed plasma thruster produces impulse bit of 23.15μNs at 1Hz. The vaporizing liquid microthruster produces a steady state thrust of 633.5μN by consuming water propellant at 1μl/s.

References

[1] Polzin K A, Markusic T E, Stanojev B J, Dehoyos A, and Spaun B 2006 *Rev. Sci. Instrum* **77** 105108
[2] Rocca S 2011 *Aerosp. Sci. Technol* **15** 148
[3] Haag T W1991*Rev. Sci. Instrum* **62** 1186
[4] Gamero-Castano M 2003*Rev. Sci. Instrum* **74** 4509
[5] Ciaralli S,Coletti M and Gabriel S B *Meas. Sci. Technol* **24** 115003
[6] Koizumi H, Komurasaki K and Arakawa Y 2004 *Rev. Sci. Instrum* **75** 3185
[7] Lun J and Law C 2014 *Meas. Sci. Technol* **25** 095009
[8] Zhou W J, Hong Y J and Chang H 2013*Rev. Sci. Instrum* **84** 125115
[9] Zhang D, Wu J, Zhang R, Zhang H and He Z 2013*Rev. Sci. Instrum* **84** 125113
[10] Yang Y X, Tu L C , Yang S Q and Luo J 2012 *Rev. Sci. Instrum* **83** 015105
[11] Cubbin E A, Ziemer J K, Choueiri E Y and Jahn R G 1997*Rev. Sci. Instrum* **68** 2339
[12] Cheah K H, Low K S, Tran Q V and Lau Z 2015*IEEE T. Instrum. Meas* **64** 3467
[13] Cheah K H and Low K S, Tran Q V and Lau Z 2015*J. Micromech. Microeng* **25** 015013

IOP Conf. Series: Materials Science and Engineering **152** (2016) 012020 doi:10.1088/1757-899X/152/1/012020

Aircraft nonlinear optimal control using fuzzy gain scheduling

I F Nusyirwan* **and Z Y Kung**

Department of Aeronautics, Automotive & Ocean Engineering, Universiti Teknologi Malaysia, Malaysia

*istaz@utm.my

Abstract. Fuzzy gain scheduling is a common solution for nonlinear flight control. The highly nonlinear region of flight dynamics is determined throughout the examination of eigenvalues and the irregular pattern of root locus plots that show the nonlinear characteristic. By using the optimal control for command tracking, the pitch rate stability augmented system is constructed and the longitudinal flight control system is established. The outputs of optimal control for 21 linear systems are fed into the fuzzy gain scheduler. This research explores the capability in using both optimal control and fuzzy gain scheduling to improve the efficiency in finding the optimal control gains and to achieve Level 1 flying qualities. The numerical simulation work is carried out to determine the effectiveness and performance of the entire flight control system. The simulation results show that the fuzzy gain scheduling technique is able to perform in real time to find near optimal control law in various flying conditions.

1. Introduction

Flight dynamics is nonlinear with the motion of an airplane subjected to inherent stability and external disturbance [1]. Hence a robust flight control system is important to stabilize and ensure favourable flight performance. Present generation aircraft has been augmented with sophisticated flight control system. This is especially true for fighter aircraft, which is designed to be inherently unstable to allow it to perform extreme manoeuvres. In view of this, optimal control and gain scheduling method are implemented to achieve the flight stability and controllability.

Linear quadratic (LQ) optimization is commonly utilised in flight control system. The general idea of LQ control is to minimize the performance index parameterized by the weighting matrices [2]. In performing stability augmented system (SAS) in flight control, the LQ optimization for tracking a command system is employed [3]. In addition, publications by [4, 5] shared a similar methodology in using LQ optimization to perform nonlinear control and they showed excellent controllability. Besides, gain scheduling method is commonly used to deal with nonlinear system and this approach reduces computational burden associated with the local linear system [6]. In agreement with research done by [7], the gain scheduling composes three important aspects: computation of linear parameter-varying models, design of linear control system and interpolation of controller gains.

In consonance with the second aspect of gain scheduling, similarity was found in using H-infinity controller together with gain scheduling method to improve the global stability [8, 9]. Meanwhile, the research study presented in [10] has highlighted the third aspect of gain scheduling, in which a global polynomial variable scheduling was examined.

In contrast, it is also stated that the conventional gain scheduling is time-consuming due to the ad-hoc nature implies more iterative work and requires more operating points [11]. Taking advantages of the research done in [12], the fuzzy logic is implemented in line with the gain scheduling. This is due to the fact that fuzzy inference system is prone to human understandable logic. Besides, fuzzy logic is widely used because of its easy realization and robustness [13]. Credible researches based on fuzzy gain scheduling could be found since the 1990's [14, 15] and similar approach was revised a decade later in [5, 11]. As presented by reference publications [5, 11, 14 and 15], they showed the capability of fuzzy gain scheduling in performing stable flight control. In addition, the stability based on fuzzy system with optimal feedback could also be found in [5, 16].

To date and to the best of author's knowledge, the work on the fuzzy gain scheduling on highly nonlinear flight dynamic is still insufficient. Previous publications showed that the application of fuzzy gain scheduling on a wide range of operating region. In fact, the most unstable performance could be observed at low speed manoeuvring flight. Therefore, the aim of this article is to perform fuzzy gain scheduling over highly nonlinear flight dynamics through examination of the eigenvalues. LQ Optimal control is applied on the particular operating point in order to search for the most suitable controller gains. The output is interfaced with the fuzzy gain scheduler to perform global flight control system. In this paper, full six-degree-of-freedom equations of motions are involved. Hence, for better illustration and understanding, Table 1 is provided to show all related abbreviations and symbols.

Table 1. List of abbreviations and symbols

Abbreviations	Definition
6DOF	six degrees of freedom
FGS	fuzzy gain scheduler
SAS	stability augmented system
LQ	linear quadratic
Symbols	**Definition**
$\dot{U}, \dot{V}, \dot{W}$	velocity derivatives in body frame with components in body system
U, V, W	body axis wind velocity components
$\dot{V}t, \dot{\alpha}, \dot{\beta}$	wind axis velocity derivatives in body frame with components in body system
ϕ, θ, ψ	Euler angles
$\dot{\phi}, \dot{\theta}, \dot{\psi}$	Euler angles derivatives
P, Q, R	body axis angular velocity components
$\dot{P}, \dot{Q}, \dot{R}$	angular velocity derivatives in body frame with components in body system
X, Y, Z	body axis x, y and z component forces
l, m, n	body axis rolling, pitching and yawing moments
Vt	absolute wind
M	aircraft mass
g_d	gravity
$J_{x,y,z,xz}$	second moment of inertia corresponding to its axis
a_x, a_y, a_z	body axis x, y, and z component accelerations
$\delta_e, \delta_a, \delta_r, \delta_T$	elevator, aileron, rudder and throttle
$\dot{x}$	state derivative variables: $\dot{x} = [\dot{V}t, \dot{\alpha}, \dot{\beta}, \dot{\phi}, \dot{\theta}, \dot{\psi}, \dot{P}, \dot{Q}, \dot{R}]^T$
x	states variables: $\dot{x} = [Vt, \alpha, \beta, \phi, \theta, \psi, P, Q, R]^T$
y	output state variables: $y = [a_x, a_y, a_z]^T$
u	control state variables: $u = [\delta_e, \delta_a, \delta_r, \delta_T]^T$
r, z	reference input state variables performance output state variables
K	controller gains: $K = [K_\alpha, K_q, K_i]^T$
Q_m, R_m	state weighting matrix and control weighting matrix
ω_n, ξ	natural frequency, damping ratio

2. Nonlinear flight dynamics

Based on [3], the F-16 jet fighter aerodynamics data is employed and the 6DOF nonlinear equations of motion are derived based on first-order differential equation. The equations are as the following.

2.1. Force equations

$$\dot{U} = RV - QW - g_d \sin\theta + X/M \tag{1}$$

$$\dot{V} = -RU + PW + g_d \sin\phi\cos\theta + Y/M \tag{2}$$

$$\dot{W} = QU - PV + g_d \cos\phi\cos\theta + Z/M \tag{3}$$

written in wind axis,

$$\dot{V}t = \left(U\dot{U} + V\dot{V} + W\dot{W}\right)/Vt \tag{4}$$

$$\dot{\alpha} = \left(U\dot{W} - W\dot{U}\right)/(U^2 + W^2) \tag{5}$$

$$\dot{\beta} = \left(\dot{V}Vt - V\dot{V}t\right)/\left(Vt\sqrt{U^2 + W^2}\right) \tag{6}$$

2.2. Euler kinematic equations

$$\dot{\phi} = P + \tan\theta\,(Q\sin\phi + R\cos\phi) \tag{7}$$

$$\dot{\theta} = Q\cos\phi - R\sin\phi \tag{8}$$

$$\dot{\psi} = (Q\sin\phi + R\cos\phi)/\cos\theta \tag{9}$$

2.3. Moment equations

$$\Gamma\dot{P} = J_{xz}[J_x - J_y + J_z]PQ - [J_z(J_z - J_y) + J_{xz}{}^2]QR + J_z\ell + J_{xz}n \tag{10}$$

$$J_y\dot{Q} = (J_z - J_x)PR - J_{xz}(P^2 - R^2) + m \tag{11}$$

$$\Gamma\dot{R} = [(J_x - J_y)J_x + J_{xz}{}^2]PQ - J_{xz}[J_x - J_y + J_z]QR + J_{xz}\ell + J_x n \tag{12}$$

where,

$$\Gamma = J_x J_z - J_{xz}{}^2$$

3. Linearization

As mentioned in the second aspect of gain scheduling, the flight control is performed under linear operating system or steady state condition, hence linearization is necessary. At equilibrium points, the trim function is constructed to determine the specific state variables and the control variables, which converge the state derivatives into zero. The constraint equations describe the relationship of 6DOF nonlinear equations and it is modelled based on analysis by [17]. To analyse with the nine equations, the numerical linearization method is utilized [3, 18]. A total number of 21 linear systems are selected according to various airspeed and altitude. The outcome of the linearized equations of motion are described in Equation 13.

$$\dot{x} = Ax + Bu \tag{13a}$$

$$y = Cx + Du \tag{13b}$$

Taking steady state condition at sea level for airspeed of 300 ft/s, the linearized matrices are determined and further decomposed into longitudinal state dependent parameters, the linearized state space matrices are as presented in Equation 14.

$$\begin{bmatrix} \dot{Vt} \\ \dot{\alpha} \\ \dot{\theta} \\ \dot{q} \end{bmatrix} = \begin{bmatrix} -0.028 & -6.2682 & -32.17 & -2.6422 \\ -0.0007 & -0.6099 & 0 & 0.9012 \\ 0 & 0 & 0 & 1 \\ 0 & -0.0710 & 0 & -0.7192 \end{bmatrix} \begin{bmatrix} Vt \\ \alpha \\ \theta \\ q \end{bmatrix} + \begin{bmatrix} 0.0112 \\ -0.0013 \\ 0 \\ -0.0628 \end{bmatrix} [\delta_e] \tag{14a}$$

$$[a_z] = [0.0066 \quad 5.6540 \quad 0 \quad 0.9238] \begin{bmatrix} Vt \\ \alpha \\ \theta \\ q \end{bmatrix} + [0.0119][\delta_e] \tag{14b}$$

4. Linear quadratic optimal control

Linearized equations of motion is constructed based on Equation 13 with respect to specific airspeed and altitude. The linear LQ optimal controller is designed based on the highly nonlinear region and this can be done by examining the root locus plot for the longitudinal mode of motion [1]. Based on the selected flying conditions, the root locus plots behave irregular pattern as shown in Figure 1 and this implies the nonlinearity of flight dynamics.

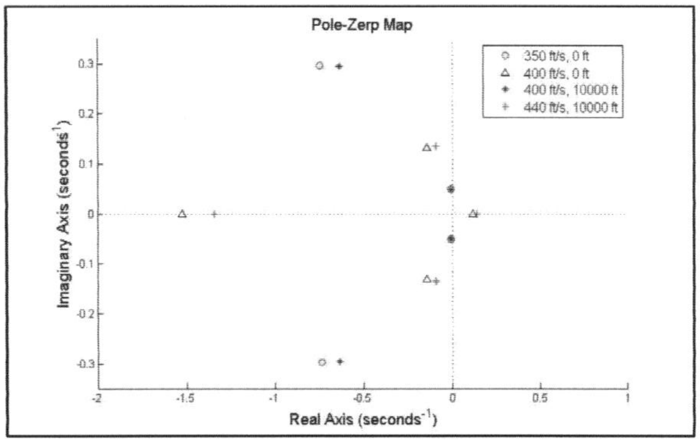

Figure 1. Root locus plots for several operating points

Taking [2, 3] as references, the optimal control for pitch rate SAS can be formulated based on Figure 2.The longitudinal SAS is performed by controlling the pitch rate. Pilot input or the reference pitch rate command is given to manoeuvre the jet fighter. In addition, the characteristics of actuator and alpha sensor are modelled in term of transfer function. This is to obtain more realistic results in performing the stability augmentation.

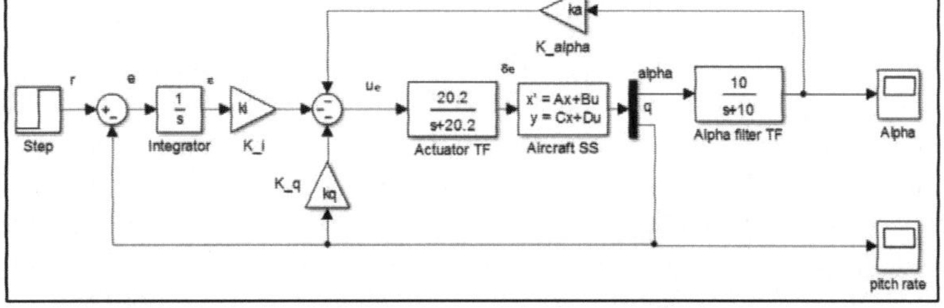

Figure 2. Pitch stability augmented system block diagram

IOP Conf. Series: Materials Science and Engineering **152** (2016) 012020 doi:10.1088/1757-899X/152/1/012020

Based on Figure 2, the LQ optimal control for command tracking is designed and derived. The mathematical formulation can be simplified in linear state space model as shown in Equation 15 and the control is static output feedback as listed in Equation 16.

$$\dot{x} = Ax + Bu + Gr \tag{15a}$$

$$y = Cx + Du + Fr \tag{15b}$$

$$z = Hx \tag{15c}$$

$$u = -Ky \tag{16}$$

where,

$$x = \begin{bmatrix} \alpha \\ q \\ \delta_e \\ \alpha_F \\ \epsilon \end{bmatrix}, \quad u = [u_e], \quad r = [q_c], \quad z = [q], \quad y = \begin{bmatrix} \alpha_F \\ q \\ \epsilon \end{bmatrix}, \quad K = \begin{bmatrix} K_\alpha \\ K_i \\ K_q \end{bmatrix}$$

Optimal control using linear quadratic for command tracking simplify the tuning parameters. By adding the time-dependent weighting component [2, 3], the general performance index or cost function is described in Equation 17.

$$J = \frac{1}{2}\int_0^\infty \left(t^k \tilde{x}^T P_m \tilde{x} + \tilde{x}^T (Q_m + C^T K^T R_m K C)\tilde{x}\right)dt \quad \text{Or} \quad J = \frac{1}{2}\text{tr}(P_k X) \tag{17}$$

where,

$$X = \overline{x}\overline{x}^T = A_c^{-1} B_c r_0 r_0^T B_c^T A_c^{-T}, \qquad A_c = A - BKC, \qquad B_c = G - BKF$$

P_m is the solution for Lyapunov equation for optimal control shown in Equation 18.

$$\begin{gathered} 0 \equiv A_c^T P_0 + P_0 A_c + P_m \\ 0 \equiv A_c^T P_1 + P_1 A_c + P_0 \\ \vdots \\ 0 \equiv A_c^T P_{k-1} + P_{k-1} A_c + P_{k-2} \\ 0 \equiv A_c^T P_k + P_k A_c + k!\, P_{k-1} + Q_m + C^T K^T R_m K C \end{gathered} \tag{18}$$

By observing the SAS presented in Figure 2, the integrator makes the whole system become type 1 control system. The natural performance index indicates the performance requirement that minimize the tracking error without huge amount of control effort [5]. Hence, the whole tuning parameter could now be simplified into single weighting value [2, 3]. This could be done by assigning the parameters into $P_m = H^T H,\ Q_m = 0,\ R_m = \rho$. Whereas the ρ term is the only tuning parameter left.

The LQ optimal control is performed with an initial guess on the feedback matrix, K and after few iterations, the cost function value, J converges into a solution that indicates the minimum performance index and results in the most appropriate controller gains. Apparently, following the LQ optimal design procedure and suitable tuning parameter, ρ, the response for pitch rate can be simulated using step input function. The result is presented in Figure 3. By checking the eigenvalues, it is found out that with the aid of the LQ optimal control, the short period poles shift to $-1.833 \pm 1.7398i$ and this gives a response of $\omega_n = 2.523, \zeta = 0.725$ that satisfies the Level 1 flying quality [19]. It is essential to highlight the advantages of using LQ optimal control. The search for the most suitable controller gains is all done by computer, which highly reduced human work load. Besides, this method enables to cope with the ad-hoc nature of fundamental trial and error gain tuning method. According to Figure 3, the LQ optimal control showed outstanding response, which avoided overshoot and settled at time less than 3 seconds.

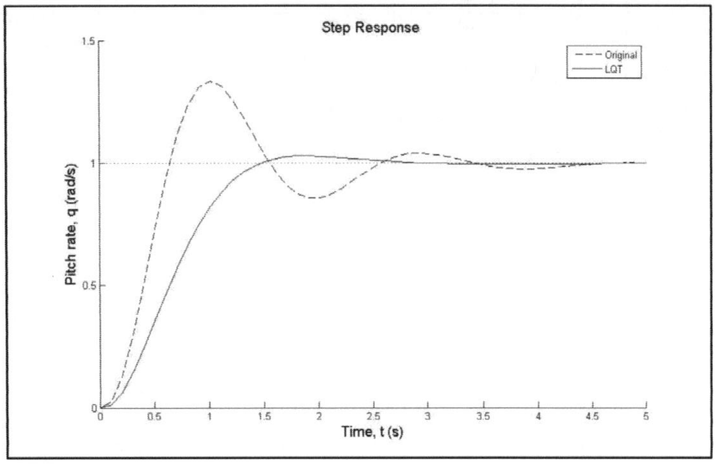

Figure 3. Pitch rate step response at airspeed of 300 *ft/s* at sea level.

5. Fuzzy gain scheduling

A series of linear systems are taken into examination and LQ optimal control is performed. Each linear controller design results in three controller gains, K_α, K_q and K_i. The results for each controller gains are fed into the FGS. Fundamental fuzzy controller is used to schedule the controller gains and the fuzzy controller design is well discussed in [20]. Design of fuzzy system is based on Model Rules, R^i.

IF $\mathrm{x}_i(\mathrm{t})$ is $\mathrm{M}_k(\mathrm{x}_i)$ **AND** $\mathrm{x}_j(\mathrm{t})$is $\mathrm{M}_l(\mathrm{x}_j)$, **THEN** $\mathrm{K} = \sum \alpha_i \, \mathrm{k}_i$

where R^i is the ith rule of the fuzzy system, $\mathrm{x}_i(\mathrm{t})$ and $\mathrm{x}_j(\mathrm{t})$ are the current inputs or states fed into the fuzzy system, M_k and M_l are the kth and i^{th} membership function subjected to inputs $\mathrm{x}_i(\mathrm{t})$ and $\mathrm{x}_j(\mathrm{t})$ respectively, k_i is the controller gain with respect to the input membership function, α_i is the weighting function.

The fuzzy system consists of four main components [20] and each component is described in detailed. First, the "rule-base" definition is determined by subjecting reference inputs to the specific outputs. Second, the inference mechanism acts as the decision maker based on the inputs parameter. Third, the fuzzification interface processes the given inputs into information for inference mechanism. The last components describe the defuzzification interface, which gives the conclusions based on the results of inference mechanism. The fuzzification inference utilises triangular membership functions for both inputs and it is shown in Figure 4. There will be only one rule involves in the whole process, which indicates four input membership functions and concludes using the **OR** defuzzification method. The actual corresponding output is weighted using the area rule given in Equation 19.

$$K = \frac{\sum_{i=1}^{4} Ar_i c_i}{\sum_{i=1}^{4} Ar_i} \qquad (19)$$

where K is the output controller gain, Ar_i is the ith triangular area of output membership function and c_i is the centre of ith output membership function.

21 linear systems are taken into consideration and these cover airspeeds for 300, 250, 400, 440, 500, 540 and 600 ft/s, and altitude for 0, 10000 and 20000 ft. The outputs from the fuzzy system are tabulated graphically and displayed as shown in Figure 5.

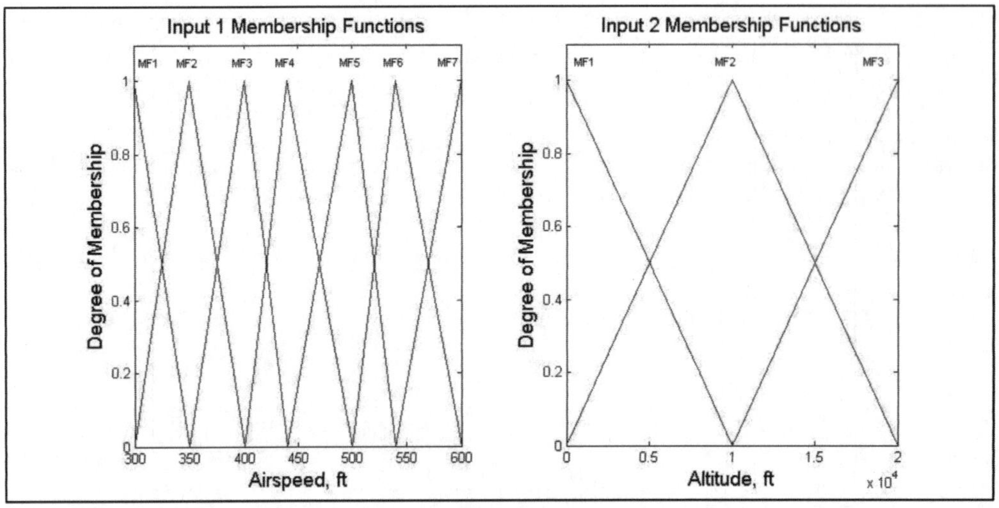

Figure 4. Airspeed (left) and altitude (right) membership function

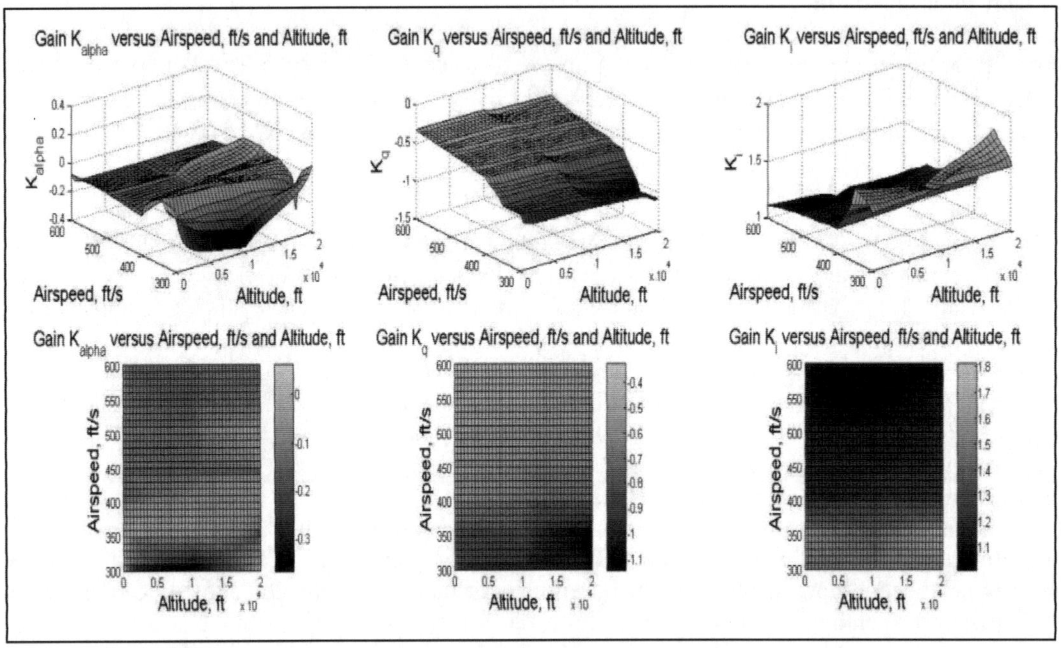

Figure 5. Controller gains K_α, K_q and K_i versus airspeed and altitude

A simple simulation on pitch rate stability augmented system is conducted by randomly selecting the flying condition. Two sets of controller gain (output from FGS and pre-set controller gain) are applied and compared. Based on the simulation work on flying condition at airspeed of 375 ft/s and altitude at 5000 ft above sea level, the step response is presented in Figure 6. The response shows non-oscillating and non-overshooting results as compared to pre-set controller gain. Besides, by checking on the eigenvalues, the short period poles are located at $-2.115 \pm 1.913i$ and this gives a response of $\omega_n = 2.852, \zeta = 0.742$ that also satisfies Level 1 flying quality. It is vital to emphasize that the LQ optimal control serve as the effective controller gain finder at a particular operating point whereas the fuzzy gain scheduling serve as the global control system for a wide range of operating range.

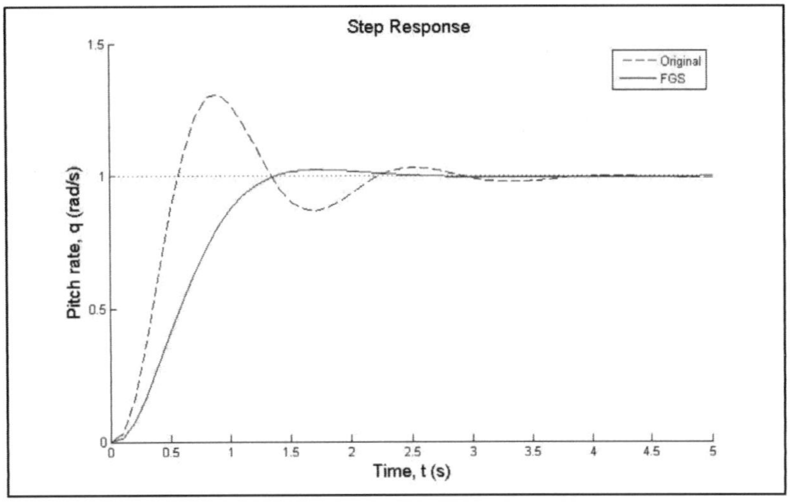

Figure 6. Pitch rate step response at airspeed of 375 ft/s and altitude of 5000 ft above sea level

6. Conclusion

In conclusion, the FGS is designed and established in longitudinal flight control system. The nonlinear control for longitudinal motion is approximated using 21 linearized models and the selection of the linearized system is based on the highly nonlinear region where the eigenvalues show irregular pattern. Using LQ optimal control, the tuning parameter is simplified into a single weighting value, ρ. Besides, this method presents the most suitable controller gain following the optimal cost function and in the meantime reduces the time required for controller gain tuning. The controller gains are determined and fed into FGS. Two inputs: airspeed and altitude are required to schedule the specific controller gains $K_\alpha, K_q$and K_i. The computer simulation demonstrates the effectiveness in using FGS and the capability in achieving Level 1 flying qualities at all time.

Acknowledgments

The authors wish to acknowledge the Department of Aeronautics, Automotive and Ocean Engineering, Universiti Teknologi Malaysia for supporting in this research. In addition, the authors also would like to present their gratitude to Universiti Teknologi Malaysia.

References

[1] Nelson R C 1998 *Flight Stability and Automatic Control* (New York: McGraw-Hill)
[2] Lewis F L and Syrmos V L 1995 *Optimal Control* (New York: Wiley)
[3] Steven B L and Lewis F L 1992 *Aircraft control and simulation* (New York: Wiley)
[4] Almutairi S H and Aouf N 2014 *Proc. of the 11th IEEE International*
[5] Bai Y, Zhuang H and Wang D 2006 *Advanced Fuzzy Logic Technologies in Industrial Applications* (London: Springer) 207-21
[6] Rugh W J and Shamma J S 2000 *Automatica* **36** 1401-25
[7] Leith D J and Leithead W E 2000 *Int. J. of control* **73** 1001-25
[8] Nichols R A, Reichert R T and Rugh W J 1993 *IEEE Trans. on control systems technology* **1** 67-79
[9] Ravanbod K M and Noll D 2012 *7th IFAC Symp. on Robust Control Design*
[10] Choi T, Jimenez H and Mavris D N 2015 *J. Aerospace Science and Technology* **46** 493-505
[11] Oosterom M and Babuska R 2006 *Control Engineering Practice* **14** 769-81
[12] Takagi T and Sugeno M 1985 *IEEE Trancs. on system, man and cybernetics* **15** 116-32
[13] Takana K and Sugeno M 1992 *Fuzzy Sets System* **45** 135-56
[14] Gonsalves P G and Zacharias G L 1994 *Proc. of the 3rd IEEE conf. on Fuzzy System*

IOP Conf. Series: Materials Science and Engineering **152** (2016) 012020 doi:10.1088/1757-899X/152/1/012020

[15] Fujimori A, Wu Z, Nikiforuk P N and Gupta M M 1997 *Proc. of AIAA Guidance, Navigation and Control Conference*
[16] Wu S J and Lin C T 2000 *IEEE Trans. on Fuzzy System* **8** 171-85
[17] Marco A D, Duke E and Nerndt J 2007 *AIAA Modelling and Simulation Technologies Conf.*
[18] Chapra S and Canale R 2005 *Numerical methods for engineers* (New York: McGraw-Hill)
[19] MIL-F-8785C1980 Flying qualities of piloted airplanes *Technical report MIL-F-8785C*
[20] Passino K M and Yurkovich S 1998 *Fuzzy Control* (Menlo Park: Addison Wesley Longman)

A blended wing body airplane with a close-coupled, tilting tail

R E M Nasir*, N S C Mazlan, Z M Ali, W Wisnoe and W Kuntjoro

Aviation Technology Research (ATR) Group, Flight Tech. & Test Centre (FTTC), Faculty of Mech. Engr. Universiti Teknologi Mara, Malaysia

*rizal524@salam.uitm.edu.my

Abstract. This paper highlights a novel approach to stabilizing and controlling pitch and yaw motion via a set of horizontal tail that can act as elevator and rudder. The tail is incorporated into a new design of blended wing body (BWB) aircraft, known as Baseline-V, located just aft of the trailing edge of its inboard wing. The proposed close-coupled tail is equipped with elevators that deflect in unison, and can tilt – an unusual means of tilting where if starboard side is tilted downward at *k* degree, and then the portside must be tilted upward at *k* degree too. A wind tunnel experiment is conducted to investigate aerodynamics and static stability of Baseline-V BWB aircraft. The model is being tested at actual flight speed of 15 m/s (54 km/h) with varying angle of attack for five elevator angle cases at zero tilt angle and varying sideslip angle for four tilt angle cases at one fixed elevator angle. The result shows that the aircraft's highest lift-to-drag ratio is 32. It is also found that Baseline-V is statically stable in pitch and yaw but has no clear indication in terms of roll stability.

1. Introduction

Blended wing body (BWB) aircraft configuration can offer reduction in fuel consumption and noise by reducing drag [1]. It has a lifting body, instead of tubular fuselage, blended smoothly to its wing, thus increases lift force. The smooth transition between body and wing reduces interference drag and its low wetted surface area compared to conventional aircraft of the same volume reduces skin-friction drag. The combination of high lift and low drag forces increases the lift-to-drag ratio (L/D) of BWB aircraft up to 20% more than conventional aircraft [2, 3]. However, high L/D of BWB aircraft design can only be practical with careful attentions to other important requirements such as flight stability and control. Despite aerodynamic advantages, BWB aircraft can become statically unstable due to short moment arm between control surface to its centre of gravity [4]. BWB aircraft have similar stability and control issues as flying-wing designs [5] and these are made worse by its sensitivity to external disturbances such as gust [6]. It is usually designed with multiple elevons located almost at full wing span because it needs large control surface area to counter its nose-down pitch moment generated by its lifting body and wing combined [7]. Just like what has been discussed in [3], a pure tail-less BWB aircraft is not only very challenging to be designed if one were to have good balance between L/D and stability, the intricate "blending" of lifting body and tail makes fabrication of aircraft structures or its wind tunnel model complex and expensive [8].

Bolsunovsky *et al.* [3] propose various plan forms and tail configurations for BWB aircraft due to inadequate control in yaw. This makes the BWB aircraft looks conventional, not to mention additional drag created by the twin vertical tails that increase interference drag. A new solution has to be found regarding inadequate yaw stability and controllability and one may look at nature to be inspired. A bird does not have vertical tail like conventional aircraft but it is still be able to control its yawing motion by tilting its tail with combination of tail pitch angle [9].

2. Design concept

A novel approach is proposed in this study to stabilize and control pitch and yaw motion via a set of horizontal tail that can act as elevator and rudder. The said tail is incorporated into a new design of BWB aircraft, known as Baseline-V, located just aft of the trailing edge of its inboard wing as shown in Figure 1. The proposed close-coupled tail is equipped with elevators that deflect in unison (port and starboard side deflect at the same angle η at the same direction), and can tilt – an unusual means of tilting where if starboard side is tilted downward at k degree, then the portside must be tilted upward at k degree too. Figure 2 shows how the elevator is deflected and the tail is tilted. Deflection of elevator angle alone can only stabilize (or trim) and control the said aircraft in pitch. Tilting the tail rotates the direction of tail lift (or down force) to an angle normal to tail plane. This produces a component of force sideways thus producing yawing moment that causes the aircraft to yaw and also roll (due to small rolling moment produced). Figure 3 shows how the combination of elevator angle deflection and tail tilt can theoretically causes this phenomenon. Increase in tilt angle shall theoretically increase yaw moment that yaw control shall be possible without using any rudder hence there shall be no need for a vertical tail.

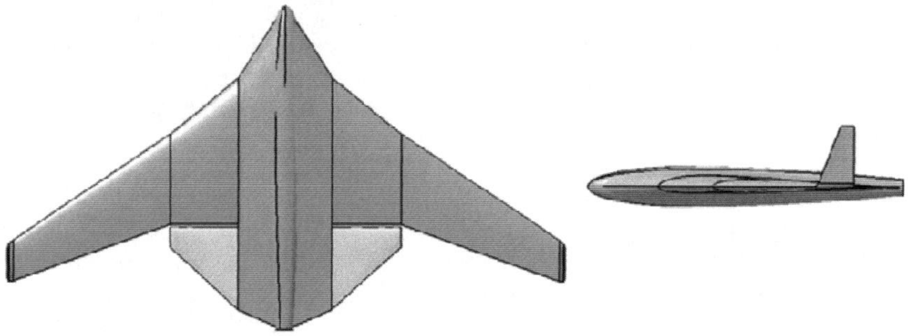

Figure 1. Baseline-V BWB with a close-coupled tail – plan and side views

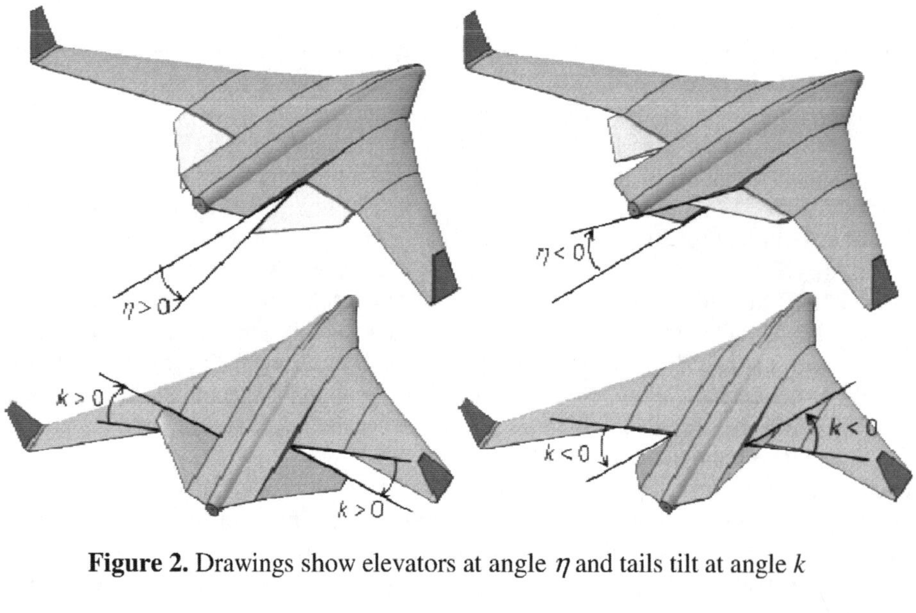

Figure 2. Drawings show elevators at angle *η* and tails tilt at angle *k*

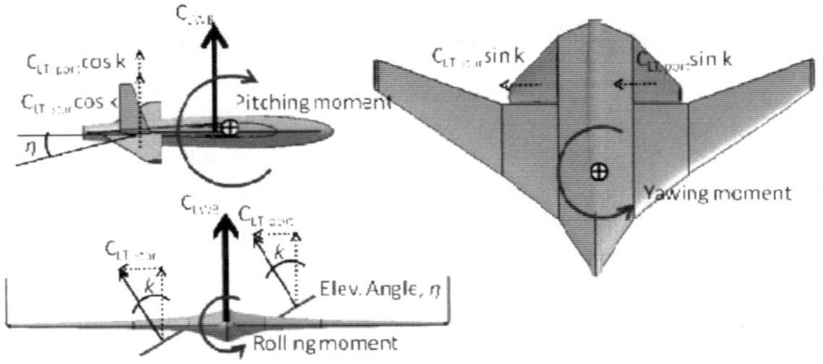

Figure 3. Schematics of theoretical effect of elevator deflection and tail tilt

The main reason for coming up with the design is to have yawing control without having a rudder on vertical tail. It is found that incorporating rudder on winglet is impractical as it requires tiny but powerful servo, which can be very expensive. Inspired by tilting tail of a bird, this solution is found where the tail can be used as pure elevator if it is horizontal, and can also act as rudder if the tail is tilted. This is impractical for big airline-sized aircraft but it is suitable for a small drone proposed here.

From plan view, the BWB seems like a tail-less aircraft just like other BWBs and hypothetically, small gap between trailing edge of the inboard wing and the nose of the tail may eliminate interference drag, thus maintaining its high L/D. How is this possible? Conventional wing-body-tail configuration has its tail located far away from the wing and the top and bottom airflow past the wing is reunited. After reuniting for quite some length, the airflow passes the horizontal tail hence; in this case, the tail becomes another "obstacle" to the airflow. On another aspect, the horizontal tail sticking out clearly from the body disrupts smooth airflow around the body and interference drag is expected. In addition, conventional tails (horizontal and vertical) add frontal area of the whole aircraft. For Baseline-V, the tail is so close that, theoretically, the top and bottom airflow on inboard wing may not be reunited and continues to flow on top and bottom of the tail separately as if the tail is integral part of the wing. It is like having a slotted flap behind the wing but in this case, it is not connected physically to the wing because it needs to be tilted to act as rudder. However, most of Baseline-V cruising time will have its tail staying horizontal at k = 0 degree with some elevator angle η for angle of attack trim. High aerodynamic efficiency (read: lift-to-drag ratio) is expected here because the wind (airflow) does not "see" the tail. However, drag is expected to increase when the tail begins to tilt because aircraft frontal area is increased thus lift-to-drag ratio is expected be reduced significantly.

3. Experiment setup

The objective of this experiment is to investigate aerodynamics and static stability of Baseline-V BWB aircraft. To achieve this, data of forces and moments in all six axes, namely longitudinal, side and vertical forces and roll, pitch and yaw moments, are collected at each angle of attack α and/or angle of sideslip β, for each elevator angle η and tilt angle k with airspeed similar to the aircraft's designed cruising/loitering speed. The experiment is conducted in a low speed wind tunnel (UTM-LST) at the Aeronautic Laboratory of Universiti Teknologi Malaysia (UTM). Raw data of forces and moments collected is then converted into coefficients that will be corrected for buoyancy, solid blockage and wake blockage. Drag, side force and lift coefficients represent the aerodynamic force to analyze their curve trends, slopes, extrema, intersections and most importantly to be converted into measurement of aerodynamic efficiency in terms of lift-to-drag ratio, L/D while pitch and yaw moment coefficients represent aerodynamic moments used to analyze static stability.

Baseline-V BWB is a small UAV for surveillance mission. The wind tunnel model is fabricated from pine woods, some lightweight plywoods, carbon fibre rods, balsa skin and plastic coverings, and is build to 1:1 scale. In fact, the wind tunnel model is build in similar manner to the actual UAV itself. Table 1 highlights the geometric specifications of Baseline-V.

Table 1. Wing-body and tail design parameters

Wing-body		Tail	
25% sweep	30^0	25% sweep	-12.05^0
Wing area, S	0.4 m^2	Tail area, S_t	0.05 m^2
Wing span, b	1.4 m	Tail span, b_t	0.36 m
Aspect ratio, AR	4.92	Aspect ratio, AR_t	2.60
Mean chord, c	0.284 m	Mean chord, c_t	0.127 m

The wing-body design can generally be divided into three main segments: body, inboard wing and outboard wing. The body is designed based on standard symmetrical airfoil section modified at the trailing edge to provide nose up pitch moment. Similarly, for inboard and outboard wing sections, the modification is also made to their respective airfoil sections near trailing edge to reduce the negative pitch moment coefficient normally found on standard 12-percent and 15-percent thickness-to-chord cambered airfoils. Again, this is done to reduce nose down moment that is prevalent on swept-back wing. The aspect ratio of this aircraft (body included since it also generates lift) is generally medium at 4.92, which is similar to some general aviation airplanes. The inboard-outboard wing combo has a planform similar to common airliners of today. The trailing edge of the inboard wing is less than five millimetres from leading edge of the tail. The tail is originally designed from standard symmetrical airfoil sections but it is then changed to a mere plate with rounded leading edge for practicality reason. The Baseline-V BWB is also equipped with vertical wing tips, mainly for reducing the trailing edge vortices' size, hence improving lift-to-drag ratio but also contribute to ensuring yaw stability especially when the tail is not tilted (horizontal). Without the wing tip, directional static stability depends on the planform shape of the BWB alone and since without it, the aircraft has small side area and couple with short length of its body relative to its wing span, the aircraft may have large tendency to spin out of control or "boomerang".

The closed circuit wind tunnel used in this study has a test section of 1.5 m x 2 m x 5.8 m. This tunnel can operate at wind speed between 3 m/s to 80 m/s. Full scale wind tunnel model of Baseline-V with control surfaces that can be deflected remotely is mounted on three struts connected to turntable on the floor as shown in Figure 4. The turntable is mounted on balance with sensors to measure forces and moments of aircraft at varying angle of attack α and sideslip angle β. The aft pitching strut was connected to the model using a single boom. Since the model tested is an actual size of the UAV, the results obtained in this experiment are assumed to be the actual aerodynamic behaviour of Baseline-V BWB. The model is being tested at actual loitering flight speed of 15 m/s (54 km/h) and Reynold's number, Re = 263,000 at its mean chord, which indicates a laminar flow. The range of Reynold's number from wing tip to the body centre section is between 120,000 to 600,000. While most of the spanwise wing-body sections are in laminar flow, there are sections especially on the body that is within transition between laminar and turbulent flow. In the meantime, the Mach number of the test is around 0.05, thus compressibility effect can be ignored. The experiment is divided into two major segments:

- varying angle of attack for five elevator angle cases at zero tilt angle ($-10° \leq \alpha \leq +10°$; $\eta = -20°, -10°, 0°, +10°, +20°$; $k = 0°$)
- varying sideslip angle for four tilt angle cases at one fixed elevator angle ($-10° \leq \beta \leq +10°$; $k = 0°, +15°, +30°, +45°$; $\eta = 0°$).

The first segment is aimed at looking for the effect of elevator angle cases to lift and drag coefficients, aerodynamic efficiency and pitch moment static stability in a simulated straight (zero sideslip angle), steady and level flight condition while the second is for the effect of tail tilt angle at fixed angle of attack but varying sideslip angle to side force, drag and yaw moment in a simulated crosswind flight condition.

IOP Conf. Series: Materials Science and Engineering **152** (2016) 012021 doi:10.1088/1757-899X/152/1/012021

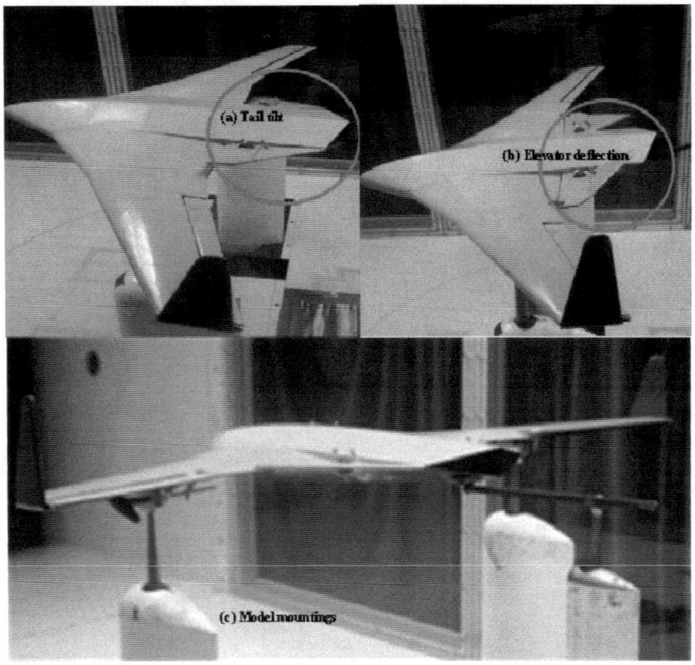

Figure 4. Baseline-V BWB (a) tilting the tail (b) elevator deflection (c) mountings

4. Results

Figure 5 shows four plots that represent the effects of elevator angle to the BWB aircraft at zero tilt angles. These are plotted to examine the effect of elevator angle to lift versus angle of attack trend (C_L vs α), drag polar trend (C_L vs C_D) and aerodynamic efficiency (L/D vs α), and also to determine pitch stability and suitable trim lift (trim angle of attack) at steady and level cruise condition.

C_L versus α plots shows almost all elevator cases have near linear trend except for $\eta = 0°$ for angles of attack within $-10° \leq \alpha \leq +10°$. There is significant increase of C_L as η increases i.e. increment of $\Delta C_L \sim 0.3$ from $\eta = -20°$ to $\eta = +20°$ roughly averaging the change of C_L with respect to the change of η to $dC_L/d\eta \cong 0.0075$ per degree. The average slope of C_L versus α plot for all cases is around 0.08 per degree or 4.58 per radian. This is 27% less than the theoretical ideal lift curve slope of 2π per radian and this can be considered steep for a BWB airplane that has aspect ratio less than 5.0. From experiences in previous studies, the lift curve slope for low-aspect ratio to medium-aspect ratio BWB airplane usually ranges from 0.04 to 0.06 per degree. Meanwhile, angles of attack at zero lift $\alpha_{CL=0}$ ranges from near $-5°$ to $-2°$, which is a normal range for a wing.

As for drag polar plot, it is expected that the lowest drag shall belong to $\eta = 0°$ case with its lowest drag coefficient C_D is lower than 0.02 and happens at $C_L \sim 0.5$, which corresponds to $\alpha = +3°$. For other η cases, the lowest C_D also occurs at near $\alpha = +3°$ but with higher magnitude of C_D than $\eta = 0°$ case. Increase in absolute elevator angle ($|\eta|$) also increases C_D, thus agreeing with the conventional understanding that larger elevator angle creates larger drag at all angles of attack. The minimum C_D for each drag polar cases does not happen at zero C_L but rather at around $C_L \sim 0.5 - 0.8$ indicating that the BWB aircraft in study is of "cambered-type". Optimal C_D (the lowest drag at a given lift, C_{Dopt}) seems to be at a point where $C_L \sim 0.5$ but one must look at L/D versus α plot to be sure.

From L/D versus α plot, the highest L/D happens at angle of attack near $\alpha = +3°$ with the highest recorded L/D ~ 32 for $\eta = 0°$ case. This means that for optimal flight C_L of 0.5 the C_D is 0.015. The drag coefficient at maximum L/D is actually rather normal for a clean, sleek airplane but the high L/D is attributed to the Baseline-V high lifting capability. The other elevator angle cases see large drop of

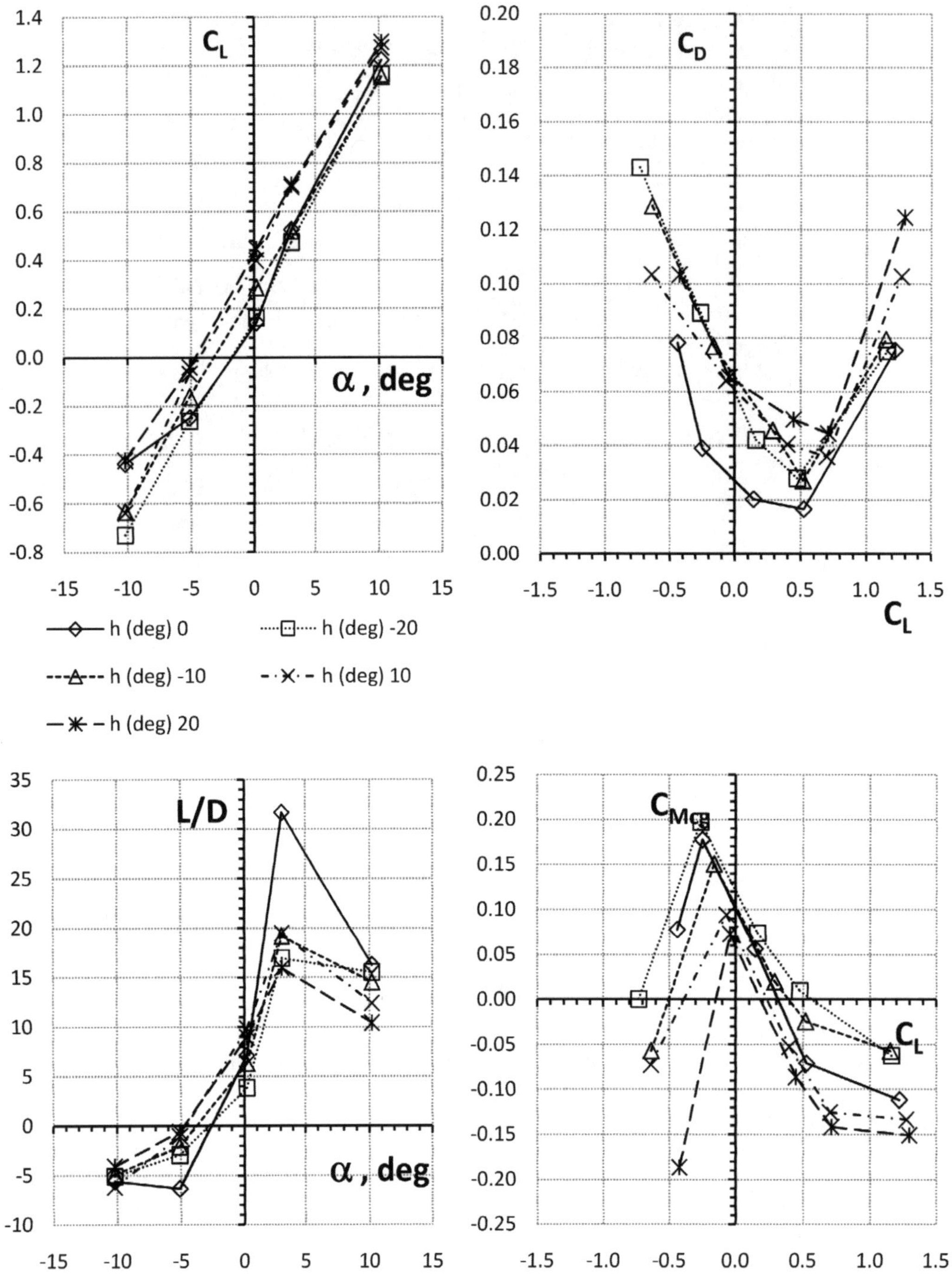

Figure 5. C_L versus α, C_D versus C_L, L/D versus α and C_{Mcg} versus C_L plots

maximum L/D to just 15-20 and even this is higher than many aircraft's L/D. As expected, the most suitable trim angle of attack α_{trim} to fly is at around $\alpha = +3°$ that give the highest L/D – i.e., if the mass of this Baseline-V BWB UAV is around 3.2 kg, then the propulsive thrust needed to overcome the drag at this trim angle is just 100 grams! However, a slight increase or decrease in angle of attack causes large drag increment, subsequently a quick drop of L/D ratio. In real flight of a remote control airplane of this design with such a small wing span, trimming at angle of attack that gives the highest L/D is one major problem. The designed UAV is susceptible to wind gust and change in airspeed – both of which change its angle of attack – hence reducing its lift or increasing its drag fairly quickly.
In the context of flow regime, it is mentioned that the Reynold's number at the mean chord is Re = 263,000 while range of Reynold's number from wing tip to centre body is Re = 120,000 to 600,000. This corresponds to the average laminar skin friction coefficient $C_f = 0.0026$ at mean chord, which is around 17% of the total drag coefficient at $\alpha = +3°$ that is 0.015. Laminar skin friction from centre body section to wing tip is $C_f = 0.0017$ to $C_f = 0.0038$ or 11.3% to 25.3% of just mentioned drag coefficient. Such a low average skin friction coefficient is one of the reasons for high L/D of 32.

However, pitch moment versus lift coefficients (C_{Mcg} versus C_L plots) show that at current centre of gravity (C.G.) location (or centre of turntable for wind tunnel experiment), the trim C_L (or C_L at C_{Mcg} = 0) for all η cases does not fall at its optimal $C_L \sim 0.5 - 0.8$ that give the highest *L/D* but rather at lower range of C_L. For instance, elevator angle of $\eta = 0°$ case has trim lift at $C_L \sim 0.3$, which is at around $\alpha \sim +2°$ with *L/D* ~ 25 only. Generally, all η cases have negative slope ($dC_{Mcg}/dC_L < 0$) with positive C_{Mcg} at $C_L = 0$ ($C_{M0} > 0$) for $-5° \leq \alpha \leq +10°$ indicating that the Baseline-V is longitudinally, statically stable. Rough average dC_{Mcg}/dC_L slope is around 0.25 – 0.30 meaning that static margin, where the aerodynamic centre of the whole aircraft (or neutral point) is approximately 25% - 30% of mean chord behind the C.G. This can further be 'relaxed' by shifting the C.G. back around 10% of mean chord to trim the aircraft at $\alpha = +3°$ where its lift-to-drag ratio is the highest.

The second part of this experiment embarks on observing the effect of tilt angle k to the drag coefficient C_D, side force coefficient C_Y, yaw moment coefficient C_{Ncg} and roll moment coefficient C_{Lcg}. These coefficients are measured with variation of sideslip angle β to observe trends and slopes $dC_D/d\beta$, $dC_Y/d\beta$, $dC_{Ncg}/d\beta$ and $dC_{Lcg}/d\beta$ that determine the hypothesis regarding increased of drag with increased in tilt angle mentioned earlier, how side force changes with sideslip angle, yaw stability and roll stability, respectively.

Figure 6 shows two plots - C_D versus k (left) and C_Y versus β (right). The drag coefficient C_D versus tilt angle k plot highlighted here is for zero elevator angle $\eta = 0°$ cases at fixed angle of attack $\alpha \sim 2°$ and zero sideslip angle $\beta = 0°$. This is to prove that the drag increases with tail tilt angle k, which increases exposed tail frontal area. At $k = 0°$ or when the tail is completely horizontal and hides behind the inboard wing, the $C_D \sim 0.011$ only. At $k = \pm 15°$ when some of the tail is exposed from frontal view the drag coefficient increases by 0.004 or 36.3%. Increasing absolute tail tilt angle by another 15 degrees to $k = \pm 30°$ sees a sudden jump in drag coefficient to 0.027 or 245% of the drag coefficient when the tail is completely horizontal, indicating at this tilt angle there is a larger frontal tail area exposed creating larger obstacle to top and bottom layers of the free airflow farther from the inboard wing surfaces. This is a sign of interference drag existence, therefore, is a proof of earlier hypothesis claiming that close-coupled horizontal tail can reduce interference drag. Increasing tail tilt angle to $k = \pm 45°$ increases drag coefficient by a slight 0.001. This can be considered insignificant because at $k = \pm 30°$, much of the frontal area of the tail has been exposed already and increased in further $\pm 15°$ in k does not changed the exposed area by much.

The right plot of Figure 6 shows the side force coefficient C_Y versus sideslip angle β for various cases of tail tilt angle k. Positive side force means that the force is directed to the starboard side or right side with positive sideslip angle is when the wind comes from the starboard or right side. General trend of these C_Y versus β curves is nearly linear with negative slopes averaging around -0.006 per degree of β. Increasing sideslip angle decreases sideslip force value, but this does not mean there is no side force at high sideslip angle. Since both sideslip angle and side force are actually vector quantities this would mean as absolute sideslip angle in one direction is increased then the sideslip force is also increased in the opposite direction. In Baseline-V case the side force coefficient is quite large at 0.08 even at only $= \pm 10°$. This is larger than drag coefficient of 0.07 at $\alpha = \pm 10°$ for $\eta = 0°$ case. This means that a sizable rudder is needed to counter side drifting of this aircraft during crosswind.

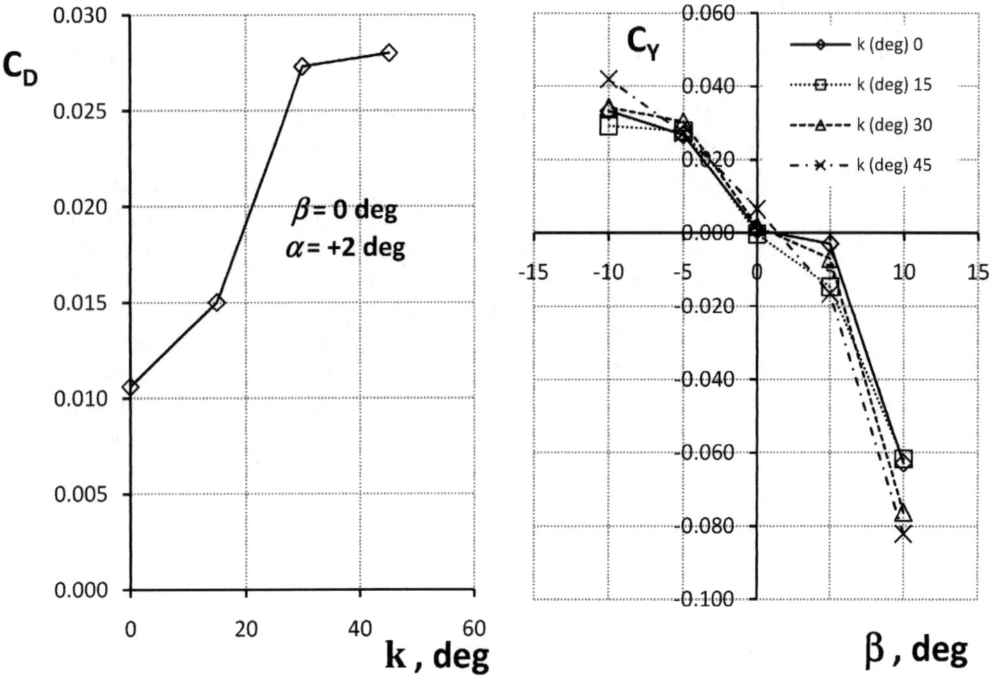

Figure 6. C_D versus k (left) and C_Y versus β (right)

Increasing tail tilt angle k is surely causing exposed side area of the aircraft to increase thus increasing the absolute value of side force. As the tail tilt angle is increased, the increase in side force (absolute value of C_Y) is 0.00033 per degree of k at $\beta = \pm 5°$ and 0.00044 per degree of k at $\beta = \pm 10°$. However, the change in side force coefficient due to changing tail tilt angle is not much larger than the effect of the aircraft's planform-vertical wing tip combo in generating side force with respect to sideslip angle at -0.006 per degree of β mentioned earlier which is more than ten times the effect of tail tilt angle to the side force. This is because the side form wing-body area and vertical wing tip area is larger than side view area of tilting tail thus they play major role in generating side force and are influential in determining how side force is enlarged with increasing sideslip angle.

Figure 7 shows two important plots related to the effect of k to C_{Ncg} and C_{Lcg} at fixed $\eta = 0°$ and fixed $\alpha = +2°$ that represent near optimal cruise condition with $L/D \sim 25$ as mentioned before. C_{Ncg} versus β plots shows that, for all k cases, the slope $\mathrm{d}C_{Ncg}/\mathrm{d}\beta$ is positive. Therefore, the aircraft is directionally, statically stable. There seems to be small but significant differences in C_{Ncg} magnitude between k cases – C_{Ncg} is larger for larger k and this seems to agree with theoretical hypothesis shown in Figure 3. Logically, as k is increased, larger tail side force is generated while lower tail lift is to be expected. The trend of C_{Ncg} versus β plots also agrees with conventional understanding that as β is increased, C_{Ncg} also increases. However, the magnitude of C_{Ncg} is small relative to C_{Mcg}.

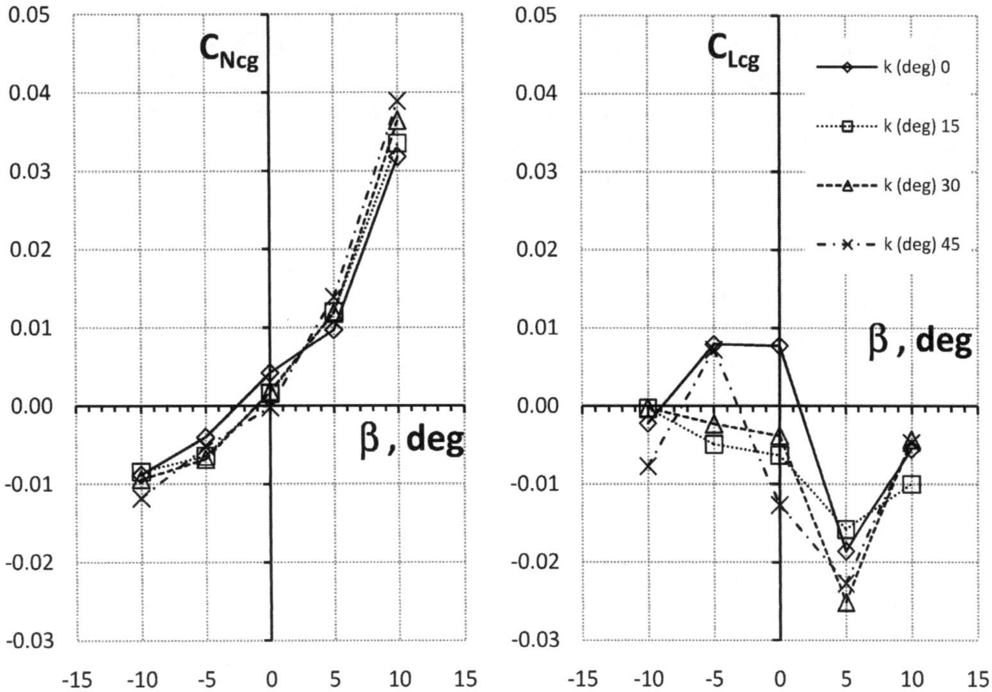

Figure 7. C_D versus k (left) and C_Y versus β (right)

Plot of rolling moment versus sideslip angle (C_{Lcg} versus β) on the right side of Figure 7 shows unclear trend. It is expected to be near linear with $\mathrm{d}C_{Lcg}/\mathrm{d}\beta < 0$ but it is not so in Baseline-V BWB case, although negative C_{Lcg} is observed at positive value of β while positive C_{Lcg} is found in most cases for negative β. The negative $\mathrm{d}C_{Lcg}/\mathrm{d}\beta$ within low β magnitude region shows some capability to return to level wing condition. However, knowing that Baseline-V has mid-wing and mid-tail setup, it is fair to say that the rolling moment slope shall be flat or $\mathrm{d}C_{Lcg}/\mathrm{d}\beta = 0$ (neutrally stable) and tilting the tail in this case may not change its rolling stability.

5. Concluding remarks

The change of lift coefficient with respect to the change of elevator angle η is approximately +0.0075 per degree. The average lift curve slope for all cases is around 0.08 per degree or 4.58 per radian of angle of attack. Angles of attack at zero lift $\alpha_{CL=0}$ ranges from near $-5°$ to $-2°$. The lowest drag coefficient is lower than 0.02 (at round 0.015) and happens at lift coefficient of approximately 0.5, which corresponds to $\alpha = +3°$. In other η cases, the lowest drag coefficient also occurs near $\alpha = +3°$ but with higher magnitude than $\eta = 0°$ case. With this BWB design, the highest L/D is recorded at 32, which is very high for such an aircraft with moderate aspect ratio. One of the reasons for high lift-to-drag ratio is its low friction coefficient, which is only around 17% of total drag coefficient due to the aircraft flying within laminar flow regime. It is proven that having close-coupled horizontal tail right behind the inboard wing reduces interference drag. As the tail is tilted the exposed frontal area of tail creates a new obstacle that increases drag. Trend of side force coefficient versus sideslip angle curves is nearly linear with negative slopes averaging around -0.006 per degree of β. As the tail tilt angle is increased, the increase in side force is 0.00033 per degree of k at $\beta = \pm 5°$ and 0.00044 per degree of k at $\beta = \pm 10°$. There seems to be small but significant differences in the yawing moment coefficient magnitude between k cases, where yawing moment is larger for larger k and this seems to agree with theory shown. Finally, there is no clear trend regarding rolling moment versus sideslip angle.

From general observation here, it is appropriate to mention that Baseline-V, a BWB aircraft with close-coupled tilting tail, is statically stable in pitch and yaw but has no clear indication in terms of rolling stability. In addition, the elevator at $\eta = 0°$ has the largest L/D. It is recommended to shift the C.G. slightly closer to aircraft neutral point such that its dC_{Mcg}/dC_L slope shall be less steep and trim at exactly $\alpha = +3°$ to fly at L/D ~ 32 where Baseline-V BWB can cruise at the lowest required thrust and saves energy, thus increases range and endurance.

Acknowledgment
The research is conducted by a team within Aviation Technology Research (ATR) Group under Flight Technology & Test Centre (FTTC), UiTM. The research presented in this paper is a part of research project under FRGS Grant (600-RMI/FRGS 5/3 (103/2014)). The authors would like to express their gratitude to Ministry of Education, UiTM's Research Management Institute and Faculty of Mechanical Engineering, Universiti Teknologi MARA for financial and administrative support.

References
[1] Liebeck R H 2004 *J. Aircraft* **41** 10-25
[2] Qin N, Vavalle A, Le Moigne, Laban M, Hackett K and Weinerfelt P 2004 *Prog. Aerosp. Sci.* **40** 321–43
[3] Bolsunovsky N, Buzoverya B, Gurevich V, Denisov A, Dunaevsky L, Shkadov O, Sonin U and Zhurihin J 2001 *Aircr. Design* **4** 193–219
[4] Nasir R E M, Kuntjoro W and Wisnoe W 2014 *Procedia Tech.* **15** 784–792
[5] Karakas H, Koyuncu E and Inalhan G 2013 *Journal of Intelligent and Robotic Systems* **69** 131-46
[6] Cai C F, Wu J H and Liang B 2014 *Adv. Mater. Res.* **1016** 359–64
[7] Li P, Zhang B, Chen Y, Yuan C and Lin Y 2012 *Chinese J. Aeronaut.* **25** 508–16
[8] Mat S, Ishak I S, Zakaria K and Khan Z A 2014 *Adv. Mater. Res.* **845** 971–4
[9] G. Sachs G 2005 *J. Theor. Bio.* **234** 471–8

IOP Conf. Series: Materials Science and Engineering **152** (2016) 012022 doi:10.1088/1757-899X/152/1/012022

Flight dynamics and control modelling of damaged asymmetric aircraft

T T Ogunwa and E J Abdullah*

Department of Aerospace Engineering, Faculty of Engineering, Universiti Putra Malaysia, Malaysia

*ermira@upm.edu.my

Abstract. This research investigates the use of a Linear Quadratic Regulator (LQR) controller to assist commercial Boeing 747-200 aircraft regains its stability in the event of damage. Damages cause an aircraft to become asymmetric and in the case of damage to a fraction (33%) of its left wing or complete loss of its vertical stabilizer, the loss of stability may lead to a fatal crash. In this study, aircraft models for the two damage scenarios previously mentioned are constructed using stability derivatives. LQR controller is used as a direct adaptive control design technique for the observable and controllable system. Dynamic stability analysis is conducted in the time domain for all systems in this study.

1. Introduction

Since the beginning of flight, aircraft control has been one of the top priorities of the aviation industry to ensure safe flight. Just like the human body, every part of an aircraft plays a role in ensuring the safe flight and its control. Air transportation is undeniably one of the safest means of transportation in recent times. However, on some occasions, accidents or incidents that involve a number of casualties do occur. Mechanical failure or damage to part(s) of the aircraft is the second most common cause of plane crashes after pilot error, accounting for about 22% of all aviation accidents [1]. Other causes of accidents also include sabotage, loss of control (LOC), weather and other human factors. In the early generation, the flight control systems were mechanical, meaning there was a direct connection between the pilot's control from the cockpit and control surfaces. Over the years, the mechanical flight control system has been replaced by the type that allows the pilot to directly control the motion of the aircraft. This digital type of flight control system uses electrical signals and is referred to by the term 'fly-by-wire'. This type of flight control system improves the stability and control of the aircraft, and also the pilot's reaction time to a flight disturbance [2].

Furthermore, in a situation where the aircraft experiences system failure of any sort, it becomes asymmetric and the workload of the pilot increases greatly. Floating trim tabs, engine fan bursts, bird strikes and frozen controls are some examples of failures that can limit the control of an aircraft. Be that as it may, in most cases, when these types of failures occur, only the control surfaces are affected while the lifting surfaces are left intact. The Sioux DC-10 crash is one of the very famous examples of a situation like this. United Airline Flight 232 was flying from Denver to Chicago when the second engine failed and rendered all hydraulic controls useless. The aircraft was then controlled by the two remaining engines and crash landed in Sioux City, Iowa. There were 111 casualties but 185 people survived [3]. This clearly demonstrates the ability of an aircraft to be controlled without the standard control surfaces. In 2003, DHL's Airbus A300B4 suffered surface-to-air missile strike to its left wing.

As a result of this, all hydraulics were disabled and thus eliminating the pilots' influence over the control surfaces completely. The pilots crash landed the aircraft after adjusting the throttle commands and moving the fuel to prevent complete drainage of the left wing's fuel storage [4]. Other significant reported accidents resulting from failures are as follows:

- Braniff Airways – Lockheed L-188 – 1968: Structural limits were exceeded in an attempt to recover from an unusual attitude caused by turbulence. This resulted in the right wing and empennage being separated during flight. The crew and passengers all died in the crash [4].
- Flying Boat Inc. Flight 101- Grumman Turbo Mallard – 2005: The aircraft experienced a right wing separation during flight due to fatigue and fracture damage in the right wing. All the passengers on board were killed after the aircraft crashed into a shipping channel [4].
- Piper PA-32 – 2006: Few minutes after a Boeing 737 had taken off, this aircraft was piloted into the same flight path. The aircraft was overstressed by the wake turbulence and it caused a section of the left wing, portions of the right wing's flap, left and right side of the stabilator and aileron to be separated. Only the pilot was killed in the crash [4].

While more examples abound, the scenarios listed above show instances of engine, wing and also control surface damage. Sadly, despite improved maintenance and inspection efforts, accidents still occur. Therefore, it is of utmost importance to develop adaptive methods to recover the aircraft from damage(s).

As it is not practical to consider all possible damage scenarios, the study focuses on two damage situations: 33% loss of the left wing from the tip and complete loss of the vertical stabilizer. In this study, a Linear Quadratic Regulator (LQR) is implemented to aid a damaged Boeing 747-200 aircraft that has a controllable and observable system even after damage in landing safely. LQR controller is an optimal controller and is chosen for this study because it is straightforward and easy to implement or use for multivariable systems, especially with the aid of computer programs. Also, for a full state feedback system, the optimal input signal can be obtained and the control output minimized.

2. Background review

This section gives reviews on past and current studies that have been carried out in the area of control of damaged asymmetric aircraft. In the first sub section, scaled models that have been used to model aircraft damage(s) are discussed. Sub sections two and three highlight studies that have been carried out to see how damage affects the aerodynamics and flight dynamics of an aircraft, respectively. Some adaptive control techniques used in previous studies are discussed in sub section four.

2.1. Scaled models

Loss of control has been a major motivation in the aerospace industry as a whole. A lot of researches have been carried out in a bid to provide solution or resolve cases involving loss of control of aircraft due to damage(s). Subscale model of a commercial aircraft known as AirSTAR was developed by the National Aeronautics and Space Administration(NASA) Langley to investigate events relating to loss of control. Cunningham *et al.* [4] discussed the details of the first subscale aircraft known as S-2 aircraft and its ground facilities. Simply described, the S-2 is an off-the-shelf aircraft model but it is fitted with further instrumentation. The aircraft was used in the development of ground facilities and procedures, allowing for more expensive scaled aircraft to be developed with less risk. The facilities in AirSTAR are able to estimate the aircraft's aerodynamic parameters in real time. This is an important feature because wind tunnel data for the S-2 can be obtained and can be used for the validation of wind tunnel data in the future. Apart from the S-2 aircraft, Jordan *et al.* [5, 6] developed two dynamically scaled aircraft known as GTM-T1 and GTM-T2. These aircraft were scaled to be about 5.5% of size of an actual commercial aircraft but the GTM-T2 was designed to have a lighter airframe so that more instrumentation and electronics could be added. As for the aircraft moments of inertia, they were scaled proportionally with geometric scaling. Also, the weight was slightly increased to make up for the increase in lift as the aircraft fly at lower altitudes, which have higher air density. One can estimate the weight of aircraft parts by comparing the available data for similar aircraft. This method was used by Beltramo *et al.* [7] to estimate the weights of the components of a commercial aircraft. It was noted that knowing takeoff weight of an aircraft is enough to estimate the weight of its wing.

2.2. Aerodynamics
The effect of the loss of wing tip on the aerodynamic stability derivatives of the aircraft, as well as the static aeroelastic effects, was explored by Woo [8]. In doing that, detailed finite element wing structure model was joined to a doublet lattice model for a fighter jet. The introduction of a rolling moment coefficient due to pitch was noted to be the most significant effect of damage among the investigated damage cases. There was also significant coupling between the pitch and yaw. For the study of the aerodynamic effects of a hole in the wing, Render *et al.* [9] used wing tunnel testing method. It was observed that this type of damage resulted in two patterns: a weak-jet and a strong-jet. In the case of the weak-jet, the flow remained attached and the hole affected the overall aerodynamic loads only a little. For the strong-jet however, the flow did not remain attached to the hole and therefore had more significant effects on the aerodynamic loads. The damage caused the drag to increase greatly for a hole close to the trailing edge and lift to be decreased as the damage progressed forward.

Extensive wind tunnel testing has also been carried out on the Generic Transport Model (GTM) for different loss of control scenarios. Foster *et al.* [10] carried out initial tests for the GTM and focused on aerodynamic properties at very high angles of attack of range -5° to 85° and sideslip angles ranging between -45° to 45°. The basis of these studies uses existing techniques for fighter aircraft and the tests carried out included that of rotary balance, static and forced oscillation. The hysteresis effects on the aerodynamic forces at large angles were observed in the studies. An attempt was made to understand the cause of these effects using flow visualisation but the results came up inconclusive. To take care of the time dependent effects, the aerodynamic forces were averaged. The tests however showed that loss of static stability can be as a result of large angles of attack. In addition, loss of rudder effectiveness at large angles of attack was also observed.

For the exploration of effects of damage on GTM aircraft, Shah [11] conducted more wind tunnel tests recently. The cases investigated include 3 horizontal tail loss cases, 6 wing tip loss cases, 3 cases of horizontal tail loss and other cases like holes in the lifting surfaces. The results of these tests showed that reduction in the lift curve slope due to reduced aspect ratio is eminent if the aircraft experiences the loss of a wing tip. The rolling moment observed by Woo [8] was also observed in these tests. Furthermore, it was noted that trimming the aircraft at negative sideslip can reduce the rolling moment experienced at trim. Rawlings [8] also carried out wind tunnel tests on the GTM to explore how the stabilator can be utilized as a speed break-brake or to recover from the loss of control. The detailed configuration of the GTM was addressed, especially the travel limits for the control surfaces. The tests result showed that there was no benefit in the use of the stabilator for recovery after loss of control but the stabilator was useful for speed brake purposes. Nonetheless, the usefulness of the stabilator was limited because of significant loading.

Thomas [12] used the lifting surface theory to explore the aerodynamics properties of a bird with an asymmetric tail or wing. He observed that the effects of asymmetry include increased turn radius and reduced lift. These effects however were not as pronounced when the asymmetry was in the tail of the bird. Disturbances to the aircraft can bring about significant aerodynamic nonlinearities. The effects of these disturbances were captured by Keller *et al.* [13] using the non-linear line theory. The wake effect was modelled using a code that was previously used for rotorcraft. This model was compared to the flight test data of a small radio-controlled aircraft and the results seemed to correlate.

2.3. Flight dynamics
In a bid to understand what causes the loss of control of an aircraft, Wilborn and Foster [14] defined a set of five quantitative metrics that includes the structural integrity envelope (airspeed and load factor) and the adverse aerodynamic envelope (angle of attack and sideslip). The study was carried out in such a way that if one of these envelopes established was exceeded, it resulted in an upset condition which was sometimes recoverable. As more envelopes were exceeded, the difficulty of recovery would also increased. Also, the aircraft experienced loss of control once three of these limits had been exceeded.

Damage to part of an aircraft causes the centre of gravity to shift and the simulation using cg-centric equations of motion becomes more complicated. Bacon and Gregory [15] succeeded in overcoming this complication by deriving the equations of motion about a fixed point. The equations of motion were then calculated using Newton's method so that the centre of gravity offset is a system parameter. This method allows for discontinuous changes of the centre of gravity's location without causing discontinuities in the system states. Although a new set of equations were employed in this study, the yielded results are equivalent to that of the traditional cg-centric equation method and even has an advantage of being simple to implement. Nguyen *et al.* [16, 17] initially used similar equations but the original equations of motion did not correctly model the change in the centre of gravity and thus were adjusted in a later erratum [18]. In order to account for damage effects, the steady level flight condition was considered as having a non-zero sideslip angle or roll angle. A finite differencing approach was then used to calculate the trim with respect to the trim of the undamaged aircraft. In the development of this model, the linear stability derivatives were obtained from a vortex lattice model of the GTM and were used to get the aerodynamic effects. An adaptive control plan based on dynamic inversion was then implemented in the resulting flight dynamics model. In the end, a recursive least squares or neural network was used to update the baseline dynamic inversion law and to achieve indirect control. To improve the performance of the flight dynamic system, a direct controller was used.

With the aim of a Model Reference Adaptive Controller (MRAC), Liu *et al.* [19] explored the flight dynamics of a damaged aircraft. The focus of this study was geared towards developing a linear model of the aircraft using the same linearized form of equations of motion as Bacon and Gregory [15]. The result showed that, for three input-output cases, the zero dynamics or relative degree were not affected and the system was minimum phased. The sign of the transfer function gain was also not affected. An adaptive controller was then implemented in the model that was able to account for the experienced changes. The effect of the wing tip loss was also investigated by Sarigul-Klijn *et al.* [4], although the stability derivatives of the undamaged aircraft was adjusted by a set of analytical corrections in order to model the loss of wing tip accurately. The reduction in aspect ratio and wing area was focused on in these methods but the coupling introduced had to be for the rolling moment only due to an angle of attack. The equations of motion were derived about a fixed point as with previous studies.

2.4. Adaptive control systems

Without doubt, asymmetric aircraft resulting from structural damage can cause the flight performance and handling quality of the aircraft to deteriorate drastically. Over the years, lots of studies have been carried out to control aircraft during failures. The control of any aircraft begins with identification of the parameters and then adaptive control. Adaptation schemes are therefore implemented directly or indirectly to improve the performance of any control and feedback systems. This study uses the direct adaptive technique where estimation of actual system parameters is avoided and instead will continue to adjust the control laws or even the system controls until the feedback errors are eliminated [20, 21]. What follows is the reviews of prior studies carried out in relation to direct adaptive controls.

Kim and Calise [22] in their work used an adaptive Artificial Neural Network (ANN) as a direct adaptive control for the inverse model of an aircraft. The ANN was modelled offline and coefficients were used to obtain the aircraft dynamics by controlling deflections using sigma-pi neurons and radial basic function. To adapt for errors, an online ANN was also used. The adaptation was actually based on input to the modelled inverse and also the aircraft output. Further research was carried out on the XV-15 Tilt Rotor [23] and involved developing an online training to show the boundedness of the weights of ANNs. The study showed that over a certain range of errors, the adaptive system began to converge to a stable solution. More analysis on bounded weights for various adaptive methods can be found in NASA's Integrated Resilient Aircraft Control (IRAC) in studies carried out by Nguyen, Huang and Bakhtiari-Nejad [21]. Furthermore, from these studies, it was found that for Lyapunov based adaptive law, when high gain occurs, the frequency of oscillations was high and other adaptive methods may be required to get rid of high gain learning.

Model-Reference Adaptive Control System (MRAC) is another direct adaptive control method that can be implemented. Shin *et al.* [24] used dynamic inversion to develop an adaptive output feedback for a modified NASA F-15. In the work they carried out, it was acknowledged that control saturation was possible for both un-adapted and adapted systems. Also, for control and damage rejection, control limitations are based around near saturation deflections. NASA and Boeing conducted simulations and flight tests on various aircraft types, from military to transport. The tests were carried out to show the performance of the flying aircraft using propulsion-only control. Propulsion Controlled Aircraft (PCA) systems tests were carried out on the F-15 and the result showed that the aircraft was able to recover from control failures and upset conditions to eventually land [25]. The PCA flight test was also carried out on the MD-11 and the similar results were achieved. In this case, the aircraft could be controlled manually, as well as make automatic landings [26]. For a Boeing-747, the PCA flight tests showed that aircraft could land successfully after recovering from gust and unusual attitudes [27].

Nguyen *et al.* [17] tested Recursive Least Squares and Lyapunov adaptive laws [21] on a damaged aircraft. The Generic Transport Model (GTM) is modelled in form of a vortex lattice code where the new flight dynamics and aerodynamic coefficient changes are developed and discussed. The new equations of motion for damaged aircraft were also derived at Georgia Tech (GT) [28] and, although the changes in the aerodynamic coefficients were discussed in the GT and NASA papers, the shift in the Center of Gravity(CG) position as a result of damage was excluded. Furthermore, some studies focused on the Flight Control System (FCS) for damaged aircraft. Tang *et al.* [29] used a Proportional Integral Derivative (PID) and an optimal control LQR system to account for the new capabilities and estimated flight envelope of the damaged aircraft, hence directing it towards the safest site to land. The further examination of the aerodynamic coefficients and equations of motion of a damaged aircraft are major contributions of this study.

3. Aircraft models

3.1. Undamaged aircraft model

The Boeing 747-200 aircraft model is used for this study. It is a long range, high capacity, wide body airliner and is the first wide-body aircraft to be built as a freighter, a combination passenger-freighter and a convertible. This aircraft model was chosen for this study primarily due to the fact that the technical specifications, aerodynamics and stability derivatives of the aircraft are widely available. The data for the nominal/undamaged Boeing 747-200 and the flight condition were obtained from [3, 4, 31, 32, 35] and are summarized in Table 1.

For symmetric/undamaged aircraft, linear equations of motion for longitudinal and lateral motions are given by Equations 1 and 2, respectively [30].

$$\begin{bmatrix}\Delta\dot{u}\\ \Delta\dot{w}\\ \Delta\dot{q}\\ \Delta\dot{\theta}\end{bmatrix}=\begin{bmatrix}X_u & X_w & 0 & -g\\ Z_u & Z_w & u_0 & 0\\ M_u+M_{\dot{w}}Z_u & M_w+M_{\dot{w}}Z_w & M_q+M_{\dot{w}}u_0 & 0\\ 0 & 0 & 1 & 0\end{bmatrix}\begin{bmatrix}\Delta u\\ \Delta w\\ \Delta q\\ \Delta\theta\end{bmatrix}+\begin{bmatrix}X_{\delta e} & X_{\delta T}\\ Z_{\delta e} & Z_{\delta T}\\ M_{\delta e}+M_{\dot{w}}Z_{\delta e} & M_{\delta T}+M_{\dot{w}}Z_{\delta T}\\ 0 & 0\end{bmatrix}\begin{bmatrix}\Delta\delta_e\\ \Delta\delta_T\end{bmatrix} \quad (1)$$

$$\begin{bmatrix}\Delta\dot{\beta}\\ \Delta\dot{p}\\ \Delta\dot{r}\\ \Delta\dot{\phi}\end{bmatrix}=\begin{bmatrix}\frac{Y_\beta}{u_0} & \frac{Y_p}{u_0} & -\left(1-\frac{Y_r}{u_0}\right) & \frac{g\cos\theta_0}{u_0}\\ L_\beta & L_p & L_r & 0\\ N_\beta & N_p & N_r & 0\\ 0 & 1 & 0 & 0\end{bmatrix}\begin{bmatrix}\Delta\beta\\ \Delta p\\ \Delta r\\ \Delta\phi\end{bmatrix}+\begin{bmatrix}0 & \frac{Y_{\delta r}}{u_0}\\ L_{\delta a} & L_{\delta r}\\ N_{\delta a} & N_{\delta r}\\ 0 & 0\end{bmatrix}\begin{bmatrix}\Delta\delta_a\\ \Delta\delta_r\end{bmatrix} \quad (2)$$

The states for longitudinal motion are forward velocity (u),vertical velocity (w), pitch rate (q) and pitch angle (θ), and the control inputs are elevator (δ_e) and differential thrust (δ_T). For lateral motion, the states are sideslip angle (β), roll rate (p), yaw rate (r) and roll angle (ϕ), and the control inputs are aileron (δ_a) and rudder (δ_r).

Table 1. The nominal/undamaged aircraft data [4, 31, 32, 35, 40]

Flight Condition Properties	h (ft) = 40,000 M_∞= 0.9 α (degrees)= 2.4 W (lbf)= 636,636	I_x (slugs.ft^2)= 18.2 × 10^6 I_y (slugs.ft^2)= 33.1 × 10^6 I_z (slugs.ft^2)= 49.7 × 10^6 V_∞(ft/s) = 871	
Longitudinal		Lateral-directional	
Derivatives	Value	Derivative	Value
C_{Lo}	0.29	$C_{y\beta}$	-0.9
C_{Do}	0.0305	$C_{l\beta}$	-0.16
$C_{L\alpha}$	5.5	$C_{n\beta}$	0.16
$C_{D\alpha}$	0.5	C_{lp}	-0.34
$C_{m\alpha}$	-1.6	C_{np}	0.020
$C_{L\dot{\alpha}}$	8	C_{lr}	0.13
$C_{m\dot{\alpha}}$	-9.0	C_{nr}	-0.033
C_{Lq}	7.8	$C_{L\delta a}$	0.014
C_{mq}	-25.5	$C_{n\delta a}$	0.0018
C_{Lu}	-0.23	$C_{y\delta r}$	0.118
C_{Du}	0.22	$C_{l\delta r}$	0.008
C_{mu}	-0.09	$C_{n\delta r}$	-0.095
$C_{L\delta e}$	0.3	C_{yp}	0
$C_{m\delta e}$	-1.20	C_{yr}	0
$C_{x\delta e}$	0	-	-

3.2. Damaged aircraft models

3.2.1. Loss of 33% of left wing from tip. In a situation where an aircraft experiences a loss to some fraction of the wing, the mass and inertia properties will be greatly affected as a result of the change in C.G. position. "The new mass of the aircraft is determined based on empirical mass build-ups for similar aircraft and the wing CG is assumed to be at 40% of the mean aerodynamic chord and is set at half span. A mass build-up of the undamaged aircraft is adjusted to produce the same CG and inertia as the production aircraft. As tip portions of the left wing are removed or damaged the wing mass lowers and its CG shifts inboard. The aircraft CG then shifts positive along body centred Y axis and positive along the body centred X axis. Any shift along the Z axis is considered negligible" [3]. The stability derivatives will also be affected and, because the whole aerodynamic structure of the aircraft is affected, the new stability derivative values have to be calculated. Since the aircraft model in [3] was assumed to be a scaled model of the Boeing 747-200, the relationship between the mass property variations and the percentage wing loss of the aircraft had to be obtained in percentage. Doing so makes it possible to adapt the changes in mass and inertia properties for the aircraft model of this study.

The aerodynamic stability derivatives were however derived from [4], where the extended vortex lattice code Athena Vortex Lattice (AVL) was utilized to model and obtain the values of stability derivatives for undamaged and 33% left wing tip loss of a Generic Transport Model Aircraft (GTM). As [4] only had sufficient information for a 33% loss of the left wing from the tip, the same damage was adapted for this study. The generic transport model aircraft was again assumed to be a scaled model for the Boeing 747-200 aircraft as they both have similar aerodynamic structure. The difference in the aerodynamic stability derivatives for the undamaged and damaged model for the GTM was then obtained in terms of percentage (%). This percentage change was applied to the aircraft model for this study using the relationship: new stability derivative value for 33% wing damage = percentage change × initial stability derivative value

The values obtained and the percentage difference are summarised in Table 2. In case of an aircraft with a portion of the wing damaged, the damage causes coupling between the longitudinal and lateral motion. It is important to note that, for this study, the effect of thrust was neglected for 33% loss of the left wing aircraft model. The linear equations of a wing damaged aircraft are [3]:

$$\begin{bmatrix}\Delta\dot{u}\\ \Delta\dot{\alpha}\\ \Delta\dot{q}\\ \Delta\dot{\theta}\\ \Delta\dot{\beta}\\ \Delta\dot{p}\\ \Delta\dot{r}\\ \Delta\dot{\phi}\end{bmatrix}=\begin{bmatrix}X_u & X_\alpha & X_q & -mgC_\theta & 0 & 0 & 0 & 0\\ M_u & M_\alpha & (M_q-m\Delta xU) & mg(S_\theta\Delta x-C_\theta\Delta z) & 0 & 0 & 0 & 0\\ Z_u & Z_\alpha & Z_q+mU & -mgS_\theta & 0 & 0 & 0 & 0\\ 0 & 0 & 1 & 0 & 0 & 0 & 0 & 0\\ 0 & 0 & 0 & 0 & Y_\beta & Y_p & Y_r-mu & mgC_\theta\\ 0 & 0 & m\Delta yU & -mgS_\theta\Delta y & L_\beta & L_p & L_r+m\Delta zU & -mgC_\theta\Delta z\\ 0 & 0 & 0 & mgC_\theta\Delta y & N_\beta & N_p & N_r-m\Delta xU & mgC_\theta\Delta x\\ 0 & 0 & 0 & 0 & 0 & 1 & \tan\theta & 0\end{bmatrix}\begin{bmatrix}\Delta u\\ \Delta\alpha\\ \Delta q\\ \Delta\theta\\ \Delta\beta\\ \Delta p\\ \Delta r\\ \Delta\phi\end{bmatrix}$$

$$+\begin{bmatrix}X_{\delta e} & 0 & 0\\ M_{\delta e} & 0 & 0\\ Z_{\delta e} & 0 & 0\\ 0 & 0 & 0\\ 0 & 0 & \frac{Y_{\delta r}}{u_0}\\ 0 & L_{\delta a} & L_{\delta r}\\ 0 & N_{\delta a} & N_{\delta r}\\ 0 & 0 & 0\end{bmatrix}\begin{bmatrix}\Delta\delta_e\\ \Delta\delta_a\\ \Delta\delta_r\end{bmatrix} \quad (3)$$

Table 2. Flight conditions and stability derivatives for 33% left wing loss [3, 4]

Flight Condition Properties	h (ft)= 40,000 M_∞= 0.9 α (degrees)= 2.4 W (lbf) = 623903.08		Δx = 0.00 Δy= 33090070 Δz = 0 V_∞(ft/s) =871		
Mass properties					
Mass property	Δ%	New Value	Mass property	Δ%	New Value
Mass (slugs)	2	19391.53	I_x (slugs.ft^2)	17.5	15097500
b (ft)	17.7	161.06	$\bar{c}$ (ft)	-4.5	28.54
I_y(slugs.ft^2)	0.03	33090070	I_z(slugs.ft^2)	6.24	46598720
Stability Derivatives					
Derivatives	Δ%	New Value	Derivative	Δ%	New Value
C_{L0}	43.1	0.29588	$C_{y\beta}$	0.34	-0.89
C_{D0}	1.43	0.0443	$C_{l\beta}$	11.6	-0.141
$C_{m\alpha}$	25.3	-1.1952	$C_{n\beta}$	-0.32	0.1605
$C_{D\alpha}$	0	0.5	C_{lp}	32.6	-0.229
C_{mq}	1.2	-25.194	C_{np}	-14.2	-0.02284
$C_{L\alpha}$	13.7	4.75	C_{lr}	6.7	0.12
$C_{X\delta e}$	0	0	C_{nr}	0.05	-0.329
$C_{Z\delta e}$	0.67	0.29	$C_{L\delta a}$	49.6	0.007056
$C_{m\delta e}$	0.08	-1.19	$C_{n\delta a}$	0	0.0018
C_{yp}	8	0	$C_{y\delta r}$	0.02	0.117
$C_{n\delta r}$	0.05	-0.094	$C_{l\delta r}$	-0.3	0.008024

3.2.2. Complete loss of vertical stabilizer. To model the complete loss of the vertical stabilizer in this study, several simplifications were made such that the lateral-directional stability derivative had to be re-examined and recalculated. It is important to note that for most conventional civil transport aircraft, the vertical tail lies on the plane of symmetry. Knowing this allows for some assumptions so that the modelling process can be simplified. The loss of vertical stabilizer of an aircraft directly results in the variation in values of some stability derivatives. The lateral-directional stability derivatives affected are [36]:

$$\{C_{y\beta}, C_{n\beta}, C_{l\beta}, C_{yp}, C_{np}, C_{lp}, C_{yr}, C_{nr}, C_{lr}\}$$

Denoting a variation in the values of the stability derivatives as Δ, one can write the unknown sets of derivative deviations as:

$$\{\Delta C_{y\beta}, \Delta C_{n\beta}, \Delta C_{l\beta}, \Delta C_{yp}, \Delta C_{np}, \Delta C_{lp}, \Delta C_{yr}, \Delta C_{nr}, \Delta C_{lr}\}$$

The relationship between the nominal values, variation and damaged values is given by:

$$\textit{New derivative value} = \textit{nominal value} - \textit{deviation value} \quad (4)$$

For the step-by-step derivation of this relationship as defined in Equation 4, please refer to [38]. The variation in some stability derivatives were obtained from [38], however, further factors had to be considered to obtain some of the other values. As a result of the complete loss of the vertical stabilizer, the vertical tail efficiency factor, volume and area will all be zero. Therefore, all stability derivatives that are dependent on these factors will be zero as well ($C_{y\beta} = C_{nr} = C_{yr} = 0$) [37]. In addition, vertical stabilizer is assumed to be primarily responsible for the weathercock stability of the aircraft, therefore bringing $C_{n\beta}$ to be equal zero as well ($C_{n\beta} = 0$) [37]. In addition, the mass and inertial properties of the aircraft will experience some changes but they are small compared to the change experienced in the wing loss scenario in the previous section [38]. The mass and inertia properties in Table 3 reflect the complete loss of the vertical stabilizer scenario. All values used for the stability derivatives deviation are summarised in Table 4. The stability derivative values used for the modelling of complete loss of vertical stabilizer for the Boeing 747-200 in this study are given in Table 5.

Table 3. Complete loss of vertical stabilizer mass and inertia data [38]

Parameter	Value
W (lbf)	629540
m (slugs)	19566.10
I_x (slugs.ft^2)	17893000
I_y (slugs.ft^2)	30925000
I_z (slugs.ft^2)	47352000

Table 4. Deviation values for complete loss of vertical stabilizer [37, 38]

Vertical Tail Contributions to Lateral-Directional Derivatives Variation		
$\Delta C_{y\beta}^{(max)} = -0.451$	$\Delta C_{l\beta}^{(max)} = -0.068$	$\Delta C_{n\beta}^{(max)} = 0.234$
$\Delta C_{yp}^{(max)} = -0.137$	$\Delta C_{lp}^{(max)} = -0.021$	$\Delta C_{np}^{(max)} = 0.071$
$\Delta C_{yr}^{(max)} = 0.467$	$\Delta C_{lr}^{(max)} = 0.071$	$\Delta C_{nr}^{(max)} = -0.242$

Table 5. Stability derivatives for complete loss of vertical stabilizer

Lateral-Directional Stability Derivatives After Complete loss of Vertical tail		
$C_{y\beta} = 0$	$C_{l\beta} = -0.092$	$C_{n\beta} = 0$
$C_{yp} = -0.137$	$C_{lp} = -0.319$	$C_{np} = -0.091$
$C_{yr} = 0$	$C_{lr} = 0.059$	$C_{nr} = 0$
Lateral-Directional Control Derivatives After Complete loss of Vertical tail		
$C_{y\delta a} = 0$	$C_{l\delta a} = 0.014$	$C_{n\delta a} = 0.0018$
$C_{y\delta r} = 0.118$	$C_{l\delta r} = 0.008$	$C_{n\delta r} = -0.095$

For an aircraft with complete loss of vertical stabilizer, the differential thrust component (δ_T) will be used as the alternate control input since the rudder has been lost due to the damage. The lateral directional state space equation for an aircraft with complete loss of vertical stabilizer is given by [37]:

$$\begin{bmatrix}\Delta\dot{\beta}\\ \Delta\dot{p}\\ \Delta\dot{r}\\ \Delta\dot{\phi}\end{bmatrix}=\begin{bmatrix}\frac{Y_\beta}{u_0} & \frac{Y_p}{u_0} & -\left(1-\frac{Y_r}{u_0}\right) & \frac{g\cos\theta_0}{u_0}\\ L_\beta & L_p & L_r & 0\\ N_\beta & N_p & N_r & 0\\ 0 & 1 & 0 & 0\end{bmatrix}\begin{bmatrix}\Delta\beta\\ \Delta p\\ \Delta r\\ \Delta\phi\end{bmatrix}+\begin{bmatrix}0 & 0\\ L_{\delta a} & \frac{I_{xz}y_e}{I_xI_z-I_{xz}^2}\\ N_{\delta a} & \frac{I_xy_e}{I_xI_z-I_{xz}^2}\\ 0 & 0\end{bmatrix}\begin{bmatrix}\Delta\delta_a\\ \Delta\delta_T\end{bmatrix} \tag{5}$$

4. Plant dynamics

4.1. Undamaged aircraft plant dynamics

The equations presented below represent the state space matrices for nominal flight condition. The matrices Alo_n and Blo_n are for longitudinal motion whereas matrices Ala_n and Bla_n are for lateral-directional motion. C and D matrices for both longitudinal and lateral motion are identity matrix and zero matrix respectively.

$$Alo_n=\begin{bmatrix}-0.02204 & 0.001422 & 0 & -32.17\\ -0.05759 & -0.3943 & 871 & 0\\ -0.000095 & -0.001793 & -0.5468 & 0\\ 0 & 0 & 1 & 0\end{bmatrix},\ Blo_n=\begin{bmatrix}0 & 0.0000191\\ -18.58 & 0\\ -1.2 & 0\\ 0 & 0\end{bmatrix} \tag{6}$$

$$Ala_n=\begin{bmatrix}-0.06399 & 0 & -1 & 0.03691\\ -2.111 & -0.5048 & 0.193 & 0\\ 0.7732 & 0.01087 & -0.1794 & 0\\ 0 & 1 & 0 & 0\end{bmatrix},\ Bla_n=\begin{bmatrix}0 & 0.00839\\ 0.1\,848 & 0.1056\\ -0.008699 & -0.4591\\ 0 & 0\end{bmatrix} \tag{7}$$

The damping characteristics for the longitudinal motion of the undamaged aircraft are given in Table 6. The results show two conjugate complex pairs of eigenvalues with negative real parts. The step input plot is shown in Figure 1 and it can be clearly seen from this plot that the system becomes stable after a while for all states and inputs. The eigenvalues and modes for the lateral-directional motion of the undamaged aircraft are shown in Table 7. Since all real parts of the eigenvalues are negative, the open loop system is stable [30]. The open loop system's response to a step input for lateral motion is shown in Figure 2 and, as in the case of the longitudinal motion, the system tends towards being stable after some time.

Table 6. Longitudinal motion characteristics for undamaged aircraft

Eigenvalue/pole location	Damping Ratio (ζ)	Frequency (ω)	Mode
-0.472 + 1.25i	0.354	1.33	Short period
-0.472 - 1.25i	0.354	1.33	Short period
-0.00988+ 0.0331i	0.287	0.0345	Phugoid
-0.00988 - 0.0331i	0.287	0.0345	Phugoid

Table 7. Lateral/directional motion characteristics for undamaged aircraft

Eigenvalue	Damping Ratio (ζ)	Time Constant (s)	Frequency (ω)	Mode
-0.0698 + 0.9i	0.0773	-	0.890	Dutch roll
-0.0698 - 0.9i	0.0773	-	0.890	Dutch roll
-0.591	-	1.69	0.547	Roll
-0.0176	-	56.8	0.0195	Spiral

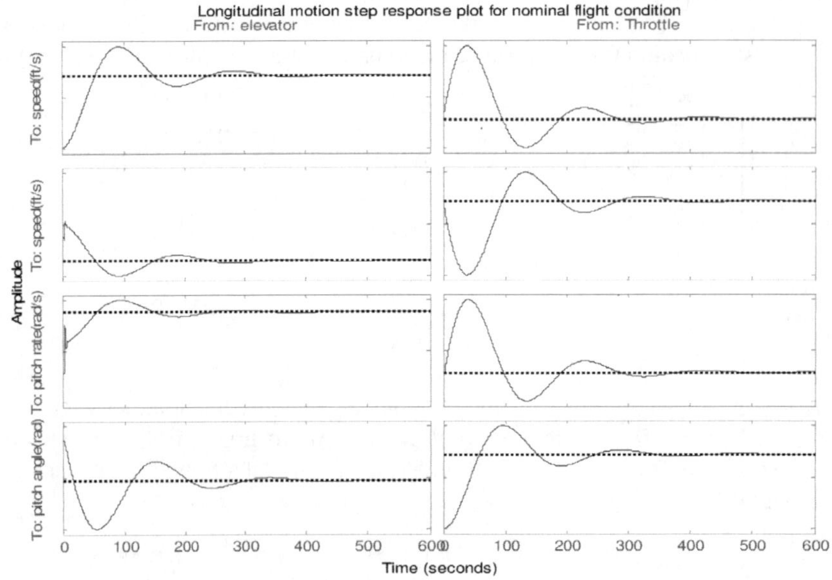

Figure 1. Step response plot for longitudinal motion for nominal flight condition

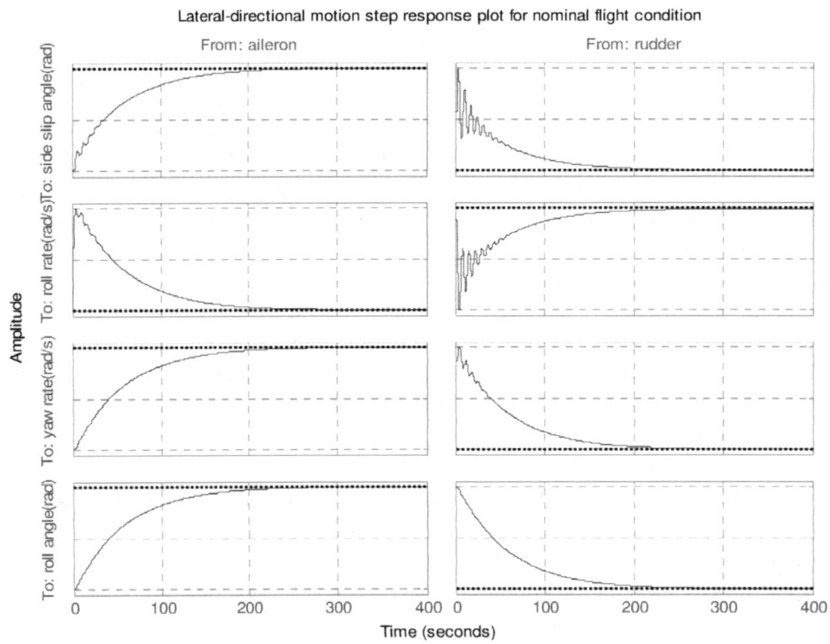

Figure 2. Step response plot for lateral/ directional motion for nominal flight condition

4.2. Damaged aircraft plant dynamics

4.2.1. Loss of 33% of left wing from tip. The first damage scenario is presented in this section. The following equations presented represent the state space matrices after the model aircraft experiences a 33% loss of the left wing from the tip. The subscript '*wd*' is to indicate that damage has been done to part of the wing.

$$A_{wd}=\begin{bmatrix} -0.0047 & -0.01077 & 0 & -623300 & 0 & 0 & 0 & 0 \\ -0.31 & -0.253 & 1.678e+7 & -28300 & 0 & 0 & 0 & 0 \\ 0 & 0 & -1.579e+5 & 264.6 & 0 & 0 & 0 & 0 \\ 0 & 0 & 1 & 0 & 0 & 0 & 0 & 0 \\ 0 & 0 & 0 & 0 & -0.047 & 0 & -1.689e+7 & 6.2e+5 \\ 0 & 0 & 4653000 & -7797 & -0.0016 & -0.2013 & 0.1055 & 0 \\ 0 & 0 & 0 & 171900 & 0.00057 & -0.0065 & -157900 & 5833 \\ 0 & 0 & 0 & 0 & 0 & 1 & 0.04541 & 0 \end{bmatrix}$$

$$B_{wd}=\begin{bmatrix} 0 & 0 & 0 \\ 13.33 & 0 & 0 \\ -0.9147 & 0 & 0 \\ 0 & 0 & 0 \\ 0 & 0 & 5.377 \\ 0 & 0.06708 & 0.07629 \\ 0 & 13.32 & -0.2895 \\ 0 & 0 & 0 \end{bmatrix} \quad (8)$$

The eigenvalues of all modes for damaged and wing damaged aircraft systems are shown in Table 8 for comparison. One of the eigenvalues for 33% left wing damage has a positive real part and for this reason, the overall system is unstable. Other damping characteristics for the undamaged and the wing damaged aircraft models are shown in Table 9.

Table 8. Eigenvalues for undamaged and wing-damaged aircraft model

Category	Mode	Undamaged	33% damage
Longitudinal	Long period	-0.00988 + 0.0331i	-0.602 + 0.861i
		-0.00988 - 0.0331i	-0.602 - 0.861i
	Short period	-0.472 +1.25i	**-15800**
		-0.472 - 1.25i	**-15800**
Lateral	Roll	-0.591	**0.896**
	Spiral	-0.017	-0.367
	Dutch roll	-0.0698 + 0.9i	-0.000547+ 0.0186i
		-0.0698 - 0.9i	-0.000547- 0.0186i

Table 9. Other characteristics for undamaged and wing-damaged aircraft model

Category	Mode	Undamaged			33% damage		
		ζd	*ωd (rad/s)*	*Time constant (s)*	*ζd*	*ωd (rad/s)*	*Time constant (s)*
Longitudinal	Long	0.287	0.0345	-	0.573	1.05	-
	period	0.287	0.0345	-	0.573	1.05	-
	Short	0.354	1.33	-	-	158000	0.00000633
	period	0.354	1.33	-	-	158000	0.00000633
Lateral	Roll	-	0.591	1.69	-	0.896	1.12
	Spiral	-	0.0176	56.8	-	0.367	2.73
	Dutch	0.0773	0.902	-	0.0293	0.0187	-
	roll	0.0773	0.902	-	0.0293	0.0187	-

The system for wing damaged aircraft turned out to not be controllable and observable. Therefore, a control system was not designed for this scenario. Observability and controllability of a system will be further explained in later section 5.

With all the derivations and equations involved in modeling a wing damaged aircraft, it is very important to validate the results obtained. Due to the lack of published information in this area of focus for this particular aircraft, other measures of validation needed to be taken. A similar study was carried out by [3] on how a 33% loss of the left wing affects the stability of a GTM aircraft. An analysis regarding the stability of the aircraft after damage was carried out. It was found that at a velocity of 160ft/s and an altitude of 625ft, the aircraft was only able to trim up to about 26% damage. It was also stated in this study that it was of a great concern that the aircraft would be unable to trim at higher damage, higher altitude and higher velocity [3]. From the results obtained, after experiencing a 33% loss of the left wing, the aircraft system was not controllable and observable. Hence a controller could not be implemented to make the system stable. This is a reasonably valid result. This could be due to the ailerons' inability to overcome the rolling moment caused by the damage. Also stated in study from [3] was that the effect of this type of damage will be greater on the roll and short period modes. A similar conclusion was reached from the results obtained in this study as the most affected modes were the short period and roll mode. The effect it has on the roll mode is due to the fact that roll mode eigenvalue is dominated by C_{lp}, which is reduced by damage [3]. Although the effects are not of exact same magnitude, this could be due to some kind of error caused by some of the assumptions made. In addition, uncontrollability of the system could be due to the fact that the effect of thrust was neglected during the modelling process because using the differential thrust from the engines in a situation like this, the aircraft is usually able to stabilise itself after suffering 33% loss of its left wing [39].

4.2.2. Complete loss of vertical stabilizer. The second damage scenario is presented in this section. The equations presented below represent the state space matrices after experiencing complete loss of the vertical stabilizer. '*vsd*' is used to indicate that damage has been done to the vertical stabilizer.

$$A_{vsd}=\begin{bmatrix} 0 & -0.001108 & -1 & 0.03691 \\ -1.235 & -0.4818 & 0.08911 & 0 \\ 0 & -0.05193 & 0 & 0 \\ 0 & 1 & 0 & 0 \end{bmatrix}$$
$$B_{vsd}=\begin{bmatrix} 0 & 0 \\ 0.1879 & 0.0142 \\ 0 & 0.6801 \\ 0 & 0 \end{bmatrix} \quad (9)$$

The open loop system response analysis was carried out on the damaged aircraft. The eigenvalues are shown in Table 10 and it clearly indicates the unstable nature of the Dutch roll mode as the real parts are positive. In addition, the pole location of the spiral mode is at the origin, which represents a very slow to unstable dynamics. The roll mode is the only stable mode of vertical-stabilizer-damaged aircraft as the eigenvalue for this mode has a negative real part (**-0.7**). The open loop step response plot in Figure 3 shows the divergent nature of the system implies that the system is unstable. Unlike the first scenario addressed, this system is observable and also controllable.

Table 10. Damping characteristics of the damaged aircraft (complete loss of vertical stabilizer)

Eigenvalue	Damping Ratio (ζ)	Time Constant(s)	Frequency (ω)	Mode
0.109 + 0.38i	-0.277	-	0.396	Dutch roll
0.109- 0.38i	-0.277	-	0.396	Dutch roll
-0.7	-	1.43	0.7	Roll
0	-	-	-	Spiral

For validation purposes, in the case of the complete loss of the vertical stabilizer of the aircraft model used in this study, a similar research was carried out on the same aircraft model but for different flight condition in [37]. The spiral and Dutch roll mode became unstable after damage. Similar results were achieved in this study as spiral and Dutch roll mode became unstable after the complete loss of the aircraft's vertical stabilizer.

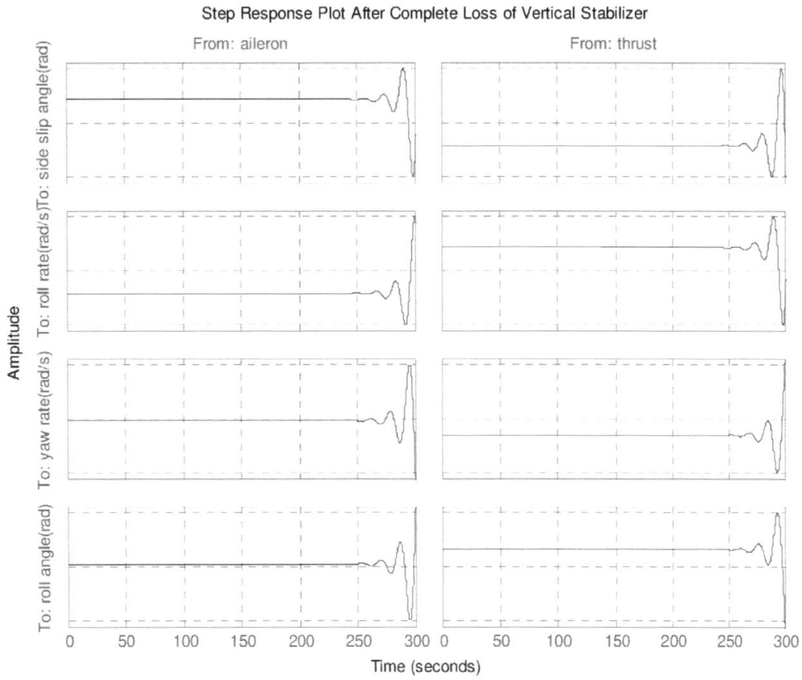

Figure 3. Step response plot for complete loss of vertical stabilizer

5. Linear Quadratic Regulator (LQR) design for controllable and observable damaged aircraft model

5.1. Controllability and observability

Controllability and observability are two very important characteristics of any system. Controllability determines the possibility of achieving the desired system states response using the input. To check the controllability of a system, the matrix given in Equation 10 is used. If the matrix is full rank, then the system is controllable [30]. Observability, on the other hand, is to know the possibility of determining all the states from the output and input signals. The system is observable if the matrix in Equation 11 is full rank. MATLAB functions *ctrb(A,B)* and *obsv(A,C)* gives the controllability matrix, ***V*** and also observability matrix, ***O*** respectively.

$$V = [B, AB, A^2B, \ldots, A^{n-1}B] \tag{10}$$

$$\boldsymbol{O} = [C^T, A^TC^T, \ldots, (A^T)^{n-1}C^T] \tag{11}$$

where n is the order of the vector $\underline{x}$ in the state space representation.

The rank of the controllability and observability matrices for 33% loss of left wing does not equal to $n((rank(V) \neq n \neq 8),(rank(O) \neq n \neq 8))$. This means that the system is not controllable or observable and due to this, the LQR controller was implemented for this damage scenario. On the other hand, after introducing the complete loss of the vertical stabilizer for the aircraft model, the system still remained controllable and observable for this case. This means the rank of the controllability and observability matrices for this damage scenario was equal to $n((rank(V)=n=4),(rank(O)=n=4))$. Hence, a full state feedback LQR controller was implemented to return the aircraft to stable condition.

5.2. Background on LQR control

Optimal control aims at producing the best results using minimum control efforts for a given set of constraints. Basically, the ultimate objective of an optimal controller is to find optimal feedback gain that can fulfil a certain performance index (PI) given a certain condition. A type of optimal controller which is the LQR is presented below. An optimal regulator problem taken from [33] is considered.

Given the equation of a system: $\dot{x}=Ax+Bu$ (12)

Determine K of the optimal vector: $u=-Kx$ (13)

Such that it minimizes the performance index: $J=\int_0^{\infty}(x^TQx+u^TRu)dt$ (14)

where Q and R are real and positive definite symmetric matrices. Q acts as a weighing matrix while R acts as the control cost matrix.

$$K=T^{-1}(T^T)^{-1}B^TP=R^{-1}B^TP \tag{15}$$

P is called a Riccati matrix and is found by solving the Riccati equation:

$$A^TP+PA-PBR^{-1}B^TP+Q=0 \tag{16}$$

MATLAB '*lqr(A,B,Q, R)*' function solves the Algebraic Ricatti Equation (ARE) and finds optimal gain '*K*' matrix. However, the dilemma of choosing the magnitude of the state weighing matrix Q and the control cost matrix R still lingers. A very simple method widely used to choose the matrices Q and R is given by Bryon's rule [34]. These matrices are selected with:

$$Q_{ii}=\frac{1}{\text{maximum acceptable value of } z_i^2} \qquad i\in\{1,2,\ldots,k\} \tag{17}$$

$$R_{jj}=\frac{1}{\text{maximum acceptable value of } u_j^2} \qquad j\in\{1,2,\ldots,k\} \tag{18}$$

Simply put, what Bryson's rule does is to scale the variables in J_{LQR} such that the maximum value is acceptable is one for each term [34]. Doing this is of great essence because, for example, if the unit of each variable differs, the full range possible for each state will also differ. So a way to normalize them is by thinking about the full range that each state can take and choosing initial guesses. The Bryson's rule usually marks the starting point for an iterative process in order to obtain the desired performance for a closed loop system [34].

5.3. Stabilizing damaged aircraft with LQR controller

After an iterative process, the magnitude of matrix Q and R chosen are shown in Equations 19 and 20. respectively. The gain '*K*' matrix found was then used in the SIMULINK model. The feedback matrix K obtained is shown in Equation 22. Figure 4 represents the block diagram used for the full state feedback LQR control system design using SIMULINK.

$$Q=50000\begin{bmatrix}1&0&0&0\\0&1&0&0\\0&0&1&0\\0&0&0&1\end{bmatrix} \tag{19}$$

$$R=\begin{bmatrix}1&0\\0&1\end{bmatrix} \tag{20}$$

$$K_{LQR}=\begin{bmatrix}1.4409&226.2909&-3.6333&223.7670\\-223.5598&3.9505&225.0357&-0.2905\end{bmatrix} \tag{21}$$

Equation 22 represents the new A matrix after implementing an LQR controller into the system. '*LQR*' is used to indicate the implementation of LQR controller.

$$A_{LQR}=\begin{bmatrix} 0 & -0.001108 & -1 & 0.03691 \\ 1.669 & -43.06 & -2.424 & -42.05 \\ 152 & -2.739 & -153.1 & 0.1976 \\ 0 & 1 & 0 & 0 \end{bmatrix} \tag{22}$$

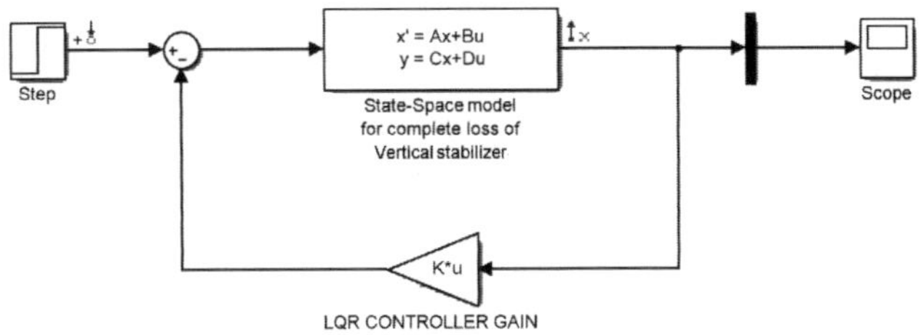

Figure 4. Block diagram used for the full state feedback LQR control system design

The new eigenvalues of the system before and after implementing the LQR controller are shown in Table 11. Using the eigenvalues, the system returns to being stable after implementation of the LQR system (all eigenvalues have negative real parts). The other damping characteristics for undamaged, vertical stabilizer damaged and LQR controlled systems are also shown in Table 12 for comparison purposes. The closed loop/feedback step response plot of the system after control is shown in Figure 5. The response plot also confirms that with the use of an LQR controller the system becomes stable in all four states after a period of time.

For validation, in reference [37], a LQR controller was also implemented for the same aircraft that suffered the same damage but for a different flight condition (lower altitude) and found to successfully stabilize all four states of the aircraft within 15 seconds. Using a Linear Quadratic Regulator (LQR) controller in this study has also yielded a similar result as the controller stabilized the damaged aircraft within 6 seconds.

Table 11. Comparison of eigenvalues for the undamaged, complete loss of vertical stabilizer and LQR controlled system

Category	Mode	Undamaged	Complete loss of vertical stabilizer	LQR controlled system
Lateral	Roll	-0.591	-0.7	-42
	Spiral	-0.0176	0	-152
	Dutch roll	-0.0698+ 0.9i	0.109 + 0.38i	-1 +0.0186
		-0.0698 - 0.9i	0.109 - 0.38i	-1 – 0.0186

Table 12. Other damping characteristics for the undamaged, vertical stabilizer damaged and LQR controlled system

Mode	Undamaged			Complete loss of vertical stabilizer			LQR controlled system		
	ζd	ωd *(rad/s)*	*Time constant (s)*	ζd	ωd *(rad/s)*	*Time constant (s)*	ζd	ωd *(rad/s)*	*Time constant (s)*
Roll	-	0.591	1.69	-	0.701	1.43	-	42	0.0238
Spiral	-	0.0176	56.8	-	0	-	-	152	0.00657
Dutch	0.0773	0.902	-	0.277	0.396	-	1	1	-
roll	0.0773	0.902	-	0.277	0.396	-	1	1	-

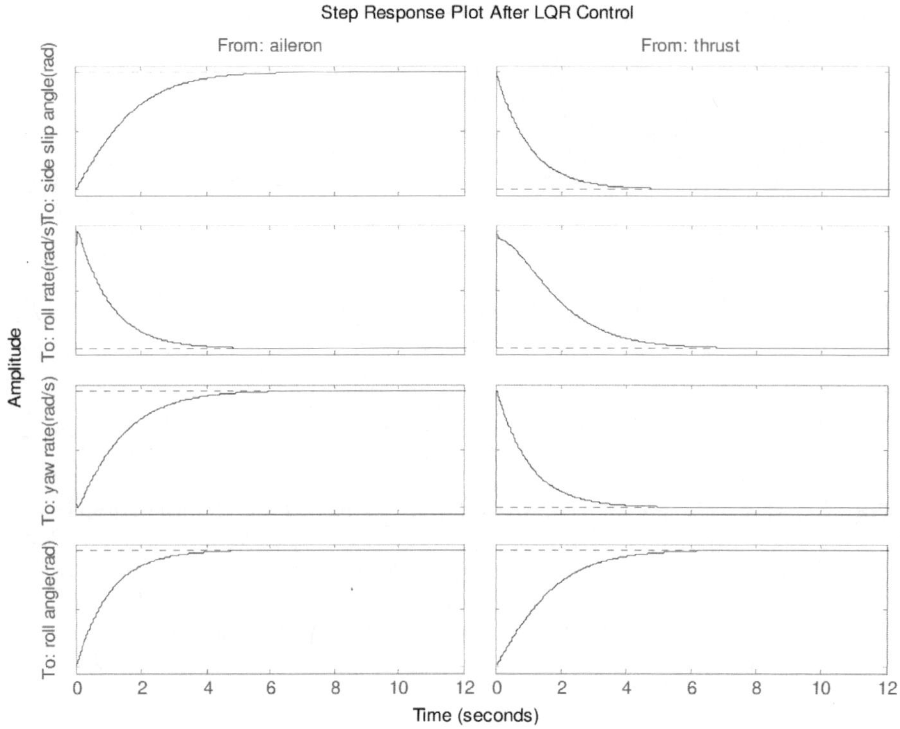

Figure 5. Step response plot after LQR controller

Finally, to know the true performance of the LQR controller, the time response characteristics (rise time, settling time, percentage overshoot) for undamaged and LQR controlled systems were compared. Table 13 shows the response characteristics due to input one (aileron) whereas Table 14 shows the response characteristics due to the second input (differential thrust). After the implementation of the LQR controller as a result of damage, the performance of the system has improved compared to its performance in nominal condition for all state variables. The controller can obviously stabilize the system within only 6 seconds, unlike the undamaged aircraft which stabilizes within 230 seconds.

Table 13. Time response characteristics due to aileron

State variable	Undamaged			LQR controlled system		
	Rise time(s)	*Settling time(s)*	Percentage Overshoot	*Rise time(s)*	*Settling time(s)*	Percentage Overshoot
Sideslip angle (β)	125	219	0	3.09	5.25	0
Roll rate (p)	0	228	-	12	4.03	0
Yaw rate (r)	125	224	0	2.76	4.97	0
Roll angle (ϕ)	125	224	0	2.2	3.94	0

Table 14. Time response characteristics due to differential thrust

State variable	Undamaged			LQR controlled system		
	Rise time(s)	*Settling time(s)*	Percentage Overshoot	*Rise time(s)*	*Settling time(s)*	Percentage Overshoot
Sideslip angle (β)	125	200	0	2.2	3.91	0
Roll rate (p)	0	207	-	-	5.81	-
Yaw rate (r)	131	222	0	-	3.97	-
Roll angle (ϕ)	125	224	0	3.09	5.21	0

6. Conclusion

The area of focus of this work is to model the dynamics of a damaged Boeing 747-200 aircraft and to implement a control system to the observable and controllable damaged system to regain the stability of the damaged aircraft. The stability derivatives for undamaged aircraft were obtained. Two damage scenarios were considered and in both cases, the stability derivative values experienced some changes and had to be re-examined and recalculated. The modelling of the 33% loss of the left wing showed that the system was not controllable or observable. On the other hand, for the complete loss of the vertical stabilizer, the system proved to be controllable and also observable. Simulations were also used to show the aircraft's response to damages. A LQR controller was implemented to return the aircraft to trim condition. The LQR controller was proven to be able to make the aircraft stable and also improve the performance characteristics.

References

[1] Romanucci and Blandin *5 Most Common Causes of Plane Crashes*
[2] Fielding C and Luckner R 2000 *IEE Control Engineering Series* 1-55
[3] Arruda M 2009 *Dynamic inverse resilient control for damaged asymmetric aircraft: Modeling and simulation* Doctoral dissertation Wichita State University
[4] Ouellette J, Raghavan B, Patil M J and Kapania R K 2009 *AIAA Atmospheric Flight Mechanics Conference* p. 6153
[5] Jordan T, Langford W, Belcastro C, Foster J, Shah G, Howland G and Kidd R 2004 *AUVSI's Unmanned Systems North America Symposium and Exhibition*
[6] Jordan T L, Langford W M and Hill J S 2005 *AIAA Guidance, Navigation and Control Conference and Exhibit*
[7] Beltramo M N, Trapp D L, Kimoto B W and Marsh D P 1977 *Parametric study of transport aircraft systems cost and weight*
[8] Woo J H 1992 *AIAA/USAF/NASA/OAI Symposium on Multidisciplinary Analysis and Optimization*
[9] Render P M, Samaad-Suhaeb M, Yang Z and Mani M 2009 *Journal of Aircraft*, **46** 997-1004
[10] Foster J V, Cunningham K, Fremaux C M, Shah G H, Stewart E C, Rivers R A, Wilborn J E and Gato W 2005 *AIAA Paper*
[11] Shah G H 2008 *AIAA Atmospheric Flight Mechanics Conference and Exhibit*
[12] Thomas A L 1993 *Proceedings of the Royal Society of London B: Biological Sciences* **254** 181-9

[13] Keller J D, McKillip Jr. R M and Wachspress D A 2008 *AIAA Paper*
[14] Wilborn J E and Foster J V 2004 *AIAA atmospheric flight mechanics conference and exhibit*
[15] Bacon B J and Gregory I M 2007 *General equations of motion for a damaged asymmetric aircraft*. American Institute of Aeronautics and Astronautics
[16] Nguyen N, Krishnakumar K, Kaneshige J and Nespeca P 2006 *AIAA Guidance, Navigation, and Control Conference and Exhibits*
[17] Nguyen N T, Krishnakumar K S, Kaneshige J T and Nespeca P P 2008 *Journal of Guidance, Control and Dynamics* **31** 751-64
[18] Nguyen N T, Krishnakumar K S, Kaneshige J T and Nespeca P P 2008 *Journal of Guidance, Control and Dynamics* **31** 751-764
[19] Liu Y, Tao G and Joshi S M 2010 *Journal of Guidance, Control and Dynamics* **33** 1500-17
[20] Lee S and Sharma M 1998 *Direct adaptive reconfigurable control of a tailless fighter aircraft*
[21] Nguyen N T, Bakhtiari-Nejad M and Huang Y 2007 *AIAA Guidance, Navigation and Control Conference and Exhibit*
[22] Kim B S and Calise A J 1997 *Journal of Guidance, Control and Dynamics* **20** 26-33
[23] Rysdyk R T and Calise A J 1999 *Journal of Guidance, Control and Dynamics* **22** 402-7
[24] Shin Y, Calise A J and Johnson M 2008 *Journal of Guidance, Control and Dynamics* **31** 1464-77
[25] Burcham Jr F W, Maine T A, Fullerton C G and Webb L D 1996 *Development and flight evaluation of an emergency digital flight control system using only engine thrust on an F-15 airplane*
[26] Burcham F W, Maine T A and Burken J J 1996 *Development and flight test of an augmented thrust-only flight control system on an MD-11 transport airplane* National Aeronautics and Space Administration
[27] Bull J, Mah R, Hardy G, Sullivan B, Jones J, Williams D, Soukup P and Winters J 1997 *Piloted simulation tests of propulsion control as backup to loss of primary flight controls for a B747-400 jet transport*
[28] Johnson E N, Calise A J and Blauwe H D 2008 *AIAA Guidance, Navigation and Control Conference*
[29] Tang Y, Atkins E and Sanner R 2007 *AIAA Guidance, Navigation and Control Conference and Exhibit*
[30] Nelson R C 1998 *Flight stability and automatic control* (New York: McGraw Hill)
[31] Caughey D A 2011 *Introduction to aircraft stability and control course notes for M&AE 5070* Cornell University
[32] Napolitano M R 2012 *Aircraft dynamics: From modeling to simulation (*Wiley)
[33] Ogata K and Yang Y 1970 *Modern control engineering*
[34] Hespanha J 2005. *Lecture notes on lqr/lqg controller design*
[35] Roskam J 2006 *Airplane Design: Part I-VIII* (Lawrence: DARcorporation)
[36] Bramesfeld G, Maughmer M D and Willits S M 2006 *Journal of aircraft* **43** 216-25
[37] Lu L 2015 *Utilization of Differential Thrust for Lateral/Directional Stability of a Commercial Aircraft with a Damaged Vertical Stabilizer*
[38] Hitachi Y 2009 *Damage-Tolerant Flight Control System Design for Propulsion-Controlled Aircraft* Doctoral dissertation University of Toronto
[39] Sarigul-Klijn N, Rapetti R, Jordan A, Lopez I, Sarigul-Klijn M and Nespeca P 2010 *Journal of Aircraft* **47** 255-67
[40] Lu L and Turkoglu K 2015 *Adaptive Differential Thrust Methodology for Lateral/Directional Stability of an Aircraft with a Damaged Vertical Stabilizer*

Optimization of a synthetic jet actuator for flow control around an airfoil

E Montazer[1, 2*], M Mirzaei[1], E Salami[2], T A Ward[3], F I Romli[4] and S N Kazi[2]

[1] Department of Aerospace Engineering, K. N. Toosi University of Technology, Tehran, Iran
[2] Department of Mechanical Engineering, Faculty of Engineering, University of Malaya, Kuala Lumpur, Malaysia
[3] Faculty of Engineering and Physical Science, Heriot-Watt University, Putrajaya, Malaysia
[4] Department of Aerospace Engineering, Universiti Putra Malaysia, Malaysia

[*] elham.montazer@gmail.com

Abstract. This paper deals with the optimization of a synthetic jet actuator parameters in the control flow around the NACA0015 airfoil at two angles of attack: 13° (i.e. the stall angle of NACA0015) and 16° (i.e. the post stall angle of NACA0015) to maximize the aerodynamic performance of the airfoil. Synthetic jet actuator is a zero mass flux-active flow control device that alternately injects and removes fluid through a small slot at the input movement frequency of a diaphragm. The movement of the diaphragm and also the external flow around the airfoil were simulated using numerical approach. The objective of the optimization process function was maximum lift-drag ratio (L/D) and the optimization variables were jet frequency, length of the jet slot and jet location along the chord. The power coefficient of the jet was considered as a constraint. The response surface optimization method was employed to achieve the optimal parameters. The results showed that the actuator is more effective for post stall angles of attack that can lead to an enhancement of 66% in L/D.

1. Introduction

Over the past decade, there has been a growing interest in small active flow control devices that have a significant impact on flow field, and modified forces and momentum around lifting surface, especially for low-Reynolds number applications such as Unmanned Aircraft Vehicles (UAVs). Extensive experimental work has shown that the synthetic jet actuator is an effective way to modify aerodynamic specifications of lifting surface. It can change the structure of vortices near the trailing edge and thus offers potential replacement of conventional control surfaces like flaps [1-5]. Practical implementation of this kind of flow control devices requires experimental development of control systems for aerial vehicles. Recently, the computational tools have been widely used to simulate the flow control. Computational fluid dynamics has provided an insight into the physics of flow control through experimental investigation, which is normally difficult to achieve.

The synthetic jet is also called Zero Net Mass Flux (ZNMF) jet since it is created by oscillating the fluid around the body periodically [3, 6-8]. The net mass flux has been zero since the air surrounding the jet orifice is sucked and blown periodically [9]. The synthetic jet generates momentum difference that changes the behavior of the flow [6, 7, 10, 11]. Figure 1 shows the most common practical form of the actuator with the output perpendicular to the stream [12, 13]. It is typically consisted of a piezoelectric disk that is attached to a metal diaphragm to form like a sealed cavity [14]. A jet slot is enclosed on one side of the cavity and the diaphragm, which is located in the lower border of the cavity, is deformed by periodic oscillation. During the time that the diaphragm oscillates alternatively, the flow goes into the cavity from the slot and it could be evacuated. In 1998, Smith and Glezer [15] proved that the interaction of the jet with the flow over the surface can displace the local streamlines and induce a virtual change in the shape of the surface.

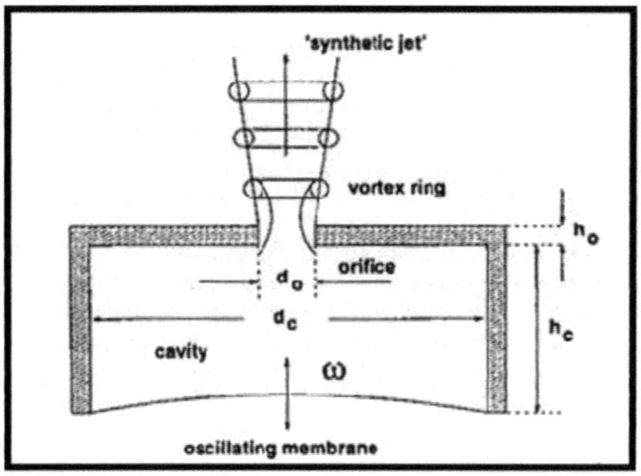

Figure 1. A view of the synthetic jet actuator [12, 13]

In 2005, Holman *et al.* [16] mentioned one of the synthetic jet actuator's ability. It has the ability to integrate the internal flow of the cavity with the external flow without any external device. This makes it an attractive device for flow control in both internal and external flows. Moreover, Ritchie *et al.* [17] demonstrated that the synthetic jet actuators have the advantage of a greater spreading rate than that of the equivalent continuous jet throughout the measured domain at the same time-averaged velocity, thus yielding a greater region of influence on the working domain.

The jet stream can be simulated like a stream with complex spatial and temporal characteristics. Flow typically separates from the edge of the slot and a vortex sheet is formed, which rotated as single and isolated vortices, and subsequently being driven by self-induced velocity. Eventually, the vortices lose their coherence and disappear as they move away from the orifice [18]. The vortices that are created never return to the orifice at suction time, if the synthetic jet can generate the strong velocity.

Recent experimental and numerical studies are testifying that use of synthetic jet actuator as a flow control device is desirable. However, there is still a limited amount of this type of flow control devices that are used on aircraft structures [9]. In recent years, the numerical simulation of the synthetic jets used for stall control around an aerodynamic body has attracted the attention of many researchers. This popularity might be because numerical simulation is a very effective way to explore the possibilities of using synthetic jet on an aircraft [3, 19]. In order to simulate flow control with the synthetic jet, Large Eddy Simulation method (LES) and Direct Numerical Solution method (DNS) for resolving small vortices induced by synthetic jet as they are very convenient [19, 20]. Simulation of the flow with high Reynolds numbers is unfeasible because of high computational cost and difficulties in the simulation. In fact, most numerical simulations for stall control with the presence of the synthetic jet were performed using Reynolds Average Navier Stokes equations method (RANS) [9].

As stated before, one of the main synthetic jet actuator applications is to modify the aerodynamic characteristics of wings and airfoils. Duvigneau and Visonneau [9] demonstrated that the synthetic jet actuator placed on the upper surface of an airfoil is able to modify the streamlines around the airfoil. Lopez *et al.* [2] have reported that using the synthetic jet actuators is an effective way to enhance the lift and modify the momentums of wings and airfoils. Effective control has been achieved using actuation frequencies with an order of magnitude larger than the natural shedding frequency of the body [2]. An automatic optimization of flow control devices is a sensitive subject due to the drastic computational time related to high accuracy analysis of unsteady flow and the possible multi-constraints of the objective function. The automatic optimization procedure joined with the flow solver is employed by Duvigneau and Visonneau [9] to optimize the actuator's parameters (momentum coefficient, frequency, angle with respect to the wall) at each incidence with the aim of increasing the time-averaged lift. Duvigneau and Chandrashekar [21] used kriging-based algorithms to optimize flow control parameters. These methods are very efficient for global optimization at a reasonable cost. The Response Surface Methodology (RSM) was employed by Tuncer and Kaya [22] for the optimization of periodically flapping airfoil parameters to maximize the thrust generation. The RSM is a collection of statistical and mathematical techniques useful for developing, improving, and optimizing processes in which a response of interest is influenced by several variables and the objective is to optimize this response [23]. It defines the effect of the independent variables, alone or in combination, on the processes [24]. The RSM is discovered to be considerably more proficient than the steepest ascent method. Akcayoz and Tuncer [6] demonstrated that the RSM allows obtaining optimum parameters with similar accuracy by performing a less number of computational evaluations.

This present work deals with the simulation of the synthetic jet actuator located on the section surface of a NACA0015 airfoil as an active flow controller. Unsteady flow over a NACA0015 airfoil is solved using a Navier-Stokes solver. The parametric study demonstrates the synthetic jet parameters must be optimized for improving aerodynamic performance. To the best of authors' knowledge, in literatures, there is no available work dealing with optimization of the jet location, jet frequency and jet slot size. The objective of the optimization is to determine those synthetic jet parameters that maximize the L/D. The RSM is used to find the optimum synthetic jet parameters.

2. Flow simulation and numerical framework

The nature of the flow around an airfoil with the synthetic jet is unsteady and turbulent. Navier-Stokes equations are considered as governing equations. Since Mach number of the free stream is less than 0.3, the flow can be considered as incompressible. The Navier Stokes equations for incompressible two-dimensional flow are:

$$\begin{aligned}
&\rho\left(\frac{\partial u}{\partial t}+u\frac{\partial u}{\partial x}+v\frac{\partial u}{\partial y}\right)=-\frac{\partial p}{\partial x}+\mu\left(\frac{\partial^2 u}{\partial x^2}+\frac{\partial^2 u}{\partial y^2}\right)+\rho g_x\\
&\rho\left(\frac{\partial v}{\partial t}+u\frac{\partial v}{\partial x}+v\frac{\partial v}{\partial y}\right)=-\frac{\partial p}{\partial y}+\mu\left(\frac{\partial^2 v}{\partial x^2}+\frac{\partial^2 v}{\partial y^2}\right)+\rho g_y\\
&\frac{\partial \rho}{\partial t}+\frac{\partial(\rho u)}{\partial x}+\frac{\partial(\rho v)}{\partial y}=0
\end{aligned} \tag{1}$$

The turbulent viscosity coefficient is modeled using the Spalart-Allmaras model. The Spalart-Allmaras is a single equation turbulence model that is designed for aerospace applications [25, 26]. For subsonic and incompressible flows, pressure-based SIMPLE algorithm was used to solve the governing equations. An approximation algorithm for velocity was obtained by solving momentum equation. The pressure gradient factor was calculated by pressure distribution in the previous stage or first prediction. The pressure equation was obtained and solved for the new pressure distribution, then the velocity was corrected and new parameter fluxes were found.

2.1. Grid study

Each RANS model requires a suitable mesh that plays an important role in the performance of the model. For this study, C-Type meshing is used. This type of mesh is very convenient for solving flow in flow field that contains vortices. Although there is no specific guideline for determining the size of domain around an airfoil, but a bigger domain as far as the boundaries of the domain do not have an

IOP Conf. Series: Materials Science and Engineering **152** (2016) 012023 doi:10.1088/1757-899X/152/1/012023

effect on the solution of the flow field around the airfoil, is the best. When the domain gets bigger, the number of cells in the flow field will also increase and this causes more computational cost. Two-dimensional view of the domain and the size is shown in Figure 2.

Reynolds number based on the airfoil chord length is 896000, the distance between the cells in the direction normal to the airfoil surface starts from 5 to 10 times smaller than the chord length. The first cell size normal to the airfoil surface is approximately two per unit area as shown in Figure 3, which is equivalent to $Y^{+}=2$.

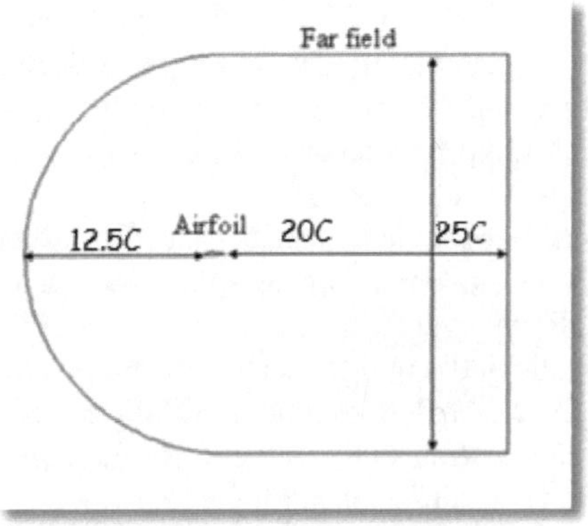

Figure 2. The schematic image of the field around the airfoil

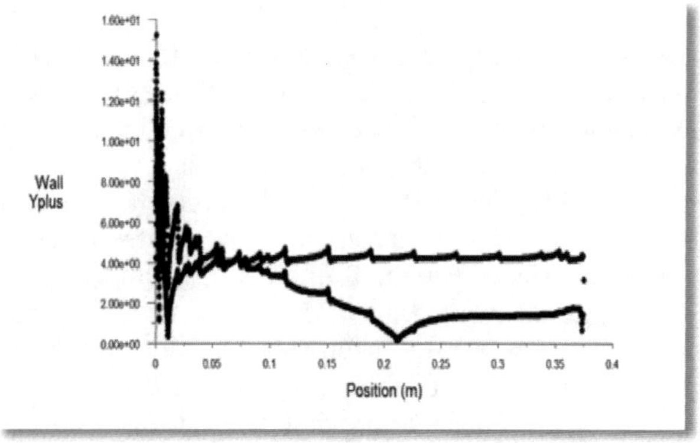

Figure 3. Y+ on the airfoil surface

In order to validate the computational results, they were compared with experimental data reported by Gilarranz *et al.* [12] for NACA 0015 airfoil. Figure 4 shows the comparison for lift coefficient at various angles of attack.

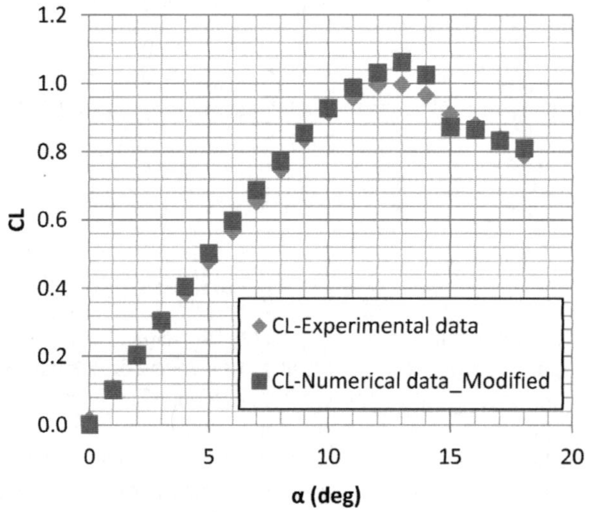

Figure 4. The lift coefficient validation for clean airfoil

As the number of cells that cover the field is increased, the result is more accurate but it should be noted that the increment of the number of small cells causes more computational efforts. Moreover, when the number of cells increases more than the suitable limit, it has no effect on the results. Figure 5 gives a view of domain meshing and clustering of the cells around the airfoil.

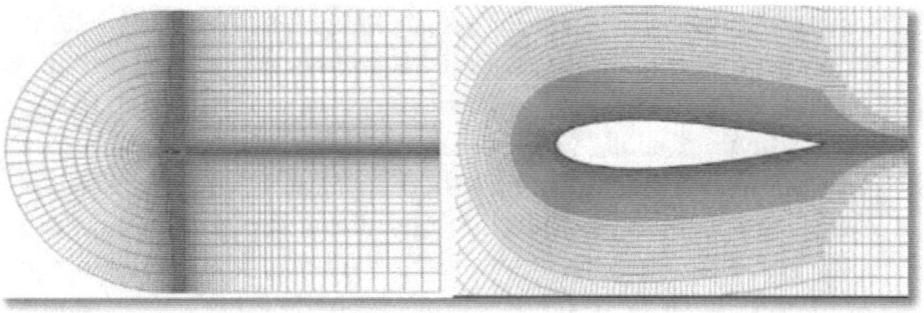

Figure 5. View of the grid in the field and near the airfoil surface

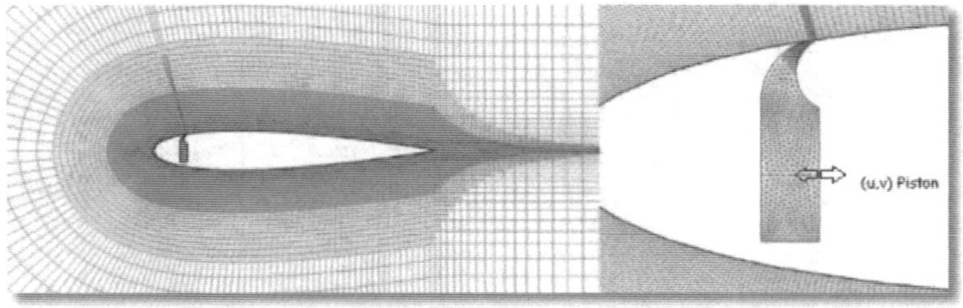

Figure 6. View of the synthetic jet simulation

2.2. Synthetic jet modeling

Previous studies have shown that the size of jet cavity does not play an important role in determining the characteristics of a synthetic jet and it is only an important issue in the modeling of a synthetic jet output flow with higher accuracy [1]. Figure 6 depicts a jet mounted on the upper surface of an airfoil. If the synthetic jet output is tangential to flow, the momentum boundary layer will directly increase. However, if the output of the jet is normal to the wall, it can increase the rate of mixing in shear layer. Gad-El-Hak [19] conducted an extensive review in this area. By utilizing Coanda effect, using a jet with blowing parallel to a wall with large curvatures is possible. He proved that this method is very useful to increase the lift force without requiring moving surfaces.

In the synthetic jet flow dynamics, not only the output flow is important but also the throat that is connecting to the cavity and the outlet is capable of playing a vital role. Based on this observation and the exact dimensions of the synthetic jet slot and the throat, the cavity size can be changed following the actual situation.

2.3. Boundary conditions

In order to simulate the oscillatory motion of the diaphragm, vertical speed set for the diaphragm is obtained as:

$$U_n = A\sin 2\pi F^{+}T \tag{2}$$

where F^+ is a dimensionless jet frequency and it depends on the size of the airfoil chord and speed of sound. The experimental jet exit velocity reaches to 85 m/s [13], This velocity is approximately 2.43 times greater than the free-stream velocity ($85 \approx 2.43U_\infty$). Considering the length of the diaphragm around 22.5 times longer than the jet slot, the maximum amount of diaphragm speed based on the experimental results is estimated $0.108U_\infty$. F^+ is calculated as:

$$F^{+} = \frac{F_{jet}C}{a_\infty} \tag{3}$$

where C is the airfoil chord length and a_∞ is the free stream speed of sound. Lastly, T represents the dimensionless time that is a function of t, U_∞ (free stream velocity) and C: $T = tU_\infty/C$.

Boundary conditions are required for finding an eddy viscosity. The eddy viscosity is dependent on the synthetic jet injected into the main flow, which also plays an important role in the downstream [27]. The turbulent velocity is assumed zero at the cavity's boundaries but still, it is a matter under investigation in numerical simulation of synthetic jet. In order to apply an oscillatory motion of the diaphragm, moving mesh based on mass-spring method was employed to all points on the diaphragm. The maximum of displacement occurred at the midpoint of the diaphragm and the two endpoints were at rest. An appropriate time step for oscillatory diaphragm motion is 0.0001 per second. For each time step, the diaphragm oscillation and cavity mesh should be updated. Figure 7 shows the displacement of the diaphragm and the moving grid points in the cavity.

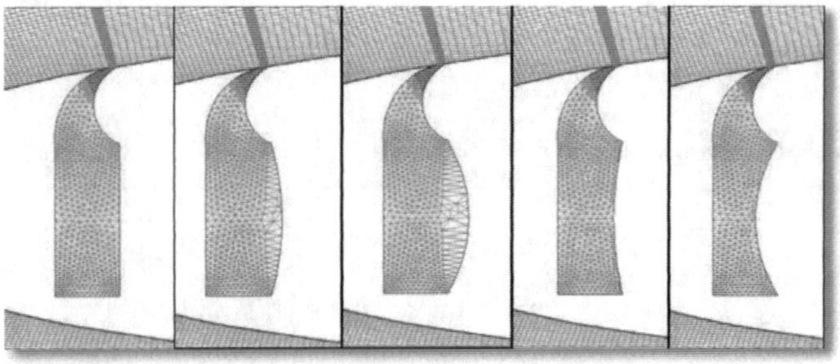

Figure 7. View of the diaphragm and remeshing at different intervals

3. Optimization

In this section, the employed optimization method, namely Response Surface Methodology (RSM), is described. The optimization is based on fitting a response surface model to the data generated by unsteady flow solver for various inputs. The response surfaces are approximated in the defined design space. The optimum design variables that maximize the value of the objective function are estimated using the response surfaces. RSM is a combination of mathematical and statistical techniques, and is based on the generation of response surfaces for a set of design variables [28]. The response can be a function of several variables and it can be obtained using experimental or numerical methods. In the generation of the response surfaces, the method of least squares is employed. Once the response surfaces are determined, the maximum or minimum values of the response and the corresponding values of optimization variables can be evaluated.

In this study, RSM is employed using Minitab software for the optimization of a problem with three optimization variables. A quadratic response surface is the approximated L/D based on the least square method. The quadratic equation for finding the best-fitted curve to numerical data points is:

$$L/D = a_0 + a_1 l + a_2 F + a_3 x_{jet} + a_{11} l^2 + a_{22} F^2 + a_{33} x_{jet}^2 + a_{12} lF + a_{13} l x_{jet} + a_{23} F x_{jet} \quad (4)$$

where a is a constant coefficient, l is related to the length of the jet slot, F is the actuation frequency and x_{jet} is the location of the synthetic jet. Once the model is created, it is required to check the goodness of the model [29]. The accuracy of the RSM is validated by calculating RSM residuals, which are the difference between the predicted and calculated responses. A perfect fit line represents the ideal design where difference between the response of the RSM model and computed responses is less than 2%.The optimization process followed throughout this study is explained in Figure 8.

An experimental design for fitting a second-order model must have at least three levels of each factor. There is numerous Design of Experiment (DoE) techniques for fitting a second order model, so it is important to choose the suitable design. The design of experiments concerns the distribution of the points in the design space. One of the best methods to distribute is Full-Factorial (FF) design. The point distribution for Full-Factorial (FF) design is shown in Figure 9. As it is illustrated the design space is covered completely. The requirement to a high number of computations and being limited low order models are the disadvantages of the full factorial design.

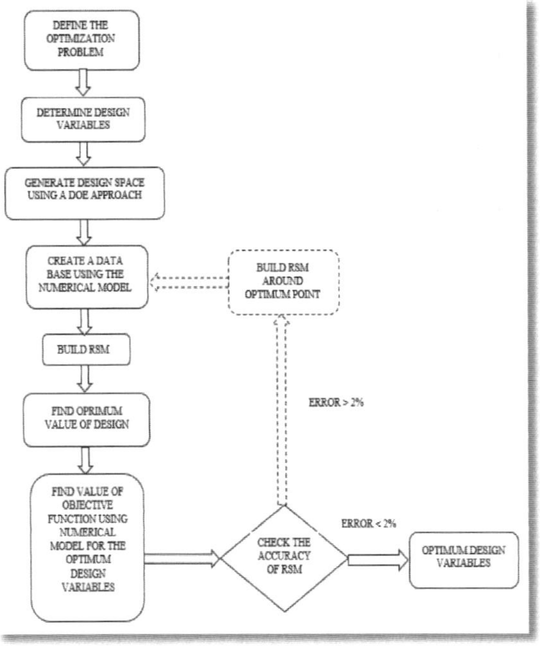

Figure 8. Optimization strategy

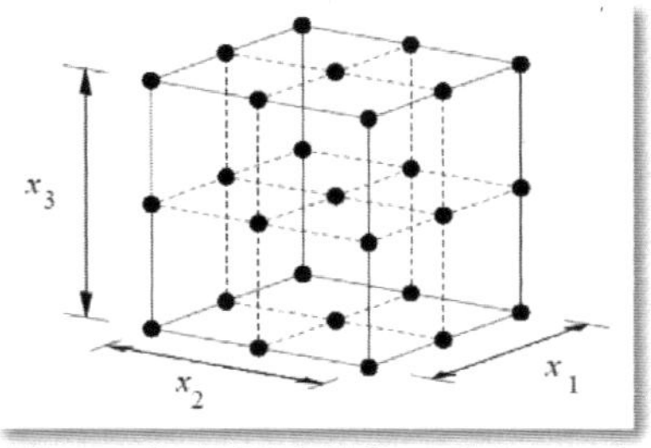

Figure 9. A 33-full factorial sample design

3.1. Optimization of synthetic jet actuator on the NACA0015 airfoil

The optimization was performed at two angles of attack, 13 degrees (that is nearly stall angle of attack of NACA0015) and 16 degrees (post stall angle of NACA0015). The objective function of the optimization was L/D and it was required to maximize this function. As stated before, the jet actuation frequency, length of the jet outlet (slot) and position of the jet slot in the upper surface of airfoil were selected as design parameters. Since the jet inlet and outlet flow are tangential to the airfoil surface, the angle of the jet relates to its location on the surface and the velocity of the jet is dependent on the actuation frequency and the length of the jet outlet. The actuation frequency was varied from 60 Hz to 130 Hz. The jet location on the upper surface of the airfoil was changed ranging from the leading edge to the trailing edge of the airfoil while the jet outlet can have three sizes of 0.6 mm, 1.2 mm and also 4.2 mm.

3.2. Constraints

Power coefficient of the jet was considered as a constraint parameter in the optimization process. Equation 5 shows a correlation between the power coefficient and other flow field parameters.

$$C_{power} = \rho_{jet} L_{jet}^2 u_{jet}^2 F_{jet} \tag{5}$$

where ρ_{jet} is the density of the free stream, L_{jet} is the length of the slot, u_{jet} is the outlet velocity of the synthetic jet and F_{jet} is the actuation frequency. Based on the numerical study conducted by Akcayoz and Tuncer [6], the power coefficient is acceptable when its value is less than 2e-7 during the optimization process.

4. Results and discussions

Optimization is carried out using the RSM. Unsteady turbulent flows over the NACA 0015 airfoil profile were computed using a Navier-Stokes flow solver over a C-grid. The flow was assumed to be fully turbulent and the Spalart-Allmaras turbulence model was employed. In unsteady calculations, the solution was initialized with the free stream conditions. The unsteady computations were carried out until a steady or a periodic behavior in aerodynamic coefficients was observed. The computed flow field was analyzed in terms of pressure coefficient distribution, aerodynamic loads and flow fields over the airfoil.

A parametric study was carried out to investigate the sensitivity of the solution to the synthetic jet parameters. The influence of the jet location, the jet frequency and the length of the jet slot on L/D was investigated. The result of the parametric study was used in the determination of the design space for the optimization study. The synthetic jet parameters were optimized to maximize the L/D. The optimization study was performed at various angles of attack. RSM was employed in the optimization study. The response surfaces for the L/D ratio were approximated using second order model based on the results of the numerical model. The optimum synthetic jet parameters and the corresponding L/D values were estimated using the approximated response surface.

4.1. Optimal control parameters at 13°

The maximum values of the data that correspond to the response surface optimization algorithm in different optimization steps are shown in Table 1. The optimization process was continued until the error gets a value below than 2%. Convergence criterion was satisfied in four optimization steps as shown in Table 1. Figure 10 shows the differences of L/D at 13° with the zero error line. In optimal conditions, the jet velocity reached up to 140 m/s while the inlet flow velocity was 35 m/s, the jet output velocity was approximately four times more than the velocity of the inlet flow. This is a good indicator of the ability of the jet on the airfoil surface and it also leads to improving aerodynamic performance where L/D is more than 28.

Figure 11 shows variation of L/D versus three optimization parameters (length of slot, jet location and actuation frequency), with 3-D surfaces approximated by RSM. Figure 11a shows variation of L/D versus the jet location and the size of the jet slot. The jet location moving between 0.12 m (32% of chord) and 0.15 m (40% of chord) from the airfoil leading edge and the size of the jet orifice varied between 0.001 m and 0.0015 m. In these range of variations, the best and maximum response surface occurred for 13 degrees. Figure 11b displays an area with the jet location moves between 0.075 m (20% of chord) and 0.1 m (26.6% of chord) from the airfoil leading edge and the excitation frequency is varied between 60 Hz to 65 Hz. The figure shows the most optimal area related to these two optimization parameters, but another important parameter that is called the length of jet outlet plays an important role in response and can change this range. Figure 11c indicates the response surface of L/D versus the actuation frequency and the size of the jet slot. The most favorable areas for these two parameters is different from the area obtained by numerical solution at 13 degrees. The difference is explained by the influence of integration of the three parameters with each other and the effect of each parameter on the response surface to define the main area.

Table 1. The optimal result for parameters at 13° in different optimization step

Step	AOA (degree)	Actuation Frequency (Hz)	Length of Slot (mm)	Jet Location (% chord)	Jet Velocity (m/s)	L/D estimated	L/D Numerical simulation	Error (%)
1	13	80	0.6	36	125	26.181	27.190	3.71
2	13	60	0.6	24	130	26.248	27.195	3.48
3	13	60	1.2	36	135	28.097	27.471	2.27
4	13	100	1.2	36	140	27.968	28.244	0.98

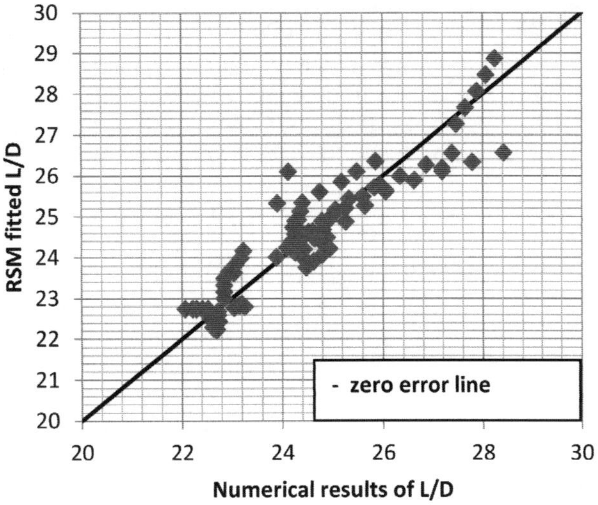

Figure 10. L/D of NACA 0015 estimated from the numerical simulation is in comparison with the fitted L/D obtained by the RSM optimization function at 13°

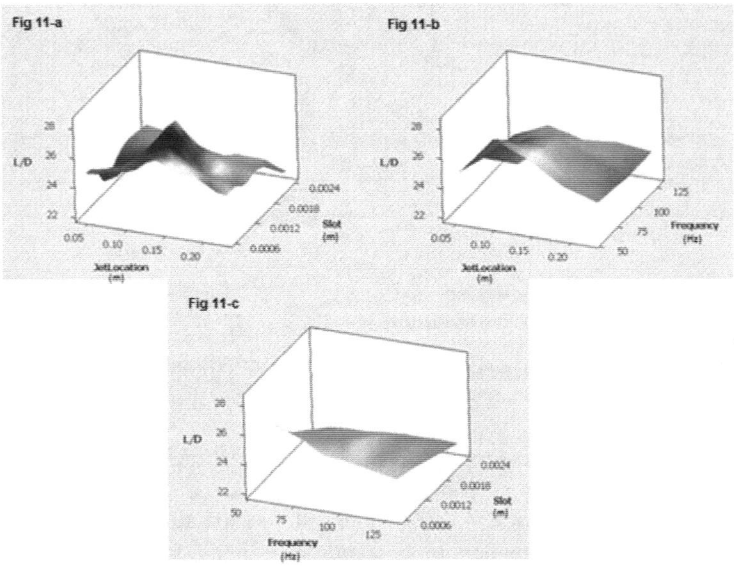

Figure 11. Response surface in 13 degrees for three optimization parameters that vary in exact range

In Table 2, the aerodynamic coefficients of the airfoil at the optimal point are compared with those of a clean airfoil (without actuator). It can be seen that the optimized actuator at stall angle of attack improves the L/D by 29.18% and it can postpone the stall conditions.

Table 2. Aerodynamic coefficients in the angle of attack equal to 13 °

AOA = 13°	C_L	C_D	L/D
Airfoil without synthetic jet	1.061	0.04853	21.864
Airfoil with synthetic jet	1.119	0.03961	28.244
Difference (%)	+5.47	-18.38	+29.18

4.2. Optimization at post stall angle of attack

The optimization study was also performed at α = 16°, which was a post stall angle of attack for the airfoil. The optimization process was ended in three optimization steps. Table 3 displays the optimum design parameters and relative errors obtained in the optimization steps. The relative difference in the third step was calculated as 1.83%, which satisfied the convergence principles. Figure 12 displays the differences of L/D at 16° with zero error line in third optimization step.

Table 3. The optimal result for parameters at 16° in different optimization step

Step	AOA (degree)	Actuation Frequency (Hz)	Length of Slot (mm)	Jet Location (% chord)	Jet Velocity (m/s)	L/D estimated	L/D Numerical simulation	Error (%)
1	16	60	0.6	12	80	9.973	9.751	2.275
2	16	75	1.2	12	55	7.228	7.684	5.93
3	16	75	0.6	12	85	10.006	9.826	1.83

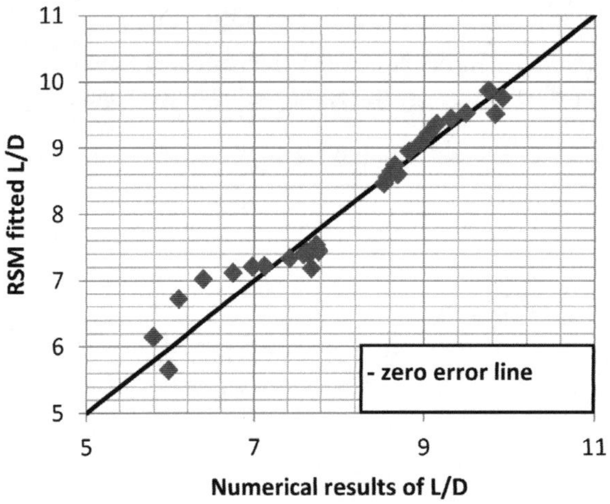

Figure 12. L/D of NACA 0015 estimated from the numerical simulation is in comparison with the fitted L/D obtained by the RSM optimization function at 16°

Figure 13 displays L/D surface approximated by RSM versus the three optimization parameters at angle of attack of 16°. Minimum and maximum points in three-dimensional surfaces were determined. In Figure 13b, an area is clearly marked with containing the jet location varied between 0.3% chord to 24% chord (0.09 m). It shows that the best input frequency range is 60 Hz-90 Hz. The best values of the data that are matching to the response surface optimization algorithm in different optimization steps are indicated in Table 3. The design parameters at optimal condition are: the jet location at 12% chord length, the outlet size is 0.6 mm and the input frequency is 75 Hz. In these conditions, the jet flow velocity reaches to 85 m/s, which is about 2.42 times more than the velocity of inlet flow.

Obviously, the jet has better performance in the post stall situation. A comparison between the aerodynamic coefficients of the optimized actuator with the coefficient of a clean airfoil is given in Table 4. It can be seen that the jet improved the aerodynamic performance of the airfoil more than 66%. This improvement can make the synthetic jet actuator as a suitable controller for active flow control for high angle of attack. According to this comparison, one can find that the effect of the jet on the flow stream in separated flow is more than the attached ones [9]. This can be found in Figure 14. In this figure, the streamlines around the airfoil at the post stall angle of attack were shown. The jet suppresses the recirculating zone and consequently changes the pressure distribution.

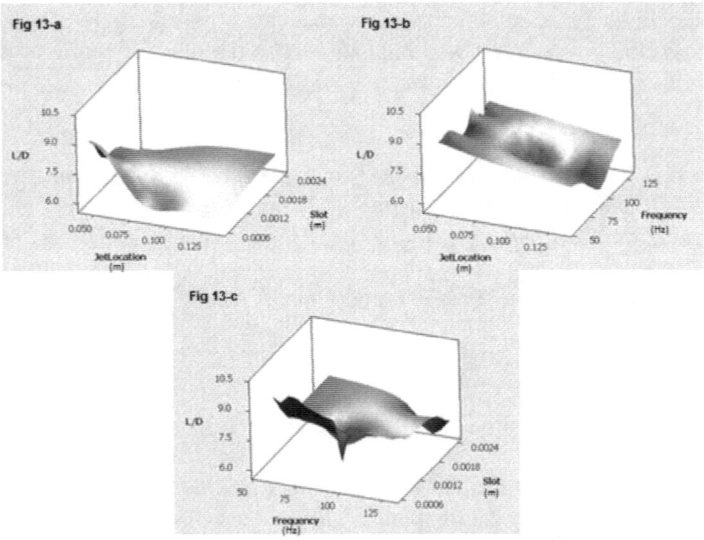

Figure 13. Response surface in 16 degrees for three optimization parameters that vary in exact range

Table 4. Aerodynamic coefficients in the angle of attack equal to 16 °

AOA = 16°	C_L	C_D	L/D
Airfoil without synthetic jet	0.8636	0.1466	5.888
Airfoil with synthetic jet	1.0118	0.1036	9.826
Difference (%)	+17.16	-29.33	+66.88

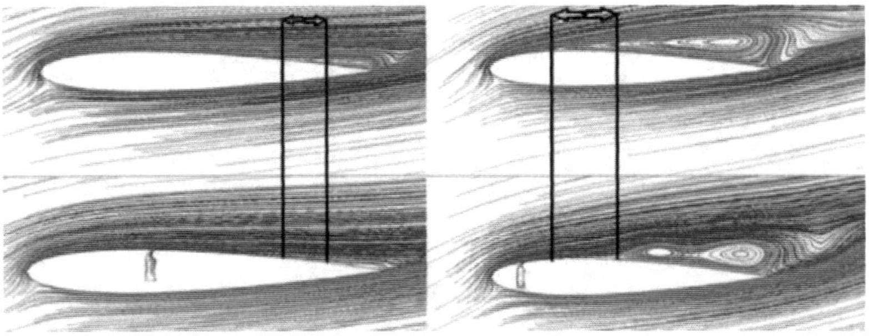

Figure 14. Effect of the synthetic jet actuator for delaying separation in optimal condition, at 13° (left) and 16° (right)

5. Conclusions

The simulation of aerodynamics active control using synthetic jet actuator was performed, solving unsteady Reynolds-Averaged Navier-Stokes equations for NACA 0015 airfoil at Reynolds number, Re = 896000. The flow solver was coupled with the response surface methodology to optimize the jet performance in order to improve the aerodynamic characteristic of the airfoil. The optimization was performed for two angles of attack, i.e. stall and post stall angles of attack. Aerodynamic performance (L/D) was increased by 29% in stall angle of attack and the stall delayed from 13°. Meanwhile, in the post stall angle of attack, L/D had 66% improvement due to reattachment of the recirculation zone on the suction side of the airfoil. The optimum length of the synthetic jet slot was detected to be decreasing as the angle of attack increases. When the angle of attack increased the optimum synthetic jet location was moving to the leading edge. The location of the separation point moved toward the tailing edge on the upper surface as the synthetic jet actuator was used. The optimum actuation jet frequency and the jet outlet velocity observed to be decreasing as the angle of attack increased from 13° to 16°. Obviously, the synthetic jet was indicated to be the most effective at post-stall angles of attack. It improved L/D significantly and postponed the flow separation.

References

[1] Lopez Mejia O D 2009 *Computational study of a NACA4415 airfoil using synthetic jet control* PhD Thesis University of Texas at Austin
[2] Lopez O, Brzozowski D, Glezer A and Moser R 2009 *19th AIAA Computational Fluid Dynamics Conference*
[3] Patel M, Kolacinski R, Prince T, Ng T and Cain A 2003 *21st AIAA Applied Aerodynamics Conference*
[4] Martin P, Tung C, Chandrasekhara M and Arad E 2003 *21st AIAA Applied Aerodynamics Conference*
[5] Budiyono A, Riyanto B and Park H 2012 *Journal of Aerospace Engineering* **25** 479
[6] Akcayoz E and Tuncer I 2009 *Ankara International Aerospace Conference*
[7] Hamdani H, Baig A and Zahir S 2003 *41st Aerospace Sciences Meeting and Exhibit*
[8] Nassirharand A 2009 *Journal of Aerospace Engineering* **23** 105-10
[9] Duvigneau R and Visonneau M 2006 *Computers & Fluids* **35** 624-38
[10] Yousefi K, Saleh S R and Zahedi P 2013 *International Journal of Engineering* **7** 10-24
[11] Mallison S, Hong G and Reizes J 1998 *13th Australian Fluid Mechanics Conference*
[12] Gilarranz J, Traub L and Rediniotis O 2005 *Journal of Fluids Engineering* **127** 377-87
[13] Gilarranz J, Traub L and Rediniotis O 2005 *Journal of Fluids Engineering* **127** 367-76
[14] Gallas Q, Wang G, Papila M, Sheplak M and Cattafesta L 2003 *41st Aerospace Sciences Meeting and Exhibit*
[15] Smith B L and Glezer A 1998 *Physics of Fluids* **10** 2281-97
[16] Holman R, Utturkar Y, Mittal R, Smith B L and Cattafesta L 2005 *AIAA Journal* **43** 2110-6
[17] Ritchie B, Mujumdar D and Seitzman J 2000 *38th Aerospace Sciences Meeting and Exhibit*
[18] Hayes-McCoy D 2012 *Direct computations of a synthetic jet actuator* PhD Thesis Brunel University London
[19] Gad-el-Hak M 2001 *Journal of Aircraft* **38** 402-18
[20] Gad-el-Hak M 1996 *Applied Mechanics Reviews* **49** 365-79
[21] Duvigneau R and Chandrashekar P 2012 *International Journal for Numerical Methods in Fluids* **69** 1701-14
[22] Tuncer I H and Kaya M 2005 *AIAA Journal* **43** 2329-36
[23] Myers R H, Montgomery D C and Anderson-Cook C M 2009 *Response surface methodology: process and product optimization using designed experiments* (Hoboken: John Wiley & Sons)
[24] Baş D and Boyacı İ H 2007 *Journal of Food Engineering* **78** 836-45
[25] Allmaras S R and Johnson F T 2012 *7th International Conference on Computational Fluid Dynamics*
[26] Spalart P and Allmaras S 1992 *30th Aerospace Sciences Meeting and Exhibit*
[27] You D and Moin P 2007 *Annual Research Briefs* Center for Turbulence Research, Stanford University
[28] Montgomery D C 2008 *Design and analysis of experiments* (Hoboken: John Wiley & Sons)
[29] Mavris D 2006 *Introduction to Design of Experiments and Response Surface Methods* Georgia Institute of Technology

Real-time monitoring system of composite aircraft wings utilizing Fibre Bragg Grating sensor

E Vorathin[1*], Z M Hafizi[1], S A Che Ghani[2] and K S Lim[3]

[1]Advanced Structural Integrity and Vibration Research (ASIVR), Faculty of Mechanical Engineering, Universiti Malaysia Pahang, Malaysia
[2]Human Engineering Group (HEG), Faculty of Mechanical Engineering, Universiti Malaysia Pahang, Malaysia
[3]Photonics Research Centre, Faculty of Science, Universiti Malaya, Malaysia

*vora.91.11@hotmail.com

Abstract. Embedment of Fibre Bragg Grating (FBG) sensor in composite aircraft wings leads to the advancement of structural condition monitoring. The monitored aircraft wings have the capability to give real-time response under critical loading circumstances. The main objective of this paper is to develop a real-time FBG monitoring system for composite aircraft wings to view real-time changes when the structure undergoes some static loadings and dynamic impact. The implementation of matched edge filter FBG interrogation system to convert wavelength variations to strain readings shows that the structure is able to response instantly in real-time when undergoing few loadings and dynamic impact. This smart monitoring system is capable of updating the changes instantly in real-time and shows the weight induced on the composite aircraft wings instantly without any error. It also has a good agreement with acoustic emission (AE) sensor in the dynamic test.

1. Introduction

Studies over the last decades have shown great interest in the use of composite structures such as fibre reinforce polymer (FRP) in engineering applications due to their high strength-to-weight ratio [1, 2]. Aerospace engineering is one of the major and critical engineering applications that implement the use of composite structures. According to [3], deployment of composite structures in Boeing 787 aircraft has increased by 38 percent compared to Boeing 777 aircraft, which utilizes 12 percent of composite structures in its manufacturing. Materials used in manufacturing the Boeing 787 aircraft are shown in Figure 1. Aircraft wings and fuselage employ the most composite materials and they are always the critical areas that prone to deformation and damage. Due to the layer by layer lamination, composite structures are prone to unforeseen damage such as cracks and delamination [4-6] after a certain period of performance. The damage must be monitored prior to further enormous deformation that can affect performance or the cause of catastrophic, which is the sudden damage of the structure.

The monitoring of a structural health condition is known as structural health monitoring (SHM). SHM is a process that involves the observation of dynamic response from an arrangement of sensors periodically over time [2]. The implementation of SHM in composites will lead to the development of smart structure in which the structure can give real time stimulation according to the environmental changes that it experienced. The topic of smart structure is relatively new and is commencing to be qualified as a distinct field of applied since in 1980 [7]. The smart structure can have distinct ways of analysis method where it can be examined as a whole structure or only focused on a specific part of it, which is a single structural element [8].

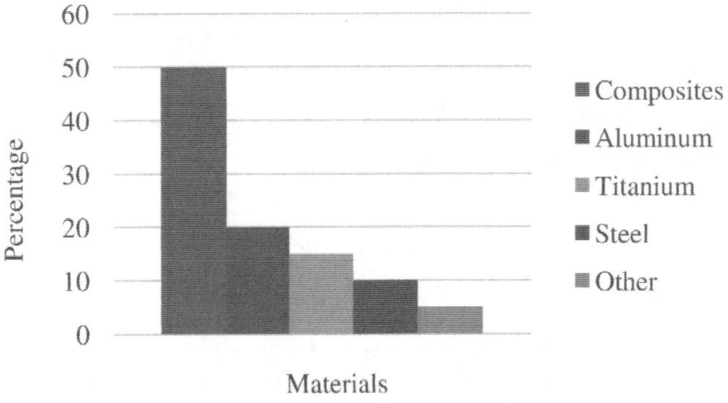

Figure 1. Materials utilization in Boeing 787 manufacturing

Aircraft wings experienced the most deflection during take-off, in which this phenomenon must be monitored before severe damage happened to the wings. Several studies [2, 3, 9-12] have been carried out with the embedment of Fibre Bragg Grating (FBG) sensors in the aircraft's composite structures, especially wings, and it is acknowledged that properties such as small diameter, multiplexing ability and immunity to electromagnetic fields [13-16] are the primary attractive characteristics of FBGs in replacing conventional sensors like piezoelectric and strain gauge for SHM in composites. However, the search for a robust and interactive real-time monitoring system that can alert the end user instantly is still far behind in achieving the idealized concept of smart structure. This drawback has caught the intention of the authors in developing a robust real-time monitoring system to monitor the deflection of the aircraft wings. At the end of this study, for a proof of concept, a fiberglass plate lamination to represent the aircraft wing was fabricated. The FBG instrumentation system and real-time MATLAB processing software are illustrated when the structure undergoes several static loads experimentation.

2. Fibre Bragg Grating (FBG) sensor instrumentation and working principle

The use of FBGs in composites has shown great interest during the late 20thcentury [17]. FBGs sensing principle enquires light source to sense the physical phenomenon.The basic working principle of FBG is by reflecting a specific wavelength of the emitted light source depending on the Bragg properties [18] as shown in Figure 2.

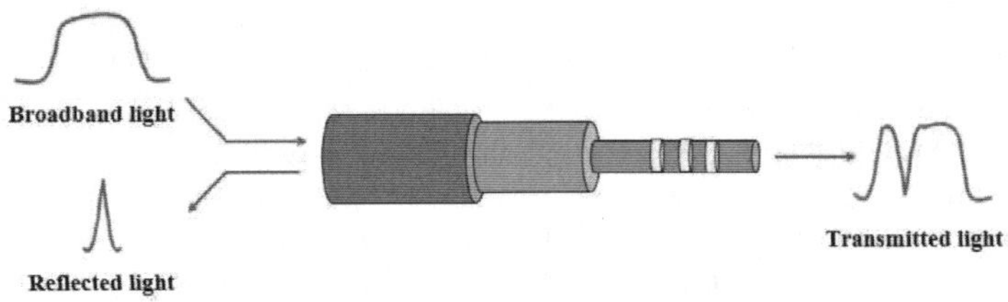

Figure 2. The working principle of FBG sensor

As the optical fibre is in tension, the gap of the Bragg gratings will be wider and vice versa when the optical fibre is in compression. The advantages of FBG as compared to the electrical sensor are shown in Table 1 [19]. The gratings area of FBG is a laser-inscribed region, which reflects the narrow band light corresponding to the Bragg wavelength, λ_B (as calculated by Equation 1) [20] where Λ is the grating period and n is the refractive index of the grating.

$$\lambda_B = 2n\Lambda \quad (1)$$

The refractive index of the grating is fixed when the FBG experienced strain, which results in the linearity between the grating period and strain. Therefore, any change in the measurement will result in the shift of the reflected wavelength. The wavelength will shift to the positive region when expanding and vice versa when contracting. The relative change in the wavelength are given as Equation 2 [21], where C_s is the coefficient of strain ε, C_T is the coefficient of temperature and ΔT is the change in the temperature.

$$\frac{\Delta\lambda_B}{\lambda_B} = C_S\varepsilon + C_T\Delta T \quad (2)$$

Table 1. Comparison between optical fibre sensor and an electrical sensor

Media	Technology	Electrical noise immunity	Measurement speed	Sensor Configuration	Mounting
Electrical	Foil gauges	Low	~ 100 kHz	Single point	Surface mount
	Vibrating wire	Moderate	~ 1 kHz	Single point	Embeddable
Optical	Fibrebragg grating	Complete	< 1 kHz	Multi sensor	Surface mount and embeddable

In order for FBG to function, an interrogation system, which is the signal transmitter and receiver, is needed. Matched edge filter detection interrogation system is the simplest and cheapest in cost as compared to other interrogation system [22]. The configuration of the matched edge filter interrogation system is shown in Figure 3 below.

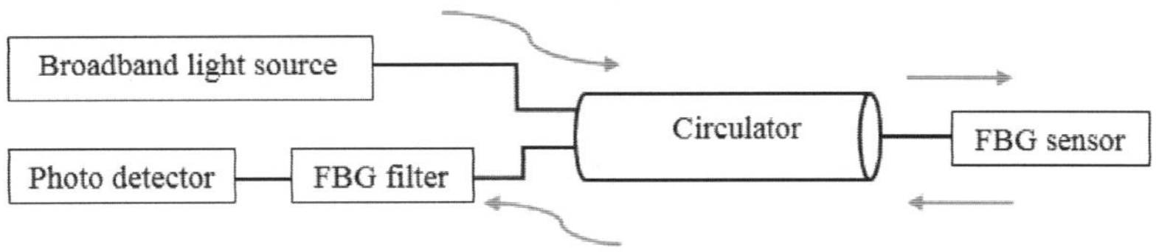

Figure 3. Matched edge filter system configuration

The system is started off with broadband light source emitted light across the optical circulator and into the FBG sensor. Gratings on the FBG will reflect the light back to the optical circular, which is functioning to circulate and convey the light passes through the FBG filter and into the photo detector. The photo detector or Photodiode (PD) functions to convert the light source into an electrical signal to be processed by the expert. The shaded area is the light intensity identified by the photo detector when the FBG sensor undergoes tension or compression [22].

3. Experimentation

3.1. Sample fabrication

A ten-layer woven fibreglass composite plate with hand lay-up lamination was fabricated to represent the aircraft wing structure. The dimension of the composite plate is 40 cm x 40 cm x 0.8 cm and it was cured with the use of glycidyl (GL) epoxy and hardener as resin. A single 1550 nm FBG sensor was positioned in the middle of the ninth and tenth layer of the composite plate. Figure 4(a) shows the woven fibreglass used and Figure 4(b) shows the cured composite plate with embedded FBG sensor.

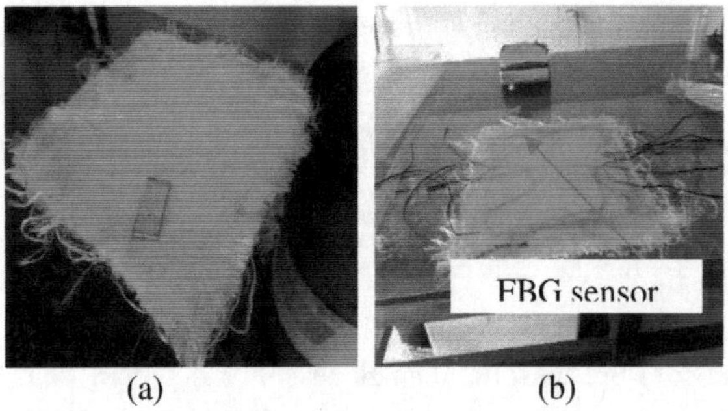

Figure 4. Fabrication process (a) Woven fibreglass, (b) Embedded FBG sensor

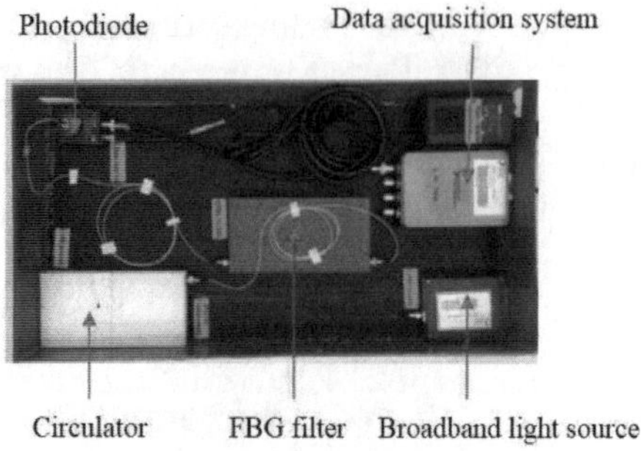

Figure 5. Matched edge filter interrogation configuration

3.2. Experimental set-up

Matched edge filter interrogation system, as shown in Figure 5, was set up for the experimentation. Broadband light source of power 20mW and 99.9% reflectivity of both FBG sensor and filter, and 3-port optical circular was used in the interrogation system. PDA10CS-EC photodiode manufactured by THORLABS was used for the conversion of light wavelength to an electrical voltage signal. NI 9234 data acquisition system from NATIONAL INSTRUMENTS was connected to PD for data acquisition. Figure 6 shows the set-up of the experimentation.

3.3. Experimental procedure

For the static loading test, a weight support stand as shown in Figure 6(a) was placed on top of the composite plate so that the distribution of loads is equal and the same at all time. 10 N and 20 N of loads as in Figure 6(b) were applied on the composite plate. The composite plate was first subjected to 10 N load, and the voltage reading and weight measurement reading were recorded and the condition of the composite plate was observed. The experiment was repeated for a load of 20 N, 30 N, 40 N and 50 N. MATLAB software was used to create graphical user interface (GUI) as virtual instrumentation tool to display the real-time deflection of the composite plate as it undergoes static point loading. Mesh grid of -10 to 10 in x- and y- axis, and 0 to 100 in z-axis artificial sample elements with four fixed locations was programmed and the prediction of strain distribution was based on the voltage reading from the FBG.

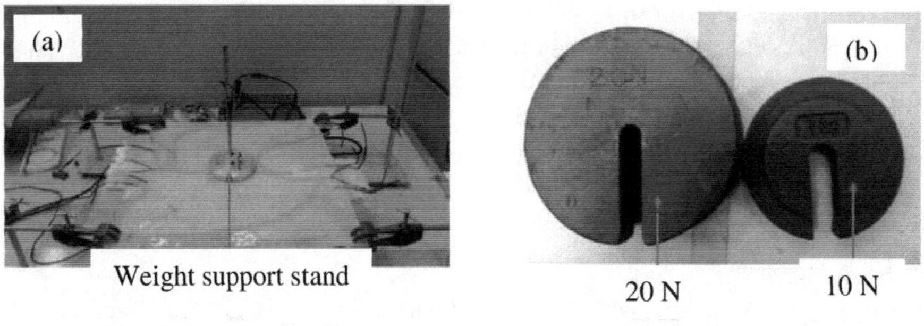

Figure 6. Experimental set-up: (a) Weight support stand, (b) Loads

For the dynamic impact test, an AE sensor was attached to the surface of the composite plate at the same position as the FBG sensor as shown in Figure 7. Three times of impact were induced with an impact hammer at a distance of 15 cm away from both the sensors' placement. The composite plate was clamped at four side corners for a better performance and away from surrounding perturbation.

MATLAB GUI with sampling rates of 1000 Hz for both FBG and AE sensors was programmed to capture the raw impact signal simultaneously. These raw signals were then transformed using Fast Fourier Transform (FFT) to the frequency domain and acquitted at frequency length of 500 Hz. The real-time natural frequency from both sensors was then captured simultaneously and compared with theoretical prediction using ABAQUS.

In Abaqus' FEA analysis, four points of the plate were fixed as the boundary conditions and the maximum frequency to be obtained was set at 500 Hz. Figure 8(a) shows the four fixed points as the boundary conditions of the composite plate and Figure 8(b) shows the meshed composite plate. Table 2 tabulates the mechanical properties of the material [23].

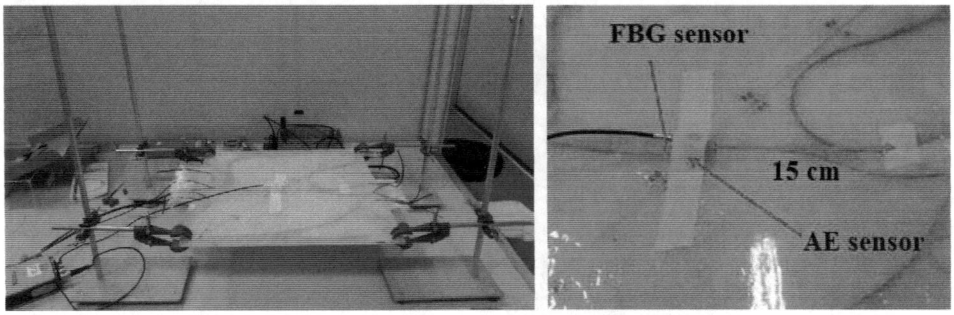

Figure 7. Dynamic impact test experimental setup

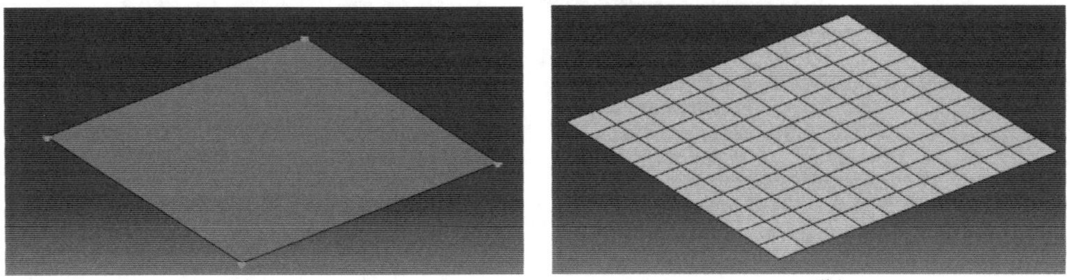

(a) (b)

Figure 8. Abaqus analysis: (a) Fixed points of the boundary conditions, (b) Meshed composite plate

Table 2. Mechanical properties of the material

$E_{xx} = E_{yy}$ [Mpa]	E_{zz}[Mpa]	$V_{yx} = V_{xy}$	$V_{yz} = V_{xz}$	G_{xy} [Mpa]	$G_{yz} = G_{zx}$ [Mpa]	ρ [kg/m^3]
60500	62356	0.075	0.4693	3900	1700	1100

4. Results and discussion

Table 3 shows the summarization of the static loading results. From the results of the experimentation, the composite plate condition shows that it is in a flat condition that implies it does not experience any strain when no load is applied. This condition can be verified from the weight measured reading where it showed 0 N. The structure condition started to deflect slightly when the 10 N load was applied to the composite.

Deflection of the composite plate sank slightly bigger and the voltage increased to 3.7 V when the load was increased to 20 N. The colour at the middle of the composite plate condition turned blue to indicate that strain experienced the most at the middle part. The maximum that can be measured by the thin composite plate is 50 N with a voltage of 4 V. The composite plate detects the load applied without any error. Linearity between the increase of weight and voltage are shown in Figure 9. The conversion of voltage to Newton was by calibration with the substitution of linearity equation obtained from the linearity graph in Figure 9.

Table 3. Summary of static loadings results

Load (N)	Structure condition	Voltage (V)	Weight measured(N)
0		3.5	0
10	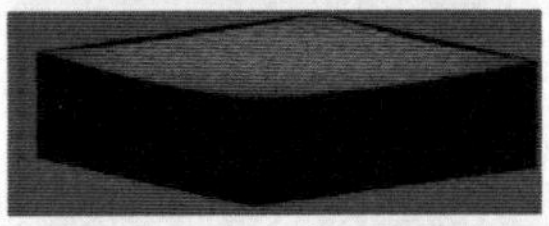	3.6	10
20	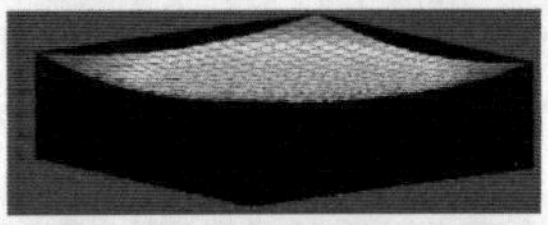	3.7	20
30	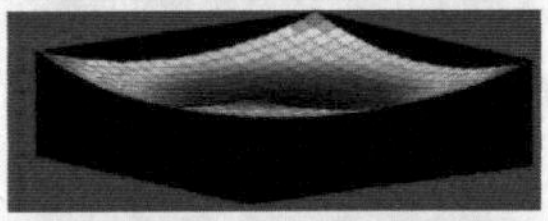	3.8	30
40		3.9	40
50		4.0	50

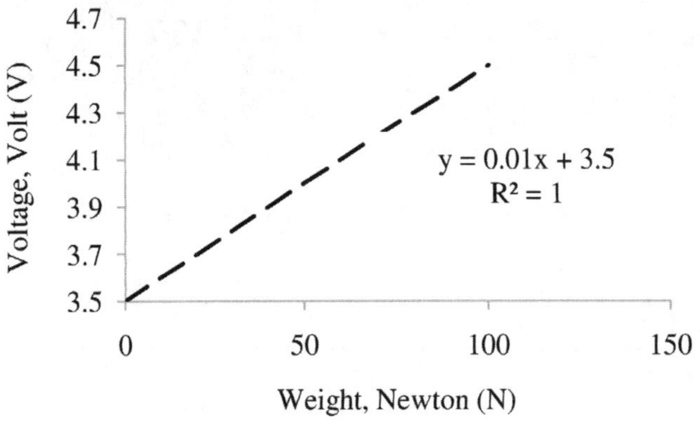

Figure 9. Voltage vs weight linearity graph

Figure 10 shows the impact seismic raw signal of FBG and AE sensor while Figure 11 shows one out of three trials comparison between FBG and AE sensor natural frequency signal. Table 4 tabulates the summarization of natural frequency results from both sensors while Table 5 shows the comparison of average natural frequency from both the sensors with theoretical Abaqus analysis.

ABAQUS FEA has predicted 4 modes of vibration for frequency below 500 Hz. Mode shape 1 (99.5 Hz) and mode shape 4 (373.81 Hz) indicate the dominance of flexural wave propagation. These wave propagation phenomena can be further verified from the high amplitude of natural frequency picked up by both the FBG (89.2 Hz, 401.3 Hz) and AE (91.2 Hz, 402 Hz) sensors. Both sensors are sensitive to flexural type wave propagation in which this type of wave induced a natural axial strain on the sensors. The FBG sensor also capable to captured slight twisting mode of wave propagation, which are mode 2 and mode 3, with a similar predicted natural frequency value of 181.02 Hz. However, the

IOP Conf. Series: Materials Science and Engineering **152** (2016) 012024 doi:10.1088/1757-899X/152/1/012024

amplitude was relatively small compared to flexural mode wave propagation. This signifies that the embedded FBG sensor is capable to capture all the important natural frequencies of various mode wave propagation. In Figure 11, the natural frequency sharp peak stood out from the rest of the noise signal.

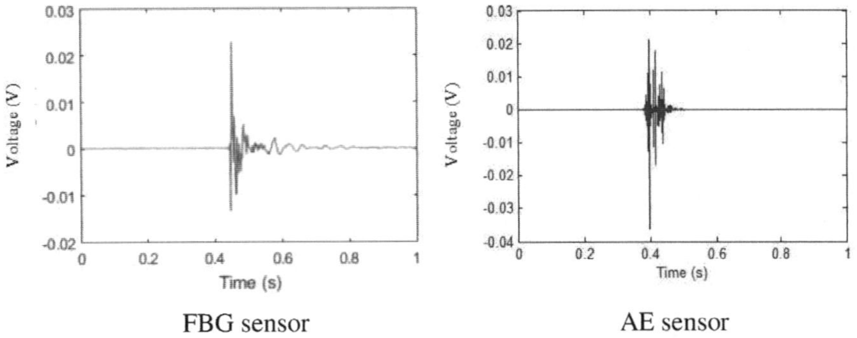

Figure 10. Comparison of impact signal between FBG and AE sensor

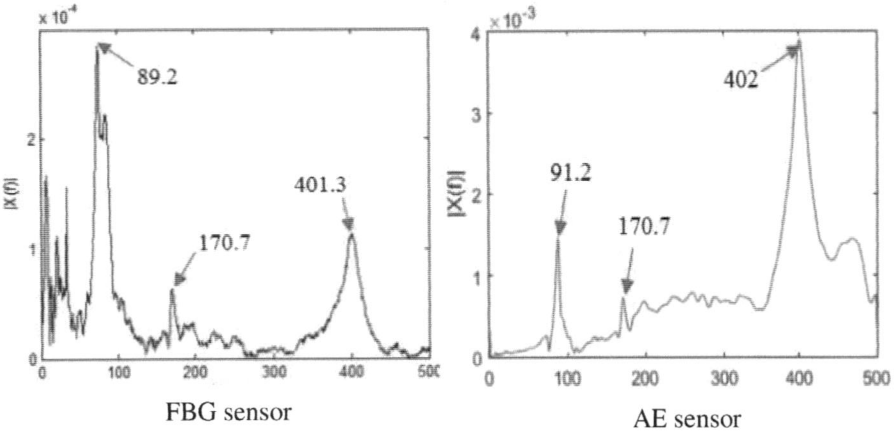

Figure 11. Comparison of natural frequency between FBG and AE sensor

Table 4. Summary of natural frequency results

Mode	Natural frequency (Hz)							
	FBG				AE			
	Trial 1	Trial 2	Trial 3	Average	Trial 1	Trial 2	Trial 3	Average
1	90.8	89.1	89.2	89.7	88.7	90	91.2	89.87
2 &3	172.7	172.7	170.7	172	198.7	172	170.7	180.47
4	404	400.7	401.3	402	401.3	398.7	402	400.7

On the other hand, the signals acquired by FBG sensor and AE sensor have revealed a very good agreement. Both sensors able to pick up all modes predicted by FEA, in which the highest percentage of natural frequency variations between both the sensors are only 4.69% for twisting mode wave propagation and 0.32% for flexural wave propagation. However, for FBG sensor, several spectra can be seen at the lower frequency (below mode 1), which is not consistent with any natural frequency content either from ABAQUS simulation or AE sensor. These unwanted spectra are simply recognized as noise signal response due to utility or power line frequency. Proper filtering can be done to screen out the undesirable signals.

Meanwhile, AE sensor shows a smoother signal compare to FBG sensor. AE sensor is a direct type of sensor where the strain signals are directly converted to electrical signal without any use of external components such as photo detector that can result in the formation of noise. However, FBG sensor has shown its superiority in term of the ability to be embedded in composite structures. All in all, from the dynamic impact results, a remark can be drawn that the embedded FBG sensor has a great potential in replacing the conventional AE sensor for SHM in aircraft composite wings.

Table 5. Summarization of dynamic impact test results

Mode	ABAQUS Mode Shape	Average natural frequency (Hz)			
		Abaqus	FBG	AE	Error (%)
1	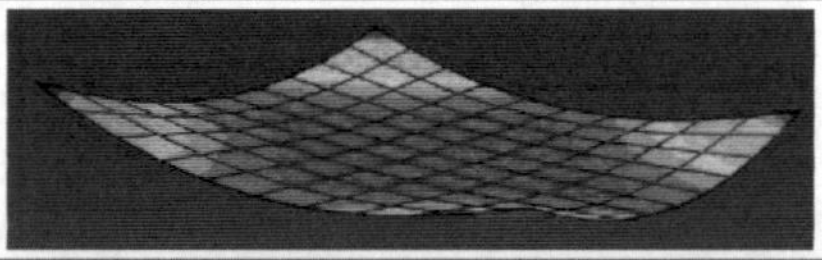	99.5	89.7	89.87	0.19
2	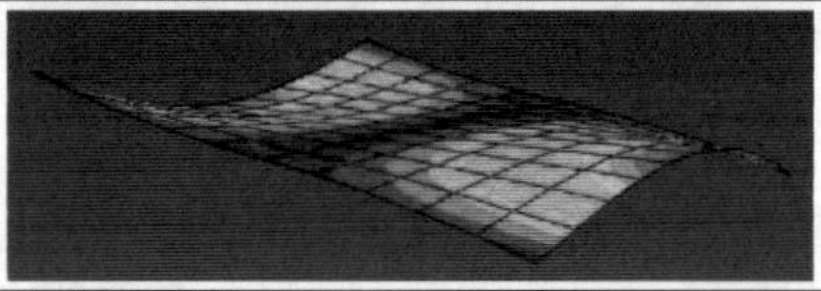	181.02	172	180.47	4.69
3	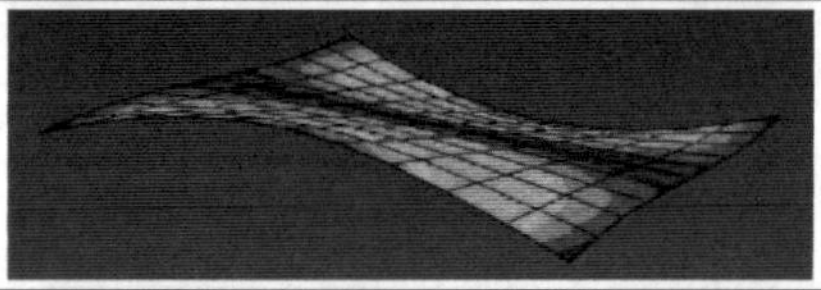	181.02	172	180.47	4.69
4	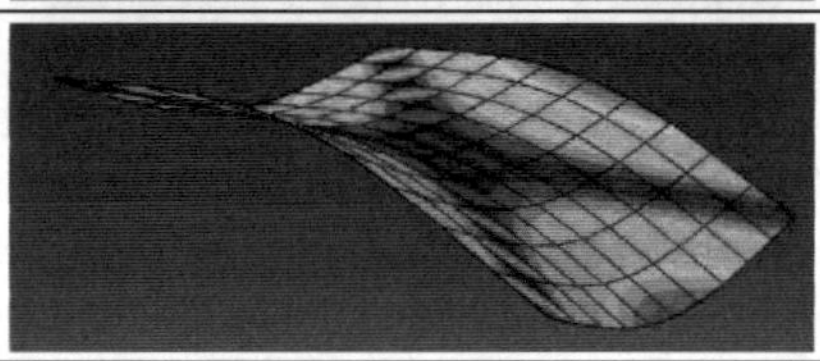	373.81	402	400.7	0.32

5. Conclusion

The study has successfully presented the application of real-time structural health monitoring system of composite aircraft wing with the embedment of FBG sensor and the use of MATLAB GUI in which the structure is capable of determining the load induced instantly in real-time without any error. The natural frequency obtained from the FBG sensor has significant similarity with AE sensor. All in all, the main objective of this paper to view real-time changes when the structure undergoes several static loadings and dynamic impact is achieved.

Acknowledgements

The authors would like to thank the Faculty of Mechanical Engineering, University Malaysia Pahang for providing the laboratory facilities and financial support. We would also like to thank the Photonics Research Centre University Malaya (PRCUM) for their support in fabricating the FBGs. Finally, special thanks to Universiti Malaysia Pahang for providing the financial assistance under Project No. RDU1403155.

References

[1] Kahandawa G C, Epaarachchi J A, Wang H, Followell D and Birt P 2013 *Sensors and Actuators A: Physical* **194** 1-7
[2] Kahandawa G C, Epaarachchi J, Wang H and Lau K T 2012 *Photonic Sensors* **2** 203-14

IOP Conf. Series: Materials Science and Engineering **152** (2016) 012024 doi:10.1088/1757-899X/152/1/012024

[3] Di Sante R 2015 *Sensors* **15** 18666
[4] Chandrashekhar M and Ganguli R 2016 *Mechanical Systems and Signal Processing* **75** 75-93
[5] Liu P F, Liao B B, Jia L Y and Peng X Q 2016 *Composite Structures* **149** 408-22
[6] de Moura M F S F and Gonçalves J P M 2004 *Composites Science and Technology* **64** 1021-7
[7] Anderson E H and Sater J M 2007 *The 14th International Symposium on: Smart Structures and Materials & Nondestructive Evaluation and Health Monitoring* pp 652702--13
[8] Gilewski W and Al Sabouni-Zawadzka A 2015 *Archives of Civil and Mechanical Engineering* **15** 469-78
[9] Guo H, Xiao G, Mrad N and Yao J 2011 *Sensors* **11** 3687
[10] Panopoulou A, Loutas T, Roulias D, Fransen S and Kostopoulos V 2011 *Acta Astronautica* **69** 445-57
[11] Mckenzie I and Karafolas N 2005 *Fiber Optic Sensing in Space Structures: The Experience of the European Space Agency*
[12] Lamberti A, Chiesura G, Luyckx G, Degrieck J, Kaufmann M and Vanlanduit S 2015 *Sensors* **15** 27174
[13] Pereira G, Frias C, Faria H, Frazão O and Marques A T 2013 *Polymer Testing* **32** 99-105
[14] Ling H-y, Lau K-t, Cheng L and Jin W 2006 *Measurement* **39** 328-34
[15] Dai Y, Li P, Liu Y, Asundi A and Leng J 2014 *Optics and Lasers in Engineering* **59** 19-24
[16] Zou H, Liang D and Zeng J 2012 *Optics and Lasers in Engineering* **50** 199-203
[17] Hafizi Z M, Epaarachchi J and Lau K T 2015 *Measurement* **61** 51-7
[18] Yang N 2011 *The Journal of Military Electronics & Computing*
[19] Instruments N 2016 *FBG Optical Sensing: A New Alternative for Challenging Strain Measurements*
[20] Zohari M H, Yusof M F, Siregar J P and Sing L K 2015 *J. Electrical Systems "Special issue AMPE2015"*
[21] Zohari M H B 2014 The Applications of Near Infra-Red Fibre Bragg Grating Sensors for Wave Propagation Based Structural Health Monitoring of Thin Laminated Composite Plates. In: *Faculty of Health, Engineering and Sciences*: University of Southern Queensland) p 183
[22] Tsuda H 2006 *Composites Science and Technology* **66** 676-83
[23] Esposito L, Pucillo G P, Penta F and Rosiello V 2016 *Procedia Structural Integrity* **2** 1870-7

Relative positioning-based system with tau control for collision avoidance in swarming application

M R Bahiki, N N A Talib and S Azrad*

Department of Aerospace Engineering, Universiti Putra Malaysia, Malaysia

*syaril@upm.edu.my

Abstract. In this paper, a relative positioning system by fusing infrared and ultrasonic range sensors data is employed to provide a more reliable relative distance data between quadcopters to achieve close proximity formation flight. This is due to lack of accuracy of positioning data from GPS due to its error of two to five meters. A leader-follower formation control strategy is used to control the distance between the quadcopters by applying data from the relative positioning system. An experiment to test the capability of the proposed strategy was done in the test platform environment equipped with Optitrack motion capture camera. Tau control was implemented as a braking system for the follower to avoid aggressive maneuvers that will make the quadcopters having high pitch along the formation control, which will affect the range detection of sensors. It has been proven from the results that close proximity formation flight is able to be achieved.

1. Introduction

The formation control or cooperative work of multiple unmanned vehicles has been proven to be more effective in terms of time consumption and overall performance. The advantages of cooperative work in unmanned vehicles including ground, aerial, space, marine and also underwater robots are of great interest to modern industries and societies. Beyond military applications, formation flights have been used widely in civilian sectors for field control and supervision, coastal surveillance and rescue, and exploration of new areas for mapping purposes. Swarm flight of unmanned aerial vehicles (UAVs) is being studied extensively by researchers as it gives a lot of advantages to not only the military forces and government agencies, but also civilian businesses and private individuals. It is promising for many applications, including search and rescue tasks for natural disaster, mobile surveillance and monitoring tasks. A high precision control of swarm flight has been made by Kushleyev *et al.* [1], which had 20 units of drones in flight. They used VICON motion capture that has a high precision positioning and a computer as a central control for each multi-copters. Although its performance is very high, it is not applicable for outdoor applications because of its dependencies to high precision camera and central computer. Meanwhile, a 10-autonomous flying robot was successfully developed by Vásárhelyi *et al.* [2] as the world's first outdoor GPS-based swarm. However, the closest achieved distance between the UAVs was around 6 to 10 m. In addition, it was only be able to fly in the velocity range of 0 to 4 m/s. Not many works are known on the capability to perform outdoor swarm flight with close proximity. This is due to limitation of the GPS-based positioning data with error of two to five meters. Therefore, a tighter swarm flight requires a better positioning accuracy to enable the sufficient 'breaking distance' between units, which is critical in collision avoidance.

There is a lack of on-board sensors technology available in the market that can provide the relative positioning between UAVs, both indoors and outdoors. In previous research by Roberts *et al.* [3], they

IOP Conf. Series: Materials Science and Engineering **152** (2016) 012025 doi:10.1088/1757-899X/152/1/012025

were able to design such sensor for indoor application. Their 3-D relative positioning sensor design is consisted of infrared emitters and receivers that are positioned in a spherical ring. However, this kind of module is only capable to operate indoor because of its limitation of high sensitivity to sunlight. Many significant works in relative position-based collision avoidance have been carried out for both individual UAVs and multi-rotor platforms. For collision avoidance, Sobers *et al.* [4] developed a quadrotor that is equipped with infrared sensors for indoor mapping and localization. Chee and Zhong [5] developed a quadrotor UAV with four infrared sensors that is capable of autonomous navigation and avoiding obstacles along the trajectory without any pilot inputs in outdoor environment. Becker *et al.* [6] presented development of active control system for quadrotor UAV to avoid collisions during the flight using four ultrasonic sensors for detecting obstacles. Gageik *et al.* [7] presented a simple approach for obstacle detection and collision avoidance of an autonomous flying quadrotor using 12 low-cost ultrasonic sensors and simple data fusion of those sensors for indoor applications.

In this work, a strategy of leader-follower formation control is proposed that integrates the relative positioning system with the Tau control by Kendoul [15]. The relative positioning system is equipped with low cost ultrasound and infrared sensors with the range data that are fused together to optimize the collected range data. This relative positioning system using sensor fusion was done in the previous work and the results were convincing enough to be implemented for close formation flight application [8]. The advantage of using this relative positioning system is the capability of the formation flight to operate in close proximity at both indoor and outdoor environment. The idea of the relative positioning system is to assist the quadcopters to have relative position data with a higher accuracy than GPS. In addition, the Tau control acts as a braking system for the follower to avoid any aggressive manoeuvres caused by limited detection range of sensors due to high-pitch-rate manoeuvre and oscillatory motion in the formation flight. The quadrotor used in this study is as shown in Figure 1.

2. Leader-follower formation control

Formation control involves a broad field of robotics that addresses the challenges in coordinating a system of mobile robots. This term applies to aerial and ground vehicles as well as to robots of varying sizes. Previous researches regarding formation flight have applied several different control techniques and methodologies in order to achieve the goal of maintaining a formation of robots while in motion. The incorporation of a "leader" into the formation is a common approach in which the motion of the lead robot is planned. In addition, the robots can have minimal sensing capabilities and communicate to at least one other robot.

The first element of the whole formation flight control scheme is a path planner as seen in Figure 2. This is where a path is constructed that represents a compromise between the risks involved in by passing obstacles and the power consumes related to the path length. Mission planning related is also implemented at this level. The second element shown in the same figure is the trajectory generator. It is used to generate a practical trajectory for the quadcopters in computing the corresponding control inputs for the formation leader while taking part for its dynamics. The next element is the formation controller of multiple quadcopters. The purpose of this part is to maintain the formation configuration during flight. The formation configuration is defined as a trajectory generator to the followers where certain desired positions and velocities are computed.

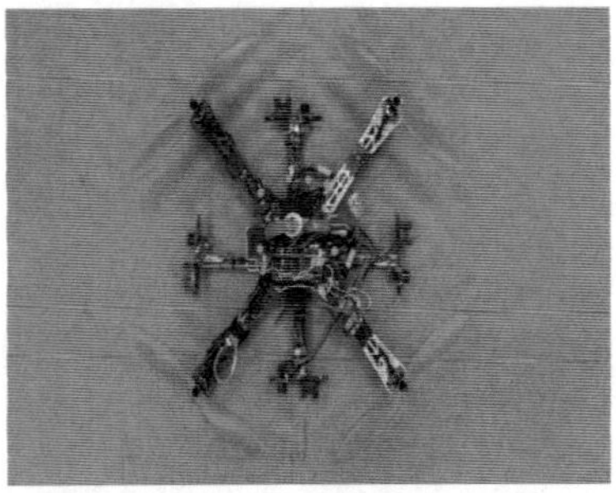

Figure 1. Quadcopter with relative positioning system installed

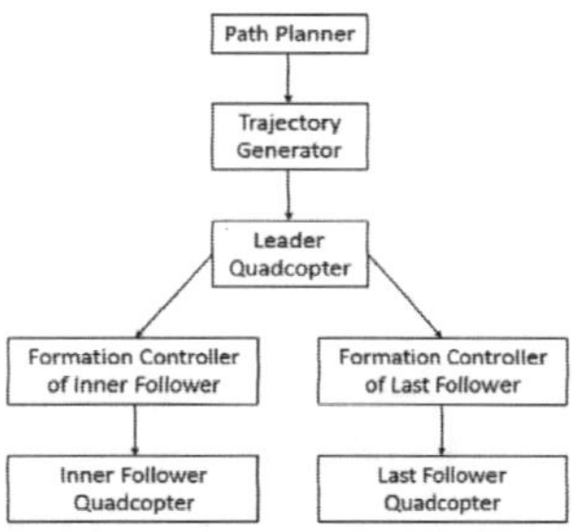

Figure 2. Leader-follower formation control scheme for multiple quadcopters

IOP Conf. Series: Materials Science and Engineering **152** (2016) 012025 doi:10.1088/1757-899X/152/1/012025

2.1. Tau theory

TauPilot by Kendoul [15] is a bio-inspired autopilot that is used to guide and control the movement of various crucial manoeuvres in unmanned aircraft system (UAS) such as docking, braking and landing. It is based on the tau theory that have been proposed by David Lee in 1976 who hypothesized that the action of a car might be visually controlled when braking at the obstacle. The approach of automated UAS is usually based on position-based guidance, navigation and control (GNC) system that requires explicit computation of position and velocity, which then limits the sensor to the position sensors like Global Positioning System (GPS). On the other hand, TauPilot can be used with a much wider range of sensors including low-cost and lightweight cameras, thus can be used in GPS-denied environment and on board of small platform with limited payload.

Existing bio-inspired autopilot is basically based on visual cues. The difference is in terms of the way the cues are used in guiding and controlling the movement, and also the type of visual cues that is used. TauPilot introduces the method that imitates the movement in humans and animals with much sophisticated guidance and control strategy, yet the strategy is simple and allows complex manoeuvres in wide range of application. TauPilot is a combination of tau information and also tau strategy in achieving various manoeuvres required for UAS to arrive at the right place in the right way within the desired time. It consists of the Tau-Navigation System and the Tau-Guidance System as described in the following sections.

2.1.1. Tau-navigation system. The system is responsible to compute taus of different gaps to achieve the desired task by identifying the gaps that need to be closed. This system generally requires the data generated from the sensing system.

2.1.2. Tau-guidance system. This system is responsible to generate the desired or reference taus, which result in task achievement if successfully tracked by the system. In this study, the focus is placed on the constant tau-dot (tauD) strategy that implemented tau law. The law posits that stopping or braking with zero velocity at the contact can be controlled by monitoring and adjusting the values of tau-dot to remain constant and positive. The law is given by Equation 1.

$$\tau_{ref}(t) = kt + \tau_0, \qquad \tau_0 = \frac{x(0)}{\dot{x}(0)} < 0 \tag{1}$$

where *t=0* and τ_0 are time and tau at initiated point, respectively. It have the advantages of avoiding the computation of $\dot{\tau}(t)$, which would give a significant noise due to the derivation of already noisy signal $\tau(t)$, and giving more stable control between control input and controlled variable $\tau(t)$ as it tracks $\tau_{ref}(t)$ instead of $\dot{\tau}_{ref}(t)$. The value of *k* has to be kept in the range of 0 to 0.5 to ensure that all trajectories (distance, velocity and acceleration) converge to zero (origin) as time goes to infinity.

2.1.3. Tau control system

This system is designed to compute and adjust the control inputs, u that are used to force actual tau in tracking the reference tau (tau-ref). The non-linear ratio control system law introduced the method of tracking error by taking the ratio between reference tau and tau. The law therefore can be defined as in Equation 2.

$$U = \begin{cases} k_p\left[1 - \frac{\tau_{ref}(t)}{\tau_i(t)}\right] & \text{if } \tau(t) \neq 0 \\ 0 & \text{otherwise} \end{cases} \tag{2}$$

This tau-controller copes well when the system operates near its singularity point when the input, u is zero, thus give a good tracking performance. The tau-controller, by considering one direction only (x-axis), can be formulated by Equation 3.

$$\mu_x = \sigma_x\left[k_p\left(1 - \frac{\tau_{ref}(t)}{\tau_i(t)}\right) + k_i\int\left(1 - \frac{\tau_{ref}(t)}{\tau_i(t)}\right)\right] \tag{3}$$

where μ is the intermediary control input and σ is the saturation function. The approximation control input is therefore given by Equation 4.

$$\theta_{ref} = \sigma_\theta\left[\mu_x \cos(\beta' - \varphi) - \mu_y \sin(\beta' - \varphi)\right] \tag{4}$$

where φ is the yaw angle and β' is the heading angle.

3. Relative positioning system

Previous works by other researchers on relative positioning are depending solely on a single type of sensors object detection. However, to have a more reliable data that can be operated both indoor and outdoor, both IR and US data are fused together using Linear Kalman Filter (LKF) based sensor fusion [10] to eliminate noises and errors from each individual sensors limitation, and increase the accuracy of the relative position (x_r, y_r) between quadcopters. The LKF provides estimation for the state vector, $\hat{x}_k$ using set of observed data,Z_k for every discrete time, k. The LKF is implemented in two steps, namely, prediction and update steps. The prediction equations determine priori estimate of the state and error covariance given Equation 5 and Equation 6.

$$\hat{x}^m_{k|k-1} = F_{k-1}\hat{x}^m_{k|k-1} + G_{k-1}u_{k-1} \tag{5}$$

$$P^m_{k|k-1} = F_{k-1}P^m_{k|k-1}F^T_{k-1} + Q_{k-1} \tag{6}$$

$$K^m_k = P^m_{k|k-1}H^T_k\left(H_k P_{k|k-1} H^T_k + R^m_k\right)^{-1} \tag{7}$$

$$\hat{x}^m_k = \hat{x}^m_{k|k-1} + K^m_k\left(Z^k_m - H_k\hat{x}_{k|k-1}\right) \tag{8}$$

$$P^m_k = (1 - K^m_k H_k)P_{k|k-1} \tag{9}$$

In the update step, the priori estimates are updated using the calculated gain matrix, K_k (in Equation 7) and *posteriori* estimates of the state and error covariance given by Equation 10 and Equation 11.

$$\hat{p}^f = \hat{p}^1 - \hat{P}^1\left(\hat{P}^1 + \hat{P}^2\right)^{-1}\left(\hat{p}^{1^T}\right) \tag{10}$$

$$\hat{x}^f = \hat{x}^1 + \hat{P}^1\left(\hat{P}^1 + \hat{P}^2\right)^{-1}(\hat{x}^2 - \hat{x}^2) \tag{11}$$

The above equations apply to both IR and US individual sensors for the estimation of range based on sensor measurement data. The *posteriori* estimates of individual sensors are fused together to obtain combined final estimates of state, $\hat{x}^f$ and error covariance, $\hat{p}^f$.

Four primary directions of the quadcopters are mounted with a pair of IR and US sensors between the quadcopter arms shown in Figure 3. In previous work [10], a relative position-based collision for swarm application using cascaded control loops made up of PI-position and P-velocity controllers was proposed. The idea was to have a formation flight using formation controller with collision avoidance override the formation controller when the distance between quadcopters was too close. However, this would increase oscillation of the follower during formation due to several act of aggressive avoidance. Therefore, in this paper, a close proximity formation flight conducted using formation controller with Tau D control as braking to avoid aggressive act and collision between quadcopters was developed. This is shown in Figure 4.

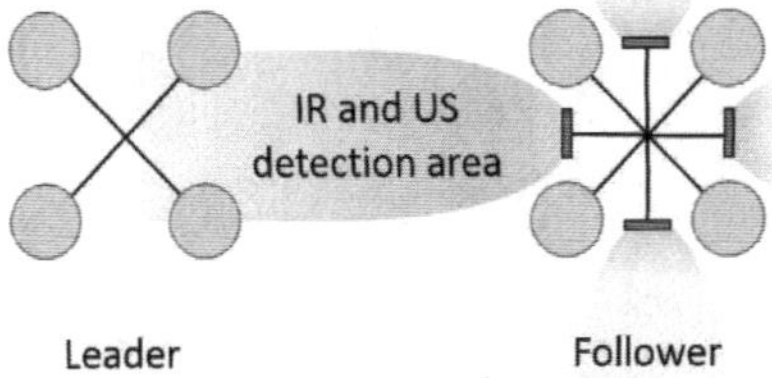

Figure 3. Visualization of IR and US leader detection using relative positioning system

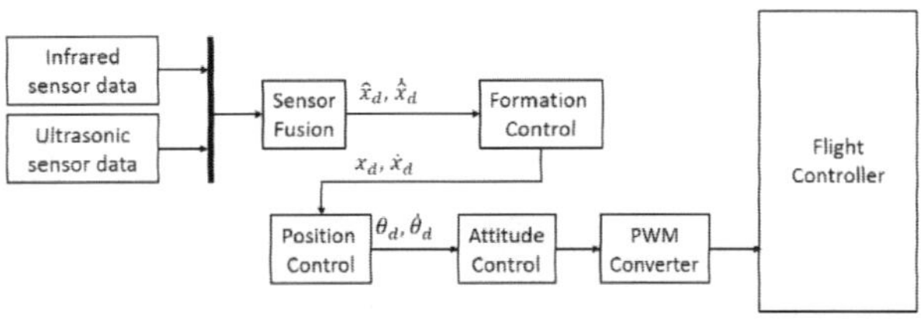

Figure 4. System block

IOP Conf. Series: Materials Science and Engineering **152** (2016) 012025 doi:10.1088/1757-899X/152/1/012025

4. Hardware setup and implementation

The development and testing of this project was done indoor environment that was equipped with the Optitrack motion capture camera for indoor positioning system. Due to slow rate of data acquisition from the flight controller to Arduino Mega, all IMU data was computed using motion capture camera to get the sampling rate of 100Hz. Quadcopter dynamics and stability control algorithm was done in computer on ground. Xbee S2 2.4Ghz was used as communication module to control the quadcopter from ground. The sensors were installed on Arduino on-board of a quadcopter platform. The Sharp GP2Y0A02YK0F infrared sensor and MaxBotix LV-MaxSonar®-EZ0 ultrasonic sensor were selected for close formation forward flight. Xbee module was also been used for sending data to the ground control. The overall setup of the hardware can be seen in Figure 5.

5. Result and discussion

An experiment was carried out in an indoor environment, fixed with Optitrack Motion Capture camera to validate the performance of the follower quadcopter to control distance with the dummy leader. The purpose of the motion capture camera is to make an indoor positioning system as testing environment to replace global positioning system (GPS) since it is easier to simulate and test the algorithm indoor repeatedly. In the experiment, there were two quadcopters fixed with at least three reflective markers onto each of them so that they can get their position and attitude data from the motion capture camera and only one of them equipped with the relative positioning system to act as the follower quadcopter.

Two quadcopters were used in the experiment, which are dummy leader and follower quadcopters. In this experiment, only the follower quadcopter was flying and the dummy leader was handled by a person. Relative position module was embedded on the follower quadcopter as the formation controller is on the follower with the PID gain setting as shown in Table 1.

Table 1. Settings for PID gain used in the system

	Position	**Velocity**
Kp	1.2	0.6
Ki	1.75	1.3
Kd	0.6	0.45

As seen in Figure 6, the dummy leader was attached with a piece of cardboard. This is necessary to help the range sensor to easily detect the follower due to the range sensors having difficulties to detect cluttered surface. The predefined desired distance between the quadcopters to stay in formation was 0.5 meter apart measured from the relative positioning system. The experiment started by hovering the follower in altitude hold flight mode at a point. The dummy leader was held 1 meter away from the follower at the same altitude. The formation controller was automatically started when the relative positioning system on the follower detected the dummy leader in a distance. Approximately around 1 meter distance of forward and backward trajectory for the dummy leader was done in linear motion at y-axis.

Figure 7 shows the results for linear formation flight response and distance between quadcopters in y-direction that were recorded. During the test, the follower was able to follow the leader in formation

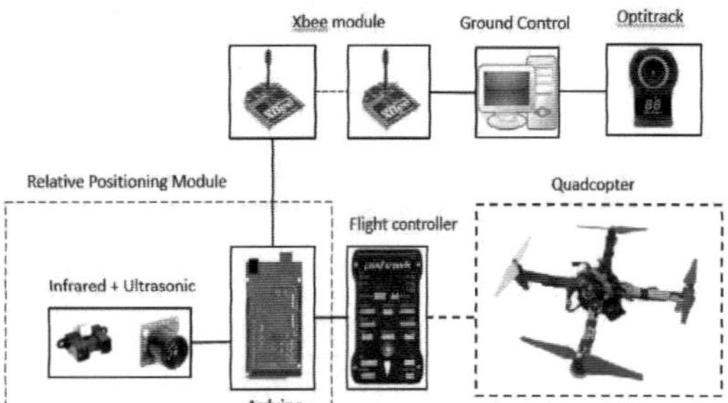

Figure 5. Hardware setup

Figure 6. Leader quadcopter with cardboard attached

as shown in Figure 7(a) and achieved to maintain its distance with the leader. As can be seen in Figure 7(b), the closest distance of the follower with the leader is approximately 0.4 meter measured from the centre of the follower to the surface of the leader recorded. Nevertheless, the pattern showed a good result from the formation controller that having just a small oscillation of controlling distance around 0.5 meter (red line in Figure 7(b)) as defined earlier.

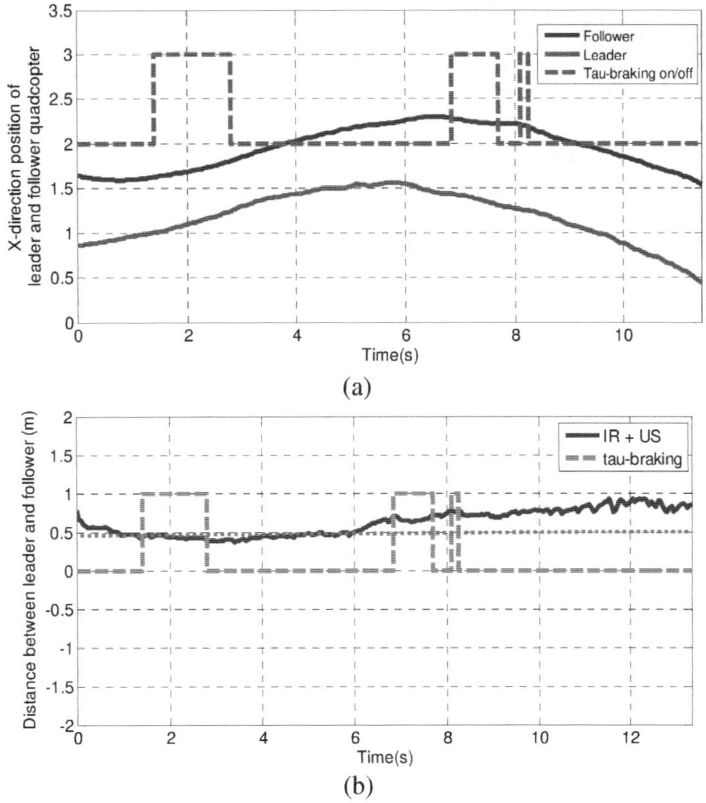

Figure 7. (a) Linear formation flight in y-direction, (b) Distance between quadcopter during formation

6. Conclusion and future work

This paper proposed a formation flight of quadcopters using relative positioning system implemented with fused sensors of IR and US to measure relative position between quadcopters by considering the inaccuracy of GPS by two to five meters for operating close formation flight control. The focus was on a close proximity formation flight that would be difficult to operate in GPS-denied environments. In view of that, the relative positioning system was considered to be embedded into the quadcopters to measure relative position between them by fusing infrared and ultrasound range sensor to gain more reliable data. As relected by the results, the close proximity flight was able to be performed by having relative positioning system with Tau control embedded into the quadcopters without fully relying on GPS in this test. However, due to range sensors that have small angle of detection, the flight can only be operated at lower velocity. This is because, to have higher velocity, the quadcopter needs to have a higher pitch angle and this will affect range detection of the sensors. Future work may also include replacing the range sensor with optical image sensor so that the quadcopter may be able to perform at higher velocity. By using image sensor, the field of view is much larger than conventional sensors and the accuracy of the sensor is higher.

References

[1] Kushleyev A, Mellinger D, Powers C and Kumar V 2013 *Auton. Robots* **35** 287–300
[2] Virágh C, Vásárhelyi G, Tarcai N, Szörényi T, Somorjai G, Nepusz T and Vicsek T 2014 *Bioinspir. Biomim.* **9** 11

[3] Roberts J F, Stirling T, Zufferey J C and Floreano D 2012 *Auton. Robots* **33** 5–20
[4] Sobers D, Chowdhary G and Johnson E 2009 *AIAA Guidance, Navigation, and Control Conference*
[5] Chee K Y and Zhong Z W 2013 *Sensors Actuators, A Phys.* **190** 66–76
[6] Bouabdallah S, Becker M and De Perrot V 2007 *XII Int. Symp. Dyn. Probl. Mech.*
[7] Gageik N, Müller T and Montenegro S 2012 *Microdrones Conference: UAVveek*
[8] Rambabu R, Bahiki M R and Azrad S 2015 *J. Teknol.* **8** 89–93
[9] Raol J R 2009 *Multi-Sensor Data Fusion with MATLAB* (Boca Raton: CRC Press)
[10] Rambabu R, Bahiki M R and Azrad S 2015 *ARPN J. Eng. Appl. Sci.* **10**
[11] Nonami K, Kendoul F, Suzuki S, Wang W and Nakazawa D 2010 *Autonomous Flying Robots* (Tokyo: Springer Japan)
[12] How J P, Bethke B, Frank A, Dale D and Vian J 2008 *Control Syst. IEEE* **28** 51–64
[13] Abas M F, Pebrianti D, Azrad S, Iwakura D, Song Y and Nonami K 2013 *Autonomous Control Systems and Vehicles* **65** 109–132
[14] Chen M 2003 *Formation and Flight Control of Affordable Quad-rotor Unmanned Air Vehicles Design* Masters Thesis The University of British Columbia
[15] Kendoul F and Ahmed B 2012 *IEEE/RSJ International conference on intelligent robots and systems*

IOP Conf. Series: Materials Science and Engineering **152** (2016) 012026 doi:10.1088/1757-899X/152/1/012026

Development of collision avoidance system for useful UAV applications using image sensors with laser transmitter

M K Cheong, M R Bahiki and S Azrad*

Department of Aerospace Engineering, Faculty of Engineering, Universiti Putra Malaysia, Malaysia.

*syaril@upm.edu.my

Abstract. The main goal of this study is to demonstrate the approach of achieving collision avoidance on Quadrotor Unmanned Aerial Vehicle (QUAV) using image sensors with colour-based tracking method. A pair of high definition (HD) stereo cameras were chosen as the stereo vision sensor to obtain depth data from flat object surfaces. Laser transmitter was utilized to project high contrast tracking spot for depth calculation using common triangulation. Stereo vision algorithm was developed to acquire the distance from tracked point to QUAV and the control algorithm was designed to manipulate QUAV's response based on depth calculated. Attitude and position controller were designed using the non-linear model with the help of Optitrack motion tracking system. A number of collision avoidance flight tests were carried out to validate the performance of the stereo vision and control algorithm based on image sensors. In the results, the UAV was able to hover with fairly good accuracy in both static and dynamic collision avoidance for short range collision avoidance. Collision avoidance performance of the UAV was better with obstacle of dull surfaces in comparison to shiny surfaces. The minimum collision avoidance distance achievable was 0.4 m. The approach was suitable to be applied in short range collision avoidance.

1. Introduction

Unmanned aerial vehicles (UAVs) hold good promises for performance in both civilian and military operations. A broad range of meaningful tasks such as aerial surveillance, simultaneous arrival and attack, search and rescue, flight formation, etc. can be accomplished using the application of UAVs. However, for UAVs to perform successfully in their respective missions and operations, collision avoidance system is necessary to be imparted into them. Without collision avoidance system, an UAV during operation would collide into unexpected objects such as wall, flying objects, flying animals, humans and others, resulting in mission failure, destruction of UAV, damage or injury to the collided objects or a combination of all. Collision avoidance for unmanned ground vehicles (UGVs) can be based on multiple choices including infrared sensors [1], image-based sensors [2], sound sensors [3] and laser scanner [4]. Image-based sensor collision avoidance on UAVs is a relatively good strategy compared to others due to the combined functions for aerial surveillance and obstacle avoidance. In other words, UAVs with other type of sensors do not provide the added ability of aerial surveillance.

For the few past years, researches on the topic of stereo-vision based collision avoidance have been carried out extensively. Most of the researches and projects were focused on determining the distance between obstacle and cameras by using the conventional method of stereo correspondence from pixel-to-pixel [5, 6, 7, 8] to obtain the disparity map, which is also known as the depth map. Apart from that, works on determining the depth information using the conventional and improved version of Scale-

IOP Conf. Series: Materials Science and Engineering **152** (2016) 012026 doi:10.1088/1757-899X/152/1/012026

Invariant Feature Transform (SIFT) are also accomplished [9, 1010]. With the aids of image filtering and computation, researchers were able to filter out the object closest to the cameras and determine its depth. However, the conventional method required massive amount of computational power and time [11], thus it is unsuitable to be applied on UAVs where high real-time processing speed is mandatory. Apart from that, high accuracy is also an important criterion that critically determines the collision avoidance performance of UAVs [12].

In this work, a strategy of depth calculation using colour-based tracking instead of the conventional method of pixel-to-pixel correspondence to obtain disparity map is proposed. A laser transmitter is used to project a laser beam onto an obstacle to create a colour-distinct point from the obstacle. The colour-distinct point acts as the reference point for image tracking and depth calculation. The depth value obtained via stereo triangulation is then integrated into the control algorithm based on non-linear model by Kendoul [13] to achieve relative-position collision avoidance. The quadrotor used in this study is as shown in Figure 1.

2. Depth estimation algorithm

2.1. Camera calibration

Camera calibration is one of the most important procedures in image processing for the purpose of obtaining accurate data from image sensors [14]. Two types of camera calibration carried out in this project were single camera calibration and stereo camera calibration, which was done to obtain the intrinsic and extrinsic parameters for the cameras. The cameras were also redefined by calibration with Optitrack motion capture system to improve the accuracy of the depth estimation algorithm.

2.2. Colour-based filtering

Colour-based filtering is used to differentiate the laser beam projection from the obstacle. The colour of each pixel on the images is defined in the HSV colour scheme where each pixel has its own HSV values. HSV is chosen over RGB colour scheme because HSV separates the image intensity from the color information, which is important. By using OpenCV, the pixels containing the laser projection point can be singled out from the rest of the image by adjusting the HSV values. Gaussian blurring method is applied to reduce the noise of the images. Morphological operators, Erosion and Dilation are also used to refine the filtered image and reduce noise. Since the orientation of the laser transmitter was fixed with the camera, the location of the laser projection was constant around a certain region of the image. The original image (640 x 480 pixels) was cropped based this region to limit the area of search to a rectangular box of size (330 x 80 pixels). Figure 2(a) and (b) show a frame of images that was captured during a test flight by left and right camera, respectively. Figure 2(c) and (d) show the cropped image from colour filtering.

Figure 1. Quadcopter with stereo cameras and relative positioning

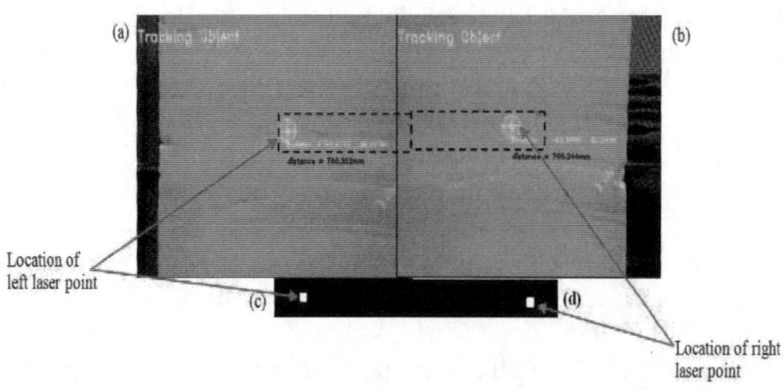

Figure 2. The original (a) left and (b) right images with track indications in 640×480. The (c) left and (d) right cropped threshold images in 330×80

2.3. *Stereo Triangulation*

The algorithm for depth calculation was based on simple Linear Triangulation method that relates the baseline, focal length and image coordinates as shown in Figure 3. This is expressed as Equation 1.

$$Z = \frac{bf}{x_l - x_r} \tag{1}$$

where Z = distance between cameras' baseline to image point, b = length of baseline, f = focal length of cameras, x_l = x coordinate of image projected on left camera and x_r = x coordinate of image projected on right camera.

3. Relative position control algorithm

The collision avoidance of QUAV is achieved mainly by using relative attitude and position control with cascade PID controller.

3.1. *Controller design and transfer function*

In this section, the vision-based controller used for target object tracking is described. In general, it is difficult to design a reliable autopilot for QUAV since the quadrotor dynamics is inherently unstable, highly nonlinear and coupled in inner-axis. The vision system, with the proposed image algorithm strategy, provides the quadrotor position and velocity with respect to the target. Therefore, the vision-based control for object tracking can be formulated as a stabilization problem. A hierarchical nonlinear controller is then proposed, which is based on the nonlinear model of the multi-rotor dynamics. The attitude and position controller by Kendoul [15] uses the non-linear controller. The outer loop of the controller is the position (x_d, y_d) controller, which runs at slow dynamics. However, the orientation (θ_d, ϕ_d) control is the inner loops controller that is running at fast dynamics. The PID structure can be expressed as in Equation 2 and Equation 3.

$$\mu_x = k_{px}(x_d - x) + k_{ix} \int (x_d - x)\, dt + k_{dx}(\dot{x}_d - \dot{x}) \tag{2}$$

$$\mu_y = k_{py}(y_d - y) + k_{iy} \int (y_d - y)\, dt + k_{dy}(\dot{y}_d - \dot{y}) \tag{3}$$

The desired attitudes are then generated as follows:

$$\theta_d = \sin^{-1}\left(\frac{\mu_x \sin\psi - \mu_y \cos\psi}{u}\right) \tag{4}$$

$$\phi_d = \sin^{-1}\left(m\frac{\mu_x \sin\psi - \mu_y \cos\psi}{u}\right) \tag{5}$$

$$u = m\sqrt{{\mu_x}^2 + {\mu_y}^2} \tag{6}$$

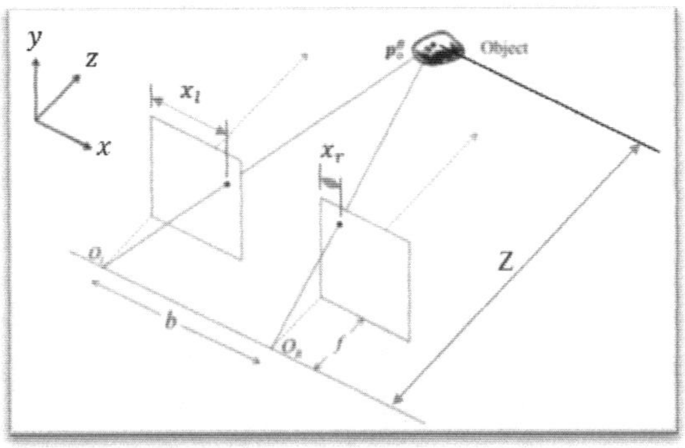

Figure 3. The basis of linear stereo triangulation

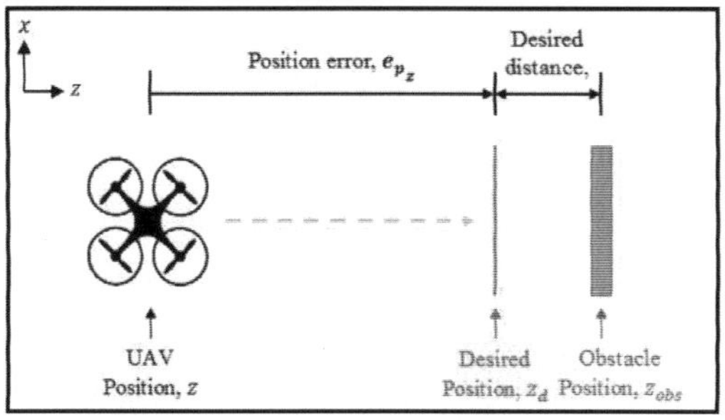

Figure 4. Illustration on linear collision avoidance scenario

Yaw, ψ is also controlled by PID controller. The output thrust of the quadcopter is shown as follow:

$$\tilde{\tau}_\theta = k_{px}(\theta_d - \theta) + k_{ix}\int(\theta_d - \theta)\,dt + k_{dx}(\dot{\theta}_d - \dot{\theta}) \quad (7)$$

$$\tilde{\tau}_\phi = k_{px}(\phi_d - \phi) + k_{ix}\int(\phi_d - \phi)\,dt + k_{dx}(\dot{\phi}_d - \dot{\phi}) \quad (8)$$

A simple control strategy was established to maintain the distance between QUAV and obstacle as shown in Figure 5.

4. Hardware set up and implementation

A pair of HD cameras, Logitech® HD Webcam C310 were calibrated before being mounted on the UAV to serve as image sensors. The purpose of the project was to demonstrate the approach for using image sensors for collision avoidance, hence the project was proceeded with on-ground processing in which the cameras were connected physically with the extended USB cables to ground computer. An Arduino compatible laser transmitter, Keyes® KY-008 was then installed at the centre between the stereo cameras to project a distinct colour spot on the obstacle for the purpose of colour-based image filtering. Evaluation of collision avoidance using image sensors was performed at indoor environment with the assistance of Optitrack motion capture system. The UAV's dynamics and attitude response was controlled based on the UAV's and the obstacle's coordinates provided by the Optitrack indoor positioning system. Overall setup of the hardware is shown in Figure 6.

5. Result and discussion

Two experiments were conducted in an indoor environment assisted by Optitrack Motion Capture system to validate performance of the UAV in static and dynamic collision avoidance. The Optitrack motion capture cameras were chosen as indoor positioning system and testing environment to replace the global positioning system (GPS) because it gives the advantage of easy conduct of repetitive flight simulations. Before conducting the flight experiments, three Optitrack markers were fixed on both the UAV and the obstacle to provide their real-time respective coordinates. The UAV was integrated with the depth estimation algorithm combined with control algorithm to achieve desired collision avoidance using image sensors. Table 1 shows the PID gain values used in the experiment. The results for static and dynamic collision avoidance performance are presented in Figure 7.

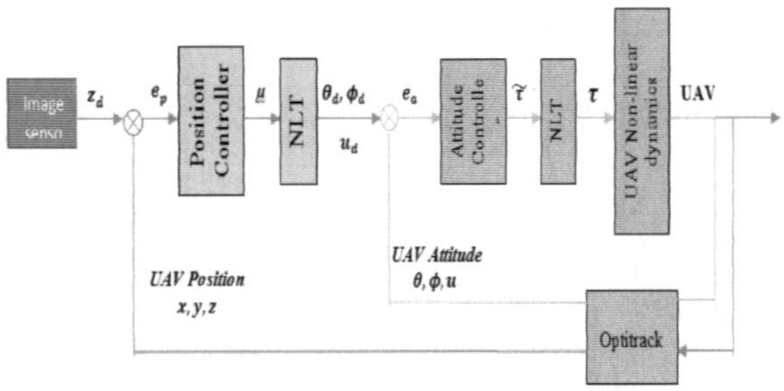

Figure 5. Detailed block diagram for attitude and position

Table 1. PID gain values

Gain	Position	Velocity
Kp	1.2	0.6
Ki	1.75	1.3
Kd	0.6	0.45

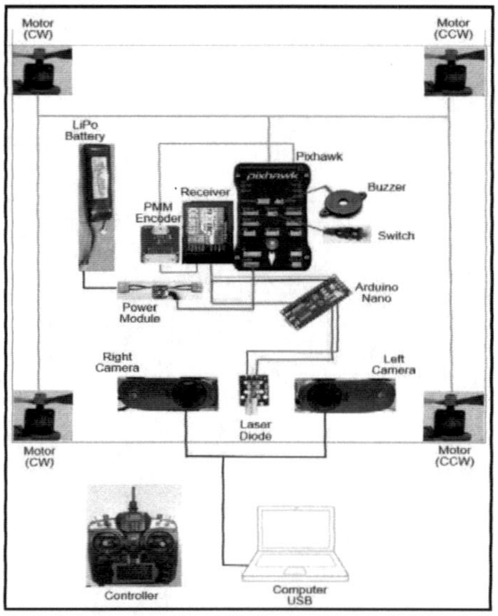

Figure 6. Full schematic drawing of quadrotor UAV hardware integration

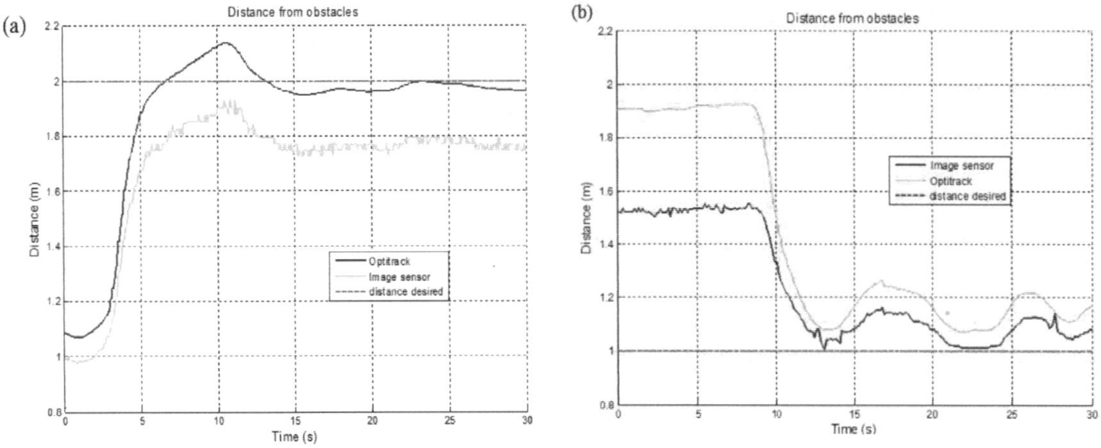

Figure 7. Comparison for calculated depth value between image sensor and Optitrack for (a) static collision avoidance, (b) dynamic collision avoidance

5.1. Static collision avoidance

The first test flight measures the UAV's collision avoidance performance in static hovering. The UAV was positioned at (0, 0) coordinate and the obstacle was positioned at (1 m, 0) coordinate. Positions of the UAV and the obstacle were tracked by Optitrack system. UAV was lifted up 1.5 m above ground by Optitrack control. The camera sensors were turned on to obtain the real distance value. Desired distance between UAV and obstacle was set to 2.0 m and response of the UAV was then recorded and comparisons were made.

As referred to Figure 7(a), the UAV was able to hover statically at the desired distance (final) with small perturbations but unacceptable error, which was greater than 10%. However, the error obtained when hovering at a shorter distance (final) was acceptable, which was less than 10%. Since the short distance is the important factor for collision avoidance, the performance of the UAV was satisfactory. The results obtained as shown in Table 2 satisfy the work by Waqar [14] where the accuracy of linear triangulation deteriorates as the distance goes higher. The reason behind the accuracy deterioration at far distance may be due to the image point not within and far away from the depth resolution at longer range [16].

Table 2. Distance values between obstacle and UAV in static flight test

	Distance from obstacles (m)		Error (m)	Error (%)	
	Optitrack	Image Sensor			
Initial	1.078	0.992	0.086	8.71	Acceptable
Final	1.962	1.751	0.211	12.08	Unacceptable

5.2. Dynamic collision avoidance

The second test flight measured the UAV's collision avoidance performance in dynamic hovering, in which the UAV was moved from a location to another location in non-linear dynamics. The UAV was initially positioned at (0, 0) coordinate and obstacle was positioned at (2 m, 0) coordinate. Positions of the UAV and the obstacle were tracked by Optitrack system. UAV was lifted up 1.5 m above ground by Optitrack control. The camera sensors were turned on to obtain the real distance value. Desired distance between the UAV and the obstacle was set to 1 m. The UAV was controlled by Optitrack to hover towards the obstacle until the desired distance was achieved. The response of the UAV was then recorded and comparisons were made.

As referred to Figure 7(b), the UAV was able to hover dynamically and stop at the desired distance (final) but with small perturbations and acceptable error, which was less than 10%. However, the error obtained when hovering from the longer distance (initial) was unacceptable, which was higher than 10%. Again, the collision avoidance performance was satisfactory because collision often occurs at small distance. The results obtained as shown in Table 3 once again satisfy that the accuracy of linear triangulation deteriorate as the distance goes higher. The depth estimation from stereo cameras is also affected by calibration errors and alignment of stereo camera parameters [17].

Table 3. Distance values between obstacle and UAV in dynamic flight test

	Distance from obstacles (m)		Error (m)	Error (%)	
	Optitrack	Image Sensor			
Initial	1.905	1.524	0.381	25.01	Unacceptable
Final	1.159	1.078	0.081	7.51	Acceptable

5.3. Colour and reflectivity test flight

Several additional test flights were conducted to test the UAV's collision avoidance performance in response towards obstacles of different colours and reflectivity to validate whether different types of obstacle's surface may degrade the quality of collision avoidance system based on image sensors. In Table 4, it is shown that the UAV's performance was degraded when a surface of high reflectivity was used. It has also proven that the background colour of the obstacle does not influence image sensing of the UAV.

Table 4. UAV's performance in colour and reflectivity flight test

Surface	Skin	Colour	Reflectivity	UAV's performance
Polystyrene		White	Low	Best
	1	Red	Medium	Good
	2	Yellow	Medium	Good
	3	Transparent	High	Degraded

5.4. Minimum distance test flight

The final section of the UAV test flight included the determination of threshold for minimum desired distance, D between the obstacle and the UAV. A few test flights were conducted to identify minimum distance that could be achieved before collision from a fixed position of 1.5 m. The minimum distance based on the dimension and geometry of UAV was measured to be 0.40 m. Hence, flying smaller than 0.4 m was undesirable for the UAV. The results were tabulated in Table 5 and it can be observed that no collision occurred when the desired distance was reduced to 0.40 m. Thus, the minimum distance that the UAV can achieved when hovering is taken as 0.40 m.

Table 5. UAV's performance in different desired range from obstacle

Desired Range (m)	Collision Avoidance
0.8	Success
0.6	Success
0.5	Success
0.4	Success

IOP Conf. Series: Materials Science and Engineering **152** (2016) 012026 doi:10.1088/1757-899X/152/1/012026

6. Conclusion and future work

This paper proposes a new approach of calculating distance using image sensor with laser transmitter to achieve collision avoidance. The laser transmitter projects a laser beam on the surface of obstacle to create a colour-distinct spot on the obstacles. OpenCV functions can be used to filter out the colour-distance spot from the rest of the images. This produces a reference point and area for colour-based tracking without the need to compare pixel by pixel as in the conventional depth map method. Control algorithm based on non-linear model developed by Kendoul [15] was applied in this work to achieve relative-positioned collision avoidance assisted by Optitrack Motion Capture System. Experiments and evaluations for the collision avoidance performance were carried out at indoor environment in range of Optitrack cameras, which provided position coordinates and velocities of the UAV and the obstacle. The results obtained showed that linear triangulation with image sensors provide satisfactory results in both static and dynamic collision avoidance. The threshold for minimum desired distance between UAV and obstacle obtained using this approach is 0.4 m. Lastly, the distance values calculated by the linear triangulation with image sensors were good when the distance was small but deteriorate when the distance increases.

The recommended future work for this project is to improve the accuracy for depth calculation by applying Kalman filter to remove highly inaccurate values. A real-time collision avoidance UAV can be developed by upgrading the UAV from on-ground processing to on-board processing by installing OpenCV compatible microcomputers such as Raspberry® Pi. The UAV can also be integrated with others sensors such as infrared (IR) and ultrasound (sonar) to collision avoidance in all four directions. Final recommendation for future work is development of control algorithm for UAV's autonomous or manual hovering at outdoor environment to achieve meaningful task, for instance high altitude surface inspections on tall buildings and structures.

References

[1] Nils G, Paul B and Sergio M 2015 *Digital Obj. Identifier* **3** 599-609
[2] Byrne J, Cosgrove M and Mehra R 2006 *IEEE Int. Conf. on Robotics and Automation*
[3] Rambabu R, Rijaluddin M and Azrad S 2015 *ARPN Jrnl. of Engin. and Appl. Sci.* **10** 12-17
[4] Matthias N, David D, Johannes S, Dirk H, Thomas L and Sven B 2013 *European Conf. on Mobile Robots*
[5] Schmidt B and Wang L H 2014 *Jrnl. of Manuf. Syst.* **33** 711-8
[6] Aydin T and Sinan Y 2010 *Pattern Recognition Lett.* **31** 2389-96
[7] Zhu L H, Cheng X H and Yuan F G 2016 *Measurem.* **77** 40-9
[8] Park J H and Kim Y D 2012 *Int. Conf. on Control, Automation and Syst.*
[9] Nunez P, Martin R V, et al. 2011 *Sensors* **11** 7262-84
[10] Chen Y and Shang L 2016 *Opt.* **127** 900-11
[11] Park J H and Park H W 2003 *Signal Processing: Image Communic.* **18** 401-16
[12] Waqar K, John M and Klette R 2009 *Int. Conf. Image and Vision Computing*
[13] Farid K 2006 *IEEE Conf. on Decision and Control*
[14] Chen G and Lingling T 2011 *Proc. IEEE I. C. on Cognitive Informatics & Cognitive* Computing
[15] Fadhil M, Pebriant D, Syaril A, et al. 2013 *Sci. and Engin.* **65** 109-32
[16] Mikko K, Mikko N and Pirkko 2011 *Proc. SPIE7864, Three-Dimensional Imaging, Interaction, and Measurement*
[17] Zhao W and Nandhakumar N 1996 *Pattern Recognition* **29** 2115-26

IOP Conf. Series: Materials Science and Engineering **152** (2016) 012027 doi:10.1088/1757-899X/152/1/012027

Defect visualization of aircraft UHF antenna radome using full-field pulse-echo ultrasonic propagation imaging system

H J Shin[1,2], S C Hong[3], J R Lee[2,3*] and J H Kim[4]

[1]LANL-CBNU Engineering Institute-Korea, Chonbuk National University, Republic of Korea
[2]X-NDT Inc., Republic of Korea
[3]Department of Aerospace Engineering, Korea Advanced Institute of Science and Technology, Republic of Korea
[4]7th R&D Institute - 3rd Directorate, Agency for Defence Development, Republic of Korea

*leejrr@kaist.ac.kr

Abstract. Most of aircraft antennas usually have various types of radome made of composite materials for protecting antenna structures. However, these antenna radome structures, which are installed on the outside of airplane, are easy to be damaged by external forces such as drag, foreign object, bird strike and others. In this study, full-field pulse-echo ultrasonic propagation imaging (PE UPI) system is proposed as the non-destructive inspection technique to visualize manufacturing defects in composite antenna radome. Based on the results of the sample case study, it is shown that the ultrasonic wave propagation imaging (UWPI) that is generated by the proposed full-field PE UPI system is able to highlight the intact internal condition of antenna structure and its defect area. Additional damage visualization techniques like ultrasonic energy mapping (UEM), variable time window amplitude map (VTWAM) and also ultrasonic spectral imaging (USI) algorithms are applied to improve the reliability of the damage visualization. It can be concluded that the proposed PE UPI system is an effective non-destructive inspection technique for the composite radome structures.

1. Introduction

Antennas for radar, communication and other wireless devices are commonly equipped on airplanes. These antennas usually have various types of radome made of composite materials for protecting them. Most of the radome structures are installed on the outside of airplane that makes it easy to be damaged by external forces such as drag, foreign object, bird strike and others. If the radome structures have internal defects induced by their manufacturing process like voids, porosity, fibre disorientation and resin rich or resin starved areas, these internal defects can give cause of damage in combination with the external impact, fatigue or load [1].

Manufacturing defects of radome structures generally can be detected by non-destructive inspection technique such as visual inspection, ultrasonic testing, radiography and others. Visual inspection is applied basically because of ease of inspection by the naked eyes with simple optical device despite its difficulty to find internal defects [2, 3]. Meanwhile, radiography is the most common non-destructive testing method to inspect the internal structure by x-rays or gamma rays [4]. However, this method has limitations due to the hazardous radiation exposure that can affect the health of the inspectors. On the other hand, ultrasonic testing method is the most effective technique to inspect composite structures. The pulse-echo technique, which is one of ultrasonic testing methods, has been widely used for aircraft structures inspection [4]. However, the typical ultrasonic testing method still has some weak points in need of improvement such as spreading coupling agents on the surface and inspecting curved surface.

In this research, full-field pulse-echo ultrasonic propagation imaging (PE UPI) system with various post-processing damage visualization technique has been applied to detect the manufacturing defects of composite antenna radome structure.

2. Full-field pulse-echo ultrasonic propagation imaging system

Figure 1 shows a concept of full-field PE UPI system that has been configured for ultra high frequency (UHF) antenna radome inspection. The full-field PE UPI system is consisted of a 1064 nm Q-switched laser, 2-axis translation stage, a 633 nm laser Doppler vibrometer (LDV), a PC and also a band-pass filter.

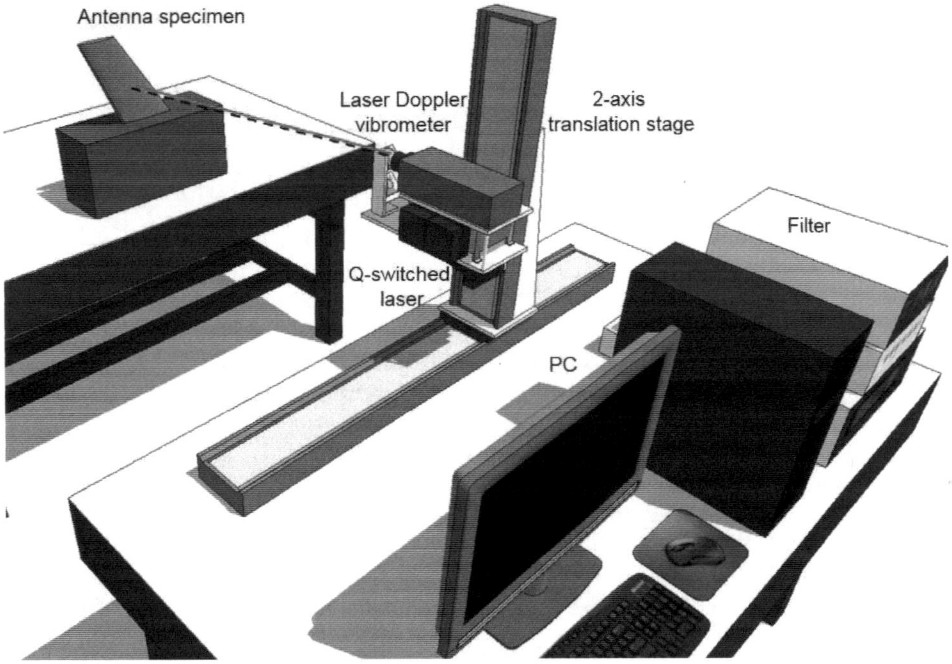

Figure 1. Concept of full- field pulse-echo UPI system

In a general UPI system, the scanning laser beam generates ultrasonic waves on the target surface and a fixed ultrasonic sensor acquires the propagated ultrasonic waves along the surface of the target [5]. On contrary, the proposed full-field pulse-echo UPI system is operating with different mechanism based on pulse-echo technique shown in Figure 2. Laser beams of the Q-switched laser and LDV are impinged along raster scanning pattern on the target surface by two-axis linear translation stage. The laser beam impinged by Q-switched laser generates ultrasonic waves from thermoelastic effect on the surface of target. These ultrasonic waves propagated through the thickness direction and are reflected by the internal structures and opposite surface [6]. The reflected ultrasonic wave signals are acquired by LDV and then processed in the PC. Finally, ultrasonic wave propagation imaging (UWPI) movie is generated along with several post image processing result.

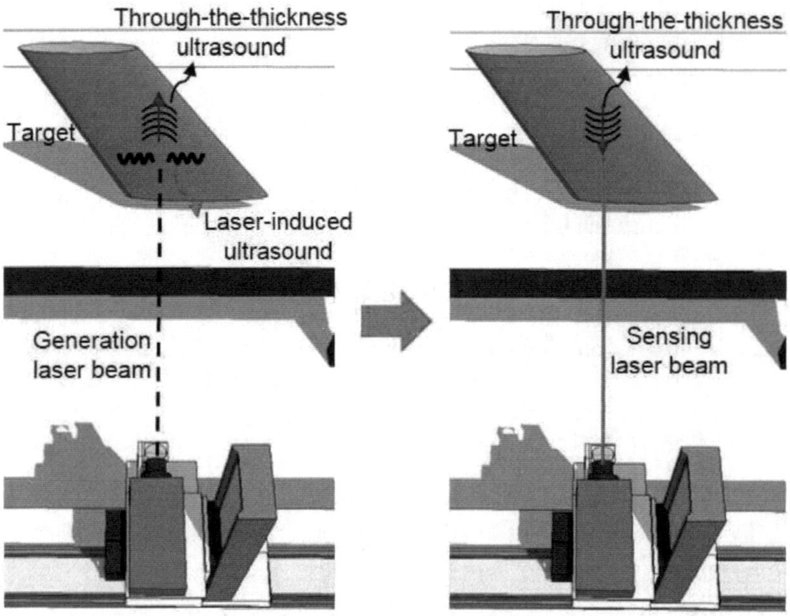

Figure 2. Mechanism of the full-field pulse-echo UPI system

In this research, ultrasonic energy mapping (UEM), ultrasonic spectral imaging (USI) and variable time window amplitude map (VTWAM), which are post image processing algorithms for previously developed UPI system, are applied to the PE UPI system for the first time. The UEM technique is a damage visualization algorithm to visualize energy of scattering waves filtered as a single wave mode (S0/A0/S1/A1, etc.) based on wave number filtering [7]. On the other hand, the USI technique is one of damage visualization algorithms to visualize damages by mapping the time domain signal that has been transformed to frequency domain using fast Fourier transform (FFT) [8]. Last but not least, the VTWAM algorithm is able to visualize the damages by mapping the averaged signal values in chosen specific time range from the UWPI data [9].

3. Defect visualization of GFRP UHF antenna radome structure

The damage visualization experiment was performed using the full-field PE UPI system to visualize internal defects of a glass fibre reinforced plastic (GFRP) antenna radome structure. As illustrated in Figure 3, the size of radome is 359.5 mm ×320.6 mm ×116.67 mm. The shape of the antenna is blade-type structure like a dorsal pin. Scan area 1 was 312 mm ×130 mm while scan area 2 was 312 mm × 52 mm, both with the same interval of 0.52 mm. The stand-off distance was 1.5 m from the full-field pulse-echo UPI system to the target surface. The energy of Q-switched laser was 3.5 mJ and the pulse repetition frequency was set to 1 kHz. The signal acquired was also filtered using band-pass frequency between 20 – 200 kHz.

The generated UWPI results are shown in Figure 4. No defect is visualized in Area 1 except for a peeling paint area of skin. Meanwhile, a defect of size of 40 mm × 2 mm was detected in the UWPI result of Area 2. Damage visualization techniques: UEM, VTWAM and USI were additionally applied to improve the reliability of damage visualization results as shown in Figure 5. As a result, the damage visualization results of Area 2 show damaged area of size of 40 mm clearly. The defect is estimated as resin-rich area formed along fibre direction of fabric. All in all, it is demonstrated that the developed full-field PE UPI system can provide damage visualization results that show defect area of the antenna structure. These results assured that the full-field PE UPI system is effective in visualizing the defect of the composite structures.

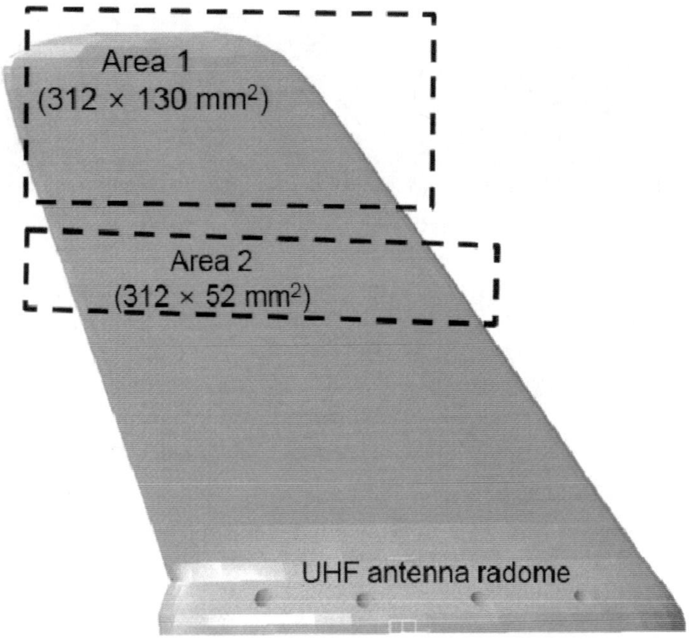

Figure 3. 3D modelling of GFRP UHF antenna specimen

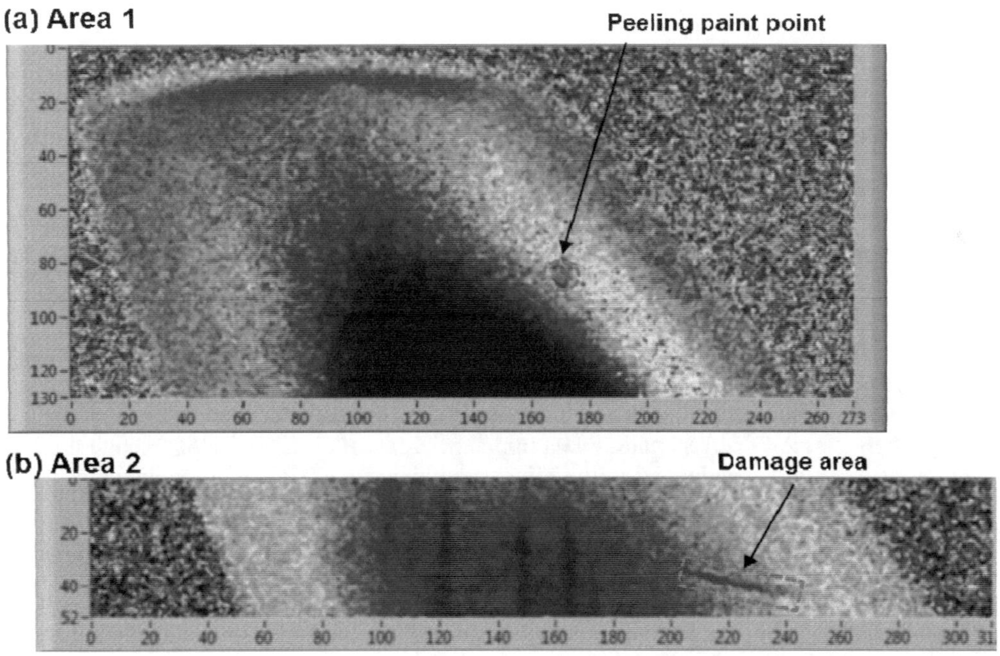

Figure 4. UWPI results (a) Area 1, (b) Area 2

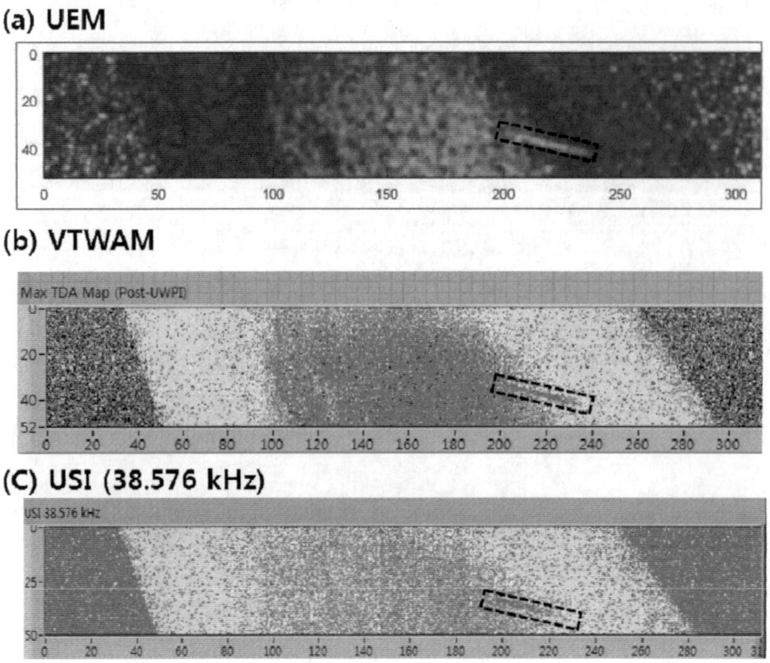

Figure 5. Damage visualization results of Area 2, (a) UEM result, (b) VTWAM result, (c) USI result

4. Conclusion

In this research, a full-field PE UPI system was proposed for actual GFRP antenna radome structure inspection. UWPI result of Area 1 that was generated by the proposed full-field PE UPI system shows an intact internal condition of the antenna structure except for a peeling paint area of skin. On the other hand, UWPI result of Area 2 shows defect area. Additional damage visualization technique such as VTWAM, UEM and USI were also applied to improve reliability of damage visualization results. As a result, the visualizations show clearly the damaged area that is believed to be a resin-rich area formed along the fibre direction of fabric. The results assured that the full-field PE UPI system is effective in visualizing defect of composite structures.

5. Acknowledgments

This research was supported by the Leading Foreign Research Institute Recruitment Program (2011-0030065) and the Space Core Technology Development Program (2013-042548) through the National Research Foundation of Korea funded by the Ministry of Science, ICT and Future Planning.

References

[1] Federal Aviation Administration 2012 *Aviation Maintenance Technician Handbook – Airframe* (Oklahoma City: Airman Testing Standards Branch)
[2] Baaran J 2009 *Visual Inspection of Composite* (Cologne: European Aviation Safety Agency)
[3] Heida J and Platenkamp D 2011 *Procs. of 16th NDT in Progress*
[4] Khan M A U 1999 *ECNDT 98*
[5] Chia C C, Lee J R, Park J S, Yun C Y and Kim J H 2008 *Proc. DEFEKTOSKOPIE 2008*
[6] Hong S C and Lee J R 2015 *20th International Conference on Composite Materials*
[7] Bae D Y and Lee J R 2015 *J. Intell. Mater. Syst. Struct.* **2015** 1–10
[8] Chia C C, Lee J R and Park C Y 2012 *Composites: Part B* **43** 2898-906
[9] Choi Y S, Jeong H and Lee J R 2014 *Shock and Vibration* **2014** 1-10

Establishing metrological traceability for radiometric calibration of earth observation sensor in Malaysia

S W Ng* and A Zulkifli

National Space Centre, Lot 2233, Kampung Sungai Lang,
42700 Banting, Selangor, Malaysia

*wayne@angkasa.gov.my

Abstract. The space borne earth observation (EO) sensor provides a continuous large spatial coverage over the earth at relatively low cost (cost-effective) and can be practically accessible worldwide. The daily synoptic view offered by instrument in earth orbit is tremendously useful in various applications, particularly long term global monitoring that needs multi-disciplinary, multi-temporal and multi-sensor data. Due to the indirect measurement nature of the EO sensor, calibration and validation (cal/val) are essentially required to establish the linkage between the acquired raw data and the actual target of interest. Ultimately, EO sensor provider must strive to deliver "the right information, at the right time, to the right people". This paper is authored with the main aim to report the process of establishing metrological traceability for radiometric calibration of EO sensor at Optical Calibration Laboratory (OCL), National Space Agency of Malaysia (ANGKASA). The paper is structured into six sections. The first section introduces the context of EO and background of radiometric calibration. The next section discusses the requirements for metrological traceability in radiometric calibration while the following third section outlines ANGKASA efforts in setting up the metrological traceability laboratory in radiometric calibration. Meanwhile, the uncertainty estimation results is reported in the fourth section and the fifth section explains some of the continuous efforts made in order to improve the current metrological traceability set up. Lastly, the summary of this paper is provided in the last section.

1. Introduction

According to [1], earth observation (EO) in narrow sense is the science of measurement of all aspects of the earth system, including its physical, chemical and biological processes by space borne sensor. Historically, EO data has been widely used in technical investigation and research by remote sensing community. The constant demand from both scientific and commercial sectors has led to continuous increment of both satellite available in orbit and satellite performance. In recent times, the use of EO data in supporting policy and regulatory decision making has gain tremendous popularity across the different countries and organisations. The ever increasing growth of EO data is also spurred by the open data policy recently adopted by United States Geological Survey (USGS) for its Landsat archive data, China-Brazil Earth Resources Satellite (CBERS) and European Union for Copernicus program, which includes constellation of Sentinel satellites [2].

However, data acquired by satellite-based remote sensing is indirect by nature, and calibration and validation (cal/val) are required to allow the retrieval of actual bio geophysical parameters of interest. In particular, quantitative analysis of EO data must take calibrated data as input to avoid "garbage in, garbage out" output. Cal/val is crucial to ensure harmonisation and interoperability of data generated from multi-source, multi-resolution and multi-temporal input at various scales. Calibration establishes confidence in EO data by providing quality indicator that allows user to readily assess "fit for purpose" of the data and the derived products. Quality indicator provides user with metrological traceability and measurement uncertainty information that represents a convenient way of seeing the big picture by just looking at a small piece of it. Evaluation of EO data suitability and adequacy is only possible with the provision of quality indicator [3].

In general, the goal of cal/val is to characterise sensor response in five major domains: radiometric, spatial, spectral, temporal and polarisation properties. Radiometric calibration characterises sensor response to known radiometric input, including radiance and/or irradiance, response linearity, detector-to-detector response uniformity, etc [4]. One particular importance of the radiometric calibration in present-day operational use of EO data is global climate monitoring. The ultimate objective of global climate change study is to distinguish human effects from natural effects that are affecting the climate change. To do that, a trusted long termed data is required by scientists to simulate sophisticated global climate model. These data are usually a combination of EO data from multiple sensors, multiple dates, multiple resolutions and different measurement techniques. Therefore, uncertainty knowledge of each input data must be obtained to quantify relative biases between different sensor outputs and temporal drift of data from the same instrument [2-6].

2. Metrological traceability for radiometric calibration

By ISO International Vocabulary of Metrology – Basic and General Concepts and Associated Terms (VIM) definition, metrological traceability is the property of a measurement result whereby the result can be related to a reference through a documented unbroken chain of calibrations, each contributing to the measurement uncertainty [7]. In addition, VIM lists down eight corresponding notes to further explain the definition. Based on the definition and the eight additional notes, the elements required for metrological traceability are summed up in Table 1. One may refers to the VIM document for detailed definition of the terminology used in Table 1, including measurement unit, measurement procedure, measurement standard and calibration hierarchy.

Table 1. Metrological traceability elements for radiometric calibration

Requirements	Remarks
Reference	A measurement unit (e.g. radiance, $Wm^{-2}sr^{-1}$) through its practical realisation. <u>or</u> A measurement procedure including the measurement unit for a non-ordinal quantity. <u>or</u> A measurement standard
Calibration hierarchy	Required
Reference information	Include the time at which the reference was used in establishing calibration hierarchy
More than one input quantity	Each of the input quantity values should itself be metrological traceable and the calibration hierarchy involved may formed a branched structure or a network.

3. Metrological traceability at Optical Calibration Laboratory

National metrology institutes (NMIs) are solely responsible for practical realization of International System of Unit (SI) quantity definition and disseminate them via primary standard. The calibration laboratory then uses the working standard, which has an unbroken record of traceability to primary standard issues by NMIs to calibrate user instrument. Unfortunately, the National Metrology Institute of Malaysia does not maintain either spectral radiance or spectral irradiance standard. The Optical Calibration Laboratory (OCL) has to source from commercial vendor for 1kw FEL-type incandescent lamp that is traceable to US National Institute of Standard and Technology (NIST) primary standard as working spectral irradiance standard. It is a pre-treated lamp and has a vertical, double-coiled, helical filament with the filament supported only at the top and bottom [8]. The NIST has utilised electrical substitution radiometry (a.k.a detector based radiometry) in assigning the spectral irradiance scale to the primary standard lamp [9].

The incandescent lamp alone is not suitable for most instrument calibration, especially the large-aperture instrument. Instead, this spectral irradiance standard lamp is applied together with a good Lambertian diffuse reflectance target. A known radiance can be computed from the reflected light off the target if the Bidirectional Reflectance Distribution Function (BRDF) of the diffuser is known. The user instrument can then be positioned to measure the known radiance reflected from the target, or alternatively, an intermediate spectro radiometer is placed such that it can then transfer the standard to another light source [9-10]. For measurement in OCL, a 99% reflectance Spectralon® target with a known BRDF is obtained from commercial vendor and it is also traceable to NIST primary standard via vendor calibration transfer. Subsequently, both measurement standards used in OCL radiometric measurement are metrologically traceable with their own calibration hierarchy. The traceability chain of this lamp-plague method is depicted in Figure 1.

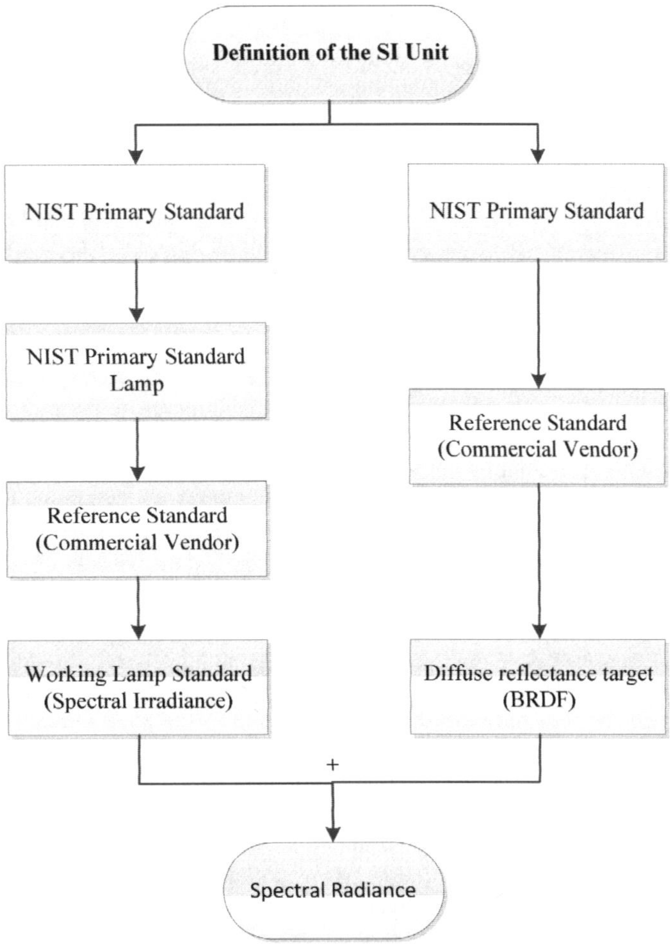

Figure 1. Metrological traceability chain in OCL for lamp-plague method

4. Uncertainty analysis

The uncertainty analysis is conducted in accordance with the Evaluation of Measurement Data — Guide to the Expression of Uncertainty in Measurement (GUM) [11]. The reflected radiance from the diffuser target is given by:

$$L_s = \frac{E_{Lamp}\beta_{\theta:\alpha}}{\pi} \quad (1)$$

where L_s is the source radiance (W.m^{-2}.sr^{-1}), E_{lamp} is the lamp irradiance (W.m^{-2}) and $\beta_{\theta:\alpha}$ is the diffuser reflectance factor. Angle θ and α represents the light incidence angle and the instrument viewing angle, respectively.

Radiance and irradiance in Equation 1 are spectral quantities but the wavelength dependence is not specifically written in the equation. However, the equation is still valid for any specific wavelength. Reflectance factor is defined as the ratio of radiant power reflected by an elemental sample surface to that which would be reflected into the same reflected beam geometry with apex at the surface element by an ideal perfectly diffuse (Lambertian) surface irradiated in exactly the same way as the sample. One should further read on [12] to have comprehensive understanding on the subject of BRDF.

From Equation 1, one can immediately identify the uncertainties sources: (a) lamp irradiance; (b) diffuser reflectance factor; and (c) reproducibility. Reproducibility is measurement precision under reproducibility conditions of measurement, which includes different locations, operators, measuring systems and also replicate measurements on the same or similar objects [7]. It is almost impossible to provide an exhaustive list of all sources of uncertainty. Nonetheless, this study has considered as many as possible sources to obtain good assessment of uncertainties that are associated with the way the measurement is set up, the values indicated by the instruments and the residual uncertainties associated with the corrections applied. In addition to the three main sources mentioned earlier, these working lamp and diffuser target may be sensitive to environmental condition, e.g. temperature, humidity and pressure. Since the calibration activity is performed in laboratory environment where temperature and humidity are controlled, the environment effects are considered negligible and hence excluded in the uncertainty analysis. Effect of stray light is also considered negligible as the room is totally maintained in "black-out" condition during the calibration activity.

The main uncertainties associated with lamp irradiance are lamp-diffuser distance, lamp alignment, lamp current control, short and long term stability. Uncertainties due to lamp irradiance output are directly read off from calibration certificate at three specified wavelength: 350nm, 655nm and 900nm. The lamp irradiance is calibrated at fixed distance 50cm, and therefore lamp-diffuser target should be set up at the same distance or else there will be irradiance output variation according to the irradiance inverse square law. The uncertainties due to lamp-diffuser distance is estimated at 0.2%, based on the accuracy of laser distance meter, i.e. 1mm over 50cm distance. Lamp alignment uncertainty is caused by the difficulty in rotating and positioning the centre of the lamp (the filament) normal to the diffuser target. To estimate lamp alignment sensitivity, measurement experiment is set up to compare standard deviation of 20 measurements with lamp re-aligned between the measurements and with lamp is not aligned between the measurements. The working irradiance lamp is operated at constant current 8.2A since lamp current fluctuation will lead to different lamp output. Similar to lamp alignment sensitivity estimation, lamp current sensitivity is determined experimentally by varying the current at (8.2 ± 0.1)A. Long term stability for lamp is not considered in this paper as the lamp used in the study is below 10 hours of operation.

Uniformity of diffuser and reflectance factor geometry are the two primary sources of uncertainties in diffuser reflectance factor. Uncertainties due to diffuser reflectance factor are directly read off from calibration certificate. The uncertainties due to diffuser uniformity arise because of the non-uniformity of diffuser surface within the instrument field-of-view and the geometry is the BRDF ambiguity as the result of uncertainty in setting up incident flux angle and instrument viewing angle. Both uncertainties are assumed to be at 0.5% and 1.45%, respectively, based on findings from literatures [13] and [14] as OCL does not have the measuring instrument to experimentally estimate the diffuser uniformity and BRDF. A list of uncertainty sources, their estimated values and their contribution to the total combined uncertainty are tabulated in Table 2. The uncertainty budget are estimated at three wavelengths (i.e. 350, 655 and 900 nm) with 0° incident flux angle and 45° instrument viewing angle. Most of the uncertainties discussed in the paper are wavelength- and angle-dependent. Therefore, re-assessment is needed to analyse uncertainty budget for other wavelength and angle.

Table 2. Uncertainty budget for radiometric calibration at OCL

	Sensitivity Coefficient	Wavelength (nm)		
		350	655	900
Standard Uncertainty (%)				
Lamp irradiance	1	2.1	1.7	1.7
Lamp-diffuser distance	2	0.4	0.4	0.4
Alignment of lamp	1	0.1	0.1	0.1
Lamp current stability	1	0.3	0.3	0.3
Lamp short term stability	1	0.1	0.1	0.1
Diffuser reflectance factor	1	0.25	0.3	0.3
Reflectance factor geometry	1	1.45	1.45	1.45
Uniformity of diffuser	1	0.5	0.5	0.5
Expanded Uncertainty (%), k=2				
Irradiance		4.33	3.56	3.56
Radiance		5.33	4.73	4.73

5. Continuous improvement

As discussed in [8], despite its popular choice as main means of spectral irradiance scale dissemination, this 1kW FEL-type incandescent lamp suffers much higher degradation rate than previously reported. The coil in the lamp will collapse in between 140 to 250 hours burn time, or after 100 cycles of turning on and off cycle. Although the commercial vendor generally recommends hundreds of hours for lamp operation, internal quality assurance program will be created to assess regularly the calibration status of the working lamp. Among others, this includes setting up three calibrated working lamp standards for inter-comparison between these lamps. Meanwhile, periodical checking of diffuser target ageing rate is also required to assess the long term stability of the target reflectance factor.

The overall combined uncertainty for spectral irradiance and radiance measurement is in the order of 2% or more. Such uncertainty may not fit for some EO sensor metrological requirements, especially those involve in mission for climate change study. There are two stumbling blocks in reducing the measurement uncertainty: the lamp irradiance and the reflectance factor geometry. One way to reduce the uncertainty associated with lamp irradiance is to source primary standard lamp directly from NIST instead of using commercial vendor working lamp. Consequently, there will be additional cost and other non-monetary resources required to purchase, maintain and transfer this primary standard. On the other hand, the plan for facility upgrade in goniometry for the angular reflectance measurement of diffuse target is currently initiated in order to reduce uncertainty in reflectance factor geometry.

6. Conclusion

The establishment of metrological traceability for radiometric calibration at the OCL, National Space Centre is reported. A details uncertainty budget has been discussed for spectral irradiance and spectral radiance measurement at OCL. Currently, the combined uncertainty is estimated in the range of 3.6–5.3 %. There are some continuous efforts in place, both in planning and on-going phases, to reduce the overall uncertainty. The improvement efforts are a continuous process implemented in OCL in refining the operation and achieving better uncertainty goals.

Reference

[1] Yang X, Blower J D, Bastin L, Lush V, Zabala A, Masó J, Cornford D, Díaz P and Lumsden J 2013 *Phil. Trans. R. Soc. A* **371**

[2] Baraldi A 2009 *IEEE Journal of Selected Topics in Applied Earth Observations and Remote Sensing* **2** 104-34

[3] Datla R, Emery W, Ohring G, Spencer R and Wielicki B 2003 *Proceedings of the International Workshop on Radiometric and Geometric Calibration*

[4] Tansock J, Bancroft D, Butler J, Cao C, Datla R, Hansen S, Helder D, Kacker R, Latvakoski H, Mylnczak M and Murdock T 2015 *Guidelines for Radiometric Calibration of Electro-Optical Instruments for Remote Sensing* (National Institute of Standards and Technology)

[5] Fox N P 2004 *Validated data and removal of bias through traceability to SI units* (London: Taylor and Francis)

[6] Johnson B C, Brown S W, Eppeldauer G P and Lykke K R 2003 *International Journal of Remote Sensing* **24** 339-56

[7] BIPM, IEC, IFCC, ISO, IUTAC, IUPAP, OIML and ILAC 2012 International Vocabulary of Metrology–Basic and General Concepts and Associated Terms JCGM 200:2012

[8] Harrison N J, Woolliams E R and Fox N P 2000 *Metrologia* **37** 453

[9] Johnson B C, Brown S W and Rice J P 2004 *Metrology for remote sensing radiometry* (London: Taylor and Francis)

[10] Fox N P, Chunnilall C J and White M G 1998 *Metrologia* **35** 555

[11] BIPM, IEC, IFCC, ISO, IUTAC, IUPAP and OIML 2008 Evaluation of measurement data — Guide to the expression of uncertainty in measurement JCGM 100:2008

[12] Nicodemus F E, Richmond J C, Hsia J J, Ginsberg I W and Limperis T 1977 *Geometrical Considerations and Nomenclature for Reflectance* (Washington D.C.: National Bureau of Standards)

[13] Bruegge C, Chrien N and Haner D 2001 *Remote Sensing of Environment* **77** 354-66

[14] Deadman A J, Behnerť I D, Fox N P and Griffith D 2011 *Geoscience and Remote Sensing Symposium*

Testing and validation of a $\dot{\mathbf{B}}$ algorithm for cubesat satellites

M Böttcher[1], S Eshghi[2*] and R Varatharajoo[2]

[1]Institute of Space Systems, University of Stuttgart, Germany
[2]Department of Aerospace Engineering, Universiti Putra Malaysia, Malaysia

*saeshghi@yahoo.com

Abstract. For most satellite missions, it is essential to decrease the satellite angular velocity. The $\dot{\mathbf{B}}$ algorithm is a common algorithm to stabilize the spacecraft by using magnetorquers. Controlling the satellite using the magnetorquers is part of the attitude control subsystem detumbling mode. Due to oscillating disturbances in the space environment, the required initial conditions needs analysis. As a consequence, the satellite stays in $\dot{\mathbf{B}}$ detumbling mode for the entire operation. In the detumbling mode, the spacecraft oscillates around its spatial axes. The purpose of this paper is to extend the $\dot{\mathbf{B}}$ algorithm with a disturbances compensation module and to achieve reduction of satellite's angular velocity. The developed algorithm is found to be able to reduce satellite's angular velocity up to 10-11 degrees.

1. Introduction

The success of a space mission depends on sophisticated attitude control subsystem (ACS). As the ACS can only be activated if the spinning about the satellite's own axes remains within certain boundaries, uncontrolled oscillation should be minimized, even if a precisely determined orientation is not required for the particular mission. Due to the increasing importance of small and very small satellites such as CubeSat satellites, which are mainly operating in a low earth orbit (LEO) under a relatively strong earth magnetic field, the relevance of magnetic attitude control systems has also grown [1, 2]. This is pushed by magnetorquers' low prices, their space saving assembly, and the independence of propellant. Therefore, they are often the only actuators in CubeSat missions. They are integrated inside the solar panels to obtain the necessary electrical power [3]. The $\dot{\mathbf{B}}$ law, a purely magnetic control scheme, was originally proposed in [4]. Since then, this control law has been adopted as the primary detumbling solution in small satellites such as Pico Satellite Solar Cell Testbed-2 (PSSCT 2) [5] and T-SAT1 [6]. While the aforementioned studies present satisfactory detumbling performances, the initial conditions are difficult to estimate in real applications. The tuning factor, for instance, is generally chosen based on trial and error experiments. However, the initial factor chosen may not reject the oscillating disturbances effectively, and the control law may fail in case the disturbances change. Therefore, the satellite is not detumbled effectively.

The aim of this work is to present a novel yet convenient $\dot{\mathbf{B}}$ algorithm to subside the effect of oscillating disturbances in the space environment. The study presents a solution to control the current by a pulse width modulation (PWM) signal and proposes a gain factor function to obtain the optimum torque to reject the disturbance. The method has been simulated and tested further on a CubeSat to validate the feasibility of the proposed scheme. The paper is organized as follows: the conventional and the novel $\dot{\mathbf{B}}$ algorithms are introduced followed by the description of the mission and the strategy.

IOP Conf. Series: Materials Science and Engineering **152** (2016) 012029 doi:10.1088/1757-899X/152/1/012029

The validation procedure using the CubeSat and the obtained results are presented later followed by the simulation results. The conclusion describes the findings of the study.

2. $\dot{\mathbf{B}}$ algorithm

The open-loop control principle for the magnetic ACS is often used in space missions to first detumble the spacecraft after its deployment and second, to force the spinning into certain boundaries defined by the main ACS as initial conditions. Since this approach comprises an open-loop control system without feedback, the correction angle accuracy of the $\dot{\mathbf{B}}$ controller is limited to less than 15 degrees.

Equation 1 shows the $\dot{\mathbf{B}}$ control law, which gives the dipole momentum $\mathbf{M}$ ($A \cdot m^2$) as a function of a gain factor K and the time derivative of the magnetic field vector. However, determination of $\dot{\mathbf{B}}$ is highly challenging, as the satellite might rotate with an uncertain angular frequency $\boldsymbol{\omega}$ (rad/s). If that frequency is known, $\dot{\mathbf{B}}$ (T) is given by Equation 2, were $\mathbf{B}_o$ denotes the magnetic field, e.g., measured with a magnetometer. However, if the angular frequency is not known, for instance for strongly spinning satellites, the time derivation of the magnetic field can only be approximated linearly with Equation 3, where $\mathbf{B}_1$ and $\mathbf{B}_0$ denote two different measurements taken within a time Δt (s).

$$\mathbf{M} = -\mathbf{K} \times \dot{\mathbf{B}} \tag{1}$$

$$\dot{\mathbf{B}} \approx \mathbf{B}_0 \times \boldsymbol{\omega} \tag{2}$$

$$\dot{\mathbf{B}} = \frac{\mathbf{B}_1 - \mathbf{B}_0}{\Delta \mathbf{t}} \tag{3}$$

Figure 1 shows a cyclic sequence of the $\dot{\mathbf{B}}$ algorithm. First, the magnetic field is measured within a control cycle of the cycle time $\mathbf{T}_{cycle}$. Second, its derivation is estimated subsequently. Then, the dipole momentum according to the $\dot{\mathbf{B}}$ control law is generated by the magnetorquers within the time $\mathbf{T}_{actuator}$. At the end of the cycle, all actuator coils have to be deactivated to avoid potential influences on the following cycle.

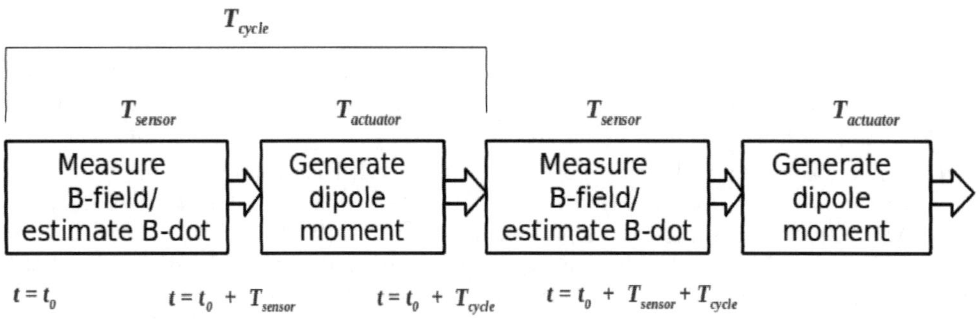

Figure 1. $\dot{\mathbf{B}}$ controller blocks

Because of both, inaccuracies in the magnetic field measurements and oscillating disturbances in the space environment, the satellite might not be stabilized properly, and its oscillation escalates. In fact, this could prevent the satellite from reaching the initial conditions for its main ACS. As a result, the satellite will try to detumble for the complete operating time, remaining in a constant oscillation.

3. Mission and strategy

There are several models for the forecast of different space disturbances. For a CubeSat mission, it can be assumed that there is a combination of disturbances with the total amount shown in Table 1 [7].

Table 1. Mission constraints

Constraints	Description
Orbit	Circular at 500 km with inclination of 0°
Size	$0.1 \times 0.1 \times 0.1\, m^3$
Mass	1 kg
Max. dipole momentum	$0.1\ \mathrm{A} \cdot m^2$
Rel. inaccuracy PWM	0.00015 %
Rel. inaccuracy magnetometer	5.888 %
External disturbance torque	$\left[5+\sin(\Omega_0 t)\right] \cdot 10^{-7}$ Nm

The parameter $\mathbf{\Omega_0}$ denotes the Earth's orbital frequency.

The gain factor K of the $\dot{\mathbf{B}}$ controller needs to be tuned to compensate for the disturbance torque. Simulations based on the modified $\dot{\mathbf{B}}$ algorithm show increasing damping behaviors. However, it is necessary to control the current driving the magnetorquers smoothly instead of only switching between maximum current and off state. Furthermore, a correlation between the gain factor and the angular position is needed as well.

4. Validation

4.1. Current control

The current, which is driven through the magnetorquers in order to generate a dipole momentum, is controlled by a pulse width modulation (PWM) signal. To use the novel $\dot{\mathbf{B}}$ algorithm, this signal must be adjusted continuously between 0 % and 100 %. It is furthermore necessary to validate the ability of the PWM signal generator to follow a given trajectory signal.

Several specific PWM duty cycles were set up and their response on the magnetic field of the earth has been measured with the satellite's magnetometers to determine a linear relation between the PWM duty cycle, DC_{PWM}, and the measured magnetic field B_{meas} presented in Equation 4.

$$\mathbf{B}_{\mathbf{meas}} = 0.1189 \cdot \mathbf{DC}_{\mathbf{PWM}}\ \mu.\mathrm{T} \tag{4}$$

Therefore, the duty cycle can be estimated within inaccuracy boundaries:

$$\mathbf{DC}_{\mathbf{PWM}} = \mathbf{B}_{\mathbf{meas}} / 0.1189\ \mu.\mathrm{T} \tag{5}$$

With this correlation, a trajectory can be specified and the response of the satellite can be plotted. Figure 2 shows the ability of the magnetorquers to follow the specified path, although a high accuracy cannot be achieved. The variations are caused by various reasons, e.g. high measurement inaccuracy of the magnetometers (see Table 1) and influences of certain magnetic disturbances on the testing environment.

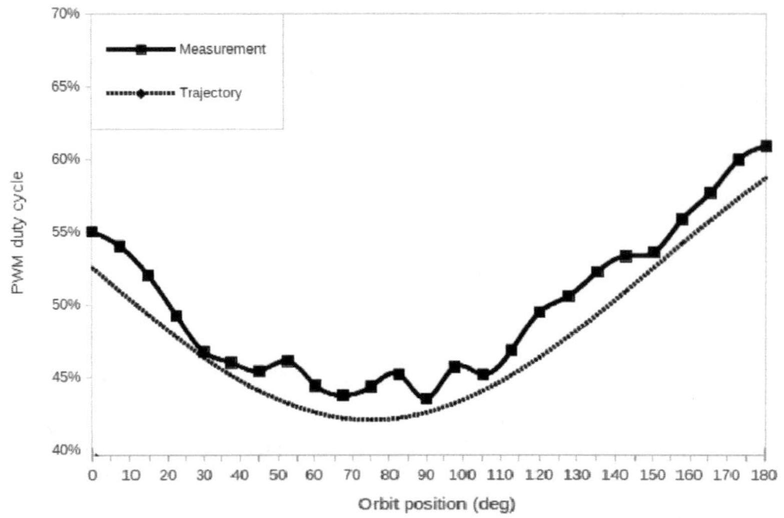

Figure 2. PWM signal setting and response

4.2. Optimal gain factor K

Assuming that the satellite is under influence of a disturbance torque, e.g. as in Table 1, and that the derivation of the magnetic field caused by the satellite's spinning remains at a low level less than *0.01* μ.T, the periodically changing controlled output torque M_{output} can be optimized to compensate for the oscillating disturbances.

$$M_{output} = \sin(\Phi)\cdot 10^{-7} \quad \text{N.m} \tag{6}$$

The controlled output torque is a function of the body position of the cubesat, and Φ denotes variation of the cubesat position along its nadir axis. Figure 3 illustrates the oscillating disturbances, the controlled output torque and the aggregate amount for different orbit positions. Basically, it also shows the effective torque induced by the magnetorquers to detumble the cubesat itself.

Assuming Equation 6 as the preferred torque output function and $\dot{\mathbf{B}}$ as fixed, the gain factor function can be set by using Equation 7.

$$K(\Phi) \approx [5.42914695\cdot\Phi - 206.318465]\cdot 10^{-7} \quad \text{N.m} / \mu.\text{T} \tag{7}$$

The result is plotted in Figure 4, including a linear approximation.

5. Testing

5.1. Simulation

According to Table 1, a CubeSat on a circular earth oriented orbit with inclination of $i = 0$ was assumed to simulate the satellite dynamics in form of the Euler angles by using MATLAB/Simulink. It is an open loop command during rocket injection/b-detumbling mode. Hereby, the pitch axis is decoupled from roll and yaw axis. As the roll and yaw axis behave similarly, it suffices to observe only one of them.

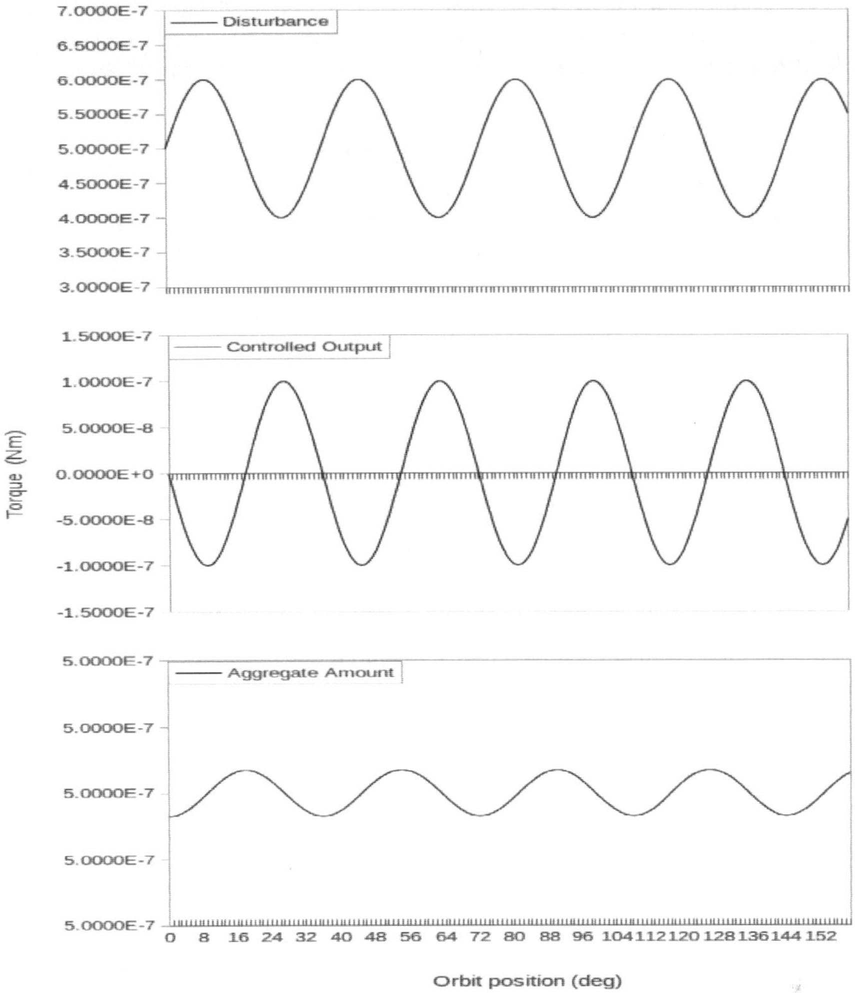

Figure 3. PWM signal setting and response

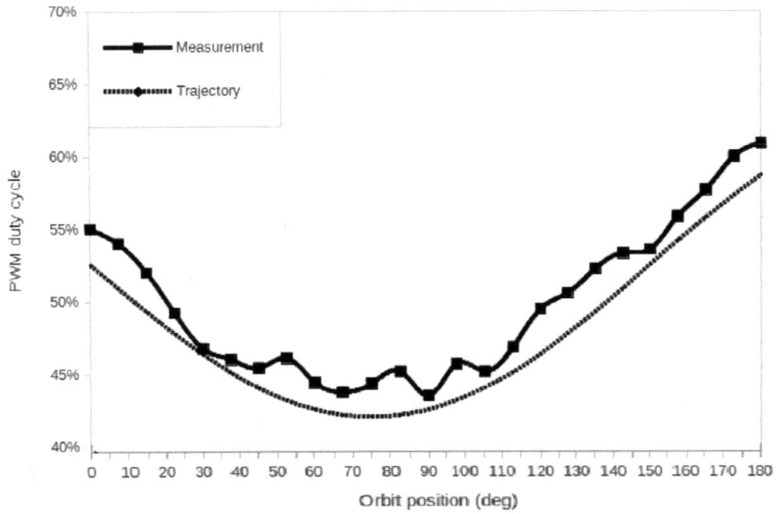

Figure 4. PWM signal setting and response

Figure 5 shows the roll angle response within 40 orbit cycles. It can be seen that, due to the oscillating disturbances, the satellite is performing an undamped oscillating movement within 90± deg. The satellite is stabilized around 0 degree; however, the accuracy of the angle control is not satisfying.

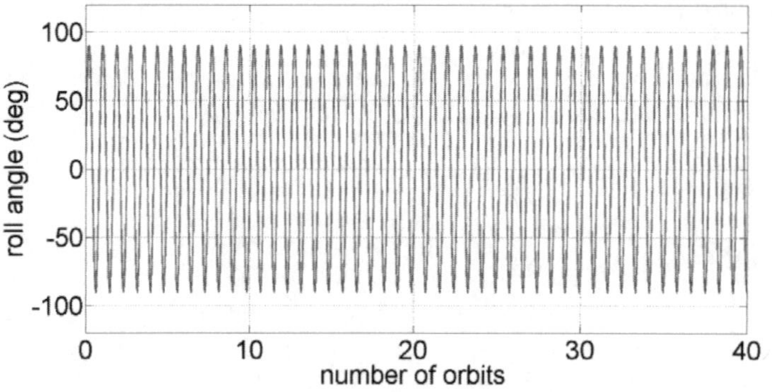

Figure 5. Satellite's roll angle simulation result with traditional $\dot{\mathbf{B}}$ controller

Figure 6 shows simulation result of the roll attitudes up to 10^{-11} degree. Even though the response exhibits more noise compared to Figure 5, the amplitude of the oscillation remains within an acceptable range during 40 orbit cycles (comment: confusing oscillation stays within 40 degree). The result is satisfactory; however, the observed oscillation is due to the inaccurately measured magnetic field and the resulting derivations in the gain factor itself. That is indeed the simulated performance. We may have variation/degradation during in-orbit flights.

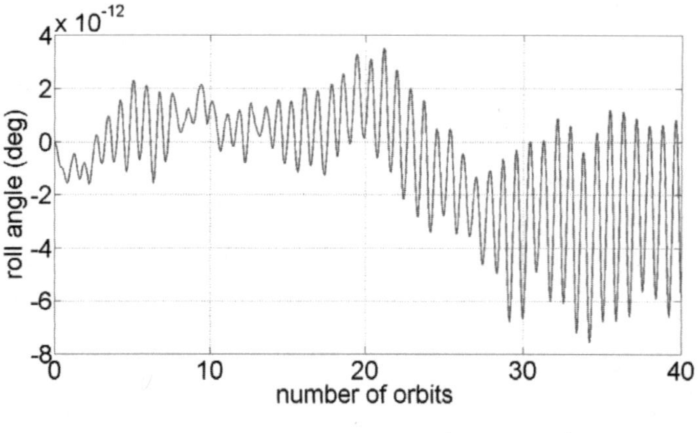

Figure 6. Satellite's roll angle simulation result with the new $\dot{\mathbf{B}}$ controller algorithm

6. Conclusion

The gain function K(ϕ) proposed for the novel $\dot{\mathbf{B}}$ control algorithm can effectively compensate for the typical space disturbances during a LEO CubeSat mission. This linear gain function only depends on the actual angular position in space in a circular orbit with 0° inclination. In addition, simulations showed that the oscillation of the satellite angles can be reduced within 40 orbit cycles compared to the existing algorithms. However, the satellite's controlled axis could not be fully stabilized to zero errors. Therefore, based on the specific mission objectives and constraints, suitable limitations of the acceptable attitude accuracies must be defined. Further work will not only focus on implementation of the $\dot{\mathbf{B}}$ algorithm in a real space environment, but also on improved magnetic field measurements, in order to decrease inaccuracies for the gain function optimization.

References

[1] Wertz J R 1973 *Spacecraft Attitude Determination and Control* (Dordrecht: Reidel Publishing Company)
[2] Sidi M J 1997 *Spacecraft Dynamics And Control: A Practical Engineering Approach* (Cambridge: Cambridge University Press)
[3] Silani E and Lovera M 2005 *Control Engineering Practice* **13** 357-71
[4] Stickler A C and Alfriend K T 1976 *J. Spacecraft and Rockets* **13** 282-7
[5] Janson S, Hardy B, Chin A, Rumsey D, Ehrlich D and Hinkley D 2012 *26th Annual AIAA/USU Conf. on Small Satellites* (Logan: Utah)
[6] Russell B, Clement L, Hernandez J, Byagowi A, Schor D and Kinsner W 2013 *26th Annual IEEE Canadian Conf.* of *Electrical and Computer Engineering*
[7] Eshghi S and Varatharajoo R 2014 *Applied Mechanics and Materials* **629** 310–7

IOP Conf. Series: Materials Science and Engineering **152** (2016) 012030 doi:10.1088/1757-899X/152/1/012030

Satellite cascade attitude control via fuzzy PD controller with active force control under momentum dumping

Z Ismail[*] and R Varatharajoo

Department of Aerospace Engineering, Universiti Putra Malaysia, Malaysia

[*]zuliana.ismail@gmail.com

Abstract. In this paper, fuzzy proportional-derivative (PD) controller with active force control (AFC) scheme is studied and employed in the satellite attitude control system equipped with reaction wheels. The momentum dumping is enabled via proportional integral (PI) controller as the system is impractical without momentum dumping control. The attitude controllers are developed together with their governing equations and evaluated through numerical treatment with respect to a reference satellite mission. From the results, it is evident that the three axis attitudes accuracies can be improved up to ±0.001 degree through the fuzzy PD controller with AFC scheme for the attitude control. In addition, the three-axis wheel angular momentums are well maintained during the attitude control tasks.

1. Introduction

Small satellites are widely used for various space missions today. Missions like weather forecasting, remote sensing and data or image collections normally require high mission capabilities. Due to that reason, the accuracy and reliability given by the attitude control of satellite system is very important [1]. Satellites equipped with reaction wheels in pyramid configuration and with a wheel momentum dumping system are studied in this paper. As the behavior of the satellite is inherently nonlinear and uncertain, nonlinear controller like fuzzy controller is recommendable for the attitude control. For flexible satellite, adaptive fuzzy sliding mode controls were explained by Liu *et al.* [2]. An adaptive fuzzy system combined with H_2/H_∞ were discussed in detail by Cheng *et al.* [3]. Fuzzy PID-controllers were investigated in [4]. The comparison between type 1 and type 2 fuzzy logic controller can be seen in [5] and fuzzy on-off controller algorithm based on Takagi-Sugeno was introduced in [6].

Besides fuzzy, AFC control scheme has been widely used in many applications; robotic controls [7], vehicle suspension system [8], vibration controls [9] and space system [10]. It has been demonstrated to be better compared to conventional methods. All those studies have proven that the dynamic system remains stable and robust in the presence of known and unknown disturbance or uncertainties. Thus, it seems that AFC and fuzzy are suitable for a dynamic system such as satellites that are exposed to the harmful space environment. This paper is organized as follows: the satellite modelling of three axis satellite attitude controls are described in Section 2 while satellite attitude controllers are presented in Section 3. Simulation and results are shown and described in Section 4. Finally, conclusions are given in Section 5.

2. Satellite modelling

2.1. Attitude dynamics and kinematics

The dynamics of a satellite as actuated by reaction wheels can be written as follows [11]:

$$\dot{\boldsymbol{\omega}} = \mathbf{I}^{-1}(\mathbf{T}_{\mathrm{d}}+\mathbf{T}_{\mathrm{m}} - \mathbf{T}_{\mathrm{w}} - \boldsymbol{\omega} \times (\mathbf{I}\boldsymbol{\omega} + \mathbf{h}_{\mathrm{w}}) \quad (1)$$

where $\dot{\boldsymbol{\omega}}$ is angular velocity of satellite, $\mathbf{I}$ is moments of inertia of the satellite's body, $\mathbf{T}_{\mathrm{w}}$ is RWs torques, $\mathbf{T}_{\mathrm{m}}$ is magnetic control torques, $\mathbf{h}_{\mathrm{w}}$ is the RWs angular momentum and $\mathbf{T}_{\mathrm{d}}$ is total external disturbance torques.

The attitude parameterization via quaternion is employed in this work. The advantage of using quaternion is that it does not require computationally intensive trigonometric function evaluations when derived from the rotation matrix, as compared to Euler Angle kinematic differential equations. Moreover, unlike Euler Angle, quaternion does not suffer from singularity problem. The kinematics of the satellite is developed by integrating the its angular velocity and the equation is as follows [11].

$$\dot{\mathbf{q}} = \frac{1}{2}\Omega(\boldsymbol{\omega}_{LVLH/B})\mathbf{q} \quad (2)$$

where, Ω is the skew symmetric matrix defined as:

$$\Omega = \begin{bmatrix} 0 & \omega_z & -\omega_y & \omega_x \\ -\omega_z & 0 & \omega_x & \omega_y \\ \omega_y & -\omega_x & 0 & \omega_z \\ -\omega_x & -\omega_y & -\omega_z & 0 \end{bmatrix} \quad (3)$$

and $\mathbf{q}$ is the attitude quaternion that represents the attitude of the satellite relative to the local-vertical-local-horizontal (LVLH) coordinate system that is defined as:

$$\mathbf{q} = \begin{bmatrix} q_1 & q_2 & q_3 & q_4 \end{bmatrix}^T \quad (4)$$

where $q_{1,}q_{2,}q_3$ are imaginary parts and q_4 is the scalar part. The quaternion can be converted into Euler angles and the conversion of quaternion vector into the Euler angles in the sequences of roll (ϕ) - pitch (θ) - yaw (ψ) are as follows:

$$\begin{bmatrix} \phi \\ \theta \\ \psi \end{bmatrix} = \begin{bmatrix} \arctan\left(\dfrac{2(q_1q_3 + q_1q_4)}{q_4^2 + q_3^2 - q_2^2 - q_1^2}\right) \\ \arcsin\left(-2(q_1q_3 - q_2q_4)\right) \\ \arctan\left(\dfrac{2(q_1q_2 + q_4q_3)}{q_4^2 - q_3^2 - q_2^2 + q_1^2}\right) \end{bmatrix} \quad (5)$$

2.2. Reaction wheels configuration

Four reaction wheels configuration is often employed to ensure that the satellite can still be controlled by the remaining wheels in case of one or two wheels are malfunctioning. The pyramid configuration is adopted in this research as shown in Figure 1 [12]. In pyramid configurations, the rotational axes of the four wheels have an inclination to X-Y plane by an angle β and each wheel also can induce torques and momentum in the Z-direction.

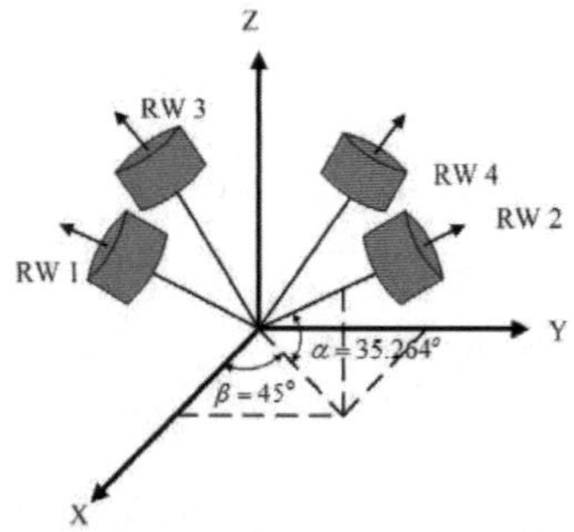

Figure 1. Reaction wheels arrangement in pyramid configuration (adapted from [13])

From the figure, the relation between each reaction wheel torque $T = \begin{bmatrix} T_1 & T_2 & T_3 & T_4 \end{bmatrix}$ and the torques developed along the three body axes of the satellite $T = \begin{bmatrix} T_x & T_y & T_z \end{bmatrix}$ can be detailed as:

$$\begin{aligned} T_x &= T_1 \cos\beta - T_3 \cos\beta \\ T_y &= T_2 \cos\beta - T_4 \cos\beta \\ T_z &= T_1 \sin\beta + T_2 \sin\beta + T_3 \sin\beta + T_4 \sin\beta \end{aligned} \tag{6}$$

Thus the wheel distribution matrix for pyramid configuration is:

$$\mathbf{A}_w = \begin{bmatrix} \cos\beta & 0 & -\cos\beta & 0 \\ 0 & \cos\beta & 0 & -\cos\beta \\ \sin\beta & \sin\beta & \sin\beta & \sin\beta \end{bmatrix} \tag{7}$$

As the pyramid configuration is rotated by an angle α around z axis, the wheel distribution matrix as related to the applied wheel torque can be written as:

$$\begin{bmatrix} T_{wx} \\ T_{wy} \\ T_{wz} \end{bmatrix} = \begin{bmatrix} \cos\beta\sin\alpha & -\cos\beta\sin\alpha & -\cos\beta\sin\alpha & \cos\beta\sin\alpha \\ -\cos\beta\cos\alpha & -\cos\beta\cos\alpha & \cos\beta\cos\alpha & \cos\beta\cos\alpha \\ \sin\beta & \sin\beta & \sin\beta & \sin\beta \end{bmatrix} \begin{bmatrix} T_{c1} \\ T_{c2} \\ T_{c3} \\ T_{c4} \end{bmatrix} \tag{8}$$

3. Satellite attitude controller

In space, the satellite will undergo many situations like detumbling phase, attitude acquisition and 3-axis stabilization state. For 3-axis satellite attitude control using reaction wheels and under momentum dumping mode, AFC is designed to act as disturbance estimator and reject the external disturbance torques while fuzzy PD controller is designed to act as attitude stabilizing controller in this work.

3.1. Active force control

The AFC algorithm is generally simple in structure, fast in computation and robust against unknown disturbances. The implementation of AFC is easy because the AFC components can be cascaded to the baseline controller [14]. Moreover, computational burden of the control system is alleviated in AFC control scheme, owing to the direct force measurements from the sensors that are readily installed in real-time [9]. The AFC idea used in satellite attitude control application relies on the measurement of angular acceleration, which can be translated to the estimated torques via crude approximation of the satellite inertia [10]. The estimated torques are then fed back to the controller to make the actual contact torques equal to the desired control torques, given by the Attitude Control and Determination

IOP Conf. Series: Materials Science and Engineering **152** (2016) 012030 doi:10.1088/1757-899X/152/1/012030

System (ACDS) specification. In space applications, constant external disturbances will cause the accumulation of momentum in the reaction wheel. If the momentum of the wheel is not actively controlled, the momentum will increase infinitely, thus saturating the wheel and hindering its ability to deliver the necessary control torque. This excess momentum must be dumped for the wheel to operate effectively. The block diagram for AFC with momentum dumping control loop is shown in Figure 2. The Fuzzy PD controller is the stabilizing controller.

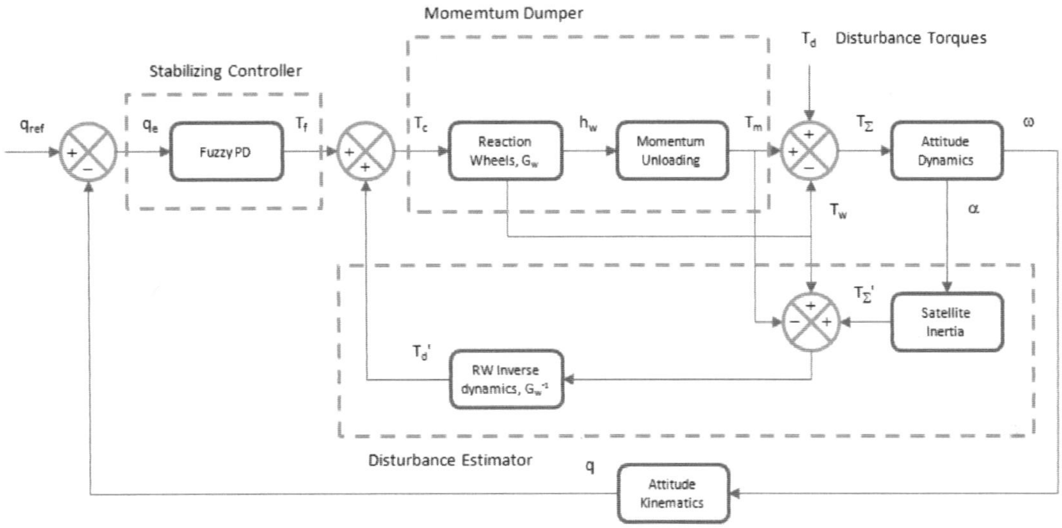

Figure 2. Block diagram of fuzzy PD controller-AFC with momentum dumping control loop

From the diagram, the estimated disturbance torque $T_d^{'}$ is given by:

$$T_d^{'} = T_\Sigma^{'} + T_w - T_m \tag{1}$$

where the total torques $T_\Sigma^{'}$ acting on the satellite can be estimated using the formula:

$$T_\Sigma^{'} = I'\alpha \tag{2}$$

and the estimated satellite inertia property, $I^{'}$ can be computed from law of conservation of angular momentum. Because the satellite inertia property is a slow-varying parameter, the ideal inertia model can be used, $I^{'} = I$ for simulation purpose [10]. The angular acceleration, α is assumed to be measured using the angular accelerator.

In the AFC scheme, the control law T_c is given by:

$$T_c = T_f + T_d^{'} \tag{3}$$

where T_f is desired control torque from the Fuzzy PD controller. Assuming perfect actuator dynamics, the actual torque delivered by the reaction wheel, $T_w = T_c$ and the disturbance torque, T_d is eliminated through feed-forward compensation:

$$\begin{aligned} T_\Sigma &= T_m - T_w + T_d \\ T_\Sigma &= T_m - \left(T_f + T_d^{'}\right) + T_d \\ T_\Sigma &= T_m - T_f \end{aligned} \tag{4}$$

The magnetic torque, T_m is necessary to remove the excess momentum accumulated in the reaction wheel.

3.2. Fuzzy PD controller
In designing the Fuzzy PD controller, basic control scheme of PD control has to be constructed first.

3.2.1. PD controller. For the 3-axis satellite attitude control, a common quaternion feedback control scheme based on PD control is employed. The control law thus can be represented as follows [11]:

$$\mathbf{T}_{PD} = 2\mathbf{K}_p\mathbf{q}_e q_{e4} + \mathbf{K}_d\boldsymbol{\omega} \tag{13}$$

where the error quaternion, $\boldsymbol{q}_e$ is the quaternion difference between the reference quaternion, $\boldsymbol{q}_r$ and the current quaternion, $\boldsymbol{q}_c$. $\mathbf{T}_{PD}$ is a $[3 \times 1]$ vector because $\mathbf{q}_e = (q_1, q_2, q_3)$ is a vector and q_4 is a scalar. The control gains $\mathbf{K}_p$ and $\mathbf{K}_d$ are matrices defined by:

$$\mathbf{K}_p = \omega_n^2\mathbf{I} \tag{14}$$

$$\mathbf{K}_\mathrm{d} = 2\xi\omega_n\mathbf{I} \tag{15}$$

These control gains are the functions of dynamic characteristics, i.e., the natural frequency ω_n and the damping ratio ξ.

3.2.2. Fuzzy controller. Numbers of research have shown the successfulness of fuzzy applications on satellite attitude control. The four main components of fuzzy are fuzzification, rule base, rule inference and defuzzification. The fuzzification involves converting the input and output data into suitable linguistic values, which may be viewed as labels of fuzzy sets. In this work, the linear Sugeno-type Fuzzy Inference System (FIS) [16] is employed to design the fuzzy controller. The inputs to the fuzzy controller are the error, $E(s)$ and its derivative change in error, $CE(s)$. The linear approximation of control torque by fuzzy controller is given as follow:

$$\begin{aligned} \mathbf{T}_f(s) &= [GE * E(s) + GCE * CE(s)] * GU \\ &= GE * GU\left[E(s) + \frac{GCE}{GE} * CE(s)\right] \end{aligned} \tag{16}$$

where the error, $E(s)$ is computed by comparing the desired attitude and actual attitude, whereas the change of error, $CE(s)$ is generated by the derivation of the error.

By assuming the error is bounded by ±100% of the step input, so $GE = 100$. As GE is known and by comparing Equation 16 with PD characteristic equations, the gain of GU can be determined by:

$$GU = \frac{K_p}{GE} \tag{17}$$

$$GCE = \frac{K_d}{GU} \tag{18}$$

In addition, as in Figure 2, the output variable of FLC is the control torque, T_f applied to the actuator. Thus the minimum fuzzy decision rules with two inputs consist of the four rules:

- IF $e(t)$ is Negative (Neg) and $\dot{e}(t)$ is Negative (Neg), THEN u is Negative Big.
- IF $e(t)$ is Negative (Neg) and $\dot{e}(t)$ is Positive (Pos), THEN u is Zero.
- IF $e(t)$ is Positive (Pos) and $\dot{e}(t)$ is Negative (Neg), THEN u is Zero.
- IF $e(t)$ is Positive (Pos) and $\dot{e}(t)$ is Positive (Pos), THEN is u Positive Big.

Figure 3 shows the Sugeno-type fuzzy inference method that has been implemented in this work.

The next step after verifying the linear Fuzzy PD controller is making FIS to achieve nonlinear control surface and fine-tune it. For non-linearization, fuzzy sets with Gaussian membership functions

(MFs) are used and it has been chosen because of their curves have the advantage of being smooth and nonzero at all points. The MFS are evenly distributed so that the tuning process of the controller can be easily done. Figure 4 shows the nonlinear control surface. The higher gain is obtained near the center of the error and change of error plane than the linear surface has. This higher gain helps quickly reduce the error when it is small.

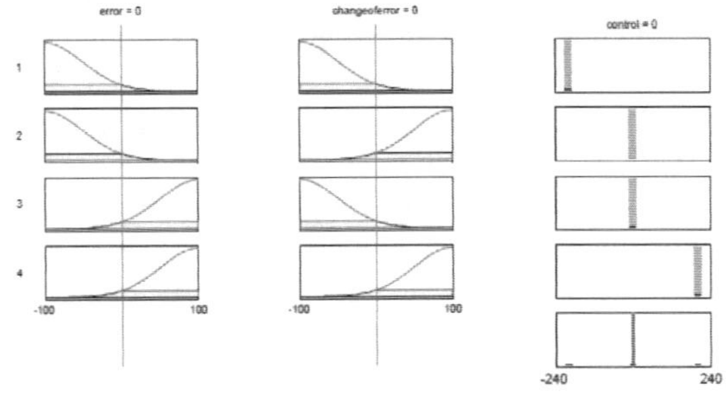

Figure 3. Sugeno fuzzy inference

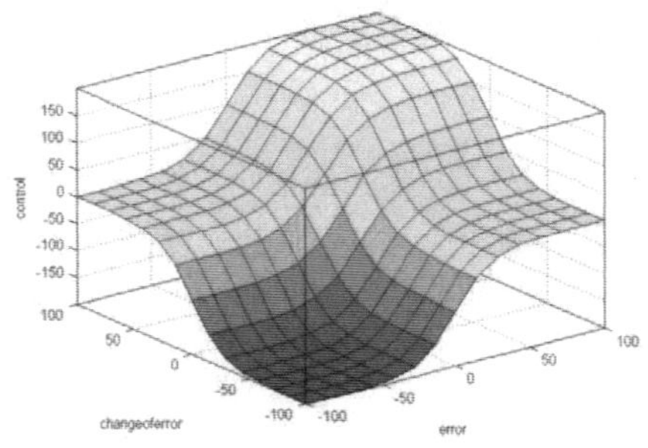

Figure 4. Nonlinear control surface

3.3. Momentum dumping control

The purpose of momentum dumping is to reject the excess angular momentum accumulated by the reaction wheels. This can be done by exerting secondary external control torques generated by the magnetic torquers. Basically, the magnetic control torques, T_m can be generated by multiplying the magnetic dipole moments, m with Earth's magnetic field strength in body coordinate systems as indicated by Equation 19.

$$\begin{bmatrix} T_{mx} \\ T_{my} \\ T_{mz} \end{bmatrix} = \begin{bmatrix} m_y B_z - m_z B_y \\ m_z B_x - m_x B_z \\ m_x B_y - m_y B_x \end{bmatrix} \tag{19}$$

By adopting the Proportional-Integral (PI) controller, the magnetic dipole moment, **m** values can be calculated as:

$$\mathbf{m} = -\frac{\mathbf{K}_p + \frac{\mathbf{K}_i}{s}}{B^2}(\mathbf{B}_{LVLH} \times \Delta\mathbf{h}) \tag{20}$$

where $\Delta\mathbf{h}$ is the excess wheel angular momentum to be removed. The proportional and integral gains can be obtained through $\mathbf{K}_p = 2\xi\omega_n$ and $\mathbf{K}_i = \omega_n^2$, respectively. On the other hand, $\mathbf{B}_{LVLH}$ represents the Earth's magnetic field strength in LVLH coordinate systems. For high inclination orbit, the magnetic moment **m** then can be re-written as:

$$\begin{bmatrix} m_x \\ m_y \\ m_z \end{bmatrix} = \begin{bmatrix} \dfrac{\sin\eta(k_y \Delta h_{wy})}{-B_z} \\ \dfrac{\sin\eta(k_x \Delta h_{wx})}{B_z} - \dfrac{\cos\eta(k_z \Delta h_{wz})}{B_x} \\ \dfrac{\cos\eta(k_y \Delta h_{wy})}{B_x} \end{bmatrix} \tag{21}$$

where η is the instantaneous satellite angle ($\eta = \omega_o t$) and note that B is the magnitude of the vector $\mathbf{B}_{LVLH}$, whereby $B = \sqrt{B_x^2 + B_y^2 + B_z^2}$.

4. Simulation and results

To have a basis of comparison, the simulation was first run for satellite attitude control system with PD controller only, second for Fuzzy PD controller and then for Fuzzy PD controller with AFC scheme. The simulations have been performed using Matlab®-Simulink™ software in order to evaluate the satellite attitude performances for the three simulations and momentum dumping control is also enabled for the all simulations. The mission parameters of satellite actuated by four pyramidical reaction wheels with the wheel momentum dumping actuated by magnetic torquers are listed in Table 1 and Table 2.

Table 1. Mission parameter

Inclination, *i*	**Altitude, h**	**Orbital period, t**	**External disturbance torques**
83 [deg]	470 [km]	22560 [s] (4 orbit)	$T_{dx} = 8\times10^{-5}\sin(\omega_o\cdot t)$ [Nm], $T_{dy} = 8\times10^{-6} + 8\times10^{-5}\sin(\omega_o\cdot t)$[Nm] $+5\times10^{-5}\cos(\omega_o\cdot t)$ [Nm] $T_{dz} = 8\times10^{-6} + 5\times10^{-5}\cos(\omega_o\cdot t)$ [Nm]

Table 2. Satellite and gain parameters

Parameters		**Values**
Moments of inertia matrix [kgm^2]		$I_{xx} = 4.2, I_{yy} = 4.4, I_{zz} = 4.2$
Wheels configuration matrix		$\beta = 45^o$, $\alpha = 35.64^o$
Limit of the reaction wheels torques [Nm]		$\mathbf{T_c} = (0.1 \quad 0.1 \quad 0.1 \quad 0.1)^T$
Gain parameters of PD	$\mathbf{K_p}$[Nm/rad]	$K_{px} = 0.458, K_{py} = 0.573, K_{pz} = 0.332$
	$\mathbf{K_p}$[Nms/rad]	$K_{dx} = 2.775, K_{dy} = 3.213, K_{dz} = 2.363$
Gain parameters of fuzzy		GU = 0.00458 , GCE = 604.8035
Saturation limit of magnetic torquers[Am^2]		$\mathbf{m} = \pm 1$
Gain parameters PI for wheel momentum dumping	$\mathbf{K_p}$[rad/s]	$K_{px} = 0.00016, K_{py} = 0.006, K_{pz} = 0.004$
	$\mathbf{K_p}\left[\frac{rad^2}{s^2}\right]$	$K_{ix} = 6.4\times10^{-9}, K_{iy} = 9\times10^{-8}, K_{iz} = 4\times10^{-8}$

The results of the simulations of the attitude control with PD, Fuzzy PD Controller and Fuzzy PD Controller with AFC are as depicted in Figure 5, Figure 6, Figure 7 and Figure 8. Comparing Figure 5 and Figure 6, the performance of Fuzzy PD controller is better than the PD controller. As can be seen in these figures, ±0.015° have been achieved by PD controller while ±0.005° have been achieved by attitude control system using Fuzzy PD Controller. Nevertheless, it can be seen in Figure 7 that the attitude accuracies for roll-pitch-yaw axes have been improved by implementing Fuzzy PD controller with AFC scheme, where ±0.001° attitude accuracies can be achieved. Figure 8 shows the graph of the wheel angular momentum. Over the longer times, the result of simulation shows that the wheel angular momentum was maintained within the acceptable limits during the attitude control operation.

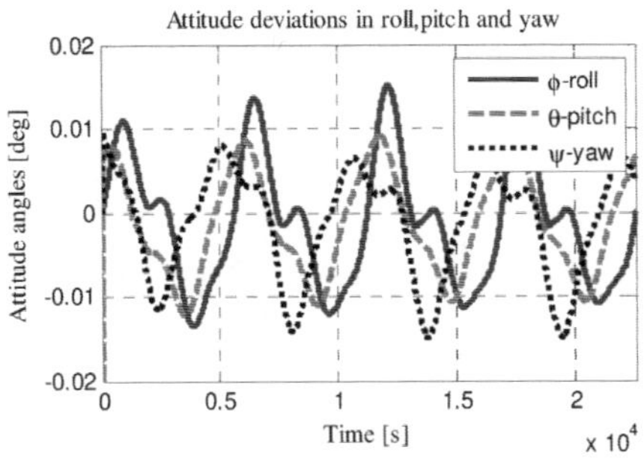

Figure 5. Attitude responses of with PD controller

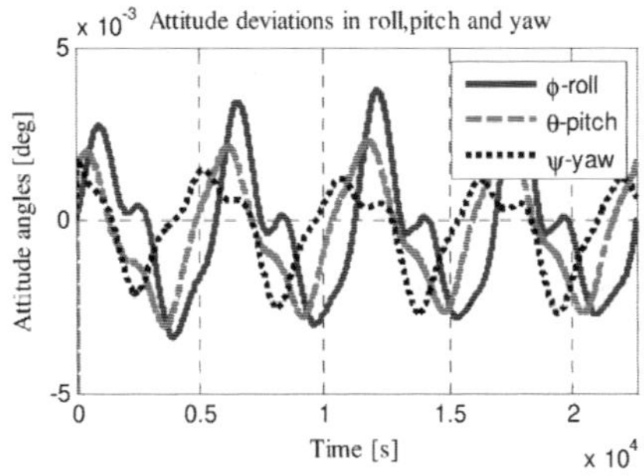

Figure 6. Attitude responses of with Fuzzy PD controller

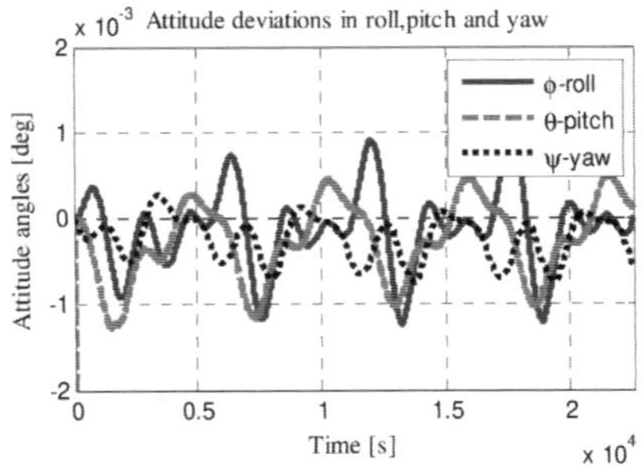

Figure 7. Attitude responses of with fuzzy PD controller with AFC

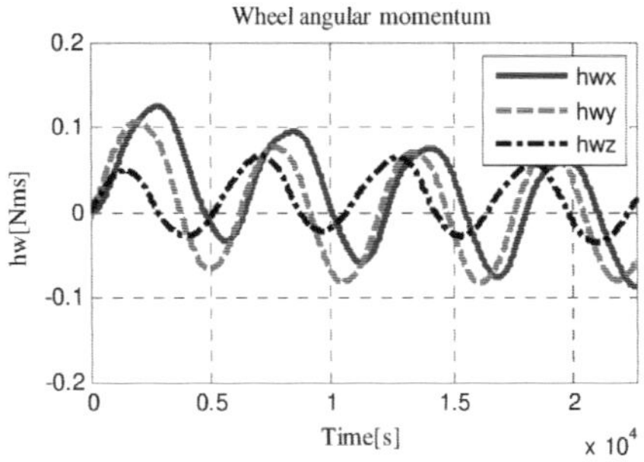

Figure 8. Wheel angular momentum responses

5. Conclusion

The satellite attitude control with reaction wheels via Fuzzy PD controller with AFC is described and tested. The results show that satellite attitude control incorporating the momentum dumping mode with Fuzzy PD controller with AFC demonstrated better attitude performances (±0.001°) in controlling all three satellite axes compared to the satellite attitude control with PD controller or Fuzzy PD controller only. For all options, the three-axis wheel angular momentums are well maintained during the attitude control tasks.

References

[1] Renuganth, V 2006 *Aircraft Engineering and Aerospace Technology* **78** 495-501
[2] Liu X J, Guan P and Liu J Z 2005 *Decision and Control, European Control Conference*
[3] Cheng C H, Shu S L and Cheng P J 2009 *Expert Systems with Applications* **36** 6613-20
[4] Gomes W and Rocco E M 2012 *Workshop on Space Engineering and Technology*
[5] Asaee A, Balochian S and Heshmati S 2013 *Buletinul Institutului Politehnic Din Iasi*
[6] Rezanezhad S 2014 *International Journal of Electrical and Computer Engineering* **4** 303
[7] Kwek L C, Wong E K, Loo C K and Rao M V C 2003 *Journal of Intelligent and Robotic Systems* **37** 143-62
[8] Priyandoko G and Mailah M 2007 *International Journal of Simulation Modelling* **6** 25–36
[9] Hashemi-Dehkordi S M, Abu-Bakar A R and Mailah M 2012 *Sadhana* **37** 637-55
[10] Varatharajoo R, Wooi C T and Mailah M 2011 *Acta Astronautica* **68** 2025-28
[11] Sidi M J 1997 *Spacecraft Dynamics and Control: A Practical Engineering Approach* (Cambridge: Cambridge University Press)
[12] Ismail Z and Varatharajoo R 2010 *Advances in Space Research* **45** 750-9
[13] Won C H 1999 *Aerospace Science and Technology* **3** 323-33
[14] Ismail Z, Varatharajoo R, Ajir R and Mohd Rafie A S 2015 *Aircraft Engineering and Aerospace Technology* **87** 546-50

Design of a terminal node controller hardware for CubeSat tracking applications

Y A Ahmad[1*], N J Nazim[2] and S S Yuhaniz[3]

[1]Kulliyyah of Engineering, International Islamic University Malaysia, Malaysia
[2]Faculty of Engineering, Universiti Selangor, Malaysia
[3]Advanced Informatics School, Universiti Teknologi Malaysia, Malaysia

[*]yasser.asrul@gmail.com

Abstract. CubeSats enable low-cost experiment and missions to be performed by universities and research institution in space. CubeSats for research use UHF and VHF communication for its tracking and telemetry applications. The current practice of a CubeSat communication is to modify radio amateur's Terminal Node Controller (TNC) to enable data to be received in the ground station. The objective of this research is to design a hardware specifically for use as a TNC for CubeSat tracking applications. A TNC is developed as an interface to the terminal and to serve as data packetization platform. The modem is integrated with a microcontroller unit (MCU) and an audio amplifier to enable the audio signals to be smoothened, amplified and interfaced with the radio. The modem, MCU and audio amplifier circuitry are designed and integrated to form a TNC platform suitable for CubeSat communication.

1. Introduction

CubeSat technology has enabled scientific research and development of space technology to be done with relatively low cost, which facilitates affordable space research and development at the university levels. In addition, many new space companies have also mushroomed based on the use of CubeSat concept for commercial applications of remote sensing, radio amateur and scientific research in space [1]. Essentially, CubeSat is a nanosatellite-class of satellites with fundamental weight of one kilogram and a volume of 10 cm x 10 cm x 10 cm. This standard is accepted worldwide and is known as 1U, where U denotes unit. The CubeSat is expandable into 2U, 3U, 6U and even 12U.

The CubeSat is basically a miniaturization of classical satellite, which is heavy, big and expensive in nature. It consists all subsystems in a satellite such as electrical power system, onboard computer, attitude determination and control system, communication system and mechanical system. CubeSats are launched into the Low Earth Orbit (LEO) as a secondary payload piggyback. Once deployed into orbit, CubeSat is tracked and communicated via a ground station.

2. CubeSat communication

CubeSat typically uses amateur radio frequencies and protocol for its packet data transmission [2, 3]. Based on the survey made by Klofas and Leveque [4], 70% of the total CubeSats launched into orbit between 2009 to 2012 used amateur radio frequencies as a part of their communication systems. Many CubeSats that are developed by universities [5, 6, 7] used slower data rates due to the ease of getting and implementing the hardware. These projects were mostly decided to build entire transceiver out of individual components that allow for tighter control of requirements and specification. Components of these custom-built transceivers also include terminal node controller (TNC), transceiver and amplifier. TNC are typically built and embedded in the satellite On Board Computer (OBC) and transceiver, but in term of TNC in the ground station, the commercial-off-the shelf (COTS) components are typically used [2, 4].

The TNC is used in the CubeSat packet radio communication, where digital signals are enabled to propagate using radio waves. It is a device used by amateur radio operators to participate in AX.25 packet radio networks [8, 9, 10, 11]. The primary usage of a TNC is as a bridge between a computer or a programmable device to radio transceiver with packet network capability [9, 10]. TNC functions as a modem and ensures the data are formatted and packetized based on packet radio requirements such as AX.25. TNC system is separated by digital part, analog part and TNC software [11]. In the CubeSat system, the experimental data will be collected and downloaded to the ground station. The telemetry data is also downloaded to the ground station. Most of the TNCs used in CubeSat missions are mostly COTS based [4]. Though the COTS TNC are implementing AX.25 packetization, the performance of each COTS is not the same because of differences and compatibility issues arises due to factors such as implementation details of AX.25 protocol, properties of the software of the TNC microcontroller and hardware properties, which affect the performance of a TNC in terms of effective transmission speed and network efficiency [11]. The use of a COTS TNC in the ground station will result in the compatibility issues and poor communication link with a CubeSat in space. Reliance on the COTS components are making the CubeSat mission difficult because the software and hardware properties of COTS components cannot be changed as it is set by the manufacturer. Therefore, it is important to have access to TNC source code for modification and capability to design the TNC hardware. A self-developed TNC for CubeSat would enable compatibility issues to be fine-tuned during the CubeSat development and modified even in the satellite operations.

A common ground station consists of an antenna system, a receiver, a transmitter, a terminal node controller, computers and operation software. TNC consists of two major components: microcontroller unit (MCU) and modem. The main function of the MCU is to implement the AX.25 packet protocol in the embedded source codes. The output of AX.25 packet encoding process will be the packet frame, which is the data in digital signal (TTL). The digital modulation must then be performed to convert the digital signal into analog signals before it that can be propagated using radio waves. The modulation process will be performed by the digital modem. The AX.25 packet protocol is a derived version of X.25 protocol that is formerly used in amateur radio packet transmission [12, 13]. It is an amateur radio specification that describes how to encode digital data to transmit it over radio frequencies. The AX.25 specification mandates a bit rate of 1200 baud and uses audio frequency shift keying (AFSK) encoding to represent binary values 0 and 1 with audio tones of 1200 Hz and 2200 Hz, respectively [14, 15]. The modem will operate in AFSK for lower baud rate and in frequency shift key (FSK) and Gaussian minimum shift keying (GMSK) for higher baud rate. Both AFSK and also FSK/GMSK modulation methods are very common in CubeSat communications [2, 3, 4]. AFSK digital modulation is implemented to convert the digital signals into analog signals before it that can be propagated using the radio waves. Figure 1 illustrates the position of a TNC and functions of its components in satellite communications.

3. The AX.25 frame

Data link layer packet radio transmissions are sent in small blocks of data called frames. Each frame is made up of several smaller groups called fields. Each field is made up of an integral number of octets (8-bit binary) and serves specific functions. All fields except the Frame Check Sequence (FCS) are transmitted low-order bit (LSB) first while FCS is transmitted high-order bit (MSB, i.e. bit-15) first [14, 15]. Specifically, the Unnumbered Information (UI) frame format of AX.25 protocol is utilized in most CubeSat communications schemes.

AX.25 UI frame shown in Table 1 starts with a flag and followed by the frame header, information (data), CRC checksum (FCS) and ends with another flag. The packet frame header consists of the destination address, source (origin) address, control bits and a protocol identifier (PID). Both of the destination and source addresses are included with Secondary Station Identifier (SSID).

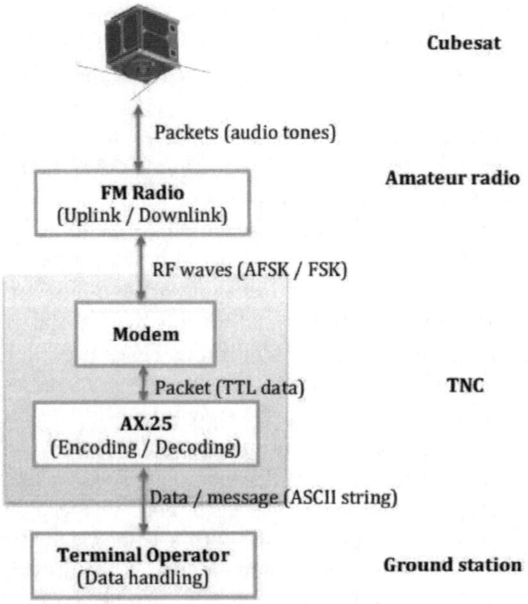

Figure 1. Application of a TNC in communication between ground station and CubeSat

Table 1. AX.25 Unnumbered Information (UI) frame [14]

Flag	AX.25 Transfer Frame Header (128 bits)				Information Field	Frame-Check Sequence	Flag
	Destination Address	Source Address	Control Bits	Protocol Identifier			
8	56	56	8	8	0-2048	16	8

3.1. Flag field

A flag is a specific number indicating start and end of a frame. It is one octet long, occurs at both the starting and the end of each frame. Two frames may share one flag, which denotes the end of the first frame and the start of the next frame. A flag consists of a zero, followed by six '1's and then another zero (b01111110 or 0x7E) [15, 16]. To ensure that the flag bit sequence mentioned does not appear accidentally anywhere else in a frame, bit stuffing is applied. The sender monitors the bit sequence for a group of five or more contiguous '1' bits. Any time five contiguous '1' bits are sent, the sending station inserts a '0' bit after the fifth '1' bit. During frame reception, any time five contiguous '1' bits is received, a '0' bit immediately following five '1' bits is discarded [15, 16].

3.2. Address field

The address field identifies both source of the frame and also its destination. The destination address consists of the callsign and the SSID of the destination, as shown in Table 2.

Table 2. Destination address frame field (56 bits) [14]

Callsign (48 bits)			SSID 8 bits
C1 (8 bits)		C6 (8 bits)	
X XXXXXX 0	....	X XXXXXX 0	0 1 1 SSID 1

The callsign is made up of six upper-case letters, numbers or space ASCII characters only (7 bits). The SSID is four-bit integer that uniquely identifies multiple stations using the same amateur callsign. The six characters of the callsign are placed in the first six octets of the field (C1 to C6). Each of the character bits are shifted one bit on the left and the LSB is set to '0'. The SSID is placed in bits 3-6, while other bits of the field are contained with fixed values [15].

Table 3. Source address frame field (56 bits) [13]

Callsign (48 bits)			SSID 8 bits
C1 (8 bits)	....	C6 (8 bits)	
X XXXXXX 0		X XXXXXX 0	0 1 1 SSID 1

The source address frame shown in Table 3 consists of the callsign and the SSID of the source. The callsign and the SSID configurations are similar to the destination callsign and SSID as stated earlier. The only bit that differs from the destination address field is the leftmost bit, which is set to '1'. This indicates that the octet to which it belongs is the last octet of the address fields [14].

3.3. Control field

The control field identifies the type of frame being passed and controls several attributes of the Layer 2 connection. For an AX.25 Unnumbered Information Frame, its value is always b00000011 (0x03) [14, 15, 16].

3.4. Protocol identifier (PID) field

The protocol identifier (PID) field appears in information frames (I and UI) only. It identifies the kind of Layer 3 protocol, if any, that is in use. The PID itself is not included as part of the octet count of the information field [16]. The definition of PIDs can be referred to Table 4. For this TNC application, the PID shall be 11110000 (0xF0).

Table 4. PID definitions of an AX.25 packet frame [15]

Hexadecimal	MSB LSB	Translation
**	yy01yyyy	AX.25 layer 3 implemented
**	yy10yyyy	AX.25 layer 3 implemented
0x01	00000001	ISO 8208/CCITT X.25 PLP
0x06	00000110	Compressed TCP/IP packet. Van Jacobson (RFC 1144)
0x07	00000111	Uncompressed TCP/IP packet. Van Jacobson (RFC 1144)
0x08	00001000	Segmentation fragment
0xC3	11000011	TEXNET datagram protocol
0xC4	11000100	Link Quality Protocol
0xCA	11001010	Appletalk
0xCB	11001011	Appletalk ARP
0xCC	11001100	ARPA Internet Protocol
0xCD	11001101	ARPA Address resolution
0xCE	11001110	FlexNet
0xCF	11001111	Net/ROM
0xF0	11110000	No layer 3 protocol implemented
0xFF	11111111	Escape character. Next octet contains more Level 3 protocol information
Escape character. Next octet contains more Level 3 protocol information	00001000	

3.5. Information field
The information field conveys the user data (message) from one end of the link to the other. The maximum (default) size of the field is 2048 bits (256 octets) [14, 15, 16].

3.6. Frame check sequence
The frame check sequence (FCS) is a 16-bit (2 octets) number calculated by the sender of a frame. It ensures that the frame was not corrupted by the transmission medium. The FCS is calculated in accordance with recommendations in the HDLC reference document, ISO 3309 [14, 16]. The FCS will be recalculated in the receiver. If the FCS does not match with the transmitted value, the frame will be aborted [16]. For this project, the FCS is a CRC calculated using polynomial x^16+x^12+x^5+1 or also known as CRC-CCITT.

4. Modulation
The term modem is a composite word that refers to the two functional entities that make up the device – a signal modulator and a signal demodulator [14]. Modulation refers to the process of transforming information of digital data into analog signals for transmission. Once the data reaches its intended destination, demodulation is performed to transform the analog signal into digital data and recovers back the original information [15]. Figure 2 illustrates the application of a modulator and demodulator at both transmitter and receiver ends.

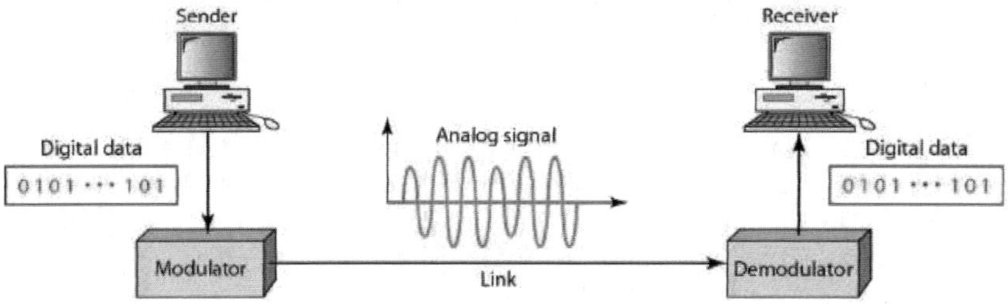

Figure 2. Digital to analog modulation [14]

A wireless modem as in the TNC converts digital data into radio signals at the transmitter and then converts the radio signal back to the digital data at the receiving end [14]. One of the most applicable digital modulation methods to transmit the radio signal using FM radio carrier is the frequency shift keying (FSK).

4.1. Frequency shift keying
In frequency shift keying (FSK), the system starts by generating a Non Return Zero (NRZ) baseband signals. The NRZ signal is where binary 1 is transmitted with square pulse of voltage +V and binary 0 is transmitted with square pulse of -V. Since FSK is based on frequency modulation, it is the carrier that varies in accordance with the information baseband signal. Since baseband signal takes only two values, it follows that frequency of modulated waveform also takes one of two values. The modulation is referred as a keying operation hence, it is known as FSK. In FSK, the carrier frequency is shifted or varied in accordance with the baseband signal. The signal frequency during each bit duration is constant and its value depends on the bit, either 0 or 1. Both peak amplitude and phase remain constant [14, 15]. The FSK signals can be characterized by feeding data into a frequency modulator. The FSK signals are represented by:

$$s(t) = A_c\left[2\pi f_c t + 2\pi k_f \int_{-\infty}^{t} m(t)\, dt\right] \tag{1}$$

where *m(t)* is the baseband signal. The serial data input is binary, hence the resulting FSK signals are called binary FSK signal. The baseband signal is bipolar 0 (-V) or 1 (+V). The integration guarantees the phase is instantaneous. The instantaneous frequency is given by:

$$f_i t = f_c \pm k_f V \tag{2}$$

The maximum frequency deviation Δf is $k_f V$. If the data signal is defined as $d_i(t)$ depending upon either 0 or 1 is being sent, the frequency of the FSK waveform is then given by:

$$f_i t = f_c \pm \Delta f d_i(t) \tag{3}$$

Figure 3 visualizes the concept of an FSK signal. FSK overcome most of the problems caused by noise that might happen in another modulation method such as in amplitude shift keying (ASK) [15].

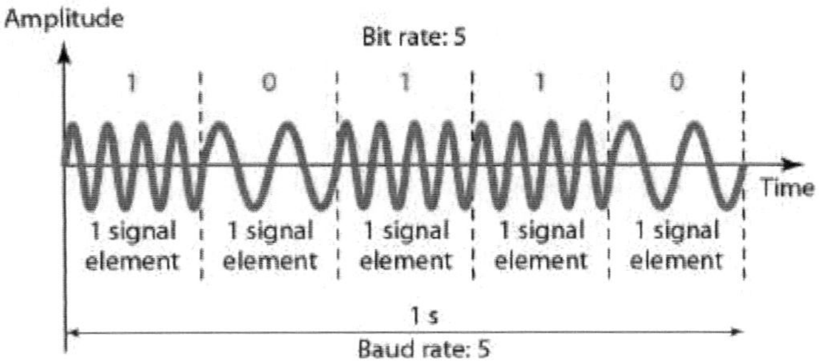

Figure 3. Frequency Shift Keying (FSK) signal [5]

4.2. FSK detection

In terms of demodulation, the non-coherent detection is preferred which does not require estimation of carrier phase. The implementation of FSK leads to orthogonal signaling where each tone is separated and cannot interfere with each other.

$$2\pi (f_1 - f_2)T = 2k_f\pi \tag{4}$$

The frequency spacing of FSK is thus k_f/T with minimum spacing of $1/T$. Simple implementation of the detector is done by using two bandpass filters tuned to each frequencies carrying bit 0's and 1's. The output of the detector is envelope-detected and then baseband-detected. This type of detector simply evaluates which of the two possible sinusoids is stronger and the receiver. The implementation of the detector is shown in Figure 4.

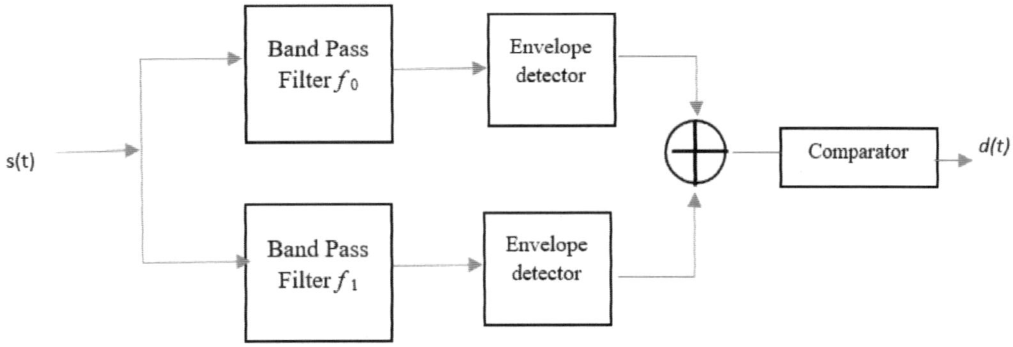

Figure 4. AFSK Non-coherent demodulator

5. Design

The integrated TNC consists of two modules: MCU module and modem module, has been designed as shown in Figure 5. The design of the TNC consists of an Atmega328P microcontroller, power supply circuitry, external clock and USB interface to terminal and PC. The modem module consists of the AFSK receiver and transmitter.

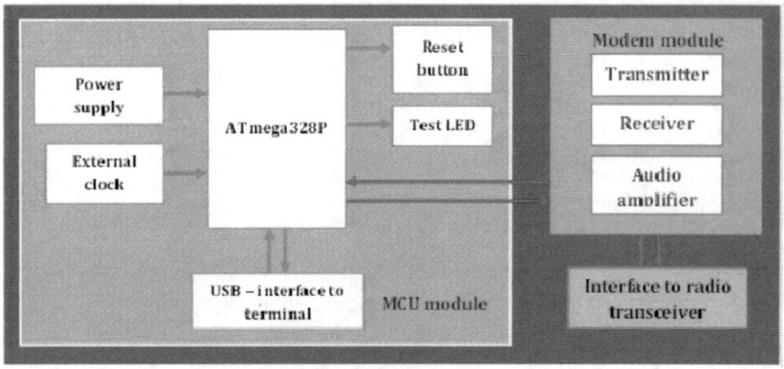

Figure 5. TNC block diagram

After the completion of the hardware design, a printed circuit board is design for the complete TNC board. The PCB is fabricated and use for testing of the modules. Figure 6 shows the printed circuit board of the TNC.

Figure 6. TNC PCB board

5.1. MCU module

The MCU module is powered by a power supply circuitry providing 5V voltages to all the circuit. The system requires a +12V DC input, which will be regulated to +5V. The MCU module interfaces with a USB to TTL board for interfacing with the terminal. The TNC MCU module prototype is connected to the terminal via USB Type-A cable. The interface enables hex file to be uploaded to the ATmega328P IC via the USB – TTL serial connection. The board pins must be correctly connected to TXD, RXD, VCC, +5V and ground pins of the ATmega328P IC. In addition to that, reset button of ATmega328P must be pressed while uploading the program's hex file from the IDE to the microcontroller IC. Note that the ATmega328P microcontroller is pre-programmed with the bootloader such that direct upload of program's hex file to the IC is possible. The external clock is a 16 MHz crystal oscillator circuitry.

The MCU module is interfaced with the modem module through the digital Pulse Width Modulation (PWM) output for modulation and the analog input for receiver modulation. Figure 7 provides the detail of the MCU module.

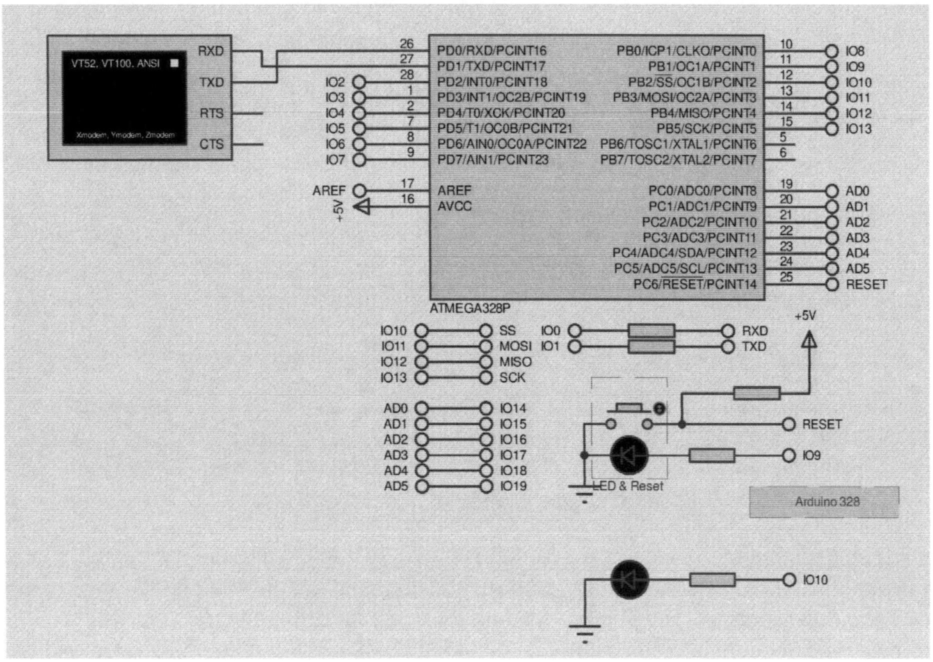

Figure 7. MCU module

The MCU module is validated by testing the software uploading process of the hex file into the Atmega328p via the USB – TTL serial connection. This is done successfully and the TNC software is able to run on the MCU module as shown in Figure 8. A serial monitor software is used to monitor data, which are able to be viewed from PC and the MCU module corresponds to the request made to the TNC.

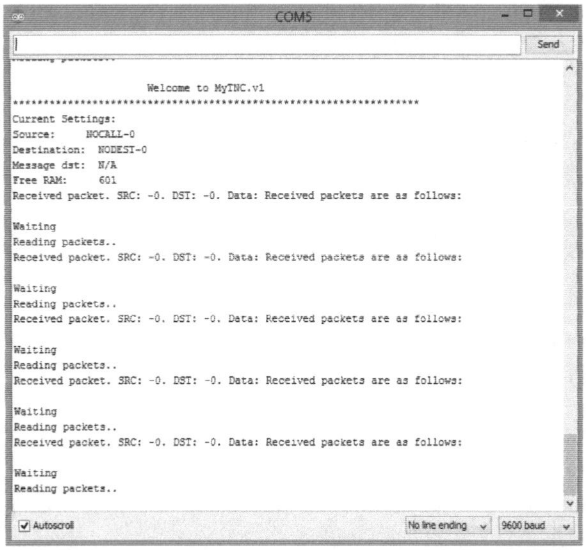

Figure 8. Serial data output from MCU module

5.2. Modem module

The modem module is made up of two parts: transmitter circuit (for modulation) and receiver circuit (for demodulation). The transmitter digital to analog conversion (DAC) circuit is shown in Figure 9 includes several resistors and capacitors.

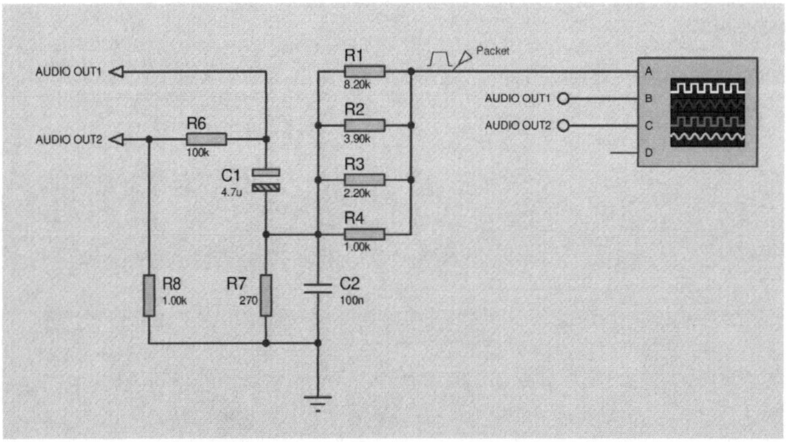

Figure 9. DAC for AFSK modulation

Software modem, which is also known as soft-modem, is implemented in the design. The software is housed in the MCU module. One of the reasons for this is to simplify the hardware modem, hence minimizing the dimensions of TNC board, leading to reduction of the overall costs. Several functions involved in the soft-modem, including digital-to-analog (DAC) in modulation module, and analog-to-digital (ADC) in demodulation module. In Figure 10, the input the digital form is compared with the DAC output.

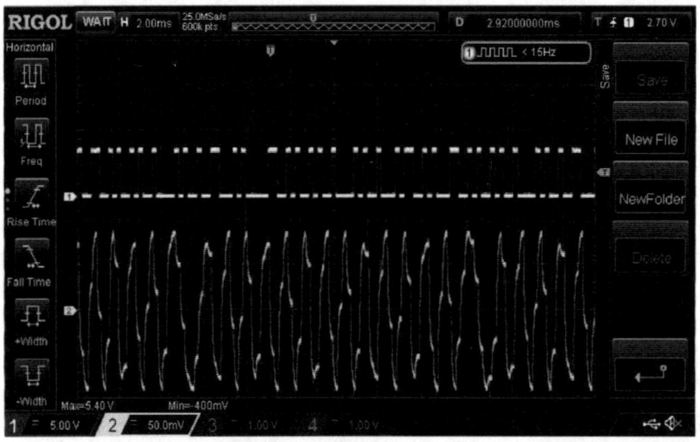

Figure 10. DAC output

Based on Figure 10, the based band data are converted into analog signals. However, the DAC is not smooth due to the small number of taps implemented. In order to smoothen the signals, the audio amplifier with bandpass filter is developed to be included in the transmitter circuit. The purpose of the amplifier is to boost up the weak signal generated at the audio output and smoothen the conversion effect for a clean conversion of AFSK signals. The band-pass filter functions to allow the AFSK signal to be transmitted in the range of band-pass frequencies from 1200 – 2200 Hz. Figure 11 provides the audio amplifier circuitry.

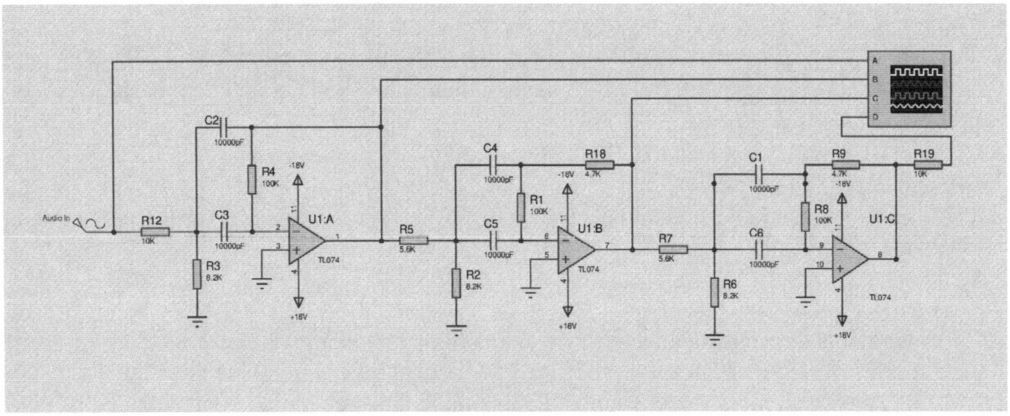

Figure 11. Audio amplifier

Figure 12 shows the output of the modem. The audio amplifier output is connected to oscilloscope to verify the output. It is observed that the modulated data has properly been smoothening by the implementation of audio amplifier and bandpass filter. The two tones of AFSK signals are observed as well.

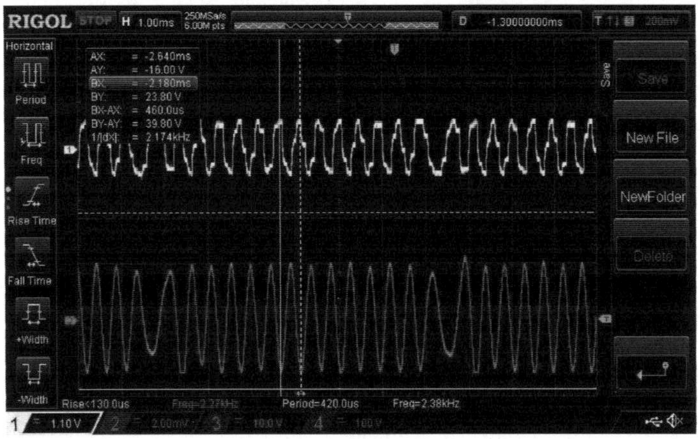

Figure 12. Modem output

6. AX.25 implementation in the TNC

The A.X25 program functions are divided into two modules: transmitter (packet encoder) and receiver (packet decoder). The source codes for each module are developed based on AX.25 UI frame format structure shown earlier. The transmitter module performs encoding messages (information) into the AX.25 packet forms while the receiver module decodes the AX.25 packets to retrieve the original message. The procedures for both modules are illustrated in the flow chart of Figure 13 and Figure 14, respectively.

6.1. Packet encoder

AX.25 transmitter module implements packet encoding functions like start and end flags insertion, bit stuffing and FCS generation for the CRC checksum. Based on Figure 13, packet encoding procedures are performed as follows:

- While the transmitter module is not reading data from input line, the flag insertion function will continue asserting flags
- The callsigns of both source and destination are asserted to start the encoding process. The SSID of the last callsign input indicates the addresses are complete and ready for the next step.

- The control bits and the PID bits are then asserted to the input line.
- The data is ready to be inserted into the input line, shifted out bit-by-bit on CLK rising edges.
- While data is being asserted, the CRC function calculates the FCS and the bit-stuffing function performs bitstuffing process.
- Once the stop signal is asserted, CRC function appends the generated 16-bits CRC into the input line and append the end flag.
- The complete packet is stored in the transmitter buffer register - queued for the next modulation process.

6.2. Packet decoder

AX.25 receiver module performs packet decoding functions including flag detection, zero un-stuffing, and CRC checking. Based on Figure 14, procedures for packet decoding can be elaborated as follows:

- The flag detection function updates the receiver module. Once a flag is detected, the receiver bit count is reset and the module is ready to read the packet header (128 bits).
- While the data reception is still going on, the zero unstuffing and the CRC functions are activated. The CRC function calculates the FCS using the same polynomial as applied in the transmitter module.
- The zero unstuffing function monitors the input data in the receiver buffer register. Once a zero that was inserted by bit stuffing process is detected, the shift register and the bit count processes are halted for one clock cycle. This is how a bit-stuffing zero is deleted from the incoming data [16].
- The procedure continues until the end flag is detected. Once detected, the calculated 16-bit (2-octects) FCS is compared with the last 2-octects that were received. If both two octets are different, the RESEND signal is asserted to appeal for a new packet.

7. Conclusion

This paper has provided a useful method in developing a terminal node controller (TNC) for CubeSat. Most of the TNCs implemented in the ground stations are COTs based and this creates an issue of data incompatibility. The incompatibilities are due to the manufacturers are using a different way of AX.25 implementation, different software properties of the TNC microcontroller and the hardware properties of TNC. As the COTs TNC are mostly designed for general radio packet network, a TNC specifically for CubeSat use in the ground station is required to overcome the issues arises with COTS TNC. The AX.25 protocol is discussed with an insight of the UI frame implementation. The fundamental of FSK modulations implemented in the design, as well as overall design and integration of microcontroller with the modem, are also discussed. Test results of the TNC development are presented and discussed. The implementation of AX.25 in the TNC is discussed at the end to complete the overall function of the TNC. Future research should address data integrity, increasing the data rates for higher bandwidth mission requirements, and focus on in-house radio development for an improved data integrity and communication performance.

Acknowledgments

The authors wish to acknowledge financial assistance from the Ministry of Science Technology and Innovation (MOSTI) for the funding of the grant no: E0620C1214 and the research facilities and opportunity provided by Jalinan Teknologi Solutions.

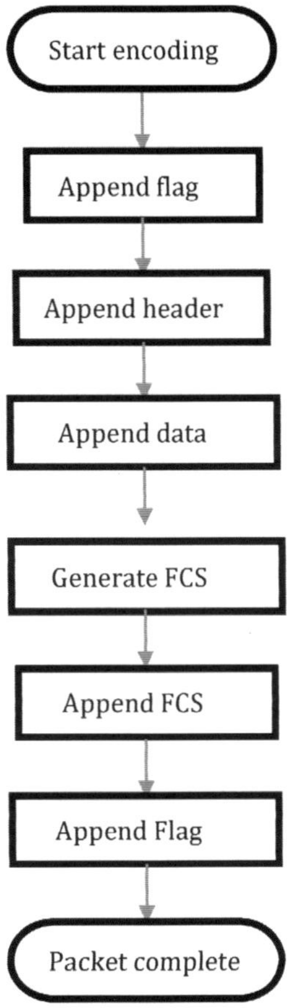

Figure 13: Program flow chart for AX.25 packet encoding

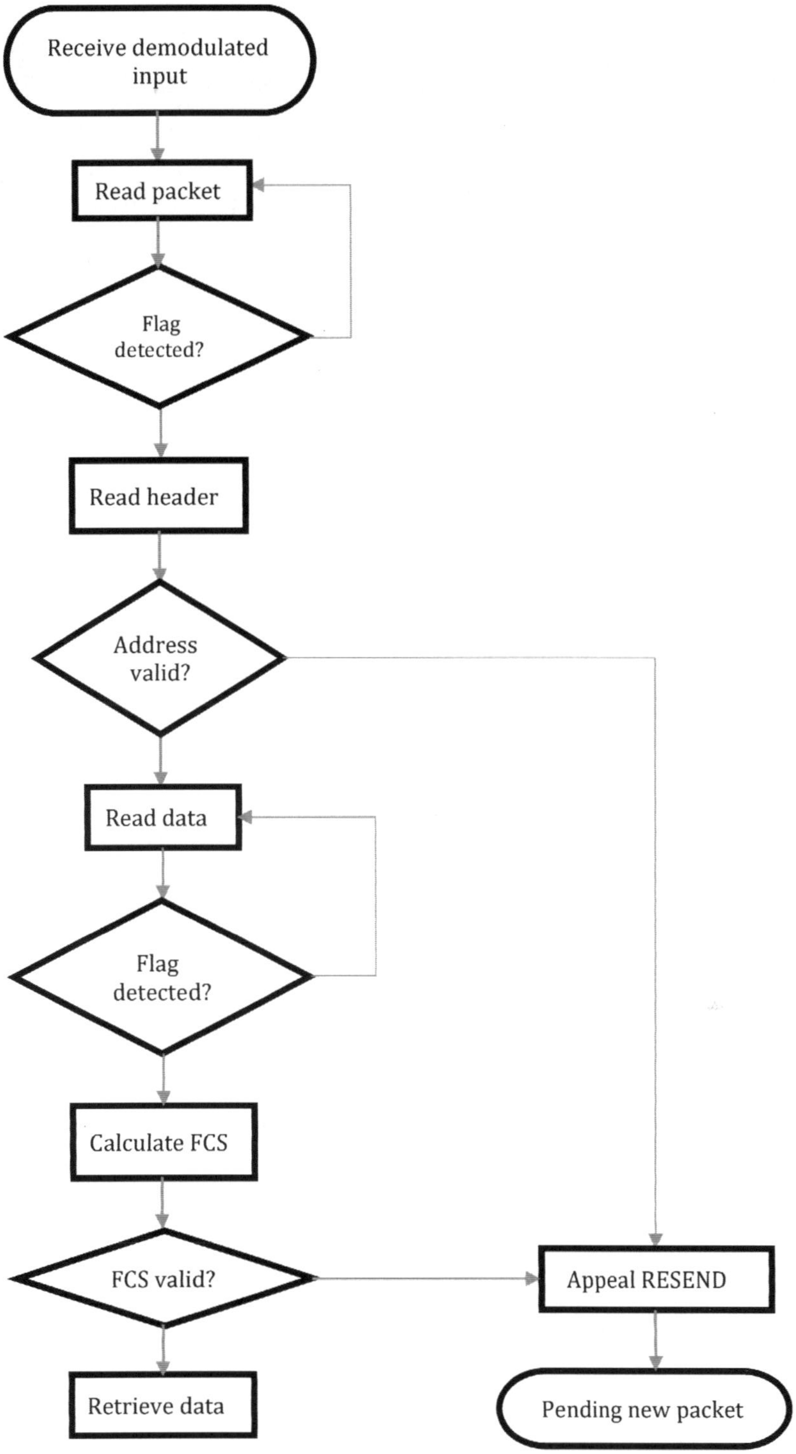

Figure 14. Program flow chart for AX.25 packet decoding

References

[1] Kramer H J and CracknellA P 2008 *Int. J. Remote Sens.* **29** 4285–337

[2] Klofas B and Anderson J 2008 *5th Annual CubeSat Developers Workshop (California)* p1-25

[3] Muri P and McNair J 2012 *J. Commun.***7** 290–308

[4] Klofas B 2013 *10th Annual CubeSat Developers Workshop (California)* p1-12

[5] Perez S, Jarrix S, Roche N J-H, Boch J, Vaille J-R, Penarier J-R, Saleman J-R, and Dusseau L 2009 *23rd Annual AIAA/USU Conference on Small Satellites (Logan)*

[6] Baumann F, Briess K and Kayal H. 2009 *Proc. of the 7th IAA Symposium on Small Satellites for Earth Observation*

[7] Cappelletti C, Martinotti G and Graziani F 2011 *Proc. of the 62nd International Astronautical Congress (Cape Town)*

[8] Dean Straw R 2006 *The ARRL handbook for radio communications* American Radio Relay League. pp. 9.14–9.15

[9] Welwarsky M *6PACK: a "real time" PC to TNC protocol* Retrieved online 2016-08-19

[10] Phil K and Chepponis M 1990 *ARRL 6th Computer Networking Conference (Redondo Beach)*

[11] Zieliński B M 2009 *Proc. Of 9th IFAC Workshop on Programmable Devices and Embedded Systems (Czech Republic)* pp 80-85

[12] Beech W A, Nielsen D E and Taylor J 1998 *AX.25 Link Access Protocol for Amateur Packet Radio* Tucson Amateur Packet Radio Corporation

[13] William A B, Douglas E N and Taylor J 1998 *AX.25 link access protocol for amateur packet radio, version 2.2* Tucson Amateur Packet Radio Corporation

[14] Parry R R 1997 *IEEE Potentials* **16** 14-6

[15] George F and Billeter S 2013 *AX.25 telemetry and telecommand transfer frames format* Swiss Space Center

[16] Huang Y 2012 *Cubesat Ground Station Implementation and Demonstration* Master's Thesis Cranfield University

Adaptive fuzzy logic controller with direct action type structures for InnoSAT attitude control system

F A Bakri[1*], M Y Mashor[2], S M Sharun[3], S N Bibi Sarpinah[1] and Z Abu Bakar[1]

[1]Universiti Teknologi MARA Sarawak, Kampus Kota Samarahan, Sarawak, Malaysia
[2]Universiti Malaysia Perlis, Perlis, Malaysia
[3]Universiti Sultan Zainal Abidin, Kampus Gong Badak, Kuala Terangganu, Malaysia

*fatimatulanis@yahoo.com.my

Abstract. This study proposes an adaptive fuzzy controller for attitude control system (ACS) of Innovative Satellite (InnoSAT) based on direct action type structure. In order to study new methods used in satellite attitude control, this paper presents three structures of controllers: Fuzzy PI, Fuzzy PD and conventional Fuzzy PID. The objective of this work is to compare the time response and tracking performance among the three different structures of controllers. The parameters of controller were tuned on-line by adjustment mechanism, which was an approach similar to a PID error that could minimize errors between actual and model reference output. This paper also presents a Model References Adaptive Control (MRAC) as a control scheme to control time varying systems where the performance specifications were given in terms of the reference model. All the controllers were tested using InnoSAT system under some operating conditions such as disturbance, varying gain, measurement noise and time delay. In conclusion, among all considered DA-type structures, AFPID controller was observed as the best structure since it outperformed other controllers in most conditions.

1. Introduction

Development of space technology is one of the main symbols of technological progress in the modern society. Hence, as a developing country, Malaysia is not left behind in becoming one of the countries involved in exploring the fields of satellite technology. With the general success of previous CubeSAT missions, it has inspired the development of new nano-satellite called Innovative Satellite (InnoSAT). Figure 1 shows the external view of InnoSAT.

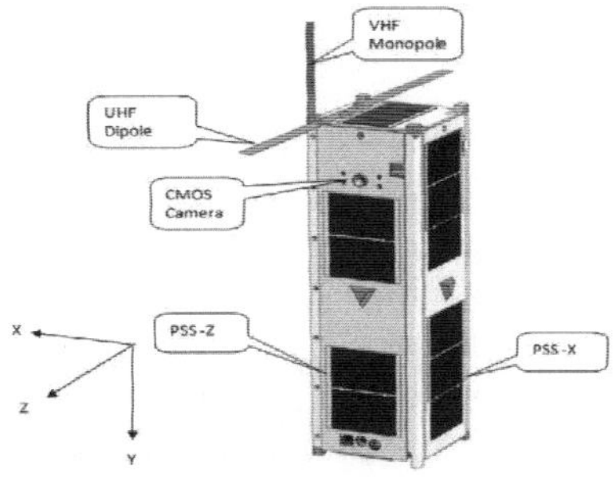

Figure 1. InnoSAT external view

IOP Conf. Series: Materials Science and Engineering **152** (2016) 012032 doi:10.1088/1757-899X/152/1/012032

Generally, the satellite receives interference from various phenomena that occurred in space. These phenomena can disturb the satellite position at any time and condition. Therefore, it is necessary to control the orientation and maintain the stability of satellite by attitude control system (ACS). In order to control satellite's ACS, there are various structures of Fuzzy PID controller that can be applied [1]. However, this paper only focuses on direct action (DA) type structures, which are Fuzzy PI, Fuzzy PD and Fuzzy PID. The DA structures have been chosen because of the simplicity of their operation and their effectiveness for most process.

2. Literature review

Proportional Integral Derivative (PID) controller is well established in classical control systems and it is often used as a benchmark against other types of controllers. Since PID controllers are linear, it is not suitable for strongly nonlinear systems. Thus combination of Fuzzy PID controller is highlighted as a substitute to the classical PID controllers in such cases [1]. There are more types of Fuzzy PID controller structure such as direct action, fuzzy gain scheduling and hybrid type [2]. The direct action type can also be divided into three categories based on the number of inputs, which are single, double and triple as depicted in Figure 2.

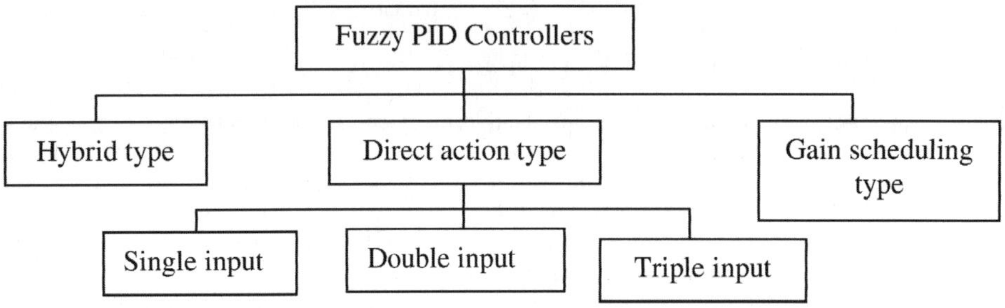

Figure 2. Classification of fuzzy PID controllers

Among Artificial Intelligent (AI) methodologies, Fuzzy Logic Controller (FLC) is the most suitable method for satellite system and accommodates the high level of human knowledge and capabilities in countering uncertain problems. It is seen possible to use FLC properties in developing an expert real time control system for the application of satellite that was exposed to non-probabilistic uncertainties such as sun flare and time dependent noises in measurement [3, 4]. Guan *et al.* [5] proposed a direct adaptive fuzzy control for stabilizing satellite attitude using Takagi-Sugeno model. The membership function and parameters of rules in fuzzy logic system were adjusted to have on-line adaptive ability. The adaptive law is derived based on Lyapunov method. The satellite also uses three-axis reaction wheel in keeping the attitude of the satellite in stable condition. Even so, the author has also stated that Fuzzy controller does not possess self-learning ability. This makes it hard to deal with uncertainty of the system as well as external disturbance effectively. Simulation results have shown that the proposed control law was able to effectively resist the perturbation and obtain better dynamic performances. The author has proven that direct adaptive fuzzy control does not require mathematical model of controlled object and has faster response as well as small steady state error. Therefore, it is concluded that direct adaptive fuzzy control is capable to deal with the real-time application of satellite attitude. Also, it is believed that this approach is robust towards uncertainty that is affecting the satellite.

In other application, Obaid [6] used different types of controllers structure to analyze and evaluate the performance of the position control for an AC motor. They are used three controllers to achieve the purpose: Classical Proportional-Integral-Derivative (CPID), Fuzzy like Proportional-Derivative (FPD), Fuzzy-like Proportional-Integral (FPI) and Fuzzy-like Proportional-Integral-Derivative (FPID). From the design and simulation results, it can be concluded that FPD provides the system response with zero overshoot. FPI, on the other hand, eliminates the error with a high overshoot and FPID gets the better response for the controlled system.

3. Methodology

For this research, Model Reference Adaptive Control (MRAC) had been applied as the control scheme while PID controller as the adjustment mechanism in the controllers' structure as presented in [7]. The proposed control scheme in Figure 3 does not estimate the plant parameters but directly estimates the controller parameters [8]. The MRAC scheme consists of two loops, which are inner and outer loops. The inner loop is the representation of a common feedback signal where the output signal from the process will be the input of the controller. The outer loop consists of the adjustment mechanisms that function to regulate error signal and signal back to the controller. In MRAC scheme, one block of stabilizer is added. It acts as a lead compensator in order to stabilize the InnoSAT plant. In addition, it can improve the stability of the system when the satellite is exposed to disturbance. In this condition, different controller structure will use different stabilizer because AFLC will be tested until the best parameters of stabilizer are found. The desired system's behavior is specified by a reference model and the parameters of the controller are adjusted based on the error model following error, e_m defined by:

$$e_m(t) = y_m(t) - y(t) \tag{1}$$

where $y(t)$ is the InnoSAT output.

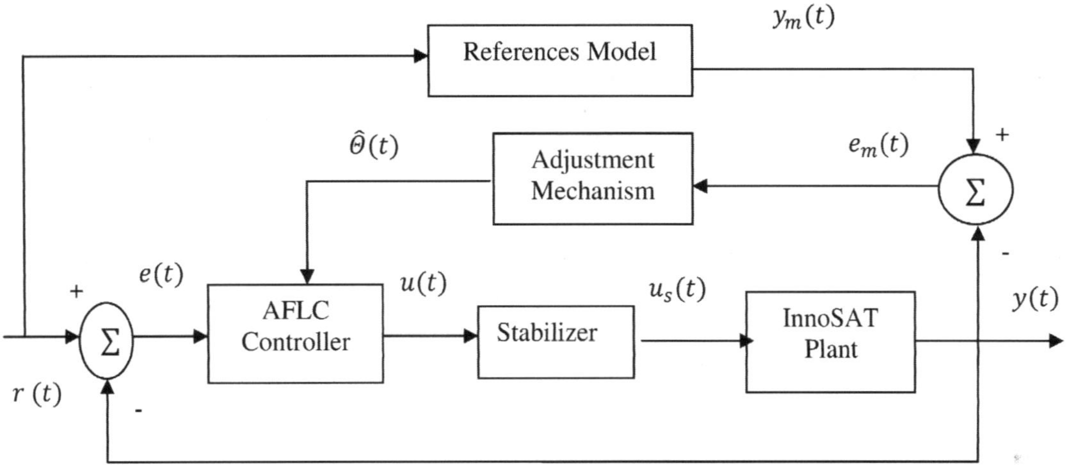

Figure 3. Modified model reference adaptive control scheme

The equation for a general second order linear continuous time reference model is given by the following differential equation [8]:

$$y_m(t) = a_{m_1}y_m(t-1) - a_{m_2}y_m(t-2) + b_{m_1}r(t-1) + b_{m_2}r(t-2) \tag{2}$$

where $r(t)$ is reference input and $y_m\ (t)$ is reference model output. In this research, parameter values for MRAC that have been chosen are $a_{m1} = 1$, $a_{m2} = 0.15$, $b_{m1} = 0.15$ and $b_{m2} = 0$. Mechanism for selecting the parameters in the MRAC can be obtained by the root locus stability.

This research aims to improve the performance of FLC by proposing a new adjustment mechanism for all controller structures, which is similar to PID error. This controller is selected because of it is well known, simple, easy to use and does not require the plant model [9, 10]. In order to obtain a good performance, PID error adjustment mechanishm will adjust the UOD boundaries accordingly based on the way that the adaptation is performed, which is governed by Equation 3.

$$\dot{\theta}(t) = 0.5e(t) + 0.01\Sigma e(t) + 0.4\Delta e(t) \tag{3}$$

For these applications, trial and error techniques are used in selecting the parameter for $\dot{\theta}(t)$. A different format and value of $\dot{\theta}(t)$ is required for different objectives. The formulation depends on the system design criteria, which includes minimizing overshoot and settling time. The combinations of these parameters are found to be adequate for most stable systems. Equations below show how the parameter UOD is adjusted by $\dot{\theta}(t)$:

$$f = f0 + \dot{\theta}(t) \quad (4)$$
$$m = m0 + \dot{\theta}(t) \quad (5)$$
$$n = n0 + \dot{\theta}(t) \quad (6)$$
$$s = s0 + \dot{\theta} \quad (7)$$
$$l = l0 + \dot{\theta}(t) \quad (8)$$

where the origin parameter of UOD is defined as [f0, m0, n0, s0, l0] in range of -4 to 4 and adjusted by $\dot{\theta}(t)$ to produce new parameter of UOD, which is defined as [f, m, n, s, l]. The parameters of the controller need to be continuously adjusted in order to force towards a stable behaviour of satellite system.

For this study, only two methods in DA type will be applied, which are double and triple inputs, since single input is known to be not practical to reduce the overshoot and settling time [11]. They are three controllers had been proposed, which are Fuzzy PI (FPI) and Fuzzy PD (FPD) for double input structure, and conventional Fuzzy PID (FPID) for triple input structure. Each controller has their own advantages and disadvantages along with their own setting parameters. The main purpose of this study is to find out which controller structure is more effective. The structure of the double input fuzzy PD controller is presented in Figure 4. The input controller with scaling gain is derived using Equation 9 and Equation 10. Fuzzy PD controller outputs, $u(t)$ function as the control command in stabilizing the attitude of the satellite. In order to gain better performance, the scaling gains of the input Fuzzy PID must be tuned properly.

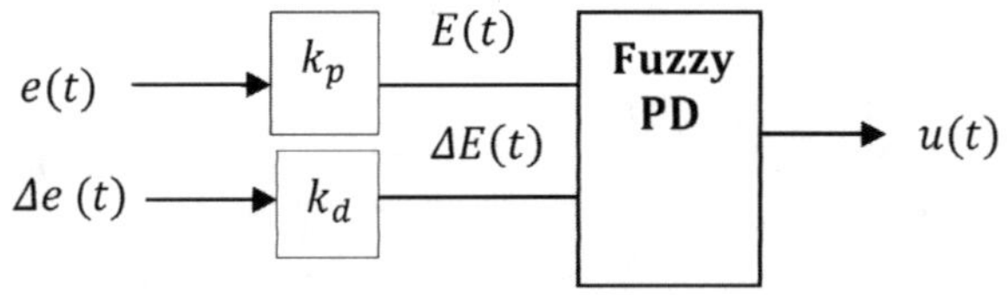

Figure 4. Structure of fuzzy PD controller

$$E = \left(k_p \times e(t)\right) \quad (9)$$
$$\Delta E = \left(k_d \times \Delta e(t)\right) \quad (10)$$

where k_p and k_d are proportional and derivative gains, respectively. Chosen value for the parameters were $k_p = 0.01$ and $k_d = 0.99$, which were suitable since the Fuzzy PD controller demonstrated the performance required by the specification. Performance of Fuzzy PD controller was greatly influenced by the changes of scaling gains of the controller inputs. The impact of setting the value of scaling gain can be summarized as follow [12]:

a) The speed response of the system will improve when the input scaling gain, k_p for $e(t)$ is increased. However, there is limitation to the amount of increment of k_p. If the value is too large, the overshoot will be large and the settling time could be longer (increase).
b) As for k_d for $\Delta e(t)$, the bigger input scaling gain, the smaller overshoot will be obtained. The input scaling gain, k_d is believed to have significant effect in reducing overshoot. However, this will cause the system to be slow in response. Each fuzzy parameter is tuned manually in suitable range of values.

It is rather difficult for a Fuzzy PD type controller to remove the steady state error and hence Fuzzy PI type controller is seen as more practical in real applications [13]. Fuzzy PI controller system also has two inputs like a Fuzzy PD controller but it has integral input $\int e(t)$ as in Figure 5. The integral input is defined in Equation 11. All parameters such as fuzzy set, UOD, membership function and rule are the same as Fuzzy PD controller.

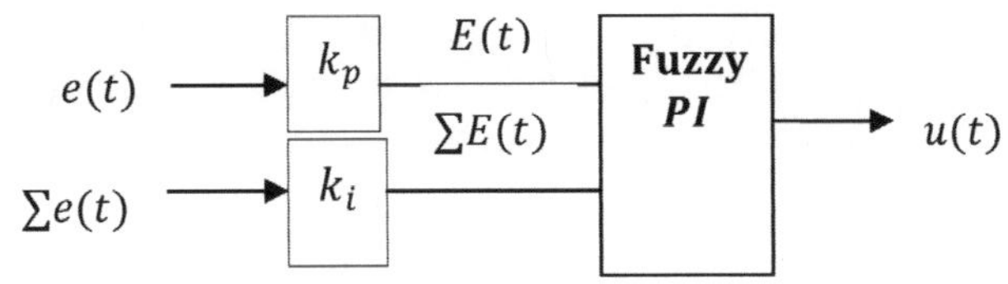

Figure 5. Structure of fuzzy PI controller

The chosen value is considered suitable since an increase in value of k_i may cause large overshoot and even make the system unstable [14]. These parameter is tuned manually with a small value to make sure the system remains stable. Then, the control signal for input controller is obtained by:

$$\sum E(t) = k_i \times \sum e(t) \tag{11}$$

where k_i is integral gain which equals to 0.01.

Although a fuzzy PI controller is able to improve steady state performance, it is known to give poor performance in transient response especially for higher order systems [15]. On the other hand, Fuzzy PD controller often finds it difficult to remove the steady state error in many cases. Hence, it is better to combine and construct a Fuzzy PID when the goal cannot be achieved by using only Fuzzy PD or Fuzzy PI type controller as shown in Figure 6. This is the main reason why the conventional fuzzy PID was built and that is to overcome this drawback. The conventional Fuzzy PID controller requires three inputs, which are error, change of error and integral of error. However, only one error is used where change of error and total of error are derived from the original error.

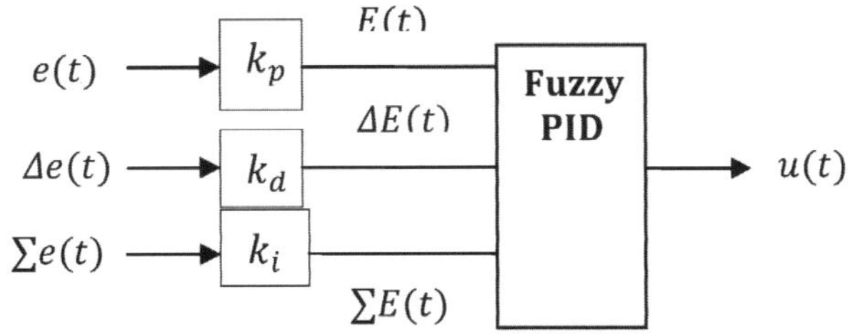

Figure 6. Structure of fuzzy PID controller

Meanwhile $\phi(t)$, $\theta(t)$ and $\Psi(t)$ are the outputs from InnoSAT plant for roll, pitch and yaw axes. The discrete model of three axes InnoSAT plants with small Euler angle such as proposed by [16]:

$$\begin{aligned}\phi(t) &= 2 * \phi(t-1) - \phi(t-2) + K_p(t) * 15.29 * \left(u_{s\phi}(t-1) + u_{s\phi}(t-2)\right) \\ &\quad + 15.29 * \left(u_{d\phi}(t+1) + u_{d\phi}(t-2)\right)\end{aligned} \tag{12}$$

$$\begin{aligned}\theta(t) &= 2 * \theta(t-1) - \theta(t-2) + K_p(t) * 10.04 * \left(u_{s\theta}(t-1) + u_{s\theta}(t-2)\right) \\ &\quad + 10.04 * \left(u_{d\theta}(t+1) + u_{d\theta}(t-2)\right)\end{aligned} \tag{13}$$

$$\begin{aligned}\Psi(t) &= 2 * \Psi(t-1) - \Psi(t-2) + K_p(t) * 15.1 * \left(u_{s\Psi}(t-1) + u_{s\Psi}(t-2)\right) \\ &\quad + 15.1 * \left(u_{d\Psi}(t-1) + u_{d\Psi}(t-2)\right)\end{aligned} \tag{14}$$

where $K_P(t)$ is a varying gain, $u_s(t)$ is controller output and $u_d(t)$ is constant disturbance torque.

4. Simulation results

The simulation results for InnoSAT plant based on direct action (DA) type for step and square input signal are shown in Figure 7. The controllers involved in the DA type are Adaptive Fuzzy PD (AFPD), Adaptive Fuzzy PI (AFPI) and Adaptive Fuzzy PID (AFPID). From the figure, it shows all controllers can follow the model reference output smoothly. Table 1 represents the numerical value for the step response analysis of all controllers. From here, it is observed that performance of AFPID controller is better than the performance of AFPD and AFPI, which indicates that AFPID is faster in time response than both controllers. Based on simulation results from adaptive neuro controller for satellite attitude control system [16], it is still acceptable for the percent overshoot to be more than 10% in the satellite system.

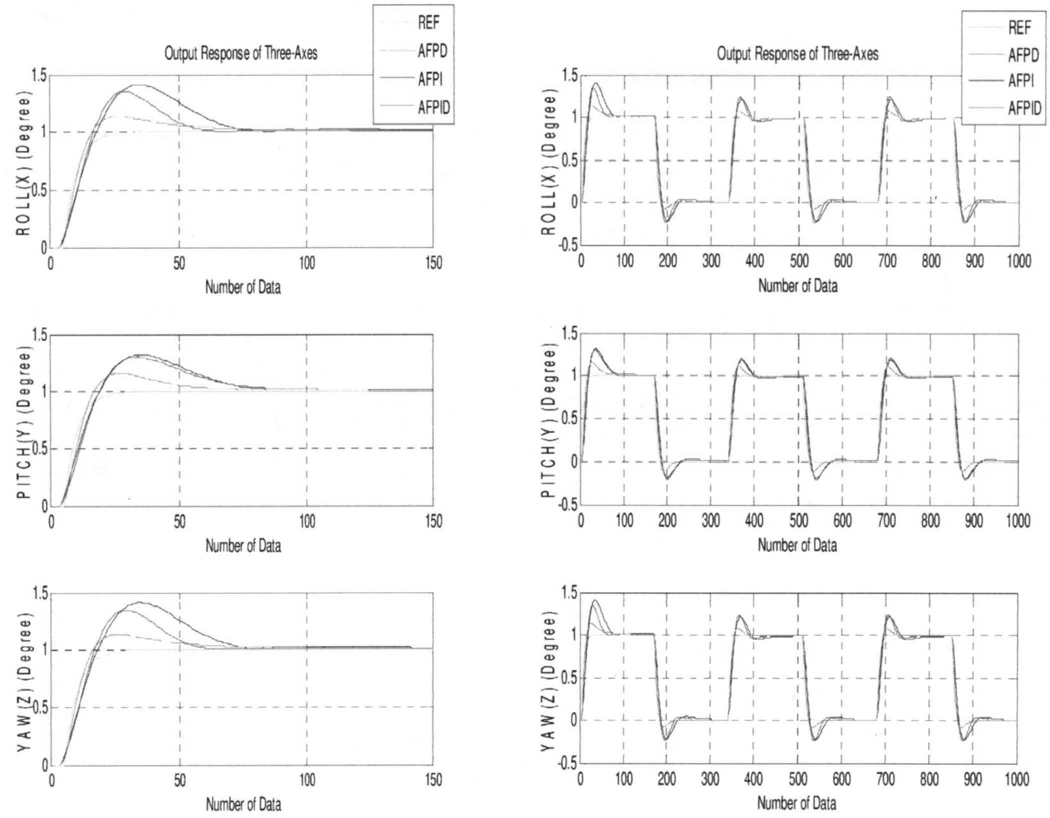

Figure 7. Performance comparison for InnoSAT Euler model with unity gain

Table 1. Step response analysis of direct action type controllers

System Characteristics	Adaptive Fuzzy PD			Adaptive Fuzzy PI			Adaptive Fuzzy PID		
	Roll (ϕ)	**Pitch (θ)**	**Yaw (Ψ)**	**Roll (ϕ)**	**Pitch (θ)**	**Yaw (Ψ)**	**Roll (ϕ)**	**Pitch (θ)**	**Yaw (Ψ)**
Rise Time (s)	9.96	11.36	10.00	10.83	11.54	10.89	8.36	8.91	8.37
Delay time (s)	6.86	7.50	6.90	6.75	6.82	6.83	5.42	6.10	5.46
Settling Time (s)	97.46	94.17	97.70	131.16	81.41	131.34	79.58	68.57	79.38
Overshoot (%)	34.73	30.20	35.04	41.12	31.91	41.42	13.69	16.04	13.87
Negative Overshoot (%)	0	0	0	0	0	0	0	0	0

The simulation results indicated that the overshoot was reduced after the first cycle of square wave input with unity gain. It was thus proven that the proposed adjustment mechanism worked efficiently to reduce the error between output reference and actual output system. Based on this result, it showed that AFPID controller had provided the best response with small overshoot for all axes as compared to AFPD and AFPI controllers. AFPD controller had presented better tracking performances than AFPI based on tuning parameter of scaling gain. AFPI, on the other hand, had the slowest response due to the presence of integral action where only small values of integral action could be used since larger value could increase the time response and cause oscillation to occur. However, the overshoot for all controllers had decreased slightly during the second cycle.

In square wave input response, the performance of the controller was evaluated based on the mean square error (MSE). MSE is defined as the average of the squared errors between actual and estimated readings in a data sample. The equation for calculating the MSE value can be written as:

$$MSE = \frac{1}{n}\sum_{t=1}^{n}\left(y_m(t) - y_p(t)\right)^2 \quad (15)$$

where $y_m(t)$ is the desired value implied by an estimator, $y_p(t)$ is the true value of the quantity being estimated and n is the number of data used for simulation. Smallest value of MSE will be considered as the best performance. Table 2 shows the result of MSE for DA type controllers with unity gain. It can be observed that AFPID controller obtained the smallest value of MSE for all axes.

Table 2. MSE for direct action type controllers with unity gain

Controller	Mean Square Error (MSE)		
	Roll (ϕ)	**Pitch (θ)**	**Yaw (Ψ)**
AFPD	0.0129	0.0141	0.0132
AFPI	0.0126	0.0099	0.0129
AFPID	0.0027	0.0042	0.0028

During the simulation process, the controllers were further tested by adding varying gain, noise, disturbance and delay as shown in Figure 8 until Figure 10. The range of signal for high gain was from 1 till 1.5 for time between 0 s and 500 s, while range of low gain from 1 till 0.5 for time between 501 s till 1000 s. Figure 11 represents the simulation results where the proposed control design methods were able to follow the reference input signal successfully even with the presence of varying gain effect. At high gain, it seems that all controllers were capable of following the reference input but at 700 s, a slight oscillation occurred due to the controllers not able to cope with the lowest valley of low gain.

Next, the controller was subjected with measurement noise as shown in Figure 12. The result points out that all of the DA type controllers were capable of handling noise signals and could maintain their stability with small oscillation. Figure 13 shows the output response of InnoSAT system with 5% of step disturbance. Here, AFPI and AFPD controllers showed the large deviation at the beginning and the ending of the disturbance where the amplitude reached up to 50% and above for three axes. As for the AFPID controller, the results showed that the deviation was a quarter than that of AFPI and tend to be consistent for all three axes.

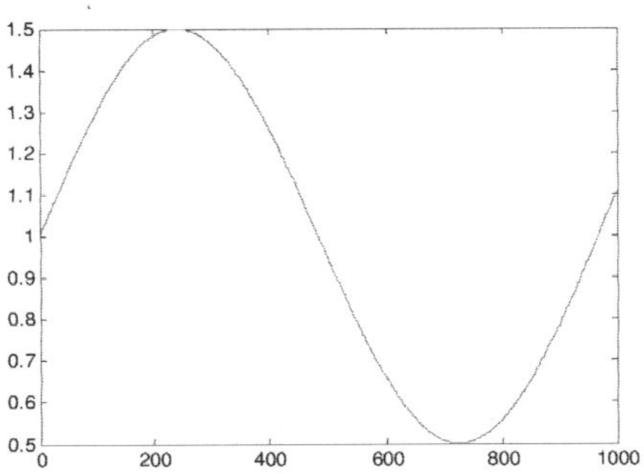

Figure 8. Varying gain

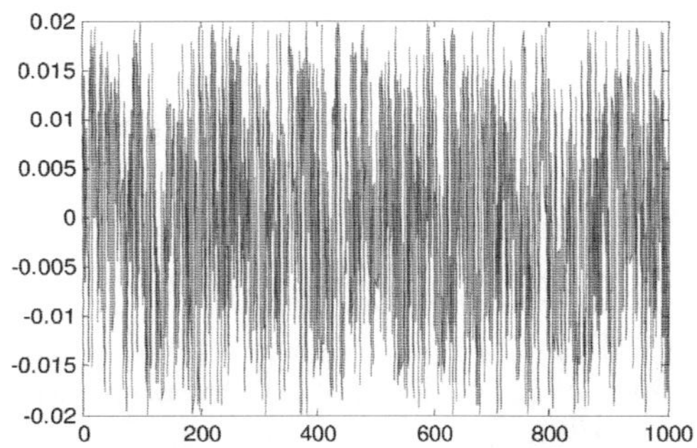

Figure 9. Measurement noise at the plant output

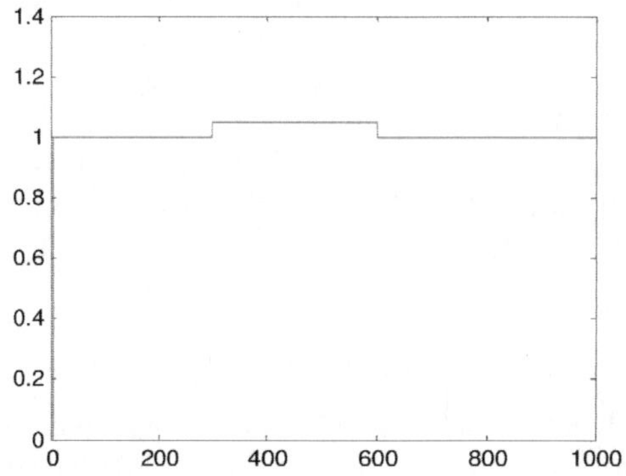

Figure 10. Step disturbance of 5% between 300s and 600s

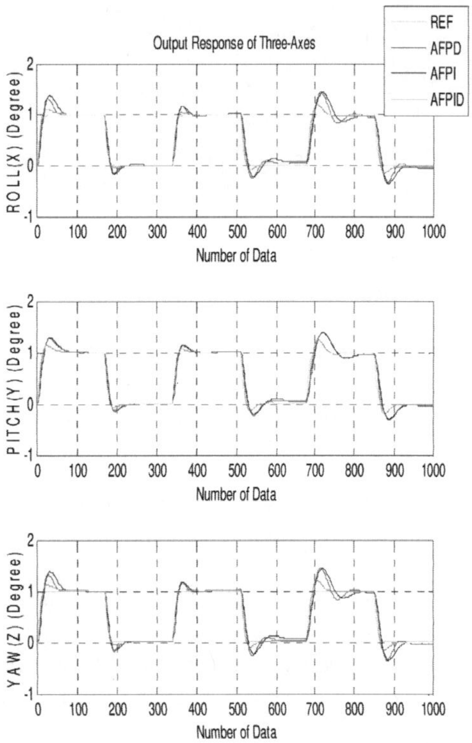

Figure 11. Performance Comparison for InnoSAT Euler model with varying gain

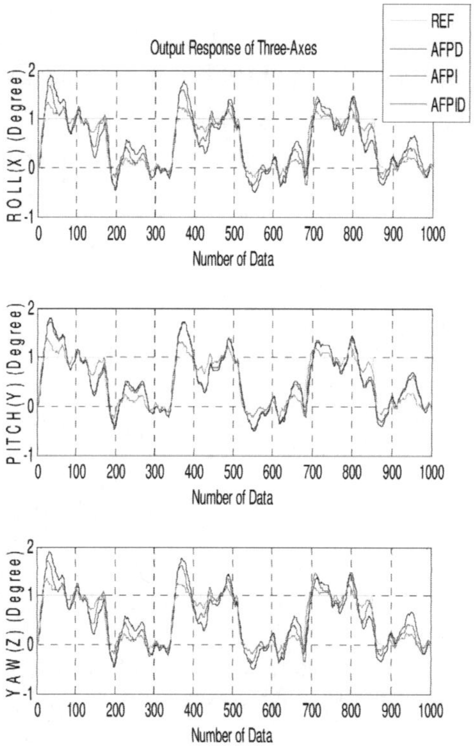

Figure 12. Performance Comparison for InnoSAT Euler model with noise

A sample time delay with value of 1 was introduced to the system in order to present the effect of delay, which normally will occur in real time. Based on Figure 14, the overall output systems tend to follow the reference input smoothly. It is somewhat difficult to distinguish the performance of the controllers in the effects of delay because the output responses only had small occurring changes where the overshoot and negative overshoot were slightly increased. Thus, MSE analysis is important to counter this situation by displaying the results such in Table 3. Based on Table 3, it clearly shows that AFPID controller outperformed the other controllers for all axes.

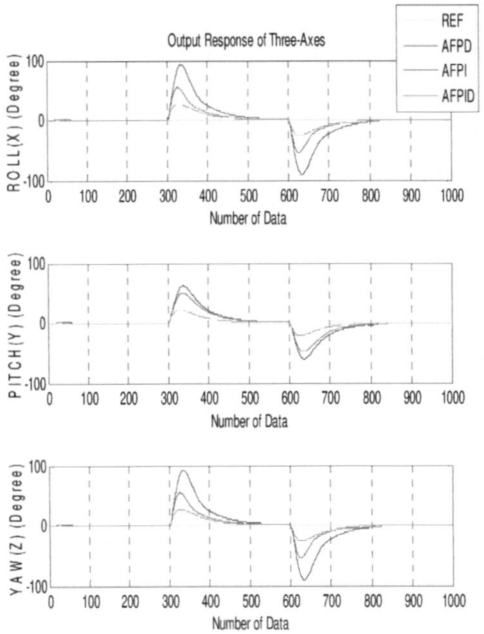

Figure 13. Performance comparison for InnoSAT Euler model with disturbance

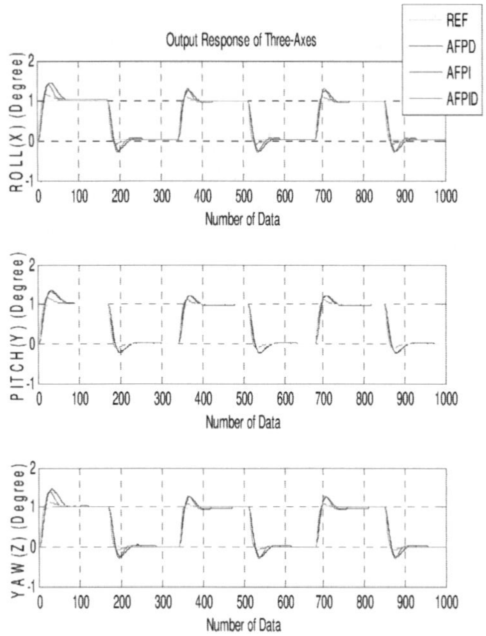

Figure 14. Performance comparison for InnoSAT Euler model with delay

Table 3. MSE for direct action type controllers with delay

Controller	**Mean Square Error (MSE)**		
	Roll (ϕ)	**Pitch (θ)**	**Yaw (Ψ)**
AFPD	0.0162	0.0167	0.0165
AFPI	0.0153	0.0120	0.0157
AFPID	0.0040	0.0057	0.0040

5. Conclusion

From the analysis that has been done, it was found that InnoSAT system with AFLC is suitable for all cases. Simulations and analyses of DA type have shown that the performance of dynamic positioning system can be substantially improved from Adaptive Fuzzy PI (AFPI), Adaptive Fuzzy PD (AFPD) and Adaptive Fuzzy PID (AFPID) where the output system has shown the capability in tracking the reference model closely. By comparing the DA type controllers, the performance of AFPID controller has shown a good result for ACS where the controller has proven its capability in tracking step and square input reference. As conclusion, the application of the AFPID controller is selected as the best controller in DA type since this controller is able to control the movement of satellite along with three axes stabilization technique that consist of roll, pitch and yaw axes.

6. Acknowledgments

The authors would like to thank Astronautic Technology (M) Sdn. Bhd. (ATSB) for providing the information and constructive guidance during the research study, as well as the Ministry of Higher Education (MOHE) and University Technology Mara (UiTM) for the scholarship.

References

[1] Passino K M and Yurkovich S 1998 *Fuzzy control* (Addison Wesley)

[2] Callai T C, Santos J E S, Sumar R R and Coelho A A R 2005 *Advances in Soft Computing* **1** 193–204

[3] Nagarajan R, Paulraj M P, Yaacob S, Zaridah M Z and Rosfariedzah R 2007 *Regional Conference on Engineering Mathematics, Mechanics, Manufacturing & Architecture*

[4] Narayana E V, Bonu V S and Rao G M 2011 *International Journal of Advanced Engineering Sciences and Technologies* **11** 190–5

[5] Guan P, Zhang W, Liu X and Li M H 2011 *Chinese Control and Decision Conference (CCDC)*

[6] Obaid Z A and Nasri S Hamidon M N 2009 *Proceeding of IEEE Student Conference on Research and Development (SCOReD)* (Serdang, Malaysia)

[7] Bakri F A, Mashor M Y, Sharun S M, Norhayati M N and Supardi N Z 2012 *8th IEEE International Colloqium on Signal Processing and its Applications*

[8] Mashor M Y 2007 *Proceeding of the International Conference on Control Instrumentation and Mechatronic Engineering*

[9] Heong K, Chong A and Yun L G 2005 *IEEE Transactions on Control Systems Technology* **13** 559–76

[10] Silva G J, Datta A and Bhattacharyya S P 2002 *IEEE Transactions on Automatic Control* **47** 241–25

[11] Asadian K, Aminifar S, Daneshwar M and Yosefi G 2005 *Turkish Journal of Electrical Engineering & Computer Sciences* **13**

[12] Jie X, Liyun L, Yanbo C and Shiyu W 2008 *Proceedings of the 27th Chinese Control Conference*

[13] Woo Z W 2000 *Fuzzy Sets and Systems* **115** 321-6

[14] Li H X and Gatland H B 1995 *Proceedings of the 1995 IEEE International Conference on Fuzzy Systems* 331-6

[15] Zhu L 1997 *International Journal Advance Manufacturing Technology*

[16] Sharun S M 2011 *International Journal of Electrical Electronic Engineering and Technology* **1**

Low-cost approach for a software-defined radio based ground station receiver for CCSDS standard compliant S-band satellite communications

M A Boettcher[1*], B M Butt[2] and S Klinkner[1]

[1]University of Stuttgart, Institute of Space Systems, Germany
[2]University of Stuttgart, Institute of Telecommunications, Germany

*boettcher@irs.uni-stuttgart.de

Abstract. A major concern of a university satellite mission is to download the payload and the telemetry data from a satellite. While the ground station antennas are in general easy and with limited afford to procure, the receiving unit is most certainly not. The flexible and low-cost software-defined radio (SDR) transceiver "BladeRF" is used to receive the QPSK modulated and CCSDS compliant coded data of a satellite in the HAM radio S-band. The control software is based on the Open Source program GNU Radio, which also is used to perform CCSDS post processing of the binary bit stream. The test results show a good performance of the receiving system.

1. Introduction

A major concern of a university satellite mission is to download the payload and telemetry data from a satellite. While the ground station antennas are in general easy and with limited afford to procure, the receiving unit is most certainly not. As experience teaches, market prices of several ten-thousand USD should be considered for a commercial transceiver system. Nowadays, flexible and low-cost software-defined radio (SDR) transceivers are available and their performance is quite convincing. In particular, since most modern SDRs use the concept of direct conversion, there is no need of up/down converting from/to an intermediate frequency, which greatly supports a cost efficient solution.

Within the university project "Flying Laptop", a small satellite has been developed that uses a 2.4 GHz HAM-Radio transmitter to downlink the payload data (data downlink system, DDS). In order to receive the CCSDS-standard [1] coded and QPSK modulated data, the SDR "BladeRF" from NUAND has been chosen [2]. The main reasons for this choice are the low market price of a few hundred USD and a requirement of minimum of 20 MS/s in order to handle the satellite's payload stream at a data rate of 10 Mbit/s [3]. The BladeRF is a fully USB 3.0 bus powered device and it can be operated autonomously using a FPGA for signal processing. It offers a RF frequency range from 300 MHz up to 3.8 GHz and an independent RX/TX with 12-bit 40MSPS quadrature sampling. With its fully integrated RF transceiver LMS6002D, the SDR is capable of achieving full-duplex 28MHz channels. The signal post processing is done by an on-board 200MHz ARM9 with 512KB embedded SRAM and an on-board 40KLE or Altera Cyclone 4 E FPGA [2]. Other options, like the HackRF One by Great Scott Gadgets [4], would have been a cheaper solution. However, with a sample size of 8 bit only, a limited performance might have to be expected. Another suitable candidate is the USRP B200 by Ettus Research [5], which is offering similar features like BladeRF but with doubled bandwidth capability and a roughly doubled price. Following the low-cost approach, this SDR has been considered as a back-up solution.

The implementation has been carried out according to the specification of DDS protocol, which can be seen in Figure 1. As a control program, open source software GNU Radio is used, including GNU Radio Companion [6] for a graphical user interface. All the functions and blocks are implemented in C++ and Python.

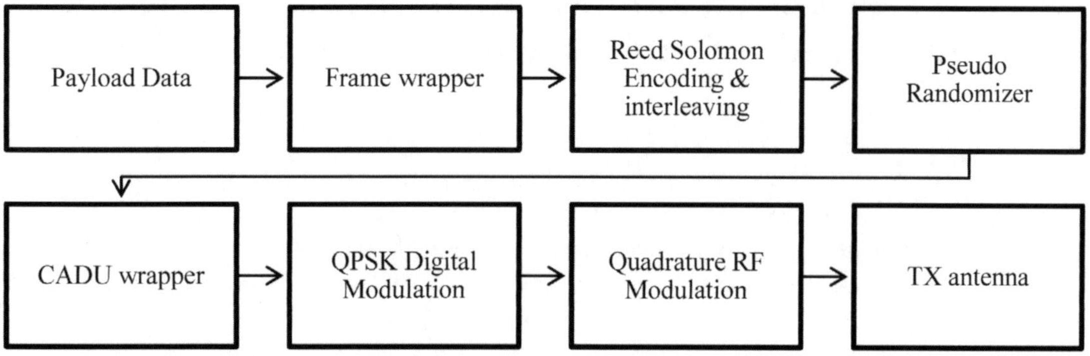

Figure 1. Signal processing chain of the payload transmitter DDS

2. Signal processing software chain

The GNU Radio Companion flow diagram is shown in Figure 2. Following this flow graph, all the steps for receiving and decoding of the DDS signal, given as an overview in Figure 3, can be taken sequentially. BladeRF offers a hardware interface via osmocom, by which the RF frequency and the filter settings can be chosen. Depending on the specific signal, the demodulation settings have to be inserted wisely. As can be seen in Figure 4, besides the wanted test signal from a signal generator, the FFT plot shows a large spike at 0 Hz. This DC offset called effect is a result from direct conversion receiving concept of the SDR. Following the solution recommended by the manufacturer NUAND [7], the DC offset can be avoided by the shifting frequency technique. As approximately 7 MHz bandwidth is occupied by the DDS signal and the BladeRF offers up to 20 MHz, the osmocom carrier frequency can be tuned at a frequency a slightly away from the original carrier frequency. In that case, as can be seen in Figure 4, the DC offset peak does not affect the input signal.

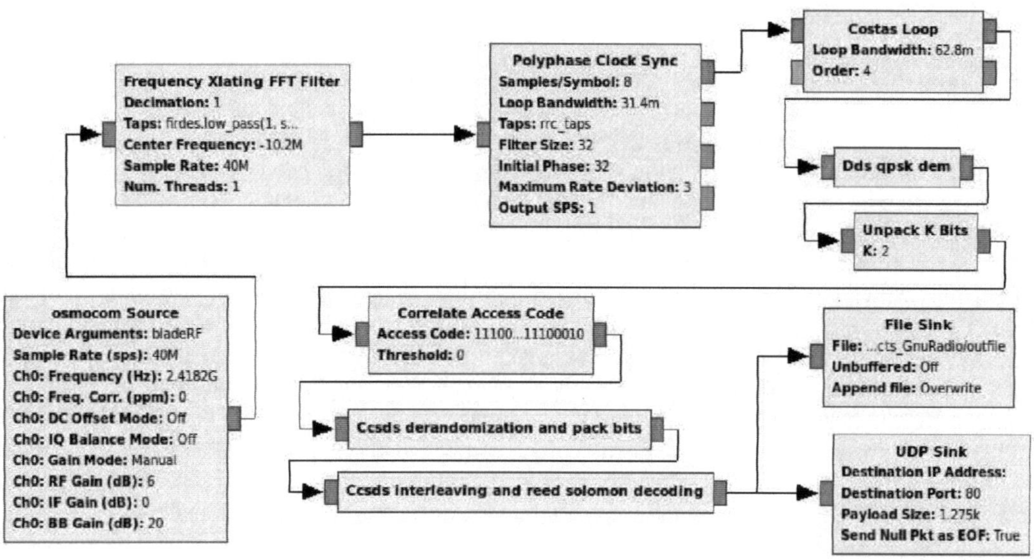

Figure 2. GRC flow graph for DDS receiver

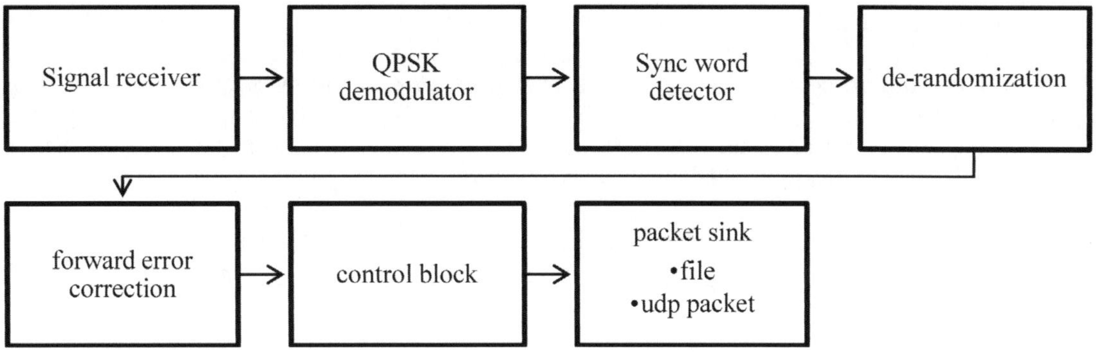

Figure 3. DDS receiver section

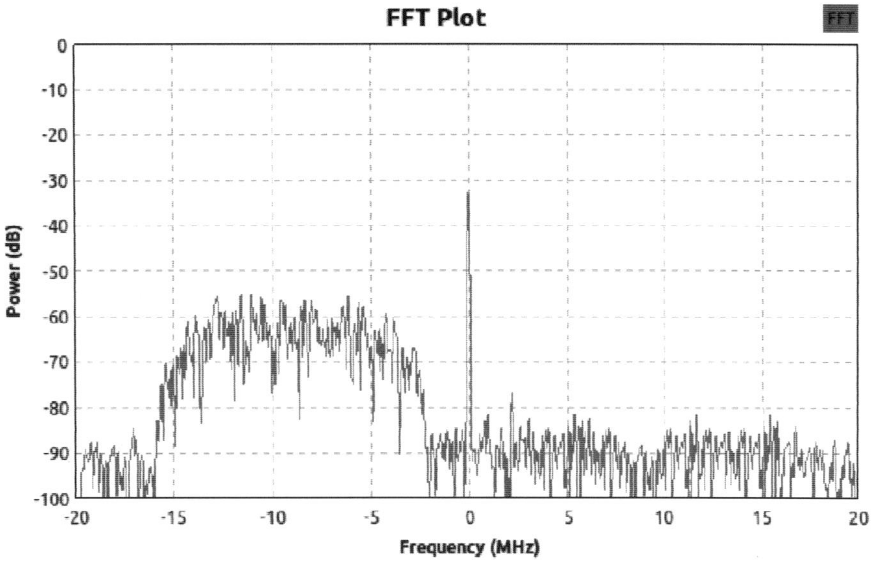

Figure 4. FFT plot of the received test signal (detuned by -10 MHz) including DC offset

In order to get only the wanted input signal, initial detuning of the SDR needs to be compensated. In GRC, a block named Frequency Xlating FFT has all the capabilities to execute two things at the same time. First, it shifts the center frequency of the signal according to the initial detune parameter, then it applies a low pass filter that helps to discard unwanted portions of the signal. Finally, the adjusted signal is handed over to the demodulation stage, consisting of clock synchronization, costas loop and QPSK demodulation. Moreover, the signal has to pass through the "Polyphase Clock Sync", which recovers the timing by trying to find the best time to sample the incoming signal. This block is able to account for the SRRC filter settings of the DDS signal (excess bandwidth of 0.4) [8]. In order to compensate for possible phase and frequency offset, the "Costas Loop" block is used. The resulting signal is then handed over to the "DDS qpsk dem" block, where the input signal is being demodulated following the QPSK scheme. Input for this block are complex numbers from the former stage and the block calculates the distance from the expected position on consolation diagram. According to the CCSDS standard, Gray coding is used and this block generates two binary bits for one point such that the data rate will double from this point. Resulting constellation diagram of the FLP DDS transmitter is shifted by 45 degrees and can be seen in Figure 5. The binary output stream is handed over to the next stage, which is the channel decoding.

With the binary data string from QPSK demodulator section, valid DDS CADU packets (CCSDS) need to be detected by searching for the "Attached Synchronization Marker" (ASM). Each packet starts with a sync pattern and after that, packets of a fixed length of 100 bytes are sent. Furthermore, the bits need to be pooled as bytes of data. Due to the fact that these bytes are scrambled before the transmission by the satellite, they need to be de-randomized for further processing. As can be seen in Figure 6, forward error correction can be implemented on the resulting bytes' string. First the data packet is de-interleaved and then the Reed Solomon (RS) decoding algorithm is applied to recover the error bytes in case of corrupted data. After recovering the original data blocks, payload information data is recovered according to the DDS protocol structure.

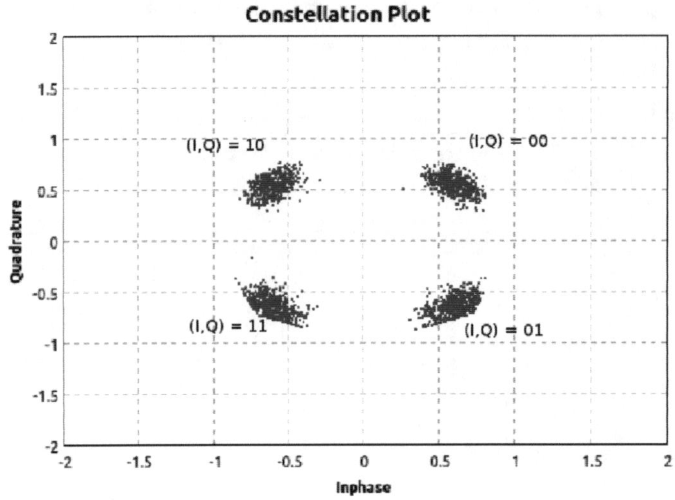

Figure 5. Constellation plot of the demodulated signal

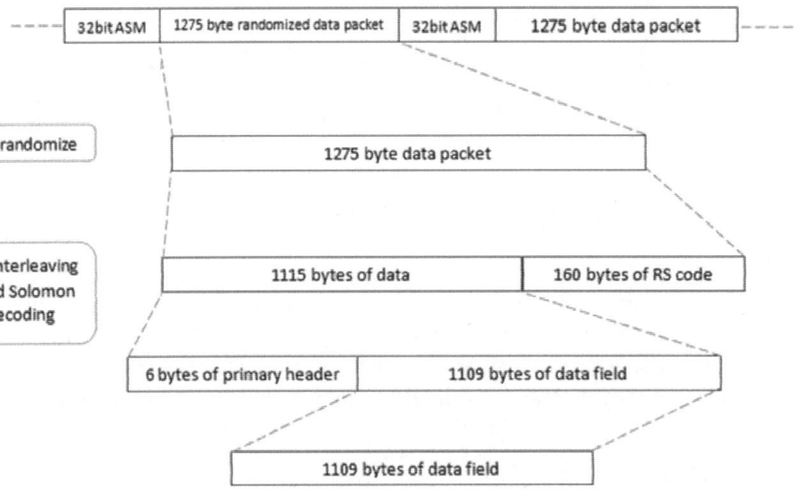

Figure 6. Example CADU packet according to CCSDS standard taken out of the binary string

2.1. Synchronization marker detection

The CCSDS Blue Book [9] recommends a 32-bit ASM to be used with the Reed Solomon and basic convolutional coding. The recommended 32-bit ASM is a hexadecimal number as 0x1ACFFC1D and the position varies as follows:

- If only Reed Solomon code is used including a bit randomizer implemented as optional, the ASM is attached at the start of each generated (randomized) code-word.
- If convolutional coding is used, ASM is attached to the start of data frame and convolutional encoding.

In the DDS protocol structure, at first the payload data are Reed Solomon coded with interleaving and an enabled pseudo bit randomizer implementation, having the ASM attached at the start of every DDS packet. The ASM is never randomized.

2.1.1. Correlation and threshold. Data frames or packets can be identified by searching for the ASM. The correlation and threshold techniques can be used to perform ASM search from the bit string in order to identify a start of packet. Here, the sync word ASM (0x1ACFFC1D) will be correlated with incoming bit stream. As bit errors may occur, which can also corrupt the ASM, a certain threshold is necessary to compensate few error bits in this sync sequence. As shown in the Figure 7, the ASM sync word is been correlated with the input stream. It generates a high peak when it is correlated with same bits that are above the threshold. Hereby, it can be assumed that an ASM is being detected in the bit stream and the start of a packet is present.

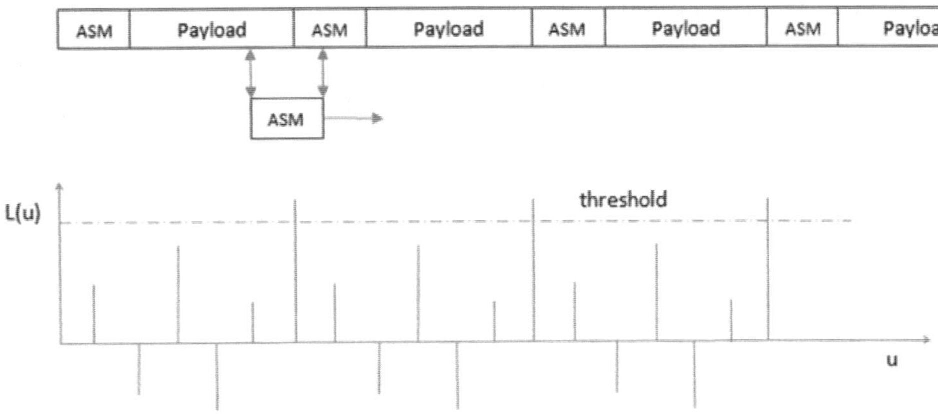

Figure 7. Correlation of ASM with the received data

2.1.2. Sync word packet detection. In GNU-radio companion flow graph, ASM detection is performed by a block named 'Correlate Access Code'. Input arguments for the block access code and threshold are needed. For access code, sync bits ASM 0x1ACFFC1D is used and for the threshold, the number of relaxation bits which will define the threshold to compare with the correlation result is inserted. Input samples for the block Correlate Access Code is the bit stream from the QPSK demodulator. It continuously correlates the input bit stream with the local argument ASM bits and compares result with the threshold. In this case, it compares how many bits are mismatched. If number of mismatched bits is less than the input argument threshold, the presence of ASM sync word in the bit stream is indicated. The immediate next bit will be marked as the start of a frame. This marker will be used by next block in GRC flow graph. Start of frame is required to perform de-randomization and further processing.

2.1.3. Pseudo randomizer. With valid detection of the start of the frame and fixed DDS packet length to 1279 bytes, 4 bytes out of 1279 are used for ASM and the remaining 1275 are the actual payload data of interest. In order to process those pseudo randomized bytes, a de-randomization is needed before further processing. In most communication systems, data is being randomized to distribute the density of 0 and 1 equally across the data string. This procedure is required especially for the receiver phased lock loop (PLL) synchronization that ensures sufficient transitions between 0s and 1s in the data bit stream. If long string of 0 or 1 is present in the data stream, the symbol recovery loop will face a difficulty to lock. CCSDS recommends a pseudo randomizer with the generator polynomial in Equation 1 [4], which is implemented in DDS transmitter.

$$h(x) = x8 + x7 + x5 + x3 + 1 \quad (1)$$

The period of this random sequence is $2^{(m-1)}$ where *m* is the degree of the generator polynomial. The sequence repeats itself after$2^{(m-1)}$ and therefore, the randomizer has a length (with *m* is equal to 8) of 255. At the start of each transfer frame, all the liner feedback shift registers are recommended in the CCSDS Blue Book [34] to be predefined as 1s. This is shown in Figure 8.

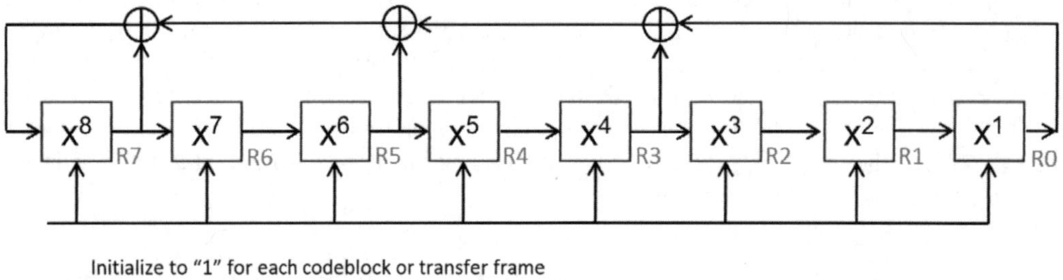

Figure 8. CCSDS driven pseudo randomizer block diagram

The output bits of the shift registers are combined to form the bytes R7, R6, R5, R4, R3, R2, R1, R0. This is the output for the pseudo random sequence bytes, which is implemented in DDS protocol. On every clock cycle, an XOR is executed between the corresponding bits from the output of the shift registers and its content is right shifted. XOR is done between the data byte and the pseudo random sequence bytes, and the output is the de-randomized transfer frames. Random sequence generation bytes are independent and constant, and they are especially independent from the input data stream. The bytes remain constant and repeat again after 255 bytes. So it is more feasible to construct a lookup table, which will optimize the speed, because bit shifting and the polynomial equation solving will not be required in real time.

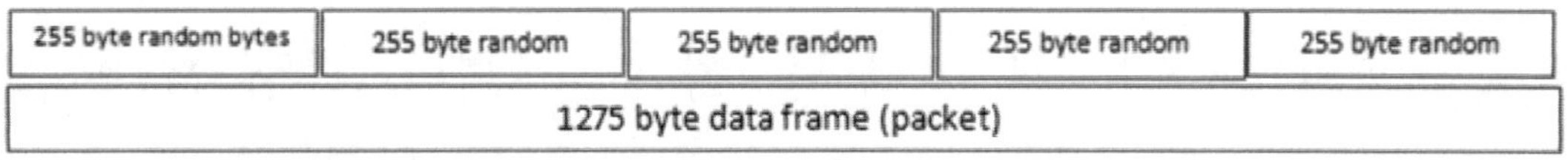

Figure 9. 255 bytes of random pattern repeats five time for one data frame

The output delivered by the Correlate Access Code is in the form of bits. It is more suitable to pack the bits into bytes by performing a packaging of bits after detection of the marker. After that, the bytes are ready for the de-randomized process. During the de-randomizing of a complete DDS frame, 255 random bytes sequence will repeat for five times within every frame. In GNU-radio companion flow graph, de-randomization is performed by a block called 'CCSD de-randomization'. The packaging of the bits is implemented as a dedicated block for the specific DDS packets. In this block, the detection marker from the previous block is needed. This marker indicates the start of the DDS frame. After detecting the frame's beginning, the bits needs to be packed to bytes and the DDS frame has fixed predefined length of 1275 bytes. These bytes are then XOR processed with pseudo random sequence bytes stored in the lookup table. In the end, a marker is placed at the beginning of the de-randomized frame to enable the next block to easily detect the start of this frame.

2.2. Forward error correction (Reed Solomon decoding)

According to the DDS protocol, the transmitter has implemented a Reed Solomon (RS) encoding with interleaving for forward error correction. Forward error correcting (Reed Solomon) is a technique of digital signal processing and it enhances the reliability of received data on the receiver side. In this technique, redundant data (overhead) is introduced and transmitted along with the original data. In the Blue Book [9], two systematic Reed Solomon codes (255,223) and (255,239) are recommended. The first one has 32 symbols of overhead and is capable for error correcting performance of 16 symbols. In the DDS protocol, the first one (255,223) is implemented by the transmitter and therefore, the receiver has a capability of correcting 16 symbols (bytes) per Reed Solomon decoding per frame.

2.2.1. De-interleaving. According to the DDS protocol, the 1275 bytes per frame contain five Reed Solomon encoded chunks. Each chunk size is 255 bytes. As RS(255,223)-coding is implemented in the DDS transmitter, 223 bytes are actual data and the rest of 32 bytes are overheads for error correction. In the DDS frame, these five RS chunks are merged with each other with a technique called interleaving. The objective of interleaving is to make the forward error correction more robust against bust error. For a whole DDS frame, up to 80 errors can be corrected because of 5 RS encoded chunks.

2.2.2. Reed Solomon error correction realization. In the GNU-radio companion flow graph, de-interleaving and Reed Solomon decoding is performed by a block named 'CCSD interleaving and Reed Solomon decoding' that is implemented as a dedicated block for specific DDS packet. It is essential for this block to use the detection maker from previous block, which indicates the start of the de-randomized DDS frame. Knowing the frame's beginning, the bytes can be de-interleaved and five different RS encoded chunks can be created. These chunks are then further fed separately to the Reed Solomon decode algorithm [10] and processed. At the end, these data chunks are organized in such a way that all the data part (first 223 bytes) of all five of the chunks placed together in the beginning, which leads to 1115 bytes of data and the RC codes for each chunk (remaining 160 bytes) placed together. The number of error corrected for each chunk is also be placed along with the data, as this information might be helpful as a signal quality indicator.

2.3. Control block and file sink
From previous stage, there are 1275 RS corrected bytes, including 1115 actual data bytes and 160RC overheads bytes. Depending on the different user needs, either all the bytes (including RS overheads) can be send to the next stage, e.g. for debugging purpose, or the overhead can be discarded and only the payload data is forwarded. The marker is detected first to ensure the start of the frame, which was marked in the previous stage. Then the data is separated or manipulated as required by using different blocks. In the last stage, the payload data stream can be transferred to a file or transmitted via UDP.
File sink requires the filename as input parameter where the samples will be stored in real time. In the case of UDP sink, the destination port, destination IP address and number of bytes per packet of UDP that is also called as payload size of UDP packet are required.

3. Test results
The tests have shown a very convincing capability of the SDR receiver. At first, the DDS channel decoding section was tested separately, showing that the whole software processing chain including the DDS protocol was working as expected. The assembly was able to decode the data package from the DDS protocol, fed by an input file. It was also successfully able to save all the data in a file and to send the packets through the network via the "UDP sink". By adding artificial bursts and bit errors of different numbers to this input file, the forward error correction section could also be successfully tested. As expected, up to 16 bit errors within one block could be detected and corrected.

In the next step, a complete system was tested to verify its functionality including RF processing chain. Therefore, the BladeRF has been connected to the FLP satellite within the clean room facilities at Institute of Space Systems in Stuttgart, Germany. In order to actually measure the radio interface, the SDR was not directly connected to the satellite's DDS transmitter. Instead, an EGSE antenna cap was attached to the FLP payload horn antenna and connected to the BladeRF. Besides, a computer connected thought an USB 3.0 interface has been used to run the GRC software.

Figure 10 shows the frequency spectrum, which was recovered from the output of the first block in GRC, the osmocom source. It can be seen that the receiver is able to receive and process the data transmitted by the satellite at a data rate of 10 Mbit/s with a corresponding bandwidth of 7 MHz. The carrier frequency is to be found around 2.408 GHz, the nominal DDS center frequency.

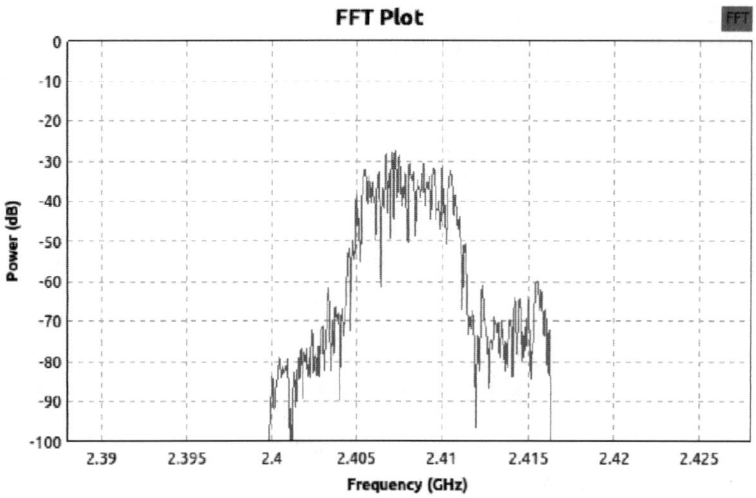

Figure 10. Received QPSK signal from satellite at 10 Mbps (7 MHz of occupied bandwidth)

During the test, most of DDS packets were received easily. Meanwhile, some of the packets were received with bit and burst errors. While most of those packets could be recovered due to the Reed Solomon coding, some contained more than 16 bit errors and were corrupted. Main reason for this behavior is to be found in actual measurement setup. Because the satellite is already fully assembled, it is not possible to get a direct access to the output port of the DDS transmitter. When using the EGSE antenna cap as a receiving antenna within the horn's near field, the SNR becomes smaller and bearing in mind that the center frequency is around 2.4 GHz, also the local Wi-Fi is accountable for some burst errors. Besides that, there were also missing packets because of instabilities in the DDS demodulation section. Displacements in the constellation plot point to the assumption that the polyphase filter bank clock recovery and SRRC filter are not working as reliable as expected, particularly in terms of the signal phase not being recovered continuously without errors. Investigation on this matter is still ongoing to improve the algorithm. Furthermore, it is intended to specify the receiver more in detail, especially to determine the sensitivity and Eb/N0 vs. bit error probability correlation.

4. Conclusion

In this work, low-cost approach for a software-defined radio based ground station receiver for CCSDS standard compliant S-band satellite communications has been shown. The SDR bladeRF by NUAND has been chosen to receive the data, which is QPSK modulated, SRRC filtered and can receive data at a rate of 10 Mbit/s and occupying a bandwidth of 7 MHz. By using this hardware, the cost could be decreased to about 10-20 % of the cost of a commercial satellite receiver. In order to configure the SDR and further process the data according to the CCSDS standard, open source software GNU Radio has been used, including GNU Radio Companion for a graphical user interface. Meanwhile, all the functions and blocks are implemented in C++ and Python. This software chain is able to demodulate the QPSK data according to the CCSDS given constellation diagram and wrap the bit stream into CCSDS packets by searching for the given synchronization marker and descrambling. Reed Solomon decoding and interleaving for forward error correction is then applied and finally, valid data packets sink via UDP over the network or into a file. The CCSDS software decoding chain has been tested successfully. During the test of data processing chain together with the hardware, the system showed very convincing results. Although most packets transmitted by the satellite were received correctly or could be recovered, some packets were found missing. Main reasons are suspected within the clock and phase recover stages because the constellation plot seemed to show some instabilities. Further investigations are still ongoing to correct this stage and to further improve the reliability of this SDR based ground station receiver. Besides, the RF parameters also need to be specified more in detail.

References

[1] *Consultative Committee for Space Data Systems* [Accessed online August 2016]
[2] *Blade RF Software defined Radio home page* [Accessed online August 2016]
[3] Thibaut C 2012 *DDS Transmitter Technical Notes and Specifications* (Stuttgart: IRS University of Stuttgart)
[4] *HackRF Software defined Radio* [Accessed online June 2016]
[5] *USRP Software defined Radio* [Accessed online June 2016]
[6] Blossom E et al. 2016 *Welcome to GNU Radio* [Accessed online June 2016]
[7] *DC offset and IQ Imbalance Correction for BladeRF* [Accessed online June 2016]
[8] Thibaut C and Boettcher M 2015 *FLP Data Downlink System Protocol Technical Note* (Stuttgart: IRS University of Stuttgart)
[9] Consultative Committee for Space Data Systems 2015 *TM Synchronization and Channel Coding. Recommended Standard CCSDS 132.0-B-2*
[10] Karn P 2016 *DSP and FEC Library* [Accessed online June 2016]

IOP Conf. Series: Materials Science and Engineering **152** (2016) 012034 doi:10.1088/1757-899X/152/1/012034

Quasi-satellite system for leader-follower spacecraft formation

A.S.M. Harithuddin

Department of Aerospace Engineering, Universiti Putra Malaysia, 43400 Serdang, Malaysia.

E-mail: a_salahuddin@upm.edu.my

Abstract. The quasi-satellite motion is where a celestial body appears to orbit around another body that is also orbiting a common focus. The recent discoveries of asteroids in quasi-satellite motion around the Earth inspired this work. This paper attempts to look into exploiting the quasi-satellite dynamics in nature to provide a new perspective in spacecraft formation flight. We consider a two-spacecraft formation flight in which the follower appears to orbit of the leader in its reference frame. We perform a parametric analysis by varying the orbital elements of leader and follower examine the initial conditions needed to simulate a quasi-satellite formation system. We illustrate the trajectories of the follower spacecraft. This serves two purposes: firstly, to demonstrate formation flight based on the idea of quasi-satellite; and secondly, to gain some insights on the shapes of the resultant "quasi-orbits".

1. Introduction

The discovery of the Earth's new "moon" [1] have inspired the work in this paper. Asteroid 2016 HO_3 is not technically a moon but one of Earth's orbital companions called quasi-satellites. It is a term first coined in a paper by Mikkola and Innanen [2] to define a celestial body that librates around the longitude of its associated planet. A quasi-satellite is different from a true satellite in that its orbit lies outside its planet's Hill sphere, meaning that its motion is not influenced by the planet's gravity. The first observed quasi-satellite system was the asteroid 2002 V $E68$ and Venus system reported in [3].

In their 1997 paper, Mikkola and Innanen [2] assert several conditions for a formation of a quasi-satellite around a planet: (i) it has to have the same mean motion and mean longitude as its associated planet; and (ii) the distance must be large enough to be outside of the planet's gravitational influence. Such celestial body will exhibit a retrograde motion around the planet of which orbits are, in general, not stable.

This observation of a natural celestial phenomenon opens up the idea of applying quasi-satellite dynamics in spacecraft formation flight. In fact, Mikkola and Prioroc [4] has very recently introduced the concept of creating a spacecraft formation flight using quasi-satellite dynamics. In their dynamics model, however, it is imposed that the spacecraft to have ion thrusters to provide inverse square-law type of force in order to imitate the gravitational force of celestial bodies like planets.

We point out here that the concept of a quasi-satellite in tandem motion with its planet can be applied in spacecraft formation flying without the need of thrusters. We can relax some of the quasi- satellite conditions proposed in [2] and [4]. The "planet" in the formation can

IOP Conf. Series: Materials Science and Engineering **152** (2016) 012034 doi:10.1088/1757-899X/152/1/012034

be a spacecraft, which has no gravitational attraction of its own or any attractive/repulsive propulsion. This simplifies the dynamics of quasi-satellite in our case in that we can reduce the dynamics into a two-body problem.

Inspired by the nature, this paper studies a way to mimic the dynamics of natural quasi-satellites in our solar system using artificial satellites. We start our investigation of this problem by modeling the relative motion of two spacecraft – a leader and a follower – using an unperturbed two-body problem model. The imposed condition for our spacecraft formation is this: the follower is to orbit like a quasi-satellite where the leader acts as the formation's planet. We conduct a parametric analysis in that we vary the orbital elements of both spacecraft to find and specify the conditions for a leader-follower formation to imitate the quasi-satellite tandem motion. We present some numerical simulations to show the trajectories of the quasi-satellite formation for different variations of orbital elements. We show that tandem flight of two spacecraft akin to a quasi-satellite system can be achieved without attractive or repulsive mutual force.

2. Related Works in Formation Flight

Satellite formation flying enables the concept of having multiple, smaller satellites cooperating to accomplish the objective of a monolithic, more expensive satellite. The type of formation can be many, e.g. a trailing formation (the *A-train* satellite formation [5]), a rigid arrangement and coordination,(typically for interferometry missions as discussed in [6]), and a constellation (GPS satellites).

Formation flight studies and analyses are mainly originates and based from terminal rendezvous studies. The Clohessy-Wiltshire equations of relative motion [7], or the CW equations, is by far the most popular method in the studies of spacecraft formation flight. There are some limitations in that the CW equations only work if the reference (leader) orbit is strictly circular and the distance between the leader and the follower has to be much less that the leader's orbit radius. A bounded, periodic solution - to find a bounded, periodic orbit of the follower spacecraft - also requires a strict initial condition which may be impractical in real application. The Tschauner-Hempel [8] equations provide a description of relative motion for an eccentric reference orbit. This model has been used to overcome the eccentricity limitation of the CW equations but its solutions are generally more complex which complicates analysis.

In finding an alternative approach to formation flight dynamics analysis, some researchers have turned into the manipulating the orbital elements. The idea of separating orbital elements to maintain safe spacecraft collocation has been applied for geostationary orbit [9]. In general, it is easier to use orbital elements approach to include orbital perturbations compared to using the state transition matrices approach (like the CW and TH equations). This is the motivation of several publications on the problem formation flight in [10], [11] and [12].

The quasi-satellite formation proposed in this paper is a formation where a spacecraft's orbit is design in such a way that it appears to orbit another spacecraft closely. We remark that this paper is not proposing any specific space mission but to introduce a type of leader-follower dynamics and present a nature-inspired innovation in formation flying design. The method presented here uses the full nonlinear model and can be applied to all closed Keplerian orbits.

3. Spacecraft Relative Motion Modeling

The analysis assumes the unperturbed two-body motion

$$\ddot{\mathbf{r}} + \frac{\mu}{r^3}\mathbf{r} = 0 \tag{1}$$

which governs the motion of a spacecraft relative to Earth with a gravitational parameter of $\mu = 398,600\ \text{km}^3/\text{s}^2$. The vectors $\ddot{\mathbf{r}}$ and $\mathbf{r}$ is expressed in the XYZ Earth-centered inertial (ECI) frame given by the X-,Y-, and Z-axes as shown in Figure 1. The X-axis points in the vernal equinox direction and the Z-axis coincides with Earth's rotational axis. The Y-axis completes the right-handed triad. We refer to the co-moving xyz frame, which originates from the spacecraft center of mass at L, as a local vertical local horizontal (LVLH) frame. We define the x-axis of the LVLH frame as pointing along the radial position vector $\mathbf{r}$ of the leader spacecraft in the ECI frame and the z-axis as pointing along its orbital angular momentum vector $\mathbf{h}$. The y-axis is aligned along its velocity vector which completes the right-handed triad. The unit vectors are therefore given as

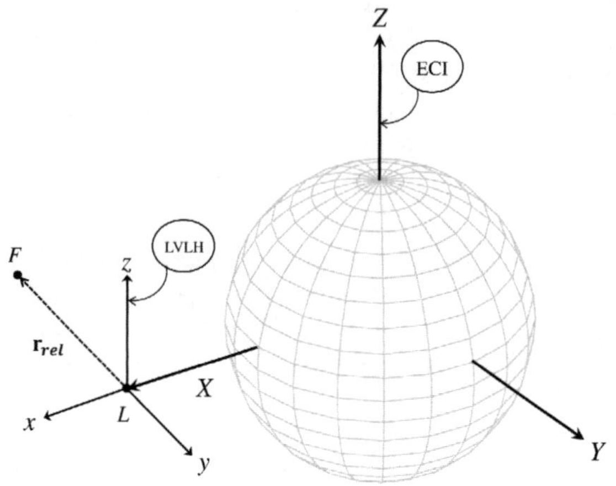

Figure 1: Position of the *leader* (L) and the *follower* (F) spacecraft in realtion to the Earth-centered Inertial (ECI) frame and the Local Vertical Local Horizontal Frame (LVLH) frame

$$\hat{\imath} = \frac{\mathbf{r}}{r} \qquad \hat{\jmath} = \hat{\mathbf{k}} \times \hat{\imath} \qquad \hat{\mathbf{k}} = \frac{\mathbf{h}}{h} \tag{2}$$

where $\mathbf{h} = \mathbf{r} \times \dot{\mathbf{r}}$.

Once we have established the systems of coordinate frames of reference, we need to define the two spacecraft: *leader* and *follower*. The leader is the spacecraft of which its center of mass is fixed at the origin o of the LVLH frame. The follower is the spacecraft which trajectory is to be design as to follow in proximity of the leader. One solution to this is, of course, to place the follower exactly in the same orbit as the leader with some distance between one another. The objective here however is to make the follower seen as moving around the leader, therefore making the follower a quasi-satellite of the leader. The quasi-satellite configuration is not intuitive and hard to visualize in the inertial ECI frame. However, if an observer is placed on the leader spacecraft, the follower spacecraft will be seen as "orbiting" the observer.

We model the two spacecraft as having no propulsive force, whether attractive or repulsive. This assumption separates from the previous studies of quasi-satellite dynamics [4], which focus more on massive celestial objects that have their own gravitational pull and satellites with ion thrusters. The apparent orbiting motion therefore is not due a central force of the leader but by a careful orbit placement. We denote $\mathbf{r}_L$ and $\mathbf{r}_F$ as the position vectors of the leader and the follower, respectively, as expressed in the ECI frame, The relative position of the follower with respect to the leader is

$$\mathbf{r}_{rel} = \mathbf{r}_F - \mathbf{r}_L \tag{3}$$

The velocity and acceleration of the follower relative to the leader can be therefore written as

$$\dot{\mathbf{r}}_{rel} = (\dot{\mathbf{r}}_F - \dot{\mathbf{r}}_l L) - \boldsymbol{\Omega} \times \mathbf{r}_{rel} \tag{4}$$

$$\ddot{\mathbf{r}}_{rel} = (\ddot{\mathbf{r}}_F - \ddot{\mathbf{r}}_L) - \dot{\boldsymbol{\Omega}} \times \mathbf{r}_{rel} - 2\boldsymbol{\Omega} \times (\dot{\mathbf{r}}_F - \dot{\mathbf{r}}_L) - \boldsymbol{\Omega} \times (\boldsymbol{\Omega} \times \mathbf{r}_{rel}) \tag{5}$$

where the angular velocity $\boldsymbol{\Omega}$ and angular acceleration $\dot{\boldsymbol{\Omega}}$ are

$$\boldsymbol{\Omega} = \frac{\mathbf{r}_L \times \dot{\mathbf{r}}_L}{r_L^2} \quad \text{and} \quad \dot{\boldsymbol{\Omega}} = -2\frac{\dot{\mathbf{r}}_L \cdot \mathbf{r}_L}{r_L^2}\boldsymbol{\Omega} \tag{6}$$

The expressions for the relative position, velocity and acceleration in this current form are expressed in the ECI frame. Therefore, the coordinates need to be rotated from the ECI frame to the LVLH frame using a rotation matrix ${}^{lvlh}R_{eci}$, which elements can be found from Equation (2). For instance, the position of a follower spacecraft expressed in the LVLH frame can be found by

$${}^{lvlh}[\mathbf{r}_{rel}] = {}^{lvlh}R_{eci}[\mathbf{r}_{rel}] \tag{7}$$

Defining orbits using the orbital elements provides more intuition of its shape and orientation in space. We employ the standard six classical orbital elements: a (semi-major axis), e (eccentricity), i (inclination), ω (argument of perigee), Ω (right ascension of ascending node), and M (mean anomaly). The last one is used interchangeably with θ (true anomaly). The definitions of these elements can be easily found in any standard reference on orbital mechanics.

Additionally, we will use another quantity, the mean longitude l, which is defined as $l = M + \omega + \Omega$ [13]. Mean longitude does not measure an angle between any vectors, but a quantity that shows the mean angular position of an object in an orbit. In our case, an indication of mean longitude is useful in showing how close, in terms of mean angular position, are two spacecraft when they are in two different orbits. Difference between the orbital elements of the leader and the follower will be denoted by the prefix δ. For example, $\delta e \triangleq e_F - e_L$.

4. Numerical Simulation

In the previous section, we have set up the referential reference frames, particularly, the Earth-centered Inertial (ECI) frame and the local-vertical local-horizontal (LVLH) frame. We have also shown the relative kinematics between the two frames. The task now is to find the state vectors, $\mathbf{r}$ amd $\mathbf{v}$, of both spacecraft at any later time Δt after their initial conditions $\mathbf{r}_0$ amd $\mathbf{v}_0$. The universal Kepler's equation is employed to solve the two-body problem. This requires the use of universal anomaly χ. Readers may consult Battin [13] for details on universal anomaly and universal variable formulation.

Initial orbital elements of the spacecraft is first converted to state vectors. From the initial state vector $\mathbf{r}$ amd $\mathbf{v}$, we define the radial component of $v_{r0} = \frac{\mathbf{r}_0 \cdot \mathbf{v}_0}{r_0}$. Using α to denote the reciprocal of semi-major axis

$$\alpha = \frac{2}{r_0} - \frac{v_0^2}{\mu} \tag{8}$$

we can estimate the initial universal anomaly χ as [14]

$$\chi_0 = \sqrt{\mu}|\alpha|\Delta t \tag{9}$$

At any given step i where $i = 0, 1, 2, \ldots$, the subsequent universal anomalies can be obtained from [15]

$$\chi_{i+1} = \chi_i - \frac{f(\chi_i)}{f'(\chi_i)} \tag{10}$$

where

$$\begin{aligned} f(\chi_i) &= \frac{r_0 v_{r0}}{\sqrt{\mu}}\chi_i^2 C(z_i) + (1 - \alpha r_0)\chi_i^3 S(z_i) + r_0\chi_i - \sqrt{\mu}\Delta t \\ f'(\chi_i) &= \frac{r_0 v_{r0}}{\sqrt{\mu}}\chi_i\left(1 - \alpha\chi_i^2 S(z_i)\right) + (1 - \alpha r_0)\chi_i^2 C(z_i) + r_0 \end{aligned} \tag{11}$$

where $z_i = \alpha\chi_i^2$. The functions $C(z)$ and $S(z)$ are Stumpff functions. For $z > 0$, the functions are defined by

$$\begin{aligned} S(z) &= \frac{\sqrt{z} - \sin\sqrt{z}}{(\sqrt{z})^3} \\ C(z) &= \frac{1 - \cos\sqrt{z}}{z} \end{aligned} \tag{12}$$

Note that the Stumpff functions shown in Equation 12 are applicable for elliptic orbit only.

With the new χ_i, one must now obtain the Lagrange f and g coefficients

$$\begin{aligned} f &= 1 - \frac{\chi^2}{r_0} C(\alpha\chi^2) \\ g &= \Delta t - \frac{1}{\sqrt{\mu}} \chi^3 S(\alpha\chi^2) \end{aligned} \tag{13}$$

to find the new position after time Δt using

$$\mathbf{r} = f\mathbf{r}_0 + g\mathbf{v}_0 \tag{14}$$

Next, from the new position we can find the new velocity. The first derivatives of the Lagrange coefficients are obtained

$$\begin{aligned} \dot{f} &= \frac{\sqrt{\mu}}{r r_0} \left(\alpha\chi^3 S(\alpha\chi^2) - \chi\right) \\ \dot{g} &= 1 - \frac{\chi^2}{r} C(\alpha\chi^2) \end{aligned} \tag{15}$$

to solve for $\mathbf{v}$

$$\mathbf{v} = \dot{f}\mathbf{r}_0 + \dot{g}\mathbf{v}_0 \tag{16}$$

This step-by-step algorithm is programmed to compute $\mathbf{r}(t)$ and $\mathbf{v}(t)$ given the initial orbital elements.

5. Quasi-satellite Formation

To simulate the orbital trajectories of the leader-follower formation, we start with a simple formation where the leader spacecraft is in a circular orbit around the Earth. For all simulations, the follower spacecraft assumes a condition that it must have the same period as the leaders orbit. This so that it has a 1:1 orbital resonance with the leader (the follower will complete one revolution around the leader in one orbital period). For this condition to be met, only the semi-major axes of both orbits need to match. In this case we consider $a = 10,000$ km for both orbits.

Difference in eccentricity

We fixed the eccentricity of the leader at zero and vary the follower's orbit eccentricity from 0.01 to 0.55. Figure 2 shows the follower trajectories in the LVLH frame when the leader is at the origin.

In Figure 3, we set the leader's eccentricity at $e = 0.1$ to and vary the follower's eccentricity from 0.01 to 0.2, with an increment of $\delta e = 0.05$. This is to see if an elliptical leader orbit would have any effect on the quasi-satellite formation. The trajectories when $e_F < e_L$ is shown in red.

All orbits are shown in the LVLH frame.

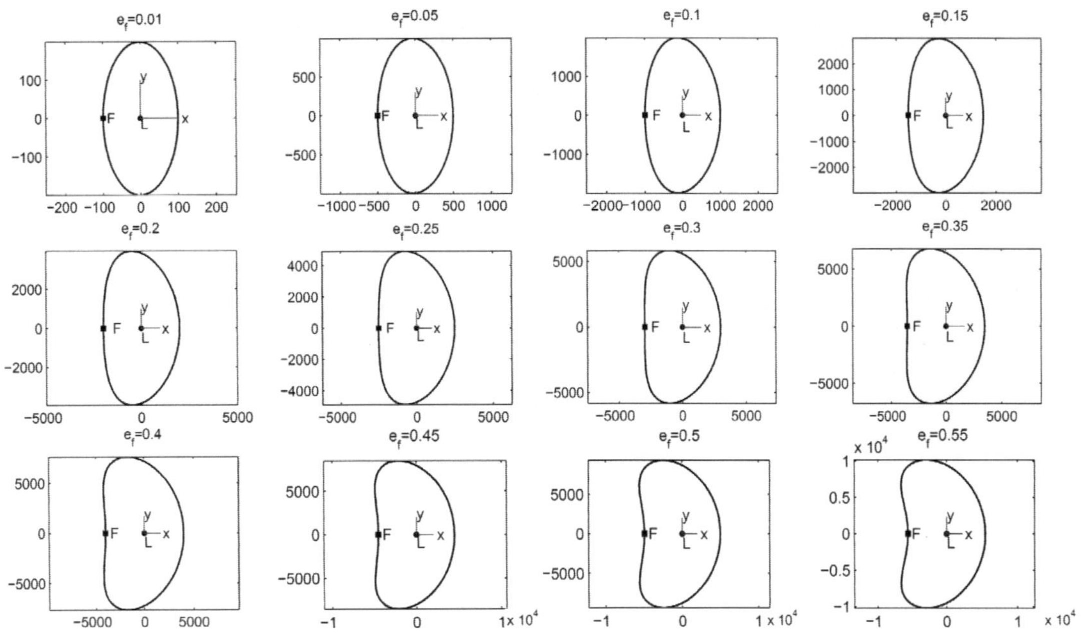

Figure 2: Quasi-satellite trajectory for increasing follower's eccentricity.

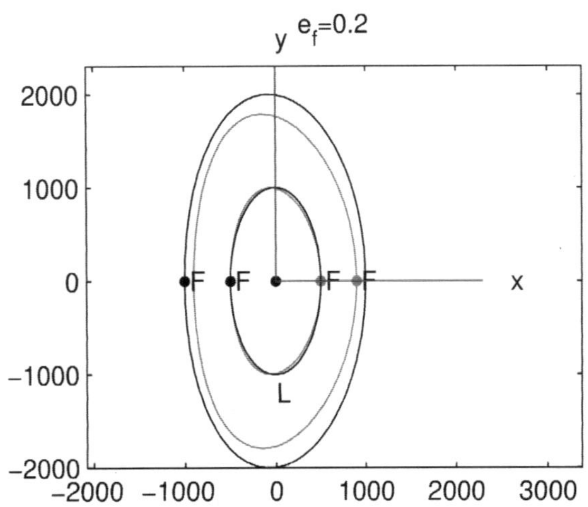

Figure 3: Negative δe orbits are indicated by red trajectories which shows that the follower's starting point is on the opposite direction.

Difference in ω, Ω, i, and θ
We start with the leader's orbit having an inclination $i_L = 10°$ and $\delta l_L = \omega_L = \Omega_L = M_L(0) = 0$. The leader is in $e = 0$ orbit and the follower is in $e = 0.2$ orbit for all cases. We vary the follower's orbital parameters -ω_F, Ω_F, i_F, and M_F - one at a time to see the effects of modifying the orbital elements of the follower in relation to the leader.

Figure 4 shows the effect of increasing the followers ω_F while the leaders ω_L is fixed. Figure 5 shows similar comparison of different $\delta\Omega$ while keeping other elements constant. The inclination δi effect is shown in Figure 6 and the $\delta\theta$ effect in Figure 7. The angles ranges from 0° to 110° for all cases.

All orbits are shown in the LVLH frame with a globe representing Earth for illustration.

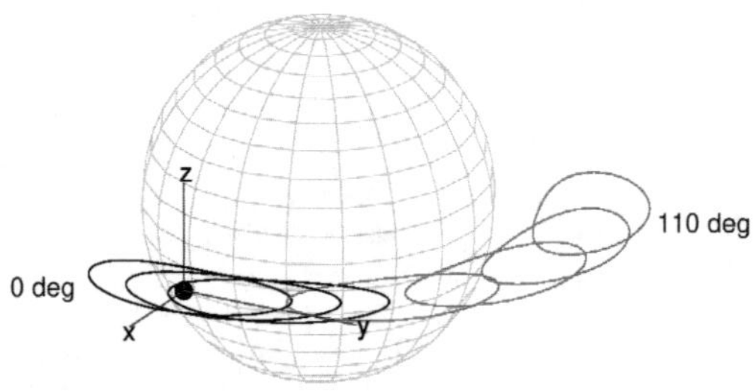

Figure 4: $i_F = 10°$; $\Omega_F = M_F(0) = 0$. Follower's *argument of perigee* is varied $\omega_f = [0°, 10°, 20°, 45°, 70°, 90°, 110°]$.
Black trajectory are when the follower is in an orbit around the leader.

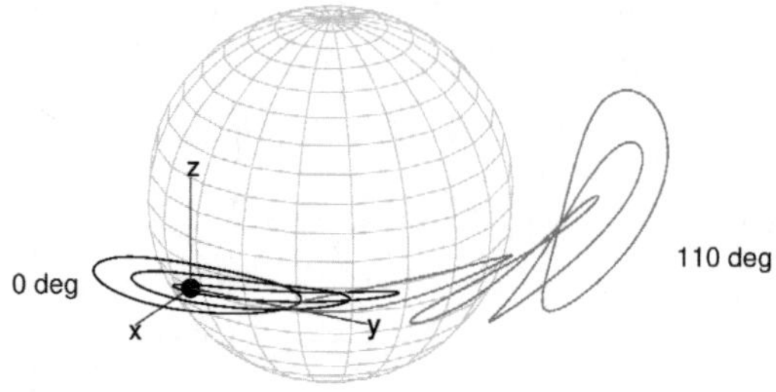

Figure 5: $i_F = 10°$; $\omega_F = M_F(0) = 0$. Follower's *right ascension node* is varied $\Omega_f = [0°, 10°, 20°, 45°, 70°, 90°, 110°]$.
Black trajectory are when the follower is in an orbit around the leader.

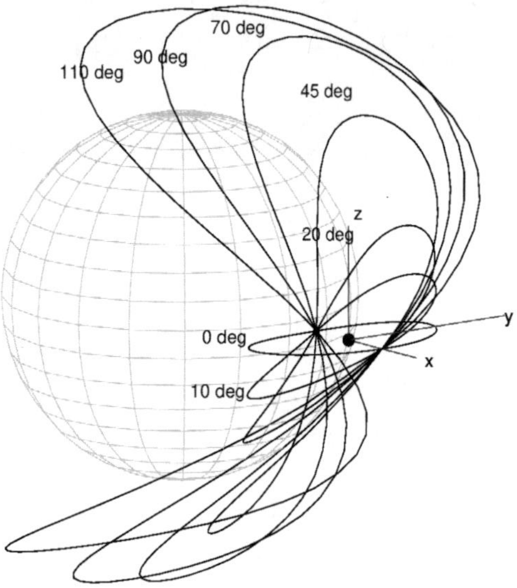

Figure 6: $\Omega_F = \omega_F = M_F(0) = 0$. Follower's *inclination angle* is varied $i_f = [0°, 10°, 20°, 45°, 70°, 90°, 110°]$.
Follower orbits the leader for all inclination.

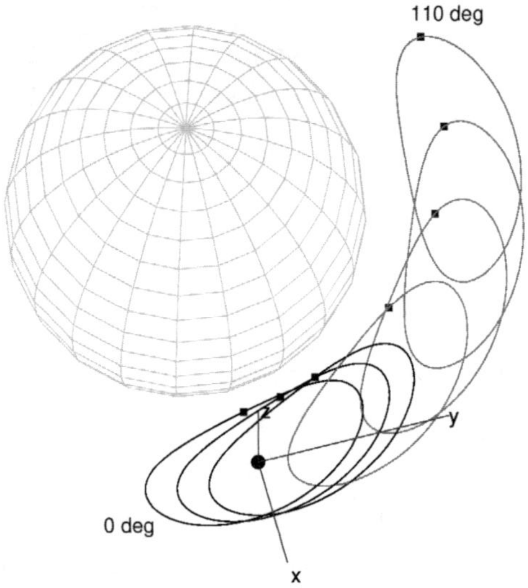

Figure 7: $i_F = 10°$; $\omega_F = \Omega_F = 0$. Follower's initial *mean anomaly* is varied $M_f = [0°, 10°, 20°, 45°, 70°, 90°, 110°]$.
Black trajectory are when the follower is in an orbit around the leader.

6. Discussion Based on the Visualized Trajectories

Our numerical simulation provides several insights in creating a quasi-satellite system for spacecraft

(i) Increasing the eccentricity difference between the leader and the follower increases the average radial distance (in the LVLH frame) of the follower from the leader. Increasing the eccentricity also evolves the followers trajectory from an oval to a kidney-shaped orbits, which are seen in horizontal Lyapunov orbit family around L2 libration point (Reader may refer to [16] for illustrations). This is more prominent when δe is large.

(ii) The follower revolves around the leader in a retrograde motion. In general (general here means if every other orbital elements, except e, is set to be the same), one can observe that the follower is the fastest when it is on the inside (negative x direction in the LVLH frame) and the slowest when it is outside. This is more intuitive if one can imagine they are the perigee and apogee points of the follower's orbit. If $e_F < e_L$, the starting point of the follower is changed 180

(iii) To get the follower to fly around the leader in the quasi-satellite fashion, the starting mean longitude difference $\delta l = \delta M + \delta\omega + \delta\Omega$ have to be close to zero. This can be seen in Figure 4,Figure 5, and Figure 7, where increasing the angular difference beyond 20° separates the leader spacecraft from the quasi-orbit. It is suggested here that for a quasi-satellite system to work, both spacecraft has to be almost co-planar. This is a good consideration especially when we incorporate perturbations later. This is consistent with the findings of [4] which had included J_2 perturbation in their analysis.

(iv) Varying inclination radically changes the shape of the follower's trajectory but still maintains the follower's orbit around the leader. We suggest here that, if an out-of-plane motion is desired, δi is to be kept small to avoid excessive maneuver to correct orbit drifting due to gravitational perturbation.

(v) The unperturbed model used here suggests that quasi-satellite dynamics is possible without having to use propulsive forces. In the presence of perturbations, however, artificial quasi-satellites cannot be achieved without control to overcome the orbital deviations. We suggest, that in order to keep the quasi-satellite configuration, the almost co-planar condition (small δl and δi) must be adhered as to minimize propulsive correction maneuver to maintain the orbit.

(vi) Recommendation based on our numerical experiment using an unperturbed two-body model is presented as such

Table 1: Quasi-satellite Requirements

Follower's Orbit Element	Recommendation Based on Numerical Simulation
semi-major axis	$\delta a \approx 0$; for a 1:1 orbital resonance
eccentricity	$\delta e < 0.05$; for close proximity formation
inclination	δi small
argument of perigee	$\delta\omega < 20°$
right ascension of ascending node	$\delta\Omega < 20°$
Mean anomaly	$\delta M < 20°$

7. Conclusion

The discovery of quasi-satellite configuration in our solar system shows that formation flight occurs in nature. Inspired by the mechanics of natural celestial bodies, we show that artificial satellites are able mimic the motion of quasi-satellites without any correcting maneuver using propulsion. We show the possible orbit shapes according to different eccentricities and, orbital orientation and position according to different orbital elements. We proposed several conditions to form a quasi-satellite for a leader-follower formation 1) both semi major axes need to be the same as to keep a 1:1 orbital resonance between the two spacecraft 2) the orbital elements relating to orientation, e.g. δi, $\delta\omega$, $\delta\Omega$ and δM, must be kept small to ensure a quasi-satellite system. To design a formation using nonlinear equations is a laborious process. This paper provides a parametric analysis based off the idea of natural quasi-satellites, like asteroids, to provide another perspective to look at the formation flight problem. The logical next step is to include perturbation to this existing model to investigate deeper the quasi-satellite formation dynamics.

References

[1] Agle DC B D and Cantillo L 2016 Small asteroid is earth's constant companion `http://www.jpl.nasa.gov/news/news.php?feature=6537`
[2] Mikkola S and Innanen K 1997 *The Dynamical Behaviour of Our Planetary System* (Springer) pp 345–355
[3] Mikkola S, Brasser R, Wiegert P and Innanen K 2004 *Monthly Notices of the Royal Astronomical Society* **351** L63–L65
[4] Mikkola S and Prioroc C L 2016 *Monthly Notices of the Royal Astronomical Society* **457** 1137–1144
[5] Stephens G L, Vane D G, Boain R J, Mace G G, Sassen K, Wang Z, Illingworth A J, O'Connor E J, Rossow W B, Durden S L *et al.* 2002 *Bulletin of the American Meteorological Society* **83** 1771–1790
[6] Beard R W, Lawton J, Hadaegh F Y *et al.* 2001 *IEEE Transactions on control systems technology* **9** 777–790
[7] Wiltshire R and Clohessy W 1960 *Journal of Aerospace Science* **27** 653–658
[8] Hempel P and Tschauner J 1965 *Astronautica Acta* **11** 104–109
[9] Eckstein M, Rajasingh C and Blumer P 1989 *International Symposium on Space Flight Dynamics* vol 6
[10] Alfriend K T, Vadali S R and Schaub H 2001 *Celestial Mechanics and Dynamical Astronomy* **81** 57–62
[11] Schaub H and Alfriend K T 2001 *Celestial Mechanics and Dynamical Astronomy* **79** 77–95
[12] D'Amico S and Montenbruck O 2006 *Journal of Guidance, Control, and Dynamics* **29** 554–563
[13] Battin R 1999 *An Introduction to the Mathematics and Methods of Astrodynamics* AIAA Education Series (American Institute of Aeronautics & Astronautics)
[14] Chobotov V 1996 *Orbital mechanics* AIAA Education Series (American Institute of Aeronautics and Astronautics)
[15] Curtis H 2009 *Orbital Mechanics for Engineering Students* Aerospace Engineering (Elsevier Science)
[16] Kolemen E, Kasdin N J and Gurfil P 2007 *New Trends in Astrodynamics and Applications III (AIP Conference Proceedings)* vol 886 pp 68–77

IOP Conf. Series: Materials Science and Engineering **152** (2016) 012035 doi:10.1088/1757-899X/152/1/012035

Current development of UAV sense and avoid system

A Zhahir[1], A Razali[1,2*] and M R Mohd Ajir

[1]Department of Aerospace Engineering, Engineering Faculty, Universiti Putra Malaysia, Malaysia
[2]School of Electrical System Engineering, Universiti Malaysia Perlis, Malaysia

*shikinrazali01@gmail.com

Abstract. As unmanned aerial vehicles (UAVs) are now gaining high interests from civil and commercialised market, the automatic sense and avoid (SAA) system is currently one of the essential features in research spotlight of UAV. Several sensor types employed in current SAA research and technology of sensor fusion that offers a great opportunity in improving detection and tracking system are presented here. The purpose of this paper is to provide an overview of SAA system development in general, as well as the current challenges facing UAV researchers and designers.

1. Introduction

Unmanned aerial vehicles (UAVs) are now used in various military fields including reconnaissance, environmental monitoring, border patrol, search and rescue operations, disaster relief, tracking and monitoring. In the next decade, UAVs' market is expected to be rapidly growing in handful of civilian and commercial industries such as agriculture, energy, utilities, mining, construction, real estate, news media, and film production [1, 2]. These applications require the small and agile UAVs to fly at lower altitude or operating inside buildings, which are exposed to many hazards and obstacles. However, current UAVs technology in automatically sensing, detecting and avoiding fixed obstacles such as power line, building, tower, tree, and moving obstacles of birds and other aircraft is still immature compared to manned aerial vehicle [3]. As UAVs' market is predicted to provide billions of dollars in economic growth especially in commercial market, researchers around the world are now in rush to develop an efficient automatic SAA system to satisfy the demands and requirements of the UAVs. Moreover, another factor to expedite this research field is due to the new regulation from aviation authorities, which will be implemented soon. The regulation requires each UAV to have this automatic SAA system to reduce collision risk and thus improve the safety flying environment in non-segregated airspace [4].

Figure 1 shows a basic diagram of how SAA system should automatically function. Several sensors are employed in the UAV to collect and record data along the flying path. During the operation, the sensors are responsible to detect and identify any obstacles or threats to the UAV depending on their specification. The sensors provide information of environmental mapping based on the implemented tracking model. The information from several sensors are gathered and processed using the collision avoidance program embedded in the main processor. There are many types of collision avoidance programs used such as nonlinear model predictive control, graph search algorithm, potentially field method and vision based neural network. Finally, the UAV executes the avoidance action depends on the processed obstacles' information.

During the avoidance action, several parameters of interest such as position, speed, size and bearing angle of UAV may change. This cycle is continuously repeating to capture next mapping environment while at the same time UAV must look ahead to pursue its operational goal without fail. The capability of the sense function is not only to detect the presence and position of intruders but also to predict the speed and trajectory of the moving intruders. On the other hand, the avoid function should provide manoeuvre path to prevent the identified collision with consideration of kinematic and dynamic aspect [2, 3, 4]

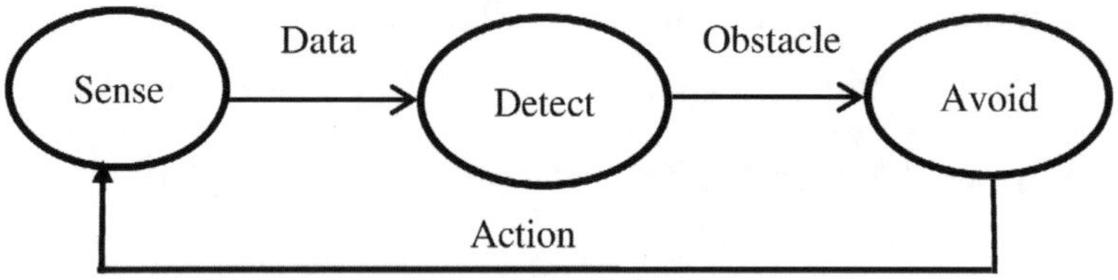

Figure 1. Diagram of SAA system

Primary challenges in implementing collision avoidance technology in UAVs are their size, weight and electrical power consumption. The limitation of payloads, especially in small UAVs, has hindered current collision avoidance technology to be integrated as the employed sensors must be lightweight, small in size and consume low power. The driving algorithm also needs to be computationally efficient and less complex as the size of main board memory is small and the processor has less speed with limited power supply. Even though UAVs' speed is slower than manned aircraft, the SAA system is expected to process the data fast enough to avoid collision with the faster manned aircraft. In addition, as UAVs' applications are widely increased, the SAA system should be capable to operate in various environment, weather and condition. Therefore, a less complex algorithm with compatible sensors at relevant cost is an important consideration in developing an efficient SAA system.

2. Sensor

Although Traffic Collision Avoidance System (TCAS) and also Automatic Dependent Surveillance–Broadcast (ADS-B) technology have already matured for collision avoidance technique in manned aerial vehicle, the system is currently satisfied only for cooperative intruders. Technology of ADS-B in terms of transmitter and receiver system is not just capable to provide data about other aircraft, but also provides information of the host aircraft's altitude and velocity. Furthermore, the system is able to deliver information of weather and terrains signal [5]. With these advantages, the requirement to integrate ADS-B to all size UAV is expected soon with current research focusing on miniaturisation of the equipment that will be beneficial to SAA system. Research on electronic sensors for ground based SAA system is also conducted for the near term solution in limited operational areas.

Many researchers have focused more on non-cooperative sensors of active and passive type as they can provide better detection for the non-cooperative obstacles especially when problem of data link loss occurs. Non-cooperative sensors are really important, particularly for the UAVs that fly inside buildings since the obstacles are mostly wall, machine, office equipment and humans. Active sensors in application include laser ranging, radar and sonar submit signal in detecting obstacles. They usually provide a very good information of the obstacles distance and it is very essential criteria in tracking operation. On the other hand, passive sensors such as electro optical, infrared, thermal imaging and motion detector depend on the signal detection emitted from the obstacle [6].

Application of electro optical sensors, combined with radar/infrared sensor/thermal imaging/motion detector, is a very popular approach in SAA system. Electro optical sensor, which is small in size and weight, has very fast scan rate and high resolution of image. They are also low in cost and consume low electric power. However, they are not capable to estimate obstacle range and sometimes generate high false alarm rate [7]. Since they need light in their operation, bad weather or cloudy condition can really affect their performance in object detection. Hence they cannot function well in low illumination [8]. The infrared sensor, which aids the electro optical sensor detection at night, is not affected by the electromagnetic interference and can measure obstacle range by calculating the infrared light radiated from objects. Thermal imaging sensor detects heat and is capable to operate in all-weather operation whereas motion detector function by sensing direction and velocity of objects [6].

Milimitre wave radar that has narrow beam width is capable to detect low altitude target without multi-path effects. The advantage of this radar is that it can work well in all-weather conditions, hence dust and lightning are less affected. However, it can be easily interfered by electromagnetic and has a bad tracking accuracy when the target is cloaked [9]. Light detection and ranging (LIDAR), which has high precision and resolution in distances measurement, is able to detect objects of different sizes and shapes by calculating the time taken for light to travel forth and back. The high-speed acquisition feature of LIDAR makes it commonly used in contour mapping but it is a high cost equipment and consumes more power. While sonar sensor is widely used in maritime for it is cheap, light weight and compactness, it provides accurate distance only to flat and upright indoor environment. Detection of irregular shaped object such as inclined roofs, trees and fences using sonar sensor will produce error to the system [10].

3. Sensor fusion technique

Sensor fusion offers a great opportunity to overcome the physical limitations of the sensing systems. It combines information from a number of different sensors to provide a robust and complete description of the flying environment that is rapidly changing. Since each sensor has its strength and weakness, a technology to integrate multiple sensors in one UAV platform is necessary to improve detection and minimize error in tracking. An efficient tracker system will result in accurate and reliable tracking and can perform fast speed avoidance manoeuvring if risky collision is identified [9, 11, 12].

The research development in sensors fusion is however still far from maturity level. The integration of the sensors into main board requires real time operation of embedded system programming, which will involve high processing power in capturing and processing various signals data from multiple sensors. Fast response algorithm without much complexity is preferable to ensure that the sense and detection system functions properly. This fast process in real time operation will increase probability of successful avoidance action in the required time frame. Another problem is the increasing UAV payload when the multiple sensors are employed. There are only few efficient lightweight sensors available in the market. As a result, some sensors with many advantages have to be neglected due to increasing payload problem since the stability and the speed of UAV also need to be considered.

Ramasamy and Sabatini (2015) have successfully shown a simulation study of sensor fusion, which combines several cooperative sensors, non-cooperative sensors and naturally-inspired sensor [11]. The integration and sequence of the sensors were made based on Boolean decision logic. The system used Kalman filter tracking at each sensor site and then communicated to a centre where a global track file was assimilated. The algorithm known as track to track fusion combines the estimation compared to observation where it can solve the robustness issues such as limited information of the environment and partial loss of information. On the other hand, Fasano *et al.* (2015) have recorded a significant improvement on the collision detection performance in multi sensor fusion of radar and electro optical sensors [12]. The effectiveness of sensor fusion in the flight experiment has successfully reduced false alarm rate, detection range sensitivity and computational burden problems. The similar result was detected in [8] for the fusion study of electro optical sensor and synthetic aperture radar, which showed the false alarm number per imaging area (in km) has significantly reduced down to 14-23%. Moreover, according to [13], the employment of infrared and ultrasonic sensors fusion has effectively reduced reading noises and errors that resulted in accurate range estimation in collision avoidance.

4. Conclusion

The overview of SAA development as well as the challenges faced by the researchers has successfully presented here. Advanced development in miniaturisation technology like MEMs and microcontroller is predicted to promote high research in the integrated sensor architecture of the SAA system. Finally, although several challenges are still unresolved for now, the industry is still young and the increased market demands for civil/commercial applications will surely attract big companies and industries to fund SAA research and thus expedites the technology growth.

References

[1] Ballve M 2015 *The Drones Report: Market Forecasts, Regulatory Barriers, Top Vendors, And Leading Commercial Application*
[2] Lin Y and Saripalli S 2013 *American Helicopter Society 69th Annual Forum (Arizona)*
[3] Alexopoulos A, Kandil A, Orzchowski P and Badreddin E 2013 *IEEE International Conference on Systems, Man, and Cybernetics (Manchester)* p1969-1974
[4] Patel A and Winberg S 2011 *IEEE Africon (Livingstone)*
[5] Cornic P, Garrec P, Kemkemian S and Ratton L 2011 *IEEE Aerospace Conference (Montana)* p1-14
[6] Muraru A 2010 *INCAS Bulletin* **2** 133–41
[7] Recchia G, Fasano G, Accardo D and Moccia A 2006 *IEEE Aerospace Conference (Montana)* p1-9
[8] Kim J and Kwag Y K 2014 *10th European Conference on Synthetic Aperture Radar (Berlin)* p632-633.
[9] Yin J, Cui B and Wang Y 2010 *Chinese Journal of Electronics* **19** 752-6
[10] Tin F E 2015 *Understanding the Sense and Avoid Challenge* EagleView Technologies, Inc.
[11] Ramasamy S and Sabatini R 2015 *International Conference on Unmanned Aircraft Systems (Denver)* p765-73
[12] Fasano G, Accardo D, Tirri A E, Moccia A and DeLellis E 2015 *Aerospace Science and Technology* **46** 436-50
[13] Rambabu R, Bahiki M R and Ali S A 2015 *Jurnal Teknologi* **76** 89-93

IOP Conf. Series: Materials Science and Engineering **152** (2016) 012036 doi:10.1088/1757-899X/152/1/012036

Temperature and heat flux measurement techniques for aero-engine fire test: a review

I Mohammed[1,2], A R Abu Talib[1,3*], M T H Sultan[1,3] and S Saadon[1]

[1]Department of Aerospace Engineering, Faculty of Engineering, Universiti Putra Malaysia, Malaysia
[2]Department of Mechanical Engineering, College of Engineering, Hassan Usman Katsina Polytechnic, Nigeria
[3]Aerospace Manufacturing Research Centre, Faculty of Engineering, Universiti Putra Malaysia, Malaysia

*abdrahim@upm.edu.my

Abstract. This review is made of studies whereby some types of fire test measuring instrument were compared based on their mode of operation, sensing ability, temperature resistance and their calibration mode used for aero-engine applications. The study discusses issues affecting temperature and heat flux measurement, methods of measurement, calibration and uncertainties that occur in the fire test. It is found that the temperature and heat flux measurements of the flame from the standard burner need to be corrected and taken into account for radiation heat loss. Methods for temperature and heat flux measurements, as well as uncertainties analysis, were also discussed.

1. Introduction

Temperature and heat flux measurements are important factors that need to be considered in fire test experiment for aero-engine applications. These two factors require accurate sensors that can easily sense and correctly read the inputs and outputs. Therefore, design and implementation of the sensors and calibration system is a complex process that requires attentive precautions to ensure accuracy in the measurements. There are many types of temperature and heat flux measuring devices that are used for fire test [1]. A good understanding of temperature and heat flux measuring devices for heat transfer to solids exposed to fire conditions is important for the fire test in building structures and aerospace industries. There are some uncertainties that arise in measuring the different forms of convection and radiation heat flux on the standard type of test that should be carried out, and on the temperature and load bearing capacity of structural components, as well as time to ignition and burning characteristics of materials and products [2].

The main issues associated with measurement of flame temperature and heat flux are the sensors in the measuring devices and the ability of the devices to measure the range of temperatures in the fire test. To the authors' knowledge, there is no available literature that reviews the temperature and heat flux measurements used for aerospace applications. The main objectives of this review are to identify devices for measuring temperature and heat flux by choosing the proper sensors according to their characteristics, to quantify the measurement errors and the problems associated with these devices, and to assess issues, methods, calibration and uncertainty on temperature and heat flux measurements in fire test techniques for aerospace engineering applications.

2. Standard fire test for aerospace applications

The standard fire test techniques used in aero-engine fire certification are explained in the ISO2685 [3] and AC20-135 [4].

2.1. ISO2685 fire test standard

ISO2685 standard specifies the test conditions applied to all components, equipment and structures located in zones designated as “fire zones” and built to satisfy the minimum level for resistance to fire. The standard does not relate to the resistance to fire outside the designated fire zones, the flammability requirements or to those conditions induced by the flame coming from the combustion chamber. Abu Talib *et al.*[5] reported that ISO propane-air burner is frequently used in UK to certify critical engine components that will be exposed to the danger of aircraft engine fires. Meanwhile, kerosene burner is widely used in the US and this is also specified in ISO 2685. The standard defines the designated fire zone as the region of an aircraft, as the compartment containing main engines and auxiliary power units, designated in accordance with the requirements of the approving authorities. The flame having the characteristic of 1100°C ± 80°C temperature and 116 kW/m^2 ± 10 kW/m^2 heat flux. In ISO 2685, the burner head must act as the flame stabilizer and prevents any flashback into the plenum chamber containing the combustible mixture. The burner is made up of 373 copper tubes, which are cooled by air flowing around them. The cooling air is supplied through 332 holes at the flame head, maintaining the burner temperature at the level required for the test. The cooling air shall be well distributed within the structure of the flame head in order to maintain the burner temperature at the required level. Air and gas are introduced and mixed in a small chamber, the mixed gas and air enter the larger plenum chamber before reaching the flame head.

Figure 1 shows the arrangement of propane-air burner, which shows the chamber where air and gas fuel are mixed at the upper base of the burner, then the mixture is passed into 1/8-inch copper tubes with a bore of 0.07 inch. The secondary air (cooling air) that is used to cool the burner and maintain the flame temperature goes through a 0.102 inch. Figure 1(a) is the main body of the burner, which is made from copper that contains fuel and an air entrance with a small chamber where fuel and air pre-mixed together and moves to plenum chamber. In addition, there is also a hole drilled at the top of the burner whereby a secondary air is introduced to the burner that provides a coolant air that stabilizes the flame temperature to be uniform. At the top of the main body is a plate that contains the burner head and burner nose as shown in Figure 1(b). The plate contains 373 copper tubes that mix the fuel and air pass through it, and another 332 holes for secondary pass through [3]. The dimensions of the stated instruments are shown in the figures.

2.2. AC 20-135 fire test standard

This is another standard that explains the methods used in determining the conformity with the power-plant fire protection prerequisite of the Federal Aviation Regulations (FAR). The standard provides procedures for the fire testing of materials and components used in the propulsion engines and APU installations, and near the fire designated zones. Although the terms are considered in classifying the material according FAR, these two terms are "fireproof" and "fire resistant," but in this standard the level of heat intensity, temperature levels, exposure time, or an appropriate wall thickness and other dimensional features were not clearly stated. When the materials are coated, composed of laminated composites and metal honeycomb for acoustically treated ducting, cowling and other components that are element of the nacelle firewall, shows the ability to withstand fire [6]. The definition of fireproof and fire resistant are as follows:

- Fireproof: the ability of a material or component to resist a flame temperature of 1093°C flame (±65.6°C) for 15 minutes minimum, and still it performed its function. When materials and parts used in an enclosed fire within designated fire zones, it means that the material or part will be used under conditions likely to occur in such zones and will resist a 1093°C flame (±65.6°C) for 15 minutes minimum.

- Fire Resistant: the ability of a material or component to resist a flame temperature of 1093°C flame (±65.6°C) for 5 minutes minimum and still it performed its function. When a materials and parts used in an enclosed fire within designated fire zones, it means that the material or part will be used under conditions likely to occurs in such zones and will resist a 1093°C flame (±65.6°C) for 5 minutes at least.

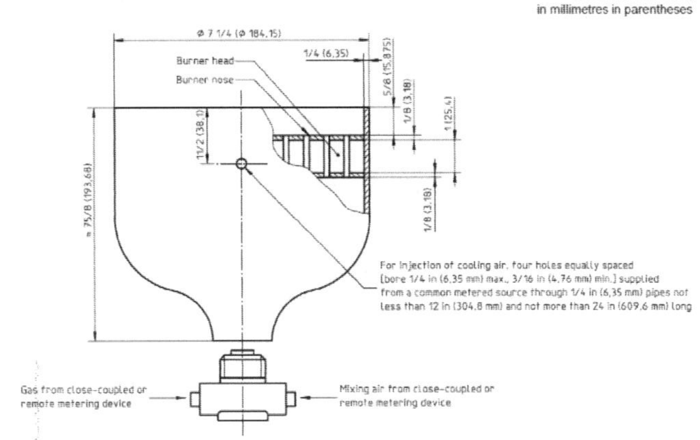

(a)

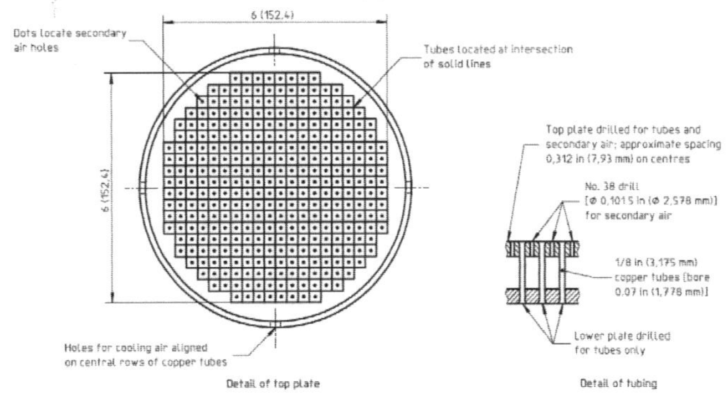

(b)

Figure 1. Configuration of propane burner (reproduced from ISO2685 [3])

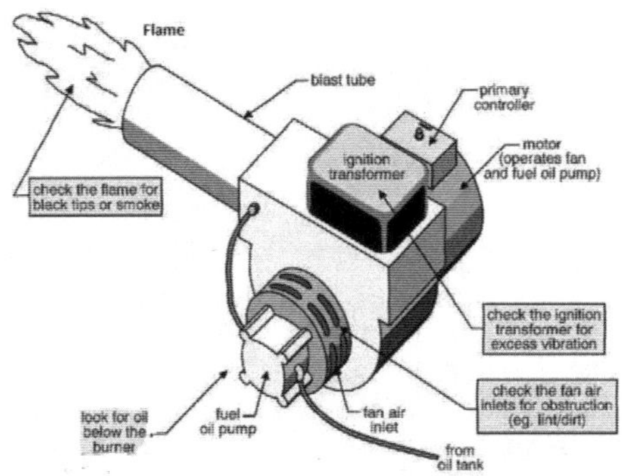

Figure 2. Oil burner [7]

3. Temperature measurement

The flame temperatures were measured at appropriate axial distance from the burner to verify that the conditions of the standard flame are achieved [8]. This section explains the methods, calibration and issues of temperature measurement.

3.1. Methods of temperature measurement

Comprehensive temperature plan was acquired if sole thermocouple over the measurement plane of the ISO burner was traversed. The measured values of temperature, heat flux and a sample in ISO 2685 propane-air burner needs to be taken at a distance of 75 mm (3 inch) from the burner face. Samir and Ryan [9] reported an experiment that was conducted on 3-inch x 3-inch samples where the temperature distribution was found to be uniform and the mean temperature reading of 1119°C were obtained. Among the types of temperature measuring devices are thermocouple, liquid-in-glass thermometers, resistant temperature detectors (RTDs), thermistors and optical pyrometers [1]. Thermocouple is the most common sensor that determines the temperature in fire test, the circuit is made up of two wires of different metals or their alloys that are joined together at one end and open at the other end, an electromotive force (EMF) is generated due to the voltage differential between the open ends when the junction is heated. The voltage differential between the open wire terminals is the outcome of the temperature differences connecting the joints. These two junctions are the hot and cold junctions that are the measuring and reference junctions respectively as shown in Figure 3(a) [1].

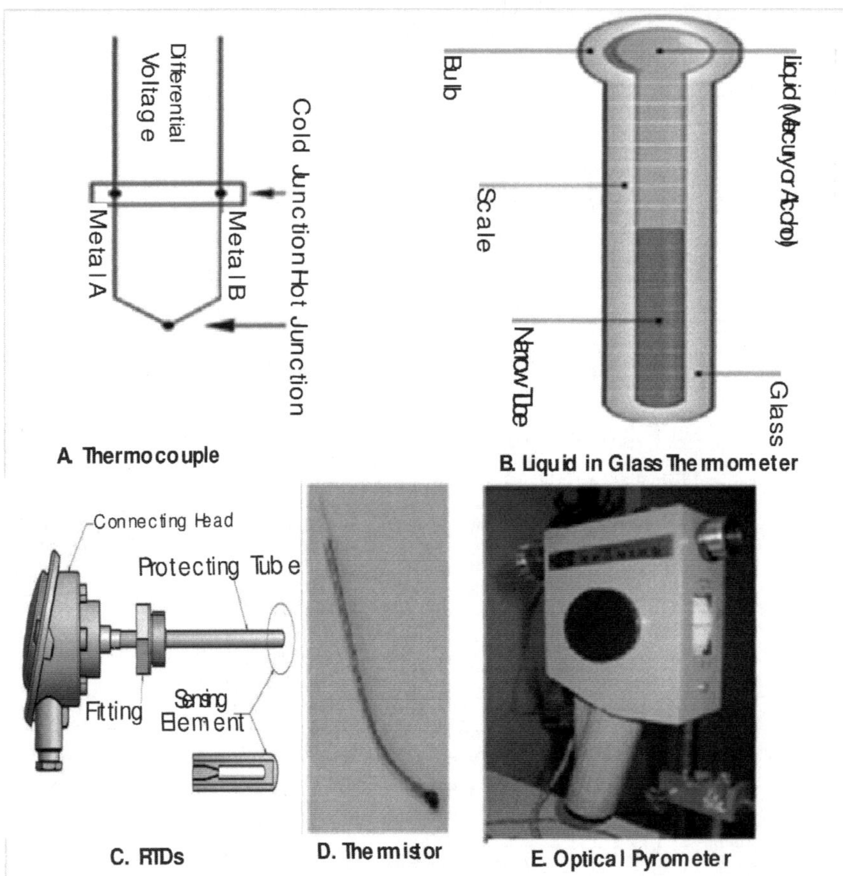

Figure 3: Types of temperature measuring devices: (A) thermocouples [10], (B) liquid in glass thermometer [11], (C) resistant temperature detector [12], (D) thermistor [13] and (E) optical pyrometer [14]

There are different types of thermocouples used in measuring the temperature and when these thermocouples are protected, they can withstand the working temperature range as listed in Table 1. Type T thermocouple is applicable to the low temperature ranging from 0 to 350°C and it is a stable thermocouple which has a very good repeatability between a temperature range of -200 to 200°C. It has a standard accuracy of ±0.75% with special limit of error of ±0.4% with typical accuracy of 1°C. Meanwhile, Type J thermocouple is the type of thermocouple that has a short life span at very high temperature. It cannot withstand high temperature, where it can only withstand moderate temperature. The temperature range is greater than the temperature ranges of T type and it is also well suited to oxidizing atmosphere. It has a standard accuracy of ±0.75% and a special limit of error of ±0.4% with typical accuracy of 2.2°C. Type E thermocouple has a moderate temperature range which is stable with good operating range in oxidizing atmosphere. It has a standard accuracy of ±0.5% with special limit of error of ±0.4% with typical accuracy of 1.7°C. Meanwhile, Type K thermocouple which is usually Chromel/constant, normally has good corrosion resistant with a good operating temperature range. It can withstand a temperature of up to 1260°C and it is not expensive when compared with other types that have higher temperature range. It is partially having radiation hardness with a standard accuracy of ±0.75% with special limit of error of ±0.4% and with typical accuracy of 2.2°C. Type N thermocouple can withstand a high temperature with good corrosion resistant and good oxidation. It has a standard accuracy of ±0.75% with special limit of error of ±0.4% and with typical accuracy of 2.2°C. Meanwhile, Type R thermocouple is used both on high and low temperature application. It has a higher percentage of Rhodium than Type S thermocouple. Hence, it is more expensive than the other types of thermocouples. This type of thermocouple has high accuracy and stability. It moderately produces higher output, and has a standard accuracy of ±0.25% with special limit of error of ±0.1% and typical accuracy of 1.5°C. Type S thermocouple can be used in inert and oxidizing atmosphere up to 1600°C. It can be used both for high and low temperature application due to its high accuracy and stability. It has a standard accuracy of ±0.25% with special limit of error of ±0.1% and with typical accuracy of 1.5°C. Furthermore, Type B thermocouple has high temperature application, with high level of accuracy and stability. It has a low output at a temperature, which is lower than 600°C. It has a standard accuracy of ±0.5%, with special limit of error of ±0.25% and with typical accuracy of 5°C at 1000°C [16].

Table 1. List of thermocouple types and its working temperature range [15]

Thermocouples	Temperature Range (°C)
T	0 – 350
J	95 – 760
E	95 – 900
K	95 – 1260
N	95 -1260
R	870 -1450
S	980 – 1450
B	871 – 1704

Liquid-in-glass thermometer is the other type of sensor used to measure the fire temperature. It has a sealed bulb connected to glass capillary in the vertical stem and because of the development of the liquid rises in the capillary; temperature growth was developed, as shown in Figure 3(B). A resistant temperature detector (RTDs) uses metals especially Nickel or Nickel alloy for its high temperature resistance as the measuring instrument. It changes when temperature increases due to its positive change in electrical resistance; it is chemically stable and has high resistivity, as shown in Figure 3(C).

Thermistor is closely related to RTDs but has semiconductors that have electrical resistance. It can be used on small temperature of maximum of 100°C as shown in Figure 3(D). On the other hand, the optical pyrometer can sense the heat temperature of solid without touching the solid, the sensor sense thermal radiation emitted for the required area surfaces [1], the diagram in shown in Figure 3(E).

Thermochromic liquid crystals have been used to measure full surface temperature distribution [17]. The TLCs are coated on the model surface in which temperature distribution is to be determined by colour-temperature relationship and imaging, colourimetry, illumination, hysteresis, film thickness and aging that are related to calibration were studied. The images of different coloured for TLCs were recorded by digital video cameras [18, 19]. Another temperature measuring device is infrared camera that is non-contact temperature measuring device. Its advantages include measuring high temperature, high speed, able to measure the temperature of moving parts, no energy was lost in measurement and the sensor must be protected from dust and condensing liquid [14].

3.2. Calibrations of temperature measurements

Thermocouples rake was used to calibrate the flame temperature with an unprotected bead. Currently, there is no uniformity on the size of thermocouples (TC) as reported in [20, 21]. AC 20-135 specifies an acceptable sheath diameter to be used in calibrating thermocouple from 1/16-inch to 1/8- inch; while the ISO2685 standard uses a thermocouple of 0.12 inch sheath diameter. Furthermore, there are differences in measuring flame temperature using the two standards for the calibration requirement, whereby ISO2685 standard requires a standard temperature of 1100°C±80°C, while AC20-135 uses an average temperature of 1093°C±83°C. The thermocouples readings were recorded in data logger and the stand of the rake were placed above the centre line of the burner for correct calibration. The average value of the number of thermocouple used is taken as the values that the standard used [22]. Samir *et al.* [20] investigated the effect of thermocouple size on flame calibration, were two different sizes were investigated, the sizes investigated were 1/8 and 1/16-inch sheath diameter as shown in Figure 4. These thermocouples had AWG 28 wires with 0.012 inch diameter bead. The thermocouples sizes used in this experiment were acceptable for the calibration under AC20-135 regulations, which specify an acceptable thermocouple wire size range of AWG 20-30, but it was found out that the one with smaller sheath diameter give more acceptable result than the bigger one [4].

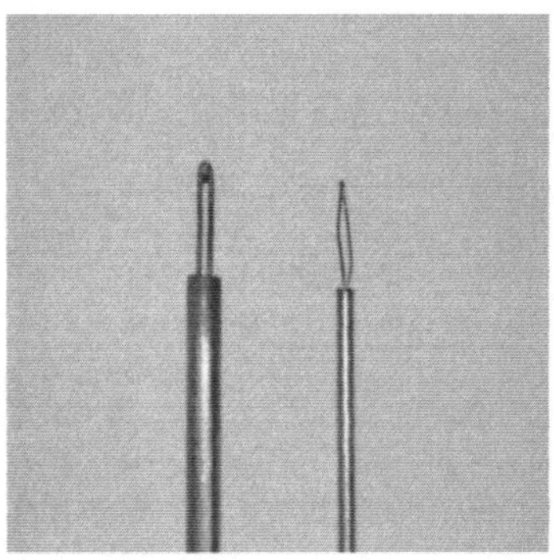

Figure 4. Baseline TC (1/8 inch) and smaller TC (1/16 inch) [20]

3.3. Issues on temperature measurement

Heat energy from convection and radiation are the main issues of heat transferred from fire that are exposed to structures. The convective heat transfer is the dominant heat transfer in liquid and gases, involve the process of conduction (heat diffusion) and advection (heat transfer by bulk fluid). The main factors that influence the convective heat transfer are the temperature differences between the target surface and the surrounding gas, and the velocity of the gas masses in the surrounding area of the exposed surface. The convective heat transfer, q_{conv} can be calculated using Equation 1.

$$q_{conv} = h(T_g - T_s)^n \tag{1}$$

where q_{conv} is convective heat transfer, h is the convective heat transfer coefficient $T_s$is the surface temperature and T_g is the gas temperature outside the boundary layer [2]. The radiation heat transfer is about the exchange of thermal radiation energy between two or more bodies. Thermal radiation is the electromagnetic radiation in the wavelength range of 0.1 to 100 microns. The radiation heat transfer, q_{rad} can be calculated using Equation 2.

$$q_{rad} = \alpha_s q_{inc} - \mathcal{E}_s \sigma T_s^4 \tag{2}$$

where q_{rad} is the net heat absorbed by radiation, q_{inc} is the incident radiation, α_s, $\mathcal{E}_s$, are the surface absorptivity and emissivity, T_s is the absolute temperature and σ is the Stefan's Boltzman constant.

The incident heat radiation on a surface arises from surrounding flames and gas masses as well as other surrounding surfaces [2]. The main issue with thermocouple measuring devices was heat loss due to radiation based on thermocouple bead size, also its sensitivity reduces accuracy as highlighted by Samir and Ryan [9]. Abu Talib *et al.* [5] have earlier reported that the effect of radiation from the thermocouple tip significantly reduces the measured flame´temperature, as the radiation losses were corrected, actual peak flame temperature was found to be approximately 800°C above the measured flame temperature. Liquid in-glass thermometer has the disadvantages of inability to measure high temperature, large sensing element, impossibility for continuous or automatic readout, long time constant, awkward dimensions, secular changes and hysteresis as reported by Wise [23]. Resistant temperature detector, thermistor and optical pyrometer are unsuitable to measure high temperature too.

4. Heat Flux Measurement

Heat flux is the rate of heat energy transfer through a given surface, per unit time measured in watt. Heat flux density is the heat rate per unit area measured in watt/metre square [24]. ISO2685 stated that the standard burner should produce 116±10kW/m^2 of heat flux value.

4.1. Method for heat flux measurement

The heat flux can be measured using different approaches such as those reported by Piccini *et al.* [25] that uses Thermographic Phosphors (TFG), Jones [26] and Shultz and Jones [27] uses optical method, and Abu Talib *et. al.* [5] and Ainsworth *et. al.* [28] uses thin film gauge (TFGs) that is more suitable for ISO 2685 standard, which can measure up to 200 kW/m^2, and endure prolonged immersion in a 1900°C flame. Different methods are used in measuring the temperature and heat flux for fire testing. The sample was immersed in flame for 15 minutes to find the material standard and this measurement is to prove the structural integration of the sample. A distance of three inches was used in between the thermocouple and the burner, so that the heat flux was measured according to the ISO 2685 standard [3]. Thin film gauges were used to measure the surface temperature due to its high thermal resistance and negligible influence on the surface flow field. In most of the practical work, radiative heat flux is normally measured with total heat flux meters of Schmidt-Boelter (thermopile) or Gardon (foil) type since they combined heat flux by radiation and convection to a cooled surface, as reported by Pullins [29]. Among the types of heat flux measuring devices are: calorimeter, Gardon gauges and Schmidt-Boelter gauges [1]. The thermocouple or thermometer are installed at the inlet and outlet of the tube of flowing water of calorimeter, which is usually a copper insulated on both sides except on the front side, to measure the inlet and outlet temperature.

Gardon gauge measuring device is in a form of copper-constantan thermocouple, as the gauge was introduced to a constant heat flux, temperature gradients develop in between the centre and at the side of the foil, the thickness of the foil determine the sensitivity of the thermocouple and is developed to produce 10mV as the output signal at the utmost heat flux. It has a water-cooled continues use, radial flow of heat in foil and parabolic temperature distribution. Schmidt Boelter gauge is another type of heat flux measuring device that the temperature difference was measured using a thermopile at the front and back side of aluminium piece, it is also like Gardon gauge, developed to generate output signal of 10mV at the utmost heat flux. It has as an axial flow of heat and a uniform temperature distribution [1]. ISO2685 [3] stated that heat flux from the standard burner must be calibrated using any approved equipment. In the process of calibration at least 180 seconds is required as warm-up time, so that a stable conditions may be obtain before taken the temperature measurements. During the test, water flows in and out of the tube and a thermocouple or thermometer is inserted at the inlet and outlet of the tube of the flowing water. The temperature is recorded of the constant flow of water at every 30 seconds for 180 seconds. This calibration process produces only a single measurement of heat flux averaged over the length of impingement of the flame onto the copper tube. The temperature difference between the inlet and outlet of the water of the surface area of the pipe exposed to heat with the volume flow rate and the density of water gives the heat flux of the flame. It was also found that like in the temperature calibration process, the two standards have different required values of heat flux calibrations. The ISO2685 standard is 116±10 kW/m^2, while AC20-135 106 kW/m^2as reported by Samir *et al.* [20].

4.2. Issues on heat flux measurement
Diller [30] reported a number of possible complications that may arises when using a differential heat flux sensor. The main issue is that, when calibration was performed on different ways of heat transfer, it may produce different results. Another issue is that the presence of the heat flux sensor may change the temperature field and heat flux result. Murthy *et al.* [31] reported heat flux calibration of Schmidt Boelter gauge using variable temperature black body (VTBB), whereby the radiating aperture and the apparent emissivity are located at fixed distance from the exit of black body with reference radiometer. The temperature was measured by sensing the radiation from one end of the furnace, which regulate and maintain the temperature within ± 0.1°C by an optical pyrometer.

In the calibration, there is heat loss due to radiation, therefore to ascertain the fraction of total heat flux due to radiation, Pullins [29] developed two different types of heat flux meter for the calibrations: total hemispherical radiometer sensitive to radiation only, and a total heat flux meter (most frequently used) sensitive to both radiant heat transfers and convective heat transfer. Samir *et al.* [20] reported a method of measuring simultaneously the heat flux and a flame temperature within the same distance as shown in Figure 5, whereby the heat flux and flame temperature measuring devices are located at four inches from the burner head.

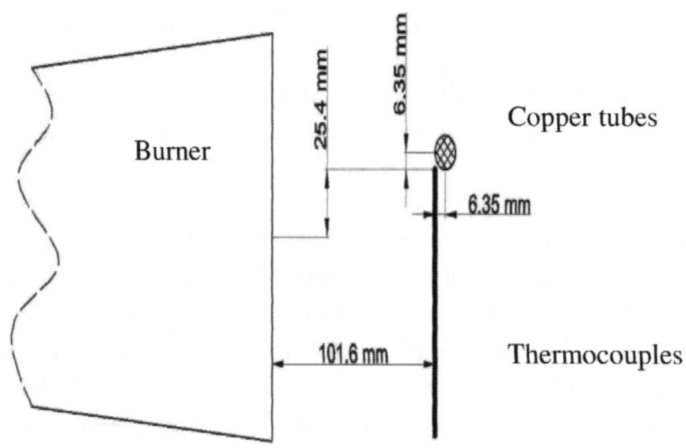

Figure 5. Simultaneous heat flux and temperature calibration for NextGen burner [20]

5. Uncertainty in measurement

There are different errors that can occur during the fire test measurement, which include human error, measurement error, standard error and equipment failure. Human error is one of the errors that brought about the dangerous conditions in the analysis of the results obtained and is prevented by designing different safety instruments in a suitable manner.

Another factor that affect the measurement is error due to measurement, the error is affected by some factors such as those affected by gas velocity over the thermometer, the emissivity and size of the thermometer body, also the radiative heat transfer between the thermometer and its surrounding may influence the temperature reading on thermometer. Abdullah *et al.* [17] reported that, supposing the actual surface temperature was reached by the junction of Type-T thermocouple attached around centre hole of the calibration plate using thermally conductive adhesive, then the general uncertainty in surface temperature measured by the thermionic liquid crystal would be approximated. Most of the general uncertainty found was due to the systematic and random uncertainty during the temperature measurement system. The uncertainty analysis was performed using the Root Sum Square method described by Coleman and Steel [32]. The systematic, random and overall uncertainty was measured to be ±0.38°C, ±0.39°C and ±1.08°C, respectively. Uncertainties that arise on thermocouple temperature measurement and plate thermometer measurement are negligible when compared to other uncertainty that arises as reported by Wickstrom *et al.* [2].

Abu Talib [33] reported the estimate of experimental method of uncertainties using the method described by Moffat [34]. The experiment was performed on flat plate test for the low-temperature analogue burner. The uncertainty level in coefficient of heat and adiabatic wall temperature depends on the initial temperature, gas temperature and surface temperature during calibration process and time. The errors obtained in the experiment were typical errors for initial temperature ±0.5, gas temperature ±0.5 surface temperature ±0.3 and time ±0.04. In the experiment first, second, third and fourth gas temperature steps for uncertainties were evaluated, and the results obtained were analysed and the coefficient of heat and adiabatic wall temperature were deducted. The root sum square (RSS) and maximum error were evaluated for the first, second, third and fourth gas temperature steps for uncertainties. The root sum square can be evaluated using Equation 3.

$$\text{Root sum square, RSS} = \sqrt{x_1^2 + x_2^2 + x_3^2} \qquad (3)$$

where x is the uncertainty on the parameters used. The maximum error is the sum of all components error, which is the worst-case error.

Murthy *et al.* [31] reported the uncertainties on heat flux measuring devices whereby some types of uncertainties are evaluated by statistical methods and the other types where evaluated by the other methods. In one of the techniques used in the calibration that is transfer techniques of heat flux sensor, the uncertainties measurement occurs in different ways, but two of these ways are the most important to know. There is uncertainty related to the calibration of the transfer standard radiometer, which is estimated to be 0.6%, and also there is uncertainty associated with calibrating the heat flux sensor with the VTBB, which has the stability of 0.1 K of the used temperature. The uncertainty obtained in this type is very negligible, which corresponds to 0.01 % at 927°C and 0.004 % at 2500°C.

6. Conclusion

From the review, it is clear that the temperature and heat flux measuring devices depends fully on the device sensors. Some temperature measuring devices do not have the ability to read high temperature measurement. It was found that, thermocouple of R, S and B types can withstand high temperature. However, the main problem with thermocouple is the bead of the thermocouple and high heat loss due to radiation, which can be reduced using an aspirated thermocouple and a smaller size thermocouple reduces heat loss by radiation. Thermometer can be used to measure the inlet and outlet temperature of heat flux measuring devices since the required temperature in that point is not as high as that recorded in the flame temperature. The sources of errors in measurement are parts of the factors that affect the measurement of flame temperature and heat flux, which can be corrected by designing a suitable safety instrument to avoid error in the measurement. Radiation and convection heat transfer are the main issues that affect the performance measurement of temperature and heat flux due to heat lost that normally encountered during the measurement and it can be reduces by using the right device.

IOP Conf. Series: Materials Science and Engineering **152** (2016) 012036 doi:10.1088/1757-899X/152/1/012036

Type R and S thermocouples are good devices used in the calibration and measurement of flame temperature that gives the higher temperature resistance and a low error of 0.1%. Likewise, heat flux calibration and measurement Schmidt Boelter gauge is the good device used in measuring heat flux that regulate and maintain the temperature within ± 0.1°C. Using these devices reduces the error due to measurement and also reduces standard error by the equipment used in calibration. The temperature should be measured using aspirated thermocouple if possible or a small size diameter bead to reduce heat loss by radiation.

Acknowledgement
The authors would like to thank Ministry of Higher Education, Malaysia for providing Fundamental Research Grant Scheme (FRGS) to support this research work (No. 5524611).

Reference

[1] Apte VB 2006 *Flammability testing of materials used in construction, transport and mining* (Sawston: Woodhead Publishing)
[2] Wickström U, Jansson R and Tuovinen H 2009 *Validation fire tests on using the adiabatic surface temperature for predicting heat transfer* SP Report
[3] ISO2685 1998 *Aircraft Environmental test procedures for airborne equipment – Resistance to fire in designated fire zones*
[4] AC 20-135 1990 *Draft Advisory Circular, Power Plant Installation and Propulsion System Component Fire Protection Test Methods, Standards and Criteria*
[5] Abu Talib A R, Neely A J, Ireland P T and Mullender A J 2005 *Journal of engineering for gas turbines and power* **127** 249-56
[6] FAA 1990 *The Materials Fire Testing Handbook* (Atlantic City: FAA)
[7] Carson D. 2016 InspectAPedia Website [Accessed on April 2016]
[8] Whitaker S 1972 *AIChE Journal* **18** 361-71
[9] Samir T and Ryan H 2015 *ISO Propane Burner, Combustion and Fire Research Laboratory* Fire Test Centre (FTC), University Cincinati
[10] Duff M and Towey J 2010 *Analog Dialogue* **44** 1-6
[11] Machin G 2012 *Measurement and Control* **45** 315-8
[12] Buenaño Andrade A M 2009 *Diseño e instalación de equipossupervisores y sensores de temperatura para las unidades de generación de la Central Hidroeléctrica*
[13] Kamat R K, Naik G M and Verenkar V M S 2001 *Synthesis and characterisation of Nickel Manganite from different carboxylate precursors for thermistors sensors* Texas Instruments Incorporated
[14] Gruner K-D 2003 *Principles of non-contact temperature measurement* Raytek Company
[15] Thermometric pts 2016 www.thermometricscorps.com/thermocouple [Accessed on May 2016]
[16] Thermocouples 2016 www.thermocoupleinfo.com/type-thermocouple.htm [Accessed on July 2016]
[17] Platzer K, Hirsch C, Metzger D and Wittig S 1992 *Experiments in fluids* **13** 26-32
[18] Abdullah N, Abu Talib A R, Jaafar A A, Salleh M A M and Chong W T 2010 *Experimental Thermal and Fluid Science* **34** 1089-121
[19] Abdullah N, Abu Talib A R, Saiah H R M, Jaafar A A and Salleh M A M 2009 *Experimental Thermal and Fluid Science* **33** 561-78
[20] Samir T, Yi-Huan K and San-Mou J 2015 *Development of Next Generation Burner Operation Setting for Fire Testing of Power Plant Components*
[21] Serge N L 2008 *Fire test on Components used in Fire Zones. Comparison of Gas Burner to Oil Burner*
[22] Horner A 2000 *Aircraft Materials Fire Test Handbook* (DTIC)
[23] Wise J A 1976 *Liquid-in-glass Thermometry* Final Report National Bureau of Standards
[24] Mullender A J, Handley B A, Coney M H, Ireland P T and Neely A J 2002 *US Patent 6,418,806*

IOP Conf. Series: Materials Science and Engineering **152** (2016) 012036 doi:10.1088/1757-899X/152/1/012036

[25] Piccini E, Guo S M and Jones T V 2000 *Measurement Science and Technology* **11** 342
[26] Jones T V 1977 *Measurement of Unsteady Fluid Dynamic Phenomena* von Karman Institute for Fluid Dynamics
[27] Schultz D L and Jones T V 1973 *Heat-Transfer Measurements in Short-Duration Hypersonic Facilities* (DTIC)
[28] Ainsworth R, Allen J, Davies M, Doorly J, Forth C, Hilditch M, et al. 1989 *Journal of Turbomachinery* **111** 20-7
[29] Pullins C A 2011 *High temperature heat flux measurement: sensor design, calibration, and applications* Virginia Polytechnic Institute and State University.
[30] Diller T 1993 *Advances in heat transfer* **23** 279-368
[31] Murthy A, Tsai B K and Saunders R D 1998 *Metrologia* **35** 501
[32] Coleman H and Steele W 1999 *Experimentation and Uncertainty Analysis for Engineers* **2** 83-115
[33] Abu Talib A R 2003 *Detailed investigation of the low-temperature analogy of an aircraft engine standard fire-test* University of Oxford
[34] Moffat R 1982 *Journal of Fluids Engineering* **104** 250-8

Capital raising of aerospace companies: equities or debts?

L Hui-Shan[1,2*], Y Taw-Onn[2] and H Wai-Mun[2]

[1]Universiti Putra Malaysia, Serdang, Malaysia
[2]Universiti Tunku Abdul Rahman, Sungai Long Campus, Malaysia

[*]hslee@utar.edu.my

Abstract. Aerospace products enhance national and economic activities, thus maintaining the sustainability of aerospace industry is crucial. One of the perspectives in ensuring sustainability of aerospace companies is expansion of firms by raising funds for research and development in order to provide a reasonable profitability to the firms. This study comprises a sample of 47 aerospace companies from 2009 to 2015 to analyze the impact of raising fund by equities or debts to the profitability of the firms. The result indicates that capital raising through equities is preferable than debts. Moreover, the study also identifies that the profit of aerospace industry is volatile and there is cyclical reduction of the net income in the first quarter of the year. The management needs to make wise decisions in raising fund to ensure a healthy growth of the aerospace company.

1. Introduction

There is a continued interest on capital raising within the increasingly competitive and global business environment, especially in aerospace industry, to remain in the market with reasonable profit. With the fact that aerospace has been a capital-intensive industry, the focus in financing activities in aerospace industry is mainly on financing aircraft with the availability of financing alternatives such as equity financing or debt financing. It was pointed out that the significance of individual options of financing may fluctuate over time as the crumple of Lehman Brothers in 2008 triggered the financial market crisis [1]. Apart from the core strategy in research and development, the development of the supply side structure in aerospace industry needs to rely on capital funding. Capital budgeting has evolved into crucial determinant for the expansion formation in the aerospace industry. Hence, the accessibility of financing sources could constitute a realistic barrier to the market entry in the aerospace industry. Wojahn (2012) highlighted that in the recent decades, the aerospace industry has not even come close to attain an adequate level of profit where losses have spread from North America to all geographical regions in the world and cyclical losses have become structural [2]. However, with the difficulty in sustaining profitability, this industry still carries on to add capacity.

Among the questions raised include whether financing the further expansion of aerospace industry through equity market or capital market is more appropriate? Due to the magnitude of debt and also declining assets values in compelling the bank crisis and private sector such as aerospace sector's de-leveraging, the recent slump in the economy activity was illustrated as "balance sheet recession" [3]. This is because the de-leveraging will lead to a crash in bank lending and also influence customers' demand. The issue of correcting balance sheet could have long lasting impacts on economic growth and by consequence on the aerospace industry demand. Pearce (2012) presented that Japan is one of the country has most previous incident of such a balance sheet recession [4]. It demonstrated that addressing the bad debts of the banking system and the impact of collapsed equity and house asset

IOP Conf. Series: Materials Science and Engineering **152** (2016) 012037 doi:10.1088/1757-899X/152/1/012037

values on private balance sheets had resulted in a decade of weak economic growth. It also highlighted that studies to achieve financial sustainability in aerospace industry are needed for future expansion purposes as well as the measures to attract new equity capital in this industry.

From Figure 1, in between the period of 2009Q1 to 2015Q4, average debt-capital financing from 47 aerospace companies worldwide is more than USD 4 billions whereas the average equity financing is less than USD 2 billions. This shows the trend that most aerospace companies rely on debt financing more than equity financing. The issue here is, which alternative is a profitable investment opportunity? Studies such as by Keuschnigg *et al.* (2013) [5], Aghion *et al.* (2007) [6], Tirole (2001) [7] and also Hubbard (1998) [8] postulated that many firms are facing dilemma in financing constraints that cause them to avoid from advancing business prospect with profitable investment opportunities. However, there is still lack of literatures that study the influence of debt financing and equity financing to the profit of the aerospace companies. In view of this, the work in this paper will assess the relationship between net incomes, revenues, total liabilities and total equities in the aerospace industry. To be more specific, this paper entails two main objectives. Firstly, it examines whether liabilities or equities will aid the aerospace companies to expand the business in return with higher profit. Secondly, it aims to investigate is there any cyclical loss in aerospace industry as suggested in [2], the loss due to structural change will influence the sustainability of the aerospace industry. This study differs from previous literatures that it endeavours to look into the finance determinants instead of technical issues related to aerospace, which will influence the profit of the aerospace firms to sustain in the market.

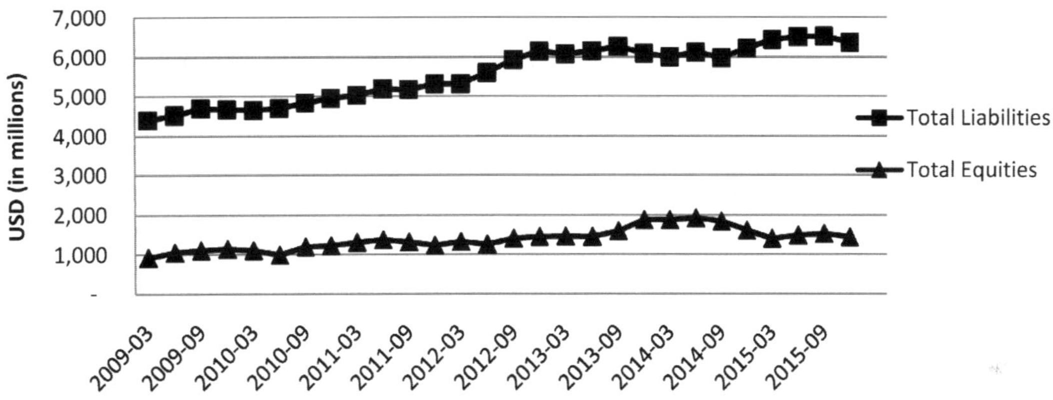

Figure 1. Average total liabilities and total equities in aerospace industry

This paper is organised as follows. In the next section, relevant literatures will be reviewed. Section 3 lays out the data and also methodology. Meanwhile, section 4 presents data and discussion on the results. Finally, section 5 provides a summary of the main findings.

2. Literature reviews

The attributes that affect business activities in airline industry had been investigated in [9]. It revealed that the airline industry is not only affected by the national characteristics but it is also influenced by internationalisation processes and international standards. The factors that had been recognised were airline alliances, national culture, implementation of business model and also consequences of market liberalization. The implication of this study served as a helpful means to potential synergies analysis. This research is focus on the operation management but not the financial management on the airline industry. Studies on the systematic risk in aviation were done by analyzing factors such as financial and operating leverage, firm size, profitability and liquidity [10, 11]. The study was later expanded by investigating the systematic financial risk exposure of the aviation industry with 5-factor asset pricing model and it was found out that the risk is positively significant to profitability and operating leverage, which resulted in negative earnings growth [12]. Nevertheless, their studies on financial management of aviation industry used financial leverage that was only focused on debts but not equities.

Chadha (2016) suggested that, when a company is incorporated, the decision on capital structure has to be taken well in advance by analysing the pros and cons of the various sources of funds before choosing the best alternative that reduces the cost of capital [13]. An optimal capital structure will maximise the company's market value, thus planning on capital structure is a permanent progression and has to be decided whenever a firm needs extra funds. The common capital structures are equities and debts. If the company raises funds by issuing equity, equity holders are the owners of the company and they will have a long term interest with the company. In other words, if the aerospace company raises fund through equity, the shareholders will have trust in the company that it will expand in the near future. In contrast, if the fund is raised through debt, the creditors of the firm are the debt holders and they do not have long term interest with the firm as their concern is the received repayment on time. Therefore, the financial managers in the aerospace company play a crucial role in deciding the capital structure choice to ensure the firm grows well without facing illiquidity trap or bankruptcy fate.

From the summary of literature reviews, aerospace industry is influenced by internationalisation processes and international standards. The study on the operation and financial management on this industry is equally important to ensure its prosperous and sustainability in the global standard. It is found from the past literatures that capital raising through equity and debt would have different impact on the firm. Nevertheless, the issue on capital structure on aerospace industry is limited. Therefore, in this study, the highlighted contributions are to assess whether equity or debt fund raising is beneficial to the aerospace firms and to identify the cyclical loss of aerospace industry in order to warrant this industry on the financial indicator that will have an effect on its growth.

3. Data and methodology

The sample of the aerospace companies is chosen based on the Bloomberg Database classification as it is one of the most powerful platforms for financial professionals who require analytical data to make business decisions. In order to eliminate structural breaks due to economic crisis, the sample period chosen is from 2009 to the most recent available quarterly data which is 2015. The quarterly period is selected since most of the regulators require public listed companies to publish the quarterly report on quarterly basis. Only public listed aerospace companies are chosen since this study is focusing on the impact of equity and debt to the profitability of the firms. There are 47 aerospace companies in total to be studied in this research. Table 1 lists all of these companies.

The dependent variable is "net income", which is the company's total earnings captivating revenues and subtracting for cost of business, depreciation, other expenses, interest and taxes. The independant variables are total revenues, total liabilities and total equities. Definition of the independent variables and the expected results are presented in Table 2.

Due to different natures of aerospace companies, i.e. airlines, aircraft manufacturers, subsystems design companies, etc., the panel data model is used to control individual heterogeneity. Unobserved characteristics may vary across individuals (known as heterogeneity). If these influence the variables of interest and are correlated with unobserved explanatory variables, then estimated effects of these variables will be biased. Panel model uses differences with regards to the individual data average as solution: $y_{it} \rightarrow y_{it} - \bar{y}_{it}$ [14]. The panel data regression model will be employed to analyse the data as it consists of 47 aerospace companies. It has been suggested that this model is practical to control the individual heterogeneity (cross-section) and implications, which are difficult to be monitored through pre-cross sections or pure time series data [15]. To run the panel regression model, Stata Software is used. Based on the model suggested in [16] and [17], the estimated equation by taking the net income of the aerospace company as the profitability determinant is as shown in Equation 1.

$$NI_{it} = \beta_i + \beta_1 Rev_{it} + \beta_2 TL_{it} + \beta_3 TE_{it} + \varepsilon_i \quad (1)$$

Table 1. List of considered aerospace companies

NO	COMPANIES	COUNTRIES	NO	COMPANIES	COUNTRIES
1	BOMBARDIER INC-B	CANADA	24	MZ ARSENAL-$BRD	RUSSIA
2	CAE INC	CANADA	25	TECHPRIBOR-BRD	RUSSIA
3	MAGELLAN AEROSPA	CANADA	26	AS YAKOVLEV DESI	RUSSIA
4	HEROUX-DEVTEK IN	CANADA	27	DUKS-RTS BOARD $	RUSSIA
5	FLYHT AEROSPACE	CANADA	28	SATURN-BRD	RUSSIA
6	AVCORP IND INC	CANADA	29	MIL MOSCOW HELIC	RUSSIA
7	IWG TECHNOLOGIES	CANADA	30	SARAPUL ELEK-BRD	RUSSIA
8	STAR NAVIGATION	CANADA	31	SINGAP TECH ENG	SINGAPORE
9	AVIC AVIATION -A	CHINA	32	A-SONIC AERO	SINGAPORE
10	AVIC AIRCRAFT-A	CHINA	33	SAAB AB-B	SWEEDEN
11	AVIC AERO-ENGI-A	CHINA	34	CTT SYSTEMS AB	SWEEDEN
12	AVIC HELICOPTE-A	CHINA	35	UNITED TECH CORP	U.S.
13	CHINA AEROSPAC-A	CHINA	36	BOEING CO/THE	U.S.
14	JIANGXI HONGDU-A	CHINA	37	ESTERLINE TECH	U.S.
15	SICHUAN CHENGF-A	CHINA	38	TRIUMPH GROUP	U.S.
16	AIRBUS GROUP SE	FRANCE	39	RBC BEARINGS INC	U.S.
17	MTU AERO ENGINES	GERMANY	40	ASTRONICS CORP	U.S.
18	BET SHEMESH ENGI	ISRAEL	41	DUCOMMUN INC	U.S.
19	JAMCO CORP	JAPAN	42	LMI AEROSPACE	U.S.
20	SHOWA AIRCRAFT	JAPAN	43	SIFCO INDUSTRIES	U.S.
21	SUMITOMO PRECISI	JAPAN	44	INNOVATIVE SOL	U.S.
22	MACROASIA CORP	PHILIPINES	45	ASTROTECH CORP	U.S.
23	KUZNETSOV-BRD	RUSSIA	46	TEL-INSTRUM ELEC	U.S.
			47	MECHANICAL TECH	U.S.

Table 2. Definition of the variables and the expected results

Variables	Explanation	Expected Result	Rationale
Total Revenue (Rev)	Revenues from the sales of commercial airplanes such as Boeing and Airbus	Positive	The higher the sales, it will generate more profit to the company.
Total Liabilities (TL)	Short term and long term debt	Negative	The higher the liabilities, the more obligation of the firm on debt settlement, reduce the profit of the firm.
Total Equities (TE)	Shareholder equities in the stock market	Positive/ Negative	The more equities invested by the shareholder, signalling that they have confident on the firm to generate more profit in the future. The more equities holders, the firm will need to pay dividend to the shareholder and reduce net income; however dividend payment is at the discretion of the firm.

IOP Conf. Series: Materials Science and Engineering **152** (2016) 012037 doi:10.1088/1757-899X/152/1/012037

First of all, pooled ordinary linear regression model (Pooled OLS) is employed to demonstrate the result based on poolability of the data but the panel structure of the data is ignored. Next, the model to be executed is the random effect model (REM), which holds the constants for each section as random parameters. The third model is the fixed effect model (FEM) where the constants are taken as group specific, which means the model allows for different constants for each firm. The Breusch-Pagan Lagrangian Multiplier Test is done to select between pooled OLS and REM, where the null hypothesis refers to the preference of Pooled OLS versus alternative hypothesis that REM is preferred. Moreover, Hausman Test is executed to decide on the choice between REM and FEM, where the null hypothesis that refers to preference of REM versus alternative hypothesis that FEM is preferred. Examination for heterogeneity and serial correlation are also performed to check for robustness of the result.

4. Data analysis

Table 3 illustrates the descriptive statistic for the data. It is noticed that there was at least one company that had negative net income in a certain quarter and this company probably was unable to cover the cost of operating. Furthermore, the minimum total equity was negative, perhaps this company was in the ramp-up stage and had spent a large amount of funds to construct infrastructure that will later yield profits [18].

Table 3: Summary of descriptive data

Variables	Observation	Mean (USD)	Std Dev	Min (USD)	Max (USD)
NI	1316	68.5 mil	2.76e+8	-1.56 bil	3.28 bil
Rev	1316	1.31 bil	4.29e+9	0	25.8 bil
TL	1316	5.57 bil	1.94e+10	282,662	111 bil
TE	1316	1.39 bil	4.34e+09	-0.87 bil	35.5 bil

Figure 2 demonstrates the average net incomes, revenues, total liabilities and total equities for the aerospace companies. The trend of the average total liabilities is increasing, which indicates that the companies were increasingly borrowing more debts. The average net income was volatile. It is noticed that for each quarter one of the year, the average net income dropped compared to previous quarter. This may signal that the management needs to discover what had happened to the aerospace industry within the quarter one of the year.

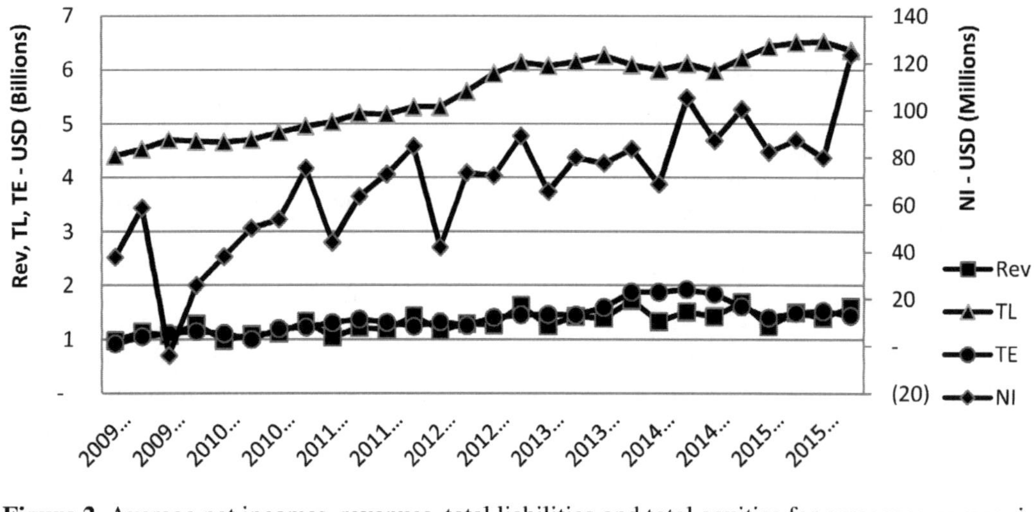

Figure 2. Average net incomes, revenues, total liabilities and total equities for aerospace companies

It is also realised that the average revenues and total equities were approximately similar. This might indicate that shareholders were willing to invest in the aerospace companies to the maximum of the revenues that the firm could earn. As the result, the firm has to opt for debt to raise further funds if there is a need for additional expansion in research and development. The higher portion of average total liabilities compared to average total equities is probably supported this likelihood.

The result in Table 4 is based on panel data model. The poolability test is used to investigate the feasibility of poolability among the independent variables in a panel although it is presumed that, in a panel, the character of heterogeneity will be controlled [19]. In model 1, two variables namely total revenues and total equities indicate positive significant impact to net income of aerospace companies. This means that the higher revenue of the firm can provide a better net profit to the firm. In addition, more capital is raised through equities will generate higher net income to the company. Conversely, total liabilities show negative significant impact to the net income, which suggests that the more debt that company hold, the income of the company will be impeded as more interest payments are obliged to be settled to the creditors. These result consistent with the recommendations in [13]. Model 2 is the robust Pooled OLS and it is executed to improve the existence of heterogeneity and autocorrelation in model 1. Similarly, total liabilities show a significant negative relationship to the net income. On the other hand, revenues and total equities indicate significant positive link to the net income. In model 3, it is assumed that the intercept is a random variable with the mean value for all sample companies. However, the result of Breusch Pagan LM Test does not reject null hypothesis and this indicates that pooled OLS is favorable than REM. Hausman test is carried out to measure the suitability between REM and FEM. The result rejects null hypothesis, which demonstrates that REM is not appropriate. In summary, model 2 is superior to the other three models as the diagnostic test Breusch-Pagan Lagrange Multiplier supported the hypothesis that pooled OLS is preferred and robust pooled OLS has corrected the heterogeneity and serial correlation issues.

Table 4. Results of the panel data

Independent Variables	Pooled OLS	Pooled OLS (Robust)	REM	FEM
Constant	-9619458** (3827255)	2.97e+7 (4.28e+7)	-9619458** (3827255)	-8.84e+7*** (7968695)
Rev	0.0821*** (0.0032)	0.0824*** (0.0086)	0.0821*** (0.0032)	0.0272*** (0.0051)
TL	-0.0119*** (0.0006)	-0.012*** (0.0012)	-0.0119*** (0.0006)	0.0226*** (0.0018)
TE	0.0269*** (0.0012)	0.0268*** (0.0047)	0.0269*** (0.0012)	-0.0034 (0.0033)
Observations	1316	1316	1316	1316
F-Test/Chi	4467.86***	8191000***	4467.86***	154.08***
Within	-	0.1031	0.0856	0.2675
R^2 Between	-	0.9875	0.9872	0.5854
Overall	0.773	0.7775	0.773	0.4866
B.P LM Test	1.89			
Hausman Test			446.97***	
Wald Test				2e+13***

Note: Figures in parentheses are standard errors.
*, **, *** indicate statistical significance at 10%, 5% and 1%, respectively

5. Conclusion

In this study, the impact of raising capital through total equities and total liabilities among aerospace companies is studied. The results show that raising fund by equities is favourable to the aerospace company as the shareholders have trust on the companies to grow well in the future with the capital invested by them. This might also suggest that they prefer to enjoy capital appreciation rather more than regular dividend payment from the funds invested. On the contrary, raising funds by debts show negative impact to the net income of aerospace firms. This suggests that if the firms are unable to raise additional fund through equities, although they could opt for debt option, they need to be aware that the consequence is the erosion of net profit that might be suffering from the illiquidity or bankruptcy issue. In addition, the quarterly net income of aerospace companies is volatile and there is also a cyclical reducing trend in the average net income of the firms in the quarter one of the year. The limitation of this study is that it is unable to identify the cause of structural decreasing of net income in quarter one. Therefore, it is recommended that the future study could focus on this area to examine on other factors like aerospace technical issues, research and development, regulations issues or economic issues that cause the volatility of net income as well as the cyclical decrease in the net income.

Acknowledgement

The authors would like to thank UTAR for subscribing Bloomberg database to enhance the process in accessing and downloading the financial data related to aerospace companies in the worldwide to work out on this research paper. Any remaining errors are our own.

References

[1] Bjelicic B 2012 *Journal of Air Transport Management* **21** 10-6
[2] Wojahn O W 2012 *Journal of Air Transport Management* **19** 1-8
[3] Koo R 2008 *The Holy Grail of Macroeconomics: Lessons from Japan's Great Recession* (Singapore: John Wiley & Sons)
[4] Pearce B 2012 *Journal of Air Transport Management* **21** 3-9
[5] Keuschnigg C and Ribi E 2013 *International Tax and Public Finance* **20** 808-26
[6] Aghion P, Fally T and Scarpetta S 2007 *Economic Policy* 731–79
[7] Tirole J 2001 *Econometrica* **69** 1–35
[8] Hubbard G R 1998 *Journal of Economic Literature* **36** 193–225
[9] Huettinger M 2013 *Baltic Journal of Management* **9** 71-90
[10] Lee J S and Jang S C S 2007 *Tourism Management* **28** 434-42
[11] Hung J H and Liu Y C 2005 *Journal of Air Transport Management* **11** 291-6
[12] Lee C H and Hooy C W 2012 *Journal of Air Transport Management* **24** 31-5
[13] Chadha S and Sharma A K 2016 *Global Business Review* **17** 411–24
[14] Law S H and Azman-Saini W N W 2012 *Economics of Governance* **13** 217–36
[15] Gujarati D N and Porter D C 2009 *Basic econometrics* (Boston: McGraw-Hill)
[16] Wang F S and Wu N H 2014 *2014 International Conference on Management Science & Engineering 21th Annual Conference Proceedings* 1393-9
[17] Margaritis D and Psillaki M 2010 *Journal of Banking & Finance* **34** 621-32
[18] Wensveen J G and Leick R 2009 *Journal of Air Transport Management* **15** 127-33
[19] Baltagi B H 2005 *Econometric Analysis of Panel Data* (Chichester: John Wiley & Sons Ltd)

A study of foreign object damage (FOD) and prevention method at the airport and aircraft maintenance area

R Hussin*, N Ismail and S Mustapa

Department of Aircraft Maintenance Engineering, Politeknik Banting Selangor, Malaysia.

*rosmawar@polibanting.edu.my

Abstract. Foreign object damage (FOD) is common risk for aviation industry since long time ago and it has contributed to many terrible incidents and fatalities. The cost of FOD cases every year is very high, which is around RM 1.2 billion. Therefore, a proper technique and strategy has to be taken by the designated organizations including airlines to further eliminate the FOD occurrences. It is not easy to control FOD due to some circumstances such as inappropriate working behaviour, poor working environment, insufficient technology and also disorganized housekeeping system. The main purpose of this research is to discuss and explain further about FOD and the techniques to prevent FOD. FOD is a universal concern in aviation industry and it is one of the reasons that contribute to aircraft failure and unwanted damages such as fatalities and causalities. Throughout this research, many information related to FOD problems and their impact on aviation industry are gathered and presented.

1. Introduction

Foreign object damage (FOD) is a big problem in aviation maintenance industry that reduces the level of safety for an aircraft. In fact, it can only be controlled and minimized properly by using the right and precise control method. Basically, FOD is known as foreign object (FO) that can cause severity and destruction to the aircraft such as engine failure and loss of human life. Nowadays, there are many cases of FOD that happened due to some reasons and this situation leads to a survey on aviation safety in the aviation industry. The result has indicated that FOD is the most potential ground base cause that contributes to catastrophic aviation failure [1].

In 1998 to 2008, there were 116 FOD occurrences reported to the Aircraft Transport Safety Board (ATSB) that affected high capacity air transport aircraft, which normally happened during the busiest hours of operations at most airport in Australia [2]. Nine of them happened at the aerodrome apron while the other 12 occurred on taxiways. Other occurrences that happened during aircraft pushback included cases when the tyre was punctured by a metal pin and the engine was damaged because FO had been sucked up into it at the aerodrome gate. Further researches done by ATSB found that FOD occurrences had no effect on aircraft operation in about 80 % of the occurrences. However, when it happened, the most typical consequential events were go–arounds and rejected take-offs [2]. This is highlighted in Figure 1.

Basically, this research will focus on FOD sources and also its prevention method at airports and maintenance areas where the most common FOD cases happened. Apart from that, this paper is aimed to classify FOD based on their specification, understand the consequences of FOD to the aircraft and identify the cost contributed by FOD. Throughout the completion of this research, several methods had been used to gather reliable information and data. Most information was obtained through research from reliable sources such as internet, technical report, books, articles and journals. Moreover, Federal Aviation Administration (FAA) and ATSB websites were really useful in searching for information, especially for FOD occurrences and the FOD Prevention Program. Beside these sources, the National Aerospace FOD Prevention (NAFPI) Conference was also very helpful in providing the most current and updated information, which also included opinions from many types of aviation organizations. Furthermore, another method that had been used in this research was analyzing the conducted surveys by authorised agencies, airline operators and aviation companies.

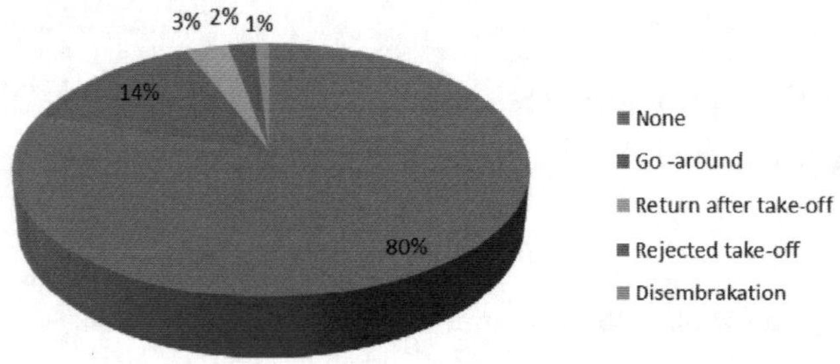

Figure 1. Consequences of reported FOD occurrences 1998-2008 [2]

2. Literature review

2.1. Definition of FOD

FOD includes debris, substances or articles that have the potential to cause damage to any vehicle or system. In other words, FOD can be defined as anything that is around or inside the aircraft and flight line operations that does not belong there. FOD varies in sizes and it has the capability to create hazard to equipment or personnel. Another definition of FOD is the damage on aircraft, helicopters, launch vehicles, engines or other aviation equipment [4], which takes place when a foreign object smashes the engine, flight controls, airframe and the other operating systems. Based on Federal Aviation Authority (FAA) [5], FOD is principally known as a hazard element that can severely harm the airport, personnel and equipment. In fact, the most serious case of FOD had involved personnel injuries or death and in most cases, it usually occurred during aircraft close-proximity taxing when the personnel was exposed to adverse effects of high velocity jet blast. The harsh blast forced FOD through the airport and often caused injuries to personnel who were working around that area.

FOD can be basically divided into two types: soft and hard bodies. According to the Research and Technology Organization (RTO)Technical Report in 2005, the impact of soft body damage can result from flexible objects such as birds, ice slabs and plastics [3]. For instance, it can usually be seen by a large radius curvature of deformation to the turbine blade fan. Meanwhile, hard body impact damage occurs with uneven appearances. For example, tear to airfoil's leading and trailing edges at the turbine blade section as a result of impact by rigid parts like metal parts, concrete and rock.

2.2. FOD sources

Through the researches and surveys done by FAA, there are many types of FOD that can be found in many forms. None of them is beneficial since they cause difficulties to airline operators, especially in maintaining safety on airfield operations. There are many types of FOD that vary in materials, colours and sizes. In general, there are four basic classes of FOD: metal, stone, miscellaneous and birds [6]. Based on the research done by the French Study on Automatic Detection Systems [5], over 60% of the collected known FOD items were made of metal, followed by 18% made of rubber.

Table 1. Types and sources of FOD [5]

Types of FOD	Sources
Personnel	It is normally caused by poor working behaviour and inappropriate housekeeping
Airport infrastructures	Sign, pavements and lights
Environment	Wildlife, snow and ice
The equipment operating on the airfield	Aircraft airport operations vehicles, maintenance equipment, fuelling and construction equipment.
Aircraft and engine fasteners	Nuts, bolts and washers
Aircraft parts	Fuel cap, oil stick, trapdoors and tyre fragments
Flight line items	Nails, personnel badges, luggage tags, soda can, etc.
Runway and taxiway materials	Concrete and asphalt chunks, rubber joint materials and paint chips

In April 2005, Delta Airlines Corporate Safety had set up forensic analysis on FOD engine events. The purpose of the analysis was to find the composition of FOD that contributed to each FOD events and as a result, Delta Airlines found the numbers of aircraft accidents that happened due to FOD. The result showed that aircraft parts were the major contributor in this analysis, as shown in Figure 2 [6]. As cited in the ATSB report [2], the most common FODs reported in Australian airports were aircraft components in the form of engine reverse thrust assemblies, blocker door and door assemblies' pins, which contributed to 33% for all FOD occurrences. Furthermore, construction activities at the airport often contributed to FOD sources. In winter conditions, FOD tends to be more predominant especially when aging pavement infrastructure may be frozen, and began to crack and break into pieces. In windy weather condition, movement of the wind can propel the FODs such as sand, papers and plastic bags from non-critical area into the runway. The knowledge of all these effects indirectly creates awareness among airline operators about the unpleasant effects of weather conditions and gives them ideas on how to solve this problem.

2.3. *FOD items location*

The UK's Civil Aviation Authority (CAA) had done a survey regarding FOD item locations [6]. CAA took 10 months to collect all data, which includes size, weight, type, location and picture of each FOD item found. Many types of FOD items with variety of sizes were found and 55% of them were located at aircraft stand area or parking area where the aircraft normally parked at the airport. Figure 3 shows the other findings from the survey.

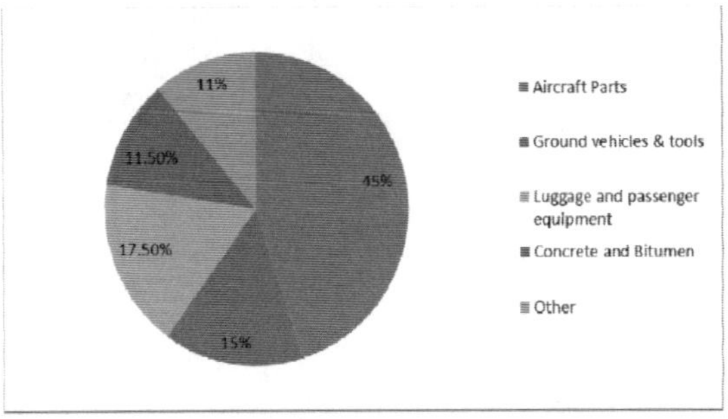

Figure 2. Sources of FOD engine damage [6]

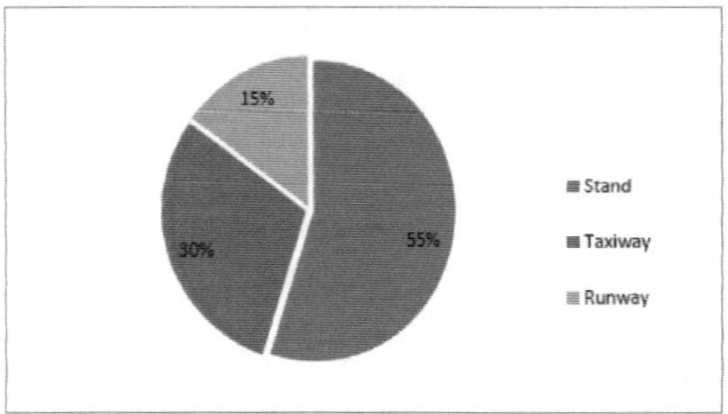

Figure 3. FOD item locations [6]

2.4. Problems associated with FOD

In the aviation industry, there are many catastrophic problems associated with FOD. Frequently, the problem is damage or destruction to the aircraft parts. The damage can be divided into two categories: minor and major damages. An example of minor damages is a skin dent while for major damages, they include control surface malfunction, jammed flight controls, electrical shots and also engine failure. According to ATSB [2], about 11% of FOD occurrences led to airframe wheel and engine damages. These damages give financial impact to the organization and contribute to a massive amount of direct and indirect costs. Direct costs involve all maintenance fees to repair the damages due to FOD. On the other hand, indirect costs include flight delays, cancellations, lost revenue, schedule disruptions and additional works by employees.

2.4.1. Engine damage. Aircraft engine has a high tendency to be ingested by FOD. Ingested objects in a form of soft and hard materials of all sizes and shapes are causing big problem once they strike rotating blades, static vanes, and other parts of the engine, thus reducing the strength of the component [7]. Furthermore, ingestion by hard body object can cause damage to the engine rotating assemblies, which leads to vibration and disruption of airflow at the early stage of the compressor section. It will cause the compressor to stall and eventually reduce the performance of the engine. A study conducted by Thompson [8] found that FOD is of great concern in the gas turbine engines because it can have deleterious effect to the rotating components in modern aircraft engines. In fact, in some severe cases, replacement of a new engine is necessary.

In years 1999-2008, the European Aviation Safety Agency (EASA) found that 44% of FOD cases in form of bird strikes had cause internal and external damages to the aircraft engine section [9], as highlighted in Figure 4 and Figure 5. Internal damages can give more significant effects since it often cannot be seen by the normal eyesight. The impact of FOD on gas turbine engine blade can be divided based on the severity of the damage as listed below:

- Minor – no more than blade blending is required
- Moderate – replacement of blade on single stage is required
- Severe – replacement of blade on more than on stage is required
- Very severe – blade replacement is required plus repair of additional damage incurred to the other engine areas

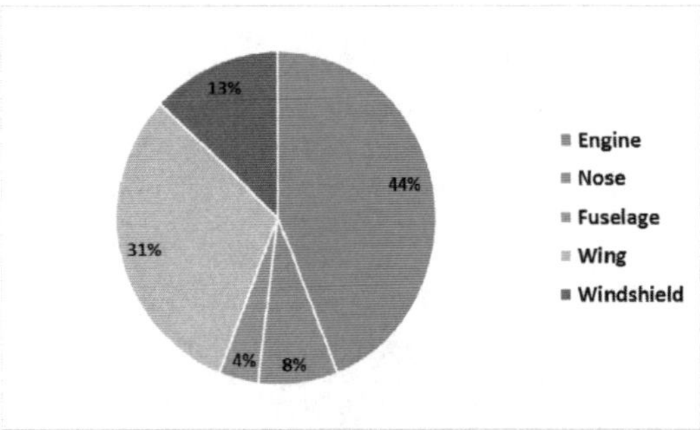

Figure 4. Location on the aircraft which was struck and damaged by bird(s), worldwide (1999-2008) [9]

Figure 5. Engine blade damage due to an ingestion of a small bird [9]

2.4.2. Tyre damage. The damage on the aircraft tyre usually happens due to the penetration of FOD into the aircraft tyre. In worst cases, it can cause tyre burst that leads to many unwanted circumstances and even loss of life. Moreover, FOD can cause the tyre treads of the nose or the main landing gear to detach. If this happen during take-off and landing, the detachment of the tyre treads will cause damage to the aircraft sections such as fuselage, wings, engine intake and compressor. The detachment of the tyre treads starts with a penetration and normally happens in take-offs and landings.

Basically, there are two types of tyre damage associated by FOD: tyres that are visibly torn by FOD and need to be replaced immediately, and tyres that have been embedded by FOD that is undetected during inspection and this will cause tyres to fail retread. According to Insight SRI Ltd [6], the net cost for tyres failing retread is USD 39,509 per 10,000 movements. In addition, it is found that the industry average for retread rejection rates due to FOD for Boeing 737 and 757 main gear tires is 2.6% [6]. The highest percentage was 4.2% for American Airlines as tabulated in Table 2.

Table 2. Retread rejection rates due to FOD for Boeing 737 and 757 main gear tyres [6]

Airline	Rejection Rate
Delta	3.7%
Air Canada	2.6%
American	4.2%
ATA	1.3%
Southwest	1.6%
Aloha	2.4%
Average	2.6%

2.4.3. Airframe damage. There are few areas that have a high potential to be damaged by FOD. These areas include fuselage, nose, radome and windshield. Besides that, FOD also has high tendency to hit the aircraft wing that will result in skin dents or serious damage at the wing spar. In most incidents that happened, these damages have led to aerodynamics loss, which have caused difficulty in controlling the aircraft and also vibration due to change in the aerofoil shape of the wing section.

In some cases, FOD may penetrate through the windshield and cause injury to the pilots. Around 13% of incidents caused by bird strike involved fatal injuries from damages to the windshield [9]. In order to avoid this problem from happening, aircraft manufacturers have to ensure the strength and durability of the windshield by making improvement in its structure. Penetration of FOD on the pressurisation area can cause rapid depressurisation. The main purpose of cabin pressurisation is to maintain a safe and comfortable environment for crew and passengers in the aircraft, which is flying at low outside atmospheric pressure. Depressurisation of the aircraft is very dangerous to everybody inside the aircraft and put the crew and passengers at risk of hypoxia, altitude sickness, barotrauma and decompression sickness.

Moreover, if FOD hit the aircraft nose radome, it can cause disruption in aircraft radar system and led to numbers of fatal incidents. Aircraft radome is an integration equipment that is equipped with the weather radar antenna. The main purpose of the weather radar is to help the pilots to identify and avoid undesirable weather formation.

2.4.4. Fuel efficiency. FOD ingestion has a tendency to drop engine's operating efficiency. This often happens when the blades are blended, which causes a slight increase in fuel consumption. Based on a survey done for the Boeing 767 aircraft, blending blades can drive an efficiency loss that increases the operating costs by USD 147.85 per flight [6]. This is more than three times the cost of normal engine running, which is USD 47.

When FOD has caused damage to the aircraft, the aircraft is normally parked at the designated area to let the maintenance personnel to inspect, and repair it if necessary. The aircraft will only return to service after the damage has been repaired. If major repair has to be conducted, the aircraft will be grounded for a long time and this causes flight cancellation. The cost of aircraft delay and cancellation fees are significantly high, hence gives big impact to airport and airline operators.

A study has found the average cost of flight delay is USD 40 per minute per aircraft [6]. It includes delays at gate waiting time, taxi-out time, airborne-time and taxi-in time. Furthermore, data from two large European airports showed that their runways were closed due to FOD and wildlife for an average of 200-240 minutes per month. The total cost of this problem was estimated about USD 26,740 per 10,000 aircraft movements and USD 1 million for large airports.

2.5. FOD prevention

FOD prevention is a method or technique to prevent FOD and promote safety in aviation world. The main purpose of FOD prevention is to reduce FOD occurrences around the airport and maintenance areas. Nowadays, most airports and airline operators have realized the importance of FOD prevention. The most important factor that contributes to the success of this method is the ongoing support and commitment from the top organisation leadership management. Without this, the effectiveness of FOD prevention cannot be achieved and it will continually suffer with lack of credibility.

A successful FOD prevention can be achieved by having procedures and implementation of Safety Management System (SMS). This system is highly beneficial because it gives major contribution for the organisation to establish decisions, attitudes, techniques of operation regarding safety culture and other related issues. Safety culture is an important element of human factors in maintaining FOD-free environment [1]. In other words, the safety culture is related to human behaviours and attitudes when working in organization. It also can be defined as habits or manners among the group of peoples in providing safety performance in an aviation organization. Through good safety culture, all duties and procedures about FOD prevention can be clearly defined and well understood. After all, designated personnel should have good personal attitudes and know their responsibilities regarding FOD hazards and how to eliminate it.

2.6. FOD prevention program

FOD prevention program is a guideline for an organization to eliminate and reduce any consequences of FOD. Normally the program is based on certain standards and guidelines that are issued by the aviation authorities such as National Aeronautics and Space Administration (NASA) and FAA. The researches that have been done clearly show that most organizations in aviation industry are practicing the same procedure of FOD prevention program such as in several aviation companies and agencies including Bell Textron Helicopter, National Aerospace FOD Prevention Inc. (NAFPI), Research and Technology Organization (RTO) and FAA. In order to achieve the ultimate goal of this program, there are three considerations to be applied: FOD designation / sensitive area, awareness and FOD airside activities preventive measure [3].

2.6.1. FOD designation / sensitive area. FOD designation area is essential to prevent FOD. This area should be designed based on maintenance activities that have been done and risks associated with FOD. However, there are many consequences and a high probability of FOD is not controlled or found in this area. Based on a study by NASA [10], FOD sensitive areas should be designed by combining two risk factors: probability and consequences as shown in Figure 6.

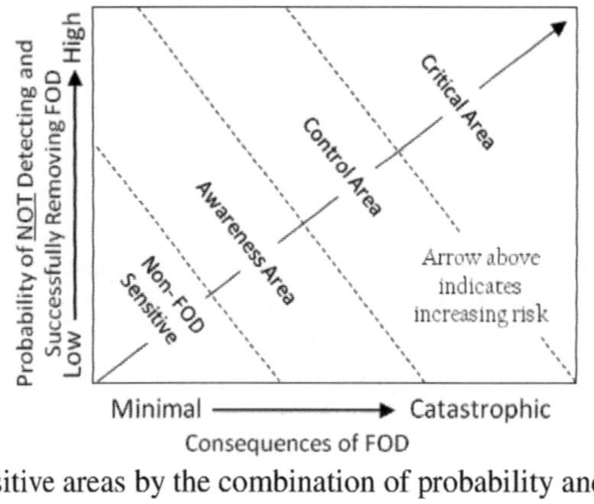

Figure 6. FOD sensitive areas by the combination of probability and consequences [10]

FOD sensitive is an area where less activities that can produce FOD is performed. Therefore, FOD risk is insignificant and no FOD control measurement is required here. On the other hand, in the FOD awareness area, although the tendency for FOD to damage the aircraft is quite low, it still cannot be ignored since the damage may become severe and disturb flight operation, as well as cause injury to the maintenance personnel. This area is divided into two parts: non-manufacturing and manufacturing areas. At non-manufacturing areas, no marking or 'FOD AWARENESS AREA' sign is required while at manufacturing areas, the sign is necessary and all personal tools must be marked properly according to their specifications. Housekeeping and 'clean as you go' standards are obligatory to be practiced within this area.

Furthermore, in the FOD control area, the potential of FOD to be set up and damage the aircraft is quite high [10]. It normally happens during the aircraft assembly and major joining operations. This area must be clearly marked with red and white stripe floor marking tapes. The other methods to mark this designated area are by using stanchion and appropriate approved signs. Besides that, housekeeping and 'clean as you go' procedures must be practiced and all maintenance personal are advised to use company-provided tools rather than their own tools. Drink, food and smokeless tobacco can only be consumed in the designated area. In fact, anybody who wants to enter this area must be FOD certified or accompanied by someone who holds a valid FOD certification. They must also wear proper dress code and jewelleries are strictly prohibited to be worn when working in this area.

Lastly, the FOD critical area contains aircraft major assembly, including assembly or disassembly of aircraft engine and other critical components [11]. Furthermore, this area can be defined as a place where the flight hardware is exposed to FOD that can cause deterioration or malfunction to the aircraft operations. 'Clean as you go' policy must be followed strictly within this area and no food and a drink is allowed except for only water. Moreover, the FOD barriers are placed on all open lines and tube, electrical connector, inlets and vent when off from work, and this is one of crucial safety precautions. Same as in FOD control area, all controlled entry and exit point must be easy to identify with clearly visible marking method and the entry is restricted to FOD trained personnel only. Any non-essential personal tools is unacceptable and any lost tool or item should be reported and investigated.

2.6.2. Awareness. The FOD prevention program is depending on awareness level among employees regarding to the program existence. The first measure of existence of the FOD prevention program is its visibility [12]. It is typically carried out within the working environment by proper communication methods such as advertisement and visual tools, which are used to remind all employees about FOD hazard. However, the methods have to be updated from time to time to ensure that all employees are aware about the program.

Moreover, design of visual tool is giving major impact in FOD prevention program visibility: it must be able to deliver a clear message that is applicable, easy to read and understand. There are many examples of visual tools to comprehend FOD message. Most of them can be a great advertisement that is usually displayed in the working areas such as FOD banners and posters. They have to be creatively designed to attract people in the working areas and to increase their awareness about the program. Additionally, the design should have an ability to convey serious messages. FOD banner is normally varied in sizes and messages, and it is permanently mounted at the hangar or in main entrance where it can easily be seen by the employees [12]. Apart from the FOD banner, posters are also able to make all employees consistently alert about FOD threats on the flight line and maintenance area.

Besides visual tools, organisation communication also can identify the potential FOD hazard and a proper way to mitigate it. FOD seminars, bulletins, case studies and electronic reporting through email and websites are examples of organisational communication. Apart from that, sharing safety related information with other airport operators also becomes other initiative of organisation communication to minimize all FOD potential hazards.

2.6.3. FOD airside activities preventive measure. This preventive measure is very important for airport owners, operators, general aviation operators and air carrier station managers to establish the FOD prevention program. Individuals in these positions have the responsibility to remind maintenances personnel, ramp crews and aircraft servicing personnel about FOD hazard. The procedures to eliminate FOD hazard must consider all airside activities [13].

For aircraft maintenance activities, all spare parts such as nuts, bolts, washers and safety wires need to be managed and disposed appropriately. The same action is also applied to hand tools used in repair jobs. After all the maintenance activities are done, those items have to be checked and put back to their original place. Moreover, they must be placed in a proper tool box or proof tote bag to prevent any lost items. To help in the this process of controlling all the spare parts and hand tools, the methods used are shadow boards, cut out tool tray liners and checklist.

Most FODs found in airport apron, service roads, baggage area and areas near the aircraft galley usually come from aircraft servicing activities. In order to avoid any unwanted incidents due to these FODs, the airport management must hire a group of people that are responsible to clean those areas. Additionally, the airport management has to set up a procedure to check Ground Servicing Equipment (GSE) for any signs of tear and wear that have tendency to create FOD hazard. Moreover, in order to reduce numbers of FOD found at the baggage area, the management should follow strict procedures to perform checking at the baggage loading and unloading areas every time when an aircraft is serviced.

Other common sources of FOD in airports are from the asphalt and concrete pavements. Towards effective pavement maintenance practices, the airport management must develop extensive resources for pavement maintenance such as pavement inspection and FOD supervision. A good supervision of FOD on airfield pavement is vital in promoting safety of aircraft operations. Pavement condition index (PCI) is developed to identify the FOD potential caused by deterioration and is incorporated in airport pavement management system [14]. The PCI is able to address any maintenance act required to reduce FOD hazards.

During windy weather conditions, debris such as plastic and cargo strapping are easily blown and they cause FOD hazards in air cargo areas. In this case, airport management should establish proper procedures on collecting the FOD, possibly by fixing fences at the right area. Since the fences trapped the blowing debris, it must be removed regularly to make sure FOD will not come back to the air cargo area.

3. Discussion

The aviation industry has a number of methods to fight against FOD hazard. In fact, guidelines from FAA A/C and NAFPI FOD prevention manual provide good and precise orientation for organizations to develop their FOD prevention program. Fortunately, the guidelines are free and easy to be accessed and adapted. Organizations that want to establish a FOD prevention program must have sufficient and adequate reliable sources before starting to emphasize on human error reduction in their FOD control measures.

Furthermore, there is no such thing of 100% FOD free at the airport and maintenance area due to the inexorable presence of debris that is impossible to be totally eliminated. However, the percentage of FOD can absolutely be reduced with the implementation of FOD control methodologies and the introduction of FOD detection technologies such as radar, camera and sensor at airport area. These technologies are really effective and convenient to be practiced. Implementation of these technologies at the airport are based on a number of factors including the type of aircraft operating, the number and size of the active runway and taxiways, and finally the location of the airport.

Technically, the airport operator should perform FOD risk assessment in order to identify the major FOD risk that has high tendency to occur. This assessment should include risk type, risk response, probability of the risk to happen, severity of the risk and lastly, effective action that should be taken in order to eliminate the risk. By having this assessment, the organizations can determine the level of the risk and construct mitigation strategies. This way, the presence of debris can be totally eliminated and this indirectly reduces the risk of having FOD at airport area.

4. Conclusion

The most effective way to eliminate FOD is by developing FOD awareness among the people who are involved in aviation industry. Due to dangers of FOD, the aviation authority, organization or company has taken appropriate action to minimize this problem by having FOD prevention program and also applying other prevention methods. The prevention program includes all techniques to eliminate FOD or everything that is prone to FOD. It is indeed one of the most effective measures to eliminate FOD in today's aviation world. The success of this program starts at the top level of an organization itself and it gets more comprehensive when there is a continuous solid support from the subordinates. Everybody in the organization must always be sensitive to the impact of FOD and needs to put an extra effort to eliminate this particular problem. After all, the most critical goal of the FOD prevention program is to promote unlimited safety level in aviation world. This is just to make sure that FOD will not become predominant and create catastrophic failures to the aircraft, as well as everyone who is flying on it.

References

[1] Kraus D C and Watson J 2001 *Guidelines for the Prevention and Elimination of Foreign Object Damage/Debris (FOD) in the Aviation Maintenance Environment through Improved Human Performance* (Federal Aviation Administration)
[2] ATSB 2010 *ATSB Transport Safety Report* (Australian Transport Safety Bureau)
[3] Wieslaw B 2005 *FOD Prevention* (NATO Research and Technology Organisation)
[4] James W C, Gorton S H and Alby F 2005 *Effects of Sand and Dust on Small Gas Turbine Engines* (NATO Research and Technology Organisation)
[5] FAA 2010 *Airport Foreign Object Debris (FOD) Management* (U.S Department of Transport)
[6] Iain M 2008 *The Economic Cost of FOD to Airlines* (London: Insight Sri Ltd)
[7] Theodore N 2006 *High Cycle Fatigue: A Mechanics of Material Perspectives* (Oxford: Elseiver)
[8] Ruschau J J, Nicholas T and Thompson S R 2001 *International Journal of Impact Engineering* **25** 233-50
[9] Maragakis I 2009 *Bird Population Trends and Their Impact on Aviation Safety 1999-2008* (European Aviation Safety Agency)
[10] NASA 2010 *FOD Prevention Program* (National Aeronautics and Space Administration)
[11] Bell Helicopter Textron 2007 *Foreign Object Debris and Foreign Object Damage (FOD) Prevention for Aviation Maintenance and Manufacturing* (Bell Helicopter Textron)
[12] Phil M 2005 *Promotion and Aware: It All Starts Here Make it FOD Free*
[13] FAA 1996 *Debris Hazards at Civil Airports* (U.S Department of Transport)
[14] Li X, Keegan K and Yazdani A 2010 *Journal of the Transportation Research Board* **2153** 81-7

Malaysian aviation technologist promotion to managerial role: an empirical overview

C L Liew[1,2*], A R Abu Talib[3,4] and R L Jacobs[5]

[1]Malaysia-Japan International Institute of Technology (MJIIT), Universiti Teknologi Malaysia, Malaysia.
[2]SR Aviation Sdn. Bhd., Taiko Hangar, Sultan Abdul Aziz Shah Airport, Subang, Selangor, Malaysia.
[3]Department of Aerospace Engineering, Faculty of Engineering, Universiti Putra Malaysia, Malaysia.
[4]Aerospace Manufacturing Research Centre, Universiti Putra Malaysia, Malaysia
[5]Human Resource Development, College of Education, University of Illinois, USA

*rickyliewcl@gmail.com

Abstract. The Malaysian aviation industry has continued to march forward. With a turnover of RM23.7 billion in 2013, it is expected to grow higher especially after the Malaysian national aerospace blueprint was launched in 2015. The aviation related organizations currently have a workforce of approximately 13500. These organizations need to be managed by competent managers who have a strong background of technologist. Aviation technologist is one of the key components in the aviation maintenance industry as they are the future managers charged with the responsibility to ensure continuation of the organization's objectives and culture. The technologist role and manager's role are somehow different. The promotion of technologist to managerial roles is quite common but whether the technologist is able to take up managerial role effectively is yet to be fully understood. It is quite common that there was insufficient training for the technologist before being promoted to take up management roles. The purpose of this paper is to give an overview of the role of technologists and managers in professional services industries such as MRO and to understand that there is a need within the industry to re-look into the perspective of a proper training to prepare them to take up management roles effectively.

1. Introduction

The Malaysian aerospace industry has continued to move forward. The industry turnover growth has gradually increased from RM21.5b in 2006 to RM32.7b in 2013 [1]. There is a clear indication that the aviation industry will continue to develop in line with the Malaysian National Aerospace Blueprint 2015 - 2030. The government has clearly listed aerospace industry as one of the key drivers of high-income economy through Economic Transformation Plan (ETP). Currently there are approximately 1800 Licensed Aircraft Engineers (LAEs), 11000 technicians and mechanics, and 600 MRO graduate engineers in this sector [2]. Due to the professionalism of aviation maintenance field, the professional body such as the Board of Engineers (BEM) Malaysia has looked into the details of the engineering profession by practise and identified the professional practitioner of engineering as technologist and technician.

After detailed studies by Board of Engineers Malaysia, the Engineering Technology Path Report [3] dated February 2003, also known as the Blueprint for a Highly Competent Engineering Technical Workforce Report, was being drafted and submitted to the Parliament for review. As a result, an Act of Malaysian Parliament was passed on 1 Dec 2014. The act is known as Bill of the Technologists and Technicians Act 2014. The bill was an act to provide for the establishment of the Malaysian Board of Technologist for recognition and registration of Professional Technologists and Certified Technicians. This may serve as a professional body that is able to support the aviation industry. According to the bill, aviation maintenance, repair and overhaul (MRO) can be determined as technology services. Both technologist and technician can provide these technology services. A technologist is one who holds a bachelor's degree that is recognized by the Board. Licensed aircraft engineers may be grouped into the technologist category due to professional license recognition by the government. On the other hand, a technician is a person who holds a certificate level of the technical education or training in practise.

This paper is focused on the aviation technologists who are one of the key components in MRO organization. It is important because this profession is a knowledge-based profession where only those with highly qualified and trained are able work in the industry safely and effectively on aircraft and its systems. Secondly, a technologist is able to perform the task to aircraft manufacturer and regulatory standards with deep insights. Finally, these technologists are needed to bring life to an organization and also to keep moving forward. It is important that this group of technologists is the future group of knowledgeable managers to ensure the continuity of the organization's objectives and culture. Without these highly qualified and competent managers, it is difficult for the MROs to operate effectively and efficiently.

2. Assessment of aviation technologist role

In order to understand the reasons why an aviation technologist is not recommended to be promoted to the position of managers without proper training, it is imperative to understand the role of technologist by nature and the role of managers in general. In addition, MRO services are a form of professional services compared to other general services such as motorcar or manufacturing sectors. Marion *et al.* state that engineers are very focussed on technical facts and do not have sufficient understanding on leadership as well as management in their research [4]. This is further suggested by Galloway [5] that it is important for engineering students to not only be furnished with latest, non-technical abilities but also with the need to change the perception of public image of these technologists. It is an indication that non-technical abilities refer to management role in an organization.

The basic technical and traditional role of aviation MRO technologist has not changed much due to the nature of the function. Although aviation technologist role is usually listed in the job scope in an MRO organization, it is important to understand technologist role in the Malaysian MRO industry and what did they study in the university or colleges. This should be contrasted against managerial role in the current aviation industry.

2.1. Technologist role

Malaysian Department of Civil Aviation (DCA) Airworthiness Notices No. 1101 [6] specifies that a licensed engineer (aviation technologist) may issue the certificates following maintenance including aircraft structure, power plant, mechanical and electrical systems. The replacement of avionic line replaceable units, requiring tests to prove their serviceability, shall also be included in the privileges. The certificates to be issued includes those for release to service in respect of the approved overhaul, inspections, repair, replacement, modification and scheduled maintenance inspection.

In simple term, aviation technologists are mainly involved in maintenance of aircraft. Maintenance seems to be a common word. Higgins [7] described maintenance as a science since its execution relies, sooner or later, on most or all of the sciences. It implies that maintenance professionals such as the aviation technologist are very much involved in the application and practical aspect of science within this profession. Kinnison and Siddiqui [8] called aircraft repairer or maintainer as aircraft mechanic in the United States (US) aviation MRO industry. Effectively, these are the aviation technologists as described by Malaysian DCA. Entropy in an aircraft is being interpreted as the degree of disorder in its design and systems. Accordingly, the aircraft design engineer designs an aircraft with minimal entropy during its operational lifetime while a technologist is to battle continuous increase in entropy during the life of the aircraft. These professionals study and involve in the practical details of aircraft systems and components. Specialization consists of the mechanical systems such as hydraulics, pneumatics, flight controls and structures of the aircrafts. The other trade is the avionics system, which covers the electronics, radios, computers, and electrical systems. They function in hangar base maintenance, line maintenance or even the workshops.

Technologists go deep into the systems and address the system or components with comprehension of its operational parameters and correct functionality. Should any discrepancy be detected, a set of established procedures for troubleshooting, discrepancy isolation and rectification is being complied. Aircraft manufacturers' procedures are complied with for removal, installation, inspection, operational test as well as for functional test. They are so specialized that they have the capability to analyze and identify fault when there is a problem with the systems or when the operational parameters of a design are out of tolerance. As aviation technologists function in the field, the skill that comes along with experience is being enhanced more and more over time. Eventually, they are able to conclude whether the problems are able to be solved or require further solution from the design engineering teams. Stevenson [9] identified the following capabilities in Table 1 that are part of a technologist.

Table 1. Capabilities of technologist [9]

Six key capabilities	To use and communicate information
	To measure and make use of measurements
	To choose materials and components and understand the processing of materials
	To understand manufacturing activities and general commercial organization and practise of their firms
	To diagnose
	To organize (but necessarily supervise) and give direction to the work of others
Other capabilities	Communication manufacturing instructions, test report, fault analysis
	Measuring: Tolerances, fits, quality and performance, checking
	Materials: strength, wear, distortion, machining, properties
	Manufacturing: assembly, limits, planning, operations
	Commercial factors: cost, standard components, customer requirements
	Diagnosed: test data, statistics, faults, value analysis
	Organize: planning work, meeting targets.

The Federal Aviation Administration (FAA) of US in its FAA-8083-30 Chapter 13 document [10] states that an aviation technician is considered the central figure in aviation maintenance. One must be fully qualified as an aviation professional in order to function to the required standards. White and Birkinshaw [11] listed a complete syllabus an aircraft maintenance technician must study and passed. The majority of the training syllabuses are of practical aspect of aviation maintenance with minimal management training. Therefore, it is clear that aviation technologist's scope is insufficient to promote them directly to take on managerial positions.

3. Assessment of manager's role

While technologist's promotion to a managerial role is common in the industry, not all promotions are effective. It is because managers play a key role in organization. The manager's role has continued to draw the attention of scholars and leaders. It was evidently written by the Chinese as early as 3500 BC and later in Sun Tzu's Art of War [12]. According to these records that were treated as one of ancient military discourse of Chinese classical works and military strategists [13, 14]. Even in the current era, business leaders and scholars have been trying to grip the roles and skills of the managers from those

times. An analysis of ancient Romans method of communication over a large empire leads to further introspection of managers' conduct and capability [15]. Historical records did not clearly spelled out how managerial roles were managed. However, work by Machiavelli in the sixteenth century provided greater contribution on understanding the managerial role. Management pioneers during the Industrial Revolution such as Adam Smith and Karl Marx also provided greater insights on the role of managers [15]. As human race continues to move along with time, classical scholars and social scientists laid out wider concepts and fundamentals in an attempt to comprehend the metamorphosis of the manager's functions. Mintzberg [16, 17] exerted an influential study in the recent era on the role a manager. His work involves detailed description of observation made on managers in his working environment. His results described 10 roles of manager functions. He divided the major dimensions of managerial role into three categories:
(1) interpersonal roles (figurehead, leader, liaison)
(2) informational roles (monitor, disseminator, spokesperson)
(3) decisional roles (entrepreneur, disturbance handler, resource allocator, negotiator)

Table 2. Mintzberg's Management functions [17]

Category	Role title	Activities
Informational	Monitor	Seek and/or receive information; scanning periodicals and reports; maintaining personal contact with stakeholders
	Disseminator	Forwarding information to organization members via e-mails, memos, reports, and phone calls
	Spokesperson	Transmitting information to outsiders via e-mails, reports, memos, and speeches
Interpersonal	Figurehead	Performing ceremonial and symbolic duties, such as greeting visitors and signing legal documents
	Leader	Directing and motivating subordinates; counselling and communicating with subordinates
	Liaison	Maintaining information links both within and outwith the organization via e-mails, mail, phone calls, and meetings
Decisional	Entrepreneur	Initiating improvement projects; identifying new ideas and delegating idea responsibilities to others
	Disturbance handler	Taking corrective action during disputes or crises; resolving conflicts among subordinates; adapting to environments
	Resource allocator	Deciding who gets resources; preparing budgets; set schedules and determining priorities
	Negotiator	Representing department during negotiations of contracts, sales, purchases, and budgets

However, Cecil *et al.* [18] challenged Mintzberg's proposition. They claimed due to the dynamism of economic transition at different levels especially in Asian countries, the new generation managers require more than the 10 roles identified by Mintzberg [17]. Even with the dynamics of organization activities and environment, it appears that minimal attention was given to re-inspect the manager's role and competency required. In addition, there are strong indications (Edwards *et al.* [19], Hornsby *et al.* [20], Kickul and Gundry [21]) that due to the swift growth associated with technology development and disposal, organization needs to adjust its organizational operations and practices. It is shows that management roles are not fixed but varies according to time. Van der Velde *et al.* [22] realize that as business environment evolves over time, work configurations appear to be more complicated with the staff clenching onto newer expectations, and together, these manifestations are focussing toward re-enacting of managerial functions. However, it appears that most management still tends to refer to the Mintzberg identified 10 manager roles. There are also concerns that the spread of industrialization from developed nation to the developing nation such as Malaysia, the Western management education concepts may result in homogenous managerial roles.

Generally, MRO managers are responsible for schedule and unscheduled maintenance checks from various aircraft operators. The managers can be classified as a form of project managers. Sommervill *et al.* [23] described that project manager is generally considered as a person accountable for ensuring that a project can be completed safely, on schedule, within budget and also to the expected quality and performance standards as specified by customers and aircraft manufacturers.

Notwithstanding the numerous discussions on the features of managerial roles, there seem to be minimal international studies with regard to managerial functions. Apparently, the Mintberg notion of the managerial roles has somehow been considered as generic roles and general management [24- 27].

4. Manager role in professional services organization

Aircraft MRO services can be considered as professional services. According to Sahin [28], expansion of professional business services is being observed as "the most profound business phenomenon of the twentieth century" [29]. In professional service organizations, it is critical that competence is the key in delivering towards customer satisfaction. These competency-based services have to be effectively managed. The specialized, highly skilled and autonomous professionals working in these firms make the manager's role challenging. Since MROs are mainly knowledge-skill intensive, it is imperative that these firms' focal point should be developing workforce quality, especially to increase the manager's competency. For this reason, it is important that technologists are trained to equip them with the right competency. Mills *et al.* [30] highlighted that in such companies, coordination and improving the competencies is the management primary task while delivering the services at the same time. This is especially true in the Malaysian MRO industry where the number of MROs are increasing as well as high number of workforce are being produced by the local training institutions and colleges each year. McKaig-Berliner [31] and Lowendahl [32] found that, top managers of professional services, unlike normal manufacturing firms that emphasise on profit and lost, investment returns and financial results, they are more immersed in activities related to non-financial purposes like new markets' development, developing workforce and expanding competitive advantages through research, development and innovation. Managers in MRO need to have similar capabilities as well.

Management scholars' interest to relate management capabilities to firm success has been going on for decades [16, 30, 33, 34, 35, 36]. The focal point was on the managers through examining their knowledge level and competency, ability to execute strategies and relationship with stakeholders. The conclusion of the studies showed that managerial resources are critical in building the competitive advantage. However, it is even more important that the managers themselves being viewed as the resources [37]. Ochel [38] highlighted that manager's education background, personality, experiences and inter-personal relationship exerts a strong impact on the firm strategies and competitiveness. An interesting note from [38], is that through combination of individual competency of its managers, an organization can improve its managerial ability. Effectively, it means to say that with strong teamwork among managers, more effective decision making can take place in practise. Kor [39], Van Den Bosch and Van Wijk [36] purported that as managers gain experience in managing over time, they establish their own management technique, the speciality is mostly tacit and hard to be duplicated, thus a source of competitive advantage itself.

Sahin [28] found that the modus operandi of a manufacturing organization is quite different to that of professional service organization in two ways. First, it is a highly customized professional service. As a result, relationship with customers itself is a feature. Secondly, these professional services require highly trained and skilled people, which in itself is a personalized service. Using MRO services as an example and concurrence with the findings of McKaig-Berliner [31], the professional service firms are grouped as knowledge-intensive that are displayed by organization core competence through customer orientated employees, empowered and autonomous members with specialized knowledge. It is clear that the key customer requirements are the specialized knowledge and specific skills possessed by the organization to be chosen by a customer. Interestingly, Ofek and Sarvary [40] claimed that knowledge constitutes the "production technology" for such services, which resulted in business knowledge on its own.

The ability to deliver professional service by attending to the needs of the customers also generates additional knowledge and experience from each of the occurring projects. As a result, there should be a capacity through effectively trained managers to capture and provide improved effective solutions over time. Eventually, an organization variable cost should decrease with improvement in efficiency that benefits organizations. Roberts [41] stated an additional benefit whereby, knowledge obtained and real experience that were captured through serving one customer can also be applied to service another customer. Therefore, a competitive advantage can be developed through increasing the list of customer portfolio, which provides an advantage to professional service organizations. The advantages could then be leveraged through use of communication technologies (like emails, electronic presentations and record system, etc.), thereby enabling effective communication within the organization teams and allowing them to retrieve the data they required [42].

Since the quality of professional services are largely dependent on individual competency, Stumpf *et al.* [43] and Ochel [38] found that inclination towards recruiting individuals equipped with a high level of education is supported by implementing a system of individual monitoring, continuous in-house training and development, and continuous improvement philosophy. Using advertising as an example of professional services, Halinen [44] identified that it is important that there is an ability to manage professionals effectively so that an organization is able to continue supporting the customer despite resignation of professionals instead of having the customer follows the individual to another organization. This also means that the ability to assign the right quality personnel who is compatible to the customer traits needs to be handled effectively. In the MRO services, multinational customers tend to source for local professional services provided that there is a well managed human capital. Hitt *et al.* [29] concluded that professional service organizations need to ensure they have well-experienced, educated, trained and skilled manager along with their employees in order to maintain a strong client relationship.

Nordenflycht [45] provided an insight on developing a "common commitment" among managers (Sanchez [46]), which has been observed to be able to keep highly mobile and skilled professionals in the professional service organization to maintain the professional value of organization. These can be in the form of prestigious recognition or shared responsibility or partnership. This leads to professional standing and trustworthiness of the managers, which is viewed as competitive abilities indicators. This team of managers is to be involved in developing and communicate the team and organizations vision, mission, objectives and values with priority setting [32]. Therefore, it is critical that the managers are able to manoeuvre within an organization so that all team members are able to share the organization strategic vision to enhance the organization performance. This tacit knowledge and experience that the manager holds can be a competitive advantage as it is unique to the organizations and managers. It is clear that a manager's ability to carry out the induction of right candidates, and develop and maintain key personnel is critical for professional service organization such as MRO to succeed. Therefore, it is acceptable when Lowendahl [32] states that the professional service organizations not only need to compete for customers but also for brilliant professionals to be more competitive.

Burke [47] considers that a successful project is achieved when quality parameters, duration and cost are equal. It is also implying that the managers, in addition to team management, need to be able to lead teams with a positive leadership, for instances in achieving trustworthiness, respectfulness, motivation, co-ordination and up keeping morale of team members. However, MRO managers need to practise other skills while taking the team lead to achieve project delivery targets. Taking example of a construction project manager, Griffith and Watson [48] identified the other skills such as conducting, controlling and administration of the project and team. Despite the manager's role needs to evolve over time, it is surprising that discussions on managerial work today still has the focal point conceived of manager that embraces the "work roles" laid out by Mintzberg [16]. From another perspective, MRO managers should possess related skills in order to competently commit to project execution. Barber [49], Fryer [50], Jha and Iyer [51], Pant and Baroudi [52], and Shehu and Akintoye [53] reasoned that these skills change as the age and experiences increase. In these contexts, managers need to continue to improve themselves to stay relevant and maintain their competitiveness. Particularly, when every project has its own uniqueness and managers must be able to identify the skills required and apply them as required to match each project requirements. This was echoed by Edum-Fotwe and McCaffer [54] that stated manager becomes competent with knowledge gained through training, together with skills sharpened as time and experience increases. Hoffman [55] described competence as what people

IOP Conf. Series: Materials Science and Engineering **152** (2016) 012039 doi:10.1088/1757-899X/152/1/012039

do and competency as how they do it. As a competent MRO manager, one must have the capability to carry out what one is supposed to do and able to do it effectively and efficiently.

The authors' observation of the managerial role for aviation technologist is being depicted in Figure 1. The left half section of the flow chart shows that the aviation technologist taking up managerial role through an unstructured root. Generally, it is through appointment or promotion due to experience and seniority within an organization. The right section of the flow shows the technologist managerial role being developed through a structured system. This is possibly the right solution that requires further study after this paper by the authors.

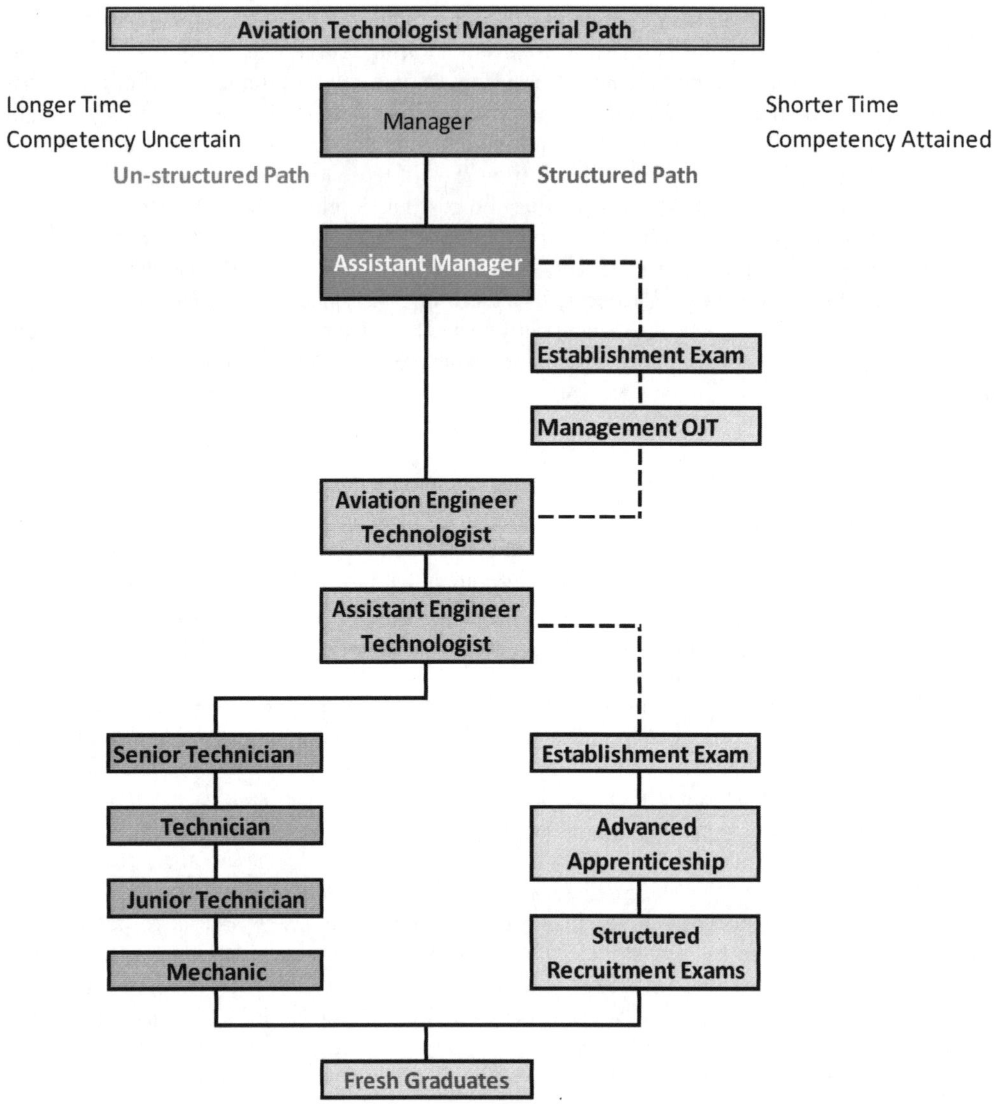

Figure 1. Aviation technologist managerial role path

5. Conclusion

It is clear that manager's role is not to be taken lightly, especially in the aircraft maintenance industry. Based on the overview presented, it is evident that the traditional way of promoting technologists to a managerial role is not simply an appointment process or fill up the position. The task of managing future work force is a great challenge. The traditional way of managing erodes due to the changing environment. This is especially true when the technologist is promoted without the understanding of the management role effectively. In other words, there were not enough training for these technologist to prepare them to take up the management roles. As the industry relies heavily on technologists and effective managers to ensure the national aerospace blueprint objectives can be attained, it is important to re-look into the issue of preparing technologists to take up the management roles. In addition, it is also a need to recognize that competitiveness-managerial capabilities and roles are a phenomenon that has several aspects. It means to say that a single view is unable to indicate the full perspective on the issue of the MRO managers. This is an initiation for future research on the issues and concerns about promoting technologists to management role in the aviation MRO industry.

Acknowledgements
The authors would like to acknowledge and thank the Ministry of Higher Education for the financial support through the scholarship for the first author's current research. Special thanks to colleagues and peers of the aviation MRO industry who have taken time to read and understand the contents of the paper after work.

References
[1] Zainal K, Abu Talib A R and Hack I 2015 *Malaysian Aerospace Industry Blueprint 2030* Malaysian Industry-Government Group for High Technology (MIGHT)
[2] Abu Talib A R and Helme N F 2015 *Aerospace Education & Training Demand and Supply in Malaysia* Malaysia's National Foresight Magazine (myForesight), Malaysian Industry-Government Group for High Technology (MIGHT) p12-17
[3] Board of Engineers Malaysia *http://www.bem.org.my/meet.pdf* Accessed on 10 August 2016
[4] Marion A, Eibl W and Kugler F 2014 *Research in Competency-Based Management* **7** 219-46
[5] Galloway P D 2004 *Leadership and Management in Engineering* **4** 127-32
[6] Malaysian Department of Civil Aviation. *Airworthiness Notices*
[7] Higgins L R 1990 *Maintenance Engineering Handbook* (New York: McGraw-Hill)
[8] Kinnison H A and Siddiqui T 2013 *Aviation Maintenance Management* (New York: McGraw-Hill)
[9] Stevenson R E 1970 *Industrial and Commercial Training* **2** 31-4
[10] Federal Aviation Adminstration 2008 *Aviation Maintenance Technician Handbook General* (Oklahoma City: US Department of Transportation)
[11] White R E and Birkinshaw J 2000 *Strategic Management: A Canadian Casebook* (Scarborough: Prentice Hall)
[12] Griffith S B 1963 *Sun Tzu: The Art of War* (Oxford: Oxford University Press)
[13] George C 1972 *The history of management thought* (Englewood Cliffs: Prentice-Hall)
[14] Seagrave D A 1995 *Lords of the Rim* (London: Cargi Books)
[15] Wren D A 1979 *The evolution of management thought* (New York: Wiley)
[16] Mintzberg H 1994 *Sloan Management Review* **36** 11-26
[17] Mintzberg H 1973 *The nature of managerial work* (Englewood Cliffs: Prentice-Hall)
[18] Cecil A L, Pearson Samir R and Chatterjee 2003 *Journal of Management Development* **22** 694-707
[19] Edwards C, Robinson O, Welchman R and Woodall J 1999 *Human Resource Management Journal* **9** 55-64
[20] Hornsby J S, Kuratko D F and Montagno R V 1999 *Entrepreneurship Theory and Practise* **24** 9-24
[21] Kickul J and Gundry L K 2001 *Journal of Management* **27** 347-61
[22] Velde V D, M E G, Jansen P G W and Vinkenburg C J 1999 *Journal of Applied Management Studies* **8** 161-74
[23] Sommerville J, Craig N and Hendry J 2010 *Structural Survey* **28** 132-41
[24] Drucker P 1995 *The practise of management* (London: Heinemann)
[25] Kotler J 1982 *The general managers* (New York: The Free Press)
[26] Bower 1970 *Managing the resource allocation process* (Boston: Havard University)

[27] Kanter R 1982 *Havard Business Review* **60** 95-105
[28] Sahin C. 2011 *Journal of Management Development* **30** 413-26
[29] Hitt M A, Bierman L, Uhlenbruck K and Shimizu K 2006 *Academy of Management Review* **40** 1137-57
[30] Mills P K, Hall J L, Leidecker J K and Margulies N 1983 *Academy of Management Review* **8** 118-31
[31] McKaig-Berliner A L 2001 *The global sourcing of competitive advantages: A study of professional business services.* Dissertation Abstracts International
[32] Lowendahl B 2000 *Strategic management of professional service firms* (Copenhagen: Business School Press)
[33] Barnard C I 1938 *The functions of the executive* (Cambridge: Harvard University Press)
[34] Prahalad C K and Hamel G 1990 *Harvard Business Review* **68** 79-91
[35] Castanias R P and Helfat C E 1991 *Journal of Management* **17** 155-71
[36] Van Den Bosch F A and Van Wijk R 2000 *ERIM Report Series: Research in Management*
[37] Penrose E T 1959 *The theory of growth of the firm* (London: Oxford University Press)
[38] Ochel V 2002 *The Service Industries Journal* **22** 1-16
[39] Kor Y 2003 *Organization Science* **14** 707-19
[40] Ofek E and Sarvary M 2001 *Management Science* **47** 1441-56
[41] Roberts J 2000 *Creativity and Innovation Management* **9** 163-70
[42] O'Farrel P N, Hitchens D M and Moffat L A R 1993 *The Service Industries Journal* **13** 40-64
[43] Stumpf S A, Doh J P and Clark K D 2002 *Organizational Dynamics* **31** 259-79
[44] Halinen A 1996 *Relationship marketing in professional services. A study of agency-client dynamics in the advertising sector* (Florence, KY: Routledge)
[45] Nordenflyct A V 2007 *Academy of Management Journal* **50** 429-45
[46] Sanchez R 2002 *Journal of Business Research* **57** 518-32
[47] Burke R 2006 *Project management, planning and control techniques* (Cornwall: John Wiley & Sons)
[48] Griffith A and Watson P 2004 *Construction management: principles and practises.* (Basingstoke: Palgrave Macmillan)
[49] Barber E 2004 *International Journal of Project Management* **22** 301-7
[50] Fryer B 2004 *The practise of construction management* (Oxford: Blackwell Publishing)
[51] Jha K N and Iyer K C 2006 *Construction Management and Economics* **24** 977-88
[52] Pant I and Baroudi B 2008 *International Journal of Project Management* **26** 124-8
[53] Shehu Z and Akintove A 2008 *The Built and Human Environment Review* **2008** 74-98
[54] Edum-Fotwe F T and McCaffer R 2000 *International Journal of Project Management* **18** 111-24
[55] Hoffman T 1999 *Journal of European Industrial Training* **23** 275-85

In-flight food delivery and waste collection service: the passengers' perspective and potential improvement

F I Romli, K Abdul Rahman and F D Ishak*

Department of Aerospace Engineering, Faculty of Engineering, Universiti Putra Malaysia, Malaysia.

*farahdiana.defy@gmail.com

Abstract. Increased competition in the commercial air transportation industry has made service quality of the airlines as one of the key competitive measures to attract passengers against their rivals. In-flight services, particularly food delivery and waste collection, have a notable impact on perception of the overall airline's service quality because they are directly and interactively provided to passengers during flight. An online public survey is conducted to explore general passengers' perception of current in-flight food delivery and waste collection services, and to identify potential rooms for improvement. The obtained survey results indicate that in-flight service does have an effect on passengers' choice of airlines. Several weaknesses of the current service method and possible improvements have been established from the collected responses.

1. Introduction

In recent years, air travel has become one of the main transport means for many people. According to the forecast made by the International Air Transport Association (IATA), the total number of airlines' passengers will reach up to 7 billion by the year 2034 [1]. In addition, Airbus has reported that more than 70% of next 20-year demand for their new passenger and freight aircraft are for single-aisle type [2], signifying progressive increase in short-haul, domestic flight routes worldwide. This development in the air transportation industry also means an increased competition among large and small airlines to draw passengers for their offered flight services. Many innovative strategies have been applied by airlines to differentiate their services from each other. Maintaining and improving the service quality has been a key competitive measure in air transportation industry, especially in attracting passengers from competing airlines [3].

One of the direct airline services that can affect passengers' evaluation towards the airlines is their in-flight services [4], which in-flight meal service has been the main central element [5]. It is observed that passengers tend to select the airlines that provide the best meals or foods for their travel and this is highlighted in a published study that links the quality of in-flight food services to passengers' intention of re-flying with the airlines [6]. It is believed that the current method used onboard for food delivery and waste collection can be further improved. Note that the waste collection service goes hand-in-hand with the in-flight food delivery service, and therefore they are considered together in this study. With this notion, the aims of this paper are to explore the general passengers' perception of the current in-flight food delivery and waste collection services, and to identify potential rooms for improvement. In order to achieve this, an online public survey has been conducted and the obtained results have been analyzed.

IOP Conf. Series: Materials Science and Engineering **152** (2016) 012040 doi:10.1088/1757-899X/152/1/012040

2. Current in-flight food delivery and waste collection

Among the earliest recorded history of in-flight food services was in the 1920s by Imperial Airways in the United Kingdom (UK), which started to serve tea and coffee during their flights along with a few variations of cold items such as ice cream, cheese, fruits, lobster salad and cold chicken [7]. Regular hot meals were then introduced in the mid-1930s and became more common in airlines after a bigger galley was first designed by Imperial Airways for their DC3 aircraft to enable extensive hot meals to be served onboard during flight [6]. Modern and more sophisticated equipment for in-flight dining, much like the ones used by airlines today, were introduced in the mid-1950s by Boeing in their B737 aircraft. The Boeing B737 aircraft's galley was the first to be equipped with several catering functions including aircraft ovens and refrigerators to improve the in-flight dining experiences [6]. This enables frozen complete meals to be kept in cold storage before they are reheated in the electric ovens prior to being served to the passengers, increasing variety of meal options that can be made available in flight.

Nowadays, most airlines are using similar method in providing their in-flight meal services, which is by means of service trolleys or carts. These trolleys are utilized to transport the food and beverages along the aisle inside the cabin. Food and beverages are prepared in the galley area before being carted on the service trolleys by the flight attendants to be distributed to the passengers. Once the passengers have finished their meals after the allocated time frame, the flight attendants will make another round inside the cabin with the service trolleys to collect the meal trays and waste. More often than not, the flight attendants will have to do few extra rounds with plastic bags or trays to collect the waste from the passengers. The waste collection service is therefore closely linked to the in-flight meal services. Overall, this general process is summarized in Figure 1.

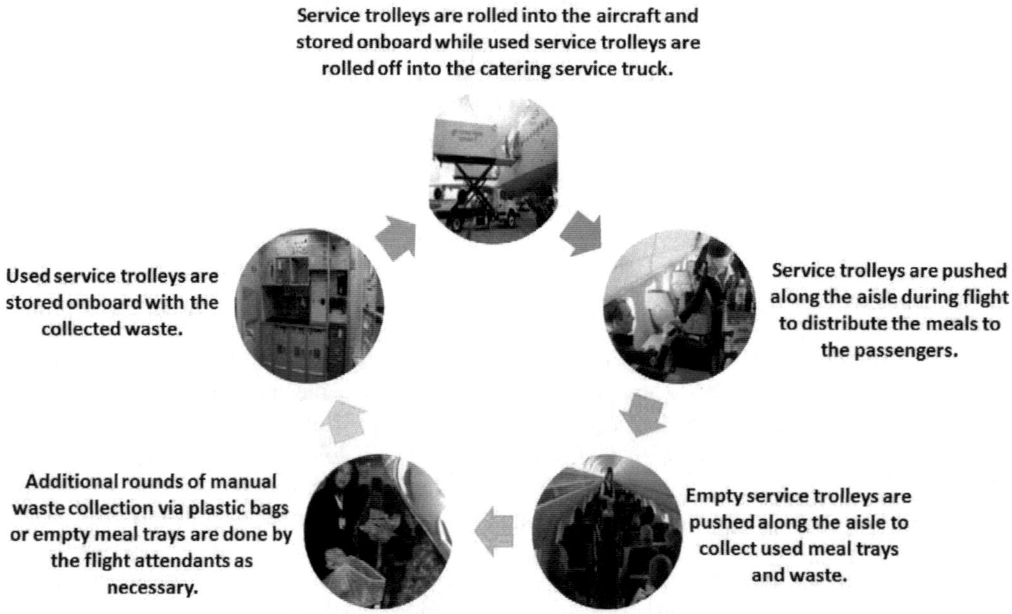

Figure 1. General in-flight food delivery and waste collection process

Similar service trolley is typically used in aircraft today, which is about 0.3m wide, 1m tall, 0.75m long and it weighs around 25kg (unladen). This is shown in Figure 2. The minimum number of cabin crew needed for a typical narrow-bodied plane is three while for a wide-bodied aircraft is 10. For some airlines, they have as many as six flight attendants on their 737s and 14 on A340s. Usually two flight attendants work together for a single service trolley in delivering food and beverages to passengers at a designated time during the flight, with the one who is pulling the trolley will hand out food while the other one who is pushing the trolley will hand out beverages.

There have also been some issues regarding this current in-flight food delivery and waste collection process. Among the common complaints are related to the long waiting time for service and also the inefficiency of the delivery and collection procedures. For instance, some passengers have to wait long for the meals because the flight attendants are serving the other rows first and sometimes, especially due to unfavourable flying conditions, the flight attendants do not have sufficient time to serve all of the passengers on short haul flights. Apart from these, during the food delivery and waste collection process, the serving cart is blocking the aisle and this makes it difficult for passengers who want to use the lavatory. Several other issues regarding the current service method are highlighted in later section based on the conducted public survey. Having highlighted some of the ongoing issues regarding the current general method used for in-flight food delivery and waste collection process, the next section will further discuss the passengers' perception and potential improvement that can be applied to this current method.

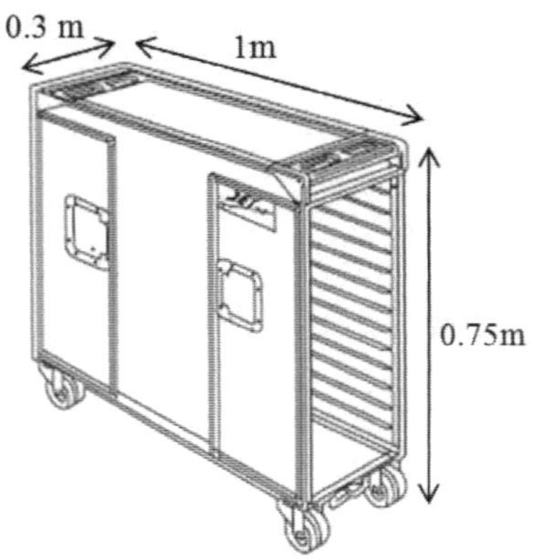

Figure 2. Typical service trolley dimensions [8]

3. The passenger survey

A questionnaire has been posted online to establish the general perception of flying passengers with regards to current method used for in-flight food delivery and waste collection services. In total, 174 people have responded to the online survey, which is comprised of questions related to their general personal and social background, flying experiences and also their comments and suggestions regarding the current in-flight food delivery and waste collection method. For this public survey, the experiences of the respondents can be from both short-haul and long-haul flights.

Of the total 174 respondents, 71% of them are female while the rest are male. Majority of them is within the age group of 21-30 years old, which makes up more than 70% of the total respondents. In addition, close to 60% of them often travel by commercial flights more than once per year. The latter characteristic makes the respondents a good sample demographics for this study as they should already have many experiences with in-flight services. Table 1 summarizes the demographic characteristics of the survey respondents in terms of their age, gender and annual income.

3.1. Passengers' perception of in-flight food delivery and waste collection

The main objective of airlines offering their in-flight services is to differentiate themselves from other competing airlines by improving their customers' flying experiences. As highlighted by the results of the conducted public survey, an overwhelming more than 90% of the respondents said that they will choose to travel with airlines that offer in-flight meals service if the flight ticket price is the same. This shows that the offering of in-flight meals service does affect the passengers' choice of airlines. In fact, in response to a question on whether they like to be served with drinks or snacks while onboard the aircraft, either for short-haul or long-haul flight, 89% of the respondents gave an affirmative answer.

Table 1. Sample demographics

Demographic characteristics	Frequency	Percent (%)
Gender		
Male	51	29
Female	123	71
Age		
Under 20	5	3
21 - 30	124	71
31 - 40	29	17
41 - 50	12	7
Above 50	4	2
Monthly income (RM)		
No specific income (student)	30	17
Below 1000	5	3
1001 - 3000	62	36
3001 - 5000	44	25
Above 5000	33	19
Total	174	100

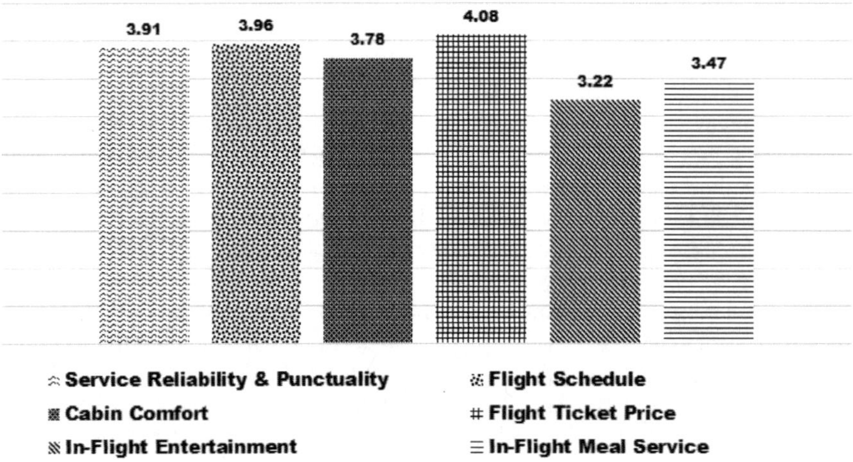

Figure 3. Considered factors when choosing airlines for travel

Figure 3 illustrates how the survey respondents rated the different factors that they consider when choosing the airlines for their travel. The rating is in a scale of 1 to 5, with 1 being the least important factor that they consider while 5 refers to the most important factor. As can be expected, flight ticket price has the highest rated importance with an average rating of 4.1, followed by the flight schedule and the airlines' service reliability and punctuality track records with an average rating of 4.0 and 3.9, respectively. Meanwhile, in-flight meal service is rated as 3.5 on average and hence can be taken to be of notable significance in relative to the other considered factors. Even though in-flight food delivery and waste collection service is not the most influential feature to be considered by most passengers while selecting their choice of airline for travel, it could still be the deciding factor. This is because in-flight meal service has a big impact on overall airline's service quality since it is directly provided to passengers through interactive encounters and occupies the largest amount of flight attendants' time with them [5]. As also argued in [6], experiences related to in-flight meal service can invoke emotional sentiments to the passengers and this can become a major factor for repeat customers. It should be noted that the questionnaire does not specify whether the selection factors are for short-haul or long-haul flights. It is believed that the factors may be rated slightly different if the type of flights has been specified. For instance, cabin comfort is expected to be rated much higher if the survey respondents are focused only on long-haul flights.

From the collected survey data, only 40% of the respondents considered the current in-flight food delivery and waste collection service to be efficient. On the other hand, almost half of the respondents felt that the service can be further improved. The respondents were asked to rate the current service in terms of eight performance characteristics as shown in Figure 4. Similar 1-to-5 rating scale is applied here, whereby a rating of 1 means that the current service is at the lowest level for the performance aspect while a rating of 5 corresponds to the current service being at the best level. Interestingly, the worst rated performance criterion is passengers' privacy, which primarily corresponds to the feeling that their privacy is disturbed when the flight attendants are serving other passengers in the same or nearby rows where they are seated. Furthermore, the respondents have also complained about the level of comfort and accident likelihood for the current in-flight food delivery and waste collection method. Nonetheless, it can be taken that most passengers are generally satisfied with the current service and this is reflected by the average rating of more than 3 for all of the measured performance criteria.

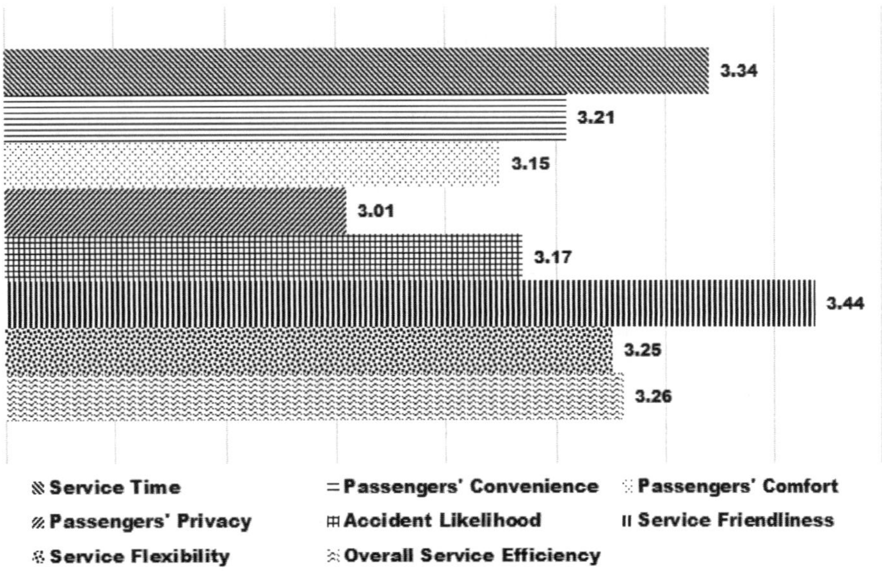

Figure 4. Performance ratings of current in-flight food delivery and waste collection service

Table 2 details out some of the negative experiences that the respondents had during their previous flight trips related to in-flight food delivery and waste collection service. It appears that the biggest comment made by the surveyed respondents is service waiting time. Depending on the rows that the passengers are seated, the waiting time can be different. More than 47% of the respondents stated that they had to wait more than 10 minutes on average for the service after it is started, which is a crucial factor when they were flying short haul flights. Secondly, in line with the rating results in figure 3 for passengers' privacy, many passengers felt that they were disturbed by the serving process, especially when they were skipping meals and wished to rest at their seat without being bothered. The situation is worse if the passenger is seated in the aisle or middle seat of the row since the service interaction with other passengers will have to go through his or her space. Other significant feedbacks from the public respondents also include missing their preferred meal options due to being serviced late and also late service by the flight attendants due to flight turbulence.

Table 2. Negative experiences regarding in-flight food delivery and waste collection service

Scenario	Percent (%)
The flight attendants could not serve your meals on-time because of flight turbulence.	28
The flight attendants served only some of the passengers since they did not have enough time to serve all passengers due to unfavourable flight conditions.	12
You had to wait long for your meals because the flight attendants started serving in other rows first.	48
There were incidents during the meal serving process such as spilled hot foods/drinks.	10
You were disturbed during your rest because other passengers in your row were being served (which you skipped to have your rest) or having their waste collected by the flight attendants.	40
You were forced to eat at the required time of meal serving even though you prefer to have it earlier or later.	20
You cannot rest comfortably after your meal because of the long wait for the flight attendants to collect the waste.	22
You had to wait to use the lavatory because the service cart was blocking the aisle during the meal serving and waste collection process.	25
The meal option that you wanted was no longer available because you were among the last passengers to be served.	30

3.2. Potential improvement of the service

With the comments regarding current in-flight food delivery and waste collection service method, the respondents were also asked several questions on the potential improvement that they would like to see implemented to the service. About 66% believe that it is more important for them to have a very efficient in-flight service instead of having plenty in-flight meal options. This indicates that the quality of service (including the quality of the food served although with limited choices) is more imperative to the passengers. For short haul flights, in which time to serve food and collect the waste afterwards is much more critical, more than 82% of the surveyed respondents were satisfied with only light meals or snacks.

When asked for their opinion on possible implementation of an automated food delivery and waste collection system onboard the aircraft in the future to make the in-flight service more efficient, 66% of the respondents agreed that it would be a great improvement effort. Only 4% were strictly against the idea since they prefer the current service method while the rest were either neutral or undecided about this revolutionary idea. Table 3 tabulates the importance rating given by the respondents for several characteristics of a new system that could improve the current in-flight service. A similar 1-to-5 rating scale is applied here, whereby a rating of 1 means that the criterion is the least important while a rating of 5 corresponds to the most important for the respondents. As can be observed from table 3, the most important factor for the passengers is safety. Improvement of the service should not expose passengers to higher accidental risk than the current method. In addition, once again the passengers' privacy issue is emphasized by the respondents with a high importance rating given on the system's capability to deliver the intended functions without disturbing other passengers.

There are indeed several inventions that have been developed for improving the food delivery and waste collection process, although none of them has made it to mass utilization onboard commercial transport aircraft. In fact, one of the earliest innovations to automate the process has been invented in 1965 by Martin Umanoff [9]. He has filed the patent of a cabinet that can move by guided cable along the aisle. The cabinet is capable of carrying a large number of separate meal trays such that in-flight meals maybe dispensed one after another. The main goal of this invention is to reduce the length of serving time and eliminate the frequent trips to the galley by the stewardess. However, the primary weaknesses of this machine are its added weight and the fact that it is still blocking the aisle during its operation. On the other hand, a recent patent filing of an automated device that can transfer meals to passengers and remove the waste is an encouraging development in improving the in-flight services. The patent describes a transport passage associated with an automated galley that is placed below the cabin floor and transports the food to the passenger cabin [10]. However, the foods and drinks that are delivered by this system still require the passengers to pass it among them in the same row.

Table 3. Importance rating of criteria for improvement

Characteristics	Rating
The system should be able to serve food at passengers' own leisure and not a fixed time.	3.18
The system should be able to deliver the food almost immediately after the passengers' request.	3.42
The system should be able to collect the waste at the passengers' own leisure and not at a fixed time.	3.43
The system should be able to deliver the food and collect the waste without discomforting other passengers.	3.72
The system should be designed for passengers' safety use by reducing the accidental risk.	3.88

4. Conclusion

Based on the results of the conducted online public survey, several conclusions regarding the in-flight food delivery and waste collection service can be drawn. It has been indicated that the service has an impact on the passengers' choice of airlines for their travel plan, which is reflected by its high rating from the survey respondents. However, the majority of respondents also points out that there are some improvements needed to increase the efficiency of current in-flight food delivery and waste collection service. The survey results highlight several potential process enhancements that could be made from the importance rating of the service criteria by the respondents. All in all, the information obtained from this survey study can be applied to either improve the current service or develop a new system for in-flight food delivery and waste collection process.

References

[1] International Air Transport Association 2015 *Air Passenger Forecasts: Global Report*
[2] Airbus 2016 *Global Market Forecast 2016-2035*
[3] Chan D 2000 *Journal of Management Development* **19** 489-514
[4] Park J, Robertson R and Wu C 2004 *Journal of Air Transport Management* **10** 435-9
[5] An M and Noh Y 2009 *Service Business* **3** 293-307
[6] Mohd Zahari M S, Salleh N K, Kamaruddin M S Y and Kutut M Z 2011 *International Journal of Social, Behavioral, Educational, Economic, Business and Industrial Engineering* **5** 1982-9
[7] O'hara L and Strugnell C 1997 *Journal of Nutrition and Food Science* **3** 105-6
[8] Jones P 2004 *Flight Catering* (Abingdon: Routledge)
[9] Martin U 1965 Patent No. US3179208 A
[10] Appolt B, Bitter M and Markus N 2016 Patent Application No. US 2016/0200439

A review on the fabrication method of bio-sourced hybrid composites for aerospace and automotive applications

M H Zin[1,2*], M F Razzi[3], S Othman[1,4], K Liew[1], K Abdan[5] and N Mazlan[3]

[1]Aerospace Malaysia Innovation Centre, German Malaysian Institute, Bangi, Selangor, Malaysia
[2]Institute of Tropical Forestry and Forest Products, Universiti Putra Malaysia, Serdang, Malaysia
[3]Department of Aerospace Engineering, Faculty of Engineering, Universiti Putra Malaysia, Malaysia
[4]Department of Mechanical Engineering, University of Malaya, Kuala Lumpur, Malaysia
[5]Department of Biological and Agricultural Engineering, Faculty of Engineering, Universiti Putra Malaysia, Malaysia

*hanafeezin@gmail.com

Abstract: Development of bio-sourced materials over the recent years has shown growing interests due to their eco-friendly characteristics. The combination of bio-sourced material such as kenaf, jute, sisal and many more into current synthetic fibres such as glass and carbon fibre, which is also known as hybrid composites, offers several significant benefits including sustainability, cost reduction, product variety and high specific mechanical properties. There are many methods used to fabricate composite parts nowadays. However, each method has its own requirement and usability. This review paper intends to focus on suitable technique to be adopted in order to fabricate bio-sourced hybrid composites. Some of the fabrication methods are customized in order to suit with the application of natural fibres. The selected methods are also highlighted with the application in aerospace and automotive industry. The process and outcomes are presented comparatively.

1. Introduction

In recent aerospace manufacturing, nearly 50% components of the aircraft are made from composites [1]. Composite material comprises of two or more different materials integrated through a certain process to obtain a resultant material with new physical and mechanical properties [2]. Resin, either from natural or synthetic polymer, is used as a binding mechanism that holds the composite together to achieve the desired strength [3]. The major benefit of composite components is strength-to-weight ratio, which is ideal for aerospace application [4]. Increasing global awareness on the environmental sustainability has led to numerous development on renewable materials such as natural fibres in the fabrication of composite parts. Hybrid composites, which is a combination of two or more materials in a common matrix [1], has been implemented since natural fibre composites do not have adequate strength to substitute conventional synthetic fibre [2]. By having hybrid composites, the advantages of one type of fibre could complement what is lacking in the other, resulting in a balance in performance and cost through proper material design [3]. The mechanical properties of natural fibre reinforced composite (NFRC) such as stiffness, strength and moisture resistant behaviour are also significantly improved by incorporation of stronger and more corrosion-resistant synthetic fibre such as glass or carbon fibre. A study by M. Ramesh *et al.* [4] indicated that hybridization of glass fibre into sisal/jute reinforced epoxy composites showed good tensile strength of 68.55MPa. Another study by M. Thwe *et al.* [5] depicted that incorporation of glass fibre up to 20% by mass with bamboo fibre increased tensile and flexural modulus by 12.5% and 10%, respectively. Sanjay [6] investigated flammability behaviour and degradation of PP/banana and glass fibre-based hybrid composites and found out that hybridization with glass improved flame retardant characteristic. T. Subash *et al.* [7] discussed about bast fibres reinforced hybrid composites for aircraft indoor structures applications. These materials provide the benefits in the making of the body panels such as in seat cushions, cabin linings, parcel shelves and many more. The natural fibres such as jute, kenaf, bagasse, bamboo, coir, sisal proved that these materials have a greater strength in aerospace and automotive industry. These composites show lower density as compared to metal composites and have a higher potential to make lightweight sustainable finished parts that can reduce tremendous amount of energy consumption in the industry.

There are various fabrication methods applicable for producing bio-sourced composites for usage in aerospace and automotive applications. This review paper will discuss processes involved using open and closed moulding technique. The correct selection of process, based on the geometry of the part, production scale, cost, and mechanical properties is vital in order to achieve optimum outcome for composite parts produced.

2. Composite fabrication method

Composite fabrication method in general can be divided into two main processes, namely open and closed moulding. For open mould technique, the top layer of the laminates and matrix are exposed to the atmosphere, resulting in uncontrolled surface condition [8]. Since the tooling fabrication process is relatively simple and low cost, rapid product development cycle is possible to be implemented using this method. This review will discuss three main processes under the open moulding method: manual hand lay-up, spray-up method and filament winding. On the other hand, closed moulding or also called liquid composite moulding (LCM) [9] is a process when fibres are laid on two-sided mould or within a vacuum bag. A liquid resin is supplied into the mould cavity to fill up the spaces within the laminates. Due to the fact that the process is normally automated and requires special equipment, it is capable of producing large part and high volume. The closed mould process that is discussed in this review include vacuum infusion process, resin transfer moulding, pultrusion, compression moulding and vacuum bag moulding. The fabrication techniques are summarized in Table 1. The selection of fabrication method depends on several factors such as material, resin system, part complexity and application. Table 2 shows suitability of fabrication method with respect to the production amount.

Table 1. Fabrication method for open and closed mould

Open Mould	Closed Mould
Hand Lay-Up	Vacuum Infusion Processing
Spray-up	Pultrusion
Filament Winding	Resin Transfer Moulding (RTM)
	Compression moulding
	Vacuum Bag Moulding

Table 2. Fabrication method according to volume production

Low Volume	Medium Volume	High Volume
Hand Lay-Up	Filament Winding	Compression Moulding
Vacuum Bagging	Resin Transfer Moulding	Pultrusion
Spray Up	Centrifugal Casting	Continuous Lamination
Vacuum Infusion Processing	Wet Lay Up Compression Moulding	Reinforced Reaction Injection Molding (RRIM)

2.1. Open mould

2.1.1. Hand lay up. Hand lay-up is the most popular fabrication process in composite manufacturing [10]. This process is widely practiced due to the low set up cost compared to other process like resin transfer moulding that requires high investment for tooling and material [8]. Hand lay-up involves application of laminate ply manually into an open mould until desired part thickness is achieved. The diagram of hand lay-up is shown in Figure 1. The use of different types of polymer is a determining factor for curing time. Epoxy resin, for instance, required 24 hours curing time at room temperature [11]. Hand lay-up is normally used with thermoset polymer based material. Although hand lay-up requires less capital investment, it is limited to relatively low production rate and low volume fraction of reinforcement [12]. Various aircraft and automotive parts can be fabricated using this method including aircraft seats, non-structural parts, car dashboards and also interior cabin parts [13]. The characterization of hand lay-up, its advantages and disadvantages are summarized in Table 3.

2.1.2. Spray up method. Spray up is a process where compressed air supplied through the spray gun is used to spray both resin and fibre onto an open mould [23]. The fibre, usually in yarn form, is cut simultaneously and mixed with the sprayed resin at the tip of the spray nozzle before both of them fall into the mould. In order to remove the air trapped in between the layers, a roller is usually applied over the sprayed material. The curing depends on the resin system used, which is normally occurs at room temperature. The cured part is detached from the mould for further manufacturing process. The process might be repeated in order to achieve desired part thickness. Due to the nature of the process, the fabrication method is efficient, hence easy to accommodate design alteration. The final quality of the composite part produced by spray up technique depends heavily on the worker's skill. In terms of application, this method is more suitable for lower load carrying components such as boats and trucks fairing. Spray-up method can accommodate a variety of part size, as well as capable of providing high volume fraction of reinforcement. Figure 2 illustrates the set up for spray-up process.

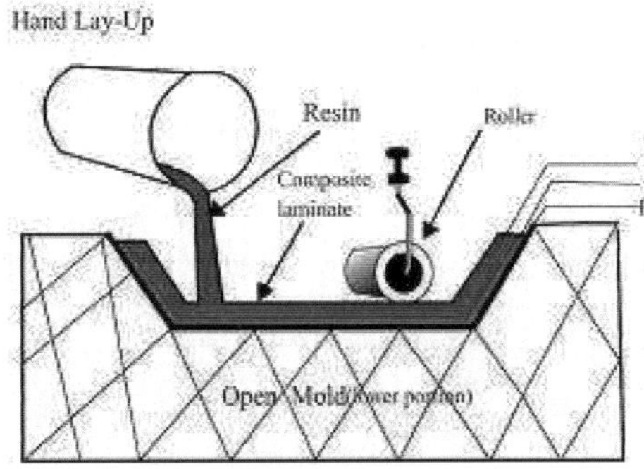

Figure 1. Hand lay-up set up

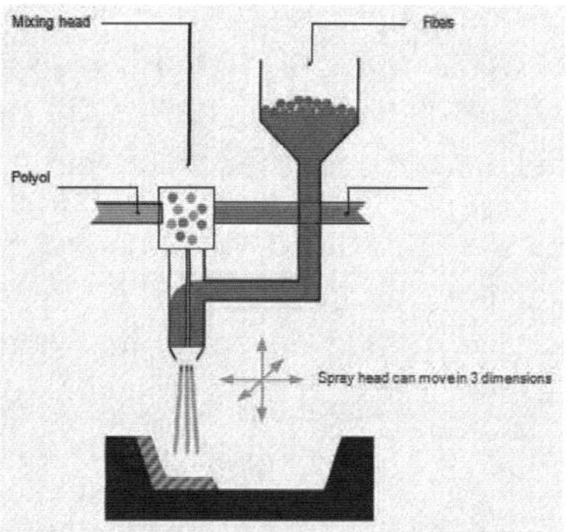

Figure 2. Schematic diagram of spray up method

Table 3. Characteristics of hand lay-up process

Resins	Fibres	Advantages	Disadvantages
epoxy, polyester, vinylester, phenolic [12, 14, 15]	Any, e.g. glass, carbon, aramid [11, 12, 16], jute [17], kenaf [11], coconut spathe [18], sisal, cotton, banana, oil palm, bamboo, wheat, rice, and bagasse [19, 20]	1. Most famous conventional method [21] 2. Relatively easy to learn process [20] 3. Low capital investment [12] 4. Variety of material types [12] 5. Higher fibre volume, and longer fibres size [22]	1. Resin mixing is very dependent on the skills of operators [23] 2. Health and safety considerations of resins [24] 3. High styrene contents of resins [22] 4. Require low-viscosity resin[10] 5. Require long curing times and the mould release characteristics are poor [25]

Although spray-up is an effective fabrication method for composites, exploration of mechanical properties for parts produced using this method is still limited. In a study by Xiao *et al.* [17], 3-point bending test was conducted to study bending properties of laminated plate with the detached acrylic resin and glass reinforced unsaturated polyester layer (GFRP). The acoustic emission, as well as thickness calculation of jute fibre reinforced plastic (JFRP), were also measured simultaneously. The experiments illustrated that JFRP laminates can obtain 75% bending modulus with the same thickness as the laminate with GFRP, suggesting possibility for JFRP to replace GFRP in industrial application. Table 4 summarizes the characteristic of spray up method.

Table 4. Characteristic of spray up method

Resins	Fibres	Advantages	Disadvantages
Matrix epoxy, polyester, polyvinyl ester, phenolic resin, unsaturated polyester, polyurethane resin	Glass fibre, carbon fibre, aramid fibre, natural plant fibres (sisal, banana, nettle, hemp, flax, coir, cotton, jute etc.) (all these fibres are in the form of chopped short fibres, flakes, particle fillers etc.)	1. Low investment for high volume capacity 2. High fibre volume	1. Inconsistent part thickness 2. Difficult to obtain thorough resin distribution

Another research done by Kikuchi *et al.* [23] on mechanical properties of jute composite by spray up fabrication reported large variation of tensile strength despite stable elastic modulus values. It was depicted that the quality of products is proportional to the operator's skill. There are several factors affecting the outcome of spray up process. Liakus *et al.* [26] investigated microstructure-property predictions for short fibre reinforced composite structures based on a spray-up process. In the study, variations in fibre length, as well as spray disposition pattern have been identified as two major factor contributing to the mechanical strength of the composite part produced.

2.1.3. Filament winding. Filament winding is mainly used for fabricating open (cylinders) or closed end structures (pressure vessel or tanks) due to high stiffness-to-weight ratios. In a filament winding procedure, the fibre tows are wetted in a resin bath before being wound onto a mandrel in different orientations. The winding process is controlled by fibre feeding mechanism and rate of rotation of the mandrel. The schematic diagram of typical filament winding process is illustrated in Figure 3.

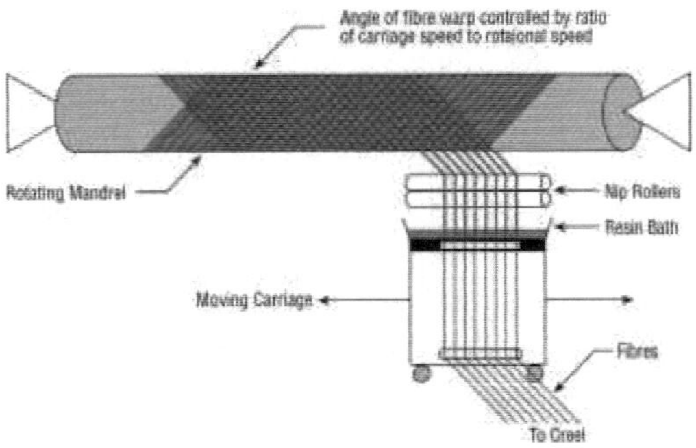

Figure 3. Schematic diagram of typical filament winding process

One of the well-known product of filament winding process is composite overwrapped pressure vessels (COPVs), which are vital to spacecraft propulsion, attitude control systems and life support applications. The COPV design requires integration between the analysis of the liner and the fibre overwrap [27]. Ductile materials are usually used as the liners, such as soft aluminium, with only minimal load-sharing capabilities. The fibre is generally applied as ribbon of multiple tows wetted in resin bath. In previous research conducted by Madhavi *et al.* [28] on design and analysis of filament wound COPV with integrated-end domes, material characterization of FRP of carbon T300/Epoxy for various configurations are determined using filament winding technique. It is found out that having alternate hoop and helical layers with hoop layers as the top and bottom most layers, gave the burst value of 12.4 MPa in the cylindrical zone. Vargas Rojas *et al.* [29], on the other hand, developed a numerical computing tool to simulate the filament winding process for various mandrel geometry, including axisymmetric mandrel. The numerical tool developed solves the general path equation that could be used as a basis for further mechanical behaviour models by integrating onto the filament winding parameters. Cherniaev *et al.* [30] conducted both experimental and numerical study of hyper-velocity impact (HVI) damage by orbital debris in the composite materials fabricated by filament winding. The result showed that the filament winding pattern with higher degree of interweaving of filament bands demonstrated better ability to hinder the dissemination of HVI induced damage, even in the presence of pre-loading. The characteristic of filament winding process is presented in Table 5.

Table 5. Characteristic of filament winding process.

Resins	Fibres	Advantages	Disadvantages
Matrix epoxy, polyester, polyvinyl ester, phenolic resin	Glass fibre, carbon fibre, aramid fibre, natural plant fibres (the fibres are used straight from a creel and not woven or stitched into a fabric form.)	1. Economic way of laying material down 2. Resin usage can be controlled 3. Minimum fibre cost 4. Good structural properties of laminates	1. Limited to convex shaped components 2. Difficult to lay fibre exactly along the length of component 3. High mandrel cost for large components

2.2. Closed mould

2.2.1. Vacuum infusion. Vacuum infusion refers to a process when the resin is driven into a laminate using a vacuum pressure [31]. Reinforcement are draped onto a mould and flow media is placed on top of it to allow equal distribution of resin throughout the laminates [32]. The vacuum is applied prior to introduction of resin. Once a complete vacuum is achieved, which is normally at 25 mmHg, resin is allowed to enter into the laminate via carefully placed tubing. Typical resin set up is shown in Figure 4.

Previous study done by Kim *et al.* [12] on hybrid glass fibre reinforce plastic (GFRP) showed that the vacuum infusion process resulted in higher ultimate strength and modulus, shear strength, and displacement over thickness value than the hand layup samples in both tension and compression tests. Another investigation carried out by Scalici *et al.* [33] on basalt fibre reinforced composites using different vacuum assisted impregnation techniques suggested that vacuum assisted resin infusion exhibited lower aptitude to thoroughly impregnate the laminates as compared to the vacuum bagging process. This results in higher fibre volume fraction, hence yielded in higher stiffness values for the composite part. However, the lack of full impregnation of the dry fabric architecture could result in under-performance outcome due to low fibre/matrix adhesion. The characteristic of vacuum infusion process is described in Table 6.

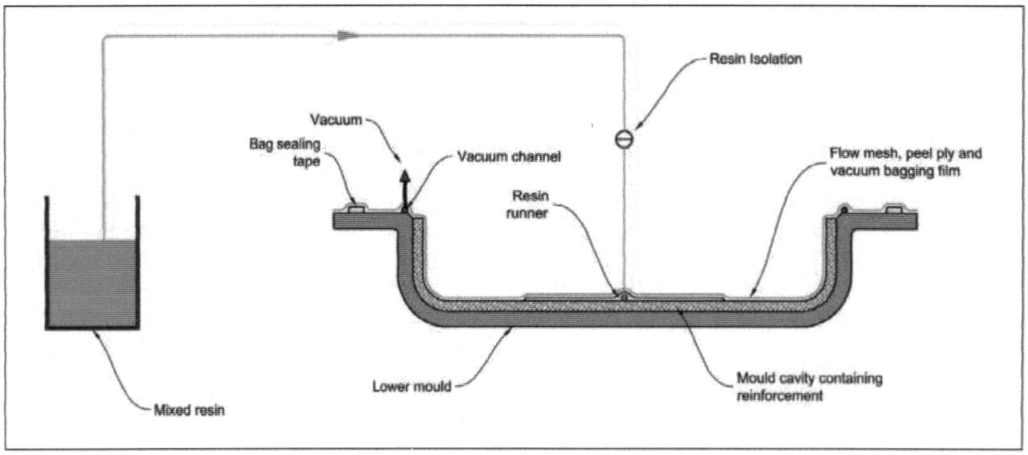

Figure 4. Schematic diagram of vacuum resin infusion process

Table 6. Characteristic of vacuum infusion process

Resins	Fibres	Advantages
Ortho, Iso, DCPD, Vinyl Ester [31], Epoxy	Any, e.g. glass, carbon, aramid, jute, kenaf, coconut spathe, sisal, cotton, banana, oil palm, bamboo, wheat, rice and bagasse	1. Better fibre-to-resin ratio 2. Less wasted resin 3. Consistent resin usage 4. Light weight, due to minimum resin contain 5. Lower tooling cost 6. Large components can be fabricated

Yenilmez *et al.* [34] investigated about variation of part thickness and compaction pressure in vacuum infusion process. Getting a consistent part thickness is a big challenge in vacuum infusion process due to the effect of varying compaction pressure on the upper mould part, a vacuum bag. Based on the experiments, it can be summarized that the final part thickness depends on the durations of initial vacuuming and gelation time for the resin, resin pressure at the location, and shrinkage ratio of the resin system used. Md Afendi *et al.* [35] studied about degassing process prior to resin infusion for reducing void formation from gasses dissolved in the resin. At 90 mbar, this method of degassing may eliminate 40-50% of the dissolved gas with the assistance of bubble thin films.

2.2.2. Pultrusion. Pultrusion is a manufacturing process for hybrid composite that offers continuous production of profile with a constant cross section [36]. In the thermoplastic pultrusion molding, resin and reinforcement fibres are pulled through a heat die. As the heat and pressure generated, resin starts to melt and impregnate into the reinforcement fibre. There are several parameters such as preheating method, die temperature and pulling speed that represent determining factors regarding a potential industrial application. The normal set up of pultrusion is shown in Figure 5 and Figure 6.

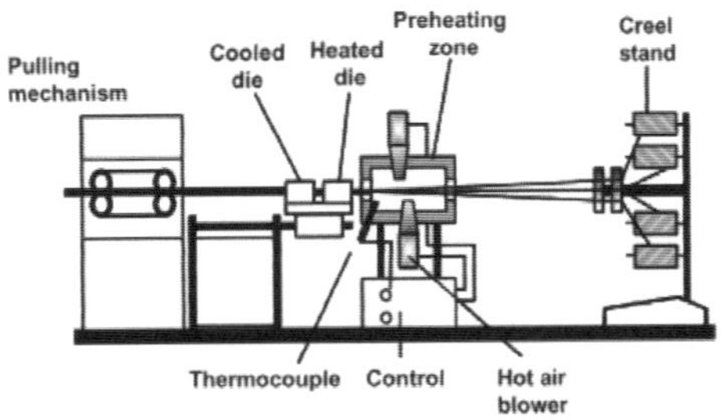

Figure 5. Schematic diagram of pultrusion process

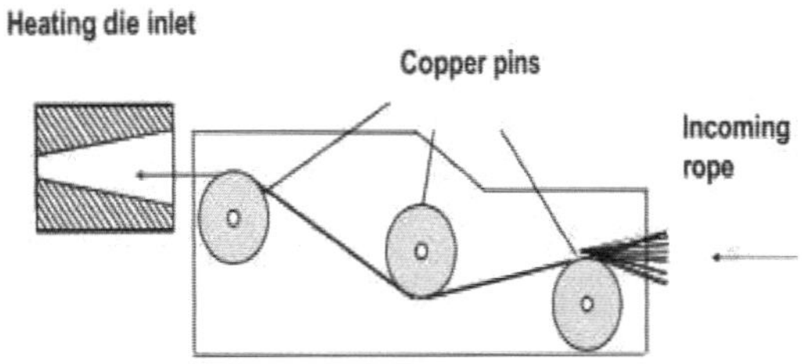

Figure 6. Pulling mechanism in pultrusion process

There are a number of studies have been carried out on pultrusion process to fabricate composite parts. Memon and Nakai [37] investigated the mechanical properties of jute spun yarn reinforced PLA tubular braided composite using pultrusion moulding. The experimental results showed that the configuration of the intermediate material and the moulding temperature influenced the mechanical properties of the part produced. It was reported that the impregnation quality increased with the size of moulding and increasing temperature reduced the mechanical properties due to deterioration of jute fibre. Another study was carried out by Peled and Mobasher [38] to investigate the use of pultrusion technique as a cost-effective method for production of thin-sheet fabric reinforced cement composites. The pultrusion process required a relatively simple set-up using low cost equipment while at the same time it was able to maintain laminates alignment, resulting in relatively smooth surface and uniform products. It was found that pultrusion had significantly improve the mechanical performance of the cement composites compared with cast composite. The mobilization of the filaments in the pultrusion process yielded in a strain hardening composite even when the yarn modulus is relatively low. The characteristic of pultrusion process is explained in Table 7.

Table 7. Characteristic of pultrusion process

Resins	Fibres	Advantages	Disadvantages
Epoxy, polyester, vinylester, phenolic	Any, e.g. glass, carbon, aramid, jute, kenaf, hemp, coconut spathe, sisal, cotton, banana, oil palm, bamboo, wheat, rice, and bagasse [39]	Fast and economic way of impregnating and curing material, minimised fibre cost, good laminates structural properties, enclosed resin impregnation area, thus limiting volatile emissions	Limited to constant or near constant cross-section components and high cost for heated die

Properties of hemp fibre composite made by pultrusion process was studied by Peng *et al.* [39]. Natural fibres generally depend on factors such as source, age, retting and separating techniques that give rise to difficulties in the property control of their composites. The diversity of compatibility to adhere to matrix required additional process such as treatments and chemical modification [40]. The result of the experiment showed that the thermosetting polymer-based composites reinforced with hemp and wool fibres have higher tensile strength compared to the common plastics at a comparable density. With the benefits of low cost, availability, and process, the polyester composite made by pultrusion process would be a good choice for manufacturing construction board, insulation board, or being used as reinforcement for other structural materials.

2.2.3. Resin transfer moulding. Resin transfer moulding (RTM) is a closed-moulding method in which reinforcement material is laid into a closed mould and resin is pumped in (through injection ports) under pressure. This process produces complex parts with smooth finishes on all exposed surfaces [41]. The lay-up of materials onto the dry mould enables any combination of different materials and orientation. The desired part thickness can be obtained by setting up the tool cavity volume and the process can achieve a fast cycle time in temperature controlled tooling. Typical set up for RTM is as illustrated in Figure 7.

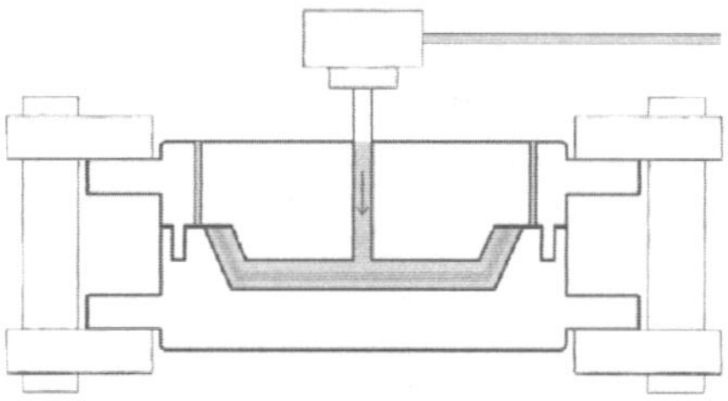

Figure 7. Schematic diagram of resin transfer moulding

Vacuum assist can be used to improve resin flow inside the mould cavity. There is a wide range of tooling variety, from low-cost composite to temperature controlled metal tooling that can be used for RTM. Mould release agent is applied prior to positioning the reinforcement (and core material) onto the mould. As the mould is closed and clamped, the resin is injected under pressure using mix/meter injection equipment and the part is cured in the mould. RTM can be done at room temperature but the heated moulds are required to achieve fast cycle times and also product consistency. Clamping can be accomplished by perimeter clamping or press clamping.

There have been many developments of hybrid composite being done using RTM process. Liu *et al.* [42] investigated the mechanical performance and failure mechanism of thick-walled composite connecting rods using RTM technique. From the study, it was recorded that the tensile failure was originated from the delamination at the round angle transition areas of the T-joints at either end of the connecting rod and the failure strength was determined by the interlaminar strength. Meanwhile, the compressive failure was due to the fracture of fibres in the main bearing beam, hence determining the failure strength of the structure. The characteristic of resin transfer moulding process is shown in Table 8.

Table 8. Characteristic of resin transfer moulding

Resins	Fibres	Advantages	Disadvantages
Epoxy, polyester, vinylester, phenolic, bismaleimides	Any, e.g. glass, carbon, aramid, jute, kenaf, hemp, sisal, cotton, banana, and bagasse, stitched material	1. High fibre volume and low void contents 2. Controlled surface on both sides of panel 3. Good health and safety, environmental control	1. Expensive tooling cost 2. Limited to smaller components 3. Unimpregnated areas can occur resulting in costly scrapped parts

Another research on mechanical and damping properties of RTM jute-carbon hybrid composites was carried out by Ashworth *et al.* [43]. The study investigated carbon (CFRP), jute (NFRP) and hybrid (HFRP) fibre reinforced polymers fabricated using RTM process. Kim and Lee [44] studied about structural behaviour evaluation of composite side beams by comparing between autoclave cure and RTM method. It was reported that the side beam made by RTM appeared to be slightly softer than the autoclave cured one due to lower fibre volume fraction. The environmental resistance of flax/bio-based epoxy and flax/polyurethane composites manufactured using RTM method was investigated by Cuinat-Guerraz *et al.* [45]. From the study, they found that for cellulosic fibres reinforced composites, the usage of polyurethane resin leads to environmental stability, durability and material's moisture resistance. It was also illustrated that the nature of the fibre/matrix interface played a key role in the moisture absorption mechanisms. Papargyris *et al.* [46] compared mechanical and physical properties of a CFRP manufactured by RTM using conventional and microwave heating and found out that there was a 9% increase in the interlaminar shear strength (ILSS) for the microwave cured composites in comparison to the conventional ones.

2.2.4. Compression moulding. Compression moulding refers to a process where composite materials are laid between two matching moulds under high pressure (up to 2,000 psi) and heat (from 120°C to 200°C) in order to cure the part. This technique is used to rapidly cure large quantities of complex fibreglass-reinforced polymer parts [47]. There are several types of compression moulding that are defined by type of moulded materials: bulk moulding compound (BMC), thick moulding compound (TMC), sheet moulding compound (SMC) and wet lay-up compression moulding. Matched metal dies are mounted in large hydraulic moulding press. Compression moulding enables part design flexibility and features such as inserts, ribs, bosses and attachments. Typical set up of compression mould is illustrated in Figure 8.

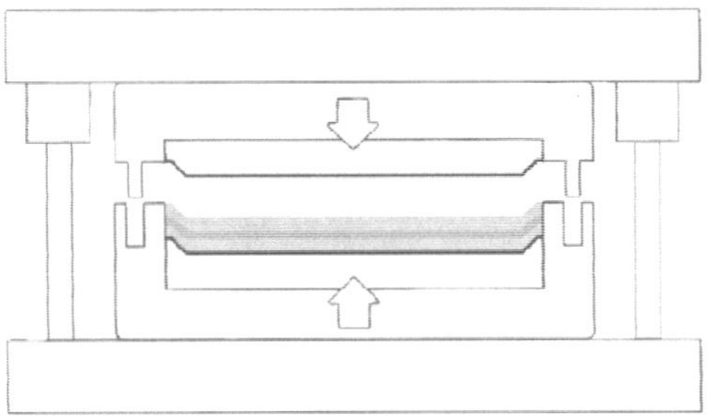

Figure 8. Set up for compression moulding

Curing time normally depends on shape of the part, size and thickness. Cured parts are removed from the mould for the next process. Compression moulding offers one of the fastest routes for the manufacturing of the composite components from thermoset matrices. This makes sheet moulding compounds (SMC) the most widely adopted material format of fibre reinforced composites within the automotive industry, accounting for 70% of composites by mass. Numerous works have been carried out to study compression moulding method as a mean to fabricate composite parts. Wulfsberg *et al.* [48] combined the carbon fibre sheet moulding with prepreg compression and found out that the combination yielded load-bearing and autoclave-quality component without being cured in autoclave.

IOP Conf. Series: Materials Science and Engineering **152** (2016) 012041 doi:10.1088/1757-899X/152/1/012041

However, each material combination has to be properly developed to comply with the requirement for particular application. There are also some parameters that need to be looked into such as flowability, moulding pressure, moulding temperature and time in order to ensure consistency in parts produced. The characteristic of compression moulding is illustrated in Table 9.

Table 9. Characteristic of compression moulding

Resins	Fibres	Advantages	Disadvantages
Epoxy, polyester, vinylester, phenolic, bismaleimides	Any, e.g. glass, carbon, aramid, jute, kenaf, hemp, sisal, cotton, banana, and bagasse, stiched material	1. Low initial set up cost 2. Fast set up time, allow intricate parts 3. Good surface finish, can apply to composite thermoplastics with unidirectional tapes, woven fabrics, randomly orientated fibre mat or chopped strand	1. Expensive tooling cost 2. Limited largely to flat or moderately curved parts with no undercuts 3. Less than ideal product consistency

In terms of the surface analysis, there is a study carried out by Serré *et al.* [49] pertaining to the morphology of the outermost layer of a compression moulded composite based on the unsaturated polyester resin. Wakeman *et al.* [50], on the other hand, conducted a systematic study of the effects of charge composition and process parameters for automotive part (door cassette) produced through the compression moulding from glass fibre and polypropylene composites.

2.2.5. Vacuum bag moulding. Vacuum bagging refers to a process that applies mechanical pressure on a laminate during its cure cycle. Vacuum bagging is an extension to the current hand lay-up. It serves important role to remove any trapped air in between the laminates, securing the position of fibre orientation during curing, reduce humidity and optimize fibre to resin ratio in the composite part. The elements consist in vacuum bagging system include vacuum pump, fitting and connectors, vacuum gauge, bagging film, sealant tape, release film, peel ply, breather, bleeder and vacuum tubing. In a typical vacuum bagging process, a release film or peel ply is draped onto the mould as first layer, followed by composite laminates, release film, bleeder, breather and bagging film as the final layer. Vacuum gauge is then installed on the bagging film and vacuum suction pressure is applied up to normally 25 mmHg. Normal set up of vacuum bagging is shown in Figure 9.

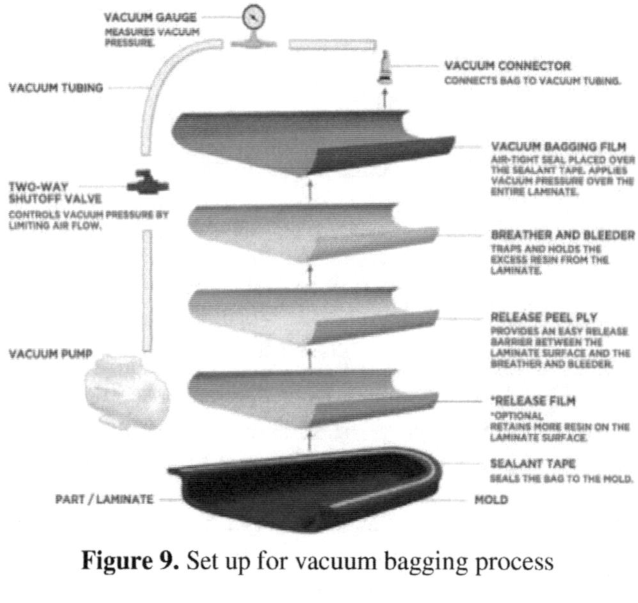

Figure 9. Set up for vacuum bagging process

Present study by Dominguez and Madsen [51] on development of new biomass-based furan/glass composites fabricated by the double-vacuum-bag technique showed that, in the optimal case of no matrix porosity, the composites yielded a well comparable stiffness with conventional thermosetting / glass composites, despite lower strength. Hamill *et al.* [52] investigated the surface porosity factors during vacuum bagging process and identified several parameters that affect air evacuation including freezer and out time, vacuum hold time, material and process modifications. The characteristic of vacuum bagging process is illustrated in Table 10.

Table 10. Characteristic of vacuum bagging process

Resins	Fibres	Advantages	Disadvantages
Epoxy, polyester, vinylester, phenolic,	Glass, carbon, aramid, jute, kenaf, hemp, sisal, cotton, banana, and bagasse, stiched material	1. Higher fibre content laminates 2. Lower void content 3. Reduce the amount of volatiles emission	1. Additional costs for bagging material 2. Higher level of skill operator 3. Resin usage depend highly on operator skill

Zahid and Chen attempted manufacturing of single piece textile reinforced riot helmet shell using vacuum bagging technique [53]. They managed to develop a technique to manufacture single-piece riot helmet shells that is usable for the development of different varieties of bi-curvature composite structures and almost any composite geometry. They also conducted a research on impact evaluation of Kevlar-based angle-interlock woven textile composite structures using vacuum bagging method [54]. The result indicated that laminate thickness plays important role in energy absorption and force transmission capabilities. Mariatti *et al.* [55] focused on investigating the properties of unsaturated polyester reinforced with banana and pandanus woven fabric. It was observed that banana woven composites show better flexural and impact properties as compared to pandanus woven fabric due to different cross section structure and higher cellulose content. As the fibre volume fraction increased, both flexural modulus and strength rose up until the maximum limit of 15 volume fraction of fibres. Sudheer *et al.* [56], on the other hand, investigated the effect of potassium titanate whiskers on the mechanical and tribological properties of glass/epoxy composites. From their study, it was reported that the addition of potassium titanate whiskers has significantly improve the property of glass/epoxy composites such as density, hardness, stiffness, wear resistance as well as friction coefficient.

3. Application

Summary of application for open and closed moulds is listed in Table 11 and Table 12, respectively.

Table 11. Application of open mould method

Hand Lay-up	Spray Lay-up	Filament Winding
Aircraft seats, non-structural parts, car dashboards and interior cabin parts	Lightly loaded structure, car bonnet, truck fairings, car bodies	Composite overwrapped pressure vessels (COPVs), pipelines

Table 12. Application of closed moulding method

Vacuum Infusion	Pultrusion	Resin Transfer Moulding	Compression Moulding	Vacuum Bag Moulding
Vehicle floor, interior cabin part, customized car body panels	Rods, tubes, interlocking floors, stiffeners, fencing, simple beams	Vehicle seat, spar, rib element, hovercraft rotor blade	Vehicle body parts, jet engine parts, nacelle structure	Helmet, door panels, vehicle body parts

4. Future overview

A significant leap in the development of composite manufacturing technologies might not be really foreseeable in the near future. However, a lot of improvement in the current manufacturing method is currently being studied and a few new manufacturing concept are in the progress. The introduction of the automated tape lay-up (ATL) and the automated fibre placement (AFP), for examples, indicates the integration of computer-guided-robotic with current composite material to perform lay-up process for large composite parts such as aircraft wing skin and fuselages. With this development, it opens up another opportunity for current manufacturing method to be integrated with automation such as spray-up and filament winding. With the view that natural fibres are becoming more substantial nowadays to replace depleting petroleum based products such as carbon fibres, emphasize should be given to the development of manufacturing technology to suit with bio-sourced hybrid composite material. This is to ensure the sustainability of the whole manufacturing ecosystem as well as the dependency on the synthetic materials.

5. Conclusion

This study sums up the current manufacturing of hybrid composites for both open and closed mould techniques. Each method is unique and some of them can be combined or integrated to achieve better results. For large size aircraft components, methods such as hand lay-up, spray-up, vacuum bagging and vacuum infusion are suitable due to their ability to suit with a big mould and working area. Meanwhile, for intricate geometry part, methods such as pultrusion, compression moulding and resin transfer moulding are more appropriate due to the pressure applied during fabrication process. The development of the composite manufacturing technologies should also take into account the growing importance of natural fibres as a substitution for current synthetic materials.

Acknowledgement

This study was carried out under joint collaboration between Aerospace Malaysia Innovation Centre, Universiti Putra Malaysia and University of Nottingham Malaysia Campus, entitled “Sustainable Aviation”, project number BF-P06.

References

[1] Boopalan M, Niranjanaa M and Umapathy M J 2013 *Compos. Part B Eng.* **51** 54–7
[2] Arthanarieswaran V P, Kumaravel A and Kathirselvam M 2014 *Mater. Des.* **64** 194–202
[3] Sanjay M R, Arpitha G R and Yogesha B 2015 *Mater. Today Proc.* **2** 2959–67
[4] Ramesh M, Palanikumar K and Reddy K H 2013 *Procedia Eng.* **51** 745–50
[5] Thwe M M and Liao K 2002 *Compos. - Part A Appl. Sci. Manuf.* **33** 43–52
[6] Nayak S K 2009 *International Journal of Plastics Technology* **13** 47–67
[7] Gupta G, Kumar A, Tyagi R and Kumar S 2016 *Int. J. Innov. Res. Sci. Technol.* **5** 6907–11
[8] Sevkat E, Brahimi M and Berri S 2012 *Polym. Polym. Compos.* **20** 321–32
[9] Sas H S, Šimáček P and Advani S G 2015 *Compos. Part A Appl. Sci. Manuf.* **78** 223–33
[10] Kozlowski R, Wladyka-przybylak M, Helwig M and Kurzydloski K 2004 *Molecular Crystals and Liquid Crystals* 418 131–51
[11] Yahaya R, Sapuan S M, Jawaid M, Leman Z and Zainudin E S 2015 *Def. Technol.* **12** 52–8
[12] Kim S Y, Shim C S, Sturtevant C, Kim D D W and Song H C 2014 *Int. J. Nav. Archit. Ocean Eng.***6** 723–36
[13] Mohd Yuhazri Y and Phongsakorn P 2012 *Int. J. Basic Appl. Sci.* **10** 54–7
[14] Saxena M and Pappu A 2011 *Adv. Compos. Mater. - Anal. Nat. Man-Made Mater.* 121–62
[15] Yuhazri M, Phongsakorn P T and Sihombing H 2010 *International Journal of Basec and Applied Sciences* **10** 54–7
[16] Hollaway L C 2010 *Constr. Build. Mater.* **24** 2419–45
[17] Xiao B, Yang Y, Wu X, Liao M, Nishida R and Hamada H 2015 *Fibers Polym.* **16** 1759–65
[18] Vijayakumar S, Nilavarasan T, Usharani R and Karunamoorthy L 2014 *Procedia Mater. Sci.* **5** 2330–7
[19] Asaithambi B, Ganesan G and Ananda Kumar S 2014 *Fibers Polym.* **15** 847–54

[20] Abdul Khalil H P S, Davoudpour Y, Saurabh C K, Hossain M S, Adnan A S, Dungani R, Paridah M T, Islam Sarker M Z, Fazita M N, Syakir M and Haafiz M K M 2016 *Renew. Sustain. Energy Rev.* **64** 823–36
[21] Sathishkumar T, Naveen J and Satheeshkumar S 2014 *J. Reinf. Plast. Compos.* **33** 454–71
[22] Quan Z, Larimore Z, Wu A, Yu J, Qin X, Mirotznik M, Suhr J, Byun J H, Oh Y and Chou T W 2016 *Compos. Sci. Technol.* **126** 139–48
[23] Kikuchi T, Tani Y, Takai Y, Goto A and Hamada H 2014 *Energy Procedia* **56** 289–97
[24] Thori P, Sharma P and Bhargava M 2013 *Int. J. Res. Eng. Technol.* **2** 350–5
[25] Deo C R 2010 *Preparation and Characterization of Polymer Matrix Composite Using Natural Fiber Lantana-Camara* PhD Thesis National Institute of Technology, India
[26] Liakus J, Wang B, Cipra R and Siegmund T 2003 *Compos. Struct.* **61** 363–74
[27] McLaughlan P B, Forth S C and Grimes-Ledesma L R 2011 *Composite Overwrapped Pressure Vessels, A Primer* National Aeronautics and Space Administration
[28] Madhavi M, Rao K V J and Rao K N 2009 *Defence Science Journal* **59** 73–81
[29] Vargas Rojas E, Chapelle D, Perreux D, Delobelle B and Thiebaud F 2014 *Compos. Struct.* **116** 805–13
[30] Cherniaev A and Telichev I 2016 *Int. J. Impact Eng.* **98** 19–33
[31] Osborne T 2014 *Reinf. Plast.* **58** 25–9
[32] Correia N C, Robitaille F, Long A C, Rudd C D, Šimáček P and Advani S G 2005 *Compos. Part A Appl. Sci. Manuf.* **36** 1645–56
[33] Scalici T, Pitarresi G, Badagliacco D, Fiore V and Valenza A 2016 *Compos. Part B Eng.* **104** 35–43
[34] Yenilmez B, Senan M and Murat Sozer E 2009 *Compos. Sci. Technol.* **69** 1710–9
[35] Afendi M, Banks W M and Kirkwood D 2005 *Compos. Part A Appl. Sci. Manuf.* **36** 739–46
[36] Wiedmer S 2006 *J. Thermoplast. Compos. Mater.* **19** 97–112
[37] Memon A and Nakai A 2013 *Energy Procedia* **34** 818–29
[38] Peled A and Mobasher B 2006 *Mater. Struct.* **39** 787–97
[39] Peng X, Fan M, Hartley J and Al-Zubaidy M 2012 *J. Compos. Mater.* **46** 237–46
[40] Corrales F, Vilaseca F, Llop M, Girones J, Mendez J A and Mutje P 2007 *J. Hazard. Mater.* **144** 730–5
[41] Pearce N R L, Summerscales J and Guild F J 2000 *Compos. Part A Appl. Sci. Manuf.* **31** 1433–41
[42] Liu G, Luo C, Zhang D, Li X, Qu P, Sun X, Jia Y and Yi X 2015 *Appl. Compos. Mater.* **22** 423–36
[43] Ashworth S, Rongong J, Wilson P and Meredith J 2016 *Compos. Part B Eng.* **105** 60–6
[44] Kim J S and Lee W G 2012 *Int. J. Precis. Eng. Manuf.* **13** 723–30
[45] Cuinat-Guerraz N, Dumont M and Hubert P 2016 *Composites Part A: Applied Science and Manufacturing* **88** 140-7
[46] Papargyris D A, Day R J, Nesbitt A and Bakavos D 2008 *Compos. Sci. Technol.* **68** 1854–61
[47] Kandar M I M and Akil H M 2016 *Procedia Chem.* **19** 433–40
[48] Wulfsberg J, Herrmann A, Ziegmann G, Lonsdorfer G, Stöß N and Fette M 2014 *Procedia Eng.* **81** 1601–7
[49] Serré C, Vayer M, Erre R and Ollive C 1999 *J. Mater. Sci.* **34** 4203–8
[50] Wakeman M D, Rudd C D, Cain T A, Brooks R and Long A C 2000 *Compos. Sci. Technol.* **60** 1901–18
[51] Dominguez J C and Madsen B 2015 *J. Compos. Mater.* **49** 2993–3003
[52] Hamill L, Centea T and Nutt S 2015 *Composites Part A: Applied Science and Manufacturing* **75** 1–10
[53] Zahid B and Chen X 2012 *J. Compos. Mater.* **47** 2343–51
[54] Zahid B and Chen X 2013 *J. Reinf. Plast. Compos.* **32** 925–32
[55] Mariatti M, Jannah M, Abu Bakar A and Khalil H P S A 2008 *J. Compos. Mater.* **42** 931–41
[56] Sudheer M, Vishwanathan K, Raju K and Bhat T 2013 *J. Reinf. Plast. Compos.* **32** 1177–87

Preliminary assessment of Malaysian micro-algae strains for the production of bio jet fuel

J T Chen[1,2*], E M Mustafa[3], V Vello[3], P Lim[3], N M Nik Sulaiman[3], N Abdul Majid[3], S Phang[3], P Md. Tahir[2] and K Liew[1]

[1]Aerospace Malaysia Innovation Centre (AMIC), Level 1, MIGHT Building, Cyberjaya, Malaysia
[2]Centre of Excellence on Biomass Valorization for Aviation, Institute of Tropical Forestry and Forest Products, Universiti Putra Malaysia, Malaysia
[3]Institute of Ocean and Earth Sciences, University of Malaya, Malaysia

*brenna@amic.my

Abstract. Malaysia is the main hub in South-East Asia and has one of the highest air traffic movements in the region. Being rich in biodiversity, Malaysia has long been touted as country rich in biodiversity and therefore, attracts great interests as a place to setup bio-refineries and produce bio-fuels such as biodiesel, bio-petrol, green diesel, and bio-jet fuel Kerosene Jet A-1. Micro-algae is poised to alleviate certain disadvantages seen in first generation and second generation feedstock. In this study, the objective is to seek out potential micro-algae species in Malaysia to determine which are suitable to be used as the feedstock to enable bio-jet fuel production in Malaysia. From 79 samples collected over 30 sites throughout Malaysia, six species were isolated and compared for their biomass productivity and lipid content. Their lipid contents were then used to derived the require amount of micro-algae biomass to yield 1 kg of certifiable jet fuel via the HEFA process, and to meet a scenario where Malaysia implements a 2% alternative (bio-) jet fuel requirement.

1. Introduction

Malaysia has long been touted as country rich in biodiversity and therefore, attracts great interests as a place to setup bio-refineries and produce bio-fuels such as biodiesel, bio-petrol, green diesel, and bio-jet fuel Kerosene Jet A-1. Numerous studies have been performed in Malaysia to determine suitable waste biomass, and among them, suitable sustainable biomass waste, as a feedstock to convert to other valued products such as bio-fuel [1]. However, the consensus of these studies seems to be in relative competition with each other for the waste biomass, especially since the industrial players' involvement to utilize a "waste to wealth" business model is only just maturing. In essence, the waste biomass in Malaysia are gaining traction as a means to create or enhance current business models. As fuel is relatively cheap and requires large amounts of volume, segregating biomass for the purpose of fuel production will not yield a high return of investment in comparison to other utilization of bio-based products [2, 3]. Therefore, a new sustainable biomass source needs to be realised to enable energy security that is not in competition with food and other bio-based economy. Micro-algae and macro-algae, as third generation feedstock, could be an alternative that provides the energy security, non-food competing and creates its own value chain rather than competing with other economic drivers [4].

The aviation industry is unique in its initiatives to be more sustainable, which was committed under the Aviation CO_2 Roadmap during the Aviation and Environment Summit organised by ATAG (Air Transport Action Group) in 2008 [5]. Major industrial aviation players have pledged to reduce their collective emissions that contributes to roughly 2% of global man-made CO_2 [6]. These pledges are: to increase fuel efficiency by 1.5 % per year (average) between 2009 and 2020, to achieve carbon neutral growth from 2020 onwards and to reduce net emissions by 50 % by 2050 (from 2005 levels) [5]. In order to achieve such ambitious targets, the aviation industry has catalysed developments across the world to search for the best sustainable biomass feedstock, its processing and conversion technology, and to support testing and certification of these bio-derived jet fuels. At present, the most widely used aviation jet fuel is Kerosene Jet A-1, which is governed by international standards such as ASTM D 1655 "Standard Specification for Aviation Turbine Fuels". In anticipation of the synthesized fuel to be widely used, ASTM D 7566 was released in 2009 that enables blending of synthetic fuel (bio-derived fuel) and conventional commercial jet fuel. ASTM D 7566 is known as "Specification for Aviation Turbine Fuel Containing Synthesized Hydrocarbon" [7, 21].

A research and development collaboration between University of Malaya, Aerospace Malaysia Innovation Centre (AMIC) and Airbus Group is aimed at searching and selecting suitable micro-algae strains in Malaysia for the production of bio-jet fuel. Micro-algae strains were screened primarily for their lipid content to determine on the viability of having mass scale micro-algae cultivation for the purpose for bio-jet fuel production. This paper assesses the selected micro-algae strains using Hydro-processed Esters and Fatty Acid (HEFA) conversion process to produce synthetic jet fuel in order to determine if this combination is able to fit Malaysia's jet fuel consumption needs.

2. Materials and methods

The Malaysian micro-algae strains were sampled from freshwater, marine, and polluted (palm oil mill factory) sources. The samples were isolated to yield the dominant strain and the isolated strains were characterized for their lipid content. Table 1 tabulates the micro-algae considered in this study.

Table 1. Micro-algae

No	Type of algae/ habitats	Enrichment media used
1.	Freshwater algae	Bold's Basal Medium (BBM) [8]
2.	Marine algae	Provasoli 50 Medium (Prov50) [9]
3.	Blue-green algae	Kosaric Medium (modified after [10])
4.	Diatoms	Diatom Medium [11]

2.1. Algae isolation

A number of isolation techniques can be employed to isolate dominant species from a cocktail of strains to obtain unialgal cultures. In order to have clonal cultures, the culture were propagated from a single cell of a few cells. Samples collected from the variety of environments (freshwater, marine, and polluted) were subjected to manipulations before isolation was carried out. The following techniques of algae isolation was used in this study: selective enrichment media technique and techniques for isolation of algae. The use of selective enrichment media was chosen to promote the growth of the particular dominant species in order to grow sufficient cells density for isolation. For isolation of algae, a number of varied techniques were used, including centrifuge-washing, streak-plating and taxis techniques [12]. Isolates were purified using antibiotic Kanamycin and Penicillin G) treatment as mentioned in [12].

2.2. Culture maintenance

Purified micro-algae cultures were maintained in sterile Bold's Basal Medium (BBM) for freshwater algae and Provasoli 50 (Prov50) Medium for marine algae. The cultures were placed on illuminated shelves, incubated at 25 ± 1 °C under 12:12 hour light-dark cycle with an irradiance of 40 – 60 μmol photon/s/m^2.

2.3. Determination of microalgae dry weight

A known sample volume containing the algae was filtered through a pre-weighed dried 0.45 μm glass fiber cellulose (GF/C-47mm) filter paper. The pre-weighed filters were heated prior to the filtration in order to remove any organic contaminants. The used filters containing algae were dried at 100±1°C in an oven (ULM-600 Memmert Oven, Schwabach, W. Germany) for 24 hours, cooled in a dessicator filled with silica gel and weighed. The algae dry weight is determined through Equation 1.

$$\text{DW (mg/L)} = \frac{\text{[weight of filters with dried algae biomass, (mg)]- [weight of blank filters, (mg)]}}{\text{Volume of algae culture (L)}} \quad (1)$$

2.4. Determination of total lipid content in micro-algae

Total lipid content of the microalgae was determined using gravimetric measurements as outlined by [13]. A known volume of filtered algae (filtered on 0.45 μm GF/C-47mm filter paper) was pounded with 5 ml Methanol-Chloroform (2:1 v/v) solution using glass hand-homoginiser and transferred into a screw-cap centrifuge plastic tube. The sample was then centrifuged (using Kubota 2100 centrifuge, Kubota Corporation, Japan) at 3000 rpm for 10 minutes. The residue was re-extracted again in 5ml Methanol-Chloroform (2:1 v/v) solution as followed in above procedures. The supernatant obtained was pooled in a new centrifuge tube and 2 mL of chloroform was added followed by 2 mL of distilled water. The tube was thoroughly shaken using a vortex mixer (S0200-230-UK Vortex mixer, Labnet International Inc, USA) until the mixture turned into a milky green colour. The sample was then centrifuged at 3000 rpm for 10 minutes (using Kubota 2100 centrifuge, Kubota Corporation, Japan). The lower layer was then removed using a Pasteur pipette and transferred into a new screw-cap glass tube. The sample was then blow-dried with stream of nitrogen gas. After drying, the extracted sample was dissolved again in 1 mL of chloroform and transferred into a 3.5 mL borosilicate glass vial (pre-weight). The extract was then blow-dried and the dry extract was placed in a dessicator containing the silica gel for 24 hours. After that, it was weighed and retained for fatty acid transesterification. The lipid content in percentage of lipid per dry weight unit was calculated using Equation 2.

$$\text{Lipid content} = \frac{\text{[Weight of lipid + vial (\textit{final weight})] - [Weight of empty vial] X 100\%}}{X \text{ (g) biomass (DW microalgae)}} \quad (2)$$

3. Results and assessment

A total of 79 samples were collected across 30 collection sites in Malaysia. They were identified and isolated, and then investigated for its lipid productivity to meet the needs of jet fuel conversion [14]. The strains of interests were narrowed down to the ones listed in Table 2.

Table 2. Micro-algae strain of interest for high lipid productivity for the purpose of bio-jet fuel conversion [14]

Location	Species	Reference
Hotspring	*Chlorella* sp.	UMACC 326
POME	*Chlorella* sp.	UMAC344
POME	*Chlamydomonas* sp	UMACC 352
Freshwater (lake)	*Chlorella* sp	UMACC 348
Marine	*Chlorella* sp	UMACC 345
Marine	*Chlorococcum* sp.	UMACC 346

A total of five species were noted to be of interest, with the majority of the species isolated is the *Chlorella* sp from various sources [14]. These five species were compared in terms of their biomass productivity potential and lipid content. This is shown in Table 3.

Table 3. Comparison micro-algae biomass productivity, lipid content and its productivity, and dry weight after 12 days of incubation [14]

Type	Biomass Productivity (mg/L/Day)	Lipid content (%wt)	Lipid Productivity (mg/L/Day)	Dry weight (mg/L)
UMACC 326	34.43 ± 8.29.	40.08 ± 2.74	13.78 ± 4.33	190.00 ± 8.00
UMACC 344	76.27 ± 11.08	40.38 ± 7.89	30.57 ± 6.63	114.00 ± 14.00
UMACC 352	32.95 ± 7.23	43.34 ± 2.44	14.03 ± 3.08	222.67 ± 12.06
UMACC 348	49.43 ± 2.07	26.72 ± 6.20	12.71 ± 3.14	506.00 ± 133.55
UMACC 345	44.90 ± 6.44	30.48 ± 5.85	13.48 ± 2.80	328.67 ± 22.03
UMACC 346	57.93 ± 14.00	39.28 ± 1.96	22.55 ± 5.70	294.00 ± 24.25

These species were grown in standard medium, autotrophic (12 hours light, 12 hours dark), in 2.5 L flasks for 12 days in the incubators, as mentioned in section 2.2 [14]. The results from Table 3 were obtained after 12 days of incubation and they can be ranked as follows:

- **Biomass Productivity:** UMAC344 > UMACC 346 > UMACC 348 > UMACC 345 > UMACC 326 > UMACC 352
- **Lipid content:** UMACC 325 > UMACC 344 > UMACC 326 > UMACC 346 > UMACC 345 > UMACC 348
- **Lipid Productivity:** UMACC 344 > UMACC 346 > UMACC 352 > UMACC 326 > UMACC 345 > UMACC 348
- **Dry Weight:** UMACC 348 > UMACC 345 > UMACC 346 > UMACC 352 > UMACC 326 > UMACC 344

As observed in the above ranking breakdown of the four criteria, the correlation between biomass productivity, lipid content and its productivity and dry weight is not apparent. Lipid content is the amount of lipid that is contained within the micro-algae cell after 12 days of incubation while lipid productivity calculates the change of lipid content within the cell across the 12 days of incubation. High lipid productivity does not mean that it also contains high lipid content. This is similar to the biomass productivity and the ensuing dry weight. Ideally, there should be high biomass productivity, lipid productivity and lipid content, which corresponds to fast growing algae that accumulates high amounts of lipid yielding plenty of biomass (dry weight) that contains large amounts of lipid. Further study can be carried out to determine the best conditions for each of these micro-algae species and they are later compared once again, including the economics.

4. Production calculation

Malaysia's consumption of aviation jet, Kerosene Jet A-1, is numbered around 3 billion litres per year (2010) and in comparison to global consumption, this is roughly 0.01% [15]. Malaysia's neighbour, Indonesia, has already introduced a biofuel mandate to introduce alternative jet fuel of 2% by 2018, increasing to 5% by 2025 [16]. IATA (International Air Transport Association) is keeping a keen eye on Indonesia and its developments, noting that other member states around the world will soon start to reflect on this move as well. In lieu of this, it can be calculated that for Malaysia to introduce a 2% alternative jet fuel, 60 million litres of sustainable bio-jet fuel has to be produced. For 5%, this number increases to 150 million litres.

HEFA has been certified in 2011 as an approved pathway to produce alternative jet fuel, which can be "dropped in" and blend up to 50% with the conventional jet fuel [17]. The HEFA process is most suitably used to process oil based feeds, such as vegetable oil, animal fat and recycled oils. In this scenario, HEFA as a process can be used as a means to convert micro-algae into jet fuel. From the literatures, it can be gathered that a maximum conversion efficiency from micro-algae oil to jet fuel is 60% (by weight) [18, 19, 20]. The process can be simply summed up in Figure 1.

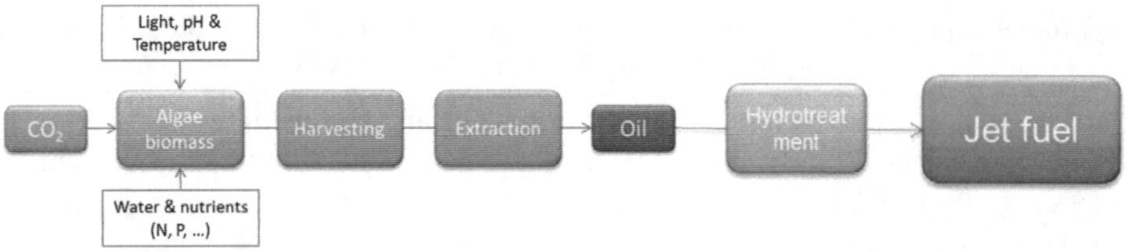

Figure 1. Simplified box-process of algae biomass to jet fuel via the HEFA process [20]

The HEFA process starts off with a hydrogenation process, after the micro-algae's lipid content has been extracted. Hydrotreatment involves oxygen removal, which achieves a paraffinic middle distillate (linear *n*C16, *n*C17, and *n*C18, and hydroisomerization) that transforms a near linear-paraffinic diesel like product into iso-paraffins. Iso-paraffins are important for low temperature performance seen in aviation fuel. After hydroisomerization, a distillation step is required to remove the gasoline range hydrocarbons to maximize the aviation jet range product. To convert micro-algae oil to jet fuel, 1.67 kg of micro-algae oil is required for process that yields 1 kg of jet fuel (based from 60% conversion). The calculated values in Table 4 are obtained by micro-algae species based on lipid (oil) percentage found per kg of micro-algae biomass.

Table 4. Calculation of lipid content to required micro-algae biomass

Type	Lipid content (%wt)	Per 1 kg of micro-algae biomass	For 1 kg of micro-algae oil
UMACC 326	40.08 ± 2.74	40.08 g ± 2.74	2.50 kg ± 0.23
UMACC 344	40.38 ± 7.89	40.38 g ± 7.89	2.48 kg ± 0.41
UMACC 352	43.34 ± 2.44	43.34 g ± 2.44	2.31 kg ± 0.13
UMACC 348	26.72 ± 6.20	26.72 g ± 6.20	3.74 kg ± 0.70
UMACC 345	30.48 ± 5.85	30.48 g ± 5.85	3.28 kg ± 0.53
UMACC 346	39.28 ± 1.96	39.28 g ± 1.96	2.55 kg ± 0.13

From Table 4, based on the lipid content, it was calculated that UMACC 326 requires 2.50 kg ± 0.23 of biomass to yield 1 kg of micro-algae oil; UMACC 344 requires 2.48 kg ± 0.41 of biomass to yield 1 kg of micro-algae oil; UMACC 352 requires 2.31 kg of biomass ± 0.13 to yield 1 kg of micro-algae oil; UMACC 348 requires 3.74 kg ± 0.70 of biomass to yield 1 kg of micro-algae oil; UMACC 345 requires 3.28 kg ± 0.53 of biomass to yield 1 kg of micro algae oil; and lastly, UMACC 346 requires 2.55 kg ± 0.13 of biomass to yield 1 kg of micro algae oil. From this tabulation, Table 5 can be calculated by determining the amount of micro-algae biomass required to have 1.67 kg of micro-algae oil to yield 1 kg of jet fuel.

Table 5. Calculation of micro-algae biomass required to yield 1 kg of jet fuel via the HEFA process

Type	For 1 kg of micro-algae oil	For 1 kg of jet fuel (via HEFA)
UMACC 326	2.50 kg ± 0.23	4.18 kg ± 0.38
UMACC 344	2.48 kg ± 0.41	4.14 kg ± 0.68
UMACC 352	2.31 kg ± 0.13	3.86 kg ± 0.22
UMACC 348	3.74 kg ± 0.70	6.25 kg ± 1.17
UMACC 345	3.28 kg ± 0.53	5.48 kg ± 0.89
UMACC 346	2.55 kg ± 0.13	4.26 kg ± 0.22

For the highest lipid containing micro-algae species, UMACC 352, it requires 3.86 kg ± 0.22 to yield 1 kg of jet fuel via the HEFA process. This means that to yield 1 kg of jet fuel, it would take approximately 17,300 L of micro-algae culture. It was mentioned that to meet the 2% alternative jet fuel, Malaysia would require 60 million litres of alternative jet fuel. Kerosene Jet A-1 having a density between 775 – 840 g/L [7], this would mean between 46.5 to 50.4 million kgs or 46,500 – 50,400 tons of alternative fuel. Ultimately this would equate to Table 6.

Table 6. Calculation of micro-algae biomass required per year to meet the expected demand of 2% alternative (bio-) jet fuel

Type	For 1 kg of jet fuel (via HEFA)	2% Bio-jet Fuel for Malaysia
UMACC 326	4.18 kg ± 0.38	ca. 194 – 212,000 tons
UMACC 344	4.14 kg ± 0.68	ca. 192 – 225,000 tons
UMACC 352	3.86 kg ± 0.22	ca. 179 – 189,500 tons
UMACC 348	6.25 kg ± 1.17	ca. 290 – 345,000 tons
UMACC 345	5.48 kg ± 0.89	ca. 254 – 296,000 tons
UMACC 346	4.26 kg ± 0.22	ca. 198 – 208,000 tons

For the highest lipid containing micro-algae species, UMACC 352, it requires 179,000 to 189,500 tons of biomass per year to meet the expected 2% alternative fuel (60 million litres) for Malaysia's consumption. For the lowest lipid containing micro algae species, UMACC 348, it requires 290,000 to 345,000 tons of micro-algae biomass per year. Typically, open pond systems are the easiest and low cost way of mass production of micro-algae. Conventionally, open ponds have a productivity of 30 – 50 ton/ha/year [22], using the data obtained from the selected micro-algae strain the following can be calculated. From Table 7, it is evident that the land size required is comparatively large. Nonetheless, compared to the entirely of Malaysia, this accounts to roughly 0.1 – 0.2 % of its land mass.

Table 7. Calculation of micro-algae biomass required to yield 1 kg of jet fuel via the HEFA process

Type	2% Bio-jet Fuel for Malaysia	Approximate Land Size Required (ha)
UMACC 326	ca. 194 – 212,000 tons	4,300 – 4,720
UMACC 344	ca. 192 – 225,000 tons	4,280 – 4,990
UMACC 352	ca. 179 – 189,500 tons	3,990 – 4,210
UMACC 348	ca. 290 – 345,000 tons	6,450 – 7,660
UMACC 345	ca. 254 – 296,000 tons	5,660 – 6,570
UMACC 346	ca. 198 – 208,000 tons	4,400 – 4,620

5. Conclusions

Malaysia has one of the highest air traffic movements in the South-East Asian region, and the need to have a blend of bio-jet fuel with conventional Kerosene Jet A-1 is just a matter of time. Having a rich biodiversity is a key enabler for Malaysia to have biomass waste to wealth economies. However, the conversion of biomass waste to bio-fuel has still yet to gain much traction in Malaysia due to many competing factors. Micro-algae is poised to alleviate certain disadvantages seen in first generation and second generation feedstock. Potential of Malaysia's micro-algae has been determined as a possible feedstock to enable bio-jet fuel production. UMACC 326, UMACC 344, UMACC 352, UMACC 348, UMACC 345, and UMACC 346 were selected from 79 samples collected over 30 sites throughout Malaysia. These species were selected based upon their high biomass productivity and lipid content. Their lipid contents were then used to derived the require amount of micro-algae biomass to yield 1 kg of certifiable jet fuel via the HEFA process and subsequently the amount of micro-algae biomass required to meet a scenario where Malaysia implements a 2% alternative (bio-) jet fuel requirement. Finally, the total amount of land size required to sustain a micro-algae cultivation plant that can cater to Malaysia's jet fuel consumption for a 2% alternative (bio-) jet fuel blend is also estimated.

Acknowledgments
This study was jointly conducted with Aerospace Malaysia Innovation Centre, AIRBUS Group and University of Malaya with the funding provided under the project "Improvement of Biomass and Lipid Productivity in Malaysian Microalgae", Grant No. PV001-2013. This activity is also a part of AMIC's industrial laboratory "Centre of Excellence on Biomass Valorization for Aviation" that is located in Universiti Putra Malaysia.

References
[1] Roda J M, Goralski M, Benoist A, Baptiste A, Boudjema V, Galanos T, Georget M, Hévin J E, Lavergne S, Eychenne F, Liew K E, Schwob C, Djama M and Tahir P M 2015 *Sustainability of bio-jet fuel in Malaysia* Centre of International Cooperation in Agronomy Research for Development (CIRAD)
[2] Ng W P Q, Lam H L, Ng F Y, Kamal M and Lim J H E 2012 *J. Cleaner Production* **34** 57–65
[3] McKendry P 2002 *Bioresource Technology* **83** 37–46
[4] Maity J P, Bundschuh J, Chen C Y and Bhattacharya P 2014 *Energy* **2014** 1-10
[5] International Air Transport Association 2013 *IATA Technology Roadmap*
[6] Air Transport Action Group 2014 *Aviation Benefits Beyond Boarders*
[7] ASTM 2015 *Standard Specification for Aviation Turbine Fuels* ASTM International
[8] Nichols H W and Bold H C 1965 *J. Phycol.* **1** 34-8
[9] Guillard R R L 1975 *Culture of Marine Invertebrate Animals* (New York: Springer US)
[10] Zarrouk C 1966 *Contribution a l'etuded'une Spirulin maxima* Thesis Paris
[11] Beakers G, Canter H M and Jaworski G H M 1988 *Can. J. Bot.* **66** 1054-67
[12] Phang S M and Chu W L 199 *University of Malaya Algae Culture Collection (UMACC) Catalogue of Strains* (Kuala Lumpur: Institute of Postgraduate Studies and Research, Kuala Lumpur)
[13] Bligh E G and Dyer W J 1959 *Can. J. Biochem. Physiol.* **37** 911-7
[14] Phang S M, Mustafa E M, Ambati R R, Sulaiman N M N, Lim P E, Majid N A, Dommange X, Schwob C and Liew K E 2015 *M. J. Science* **34** 148-74
[15] US Department of Energy 2012 *Energy Information Administration, Independent Statistics & Analysis, Consumption of Jet Fuel, Malaysia* U.S. Energy Information Administration
[16] IATA 2016 *Alternative Fuels*
[17] Radich T 2015 *The Flight Paths for Biojet Fuel, Independent Statistics & Analysis, U.S Energy Information Administration* U.S. Energy Information Administration
[18] Endres C, Falter C, Roth A, Riegel F and Sizmann A 2012 *Deutscher Luft-und Raumfahrtkongress 2012*
[19] Stratton R W, Wong H M and Hileman J I 2010 *Life cycle greenhouse gas emissions from alternative jet fuels* Massachusetts Institute of Technology
[20] Lundquist T J, Woertz I C, Quinn N W T and Benemann J R 2010 *A realistic technology and engineering assessment of algae biofuel production* Energy Biosciences Institute University of California
[21] ASTM 2016 *Standard Specification for Aviation Turbine Fuels Containing Synthesized Hydrocarbons* ASTM International
[22] Thangavel P and Sridevi G 2015 *Environmental Sustainability: Role of Green Technologies* (New York: Springer)

IOP Conf. Series: Materials Science and Engineering **152** (2016) 012043 doi:10.1088/1757-899X/152/1/012043

The adaptation of sustainable biojet fuels and its effect on aircraft engine maintenance

H Mohd Noh[1*], G A Rodrigo[2] and N A Abdul Rahman[1]

[1]Universiti Kuala Lumpur- Malaysian Institute Aviation Technology, Malaysia
[2]Universidad Politecnica De Madrid- Escuela Tecnica Superior de Ingenieros Aeronauticos, Spain

*hazariah@unikl.edu.my

Abstract. Aviation industries are looking into several resources for renewable and sustainable energy. Among those attention is focused in biojet fuel. This paper engages the issue of biojet fuel emissions that increase the environmental concern in the air transport sector. The paper presents the use of biojet fuel and its effect on aircraft engine maintenance through preliminary data collections, and a review of its development process in operations for time and goal. As conclusion, airlines management needs to adapt and adopt the transition to alternative fuels, especially given the global biofuel trend emerging due to the authority approval.

1. Introduction

In the recent years, there has been an increasing research interest in alternative energy due to aviation environmental pollution from the usage of fossil fuel [1, 2]. Among the many anticipated benefits of alternative energy, sustainability is considered as a global need that has to be measured. This is due to its reliability, economics and supply chain optimism [3]. According to other past researches, in Europe itself, 216 million tons of $C0_2$ emissions were produced in the year 2010. This is projected to be 334 million tons for optimistic scenarios in the year 2030 [4]. The reduction of emissions is considered to be more a matter of engine design than fuel technology. The adverse effects of the general degradation of combustion performance in kerosene for over 25 years cause emissions and engine life span to be in need of consideration, and this has led to a rapid demand for new and alternative fuel [5]. These rapid changes bring a serious effect in creating global carbon neutrality, where in this case, improvements in carbon engine outputs alone will not work. Adaptation and current work with new 'alternative energy' with zero carbon distributions will help to reduce air pollution that directly impacts the atmosphere [6]. Alternative energy in commercial aviation can be defined as using biojet fuel, gas to liquid, and hydrogen cells. The use of battery and hydrogen cells addresses some issues such as weight limitation, and thus it will not be discussed in this paper. Similarly, GTL or gas to liquid will not be considered in this article as it has an adverse feedback on the environment [7].

When alternative energy approaches sustainability, factors such overall efficiency of combustion process and level of the emissions produced are its crucial assets. A significant increasing amount of research has been carried out in providing the definitive evidence of performance for these alternative fuels in relationship to combustion chamber [8, 9, 10]. Nevertheless, this research area is still lacking. Essentially, air transportation has to ensure the future forecasting of 27,200 new airplanes worth 2.7 trillion that will be needed by 2025 to be embarked towards sustainability of the aviation business entities across the world. In fact, it is to be forecasted that the total CO_2 emissions due to commercial aviation may reach 1.2 billion to 1.5 billion tonnes. It is becoming an important part of the airlines industry and associate to be part of a sustainability transition towards the environment.

Therefore, this study will present the knowledge of biofuel usage in the engine maintenance and also a review of its development in operations, specifically for time and goal. These clearly help the airlines to ensure the adaptation of the biojet fuel to its application in civil aviation. In fact, this study will also benefit to the policy maker or aviation authority in preparing the aviation industries to this new alternative energy resources.

2. The research method

This research reviews current maintenance programs as an approach to planning and constructing the way ahead in adaptation for the use of new alternative energy. Therefore, this study supports airlines in economic view and maintaining the safety of aircraft components for the engine. Doing so provides a reliable yet efficient way to examine and identify maintenance perspective. Information obtained was collected with deeper document analysis including report analysis from biojet fuel flight test and aircraft engine and maintenance manual. The approach implemented in the research involves acquiring an expert from the industries such License Aircraft Engineer in both airframe and engine, and the fuel workshop approval holder to form a line of reasoning to achieve the solution and provide preliminary result.

3. Aircraft engine emissions

Higher efficiencies of the engine provide an increase rate of performance. However, negative rate of performance of the turbine engine can also occur due to different level system losses, which can take many forms. The simplified internal flow shows the relevance, description and products from the gas turbine engine operation in an adaptation from research article in Table 1 [11].

Table 1. Description of simplified internal flow systems in turbine engine combustion systems

Relevance	Description	Products
Energy source	System delivering super-heated steam to represent combustion and compression process	Air $N_2 + 0_2$
Energy transfer	System with heat transfers possible	$C_NH_M + S$
Energy conversion	System converting energy from super-heated steam to work	$CO_2 + H_20 + N_2 + SO_2$ = *complete combustion* $C0_2 + H_2O + N_2 + O_2 + NO_X$ + SO_X + HC + Soot Particles = *Actual combustion*

The product of the turbine engine is creating emissions. The exhaust gas is roughly about 70% CO_2. A little less than 30% non-CO_2 emissions include water vapor (H_2O), and less than 1% each of nitrogen oxides (NO_x), sulfur oxides (SO_x), hydrocarbons (HC), and black carbon (or soot) particles [12]. These emissions from standard Jet Fuel for aviation such CO_2, H_2O and soot emissions will contribute directly to the environmental impacts, where these emissions are released directly into the upper troposphere and lower stratosphere. This is where the effects of the climate change are discussed in detail [13, 14, 15].

4. Biojet fuel for commercial engine

According to NASA research center on propulsion technology, their primarily focus will be on high-pressure engine core components and propulsion technologies, with the goals of lowering harmful emissions and fuel burn [16]. As mentioned earlier, improvements in the engine itself will not lead to zero carbon outputs. According to European Commission [17], in the biofuel flight path 2050, the target of European countries is to be established as a centre of excellence on sustainable alternative fuels, including those for aviation, based on a strong European energy policy. This will lead to more biofuel projects and research. The decisive evidence shows biojet fuel need to come into the depiction. It is certainly true that advantage of biojet fuel because it can be a direct drop-in fuel. It is complimentary and compatible with the conventional jet fuel. With biojet fuels to displace the conventional jet fuel without the need to change the engine technology. Currently, there are over 300,000 commercial aircraft in the world. Since biojet fuel is complimentary and compatible with the conventional jet fuel, it will have a ready market of over 300,000 commercial aircraft [18].

The future base of aviation fuel research conducted by IATA described chemical and physical characteristics of the fuel and reviewed the contents of its requirements, which include heat content, fluidity, corrosion protection, and stability [19]. In 2008, the ground test was performed for CFM 56-7B for new alternative energy using biomass Algae and Jatropa for the bioJet fuel as a 50% mix to current jet fuel. It is the continuation of the test by Bio Spk that blends Jatropa-Jet A-1, Jatropa/Algae, Jet A and Camelina/Jatropa – Algae/JetA. The detailed result of composition, volatility, fluidity, combustion, thermal stability and contaminants can be seen in Table 2 from the Evaluation BioSPk report 2009 [20].

Table 2. Evaluation BioSPk report 2009 for different types of biomass versus traditional jet fuel

Property	Jet A/Jet A-1	Jatropha	Jatropha/Algae	Camelina/ Jatropha/Algae
COMPOSITION				
Acidity, total mg KOH/g Aromatics	0.10	0.002	0.001	0.003
1. Aromatics, volume%	25	0.0	0.0	
2. Aromatics, volume%	26.5	0	0	0
Sulfur, mercaptan, mass%	0.003	<0.0001	<0.0001	<0.0001
Sulfur, total mass%	0.30	<0.0001	<0.0001	<0.0001
FLUIDITY				
Freezing point, °C	-40 Jet A	-57.00	-54.5	-63.5
Viscosity −20°C, mm2/s	8	3.663	3.51	3.353
COMBUSTION				
Net heat of combustion, MJ/kg	**42.8**	44.3	44.2	44.2
Hydrogen content, mass %				
One of the following requirements shall be met:				
1. Smoke point, mm, or	25	0	0	0
2. Smoke point, mm, and	18	0	0	0
Naphthalene's, volume, %	3.0	<0.01	<0.01	<0.01
THERMAL STABILITY				
JFTOT (2.5 h at control temperature)				
Temperature, °C	260	340	340	300
Tube deposits less than	3	1	1	1
CONTAMINANTS				
Existent gum, mg/100 mL	7	0	<1	<1

The result from biojet fuel being used in all three flight tests can be concluded as positive outcomes as the reading is not just meeting the lower requirements aviation fuel needs but supersede the current jet fuel composition benefit. Enhanced composition in biojet fuel to be compared with conventional fuel creating more advantages in term of carbon emissions, fuel efficiencies, components wear and tear and reduction in fuel cost and maintenance in prolong future.

Different types of biojet fuel generation being used and each of this generation was analysed and were identified for Air transportation use by giving the potential to deliver in large quantities, for greener, cleaner, and cheaper fuel. This can be categorised as Jatropha and Camelina for the short term feedstock while algae and halophytes as long-term feedstock [21].

5. Engine maintenance

Current practice in aircraft operations and maintenance are being measured for aircraft's preventative maintenance. Future biofuel to be introduced into the aircraft fuel system requires no modification, but it demands further analysis on predicting maintenance schedules. The utilization of current standard fuel Jet, A-Jet A-1 to ASTM D7566 drop in bio jet fuel, such as Camelina, Jatropa and Algae types of oils, is analysed [22]. To develop quantification of maintainability for the new alternative energy being used in the fuel system components, two affected maintainability prediction techniques are considered: primary procedures and secondary procedures.

Primary procedures take into account the early development and the entire separate processes of design and diagnostic. This will give a very accurate list of data for predictions limited throughout the engine response [23]. The secondary procedure will be the earliest procedure which is applicable earlier in the components diagnostic routines inspection and troubleshooting procedures, where the failure rates of the components can be recorded and measured. Since there is no complete data analysis for the real development of the biojet fuel flow through the pump itself, this identical value can be used to project into the maintenance schedule and maintenance planning data in MRO similar to overall fuel flow throughout the entire system [24]. This can be seen in Figure 1, where fuel reflects design elements.

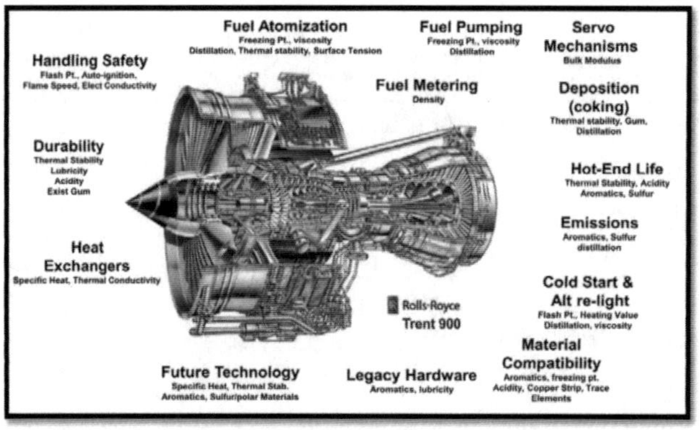

Figure 1. Fuel effects on engine design elements (Source: FAA)

6. Biojet fuel development – time and goal

In the aviation biojet fuel development context, it is acknowledged that the development needs to be focused in two different sets of scenarios namely time and goals. In term of time, more deliberate ideas can be developed when the process is focused on short term, medium term, and long term. This can be demonstrated in the adaptation of the elements of a featuring process for sustainable effectiveness in Table 3 [25]. As shown in Table 3, the development process in operations can be divided to six categories namely adaptation to MRO for biojet fuel, the purpose, the key outputs, tools, parties that involved and number of scenarios.

Table 3. Development process in operations time and goal adaptation [25]

	Short term	Medium term	Long term
Adaptation to MRO for biojet fuel	Observation through biojet fuel trending through test and schedule flight testing	Developing the gap for future biojet fuel and current jet fuel.	Identifying the business assumption for new fuel and disaggregating the development and strategic aims
Purpose	Identify adjustments to current strategic intent to optimize sustainable effectiveness	Develop boundaries for next version of robust strategic intent	Identify business assumptions and develop a range of possible future strategic plans
Key outputs	Operating decisions that change aggressiveness and differentiation	Identify capability investments	Identify potential capabilities to build team responses
Tools	Traditional environmental scanning tools, prediction markets, extreme strategizing	Ecosystem mapping base, best and worst case forecasts	Future scenarios
Who is involved	Top management -20 % Any boundary- spanner, current customers and relevant stakeholders	Top management 50%, Network and ecosystem community	Top management 30%, Board 75%, High-level government, NGO and community representatives, futurist
Number of scenarios	Zero, use goals instead	Three in current trajectory	Three of five scenarios of diverse conditions

The economic analysis in identifying the incentive and mechanism in understanding further the implication in biojet fuel usage will be vital. Subsidies, on the other hand, are forms of government financial support for activities believed to be environmentally friendly. Rather than charging a polluter for emissions, a subsidy rewards a polluter for reducing emissions [28]. The cost benefits of using bio jet fuel remain unclear; this is often taken to support the view why Airlines is hesitant and reluctant to invest and adopt this new fuel. Some aspects derived from the resistant required further investigation. A more promising approach would be through risk assessment. At the risk offending the reluctant, the risks classification, must be examined more carefully [26]. It does suggest that the other factors will be a balance that can be drawn between risk and opportunity.

7. Preliminary data collection

This questionnaires Phase 1 discusses the data analysis and findings from 40 questionnaires completed by individuals who are working in Aviation industries in Malaysia, Spain, Saudi Arabia and Singapore. The result can be seen in all figure and table below. Details explanation can be concluded in discussions and analysis. Four sections of questionnaires has been handed out consisting of:

- Section A: Personal (biographical) data
- Section B: General knowledge on biojet fuels
- Section C: The feasibility of biojet fuels in aviation
- Section D: Impact of the usage of biojet fuels

7.1. Section A

The targeted group was Aviation personnel whose has more than four years' experience in handling aircraft and more accurate towards Engine or Aircraft systems. The data was collected from various Aviation company included Malaysia, Saudi Arabia, Singapore. The random sample of 40 participants provided an in-depth understanding of the general knowledge on alternative biofuel, the feasibility of biojet fuels in aviation and its impact. The selection process was identified through each participant's role in the use or testing of alternative fuel.

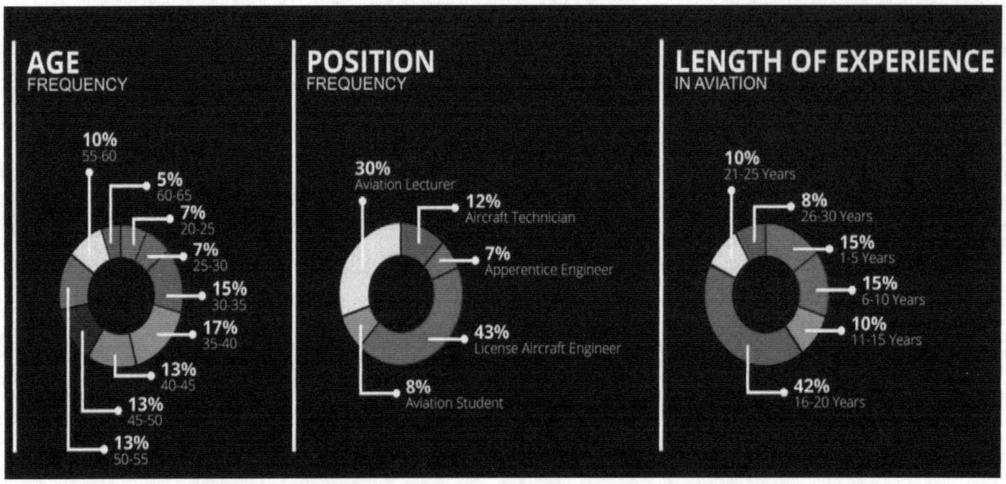

Figure 2. Respondents age, position and length of experience

7.2. Section B

When addressing the question in this section B, it focuses in how good biojet fuels to the environment, to aircraft engines, and reducing fuel cost by using the biojet fuels. The questions (B1, B3, and B5) have the highest value of this section - Moderate by 50%. This is shown in Figure 3 that average respondents know about biojet fuel, it is used in aircraft and airlines. With the second highest with 20%, the respondents have very low knowledge in biojet fuel. Most of the respondents rated the second question until fourth question, (B2-B4) as very high.

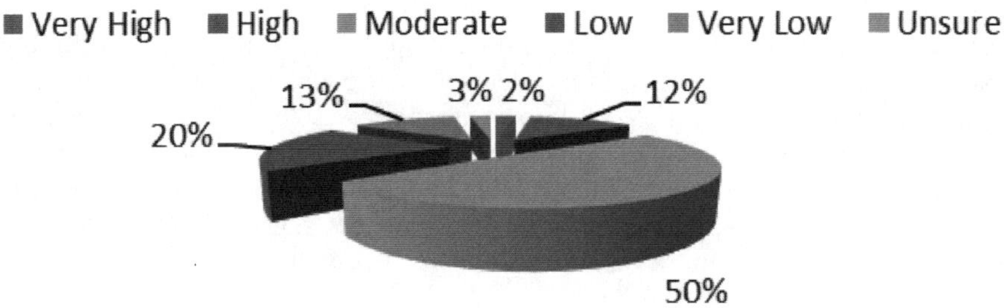

Figure 3. Section B: personal knowledge about biojet fuels in aviation

7.3. Section C

The first question (C1) probed if the biojet fuel is safe for the aviation industry. Most of the respondents agree with that statement only nine respondents (23%) are not sure. For the second question until fourth question, (C2, C3, and C4), the respondents averagely answered as moderate. The questions are more on the technical side if the biojet fuels will be used in aircraft engines. The questions asked about the modification is required to the engine if using biojet fuels, level adaptation to biojet fuel will be easier if that happened, and low technical feasibility changing from normal fuel (Jet A/Jet A-1 to biojet fuel) is needed. The last question (C5) however, they agree that biojet fuels will be widely used in future leaving three respondents unsure.

7.4. Section D

Figure 5 identifies the impact of biojet fuel usage. First question (D1) asked the respondents if biojet fuels lead to a better environment, most of them rated as very high and high. The second question (D2) asked if biojet fuels provide better fuel costs, the common answer is moderate. For questions (D3) and (D4), the respondents averagely answered as moderate. The questions are about biojet fuels will increase engine performance, and saving airlines fuel cost. The last question (D5) asked if sustainable biojet fuels can be achieved. Most of the respondent rated as moderate.

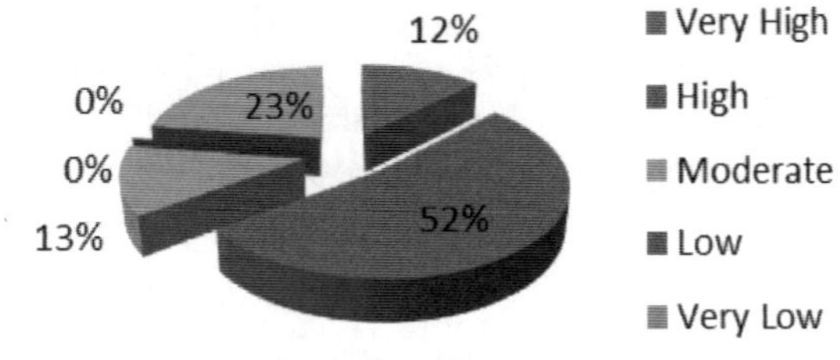

Figure 4. Section C: the feasibility of biojet fuels in aviation

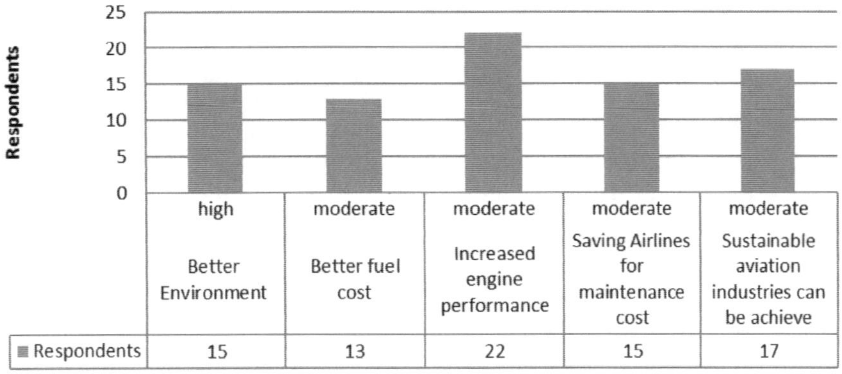

Figure 5. The impact on using biojet fuel

Questionnaires forms that have been conducted have been major contributors to this research. As preliminary data research, few important of conclusions can be made from this finding. It can conclude that the level of knowledge of new green technology, especially biojet fuels are still at a moderate level, among the personnel in the industry. Also from the results, it shows that biojet fuels knowledge is at 25% at the lower rate or unsure. The airlines or aviation personnel is aware and opened to the application of the biojet fuels in the aviation industry. From this research, respondents aware that biojet fuels are safe to be used in aviation. This is because more than 25 airlines already tested or conducted the biojet fuel flight as Figure 7 and Table 3. Besides, biojet fuels release zero gas emission compared to current fossil fuels. Even more, safety and fuel quality specifications are of tremendous importance in the aviation sectors. The technical requirements for aviation biojet fuels are; a high-performance " fuel (that can withstand a range of operational conditions)", a fuel that does not compromise safety, a fuel that can directly substitute traditional jet fuel aviation, and also a fuel that meets the required performance target. The feasibility of biojet fuel adaptation on Aviation industries is also giving results by 52% at the higher rate.

As conclusion, this researched questionnaires shows that awareness led to where the safety in biojet fuel usage is very crucial. The cost has a high impact when coming to new alternative fuel, as increasing in cost will create a huge hurdle to the Aviation industries. The policy is creating a medium resource that can lead to projection in the adoption of biofuel in aviation industries. From this questionnaires its objectives; as being a major contribution to the feasibility biojet fuels and its impact on the aviation industry in the context of Malaysian environment has been achieved.

8. Biofuel drop in and successful roadmap flight (test and schedule)

Airlines need to adapt to the new energy, but the most necessary part is ensuring that it can be used effectively without any alteration to the systems, and it is affordable to maintain it. The ASTM D4054 was developed as a guide related to testing and procedures ensuring certification of drop in alternative fuels have remained unchanged to the limitation of engine operating and aircraft operations, as well as

IOP Conf. Series: Materials Science and Engineering **152** (2016) 012043 doi:10.1088/1757-899X/152/1/012043

aircraft flight manual as in Figure 6 below. Besides, this indicator needs to relate and ensure that the transition into the prolonged future towards engine systems can continually perform its function in the levels of reliability and safety [27].

The development of biofuel technology for commercial aviation has made extraordinary advances in the last eight years. On February 23, 2008, a Virgin Atlantic B747/400 with General Electric CF6-80C2 engines became the first commercial aircraft in making a flight with a mix of 80% conventional kerosene and 20% first generation biofuel, produced from coconut oil and seeds of babassu palm tree. Since then, more than fifteen airlines have tested the operational possibilities of different aviation biofuels with satisfactory results from the technical point of view. The current testing in May 2016, was by Cathay Pacific using Airbus 350 with the mix of 10% blend fuel by Farnasene. Details world map on biojet fuel test in three different continental can be seen on Figure 7 below. Details summary on biojet fuel flight test in Asia Pasific region since 2008 until the year of 2016 can be seen in Table 3.

Figure 6. ASTM D4054- Drop in alternative fuel with unchanged operating limitation (Source: ASTM)

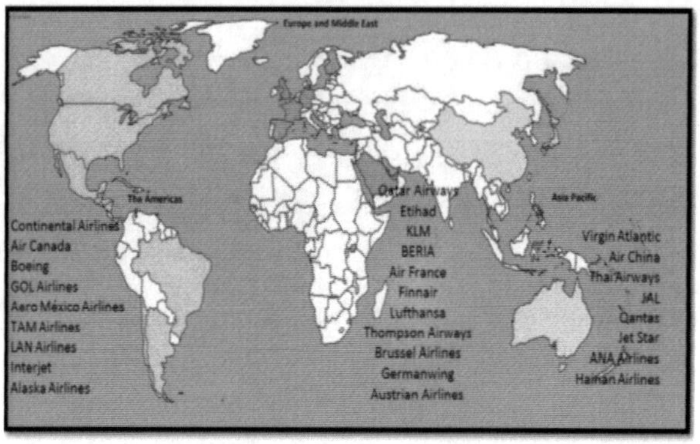

Figure 7. Biojet fuel flight test and scheduled flight according to the region

Table 3. A summary of biofuel tests performed on commercial flights in Asia Pacific

CARRIER	AIRCRAFT	PARTNERS	DATE	BIO-FUEL	BLEND	DESTINATION
Virgin Atlantic	B747-400	Boeing, GE Aviation	23.02.2008	Coconut Babassu	20% one engine	Amsterdam-London
Air New Zealand	B747-400	Boeing, Roll Royce	30.12.2008	Jatropha	50% one engine	Two-hour flight test
JAL	B 747-300	Boeing, P&W, Honeywell UOP	30.01.2009	Camelina, Jatropha, Algae blend	50% one engine	Demo flight
Air China (CA)	B747-400	Boeing, Petro China, Honeywell UOP	June 2011	Jatropha	50% of one engine	Test Beijing airspace
Thai Airways TG	B777-200	Rolls Royce, Boeing, SkyNRG	21.12.2011	Castor seed	One engine	Bangkok-Chiang Mai
Qantas	A330	Airbus	April 2012	Cooking Oil	50%	Sydney-Adelaide
ANA	B737 Dream-liner	Boeing	April 2012	Cooking Oil	10% blend	Evertt's Paine Field (KPAE) to Haneda Airport (HND)
Jetstar	A320	Airbus	April 2013	Cooking Oil	50% Blend	Melbourne - Hobart
Hainan Airlines	737-800	Boeing	March 2015	Cooking Oil	50% Blend	Shanghai to Beijing
Cathay Pacific	A350 -900	Airbus/ Amyris-Total	May 2016	Farnesane	10% Blend	Toulouse to Hong Kong

9. Conclusions

Biofuels developments in an advance generation are a crucial evaluation in ensuring the sustainability of the aviation industries. Detailed information on the adaptation and adoption towards biojet fuel in the turbine aircraft fuel system suggests that viability of all the process regarding minimising the costing, especially in reducing the maintenance and operating costs. The reliability and maintainability will be the key factors for the success of deployment of commercialization of biojet fuel for aviation industries, especially in commercial aviation. The urgencies, risks, and opportunities of this new alternative energy need to be broadened in various aspects for sustainability to reach and achieve its destination. Maintainability in the aircraft system, on the other hand, measures the right direction in supporting the quest to meet high sustainability goals by implementing biojet fuel. Moreover, biofuels may provide valuable economic opportunities to communities that can develop new sources of income – including in many developing nations.

9.1. Maintenance view

A direct impact of the maintenance program of this new alternative energy can easily be focused on economics cost benefit. As long as the price of petroleum remains high, the fossil source continues to be sufficient, the concerns about the environment persist, and then the interest in alternative jet fuels derived from non-petroleum sources will continue to grow. Despite the presence of technical hurdles and the growing necessity for further measurement and study on the costs and benefits, alternative jet fuels appear to be both technically and economically feasible in engine maintenance system because considerable momentum and coordinated work have resulted in recent years.

Most of the turbine maintenance is driven by the highest temperature against the average temperature which the parts is subjected. If the biojet fuel can improve and provide uniformly burnt in the combustor section, the engine will last longer on the wing, and it is proving to have an immense maintenance cost benefit. The fuel producers and fuel researchers need to ensure that the new fuel complies with the material to the standard of engine hot sections. Throughout many biofuel flight tests being conducted since 2008, most of the engines using biofuel have more optimum temperature compared to engines that used normal fuel. With this evaluation, it is acknowledged that in the long run, without a doubt, the use of biofuel will offer the benefit to the maintenance. With the reduction of fuel consumption and the specific requirements meet, prevention maintenance can safely be attained, and the components can reach beyond standard life hours/cycle or even more. Above all, maintenance department will be able to allow the operator or airlines to identify whether using biojet fuel is beneficial and sensible in term of overall operating cost.

9.2. Sustanability of biojet fuel

The economic analysis in identifying the incentive and mechanism in understanding further the implication in biojet fuel usage will be vital. Subsidies, on the other hand, are forms of government financial support for activities believed to be environmentally friendly. Rather than charging a polluter for emissions, a subsidy rewards a polluter for reducing emissions [28]. Although knowledge of biojet fuel is considered a necessity for sound decision making in aviation, the airlines company and the aviation authority needs to reset the action in taking this new alternative energy to the next level to create the sustainable growth. The value required to interconnect increasingly with various departments of the airline, the organizational strategy, structure, systems, final results and feedback (increasing or decreasing the cost). The global avation world need to be underpinned by the awareness of the good effect of the usage of biofuel on engine process and procedure. In fact, the support by the policy and regulatory framework also is recognised as a pre requisite for a successful biojet fuel implementation across the globe.

Referring to the previous study on biojet fuel in in other research streams, there is a white space in the aviation biojet fuels that looks into economic perspectives through the use of biojet fuel in aviation, which needs further attention. Future research should look into how, when and by whom activities to strengthen the importance of biojet fuel for aviation sustainability in the future. This study ideally shall enable us to create new theoretical insights of the usage of biojet fuel and extend the knowledge of its effect on aircraft engine maintenance which is infrequently explored.

Acknowledgments
Thanks to Universiti Kuala Lumpur and Universidad Politecnica De Madrid for continuous support. Gratitude is also extended to various airlines engineering and maintenance personnel from Malaysia, Saudi Arabia and Spain, whose help contributed and generated information for this research.

References

[1] H Gorter and D R Just 2010 *Applied Economic Perspectives and Policy* **32** 4–32
[2] Noh H, Alonso G and Rahman N A A 2015 *Applied Mechanics and Materials* **747**
[3] Janic M 2002 *J. Air Transp.* **7** 1999–2001
[4] Alonso G, Benito A, Lonza L and Kousoulidou M 2014 *J. Air Transp. Manag.* **36** 85–93
[5] Goodger E M 1995 *Proc. Inst. Mech. Eng. Part G J. Aerosp. Eng.* **209** 147–56
[6] Whitfield R 2011 *Proc. Inst. Mech. Eng. Part G J. Aerosp. Eng.* **225** 619–30
[7] ICAO 2009 *A summary of research and perspective ICAO on alternative fuels*
[8] Jenkins B M, Baxter L L, Miles Jr. T R and Miles T R 1998 *Fuel Processing Technology* **54** 17-46
[9] Shah A N, Yun-shan G E, Chao H and Baluch A H 2009 *Pak. J. Engg. Appl. Sci.* **4** 34–42
[10] Uryga-Bugajska I, Pourkashanian M, Borman D, Catalanotti E and Wilson C W 2009 *European Air and Space Conference*
[11] Doty J H, Camberos J A and Moorhouse D J 2009 *Int. J. Aerosp. Eng.* **2009** 1–11
[12] Blakey S, Rye L and Wilson C W 2011 *Proceedings of the Combustion Institute* **33** 2863-85
[13] IATA 2009 *IATA Environment Report*
[14] ICAO 2010 *ICAO Environment Report*
[15] Graham B and Guyer C 1999 *Journal of Transport Geography* **7** 165-80
[16] Withers M R 2014 *Environmental and economic assessment of alternative transportation fuels* Master Thesis Massachusetts Institute of Technology
[17] European Commission 2011 *Flightpath 2050: Europe's Vision for Aviation*
[18] IATA 2014 *Overview of Alternative Jet Fuels in 2014*
[19] IATA 2012 *IATA 2012 Report on Alternative Fuels*
[20] Rahmes T, Kinder J D et al. 2009 *9th AIAA Aviation Technology, Integration, and Operations Conference (ATIO)*
[21] Noh H and Zulkifli N 2012 *Applied Mechanics and Materials* **225** 572–7
[22] Holmgren J 2007 *ERTC 12th Annual Meeting*
[23] Lipa J F 1981 *IEEE Transactions on Reliability* **30** 218-21
[24] Rumizen M 2012 *Public Meeting for Center of Excellence for Alternative Jet Fuels and Environment* FAA
[25] Lawler E E, Worley C G and Creelman D 2011 *Management Reset: Organizing for Sustainable Effectiveness* (San Francisco: Jossey-Bass)
[26] Noh H, Alonso G, Nair S and Dahdi Y 2015 *WIT Transactions on Ecology and the Environment* **206** 103–11
[27] Kinnison H A 2004 *Aviation Maintenance Management* (New York: McGraw-Hill)
[28] Noh H M, Benito A and Alonso G 2016 *Transportation Research Part D: Transport and Environment* **46** 298–316

The potential risk of communication media in conveying critical information in the aircraft maintenance organisation: a case study

S Ahmad Shukri[1*], R M Millar[2], G Gratton[3] and M Garner[4]
[1]Universiti Kuala Lumpur - Malaysian Institute of Aviation Technology, Malaysia
[2]University of Aberdeen, Scotland
[3]Brunel University, West London
[4]Whitelands College, University of Roehampton, London

*suhailah@unikl.edu.my

Abstract. In the world of aircraft maintenance organisation, verbal and written communication plays a pivotal role in transferring critical information in relation to aircraft safety and efficiency. The communication media used to convey the critical information between departments at an aircraft maintenance organisation have potential risk in misunderstanding of the information. In this study, technical and non-technical personnel from five different departments at an aircraft maintenance organisation were interviewed on the communication media they normally utilised to communicate six different work procedures that are closely related to aircraft safety and efficiency. This is to discover which communication media pose higher risk in misunderstanding critical information. The findings reveal that written communication pose higher risk of misinterpretation compared with verbal communication when conveying critical information between departments.

1. Introduction

Civil aviation is a highly regulated industry controlled by numerous national regulatory bodies such as the UK Civil Aviation Authority (UKCAA), the US Federal Aviation Administration (FAA) and the Department of Civil Aviation Malaysia (DCAM). In order to ensure the safety of aircraft and the people on board, pilots must be trained and licensed. For the same reason, Aircraft Maintenance Technicians (AMT) and Aircraft Maintenance Engineers (AME) must be trained and certified to maintain and repair the complex system of aircraft so that they are airworthy. Research into human factors has contributed significantly to understanding what contributes to aircraft failures. The International Civil Aviation Organization (ICAO) and National Aviation Authorities (NAA) are very much concerned with the crucial human contribution to aircraft safety, and have undertaken a great deal of research into human factors, particularly the importance of communication in aviation maintenance organizations. As such, many studies are undertaken aiming to understand the causes of communication errors and find solutions to mitigate incidents that might affect safety. By analysing areas of potential inefficiencies in communication in a critical organisational communication system, the study reported here reveal the type of communication media used in the aircraft maintenance organisation and the work procedures that have higher risk in misunderstanding critical information.

1.1. Communication theories

Communication system consists of five parts; a message, a transmitter which operates on the message to produce a signal channel, the receiver and the destination [1]. Ma *et al.* (2008) explained that the basic communication process consists of speaker (people, documents, machines etc), the message (information, feelings, etc) the medium (telephone, speech, text, etc) barriers (psychological, physical & physiological noise) and feedback [2]. Communication in an organisation involves media choices that can help to convey the message effectively. The media that are used for transferring a message; critical and non-critical information, have evolved tremendously over the years; from letter writing, phone calls, face to face discussion to advanced technology; electronic mail, teleconferencing, video chat and voice calls [3]. One theory that highlights the effectiveness and capability of the media in transferring information is the Media Richness Theory (MRT) [3, 4]. The theory is about the choice of the medium made by managers and subordinates in order to accomplish a task that may carry ambiguous information [5]. Richness of a medium is defined when four features below are fulfilled in getting a message across:

- Immediate feedback of the message
- Use of various cues (body language, voice, pictures, and so forth)
- Use of natural language
- Personal focus on the medium

A message is considered lean when one or more of the features is absent. Failure of communication therefore occurs due to misunderstanding or the message. The view of communication failure taken here is inspired by Cushing's (1994) definitions of communication breakdown between a pilot and a traffic controller [6], with the definition being modified to account for the reasons for communication failures during the performance of maintenance, repair and overhaul on aircraft:

- A message that is unavailable
- A message that is available but incomplete
- A message that is available and complete but incorrect
- A message that is available, complete and correct but not clear
- A message that is available, complete, correct and clear, but not understood
- A message that is available, complete, correct, clear and understood but mistakes still happen due to human factors

The involvement of human factors cannot be denied as the act of reading and interpreting documents comprises of the skills and experience of both the writer and the reader of the documents. Next, the errors that are commonly found in documents that lead to aircraft maintenance errors are highlighted.

1.2. Aircraft maintenance errors

To err is human, to forgive divine. The proverb illustrates that it is human nature to never be free from making mistakes. Knowing the fact that humans are susceptible to errors, it is important to identify the contributing factors of errors and the type of errors that may lead to aircraft incidence and/or accidents, also to understand and relate the causes of good practices that leads to aircraft safety and efficiency. It is stressed that one of the factors of maintenance errors is confusion, misunderstanding or differences of opinion about the procedures [7]. Apart from human and organizational factors, documentation error can also cause technicians to make mistakes as listed below [8]:

- Contributing factors or procedural error that cause technician to err
  - Information not understandable
  - Information incorrect
  - Information not enough
  - Information not used
  - Information unavailable
- Maintenance error type-unexpected discrepancy made by technician
  - Improper documentation
  - Wrong procedure

With regards to procedural compliance, there are several reasons for maintenance personnel to not following the procedures given. Civil Aviation Authority (CAA) highlights them as follow [9]:

- The inaccuracy of the procedures
- The procedures are not practical to perform
- Difficult to find information within the written instructions
- Unclear policy on the usage of the procedures

- Rely on skills and experience when performing task
- Unaware of the existence of procedures for certain task

In relation to effects of errors in the written media, maintenance errors were predominantly due to maintenance personnel forgetting a step (60%) during the maintenance work, followed by incorrect installation with 32% and installing using wrong components with 8% (see Figure 1) [10]. However, according to Chaparro et al (2001), the failure to comply with maintenance documents was the main reason for system malfunction, especially when the content of the documents comprise of unclear, insufficient and false information [11]. Confidential Human Factors Incident Reporting Programme (CHIRP) in 2007 reported that mistakes most repeatedly made are not following procedures and not using information that is readily available and to rectify these problems, the producers of the documents are recommended to make sure that the procedures are user friendly and to have an association of the company task cards with the Aircraft Maintenance Manual [12].

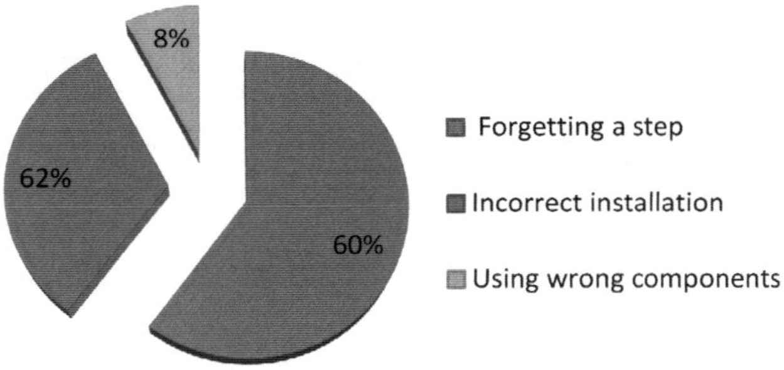

Figure 1. Maintenance errors

2. Methodology

2.1. Data gathering

This case study started with an exploratory research by interviewing technical and non-technical employees from an aviation maintenance organization. The exploratory research is conducted to investigate a new field of which little is known and in this case study, there is no study to the best of the researcher's knowledge that has been carried out in discovering the inter-department and inter-organization communication media in the aircraft maintenance organization and the possible risks the media may carry in conveying the critical information. To explore something new, the researcher employed a semi-structured interview as it is a widely used method for gathering data in social research and can help discover another person's knowledge and experience depending on the breadth and depth of the questions asked. The semi-structured interview was then used to allow participants to provide their own account of daily operations in communicating the critical information between departments. Individual interviews were conducted to let each participant feel comfortable in expressing his/ her thoughts especially when they were talking about the communication issues they had with their colleagues. Interviewing professionals will help to differentiate between process knowledge and context knowledge. Process knowledge is designed to discover a certain type of process, for example, in this case, study, the communication process between departments in GrA, the communication problems they face and the process of addressing those issues. Context knowledge, on the other hand, focuses on the frequency of those communication problems and the role played by individuals or team responsible in helping to solve those issues.

The interview data was gathered at a regional (Malaysian) subsidiary of a major aircraft manufacturer providing sales and maintenance support on behalf of the Original Equipment Manufacturer (OEM). For the purpose of confidentiality as requested by the organisation's top management, the aircraft maintenance organization being studied is fictitiously named as Grand Air (GrA). It consisted of six core departments and one unit; Quality Assurance (QAD), Aircraft Management (AMD), Planning (PD), Supply Chain and Logistics (SCLD), Engineering (ED),Technical Publication Unit (TPU) that is under the Engineering Services (ESD) and Flight Operation (FO). Only thirty three personnel were interested to be interviewed. The interviews were conducted in the respective departments, only when the employees were available. The data was audio recorded using a digital recorder, with the permission of the informants, and transcribed. Analysis of the data identified themes relevant to the research questions was then paraphrased and summarized for quoting.

2.2. Departmental procedures

Once all the interviews were completed, the next step was to list the critical procedures of each department that linked up to the engineering department. This could only be done after the researcher obtained a full view based on how each department communicated between departments and between organisations. It was essential to divide these procedures into two organizational levels: inter-departmental and inter-organizational communication for the purpose of exploring which communication medium, through these procedures, poses a high risk of conveying critical information that may affect aircraft safety and efficiency. Table 1 below is the list of GrA Inter-departmental communication, which consists of seven core procedures. The communication media and their frequency used in each of the inter-departmental procedures will be further discussed below.

Table 1: GrA Inter-departmental communication

Department	Procedures
Quality Assurance department communication with all departments	Audit
Aircraft management department communication with Planning department	Aircraft Status List & Maintenance Forecast
Planning department communication with Aircraft Management department	Return of Completed Work Pack of Scheduled Maintenance
Planning department communication with Engineering department	Work Package Delivery
Engineering department communication strategies with supply chain/ logistic department	Request for Aircraft Spares
Engineering department communication with Flight Operation	Flight Test Procedures
Technical Publication Unit communication with core departments	Technical Documentations from Aircraft Manufacturer (Service Bulletin & Alert Bulletin) and European Aviation Safety Agency (Airworthiness Directives)

3. Data analysis

Data were gathered during the first phase of exploratory research. Selection of data of what was deemed essential in answering the research questions was necessary. Even though employees from eleven departments were interviewed, however, only six departments and one unit, namely the Quality Assurance Department, Aircraft Maintenance Department, Planning Department, Supply Chain and Logistics Department, Engineering Department, Technical Publication Unit and Flight Operation communication strategies were chosen to be analysed. This was because the departments had a closer link to the engineering departments in terms of communicating critical information that had potential effects on aircraft safety and efficiency.

The findings from the interview data relating to each department and communication media was divided into two: verbal and written media. The verbal media consisted of scheduled and unscheduled meetings, telephone calls and face to face communication whereas the written media consists of System Software, Email, documents and fax. The legend (from never (0) to always (4)) helped the researcher to insert the frequency of communication used in the respective departments gathered based on the interview data and on the researcher's interpretation. Once the data were inserted, further analysis of the data was displayed through the risk assessment model. Risk assessment was introduced in this study as an additional analysis to the qualitative data to find out which communication media presented the highest risk of transferring critical information. Risk assessment is widely used in the engineering field to predict component failures to system failures of in an engineering product, for example in nuclear power, oil and gas, health science, aerospace industries, etc. Table 2 below is the risk assessment matrix developed to indicate GrA communication media used in transferring critical information based on its inter-departmental communication procedures.

Table 2. GrA inter-departmental communication risk assessment matrix

1 Very High	5 Low	20 Medium	45 High	80 Extreme	125 Extreme
2 High	4 Low	16 Low	36 Medium	64 High	100 Extreme
3 Medium	3 Low	12 Low	27 Medium	48 High	75 High
4 Low	2 Low	8 Low	18 Medium	32 Medium	50 High
5 Very Low	1 Low	4 Low	9 Low	16 Low	25 High
Scale of Impact	1 Insignificant	2 Minor	3 Moderate	4 Major	5 Catastrophic

It is important to note that in measuring risk, there are different levels of risk measurement that are motivated by various circumstances. Table 3 indicates the risk rating for the assessment matrix. The reason for the rating is being classified as such indicates the level of risk of both verbal and written media used in communicating critical information based on the critical procedures of each GrA core department with other departments. The low level shows that the critical information being conveyed by the media used present very little indication of miscommunication between departments and organisations and it does not have a serious impact on aircraft safety and efficiency. However, the medium level of risk though does not show an alarming sign of risk but does need attention so that the media used can be improved to prevent further escalation of conveying misunderstood information. The high level of risk rating demonstrates a worrying indication that the media used to communicate critical information can possibly affect aircraft safety and efficiency and needs to be attended to, to eliminate misunderstanding. The final level is the extreme risk rating. When a procedure being conveyed through the media and they are identified as delivering the highest level of misunderstood information, to a point that the outcome is catastrophic to aircraft safety and efficiency, then the media needs to be re-evaluated and enhanced so that unwanted incidents can be avoided.

Table 4 indicates the classifications for the frequency of communication media being used between departments and between organisations in GrA. 'Never' represents the verbal and written media are not used at all to convey critical information. 'Seldom' demonstrates that the verbal or written media are rarely used to convey the critical message while 'Sometimes' indicates that the media are occasionally used to transfer critical information. 'Often' informs us that the verbal and written media are used regularly, however 'Always' illustrates that the media are used constantly to convey the critical information. Table 5 shows the frequency, probability and consequences classification of GrA communication media

4. Results

4.1. The participants' demographic background

Out of 70 technical and non-technical personnel from the eight departments, only thirty three employees volunteered to participate in the interview. Table 6 below shows the distribution of the interview subjects based on gender, age, nationality and position. There were twenty seven (82%) males and six (18%) females that were interviewed and ten of them (30%) were between thirty six and forty years old, while nine (27%) represented between thirty one and thirty five and another eight

(24%) were between forty and above. However, only six (18%) were between twenty five and thirty years old. Most of the subjects were Malaysians with 32 (97%) while only one French (3%). The subjects interviewed held different positions. Eight (24%) of them were managers from eight departments, five (15%) were assistant managers, three (11%) executives, one senior officer, inspector and a pilot (3%) each, seven senior technician (21%), three (11%) junior technicians, with another two being (6%) officers.

Table 3. Risk rating and action based on calculation

Risk Calculations Frequency*Probability*Consequences	Risk Rating	Actions
Frequency (3) x Probability (2) x Consequences (2) = 12	Low 1 - 16	Even though the risk is low, small errors need correction
Frequency (3) x Probability (3) x Consequences (3) = 27	Medium 17 - 36	Close monitoring of work process needed
Frequency (5) x Probability (3) x Consequences (3) = 45	High 37 - 79	Fast action needed to rectify problems
Frequency (5) x Probability (5) x Consequences (5) = 125	Extreme 80 - 125	Discontinue of work immediately to prevent serious risk of injury

Table 4. Frequency classification of GrA communication media

Frequency	**Never (1)** Verbal or written media are not used to convey critical information	**Seldom (2)** Verbal or written media are rarely used to convey critical information	**Sometimes (3)** Verbal or written media are occasionally used to convey critical information	**Often (4)** Verbal or written media are regularly used to convey critical information	**Always (5)** Verbal or written media are constantly used to convey critical information

Table 5. Frequency, probability and consequences classification of GrA communication media

Consequences	Frequency	Work Performance	Probability
Insignificant (1)	Never (1) Verbal or written media are not used to convey critical information	Trivial maintenance error done due to outdated reference	Very low (1) May occur only in exceptional circumstances
Minor (2)	Seldom (2) Verbal or written media are rarely used to convey critical information	Small maintenance errors due to misunderstanding of instructions	Low (2) Could occur at some time
Moderate (3)	Sometimes (3) Verbal or written media are occasionally used to convey critical information	Tolerable delay of maintenance work due to non-conforming aircraft products from manufacturer	Medium (3) Possible to occur at some time
Major (4)	Often (4) Verbal or written media are regularly used to convey critical information	Major maintenance errors made due to incorrect or misunderstanding of maintenance manual	High (4) Likely to occur at some time
Catastrophic (5)	Always (5) Verbal or written media are constantly used to convey critical information	Severe maintenance errors made due to misunderstanding of instructions and failure to comply with standard operating procedures and regulations	Very High (5) Expected to occur in most circumstances

Table 6. Participants' demographic background based on gender, age, nationality and position (n=33)

Characteristics			
Gender:			
Male	27 (82%)		
Female	6 (18%)		
Age:			
25 – 30	6 (18%)		
31 – 35	9 (27%)		
36 – 40	10 (30%)		
40 and above	8 (24%)		
Nationality: Malaysian	32 (97%)		
French	1 (3%)		
Position:			
Manager	8 (24%)	Senior Technician	7 (21%)
Assistant Manager	5 (15%)	Junior Technician	3 (11%)
Executive	3 (11%)	Pilot	1 (3%)
Senior Officer	1 (3%)	Officer	2 (6%)
Inspector	1 (3%)		

4.2. GrA inter-departmental communication

The results of the analysis were first presented by communication media used within each department. The core departments, namely the Quality Assurance (QAD), Aircraft Management Department (AMD), Planning Department (PD), Supply Chain and Logistics Department (SCLD), Engineering Department (ED) and the Technical Publication Unit (TPU), which is located under the Engineering Services Department (ESD). The communication media used to communicate with each other with regards to daily aircraft maintenance operations were:-

- 'Unscheduled Meeting' (meetings called when necessary to discuss important or urgent matters)
- 'Face to face' (regular social interaction between personnel discussing work)
- 'Documents' (written sources of critical information used to refer to and act upon)
- 'Scheduled meetings' (formal meeting with set dates, agendas and minutes)

From the interview data and observation, it was found that scheduled meetings were very rare at GrA, especially at the executive level: they took place between once a month and once in every three months. At unscheduled meetings, the manager of the department briefed the staff on the critical nature of certain issues. When asked why this approach was used, managers commented that they preferred to deal with daily issues as they arise, talking to their staff face to face about what problems they encountered, discussed and finding solutions. This medium of communication was observed in almost all departments. Most of the staff indicated that they were comfortable with this approach as an efficient way of solving many issues rather than spending a lot of time in meeting rooms. The document was essential when dealing with identification and traceability of critical information. In all departments, these three communication media were supplemented by email as records to keep track of information that was passed on and/ or needed urgent attention.

'System software' was used mainly by Aircraft Maintenance Department, Planning Department, Engineering Department, Supply Chain and Logistics Department. This was used based on the Aircraft Status List & Maintenance Forecast Procedures that required these four departments to communicate with each other using the System software. Phone calls were used much less frequently than face to face communication as the departments were in close proximity. Staff members were frequently observed to walk to the other department and had face to face contact in order to question, reaffirm, show or discuss issues on certain documents that may seem difficult to explain on the phone.

The results of the risk assessment focused on the communication media used in inter-department and inter-organizational at GrA. The data multiplications of risk assessment were designed to identify those communication media that presented the greatest potential for conveying incorrect information. The findings were categorised into verbal and written media. The former include scheduled and unscheduled meetings, telephone calls and face to face communication; the latter includes software, email, controlled documentation and faxes. The frequency of use of the media was derived from the interview data. Next, we shall look at the level of risk in conveying incorrect information through the analysis of communication medium from various work procedures.

4.2.1. Quality assurance department communication strategies with all departments on audit procedures. As shown in Table 7, it was found that the Quality Assurance Department scored 20, low in risk, in conveying incorrect information by using face to face meetings and email to communicate the audit. The incorrect information coming from this department to other departments would create a delay of audit preparation and might affect the audit plan. However, it did not have a direct impact on aircraft safety and efficiency, but would rather slow down audit plan activities.

Table 7. Quality assurance department communicating audit procedures

	Verbal block				Written block			
	Meetings		Phone call	F2F	Software	Email	Controlled	
	Scheduled	Unscheduled					Document	Fax
Frequency	5	5	4	5	1	5	5	1
Probability of miscommunication	2	2	2	2	1	2	2	1
Consequences of miscommunication	2	2	2	2	1	2	2	1
Triple multiple	**20** MEDIUM	**20** MEDIUM	**16** LOW	**20** MED	**1** LOW	**20** MED	**20** MED	**1** LOW

4.2.2. Aircraft management department communication strategies with planning department on aircraft status list and maintenance forecast procedures. The outcome scales in Table 8 clearly indicates that the written media (Software and Controlled documentation) used between Aircraft Management Department and Planning Department in relation to the Aircraft Status List and Maintenance Forecast Procedures present the highest score of 45. This provides a serious miscommunication indication that may negatively affect maintenance work in relation to these media. This was due to the wrong information from the documents being keyed into the software system where the aircraft status list and maintenance forecasting information passed the planners. The same information was then transferred to the Engineering department and incorrect maintenance work took place.

Table 8. Aircraft management department communicating aircraft status list and maintenance forecasting procedures

	Verbal block				Written block			
	Meetings		Phone call	F2F	Software	Email	Controlled	
	Scheduled	Unscheduled					Documentation	Fax
Frequency	2	2	2	1	5	2	5	1
Probability of miscommunication	1	1	2	1	3	2	3	1
Consequences of miscommunication	1	1	2	1	3	2	3	1
Triple multiple =	**2** LOW	**2** LOW	**8** LOW	**1** LOW	**45** **HIGH**	**8** LOW	**45** **HIGH**	**1** LOW

4.2.3. Planning department communication strategies with aircraft management department on return of completed work pack of scheduled maintenance. The incorrect information conveyed through software and document is rather low here with a score of 16 (Table 9). This was due to the insertion or update of information that was written in the work pack for the Aircraft Maintenance Department personnel to check and verify.

Table 9. Planning department communicating the return of completed work pack of scheduled maintenance

	Verbal block				**Written block**			
	Meetings		**Phone call**	**F2F**	**Software**	**Email**	**Controlled**	
	Scheduled	**Unscheduled**					**Documentation**	**Fax**
Frequency	2	2	1	4	4	2	4	1
Probability of miscommunication	1	1	1	1	2	1	2	1
Consequences of miscommunication	1	1	1	1	2	1	2	1
Triple multiple =	**2**	**2**	**1**	**4**	**16**	**2**	**16**	**1**
	LOW	LOW	LOW	LOW	LOW	LOW	LOW	LOW

4.2.4. Planning department communication strategies with engineering department on work package delivery. Table 10 shows that this procedure scored 20. This indicates a low risk of conveying incorrect information as the information was mainly on incorrect dates, signature, aircraft part numbers or reporting the procedures of the maintenance work being done on the aircraft.

4.2.5. Engineering department communication strategies with supply chain/logistic department on request for aircraft spares. Table 11 indicates that in communicating this procedure with the Supply Chain and Logistics Department, the Engineering Department scored 16 in the controlled documentation. This was due to the wrong request of aircraft spares when ground crew referred to incorrect aircraft manuals or illustrated parts catalogue prior to maintenance work on aircraft. When the Aircraft Maintenance Technicians realized they had wrongly requested the parts, requesting a new one might take a longer amount of time if it was not available at the Supply Chain and Logistics Department. The Engineering Department personnel felt that this did not affect aircraft safety and efficiency as it would be detected before commencing the work by double checking the aircraft manual and work packages.

Table 10. Planning department communicating work package delivery

	Verbal block				**Written block**			
	Meetings		**Phone call**	**F2F**	**Software**	**Email**	**Controlled**	
	Scheduled	**Unscheduled**					**Documentation**	**Fax**
Frequency	3	4	3	4	2	3	5	1
Probability of miscommunication	1	1	1	1	1	1	2	1
Consequences of miscommunication	1	1	1	1	1	1	2	1
Triple multiple =	**3**	**4**	**3**	**4**	**2**	**3**	**20**	**1**
	LOW	LOW	LOW	LOW	LOW	LOW	MEDIUM	LOW

Table 11. Engineering department communicating requisition of aircraft spare

	Verbal block				Written block			
	Meetings		Phone call	F2F	Software	Email	Controlled	
	Scheduled	Unscheduled					Documentation	Fax
Frequency	2	3	4	4	1	3	4	1
Probability of miscommunication	1	1	1	1	1	1	2	1
Consequences of miscommunication	1	1	1	1	1	1	2	1
Triple multiple =	**2**	**3**	**4**	**4**	**1**	**3**	**16**	**1**
	LOW	LOW	LOW	LOW	LOW	LOW	LOW	LOW

4.2.6. Engineering department communication strategies with flight operation on flight test procedures. Another indication of the low risk of incorrect information is when the Engineering Department communicated with Flight Operation on flight test planning (Table 12). The wrong information was caused by incorrect information on the dates and time for flight test and which aircraft to flight test on, which did not have any effect on flight safety and efficiency.

4.2.7. Technical publication unit communication with core departments on technical documentations received from aircraft manufacturer (Service Bulletin & Alert Bulletin) and EASA (Airworthiness Directives). It was found that the written media (email and controlled documentation) used between the technical publication unit and the other core departments on technical documents scored 45 as shown in Table 13, and thus posed a high risk of miscommunication. This would happen especially if the technical documents were not being updated. The wrong information found in the technical document, if not corrected, could be used as a reference during maintenance work and eventually lead to serious maintenance error done on the aircraft.

5. Discussion

The results show that out of seven core inter-departmental procedures, only two procedures indicate a high risk of conveying incorrect information connected to written communication which is considered to be a less rich media as it has the potential to convey an equivocal message. In this case, the lean media could potentially affect aircraft safety and efficiency due to its heavy usage of written media in transferring the critical information.

Table 12. Engineering department communicating flight test procedures

	Verbal block				Written block			
	Meetings		Phone call	F2F	Software	Email	Controlled	
	Scheduled	Unscheduled					Documentation	Fax
Frequency	2	3	3	4	1	2	4	1
Probability of miscommunication	1	1	2	2	1	1	2	1
Consequences of miscommunication	1	1	2	2	1	1	2	1
Triple multiple =	**2**	**3**	**12**	**16**	**1**	**2**	**16**	**1**
	LOW	LOW	LOW	LOW	LOW	LOW	LOW	LOW

Table 13. Technical publication unit communicating technical documents

	Verbal block				Written block			
	Meetings		Phone call	F2F	Software	Email	Controlled	
	Scheduled	Unscheduled					Documentation	Fax
Frequency	4	3	3	4	1	5	5	3
Probability of miscommunication	1	1	1	1	1	3	3	1
Consequences of miscommunication	1	1	1	1	1	3	3	1
Triple multiple =	**4**	**3**	**3**	**4**	**1**	**45**	**45**	**3**
	LOW	LOW	LOW	LOW	LOW	**HIGH**	**HIGH**	LOW

There are two inter-department and three inter-organization communication procedures that are found to pose a high risk in conveying critical information through written media as listed below:

- Aircraft Maintenance Department with Planning Department on Aircraft Status List
- Technical Publication Unit with GrA core departments on technical documents

The case study clearly indicates that using software and Controlled documentation between Aircraft Management and Planning departments to communicate on aircraft status list and maintenance forecasting procedures result in miscommunication that may negatively affect maintenance work. Based on the interviewees' comments, the danger of miscommunication arises from the manual insertion of critical information in relation to the aircraft maintenance forecasting into the software. The information may lead to misinterpretation by the planners which is then transferred into the Work Package (WP). Unless the planners realize the incorrect information and inform the Aircraft Management Department to amend them, incorrect maintenance work ensues and this may affect aircraft safety and efficiency. The wrong information passed to two written media (email and controlled documentation) from the aircraft manufacturer to the Technical Publication Unit would eventually transfer to other core departments in GrA. According to the technical library, officer interviewed, this may occur when the technical documents namely flight manuals and maintenance manuals that accompanied the Illustrated Parts Catalogue (IPC) in a compact disc from the manufacturers, were outdated. This would affect inefficiency of work throughout the departments and it was plain that Engineering Department would usually detect the incorrect information and would inform Technical Publication Unit to seek clarification. The wrong revision of technical documents produced by the manufacturer posed a high risk of misinformation. Hence, aircraft safety and efficiency might be compromised, especially when Licenced Aircraft Engineers and Aircraft Maintenance Technicians performed the maintenance work on aircraft based on outdated procedures.

The findings of this study were similar to those in [8] where the document deficiencies was found to be one of the two major causes (the other being user errors) of incorrect or inefficient maintenance procedures. All three verbal media, meetings, phone call and face to face communication, however, scored low (1 to 16). This is reflected in the interview and observational data, which showed that GrA personnel preferred face to face meetings to solve critical or non-critical issues immediately as they arose. Table 14 below is the summary of two departments' communication procedures that pose a high risk in conveying incorrect, insufficient and unavailable critical information. It suggests that the two procedures that carried critical information using written media had the potential to convey incorrect information. These procedures had relation to aircraft maintenance work and were likely to jeopardise aircraft safety and efficiency if they were not detected and improved promptly.

Table 14. Summary of departments with high risk in communicating critical information using written media

Department	Procedures	Written Media	High Risk
Aircraft Maintenance Department with the Planning Department	Aircraft Status List & Maintenance Forecasting	Software & Documentation	Incorrect information
Technical Publication Unit with GrA core departments	Flight Manuals, Aircraft Maintenance Manual, Service Bulletin & Airworthiness Directives	E-mail & Documentation	Incorrect, Insufficient & Unavailable information

The overall results also agree that the Media Richness Theory [3, 4] has proven that written media such as documentation, electronic mail and software are less rich and pose a higher risk to due to incorrect, insufficient and unavailable critical information that may lead to maintenance errors and aircraft delay. On the other hand, verbal media, for example, face to face communication, phone calls and meetings were suggested as carrying a low risk of misinterpretation as any incorrect or unclear critical information was thoroughly discussed and clarified as they arose. However, the study has also shown that instead of relying solely on the written media to interpret the message, both technical and non-technical personnel did take the initiative to confirm any information that was considered to be equivocal by utilising the rich media (face to face, phone call and meetings).

6. Conclusion

The study reveals a scenario of verbal and written communication in conveying critical information in an aircraft maintenance organisation. Looking at the bigger picture, written media in the maintenance organisation can potentially affect aircraft safety and efficiency in the following situations.

Aircraft safety and efficiency can be affected due to incorrect, insufficient and unavailable information and if it was not detected earlier, it can be taken and used as true information. It would be more harmful if the same information was passed through from one person to the other; this could create a greater problem, especially when maintenance, repair and overhaul work on aircraft is carried out. Early discovery of errors could save time and cost. Consequently, maintenance work on aircraft can be carried out accordingly. Another factor in communication failure is the misinterpretation of the content by the user. Even though the information conveyed in the written media carries complete, correct and clear information, misunderstanding of the message leads to the creation of misunderstood information in the reader's mind. This could also jeopardise aircraft safety and efficiency when the wrong interpretation of the initially correct information being passed from one department to the other thus result in maintenance errors.

Written media such as the technical manuals are documented and are referred to by aircraft maintenance personnel, to be used in daily maintenance operations. How it is written and presented plays a vital role in making sure every user understands the content; the outcome of the work produced should be as it was intended to be. However, one person may have a different interpretation and understanding of the message most probably because of his/ her knowledge and experience. If a common understanding of the written media is not achieved through open discussions among Licensed Aircraft Engineers and Aircraft Maintenance Technicians, then mistakes will happen. The verbal media of communication, however, posed a lower risk in conveying critical information in the maintenance organisation. This was because verbal clarification took place whenever misunderstanding of critical information was detected. The confirmation and verification of uncertain or potentially incorrect or insufficient information were carried out through face to face discussion. Therefore, verbal media carried a low risk in affecting aircraft safety and efficiency.

Acknowledgements
Thank you to Universiti Kuala Lumpur Malaysian Institute of Aviation Technology and University of Aberdeen, Scotland for the strong support.

References

[1] Shannon C E 1948 *The Bell System Technical Journal* **27** 379-423
[2] Ma J, Drury C G and Marin C V 2009 *The International Journal of Aviation Psychology* **20** 25-47
[3] Daft R L and Lengel R H 1986 *Management Science* **32** 554-71
[4] Daft R L and Lengel R H 1984 *Research in organizational behavior* **6** 191-233
[5] Sheer V C and Chen L 2004 *Management Communication Quarterly* **18** 76–93
[6] Cushing S 1994 *Fatal Words: Communication Clashes and Aircraft Crashes* (Chicago: University of Chicago Press)
[7] Hobbs A 1999 *Asia-Pacific Air Safety* **21** 2-7
[8] Lattanzio D, Patankar K and Kanki B G 2008 *The International Journal of Aviation Psychology* **18** 17-29
[9] CAA 2003 *CAP 716: Aviation Maintenance Human Factors (EASA/JAR145 Approved Organisations) Guidance Material on the UK CAA Interpretation of Part-145 Human Factors and Error Management Requirements*
[10] ICAO 2010 *Annual Report of the Council*
[11] Chaparro A and Groff S L 2001 *Human Factors Survey of Aviation Technical Manuals* Federal Aviation Administration
[12] William L R, Shappell S and Wiegmann D 2008 *Human Factor Guide for Aviation Maintenance and Inspection* Federal Aviation Administration

IOP Conf. Series: Materials Science and Engineering **152** (2016) 012045 doi:10.1088/1757-899X/152/1/012045

An experimental study of non-destructive testing on glass fibre reinforced polymer composites after high velocity impact event

N Razali, M T H Sultan* and F Cardona

Aerospace Manufacturing Research Centre (AMRC), Faculty of Engineering, Universiti Putra Malaysia, Malaysia.

*thariq@upm.edu.my

Abstract. A non-destructive testing method on Glass Fibre Reinforced Polymer (GFRP) after high velocity impact event using single stage gas gun (SSGG) is presented. Specimens of C-type and E-type fibreglass reinforcement, which were fabricated with 6mm, 8mm, 10mm and 12mm thicknesses and size 100 mm x 100 mm, were subjected to a high velocity impact with three types of bullets: conical, hemispherical and blunt at various gas gun pressure levels from 6 bar to 60 bar. Visual observation techniques using a lab microscope were used to determine the infringed damage by looking at the crack zone. Dye penetrants were used to inspect the area of damage, and to evaluate internal and external damages on the specimens after impact. The results from visual analysis of the impacted test laminates were discussed and presented. It was found that the impact damage started with induced delamination, fibre cracking and then failure, simultaneously with matrix cracking and breakage, and finally followed by the fibres pulled out. C-type experienced more damaged areas compared to E-type of GFRP.

1. Introduction

When composites were introduced into aircraft components and aerospace industries, unexpected impacts occurred. These might be due to damages during flight operations such as runway debris on composite airframes, bird strike during flight operation and dropping of hand tools during maintenance work. For this research, the impact velocity is set to be less than 31 m/s, which indicates low velocity testing. For high velocity testing, the impact velocity is in the range of 31 m/s to 240 m/s. In this study, high velocity impact test has been chosen to be conducted on Glass Fibre Reinforced Polymer (GFRP) since this material has been used widely in many applications. The study of a high velocity impact comparison between two types of GFRP: durability (Type C) and electrical conductivity (Type E) has never been done before. This study is necessary to analyse the impact behaviour of GFRP in the form of chemical. When these materials are subjected to the high velocity impacts, their structural integrity, stiffness and also toughness are significantly reduced and this may result in a catastrophic failure to the structure in extreme scenarios. Matrix cracking, fibre fracture, fibre pullout and delamination are examples of major undetected hidden damages in composite materials after an event of impact. There is a need to study the behaviour of the composite materials under impact loading as the impacts occur mainly during maintenance and work manufacturing.

Optical microscopy (OM) as well as scanning electron microscopy (SEM) are classical techniques for composite surface studies [1–3]. From the observation, the types of failure can be determined and analyzed. A variety of failure modes after high impact velocity tests may lead to perforation of the composite panels. Failure modes of the target will vary depending on the properties of the materials, impact velocity, projectile nose shape, target geometry, support conditions, relative mass of projectile and target, etc. The most common failure modes are illustrate as in Figure 1.

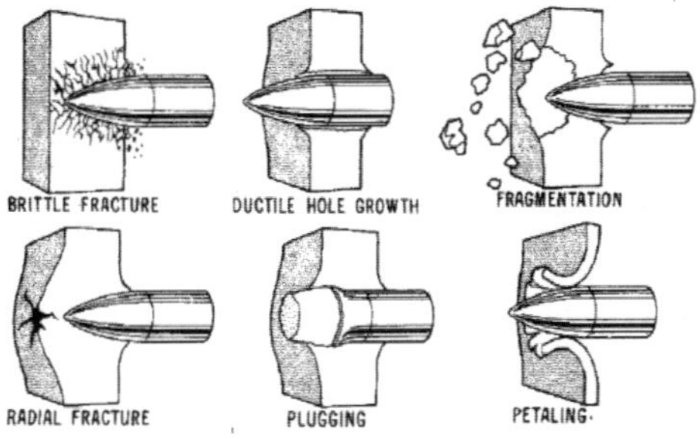

Figure 1. Common failure modes on the different type of materials [4]

The most significant failure modes in the present exploration are brittle fracture and petaling. A brittle fracture happens as a consequence of tensile stress acting typical to the crystallographic planes with low bonding when the bullets hits the surface material. Petaling failure happens when the rigidity is surpassed at the back side of the objective and a star-moulded break creates around the tip of the bullets. The parts shaped are then pushed back by the movement of bullets, framing petals as appeared in Figure 1. Matrix damage is the main sort of failure instigated by a transverse low-velocity impact. Typically, this appears as matrix breaking and de-bonding between the fibre and the matrix. Matrix breaks happen because of property confounding between the fibre and the matrix, and are typically situated in planes parallel to the fibre heading in the unidirectional layers. Delamination is a split that keeps running in resin-rich range between various fibre orientation and it is not between lamina in the same ply bunch. Delamination is an after-effect of bending mismatch coefficient between two nearby laminates such as diverse fibre orientations between the layers. The more the mismatch (0/90 degrees), the more delamination range will be [5]. The delamination will influence the materials' properties and it disrupts the stacking arrangement and the laminates thickness. Both bending splits and shear breaks could start delamination. Delamination initiated by shear breaks is precarious while the bending break prompts delamination to develop in a steady way in correspondence to the applied load. This damage mode generally happens much later in the break procedure than matrix splitting and delamination.

Fibre failure happens under the penetrators because of local high stresses and the space impacts of shear forces. It also occurs on the non-affected face because of high bending stresses. The fibre failure is a forerunner to the disastrous infiltration mode. Penetration is a plainly visible method of failure. It happens when the fibre failure achieves a basic degree, empowering the penetrators to totally infiltrate the material. It is anything but difficult to anticipate the introduction of matrix cracking fibre of the unidirectional layers while the break patterns of haphazardly situated layers are less simple to set up [6-8]. The aims of this research are to analyse the damage area and the effect of high velocity impact, and to observe the types of failure using non-destructive technique corresponding to the different types of bullets and impact energy level between the two types of GFRP using single stage gas gun (SSGG).

Various NDT techniques are used for damage detection in composite structures. Examples include acoustic emission technique (AE), ultrasonic testing, structural health monitoring, visual inspection technique, eddy current testing, holography and shearography imaging technique, vibration-based technique and lamb wave techniques. AE is a passive method that has been widely used as the NDT tool for damage detection in composite structures. AE analysis is a useful method for the investigation of local damage in materials. One of its advantages in comparison to the other NDE techniques is the possibility to observe the damage processes during the entire load history without any disturbance to specimen. AE analysis is used successfully in a wide range of applications including for detecting and locating faults in the pressure vessels or leakage in storage tanks or pipe systems, monitoring welding applications, corrosion processes, partial discharges from components subjected to high voltage and the removal of protective coatings [9]. Ultrasonic testing uses high frequency sound energy to conduct examination and make measurements. The ultrasonic inspection can be applied for flaw detection or evaluation, dimensional measurements, material characterization and more [10].

Standard eddy current testing is essentially a near-surface technique, which is useful for detecting surface breaking or near-surface cracking and variations in material composition. It can also be used to measure the thickness of non-electrical conductive coatings on electrically conductive substrates. In general, standard eddy current methods are only used in plant inspection for nonferritic material where the eddy current penetration is deeper or for special applications such as inspection of heat exchanger tubing for cracking or corrosion thinning [11]. The vibration based model dependant methods with piezoelectric sensor and actuator incorporated into composite structures offer a promising option to fulfil such requirements and needs. These methods utilize finite element analysis techniques, together with experimental result, to detect damage. They locate and estimate the damage events by comparing dynamic responses between the damaged and undamaged structures. Meanwhile, holography is 'lens-less photography' in which the image is captured not as an image focused on film but instead as an interface pattern at the film. Typically, coherent light from a laser is reflected from an object and combined at the film light from a reference beam. This recorded interference pattern actually contains more information than a focused image and enables viewers to look at the appearance from several different angles, just as if looking at a real 3D object [12].

On the other hand, structural health monitoring is an upcoming technology in civil, mechanical and aerospace engineering. It is an approach used to monitor structure physical properties such as loading, stresses, strains, accelerations, cracks and etc. It is also a maintenance approach that can detect and identify global structural damage through automated continuous or periodical monitoring. Moreover, visual inspection is the most cost-effective method used for damage detection in aircraft structures. Visual inspection requires little equipment. Aside from good eyesight and sufficient light, all it takes is pocket rule, weld size gauge, magnifying glass, and possibly straight edge and square for checking straightness, alignment and perpendicularity [13]. Liquid penetrant inspection technique is a widely applied and low-cost inspection method used to locate the surface breaking defect in all non-porous material including metals, plastics, ceramics and composites. Penetrant may be applied to all non-ferrous materials but magnetic particle inspection is preferred for inspection of ferrous components due to its subsurface detection capability. Liquid penetrant inspection is used to detect casting and forging defects, cracks and leaks in new products, and fatigue cracks in service components. In this research, visual inspection techniques using a dye penetrant was used to observed the damage area and the mode of failure after high velocity impact.

2. Methodology

2.1. Specimens fabrication

GFRP of type E-glass and type C-glass are chosen for this research since they are easy to obtain and of low cost. Thermal properties of the fibre is one of the criteria that affect the material selection. Since this research is more concerned with high velocity impact, the material that can perform well in high application should be considered. Thus, the same materials are used for low velocity impact testing. Fibres made primarily from silica-based glass containing several metal oxides offer excellent thermal and impact resistance, high tensile strength, good chemical resistance and outstanding insulating properties. E-glass has a low thermal conductivity and is resistant to thermal shock. Fibreglass woven roving fabric also easier to fabricate compared to the chopped strand mat, giving it design flexibility. The manufacturing procedure for this GFRP consists of material preparation, a lay-up process, curing and CNC machine cutting. All specimens are oriented in the same 0^0 direction.

The GFRP woven cloth was first cut into large panels of 320 mm long and 320 mm wide. After the cutting process, the next step was to perform the lay-up preparation using a heat blanket machine. The laminates were fabricated for the impact tests with different thickness, i.e. 6, 8, 10 and 12 mm. The thickness of each plies of fibre had been measured before the fabrication. The estimation of number of layers had been made after a few sample had been fabricated. All of the fabricated specimens had the same fibre loading and matrix weight fraction. Epoxy resin was used as the matrix by the mixing of epoxy and hardener in 2:1 weight ratio. Both the epoxy and the hardener used were of type Zeepoxy HL002TA and Zeepoxy HL002 TB, respectively. The mixed epoxy-hardener had a low viscosity that allowed easy handling and provided good wetting of the glass reinforcements. The curing process for the high velocity impact test specimens involved the use of heat blanket machine Heatcon HCS9000B and vacuum bagging process. This method accelerated the curing process and also resulted in better strength of the composite material due to the pressure applied by the metallic plate that properly compacted the panel. The material was cured in the vacuum bagging for two hours at 150 °F (65.56°C) to ensure the resin flowed uniformly to the whole glass layers and that the resin fully cured inside the laminates. Each panel then was cut into nine specimens with the required dimensions of 100 ×100 mm as shown in Figure 2 using a CNC router machine.

2.2. Impact testing

The impact tests were carried out using a single stage gas gun as shown in Figure 3. The test machine consists of five main components: catch chamber, ballistic data acquisition system, pressure reservoir unit, firing mechanism unit and the launching unit. The catch chamber provides room for holding the specimen to be tested. The impact tests were performed by clamping the specimens between two steel frames at the catch chamber and striking them at the centre with a bullet.

The ballistic data acquisition system collected the data from the test such as the impact energy and the impact force. The pressure reservoir unit consists of a cylindrical gas tank, pressure regulator, a pressure vessel and a pressure control valves. This unit is used to control the pressure of the gas gun. The pressures were varied from 6 bar to a maximum of 60 bar. Varying the gas pressure in the firing chamber will proportionally vary the bullet velocity. Table 1 shows the details of gas pressures used according to the test panel's thickness. The applied gas pressure was varied due to the fact that thin panel samples were not able to withstand the high velocity impact (given by high gas pressure) and therefore, it was necessary to reduce the gas pressure to avoid complete damage of the panels by the total penetration of the bullets on impact. Initial impact pilot tests established the gas pressure to apply in accordance with the thickness of the test panels.

Figure 4 shows three different types of bullets used in this experiment: blunt, hemispherical and conical shape. Each bullet had different cross section area that would affect the impact test result. The bullets that had been used were made from stainless steel. All bullets had the same weight of 5 g. The dimensions of projectile used were 15 mm in length and 10 mm in diameter. The firing mechanism unit influenced the gun's performance by insuring that the bullet slid smoothly along the chamber and shot the target specimen.

Figure 2. GFRP specimens for high velocity impact testing

Figure 3. Single Stage Gas Gun (SSGG)

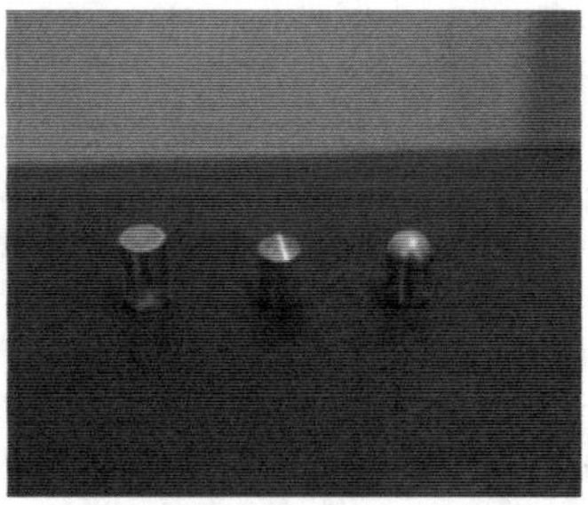

Figure 4. Type of bullet

IOP Conf. Series: Materials Science and Engineering **152** (2016) 012045 doi:10.1088/1757-899X/152/1/012045

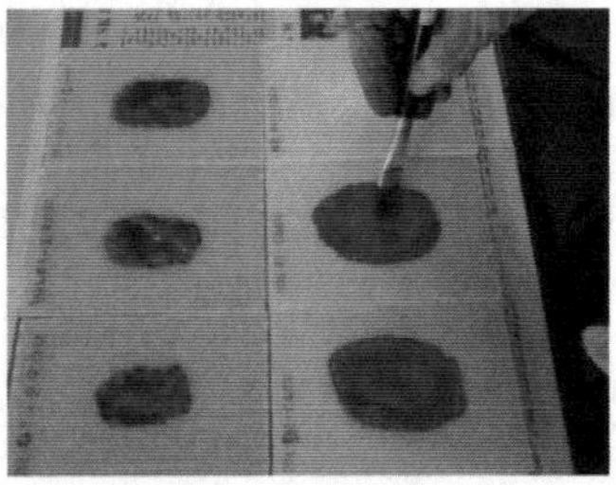

Figure 5. Dye penetrant applied on the surface of specimens using brush

Table 1. The pressure varied for each thickness of specimens

Thickness (mm)	Pressure (bar)		
	P_1	P_2	P_3
6	6	12	18
8	10	20	30
10	15	30	45
12	20	40	60

The test was carried out by varying the pressure of the gas gun from 6 bar to 60 bar. Three projectiles of different geometries, which were blunt, hemispherical and conical, were impacted to the laminated specimen of 100 mm x 100 mm with 6 mm, 8 mm, 10 mm and 12 mm thickness. Each specimen was subjected to a single impact event.

2.3. Non-destructive technique

There are many non-destructive techniques that can be used to examine impact damage of composite materials. One of the non-destructive techniques is by using a dye penetrant process. In this particular work, this technique was selected to examine the damage. In this experiment, Spotcheck SKL-SP2 dye penetrant was used. It is a red-colour contrast penetrant that is removable by solvent and with an outstanding penetrating characteristics. Thinner was used to clean the surface of the specimens from small debris and dirt. A synthetic cloth was used to wipe the surface. The dye penetrant was applied on the surface of the test specimens using a brush as shown in Figure 5, which was then left to dissipate through the damage area for 20 minutes. After 20 minutes, the specimens once again were wiped with the synthetic cloth and thinner to clean the surface. The damage area could be seen as the red colour of the dye contrasting with the colour of the specimens. As the damage areas were much clearer after the dye penetrant process, they will be easily examined using optical microscope. The optical microscope used in this examination was Olympus BX51 microscope, which can observed failure modes of the specimens over the impact zone. The advantages of using this optical microscope are its superb optical performance, optimised contrast and resolution, and outstanding fluorescent capability. The damage area was then calculated manually by grid paper.

3. Results and discussion

After the high velocity impact test was completed, the same non-destructive procedure techniques were perform on the tested specimens, which were the dye penetrant process and optical microscopic examination. Figure 6 shows the failure mode of the specimens after impact for both types of GFRP. Meanwhile, Figure 7 to Figure 19 show the surface of the GFRP laminates after impact by different types of bullets and with different applied gas pressure. The actual penetration of the bullets into the GFRP laminates was 70~80% maximum of the panel thickness, without going through the whole panels. After high velocity impact, all specimens underwent fibre breakage and fibre pullout as shown in Figure 6. Due to the high velocity impact event, the bullet penetrated through the surface of the specimens and broke the fibres, inducing a delamination, matrix cracking and matrix breakage on the side of the test specimens. The figures highlight that the bullets also created a hole on the impacted surface of the test specimens. The diameter of the damage area was increased as the applied gas gun pressure was increased. The impact damage started with induced delamination, fibre cracking and failure, simultaneously with matrix cracking and breakage, and finally followed by the fibres pulled out (see Figure 6). From the observation, Type E-glass/epoxy 600 g/m^2 laminates experienced the same failure as the type C-glass/epoxy 600 g/m^2 panels. The only main difference was that type C-panels suffered much larger and wider crack damage, and also much deeper bullet penetration than type E-panels. This shows that E-glass is tougher and stronger in comparison to C-glass. The visual observation of the induced damage was much clearer after the dye-penetrant process. The recorded images of the impact areas showed similar damage conditions (impact area and penetration hole) for panels hit by the same type of bullets and with the same gas pressure, indicating a high grade of reproducibility and consistency on the high velocity impact tests. The micrographs also revealed that the increase in the gas gun pressure enhanced the damage sustained by the fibreglass and the whole composite structure. The results demonstrated that for the same applied gas pressure, the conical shape bullets penetrated deeper through the surface specimens than both hemispherical and blunt shaped bullets due to its smaller surface area.

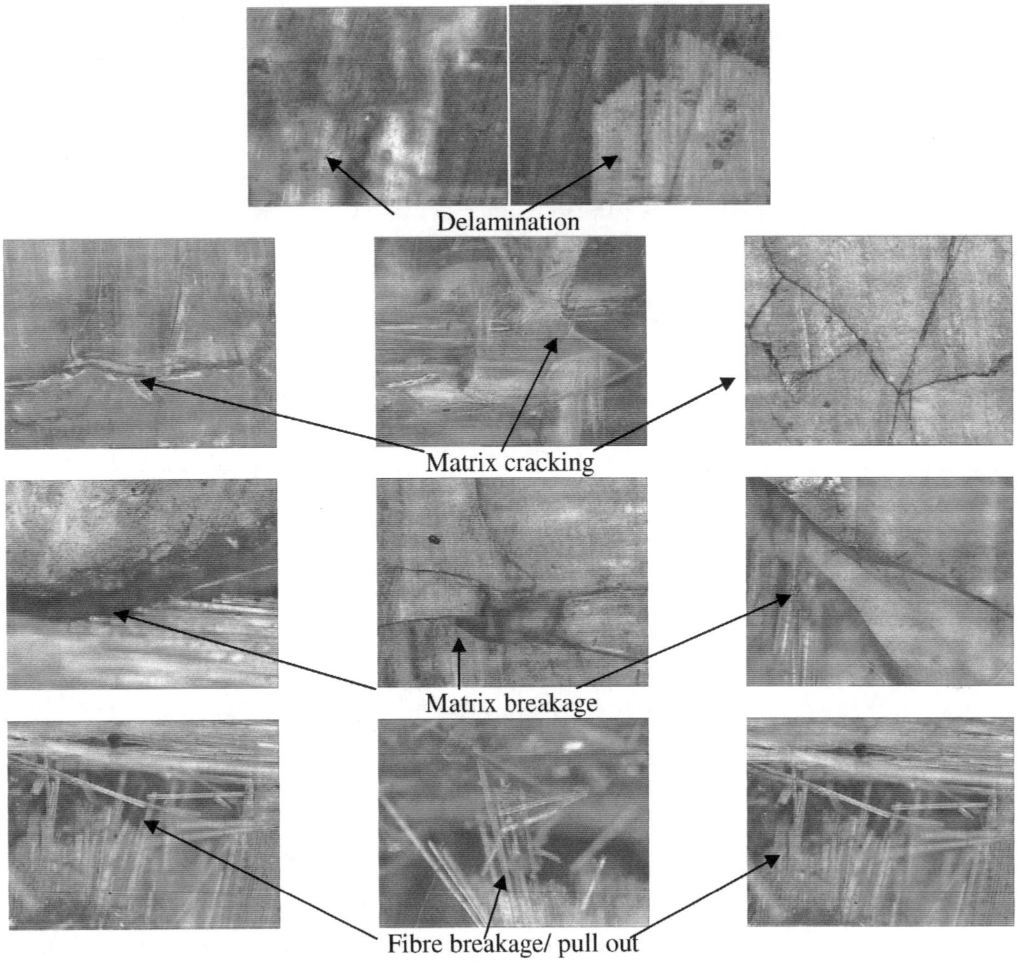

Figure 6. Failure modes of the test specimens (magnification power is 5 times)

Table 2 until Table 5 shows the data of damage area for all specimens impacted by the three types of bullets with different levels of impact energy. It also shows the failure mode experienced by the impacted specimens: fibre crack and fibre pull-out. The impacted specimens experienced delamination and matrix cracking as shown in Figure 6 before it turned to fibre breakage and fibre pull-out. Multiple matrix breakage extended at the back of the impacted area, and the fibre broke just beneath the impact point due to the high impact velocity. Extensive fibre cracks, which appeared as catastrophic failure, were generated beneath the impact point. Outside the ply failure zone, delamination occurred and extended from the tips of the matrix cracks. Most of the specimens suffered a fibre crack at the low impact energy while the higher impact energy caused the specimens to undergo fibre pull-out failure.

Figure 7 until Figure 10 illustrate the differences of impact area between type C and type E specimens. The graph also shows the differences of damage area using a different type of bullets. As can be seen in the figures, C600 had larger impact area compared to E600. From this results, it was concluded that the C-type suffered more damage than E-type. Blunt type of impactor had larger impact area, followed by hemispherical and conical. This is because of blunt type has larger cross section area compared to the other two. Conical type of bullet penetrated much deeper than the other bullets, hence its surface area had smaller damage.

Table 2. Damage area and mode of failure for 6 mm specimens

GFRP	Impact energy / failure	Type of bullets		
		Blunt	Hemispherical	Conical
C600	6	0.0007	0.0007	0.0004
		Fibre crack	Fibre crack	Fibre crack
	12	0.0012	0.0011	0.0006
		Fibre crack	Fibre crack	Fibre crack
	18	0.0015	0.0013	0.0011
		Fibre crack	Fibre crack	Fibre pull-out
E600	6	0.0006	0.0005	0.0003
		Fibre crack	Fibre crack	Fibre crack
	12	0.0008	0.0007	0.0005
		Fibre crack	Fibre crack	Fibre crack
	18	0.0014	0.0013	0.0011
		Fibre crack	Fibre crack	Fibre crack

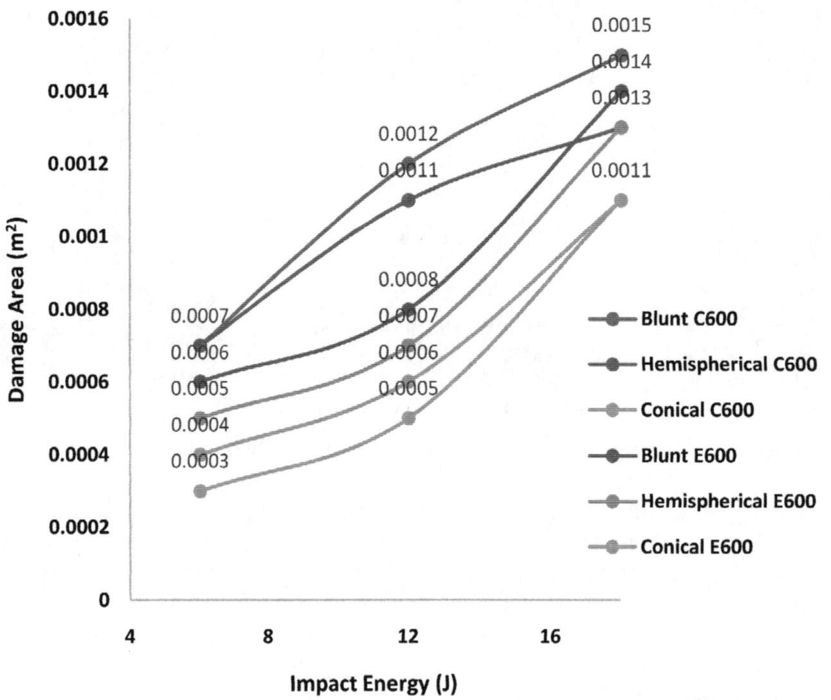

Figure 7. Graph of damage area against impact energy for 6 mm specimens

Table 3. Damage area and mode of failure for 8 mm specimens

GFRP	Impact energy / failure	Type of bullets		
		Blunt	Hemispherical	Conical
C600	10	0.0012	0.0008	0.0007
		Fibre crack	Fibre crack	Fibre crack
	20	0.0014	0.0013	0.0013
		Fibre crack	Fibre crack	Fibre pull-out
	30	0.0016	0.0015	0.0013
		Fibre crack	Fibre pull-out	Fibre pull-out
E600	10	0.0010	0.0008	0.0007
		Fibre crack	Fibre crack	Fibre crack
	20	0.0012	0.0011	0.0011
		Fibre crack	Fibre crack	Fibre crack
	30	0.0014	0.0012	0.0011
		Fibre crack	Fibre crack	Fibre pull-out

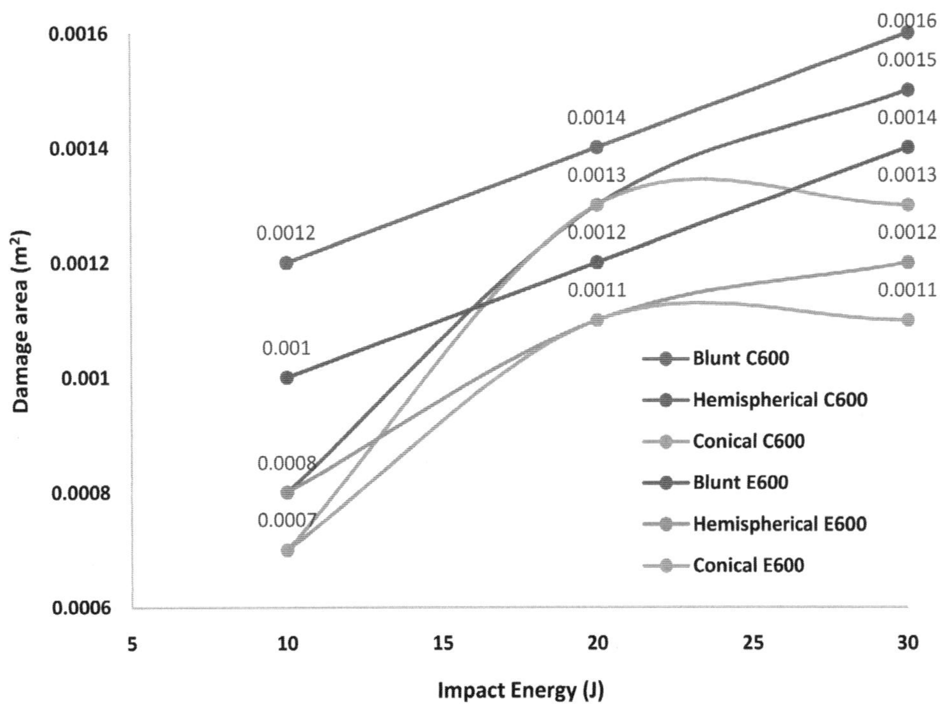

Figure 8. Graph of damage area against impact energy for 8 mm specimens

Table 4. Damage area and mode of failure for 10 mm specimens

GFRP	Impact energy / failure	Type of bullets		
		Blunt	Hemispherical	Conical
C600	15	0.0008	0.0007	0.0006
		Fibre crack	Fibre crack	Fibre pull-out
	30	0.0009	0.0008	0.0007
		Fibre crack	Fibre pull-out	Fibre pull-out
	45	0.0011	0.0010	0.0008
		Fibre crack	Fibre pull-out	Fibre pull-out
E600	15	0.0006	0.0005	0.0004
		Fibre crack	Fibre crack	Fibre pull-out
	30	0.0007	0.0007	0.0006
		Fibre crack	Fibre pull-out	Fibre pull-out
	45	0.0009	0.0007	0.0007
		Fibre crack	Fibre pull-out	Fibre pull-out

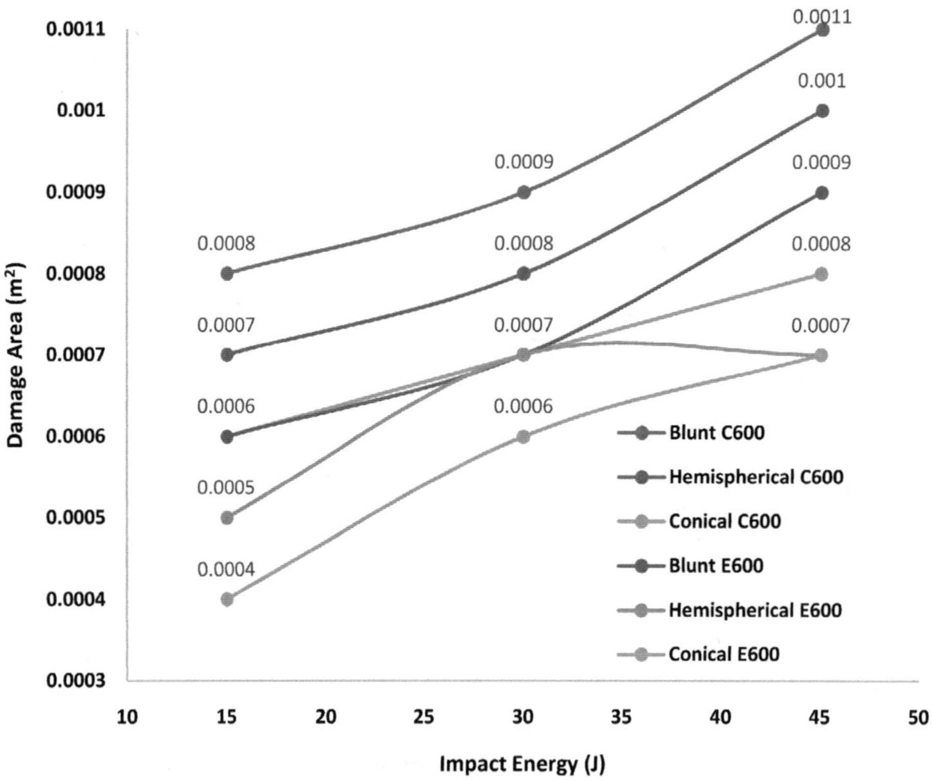

Figure 9. Graph of damage area against impact energy for 10 mm specimens

Table 5. Damage area and mode of failure for 12 mm specimens

GFRP	Impact energy / failure	Type of bullets		
		Blunt	Hemispherical	Conical
C600	20	0.0006	0.0006	0.0006
		Fibre pull-out	Fibre pull-out	Fibre pull-out
	40	0.0008	0.0007	0.0007
		Fibre pull-out	Fibre pull-out	Fibre pull-out
	60	0.0009	0.0008	0.0008
		Fibre pull-out	Fibre pull-out	Fibre pull-out
E600	20	0.0005	0.0004	0.0004
		Fibre pull-out	Fibre pull-out	Fibre pull-out
	40	0.0007	0.0005	0.0005
		Fibre pull-out	Fibre pull-out	Fibre pull-out
	60	0.0008	0.0007	0.0007
		Fibre pull-out	Fibre pull-out	Fibre pull-out

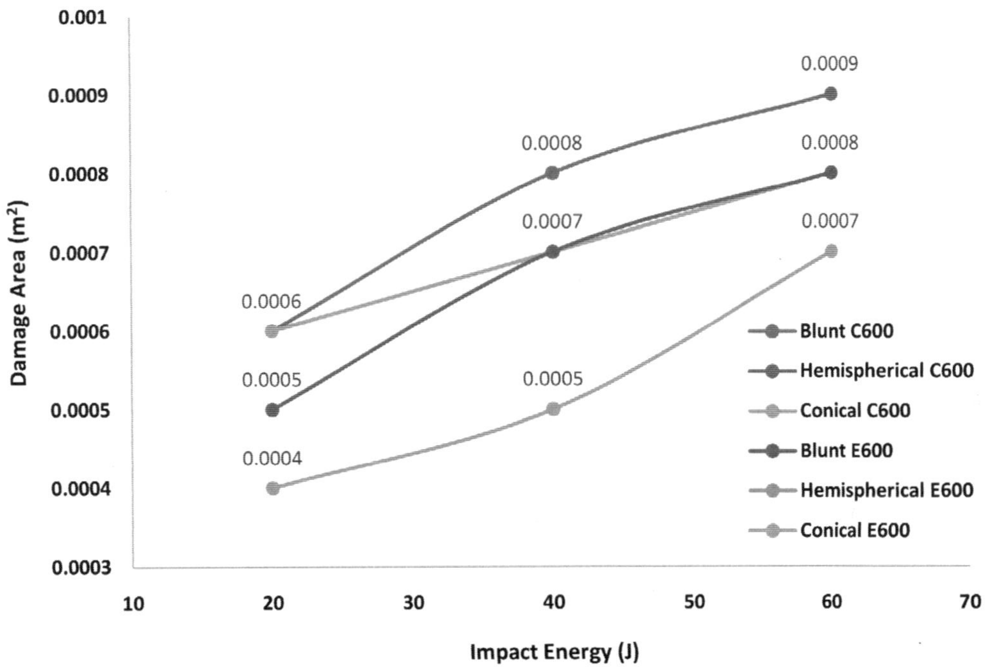

Figure 10. Graph of damage area against impact energy for 12 mm specimens

4. Conclusion

A non-destructive technique was used to evaluate impact damage and type of failure experienced by the C-glass and E-glass types of reinforced composite panels. The identified damage mechanism of the composite laminates after the high velocity impact tests involved delamination, matrix cracking and breakage with simultaneous fibre cracking and fibre breakage, and finally followed by the fibre pull-out. The increase in the applied gas gun pressure resulted in an increasing impact damage area of the panel's surface. A general trend was observed. As the gas gun pressure was increased, initial velocity of the projectile also increased, the projectile kinetic energy increased, the maximum force exerted to specimens increased and the energy absorbed by the specimens also increased. Most of the impacted specimens showed that the E-types experienced smaller damage area compared to C-types of GFRP. In this analysis, the character of impact damage can be divided into two parts: intra-laminar damage and inter-laminar damage. The intra-laminar damage corresponds to matrix cracking, fibre or matrix de-bonding and fibres breakages whereas the inter-laminar damage corresponds to delamination effect. Furthermore, another non-destructive technique such as ultrasonic C-scans and X-rays can be used to determine the failure in the damaged area. To investigate further the damage in the composite, the specimens can undergo destructive techniques to see the extent of the damage. Another material can be tested to see the damage behaviour. The same procedure discussed in this research work can be used for different types of material like carbon fibre, Kevlar, natural fibres etc. A better understanding about fibreglass with a different type of fibreglass and a different orientation should be performed in the future to examine the differences between impact energy and force with the existing specimens discussed in this research. It can be concluded that E-glass/Epoxy 600 g/m^2 is stronger and tougher than the C-glass/Epoxy 600 g/m^2 since it has greater mechanical properties. Therefore, type E-glass is recommended to be used in structural applications as compared with Type C-glass.

Acknowledgement

This work is supported by UPM under GP-IPB grant, 9415402. The authors would like to express their gratitude and sincere appreciation to the Aerospace Manufacturing Research Centre (AMRC) and the Department of Aerospace Engineering, Universiti Putra Malaysia (UPM).

References

[1] David C, Sublet P, Auriol A and Rappeneau J 1964 *Carbon* **2** 139–48
[2] Thornton P R 1967 *Scanning electron microscopy* (London: Chapman and Hall)
[3] Pfrang A, Reznik B, Gerthsen D and Schimmel T 2003 *Carbon* **41** 179–98
[4] Zukas J A, Nicholas T and Swift H F 1992 *Impact dynamics* (Malabar: Krieger Publishing)
[5] Drzal L T, Herra-Franko P J and Hoe H 2000 *Comprehensive Composite Materials* (Amsterdam: Elsevier)
[6] Sultan M T H and Rafie A S M 2010 *International Journal of Mechanical and Material Engineering* **5** 260-7
[7] Sultan M T H 2011 *Impact Damage Characterisation in Composite Laminates* PhD Thesis University of Sheffield
[8] Sultan M T H, Worden K, Staszewski W J and Hodzic A 2012 *Composites Science and Technology* **72** 1108-20
[9] Duke J C and Henneke E G 1980 *The Fifth International Acoustic Emission Symposium*
[10] Rose J L 1999 *Ultrasonic Waves in Solid Media* (Cambridge: Cambridge University Press)
[11] The Welding Institute Website [Accessed on March 2016]
[12] Ambu R, Aymerich F, Ginesu F and Priolo P 2006 *Composite Science and Technology* **66** 199-205
[13] Pang J W C and Bond I P 2005 *Composite: Part A: Applied Science and Manufacturing* **36** 183-8

Spring-back simulation of flat symmetrical laminates with angled plies manufactured through autoclave processing

M N M Nasir[1], Y Aminanda[1*], L Mezeix[2], M A Seman[3], A Rivai[3] and K M Ali[4]

[1]Department of Mechanical Engineering, International Islamic University Malaysia, Kuala Lumpur, Malaysia
[2]Faculty of Engineering, Burapha University, Chonburi, Thailand
[3]Faculty of Mechanical Engineering, Universiti Teknikal Malaysia Melaka, Malaysia
[4]Aerospace Malaysia Innovation Centre, Cyberjaya, Malaysia

*yulfian@iium.edu.my

Abstract. This paper presents a non-linear finite element analysis (FEA) method to predict the spring-back deformation for composite structures manufactured using autoclave processing. It is a progression from previous accompanying study on flat unidirectional samples and the aim is to observe spring-back warpage on laminates consisting of angled ($\pm 45^o$) plies compared to unidirectional (0°) laminate. Three samples for each of the symmetrical laminates with angled plies $[45/0]_S$ and $[45/\text{-}45]_S$ are manufactured and the warpage form is observed. FEA model that was utilized in the previous study, along with the physical mechanisms of spring-back such as the first ply stretching and tool-part interaction mechanisms, are maintained with only changes in material orientation for the part and the tool-part interface components. Upon comparison, the data shows that the spring-back form for symmetrical laminate becomes more complex to predict.

1. Introduction

Composite usage within the aerospace industry has increased significantly over the past 30 years, led by original equipment manufacturers (OEMs) such as Airbus that have been incorporating composite structures into their aircraft designs. The recent A350 project has seen more than 50% application of composite materials due to its higher specific stiffness compared to metals. Like any other material, composites induce residual stress as a result of the manufacturing process, which in this case involves curing at high temperatures in an autoclave. Those residual stresses pre-stress the composites, resulting in strength degradation. The apparent consequence of this is the deviation of the final product from what is initially designed. This phenomenon is known as spring-back deformation. Other weaknesses of laminate composite structures also include impact strength and hygrothermal loading [1].

Spring-back deformation can be attributed to a multitude of factors. One of them is the change in mechanical properties of the laminate during curing process whereby fibre reinforcement properties essentially remain constant but the matrix/resin properties evolve with respect to the curing process. Another source of the deformation is the difference in fibre orientation between individual plies, which results in non-uniform in-plane stresses through the laminate thickness. The severity of the warpage is more for an asymmetrical and unbalanced layup due to the multiple constraints that has been imposed [2]. Several studies have highlighted the contribution from the autoclave curing cycle. For example, a study showed that a 2-hold cure cycle yields significantly higher spring-back than a 1-hold cure cycle if there is no release film sheet between the laminate part and tool [3]. This is because in a 2-hold cure cycle, temperature of the autoclave is raised after a period of time that induces more resin shrinkage inside the laminate, hence greater thermal stresses. Other studies have also stated that a 2-hold cure cycle, with a first lower temperature hold and a second temperature hold, is designed such that the resin gels during the first hold. As a result, spring-back is more severe compared to a 1-hold cure cycle [4, 5]. Another study has demonstrated that by optimizing the cure cycle, the quality and strength of the part can be enhanced [6]. Nevertheless, tool-part interaction is seen as the most critical mechanism [7, 8, 9]. The difference in stretching between the tool and laminate during heat-up creates interfacial shear stresses, which generate a stress gradient through the laminate thickness that creates the bending moment [10]. The varied stacking sequence of a laminate results in in-plane stresses within the plies due to the disparity in thermal expansions in both the fibre and resin directions at the ply level [11, 12]. With cross-ply composites [0/90], the 90° fibres will impose a mechanical constraint on the 0° fibres during the processing phase [13]. The geometry after curing for the non-unidirectional laminate varies depending on the stacking layup as shown in Figure 1.

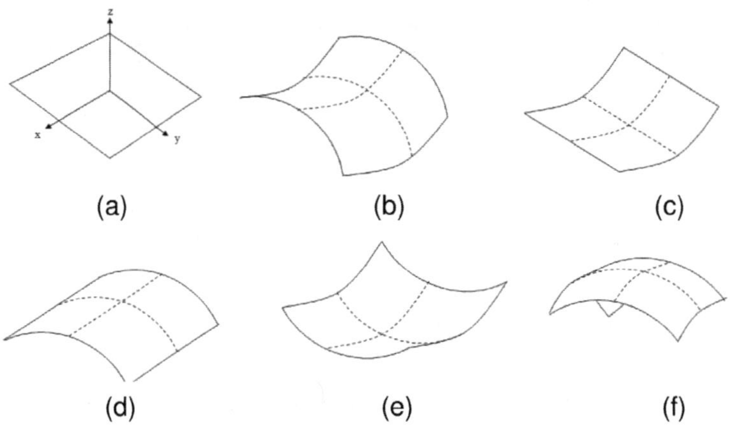

Figure 1. Part warpage due to asymmetrical layups: (a) original, (b) saddle, (c) and (d) cylindrical, (e) and (f) spherical [2]

The goal of this study is to develop a robust numerical method that can accurately predict spring-back warpage on flat symmetrical laminates via finite element analysis (FEA). Previous accompanying study has managed to predict spring-back warpage by utilizing the same FEA model for unidirectional flat laminates [14]. In that study, the two critical parameters in providing accurate results were the in-plane stress generated collectively by the first ply stretching and fibre volume fraction gradient, and the out-of-plane shear stress via tool-part interaction. In real life manufacturing, the in-plane stresses are due to resin polymerization that occurs during cure, so the composite law behaviour of the FEA model is modelled as a function of the degree of cure of the laminate. The law behaviour is obtained by integrating the elastic-linear law behaviour with respect to the three different resin polymerization stages: viscous, rubbery and glassy. Meanwhile, the tool-part interaction behaviour is modelled with a layer of solid elements associated with isotropic elastic property. For the current symmetrical laminate study, the mechanisms established for unidirectional are retained with the variables being the layup configuration and the corresponding material modulus.

2. Experimental

The current study employs a tool machined from S275JR carbon steel to manufacture the laminate composites made from IMA/M21E, which were symmetrically configured. Both were fabricated as per industry standards by CTRM, the leading supplier of aerospace composite structures in Malaysia. For proprietary reasons, the material property values of the steel tool and laminate together with its curing profile will not be provided in this paper. The laminate has an area of 300 x 300 mm^2 with a thickness of four plies and laid up at $[45/0]_S$ and $[45/-45]_S$ configurations. For repeatability, three samples were manufactured for each layup.

The resulting warpage for both $[45/0]_S$ and $[45/-45]_S$ laminates was different compared to that of the unidirectional, i.e. $[0]_4$ laminate from previous study [14], such that the warpage form is maximum in two diagonal corners as opposed to the symmetrical distortion of the unidirectional stacking. This is shown in Figure 2. Additionally, the direction of the warpage followed the fibre orientation, i.e. longitudinal direction, of the first ply. The maximum points were measured from the respective vertices of the warped sample to the reference plane and averaged as in Table 1. If based purely on the yielded spring-back warpage, the best configuration is the unidirectional layup. However, in real life aerospace applications, load path of the structure needs to be accounted for and this may require the use of symmetrical laminates with angled plies, especially ±45°, to account for the shear stresses [15].

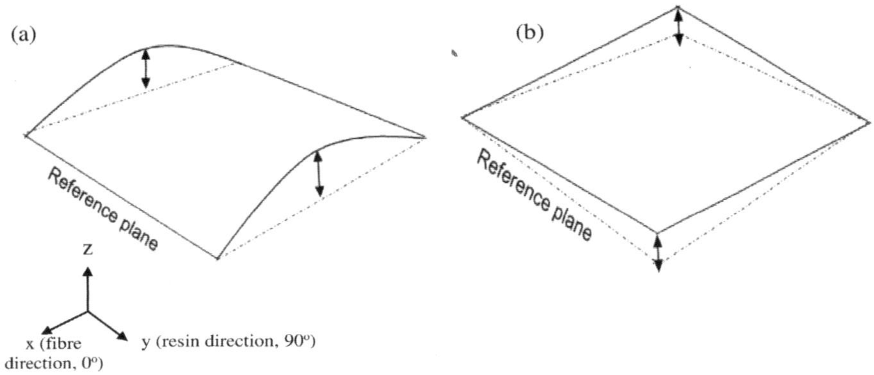

Figure 2. Spring-back warpage for: (a) $[0]_4$, (b) $[45/0]$ and $[45/-45]_S$

Table 1. Averaged maximum warpage of the 300 x 300 mm^2 with 4 plies thickness laminate samples

Laminate layup	Averaged maximum warpage
$[0]_4$	-5.2 [mm]
$[45/0]_S$	-13.3 [mm]
$[45/-45]_S$	-6.2 [mm]

At the laminate level, having a symmetrical layup uncouples the in-plane (membrane) and out-of-plane (bending) responses should theoretically eliminate the spring-back warpage due to the difference in thermal expansion between longitudinal and transversal directions. This is not the case as evident from Table 1. Nonetheless, the warpage can be minimized if the layup configuration is balanced aside from being symmetrical through two mechanisms. Firstly, the membrane coupling between the in-plane normal and shear behaviours is removed as a result of the membrane stiffness coefficients A_{16} and A_{26} being zero. Secondly, the laminate bending response is simplified when the bending stiffness coefficients D_{16} and D_{26} that couple bending and twisting are non-zero for symmetrical configuration, e.g. $[45/0]_S$. Of course, the application of a bending moment will generate twisting and the twisting moment produces a bending curvature. However, if the angled plies are dispersed in $\pm$ orientation, the D_{16} and D_{26} will be relatively less compared to the other terms in the D bending matrix [16].

Having a balanced configuration still does not eliminate the spring-back warpage, which brings into question the fabricating tool. There exists a great disparity (several orders of magnitude) of the thermal expansion between the tool and the laminate. It is this manufacturing reality that forms the basis of the first ply stretching that creates the in-plane through thickness stress and the out-of-plane shear stresses from the tool-part interaction mechanism. Both mechanisms are detailed in the following section.

3. Finite element analysis

3.1. Hypothesis and model characteristics

As mentioned earlier, this study is a progression from previous study that was done for unidirectional flat samples in [14]. Therefore, the methodology is maintained with only differences in the material orientation to reflect the various fibre orientations in a symmetrical laminate. The FEA model for the unidirectional study was constructed based on several assumptions. Considering that the goal was to develop an efficient but robust model, only the cool-down phase (180°C to 25°C) was simulated based on a study that observed inconsequential effects in terms of spring-back deformation from the strain gage readings during cure and only seeing an exponential rise in the contribution during the cool-down phase [17]. The temperature loading started at 180°C, which was the maximum curing temperature that the samples underwent during the cure cycle and ended at 25°C, the room temperature at which the samples were cooled to when it had achieved full vitrification. In tandem with that, the autoclave pressure was also neglected. Like in the unidirectional study, the FEA model for the symmetrical study comprised of the tool, interface and laminate part modelled using solid elements (refer to Table 2 and Figure 3). Both the tool and laminate components were respectively assigned isotropic and orthotropic thermal-mechanical elastic laws while the interface properties were input via a user-defined subroutine file (VUMAT) compiled using FORTRAN.

Table 2. Material property input for the FEA model components

FEA Model Component	Material Properties
Tool	As per for S275JR carbon steel
Interface	User defined subroutine (orthotropic linear with failure criterion)
Laminate part	Rest of the laminate: As per for carbon fibre IMA/M21E prepreg and the properties oriented in the ply angle direction
	First ply: As per for carbon fibre IMA/M21E prepreg except the coefficient of thermal expansion in the longitudinal direction modified to simulate the ply stretching effect and the properties oriented in the ply angle direction

(a) (b)

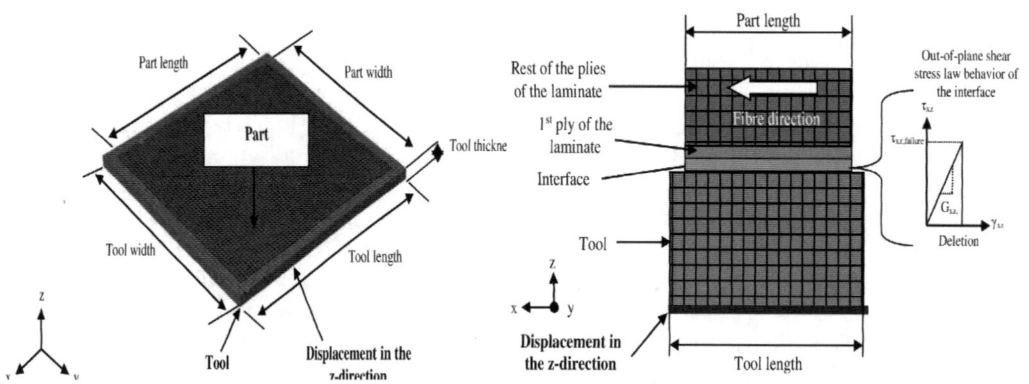

Figure 3. FEA model, 300 x 300 mm^2 (a) 3D view and (b) cross-sectional view and the out-of-plane shear stress law behaviour of the interface

3.2. Laminate part (IMA/M21E) properties

As mentioned earlier, ply stretching, fibre volume fraction gradient and resin shrinkage create an in-plane stress gradient through the laminate part thickness during the cure cycle [3]. Integrating all three phenomena would condense the FEA model and complicate the simulation, not to mention that the required material data would be difficult to obtain. The resin shrinkage is considered to have a minimal effect on the warpage [7]. Moreover, laminates made out of carbon fibres and M21 epoxy resins were observed to not exhibit any fibre volume fraction gradient effects [8]. Thus the in-plane stress gradient was initiated through the ply stretching component only. As only the cool-down phase was modelled, the in-plane stress created by the stretching of the laminate plies during the heat-up phase of the cure cycle was initiated by assigning a specific longitudinal coefficient of thermal expansion (CTE), α11 to the first ply (see Figure 3b) as it is hypothesised that only the closest ply to the tool bears most of the stress transfer from the tool [3].

For both $[45/0]_S$ and $[45/-45]_S$ layups, the longitudinal coefficient of thermal expansion of the first ply that was determined for the unidirectional laminate (3.06×10^{-6} °C^{-1}) in [14] to accurately capture the spring-back form and magnitude for 300 x 300 mm^2 and 4 plies, was maintained with only differences in the material orientation to reflect various fibre orientations in the symmetrical laminate. Table 3 shows the resultant modulus and principal material direction of the laminate samples relative to the unidirectional (0°) laminate as calculated using Classical Laminate Analysis (CLA). Examining both Table 1 and Table 3, the $[45/0]_S$ layup possesses a longitudinal modulus, E_{11} of approximately 0.5 times smaller but has a displacement value of approximately 2.6 times larger than the unidirectional laminate. Meanwhile, the $[45/-45]_S$ layup possesses a longitudinal modulus 12.6 times smaller but still yields a maximum displacement greater only by 1 mm to the unidirectional laminate. This situation is most probably due to the quasi-isotropic nature of $[45/-45]_S$ layup compared to $[45/0]_S$ (see Figure 4).

Table 3. Resultant modulus and principal fibre direction of IMA/M21E laminate samples with 300x300 mm^2 and 4 plies thickness

Laminate layup	Longitudinal modulus	Transversal modulus	Principal material direction relative to the unidirectional laminate
$[0]_4$	154000 [MPa]	8500 [MPa]	0 [°]
$[45/0]_S$	83100 [MPa]	10300 [MPa]	10 [°]
$[45/-45]_S$	12200 [MPa]	12200 [MPa]	45 [°]

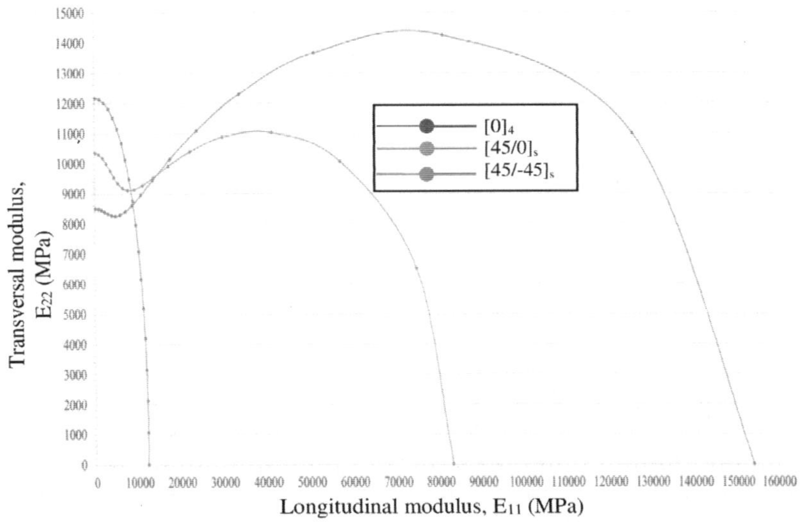

Figure 4. Evolution of the laminate material modulus

3.3. *Out-of-plane shear (interface) properties*

The interface component in the FEA model represents the polytetrafluoroethylene (PTFE) release film and has significantly less rigidity compared to other two components above and below it. Therefore, to ensure that the laminate component does not penetrate the interface and tool, i.e. effective out-of-plane shear stress transfer, the transversal through thickness Young's modulus of the interface was assigned the same value as the tool while the in-plane modulus was maintained from the release film (PTFE) properties. Therefore, a single layer of elements was assigned with orthotropic linear coupled with an out-of-plane shear stress failure criterion, $\tau_{xz,failure}$, which was developed from previous unidirectional study [14]. The transferred stress from the interface to the laminate part can be controlled via the out-of-plane shear modulus, G_{xz} and the out-of-plane shear stress failure, $\tau_{xz,failure}$. The FEA model maintains the same interface properties as before, using the same values of 56 MPa and 3.8 MPa for the out-of-plane shear modulus and out-of-plane shear stress failure, respectively, with the exception of the orientation being parallel to the principal material direction (refer to Table 3).

4. Result and discussion

Table 4 shows the comparison between the FEA results and the experimental data of the $[45/\text{-}45]_S$ and $[45/0]_S$ laminates (refer also to Figure 5). The FEA results illustrate the complexity in predicting the geometry and modelling the non-linear analysis that is instrumental in capturing spring-back behaviour using a numerical model. The deviation is correlated to the severity of the warpage. From the results, having a 45° fibre direction at the first ply will generate more warpage compared to a layup having 0° fibre direction at the first ply adjacent to the tool. The significant difference of the maximum warpage magnitude between the two layups is most probably due to the balanced laminate of $[45/\text{-}45]_S$, which provides the laminate with more uniform properties compared to the $[45/0]_S$ layup (see Figure 4).

Table 4. FEA simulation vs Experimental data

Laminate layup	Longitudinal coefficient of thermal expansion	Spring-back warpage		Error deviation
		Experimental	FEA	
$[0]_4$		-5.2 [mm]	-5.2 [mm]	0 [%]
$[45/0]_S$	$3.06\text{x}10^{-6}$ [°C^{-1}]	-13.3 [mm]	-7 [mm]	47.37 [%]
$[45/\text{-}45]_S$		-6.2 [mm]	-5.74 [mm]	7.42 [%]

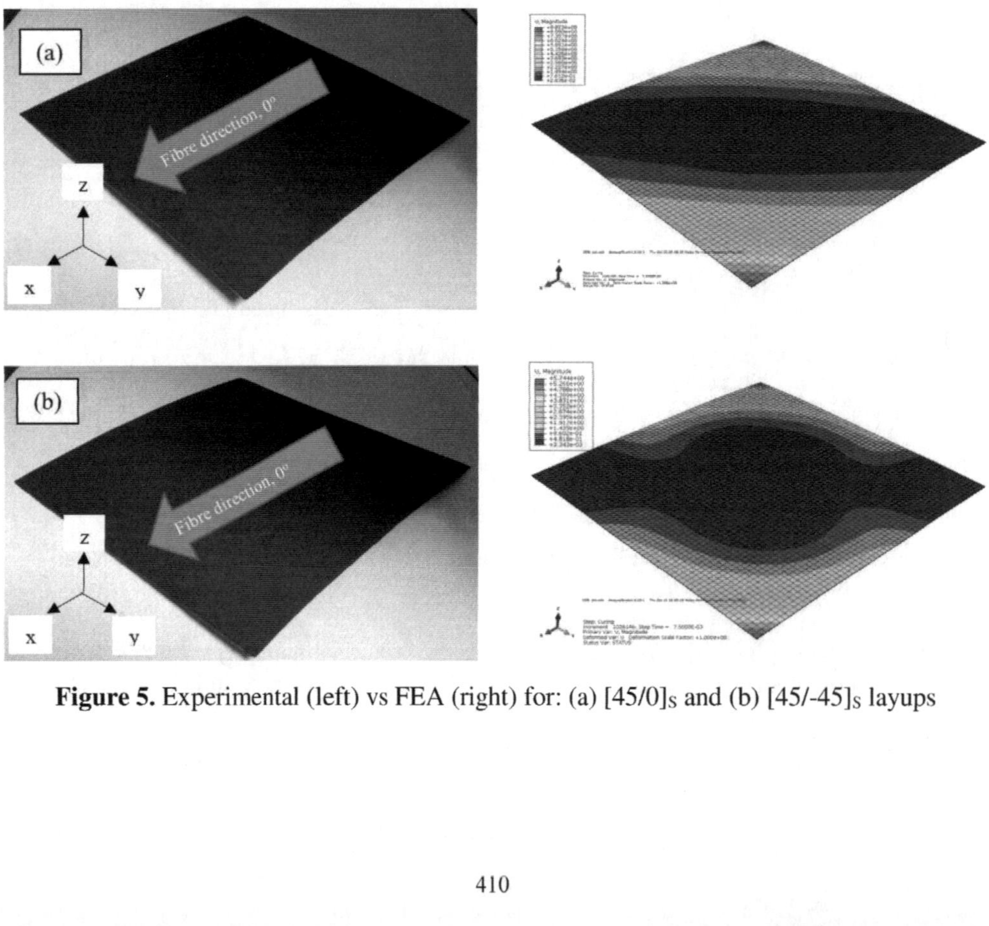

Figure 5. Experimental (left) vs FEA (right) for: (a) $[45/0]_S$ and (b) $[45/\text{-}45]_S$ layups

5. Conclusions and future work

The objective of this study is to develop an FEA model capable of predicting spring-back warpage for symmetrical laminates, i.e. $[45/0]_S$ and $[45/-45]_S$. Several main conclusions that can be made are listed as follow:

- Having an angled ply as the first ply in contact with the tool will significantly increase the severity of the warpage
- The FEA model has close proximity with the experimental data for the [45/-45]s layup, which is balanced and quasi-isotropic in behaviour
- For the unbalanced [45/0]s layup, the disparity between the FEA simulation and experimental measurement is great but the spring-back form is accurate in the simulation. With that, an FEA model refinement study which involves the bending strength properties of the laminate and its first ply specific longitudinal CTE might be performed to close the gap

This study has increased the understanding on the part design parameters affecting spring-back and illustrates the many forms of spring-back deformation. However, the conclusions drawn should be used with caution outside the scope of this study. There is also a need to study the various symmetrical configurations and their corresponding warpage patterns.

Acknowledgement

Much appreciation goes to Aerospace Malaysian Innovation Centre (AMIC) for financial, operational and other related supports that ensure the progress of this research. Credit also goes to CTRM Sdn. Bhd. for its support in the manufacturing of the samples.

References

[1] Mokhtar H, Sicot O, Rousseau J, Aminanda Y and Aivazzadeh S 2011 *Procedia Engineering* **10** 2615-20
[2] Rapp S 2011 *Shape memory polymers in fibre composite structures for shape adjustment* Technical University of Munich
[3] Albert C and Fernlund G 2002 *Compos. Sci. Technol.* **62** 1895-912
[4] Fernlund G, Rahman N, Courdji R, Bresslauer M, Poursartip A, Willden K and Nelson K 2002 *Compos. Part* A **33** 341-51
[5] White S R and Hahn H T 1992 *J. Compos. Mater.* **26** 2433-53
[6] White S R and Hahn H T 1993 *J. Compos. Mater.* **27** 1352-78
[7] Nelson R H and Cairns D S 1989 *34th Int. SAMPE Conf. on Symp. & Exhibition* (Nevada: USA)
[8] Ridgard C 1992 *Accuracy and distortion of composite parts and tools: Causes and solutions*
[9] Wisnom M R, Ersoy N, Potter K and Clegg M J 2005 *Compos. Part A* **36** 1536-44
[10] Pagliuso S 1982 *4th International Conference on Composite Materials*
[11] Hyer M W 1981 *J. Compos. Mater.* **15** 175-94
[12] Kim K S and Hahn H T 1989 *Compos. Sci. Technol.* **36** 121-32
[13] Harris B 1999 *Engineering composite materials* (London: IOM Communications)
[14] Mezeix L, Seman A, Nasir M N M, Aminanda Y, Rivai A, Castanie B, Olivier P and Ali K M 2015 *Compos. Struct.* **124** 196-205
[15] Federal Aviation Administration 2012 *Advanced Composite Material Aviation Maintenance Technician Handbook – Airframe* (Washington: FAA)
[16] Northtrop Grumman 1997 *A Summary and Review of Composite Laminate Design Guidelines* (California: Northrop Grumman)
[17] Olivier P A 2006 *Compos. Part A* **37** 602-16

IOP Conf. Series: Materials Science and Engineering **152** (2016) 012047 doi:10.1088/1757-899X/152/1/012047

Numerical simulation of flow around a simplified high-speed train model using OpenFOAM

I A Ishak*, M S M Ali and S A Z Shaikh Salim

Wind Engineering, Malaysia-Japan International Institute of Technology, UTM Kuala Lumpur, Malaysia

*izuan.amin@gmail.com

Abstract. Detailed understanding of flow physics on the flow over a high-speed train (HST) can be accomplished using the vast information obtained from numerical simulation. Accuracy of any simulation in solving and analyzing problems related to fluid flow is important since it measures the reliability of the results. This paper describes a numerical simulation setup for the flow around a simplified model of HST that utilized open source software, OpenFOAM. The simulation results including pressure coefficient, drag coefficient and flow visualization are presented and they agreed well with previously published data. This shows that OpenFOAM software is capable of simulating fluid flows around a simplified HST model. Additionally, the wall functions are implemented in order to minimize the overall number of grid especially near the wall region. This resulted in considerably smaller numbers of mesh resolution used in the current study compared to previous work, which leads to achievement of much reasonable time simulation and consequently reduces the total computational effort without affecting the final outcome.

1. Introduction

The study of flow around high-speed train (HST) has been, and continues to be, the subject of intense research [1-3]. Rapid developments of HST, especially the ones currently running in Europe and Japan and some other countries, have attracted people's attention. Aggressive improvement in the context of the train's technology expansion has shown a trend to be faster, yet more energy efficient, from the last three decades, hence shaping a new interest among travellers. Many of the latest generation of HST such as Japanese Shinkansen E6, Spanish AVE class 103, China CRH380B, Italy ETR1000, German ICE-3, South Korea KTX Sancheon and France TGV POS reach speeds more than 300 km/h in regular operation. However, modern technologies nowadays are only focusing on the electric motor to speed up the train instead of understanding the flow field around it, which in the end leads to a decreased performance of HST due to large energy losses [4]. At these speeds, the aerodynamic loads become significantly important for the running performance of the train [3]. For instance, turbulence that occurs due to increased speed leads to disturbance of flows around the train, which will then cause the flow energies to be converted to aerodynamic drag, noise and vibration [4].

The alertness on the safety factors concerning the operation of HST, especially the one that relates to crosswind stability for railing vehicles, has grown significantly in the HST community. It has also become an alarming sign to those countries with high-speed rail networks [5]. The lateral stability of a train is an important safety issue as it is largely dependent on aerodynamic loads, which are deduced from flow field phenomena surround the vehicle. As trains nowadays are often built using lightweight materials in order to achieve higher speed and they are shifting towards more comfortable vehicles in order to cater for high passengers capacity, these trends conclude that it is essential for HST to have a much better handling stability. However, the demands for considerably lighter weight construction and higher driving velocities have shown a conflict with the crosswind stability of a particular HST. The fact that there have been several serious incidents that occurred in recent years where strong winds were thought to be the main cause of the derailment has intensified the importance of further research on flow over a HST [6].

In the context of investigating the flow field around a HST using numerical analysis, the factor that may lead to result discrepancies is due to different shapes of HST that are modelled. Some analysed a more complex geometries with consideration of additional structures such as front spoiler, bogies and pantograph while others define the structure in much more simplified train shapes. Recent study by Hemida *et al.* [1] used a HST model without underbody complexities or inter-carriage gaps between coaches, and still showed a good comparable result with other previous studies that used complex geometries. Furthermore, shortened train length was also implemented in this study since the execution of numerical simulation for a complete train length requires more advanced computational resources. Based on Khier *et al.* [7], the reduction in length will not change the important physical features of the flow since the flow characteristic downstream of a certain distance from the nose of the train (less than one coach length) is comparatively constant. Therefore, HST model that is presented in this paper is in the form of a simplified and shortened model in order to optimize the simulation time.

As the external flow of fluid moves around bodies, it is anticipated that there would be changes in pressure and velocity fields. It is important to take into consideration the characteristics of the flow when designing and constructing bodies that are exposed to external flows and computational fluid dynamics (CFD) is a useful tool in understanding these conditions. CFD method is gaining importance nowadays as an option for conducting investigation related to wind engineering. The technique is widely used due to its advantages that include low cost and time consuming, as well as flexibility in measuring various parameters. There are numerous validation and verification studies based on wind tunnel and on-site measurement that have already been carried out to justify the accuracy of the results provided by CFD application. Most of the comparison studies confirmed that the method is reliable enough to accurately predict the flow over a HST.

In this study, the OpenFOAM CFD software package is used to numerically analyse the fluid flow characteristics surrounding the HST. The advantages of OpenFOAM computational toolbox [8] are that it is a free and open software package that is capable of simulating a wide variety of fluid flow processes. With an extensive range of features that are contributed voluntarily by the CFD community around the world, OpenFOAM has the capabilities to simulate a wide range of flow problems. In total, there are over 170 utilities available for grid generation, and pre- and post-processing [9]. This open source CFD package is the alternative to the commercial CFD packages and it is increasingly popular nowadays. In the past few years, several notable studies have been published by OpenFOAM users worldwide in the fields of CFD, computational heat transfer, fluid structure interaction and multiphase flow [9]. The objective of this study is to present a numerical simulation of flow around a simplified HST using the OpenFOAM free software package. The investigation method applied in this work is unsteady simulation using numerical method of unsteady RANS combined with SST k-ω turbulence model. Two grid resolutions, which represent the coarse grid and the fine grid, are carried out in order to investigate the dependency of the results to the flow simulation. As the grid is refined, grid cells will become smaller. In the meantime, the number of cells in the flow domain increases and the time step is refined (reduced).

This paper is organized as follows. Firstly, the initial set-up inclusive of the model, domain and mesh are described. Next, the solution methodology including the equations used to solve the fluid flow problem is mentioned in details. Then, the result obtained from both coarse and fine meshes are compared with previously published simulated and experimented data. The computing machine for these simulations are then reported to evaluate the computational cost required to simulate both cases. In the end, the conclusion is given in the last section.

2. Model description

The model used in this study is a simplified HST model, which replicates a similar model investigated in experiment by Sakuma *et al.* [11] and in simulation by Osth *et al.* [10], as can be seen in Figure 1, Figure 2 and Figure 3. Design configurations for the train model are as listed below:

- Leading top and side edges on the front are rounded using an elliptical profile with major axis in the ellipse length of 0.07H and the minor axis length of 0.04H as can be seen in Figure 3(a).
- The top and side edges on the rear end of the bluff body are rounded with a circular radius 0.107H as can be seen in Figure 3(b).
- Both front and rear bottom edges are not rounded at all and thus sharp.
- The model is placed on two egg-shaped supports and is lifted 0.41H above the ground in order to replicate the same condition as in the wind tunnel Sakuma *et al.* [11].
- The length of the train is 7H while the width and height are both equal to H (W = H = 0.56m).

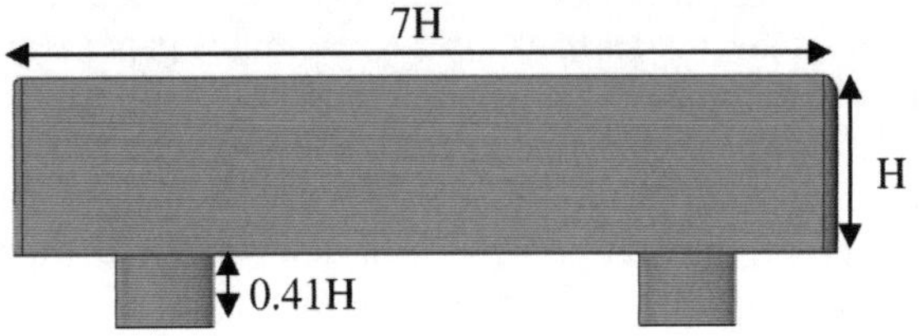

Figure 1. Side view of the train model

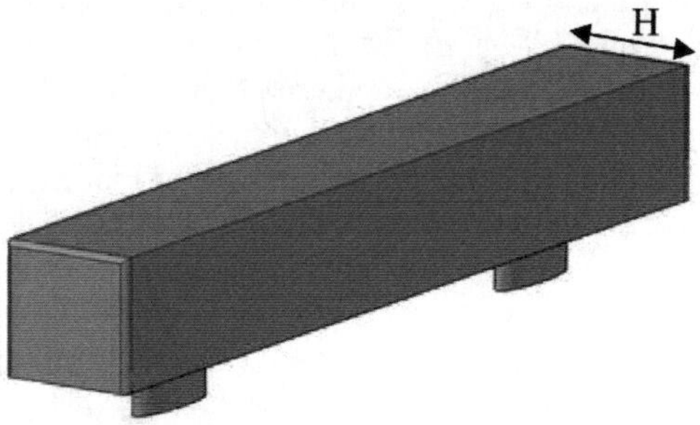

Figure 2. Isometric view of the train model

(a) (b)

Figure 3. Details of front and rear corners rounding (a) front corner elliptic rounding, (b) rear corner circular rounding

3. Domain description

The dimension of domain cross section is 36H x 21H x 11.41H (length x width x height) whereby H is the height of the train model that equals to 0.56m. The distance from the inlet to the bluff body is 8H and the distance from bluff body to the outlet is 21H. These lengths were found to be sufficient in the previous simulation of flows around a simplified train models by Hemida *et al.* [12], Krajnovic and Davidson [13], and Hemida and Krajnovic [14]. As for the boundary condition on the inlet, a uniform velocity, U_∞ is set in the x-direction. On the ground plane, the slip condition is used together with the velocity component in the x-direction equals to inlet velocity, U_∞ in order to prevent the development of the boundary layer on the ground plane. The homogenous Neumann boundary condition is applied at the outlet. Lastly, on the lateral side and roof, the slip condition is used similar to the ground plane. In this study, the boundary condition setup is similar as in simulation works by Osth *et al.* [10]. Details of the domain size and its boundary conditions can be seen in Figure 4 and Figure 5.

The Reynolds number (*Re*) used in this paper is 3.7 x 10^5 ($Re = U_\infty H/v$) based on height of the train H, kinematic viscosity v, and free stream velocity U_∞, respectively. This particular Reynolds number is selected in order to validate the simulation with previous work by Osth *et al.* [10]. Based on Alam *et al.* [15], the effects of Reynolds number can actually be neglected when $Re > 3 \times 10^5$. There were also several other researchers who claimed that the effects of Reynolds number on aerodynamic coefficients are indeed very little. Among them are Orellano and Schober [16], who experimentally studied the influences of changing Reynolds number on aerodynamic coefficients of trains subjected to side winds. They measured the aerodynamic coefficients at three different Reynolds numbers and it was found that the Reynolds number had almost negligible influence. Per-Ake [17], who carried out similar experiments on a bus subjected to crosswind, also arrived at the same conclusion that Reynolds number has insignificant effect on the aerodynamic coefficients.

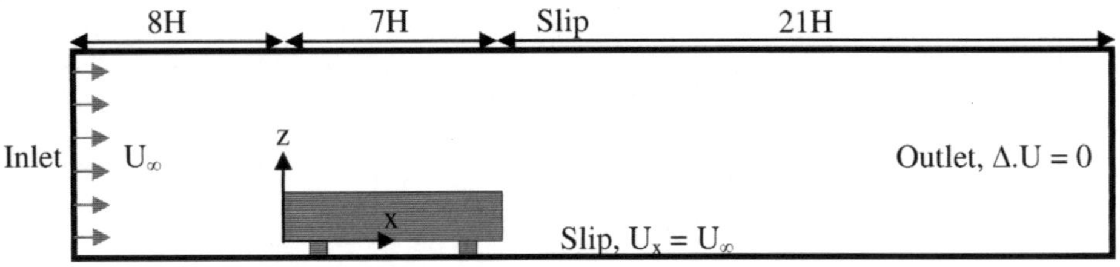

Figure 4. Computational domain used in the numerical investigation: side view (not to scale)

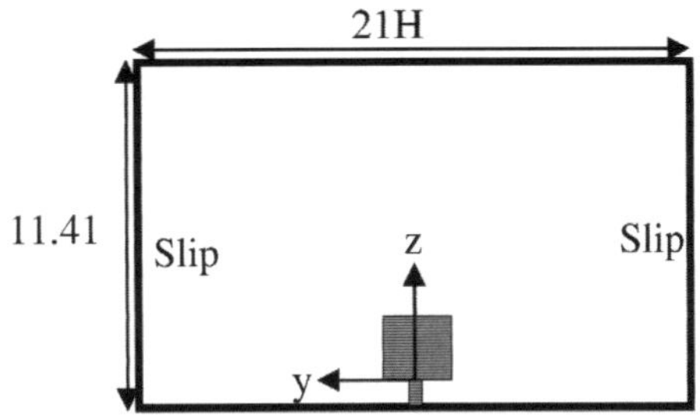

Figure 5. Computational domain used in the numerical investigation: front view (not to scale)

4. Mesh description

The cells in the computational domain are constructed using a structured non-uniform Cartesian mesh. Mesh refinement is applied on the train surface and its surrounding areas. Two grid resolutions are carried out, which indicate coarse and fine grids. These different grid resolutions are selected based on grid refinement ratio (r). Based on Celik *et al.* [18], it is desirable that the grid refinement ratio to be greater than 1.3. In this case, $r = 1.4$ is selected. Since the meshes are not uniform, the grid refinement ratio is calculated based on average grid size (h_{ave}) as shown in Equation 2 and Equation 3. Details of the grid resolutions can be seen in Table 1. The grid refinement ratio, *r* and is calculated as follows:

$$r = \frac{h_{ave\,(\text{coarse})}}{h_{ave\,(\text{fine})}} \tag{1}$$

where the average cell size (h_{ave}) is obtainable using below formulation:

- Three dimensional calculation: $h_{ave} = \left[\frac{1}{N}\sum_{i=1}^{N}(\Delta V_i)\right]^{\frac{1}{3}}$ (2)
- Two dimensional calculation: $h_{ave} = \left[\frac{1}{N}\sum_{i=1}^{N}(\Delta A_i)\right]^{\frac{1}{2}}$ (3)

where ΔV_i is the volume, ΔA_i is the area of the ith cell and *N* is the total number of cells used for the computations.

Table 1. Grid parameter for case A and case B

CASE	A (Coarse)	B (Fine)
Total no. of cells	844,148	2,114,715
Cell size, h_{ave}	0.1215	0.08895
Averaged y^+	100.34	81.76
Refinement ratio, r	1.4	

On the other hand, Figure 6 and Figure 7 show detailed view of grid construction for both Case A (coarse) and Case B (fine) across the simulation domain.

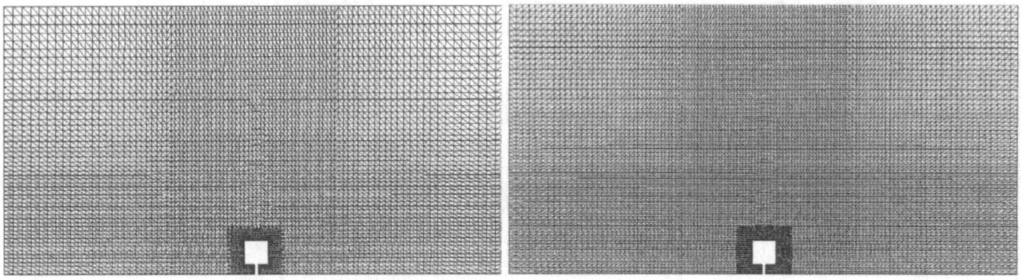

(a) (b)

Figure 6. Overall grid structure from front view: (a) Case A (coarse), (b) Case B (fine)

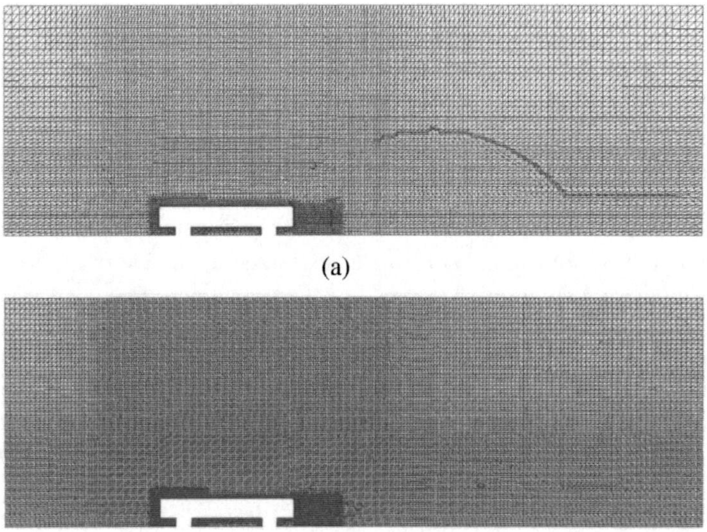

(a)

(b)

Figure 7. Overall grid structure from side view: (a) Case A (coarse), (b) Case B (fine)

Wall function is used in all the cases to reduce the computational cost by properly treating the cell size near the surface [19]. The distance of the first cell layer to the model surface should be located within the requirement of y^+ ($30 < y^+ < 300$). Figure 8 shows the details of y^+ value acquires for both cases A (coarse) and B (fine) along the mid plane of the train model. As can be seen, the result of y^+ value was distributed well in the requirements range.

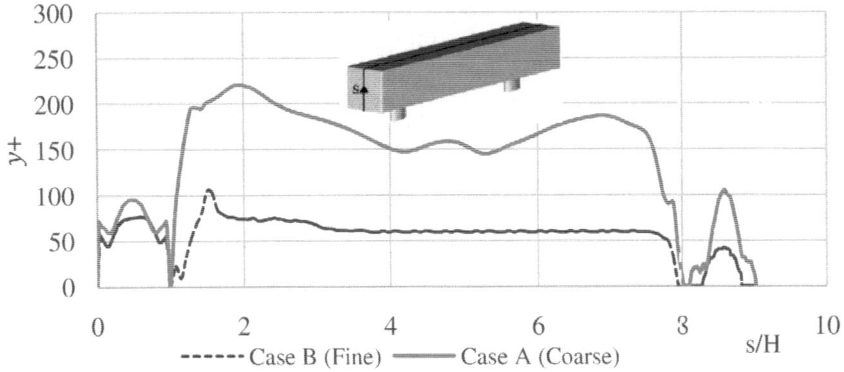

Figure 8. Value of y^+along a line 's' starting at origo along the mid plane of the train model

5. Solution methodology

Primitive variables are numerically calculated based on the three dimensional unsteady incompressible Navier-Stokes and continuity equations. The pressure implicit split operator (PISO) solution algorithm [20] with one predictor step and two corrector steps for the pressure-velocity coupling is used in this study. For the discretization scheme, 2nd order backward scheme is used for temporal discretization, 3rd order QUICKV scheme is used for convection term and 2nd order unbounded Gauss linear differencing scheme is used for viscous term. Two different time steps have been used in correspondence to the two different grid cases, such that the Courant-Fredichs-Lewy (CFL) number is always below unity. Table 2 shows time step for pressure, convection and diffusion term, Δt for each case.

Table 2. Time step for case A and case B

CASE	A (Coarse)	B (Fine)
Δt	0.005	0.002

Flow around the train was considered as incompressible and obtained by solving incompressible form of the Unsteady Reynolds Averaged Navier-Stokes (URANS) equation combined with the help of $k-\omega$ based Shear-Stress-Transport (SST) turbulence model. Actually, there are several selections of turbulence models that can be chosen to solve the fluid flow problems. The simplest is one-equation model such as Baldwin-Barth's, Spalart-Almaras' and Prandtl's models. The solution for this type of models is quick but it does not take into account the convection and diffusion of the turbulent energy. To include these variables, two-equation models can be used [21]. Two most predominant models in two-equation models are $k-\varepsilon$ model and $k-\omega$ model.

Menter [22] introduced the SST model that combines positive features of both $k-\varepsilon$ and $k-\omega$ models. The idea is to employ the wall-bounded advantages through viscous sub-layer of the $k-\omega$ model at near the wall region and the $k-\varepsilon$ model's free stream advantages near the boundary layer edge. This avoids the common $k-\omega$ problem that the model has, which tends to be too sensitive to the inlet free-stream turbulence properties. By taking these benefits, the model was selected in this study due to the fact that severe separation region near the wall bounded area and the wake structure of flow over the HST model in high Reynolds number was anticipated in the simulation. In addition, there was also a similar study in the past by Prime *et al.* [19] that showed comparable results with the experiment when the SST $k-\omega$ turbulence model was imposed in the simulation method carried out. In this numerical study on flow modelling of interacting prisms, both the flow velocity and turbulence intensities in various positions in the wake were compared with experimental hotwire data measured and the results showed a good agreement for both approaches.

5.1. Governing equations

The URANS equations are principally obtained from the usual RANS equation but the unsteady term has been maintained [22]. Calculations are carried using two equations: continuity and Navier-Stokes equation for the incompressible flow as follows.

$$\frac{\partial u_i}{\partial x_i} = 0 \tag{4}$$

$$\rho\frac{\partial u_i}{\partial t} + \rho U_j\frac{\partial u_i}{\partial x_j} = -\frac{\partial p_i}{\partial x_j} + \frac{\partial}{\partial x_j}\left[\mu\left(\frac{\partial u_j}{\partial x_j} + \frac{\partial u_i}{\partial x_j}\right)\right] + \rho f_i \tag{5}$$

The velocity components, u_i and the pressure, p_i are both nonlinear partial differential equations. This means that there is no analytical solution for the problem with arbitrary boundary conditions. The unsteadiness of the flow variables (i.e. velocity and pressure) are decomposed into mean value and fluctuations as follows:

$$u_i = U_i + u_i' \tag{6}$$

$$p_i = P_i + p'_i \tag{7}$$

where U_i and P_i are the time averaged terms, u_i' is the fluctuation terms of velocity, and p_i' is the fluctuation terms of pressure. Substituting these Reynolds decomposed velocities and pressures into the equation of continuity and Navier-Stokes equation yields the Reynolds Averaged Navier-Stokes equation of motions as shown below:

$$\frac{\partial U_i}{\partial x_i} = 0 \tag{8}$$

$$\frac{\partial U_i}{\partial t} + U_j\frac{\partial U_i}{\partial x_j} = -\frac{1}{\rho}\frac{\partial P_i}{\partial x_i} + \frac{\partial}{\partial x_j}\left(\mu\frac{\partial U_i}{\partial x_j} - \rho\overline{u_i'u_j'}\right) \tag{9}$$

5.2. Turbulence model

For a URANS simulation, in order to solve the governing equations, Reynolds stress tensor $-\rho\overline{u_i'u_j'}$ must first be determined. The Reynolds stress can be modelled either by additional equations or from the known quantities in order to achieve "closure" for the governing equations. Closure denotes that there is a sufficient number of equations for all the unknowns, including the Reynolds stress tensor resulting from the averaging procedure. The equation used to close the system is depending on the turbulence model being used.

A turbulence model is a computational procedure to close the system of flow equations as derived earlier. Most of them are based on the Boussinesq hypothesis. Based on Boussinesq, Reynolds stress tensor could be linked to the mean rate of deformation. The most widely used concept is as below:

$$-\rho\overline{u_i'u_j'} = \mu_t\left(\frac{\partial U_i}{\partial x_j} + \frac{\partial U_j}{\partial x_i} - \frac{2}{3}\frac{\partial U_k}{\partial x_k}\delta_{ij}\right) - \frac{2}{3}\rho k\delta_{ij} \tag{10}$$

where the turbulent kinetic energy (k) and specific dissipation rate (ω) are solved using the following equations:

Turbulence Kinetic Energy, k: $\frac{\partial k}{\partial t} + U_j\frac{\partial k}{\partial x_j} = P_k - \beta^* k\omega + \frac{\partial}{\partial x_j}\left[(\nu + \sigma_k\nu_T)\frac{\partial k}{\partial x_j}\right]$ (11)

Specific Dissipation Rate, ω :

$$\frac{\partial \omega}{\partial t} + U_j\frac{\partial \omega}{\partial x_j} = \alpha S^2 - \beta\omega^2 + \frac{\partial}{\partial x_j}\left[(\nu + \sigma_\omega\nu_T)\frac{\partial \omega}{\partial x_j}\right] + 2(1 - F_1)\sigma_{\omega 2}\frac{1}{\omega}\frac{\partial k}{\partial x_i}\frac{\partial \omega}{\partial x_i} \tag{12}$$

where ν_T is the kinematic eddy viscosity and can be defined as follows:

$$\nu_T = \frac{a_1 k}{\max(a_1\omega, SF_2)} \tag{13}$$

The following closure coefficient is used in this study

$$F_2 = \tanh\left[\left[\max\left(\frac{2\sqrt{k}}{\beta^*\omega y}, \frac{500v}{y^2\omega}\right)\right]^2\right] \tag{14}$$

where y is the distance to the next surface.

$$P_k = \min\left(\tau_{ij}\frac{\partial U_i}{\partial x_j}, 10\beta^* k\omega\right) \tag{15}$$

$$F_1 = \tanh\left\{\left\{\min\left[\max\left(\frac{\sqrt{k}}{\beta^*\omega y}, \frac{500v}{y^2\omega}\right), \frac{4\sigma_{\omega 2}k}{CD_{k\omega}y^2}\right]\right\}^4\right\} \tag{16}$$

$$CD_{k\omega} = \max\left(2\rho\sigma_{\omega 2}\frac{1}{\omega}\frac{\partial k}{\partial x_i}\frac{\partial \omega}{\partial x_i}, 10^{-10}\right) \tag{17}$$

$$\phi = \phi_1 F_1 + \phi_2(1 - F_1) \tag{18}$$

$$\alpha_1 = \frac{5}{9}, \alpha_2 = 0.44 \tag{19}$$

$$\beta_1 = \frac{3}{40}, \beta_2 = 0.0828, \beta^* = \frac{9}{100} \tag{20}$$

$$\sigma_{k1} = 0.85, \sigma_{k2} = 1, \sigma_{\omega 1} = 0.5, \sigma_{\omega 2} = 0.856 \tag{21}$$

6. Results and discussions

6.1. Pressure coefficient

For comparison purposes, results obtained from the two grid resolution were compared with numerical data and experimental data studied by [10, 11]. Pressure coefficient is selected and defined as follows:

$$C_p = (\rho_F - \rho_\infty / 0.5\rho U_\infty^2) \tag{22}$$

The pressure coefficient along a line in the mid plane surface on train model is shown in Figure 9. In general, the pressure follows the same pattern with the result in [10, 11]. However, some discrepancies occurred at separated flow region on the front roof of the train. Recent work seems to underestimate the negative pressure at this specific point. From the simulation, the flow reattached at a bit further downstream as compared to previous simulation and experimental results. Fine mesh resolution (Case B) appeared to produce more consistent result in the separated region compared to coarser mesh (Case A). However, Case B and Case A showed significant similarity in the reattachment region as the flow pass through the roof. In the wake region, the simulation slightly overestimated the value of pressure coefficient. Nonetheless, both meshes still captured the lowest peak value of the pressure coefficient similar to Osth *et al.* [10].

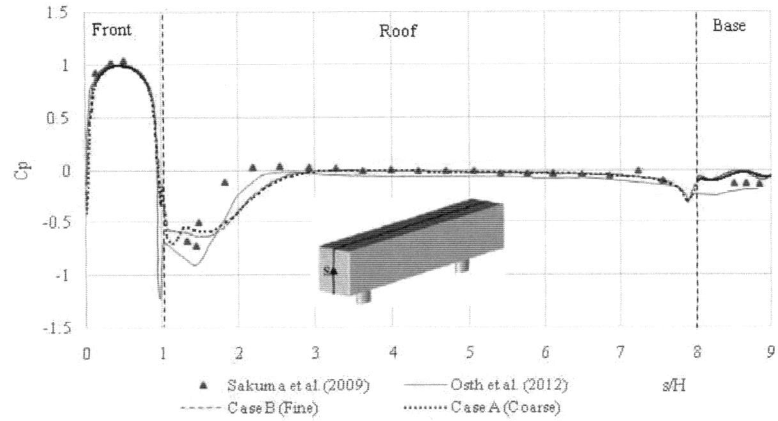

Figure 9. Pressure coefficient starting at origo along the mid plane of the train model

The discrepancies on the results obtained with previous simulation work by Osth *et al.* [10] are mainly due to the differences in numerical approach for the simulation model used, as well as the mesh resolution covered. Table 3 shows the differences of the computational setting for Case B (fine mesh) and Osth *et al.* [10].

Table 3. Comparison of simulation set-up with the validation paper

CASE	Osth et al. [10]	Case B (Fine)
Simulation model	PANS (Partially Averaged Navier Stokes)	URANS (Unsteady Reynolds Averaged Navier Stokes)
No. of extra transport equations	Four model equations $(k-\epsilon-\zeta-f)^{a}$	Two model equations $(k-\omega)^{a}$
Algorithm	SIMPLE (Semi-Implicit Method for Pressure Linked Equations)	PISO (Pressure Implicit with Splitting of Operator)
Time step	0.00038 (CFL < 1)[b]	0.002 (CFL < 1)[b]
No. of cells	12 million	2 million

[a] where k is the turbulence kinetic energy, ϵ is the dissipation in the flow, ζ is the scale ratio (e.g. ζ_u is the velocity scale ratio), f is the ratio (e.g. $f_\epsilon$is the ratio of unresolved dissipation to resolved) and ω is the specific dissipation rate.
[b] CFL is the Courant-Fredichs-Lewy condition which must be kept below unity

6.2. Global quantity

Time-averaged drag force coefficient from simulations was compared with that obtained numerically by Osth *et al.* [10] and experimentally by Sakuma *et al.* [11]. The result is presented in Table 4. The drag coefficient is defined as follows:

$$C_D = \frac{2F_D}{\rho U_\infty^2 A_x} \tag{23}$$

where ρ is the density of air at 20°C and $A_x = H^2$.

Table 4. Drag force coefficient for comparison

CASE	C_D
Experiment by Sakuma et al. [11]	0.86
Numerical by Osth et al. [10]	0.78
Case A (Coarse)	0.71
Case B (Fine)	0.73

The drag force coefficient obtained is relatively underestimated compared to numerically attained by Osth *et al.* [10] by 6% for Case B and 9% for Case A. The under-prediction of the negative pressure that occurred in separated region on the roof might be one of the reasons that resulted to discrepancies in the results of drag force coefficient with the previous works. This is due to the fact that separation region that started from the front leading edges extended much longer, hence deduced much negative pressure that contributed to a decrease in the total drag coefficient of the train model.

6.3. Streamlines of the time-averaged flow

Figure 10 and Figure 11 show a side view streamlines of the time averaged velocity field comparison with the previous data obtained by Osth *et al.* [10]. Only Case B (fine) is presented here. Generally, the current simulation was able to replicate the flow phenomena as obtained by Osth *et al.* [10].

IOP Conf. Series: Materials Science and Engineering **152** (2016) 012047 doi:10.1088/1757-899X/152/1/012047

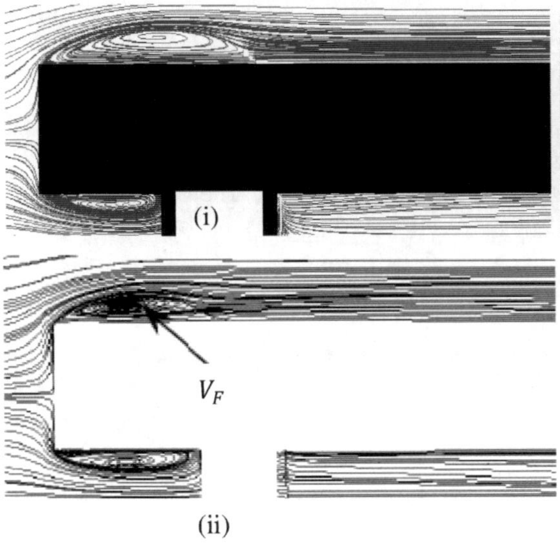

Figure 10. Side view of streamlines of the time-averaged velocity field around the first half of the train: (i) Case B and (ii) Osth *et al.* [10]

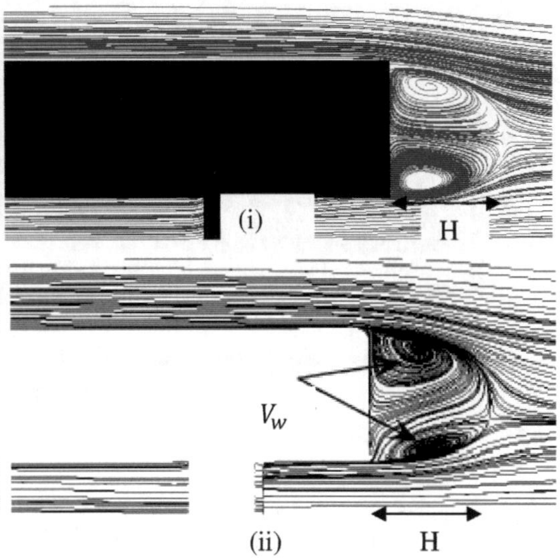

Figure 11. Side view of streamlines of the time-averaged velocity field around the second half of the train: (i) Case B and (ii) Osth *et al.* [10]

In Figure 10, the flow separation occurred at the leading edge at the top and bottom sides, and the vortex produced due to this is denoted as V_F. Only a small discrepancy in the size of vortex appears on the roof of the HST model, which was a bit lower than attained by Osth *et al.* [10]. This confirmed the earlier argument that the flow reattached further back, which made the drag force coefficient, C_D larger than the one obtained by Osth *et al.* [10]. On the other side, a similar flow structure produced at the second half of the train as can be seen in Figure 11(i) and (ii). There were two vortices at the upper and bottom sides that occurred in the wake, denoted as V_w. Simulation result agrees well with previous simulation data with the vortex V_w extends a distance H in the streamwise direction from the base.

Figure 12 and Figure 13, on the other hand, illustrate top view streamlines of the time averaged velocity field comparison. In Figure 12, flow separation occurred at the both of side leading edges and produced vortex similar in size. Same as previously stated, the vortex appeared to be slightly extended in the streamwise direction as compared to Osth *et al.* [10]. For the flow structure formed at the second half of the train as can be seen in Figure 13(i) and (ii), it similarly resembles the same shape as in the previous paper. There were two vortices occurring sideways in the wake. The vortex extended with a distance H in the streamwise direction from the base.

7. The computing machine

The simulations using OpenFOAM software in this recent work utilized parallel processors. To run the simulation job in parallel, the computational domain needed to be decomposed into several smaller domains. This was set by using *decomposePar* utility, where the total number of processors to run a particular job was selected. With the use of parallel processors, running any OpenFOAM job could be made faster. Table 5 shows the performance of the computer in simulating case A (Coarse) and case B (Fine).

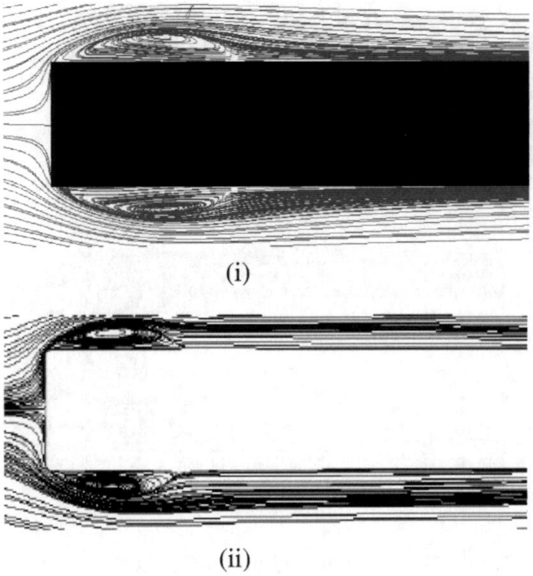

Figure 12. Top view of streamlines of time-averaged velocity field around first half of the train: (i) Case B (ii) Osth et al. [10]

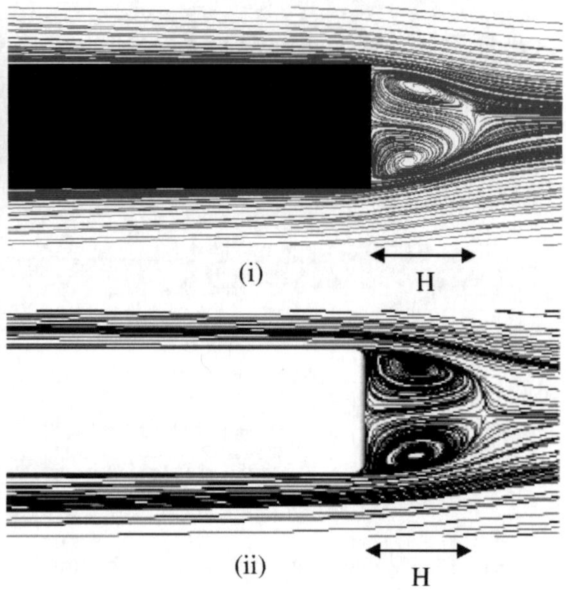

Figure 13. Top view of streamlines of time-averaged velocity field around second half of the train: (i) Case B (ii) Osth et al. [10]

Table 5. Computer performance for different number of processors used in the simulation

CASE	A (Coarse)	B (Fine)
No. of processors	8	32
Clock time (hour)	265	325
Simulation time $\left(U_\infty/H\right)$	1000	1000
Time step $\left(\Delta t U_\infty/H\right)$	0.005	0.002

8. Conclusion

The flow around a simplified HST has been simulated numerically with OpenFOAM framework by utilizing Unsteady Reynolds Navier-Stokes (URANS) equation combined with SST k-ω turbulence model for two different mesh resolutions. In this paper, it is well confirmed that, with the usage of OpenFOAM, flow structure and pressure distribution can be captured correctly and the outcomes have a good agreement with other similar studies. Study on two different mesh resolutions shows that the fine grid results in a more accurate data when compared to previous work. With the smaller mesh resolution used in this study compared to that previously used by other examiners, much reasonable time simulation can be achieved and this subsequently reduces the computational cost without effects on the final result.

Acknowledgement

This research was financially supported by Malaysia Ministry of Higher Education (MOHE) under The Fundamental Research Grant Scheme (FRGS PY/2015/05383) and also from the COMSTECH-TWAS 13-272. The authors also acknowledge the High Performance Computer (HPC), Universiti Teknologi Malaysia for the use of their supercomputer facilities.

References

[1] Hemida H and Krajnovic S 2009 *Engineering Applications of Computational Fluid Mechanics* **3** 28-41
[2] Krajnovic S, Ringqvist P, Nakade K and Basara B 2012 *Journal of Wind Engineering and Industrial Aerodynamics* **110** 86-99
[3] Biadgo A M, Simonovic A, Scorvan J and Stupar S 2012 *Addis Ababa University, Technology Faculty of Mechanical Engineering Transaction, Ethiopia* **42** 10-8
[4] Raghunathan R S, Kim H-D and Setoguchi T 2002 *Progress in Aerospace Sciences* **38** 469-514
[5] Baker C J, Jones J, Lopez-Calleja F and Munday J 2004 *Journal of Wind Engineering and Industrial Aerodynamics* **92** 547-63
[6] Ekeroth F, Kalmteg N, Pilqvist J and Runsten J 2009 *Crosswind flow around a high-speed train on embankment – An experimental and numerical study.* Bachelor thesis Chalmers University of Technology (Sweden)
[7] Khier W, Breur M and Durst F 2000 *Computers & Fluids* **29** 179-95
[8] OpenFOAM© 2011 *The Open Source CFD Toolbox* (Paris: ESI Group)
[9] Robertson E, Choudhury V, Bhushan S and Walters D K 2015 *Computers & Fluids* **123** 122-45
[10] Osth J and Krajnovic S 2012 *The 15th International Conference on Fluid Flow Technologies*
[11] Sakuma Y and Ido A 2009 *Aerodynamic Laboratory, Environmental Engineering Division, Japan* **50** 20-5
[12] Hemida H, Krajnovic S and Davidson L 2005 *17th AIAA Computational Fluid Dynamics Conference*
[13] Krajnovic S and Davidson L 2004 *ASME Heat Transfer/Fluids Engineering Summer Conference*
[14] Hemida H and Krajnovic S 2010 *Journal of Wind Engineering and Industrial Aerodynamic* **98** 34-46
[15] Alam F and Watkins S 2006 *Proceeding of the 3rd BSME-ASME International Conference on Thermal Engineering*
[16] Orellano A and Schober M 2006 *Proceedings of the 4th WSEAS International Conference on Fluid Mechanics and Aerodynamics*
[17] Per-Ake T 1999 *Report (in Swedish) FFA TN 2000-05*
[18] Celik I, Ghia U, Roache P J and Freitas C 2008 *Journal of Fluids Engineering-Transactions of the ASME* **130**
[19] Prime Z, Moreau D, Doolan C J and Mat Ali M S 2014 *20th AIAA/CEAS Aeroacoustics Conference*
[20] Barton I E 1998 *International Journal of Numerical Methods in Fluids* **26** 459-83
[21] Wilcox D C 1993 *AIAA Journal* **32** 1414-21
[22] Menter F M 1994 *AIAA Journal* **32** 1598-1605

IOP Conf. Series: Materials Science and Engineering **152** (2016) 012048 doi:10.1088/1757-899X/152/1/012048

Experimental study on energy absorption of foam filled kraft paper honeycomb subjected to quasi-static uniform compression loading

N Abd Kadir, Y Aminanda*, M S Ibrahim and H Mokhtar

Mechanical Engineering Department, Faculty of Engineering,
International Islamic University Malaysia, 53100 Kuala Lumpur, Malaysia

*yulfian@iium.edu.my

Abstract. A statistical analysis was performed to evaluate the effect of factor and to obtain the optimum configuration of Kraft paper honeycomb. The factors considered in this study include density of paper, thickness of paper and cell size of honeycomb. Based on three level factorial design, two-factor interaction model (2FI) was developed to correlate the factors with specific energy absorption and specific compression strength. From the analysis of variance (ANOVA), the most influential factor on responses and the optimum configuration was identified. After that, Kraft paper honeycomb with optimum configuration is used to fabricate foam-filled paper honeycomb with five different densities of polyurethane foam as filler (31.8, 32.7, 44.5, 45.7, 52 kg/m^3). The foam-filled paper honeycomb is subjected to quasi-static compression loading. Failure mechanism of the foam-filled honeycomb was identified, analyzed and compared with the unfilled paper honeycomb. The peak force and energy absorption capability of foam-filled paper honeycomb are increased up to 32% and 30%, respectively, compared to the summation of individual components.

1. Introduction

Honeycomb sandwich structure has been used extensively as an energy absorber or a cushion to resist external loads due to its lightweight and high energy absorbing capability. However, its weakness due to impact loading, hygro-thermal and others becomes a significant preoccupation for its application in the industry. Extensive work has been carried out on the structural and failure behavior of honeycomb sandwich structure. Aktay *et al.* [1] developed several numerical techniques for modeling transverse crush behavior of honeycomb core materials and compared with test data on aluminum and Nomex honeycomb. Ajdari *et al.* [2] investigated in-plane dynamic crushing of two dimensional honeycombs with both regular hexagonal and irregular arrangements. Furthermore, the critical buckling loads for various core densities and materials of honeycomb composite panels were studied both experimentally and analytically by Kaman *et al.* [3]. In addition, Zhang *et al.* [4] investigated the crashworthiness of kagome honeycomb sandwich cylindrical column under axial crushing loads and presented the new type of kagome honeycomb sandwich in order to expand the plastic deformation zones and improve the energy absorption efficiency.

Petrone *et al.* [5] reported that energy absorption in honeycomb cores is minimally dependent on the height of the cell walls but predominantly dependent on geometrical parameters of the honeycombs, essentially the cell wall thickness-to-length ratio (*t/l*) or the relative density of the core. In the case of compressive loads, the critical buckling load is dependent on the bending of the cell wall that is related by (t/l^3). Therefore, by increasing the relative density of the core, an increase of energy absorption in these cores can be expected. In addition, Xu *et al.* [6] reported that cell wall thickness to edge length ratio, *t/l* also plays the dominant role in the dynamic properties of honeycomb materials. In their work on the experimental study of out-of-plane dynamic compression of hexagonal honeycombs, they found that the tangent modulus increased towards the end of the crushing process, especially for honeycombs with small values of wall thickness to edge length ratio (*t/l*). For large cell size honeycombs, the plastic buckling is more likely to occur in the middle region of the specimens under quasi-static and dynamic compression, indicating that the deformation pattern is influenced by the dimension of the honeycomb cells. Other noteworthy studies on the effect of parameters of honeycomb on the energy absorption capability include Zhang *et al.* [4], Alavi and Sadegia [7] and Dongmei Wang [8].

The compression deformation behavior of polymer foam is one of the characteristics that make the material desirable in energy absorption applications. In particular, the ability of the foam to undergo large amounts of plastic deformation at nearly constant stress is advantageous. Therefore, polymeric foams have been investigated extensively as fillers for hollow cores [7, 9–11]. The results have shown that these low density foams have positive effect on strengthening the cell walls of panel and also has been credited with improving energy absorption capability and damping properties of the honeycombs [10, 12–13]. Alavi and Sadega [7] found that foam filling of hexagonal cell can increase their mean crushing strength and energy absorption capability up to 300%. In addition, the mean crushing strength of foam filled panel is always greater than summation of individual component. Aktay *et al.* [14] also reported that the restraining effect of filler shifted the deformation mode of Al tube from diamond to concertina in larger diameter tube. Besides, energy absorptions in foam-filled tube were increased with increasing filler density and higher than the summation of the energy absorption of empty tube (alone) and filler (alone).

In this research, honeycomb structure made from Kraft paper for load bearing application was used and manufactured manually in order to obtain the desired thickness, density and cell size. Optimum configuration of the unfilled Kraft paper honeycomb on both specific energy absorption and specific compressive strength was determined through statistical analysis of variance (ANOVA). Then, Kraft paper honeycomb with optimized parameters was filled with low density polyurethane foam in order to strengthen the cell walls. Series of quasi static experimental tests were carried out to investigate the force and specific energy absorption capability of foam filled kraft paper honeycomb.

2. Experiment materials and methods

2.1. Materials

The kraft paper honeycomb was fabricated from commercial kraft paper with three different densities (80gsm, 120gsm, 175 gsm) by using handmade honeycomb maker. The kraft paper honeycomb were coated with resin and cured at room temperature for 24 hours. Polyurethane foam were used as filler in the kraft paper honeycomb structure and it was made by mixing polyol and isocyanate in liquid foam. Five different types of polyol were used to produce five different density of polyurethane. However, only one type of isocyanate was used namely Maskiminate 80. The type of polyol and density of foam are shown in Table 1.

2.2. Design of experiment method

The full factorial design using Design-Expert Software was used to evaluate the effect of factors and to obtain the optimum configuration of Kraft paper honeycomb on specific energy absorption (SEA) and specific compressive strength(SCS). Full factorial design is a collection of statistical and mathematical techniques that are useful for the modeling and analyzing problems in which a response of interest is influenced by several variables and the objective is to optimize this response [15].

Table 1. Foam density and polyol type of polyurethane foam

Polyol type	Foam density (kg/m^3)
788B/4x2	31.8
788B/8	32.7
788B/2	44.5
788B/9/45	45.7
788B/9/68	52.0

Design-Expert software fitted four models; linear, two-factor interaction (2FI), quadratic and cubic polynomials to the responses and displayed a measure of progress during calculation. Linear models are generally used in most studies to assess the independent and dependent factors. In this model, the behavior of the dependent variables (response) can be expressed as Equation 1 [15].

$$y_i = \beta_o + \sum_{j=1}^{n} \beta_j X_{ij} + \varepsilon_i \tag{1}$$

where ε_i is independent random variables, β_o is the mean of observations, β_j is unknown constant, J is the factor and n is the number of observations.

The non-linear models are important and necessary to consider an experimental design that allows one to fit the experimental data to quadratic model. The general model presented in Equation 2 is used to describe the non-linear model [15].

$$y = \beta_o + \sum_{j=1}^{n} \beta_i X_i + \sum_{j=1}^{n} \beta_{ii} X_i^2 + \sum_{i=1}^{n} \sum_{i<j=2}^{n} \beta_{ij} X_i X_j \tag{2}$$

2.3. Foam filled kraft paper honeycomb

The Kraft paper honeycomb with the optimized configuration is used to fabricate the foam-filled Kraft paper honeycomb by using dipping process. The dipping process gave a uniform foam-filling inside the honeycomb cell for bigger cell size, hence this method is more practical manufacturing process in comparison to foam injection as proposed by [12]. The foam-filled paper honeycombs were subjected to quasi-static compression loading by using SHIMADZU Autograph AG-X 250 compression machine. The bottom plate was fixed and the upper plate was subjected to vertical downward displacement up to 22.5mm (50% of the honeycomb height) with a displacement rate of 0.5 mm/min. The force acting on the specimens was measured by load cell and force-displacement curve was plotted for further analysis.

3. Results and discussion

3.1. Analysis of design of experiment of unfilled kraft paper honeycomb

Analysis of variance (ANOVA) was conducted on the collected data to investigate the main effects of density of paper (*A*), thickness of cell (*B*) wall and also cell size of honeycomb (*C*), with three level interaction effects on the specific energy absorption (SEA) and specific compression strength (SCS). Table 2 shows the factors and levels of kraft paper honeycomb. The height for all specimens was 45mm. According to full factorial design method, 27 configurations of kraft paper honeycomb are generated considering three factors with three levels and the experimental results are shown in Table 3.

Table 2. The three factors and their levels

Factor	Code	Unit	Level 1	Level 2	Level 3
Density of paper	A	gsm	80	120	175
Thickness of cell wall	B	ply	1	2	3
Cell size of honeycomb	C	mm	10	15	20

Table 3. The experimental results obtained based on full factorial design

Run	*A*	*B*	*C*	Specific compression strength (Mpa/kg)	Specific energy absorption (J/kg)
1	175	2	10	16.79	1128.41
2	80	2	20	9.25	657.42
3	80	2	10	23.10	535.93
4	120	3	15	6.69	203.55
5	175	3	20	7.64	808.94
6	80	3	15	3.50	71.69
7	120	2	10	12.79	325.15
8	120	1	20	7.87	436.98
9	120	2	15	14.48	458.22
10	120	3	10	15.59	270.64
11	80	1	20	1.70	81.09
12	80	3	20	6.06	426.57
13	175	2	20	4.09	452.96
14	80	1	10	7.59	97.67
15	175	1	20	7.09	560.30
16	175	2	15	11.16	647.34
17	120	1	10	14.91	266.13
18	120	1	15	16.69	421.35
19	80	2	15	16.44	595.467
20	80	3	10	22.15	569.33
21	175	1	10	17.50	559.68
22	120	3	20	4.87	332.11
23	175	3	20	9.50	661.96
24	175	1	15	8.02	337.70
25	120	2	20	6.06	472.63
26	175	3	10	13.422	601.65
27	80	1	15	7.29	335.18

3.1.1. Analysis of specific energy absorption. According to the sequential model sum of squares, the models were selected based on the highest order polynomials where the models were not aliased and the additional terms were significant. From the analysis, it found that the two-factor interaction model come out the best for specific energy absorption (SEA). This is because it exhibits a high R-square values of 0.8496. The predicted R-square of 0.5717, is in reasonable agreement with adjusted R-square of 0.7556. Adequate precision measures the signal to noise ratio and the ratio greater than 4 is desirable. The ratio for SEA is 11.819 and this indicates that an adequate signal and model can be used to navigate the design space.

The result of ANOVA analysis for SEA is presented in Table 4. For each response, the probability (Prob>F) was examined to check if it falls below 0.05. The model for SEA was developed with 99% confident level and P-value less than 0.001, which indicates that this model are highly significant. *A*, *B* and *C* refers to density of paper, thickness of cell wall and cell size of honeycomb, respectively. Based on the P-value, density of paper, thickness of cell wall and cell size of honeycomb were found to have significant effect on SEA. Values greater than 0.1000 indicate the model terms are not significant.

The F-value for each factor is simply a ratio of the mean of squared deviations to the mean of the squared errors. A larger F-value means that the factor has greater significance for SEA. From Table 4, it reveals that thickness of cell wall, *B* as most significant factor with F-value of 24.10, followed by density of paper, *A* and cell size of honeycomb, *C* with F-value of 12.13 and 4.77, respectively. The F-value for interaction between density of paper, *A* and cell size of honeycomb, *C* is 2.10, which implies a weak influence of the two factor interaction to SEA value. The two-factor interaction model for SEA is given in Equation 3.

$$SEA= 426.33 - 105.03A_1 + 27.55A_2 -145.94B_1 + 31.68B_2 + 62.86\, C_1 - 8.35C_2+ 6.10A_1C_1 - 49.93A_2C_1 - 55.27A_1C_2 + 21.03A_2C_2 \quad (3)$$

Table 4. Analysis of variance (ANOVA) for SEA

Source of data	Sum of square	Degree of freedom	Mean square	*F* value	Prob >*F*	Comment
Model	5.967E+005	10	59669	9.04	<0.0001	Significant
A	1.602E+005	2	80077.86	12.13	0.0006	
B	3.182E+005	2	1.591E+005	24.10	<0.0001	
C	62932.46	2	31466.23	4.77	0.0238	
AC	55395.17	4	13848.79	2.10	0.1287	

3.1.2. Analysis for specific compressive strength. Similarly, the best model for specific compressive strength (SCS) is two-factor interaction (2FI). This is due to low standard deviation (Std. Dev) of 0.75, high R-square value of 0.9390 and a low PRESS value of 34.61. The predicted R-square of 0.6912 is in reasonable agreement with adjusted R-square of 0.8678. The adequate precision is 15.95, which indicates that a satisfactory signal and model can be used to navigate the design. Table 5 summarizes the result of ANOVA analysis for SCS. The same parameters have been used for SEA analysis previously. Density of paper, thickness of cell wall and cell size of honeycomb were found to have significant effect on SCS with P-value 0.05. Thickness of cell wall resulted as most significant parameter for SCS with F-value of 48.91, followed by density of paper and honeycomb cell size with the F-value of 17.03 and 16.66, respectively. The lower F-value for interaction between AB and BC implies weak interaction between these parameters. The proposed models for SCS is expressed in Equation 4.

$$SCS = 6.08 - 1.19A_1 + 0.49A_2 - 1.92B_1 + 0.40B_2 + 1.12C_1 - 0.23C_2 - 0.100A_1B_1 - 0.54A_2B_1 + 0.15A_1B_2 - 0.15A_2B_2 - 0.18B_1C_1 - 0.38B_2C_1 + 0.66B_1C_2 + 0.20B_2C_2 \quad (4)$$

Table 5. Analysis of variance (ANOVA) for SCS

Source of data	Sum of square	Degree of freedom	Mean square	*F* value	Prob >*F*	Comment
Model	105.23	14	7.52	13.19	<0.0001	Significant
A	19.43	2	9.73	17.08	0.0003	
B	55.73	2	27.86	48.91	<0.0001	
C	18.98	2	9.49	16.66	0.0003	
AC	4.94	4	1.23	2.17	0.1348	
BC	6.12	4	1.53	2.69	0.0827	

3.1.3. Optimization parameter of kraft paper honeycomb. The density of paper, thickness of cell wall and cell size of honeycomb were found to have significant effect on SEA and SCS. Both SEA and SCS were found to increase with decreasing of honeycomb cell size and increasing of density of paper and thickness of cell wall. The trend was agreed with other researchers who found that the higher cell wall thickness and smaller cell size increase the energy absorption and crush strength [6, 16]. The highest value of responses were obtained when honeycomb cell size at minimum point, while density of paper and thickness of cell wall at maximum point within the range of study. As a result, it was found that the optimum configuration of the unfilled Kraft paper honeycomb on SEA and SCS within the range of this study is: kraft paper density of 175 gsm, cell wall thickness of 3 ply and honeycomb cell size of 10mm.

3.2. Unfilled kraft paper honeycomb

The force-displacement curves of selected kraft paper honeycomb compressed to the 50% deformation in quasi static compression are shown in Figure 1. The graphs show that there is elastic behavior at the beginning of the indentation until a critical load is reached. Then, sharp drop is observed after the peak load, which corresponds to the beginning of the vertical edge deformation. The force then decreases to reach a plateau, which corresponds to the succession of fold forming and ends up by condensation of the honeycomb [17]. The deformation of selected kraft paper honeycomb of paper can be observed in Figure 2.

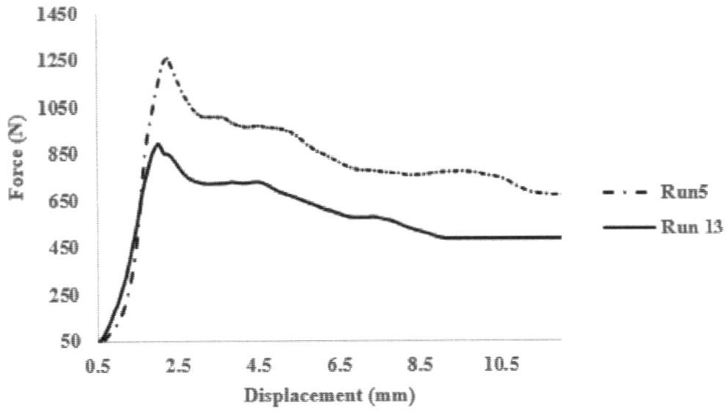

Figure 1. Force-displacement curves of selected kraft paper honeycomb

Figure 2. Condition of selected kraft paper honeycomb before (top) and after (bottom) quasi static compression test

The failure mechanism of kraft paper honeycomb consists of buckling of the cell walls and vertical edges, wall buckling on the top or bottom surfaces and the tendency of unstable outer cells to buckle towards inward or outward direction. Initially, as kraft paper honeycomb was compressed, the lower part of the cell walls started to buckle. The cell walls started to buckle locally at the weakest point along the height, which is the bottom part of the structure. This lower part of cell walls continued to buckle as the kraft paper honeycomb was compressed further. Then, the number of folds of walls and vertical edges increased until reaching the densification phase. It is also can be observed from Figure 2 that the kraft paper honeycomb with 3 ply of thickness of paper (c) buckled only on the top surface compare to 1 ply (b) and 2 ply (a), where the wall buckling on both side of honeycomb. This condition happened due to the imperfections of the cell alignment during the manufacturing process. However, the fold shape is same for all samples. The folding mechanism is similar to Nomex honeycomb that reported by Aminanda *et al.* [1].

3.3. Foam-filled kraft paper honeycomb structure

The Kraft paper honeycomb (Figure 3a) with optimized configuration (175gsm, 3ply, 10mm) was used to fabricate the foam-filled Kraft paper honeycomb (Figure 3b). Five different densities of foam was used with its identification using FFP1 to FFP5 respectively with increase of foam density. The list of the five different densities of foam and its designation are shown in Table 6.

Figure 3. Condition of (a) unfilled kraft paper honeycomb (b) foam-filled kraft paper honeycomb

Table 6. Foam density and designation of foam-filled kraft paper honeycomb

Designation	Foam density (kg/m^3)
FFP1	31.8
FFP2	32.7
FFP3	44.5
FFP4	45.7
FFP5	52.0

The force in function of the compression displacement for all specimens are drawn in Figure 4. The shape of the curves is almost similar with that of unfilled kraft paper honeycomb. However, the peak load of the foam-filled kraft paper honeycomb is broader than unfilled honeycomb, which is due to the foam occupied at the fold angle that gives a higher residual resistance and less abrupt behavior. From the graph, the energy absorption were determined by measuring the net area under force-displacement curve and it was measured until 22.5mm. Meanwhile, peak force was obtained by taking the maximum force of each curves. Table 7 and Table 8 represent the comparison of energy absorption capability and peak forces between the foam-filled Kraft Paper honeycombs and the summation of their individual components. The specific energy of individual component were calculated using Equation 5.

$$SEA_{f+h} = \frac{E_f + E_h}{m_f + m_h} \quad (5)$$

where E_f and E_h, are energy absorption for foam and honeycomb, while m_f and m_h are mass for foam and honeycomb, respectively. Meanwhile, specific energy absorption of foam-filled were calculated using Equation 6.

$$SEA_{ff} = \frac{E_{ff}}{m_{ff}} \tag{6}$$

where E_{ff} is energy absorption for foam-filled, while m_{ff} is mass for foam-filled. Since the total mass of individual element is almost the same as compared to that of the foam-filled honeycomb, the energy absorption increment relates also to the specific energy absorption.

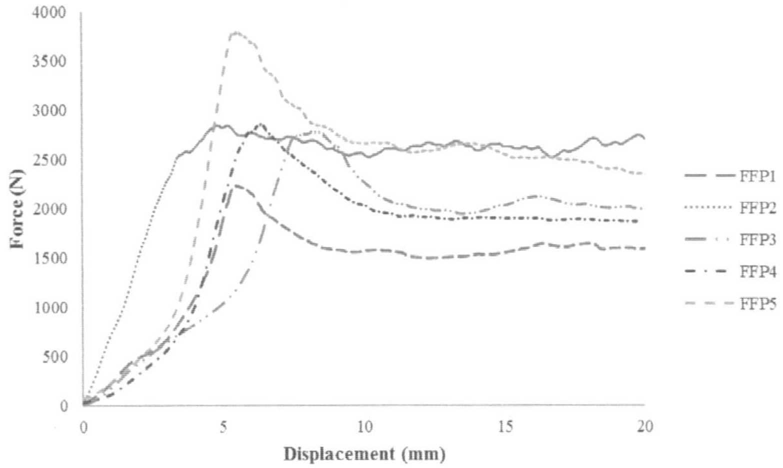

Figure 4. Force-displacement curves of foam-filled with five densities of foam

Table 7. Comparison of energy absorption capabilities between foam filled paper honeycomb and summation of its individual component

Foam-filled honeycomb	Energy absorbed (J)				
	Unfilled paper honeycomb (a)	Polyurethane foam (b)	Summation (a) + (b)	Foam-filled paper honeycomb	% Increase
FFP1		5.87	30.43	32.19	5.78 %
FFP2		7.81	32.37	38.75	19.70 %
FFP3	24.56	9.00	33.56	33.92	1.07 %
FFP4		12.97	37.53	48.93	30.38 %
FFP5		17.85	42.41	55.03	29.76 %

Table 8. Comparison of peak forces between foam filled paper honeycomb and summation of its individual component

Foam-filled honeycomb	Peak Force (kN)				
	Unfilled paper honeycomb (a)	Polyurethane foam (b)	Summation (a) + (b)	Foam-filled paper honeycomb	% Increase
FFP1		0.121	2.502	2.235	-10.67 %
FFP2		0.191	2.572	2.844	10.58 %
FFP3	2.381	0.247	2.628	2.780	5.78 %
FFP4		0.387	2.768	2.852	3.03 %
FFP5		0.490	2.871	3.796	32.21 %

From Table 7, it can be seen that energy absorbed of foam-filled honeycombs is always higher than the summation of their individual components. Foam-filled Kraft paper honeycomb with foam density 45.7 kg/m^3 (FFP4) shows the highest increment of energy absorbed with 30.38 % increase. It is also found that the percentage increase in SEA for higher density of foam filling (FFP4 and FFP5) are much higher compared to low density of foam (FFP1, FFP2, FFP3). This results also show agreement with study of Aktay *et al.* [14] that the energy absorption of foam-filled tube increases with increasing filler density and higher than the summation of the energy absorption of empty tube (alone) and foam filler (alone). Increment of energy absorption capability was due to the change of boundary condition on the top and bottom surfaces of the honeycomb from free-free to fixed-fixed boundary condition. The foam densification was found to occur at the buckled cell walls. This foam densification occurred due to the compression between one buckled wall and its neighboring walls. Therefore, after the sharp drop of force, the force fluctuated with number of wall buckling and densification of foam. As a results, the energy absorption capability of the foam-filled kraft paper honeycomb was increased and its values were higher than the summation of individual components.

Comparison of peak forces between foam-filled paper honeycomb in Table 8 shows that almost all peak force values of foam filled are higher than the summation of their individual component except for the peak force of foam-filled Kraft paper honeycomb with 31.8 kg/m^3 density of foam (FFP1). The Kraft paper honeycombed filled by the highest density of polyurethane foam (52 kg/m^3) displays the highest increment of peak force with 32.21% increases. The foam infill has increased the bending and buckling resistivity of the honeycomb core cell walls. Due to this support, the peak force is increased and the transmitted load from the front face to the back face is increased, although the foam inside the cell has not compressed to its maximum value.

The specific energy absorption of the foam-filled kraft paper honeycomb with different densities is plotted in Figure 5. From the graph, the optimum value of specific energy absorption is obtained at the maximum point of the specific energy absorption curve. It is found that the optimum specific energy absorption can be divided into two regions within the range of density studied for this work. For foam density less than 44.5 kg/m^3, the optimum SEA is found for foam density of 33 kg/m^3. As for density higher than 44.5 kg/m^3, more experiments needs to be performed to reach an eventual optimum SEA as shown in Figure 5. It is also noted that for lower density foam filling (less than 44.5 kg/m^3), it is almost ineffective to increase the SEA of foam-filled kraft paper honeycomb over unfilled kraft paper honeycomb.

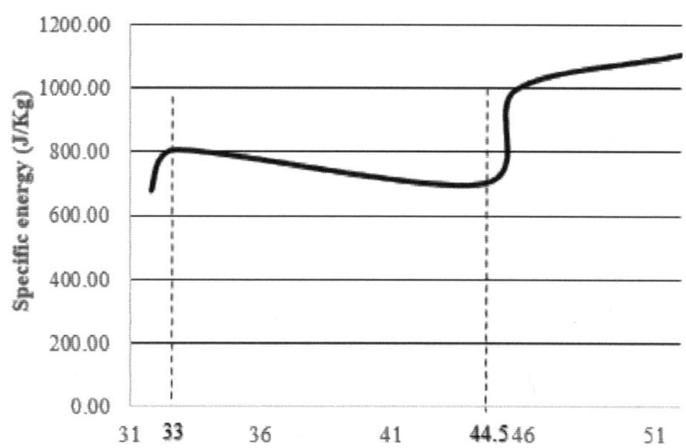

Figure 5. Specific energy absorption of foam-filled kraft paper honeycomb with different densities of foam

Failure mechanism of foam-filled paper honeycomb is comprised by interaction between cell walls and foam, where the foam filled up the folds. The condition of the foam filled kraft paper honeycomb before and after compression can be observed clearly in Figure 6 and Figure 7.

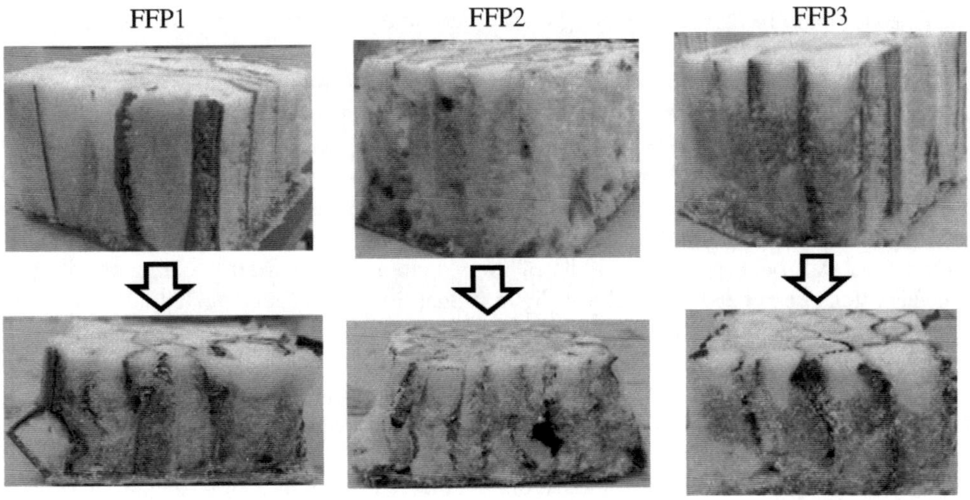

Figure 6. Condition of foam-filled kraft paper honeycomb before (top) and after (bottom) quasi-static compression test for FFP1, FFP2 and FFP3

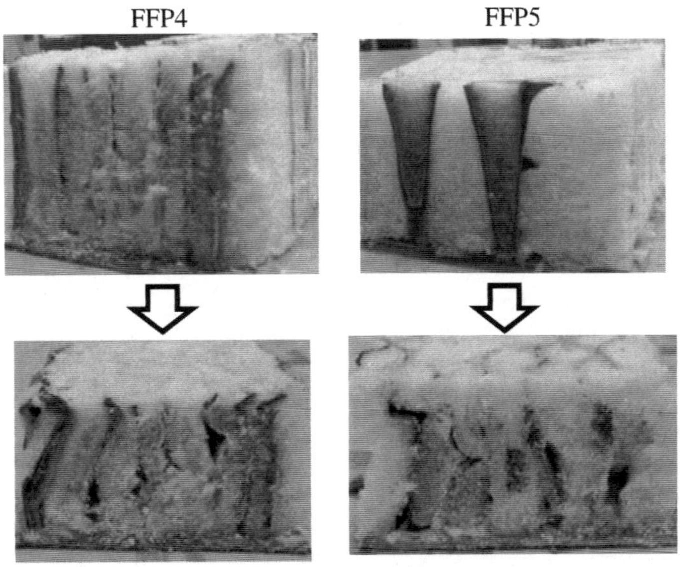

Figure 7. Condition of foam filled kraft paper honeycomb before (top) and after (bottom) quasi-static compression test for FFP4 and FFP5

The buckling of cell walls starting at the middle of the structure and followed by foam densification at the same location. As illustrated in the figures, the most obvious observation if compared to unfilled kraft paper honeycomb is the condition on the top and bottom surface of the structure. There is no buckling at the wall on the top and the bottom surfaces of kraft paper filled with polyurethane foam even though the specimens were compressed until densification phase. This is in contrast with unfilled kraft paper honeycomb, where the numbers of buckled wall on the surface increase as displacement increases. On the surface of foam-filled kraft paper honeycomb, the shape of multiple hexagonal tubes remains as hexagonal even though they were compressed until the densification phase. Therefore, it is found that the failure mode was changed from wall buckling on the top and bottom surface for unfilled kraft paper honeycomb to the wall buckling on the middle of the structure for foam-filled kraft paper honeycomb.

The failure mechanism of foam-filled kraft paper honeycomb was divided into three stages: before plateau region, plateau region and end of plateau region. Before plateau region, the buckling of cell wall and vertical edge is less severe compared to unfilled kraft paper honeycomb and there is still no tearing of double thickness wall. In plateau region, the foam starts to push the bucked wall and filling the void between the folds. Tearing of cell wall occurs but pushed against each other. At the end of the plateau region, the whole cell wall bucked with bigger fold length compared to unfilled kraft paper honeycomb. The wall buckled at the middle of the structure, and no wall buckling at the upper and lower surface. It also can be seen clearly in Figure 6 that few outer cells buckled out since they were not supported from the outside and because they were only supported by three neighboring cells and only two for the cells at the end corners.

The failure behavior of the kraft paper honeycomb was found to be similar to foam-filled Nomex honeycomb that was reported by Wan *et al.* [12]. Based on the figures, it also shows a change of the boundary condition from free-fixed type (local buckling) of unfilled kraft paper honeycomb to fixed-fixed (global buckling) for foam-filled honeycomb subjected to compression loading. The change of the boundary conditions explains the peak force and energy absorption increment of the foam-filled honeycomb, together with the fact that the foam strengthens the honeycomb cell walls and cell walls confined the foam during the compression loading.

4. Conclusion

In conclusion, experimental design methodology has been shown to be a valuable tool to obtain the optimum configuration for kraft paper honeycomb and to explore the influences of density of paper, thickness of cell wall and cell sizes of honeycomb on the specific energy absorption and the specific compression stress. The results reveal that all factors give significant effect, however the thickness of cell wall has the most influence on the performance of kraft paper honeycomb. It is also found that the optimum configuration for kraft paper honeycomb is paper density of 175gsm, cell wall thickness of 3 ply and honeycomb cell size of 10mm. Besides, the kraft paper honeycomb is found to collapse by the buckling of cell walls and vertical edges on the top and bottom surfaces when they were compressed under quasi-static compression. The kraft paper honeycomb with the optimum configuration was filled up with low density polyurethane. The results reveal that filling the kraft paper honeycomb up with polyurethane foam does help strengthening its cell wall, hence improves the energy absorption and peak force of kraft paper honeycomb. This was proven experimentally where the energy absorption and peak force were found to be larger for foam-filled honeycomb structure in comparison to the summation of individual component while the total weight of components remains the same compared to foam-filled honeycomb. From the experimental observation, the increment of peak force and energy absorption is due to change of boundary conditions from free-fixed on honeycomb structure producing local buckling to fixed-fixed on foam-filled producing global buckling. In addition, the effect of foam in strengthening the cell walls and the effect of cell walls in confirming the foam contribute to the higher value of peak force and energy absorption of foam-filled honeycomb structure. Furthermore, the change of failure deformation from wall buckling on the top and bottom surface for unfilled kraft paper honeycomb to the wall buckling at the middle of the structure for kraft paper honeycomb filling with polyurethane foam is also noted.

Acknowledgements

This research was supported by Ministry of Higher Education Malaysia and Research Management Centre of the International Islamic University Malaysia (Grant no: FRGS14-155-0396)

References

[1] Aktay L, Johnson A F and Kröplin B-H 2008 *Eng. Fract. Mech.* **75** 2616–30
[2] Ajdari A, Nayeb-Hashemi H and Vaziri A 2011 *Int. J. Solids Struct.* **48** 506–16

[3] Kaman M O, Solmaz M Y and Turan K 2010 *J. Compos. Mater.***44** 2819–31
[4] Zhang Z, Liu S and Tang Z 2010 *Thin-Walled Struct.* **48** 9–18
[5] Petrone G, Rao S, De Rosa S, Mace B R, Franco F and Bhattacharyya D 2013 *Compos. Struct.* **100** 356–62
[6] Xu S, Beynon J H, Ruan D and Lu G 2012 *Compos. Struct.* **94** 2326–36
[7] Alavi Nia A and Sadeghi M Z 2010 *Mater. Des.* **31** 1216–30
[8] Wang D 2009 *Int. J. Impact Eng.* **36** 110–4
[9] Burlayenko V N and Sadowski T 2010 *Compos. Struct.* **92** 2890–900
[10] Niknejad A, Liaghat G H, Moslemi Naeini H and Behravesh A H 2011 *Mater. Des.* **32** 69–75
[11] Vaidya U K, Ulven C, Pillay S and Ricks H 2003 *J. Compos. Mater.* **37** 611–26
[12] Wan Abdul Hamid W L H, Aminanda Y and Shaik Dawood M S I 2013 *Appl. Mech. Mater.* **393** 460–6
[13] Zhang G, Wang B, Ma L, Wu L, Pan S and Yang J 2014 *Compos. Struct.* **108** 304–10
[14] Aktay L, Toksoy A K and Gü Den M 2006 *Mater. Des.* **27** 556–65
[15] Douglas C. Montgomery 2001 *Design and analysis of experiments* (Hoboken: Wiley)
[16] Khoshravan M R and Najafi Pour M 2014 *Thin-Walled Struct.* **84** 423–31
[17] Aminanda Y, Castanié B, Barrau J-J and Thevenet P 2005 *Appl. Compos. Mater.* **12** 213–27

IOP Conf. Series: Materials Science and Engineering **152** (2016) 012049 doi:10.1088/1757-899X/152/1/012049

Stress intensity factors of double edge cracks in large groove plates under mode I tension

A E Ismail[1*], M A Azlan[1], M L M Tobi[1] and M H Ahmad[2]

[1]Faculty of Mechanical and Manufacturing Engineering, Universiti Tun Hussein Onn Malaysia, Batu Pahat, Johor, Malaysia.
[2]Faculty of Civil and Environmental Engineering, Universiti Tun Hussein Onn Malaysia, Batu Pahat, Johor, Malaysia.

[*]emran@uthm.edu.my / al_emran@hotmail.com

Abstract. The solutions of stress intensity factors (SIFs) of double edge cracks in large groove plates are investigated. The plate is assumed to fulfil the plain strain condition and it is axially stressed. According to the literature, the SIFs of the cracks in circular notches are available. However, the roles of cracks in large groove are difficult to obtain. Therefore, different grooves and crack geometries are modelled numerically using the ANSYS finite element software. The present model is first validated using an existing model of double edge cracks without notches and it is found that both models are in close agreement. The large grooves are then introduced and the cracks are positioned in the mid-point. It is found numerically that the SIFs are strongly affected by the relative crack depth and the groove geometries. It is also found that the large groove is capable of reducing the SIFs in comparison with the circular notched due to lower stress concentration factors.

1. Introduction

In modern engineering, discontinuities in engineering components are unavoidable due to the design complexities. Under certain circumstances, cracks might appear on the surface of such discontinuities due to corrosion or fatigue [1]. Generally, stress intensity factor (SIF) approach is used successfully to characterize the cracks. On the other hand, other different methods are also available for the purpose of crack characterizations [2]. According to the recent literature survey, the solutions of SIFs for the circular holes or notches containing cracks are available [3-5]. The solutions of SIFs without notches are also vastly available [6-10]. However, the roles of cracks in large grooves are rarely investigated and not fully understood. Therefore, this present paper studies the effect of notches and large grooves containing different crack depths. In this study, four groove depths and eight crack lengths have been numerically modelled using ANSYS finite element software. The SIFs are based on the displacement extrapolation method and they are then normalized and plotted against the relative crack depth.

2. Theoretical background

The evaluation of J-integral around the crack tip is based on the domain integral method initially introduced by Shih *et al.* [11]. This integral formulation uses area integration for 2-dimensional and volume integration for 3-dimensional problems which offer much better accuracy than contour integral and it is also much easier to implement numerically. Equation 1 represents the 2D domain J-integral taking accounts the absence of thermal strain, path dependent plastic strain, body forces occur within the integration area [12].

$$J = \int A \left[\sigma_{ij} \frac{\partial u_j}{\partial x_1} - w\delta_{1i} \right] \frac{\partial q}{\partial x_i} dA \tag{1}$$

where q_{ij} is the stress tensor, u_j is the displacement vector, w is the strain energy density, δ_{ij} is the Kronecker delta, x_i is the coordinate axis and q is referred to as the crack extension vector. The direction of q is similar with x-axis of the local coordinate specified at the crack tip and it is normally chosen as zero at nodes along the contour Γ. It is also a unit vector for all nodes inside Γ except the mid-side nodes and known as virtual crack extension nodes. For 3-dimensional problems, the principal is similar to 2-dimensional problems. However, domain integral representation of J-integral becomes volume integration [13]. Two approaches for calculating stress intensity factor (SIF) are available in ANSYS software: interaction integral and displacement extrapolation methods. The first method is used because it is much easier to implement numerically, and it also offers better accuracy and fewer mesh requirement. This method is similar to domain integral method for J-integral evaluation that is described previously. Discussion on the domain integral methods can be found in [13]. The interaction integral is defined as in Equation 2.

$$I = -\int_V q_{ij} \left[\sigma_{k,l} \varepsilon_{k,l}^{aux} \delta_{i,j} - \sigma_{k,j}^{aux} u_{k,i} - \sigma_{k,j} u_{k,i}^{aux} \right] dv \Big/ \int_s \delta q_n ds \tag{2}$$

where σ_{ij}, ε_{ij} and u_i are the stress, strain and displacement, and $\sigma_{k,j}^{aux}$, $\varepsilon_{k,l}^{aux}$ and $u_{k,i}^{aux}$ are the stress, strain and displacement of the auxiliary field, and q_i is the crack extension vector.

$$\sigma_{ij}^{aux} = \frac{K_I^{aux}}{\sqrt{2\pi r}} f_{ij}^{I}(\theta) + \frac{K_{II}^{aux}}{\sqrt{2\pi r}} f_{ij}^{II}(\theta) + \frac{K_{III}^{aux}}{\sqrt{2\pi r}} f_{ij}^{III}(\theta) \tag{3}$$

$$u_{ij}^{aux} = \frac{K_I^{aux}}{2\mu}\sqrt{\frac{r}{2\pi}} g_j^{I}(\theta,\nu) + \frac{K_{II}^{aux}}{2\mu}\sqrt{\frac{r}{2\pi}} g_j^{II}(\theta,\nu) + \frac{2K_{III}^{aux}}{\mu}\sqrt{\frac{r}{2\pi}} g_j^{III}(\theta,\nu) \tag{4}$$

$$\varepsilon_{i,j}^{aux} = \frac{1}{2}\left(u_{i,j}^{aux} + u_{j,i}^{aux} \right) \tag{5}$$

An expression for the energy release rate in terms of mixed-mode SIFs is defined as for plane strain condition as in Equation 6, Equation 7 and Equation 8.

$$J = \frac{\left(K_I^2 + K_{II}^2\right)\left(1-\upsilon^2\right)}{E} + \frac{K_{III}^2 (1+\upsilon)}{E} \tag{6}$$

$$J = \frac{\left[\left(K_I + K_I^{aux}\right)^2 + \left(K_{II} + K_{II}^{aux}\right)^2 \right]\left(1-\upsilon^2\right)}{E} + \frac{\left(K_{III}^2 + K_{III}^{aux}\right)(1+\upsilon)}{E} \tag{7}$$

$$J = J + J^{aux} + I \tag{8}$$

The interaction integral can be associated with the SIFs as Equation 9.

$$I = \frac{2(1-\upsilon^2)}{E}\left(K_I K_I^{aux} + K_{II} K_{II}^{aux} \right) + \frac{1}{\mu} K_{III} K_{III}^{aux} \tag{9}$$

By setting $K_{II}^{aux} = 1$ and $K_I^{aux} = K_{III}^{aux} = 0$,

$$K_I = \frac{E}{2(1-\upsilon^2)} I \tag{10}$$

By setting $K_{II}^{aux} = 1$ and $K_{I}^{aux} = K_{III}^{aux} = 0$ and $K_{III}^{aux} = 1$ and $K_{I}^{aux} = K_{II}^{aux} = 0$ leads to the relationship between modes II and III SIFs with *I*, respectively.

$$K_{II} = \frac{E}{2(1-\upsilon^2)} I \tag{11}$$

$$K_{III} = \mu I \tag{12}$$

where K_i is a SIF with *i* is a loading mode, *i* = 1, 2 and 3. J-integral can be represented as *J* and *I* is a interaction domain integral. Furthermore, *E* and μ is a modulus of elasticity and modulus of rigidity, respectively.

3. Methodology

ANSYS finite element software is applied to model the cracks in large groove plates. The plate is assumed to fulfil the plain strain condition and it has two grooves at both sides and the cracks are located in the middle as shown in Figure 1. Due to the symmetrical geometry, only a quarter model is used and modelled numerically. A quarter finite element model is developed using ANSYS software and a special attention is given to the crack tip by employing 20-node iso-parametric elements. The square-root singularity of stresses and strains at the crack tip is modelled by shifting the mid-point nodes to the quarter-point location close to the crack tip, as shown in Figure 2. The bottom and right lines, except the crack face, are symmetrically restrained while the pressure is applied onto the left line. In order to calculate the stress intensity factors (SIFs), J-integral approach is used and it is then converted to SIFs according to Equation 13. The SIFs are normalized using Equation 14 and plotted against the relative crack depth, *a/d*.

$$K_I = \sqrt{\frac{J_e E}{1-v^2}} \tag{13}$$

$$F_{I,a} = \frac{K_I}{\sigma\sqrt{\pi a}} \tag{14}$$

where K_I is a mode I stress intensity factor, $F_{I,a}$ is a normalized stress intensity factor, J_e is an elastic J-integral, *E* is a modulus of elasticity, *v* is a Poisson's ratio, σ is a tension stress and *a* is a crack length. Mode I is a condition where the crack is opened. Large groove is defined when $r >> c$.

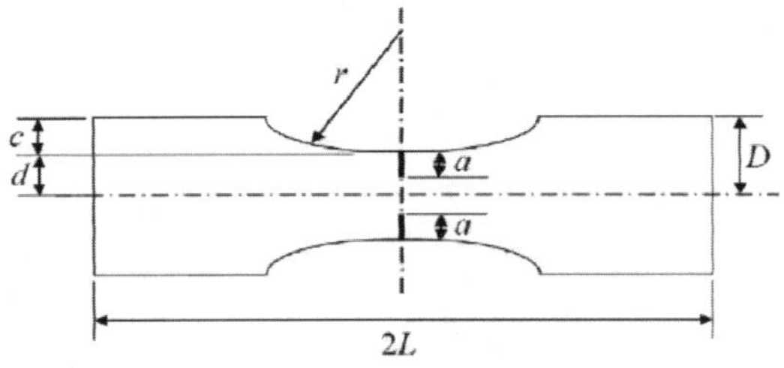

Figure 1. Large groove plate containing double edge crack in the middle location

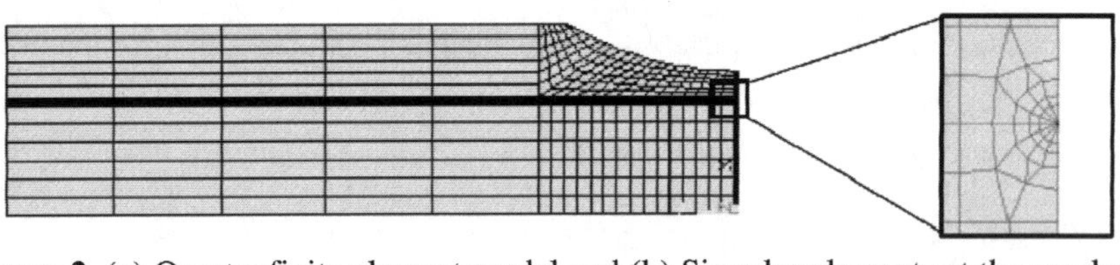

Figure 2. (a) Quarter finite element model and (b) Singular elements at the crack tip

In this paper, the range of values of *a/d* used is between 0.1 to 0.8. Meanwhile, the relative groove depth, *c/D* are 0.125, 0.25, 0.375 and 0.50 where the plate width, $W = 2D$. In order to study the role of groove geometries, there are fifteen ratios of *c/r* that are used depending on the radiuses and depth of the grooves. Before the model is further used, it is an important task to validate and compare the present model with an existing model. Since there is no cracks available in large grooves, then the crack without notch is used. It is found both models are in a good agreement as shown in Figure 3.

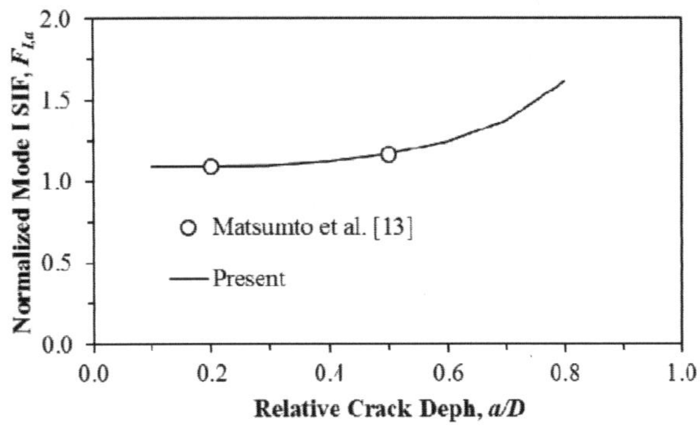

Figure 3. Finite element validation

4. Results and discussion

Figure 4 shows the effect of *a/d* on the SIF for various groove geometries. The distributions of the SIFs are almost identical where the SIFs have reduced as longer cracks are used. However, when the crack length reached a certain crack depth depending on the groove geometries, the SIFs are increased. The turning point of the SIFs is approximated half of the plate width. This behaviour of crack is quite different with the cracks without notches where the SIFs increased as the crack length is increased [9, 10]. Figure 4 also reveals that the SIFs distributions can be divided into two regions. The first region shows that the SIFs are strongly depend on the crack depth and radiuses of the grooves. It is revealed that SIFs for circular notch are greater than the SIFs obtained from the large grooves. The dispersions of SIFs increased in magnitude if the groove depths are increased. In the second region where the cracks become deeper, all the SIFs converged into a single value indicating that varying the groove geometries are not affected the SIFs. Table 2 tabulates the regression statistics of the Equation 15.

$$\begin{aligned} F_{I,a} &= 2.0324 - 0.5200\left(\frac{r}{W}\right) - 1.491\left(\frac{a}{L}\right) - 0.7810\left(\frac{c}{r}\right) + 0.2751\left(\frac{r}{W}\right)^2 - 0.6060\left(\frac{a}{L}\right)^2 \\ &- 0.0840\left(\frac{c}{r}\right)^2 - 0.0920\left(\frac{r}{W}\right)^3 + 2.3318\left(\frac{a}{L}\right)^3 - 0.4130\left(\frac{c}{r}\right)^3 + 0.1591\left(\frac{r}{W}\right)\left(\frac{a}{L}\right)^2 \\ &+ 5.2377\left(\frac{r}{W}\right)\left(\frac{c}{r}\right)^2 + 0.1247\left(\frac{a}{L}\right)\left(\frac{r}{W}\right)^2 - 2.1930\left(\frac{a}{L}\right)\left(\frac{c}{r}\right)^2 + 0.7629\left(\frac{c}{r}\right)\left(\frac{r}{W}\right)^2 \\ &+ 0.7629\left(\frac{c}{r}\right)\left(\frac{r}{W}\right)^2 \end{aligned} \tag{15}$$

The results conforms that by using Equation 15, the SIFs values can be estimated appropriately. In order to show the performance of the expression, a certain crack condition is selected for example for the case of $r/W = 0.6$, $r/c = 4.8$ and $a/L = 0.1$ to 0.8. The comparisons between two methods are shown in Figure 5 where both results are in good agreement.

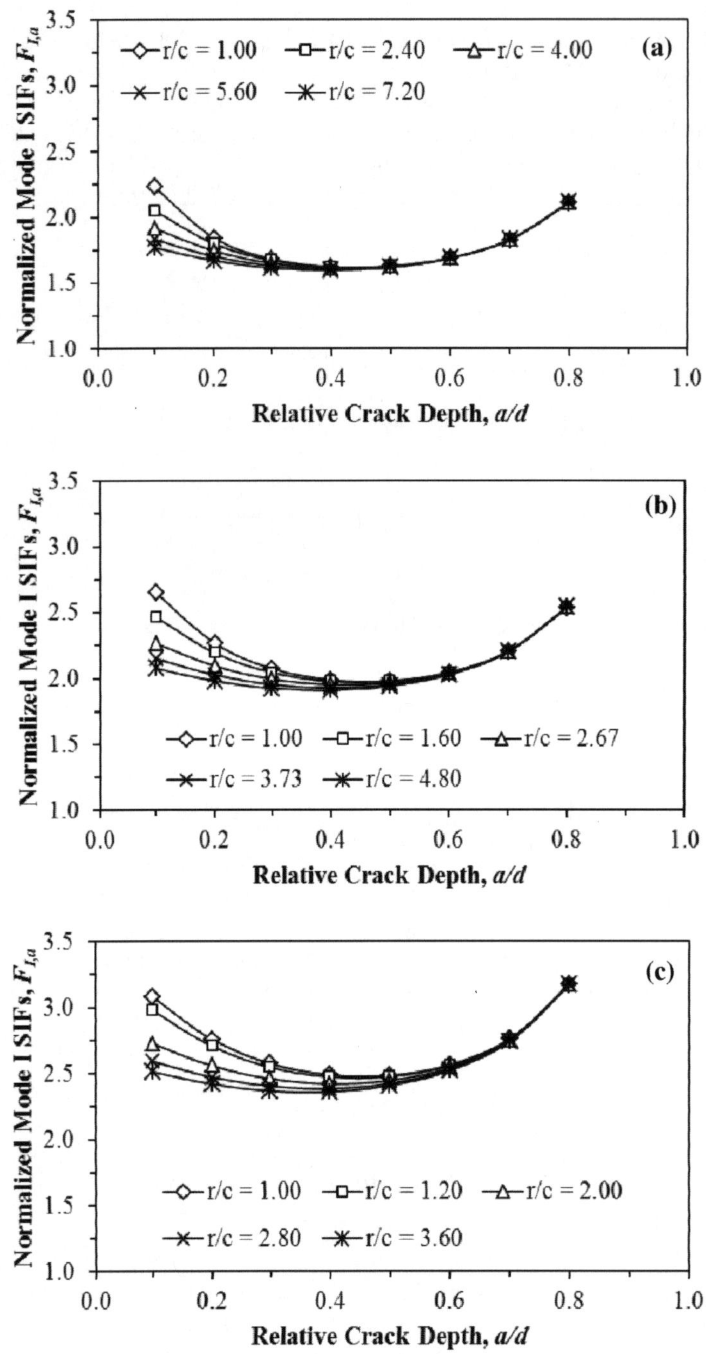

Figure 4. The effect of relative crack depth, *a/d* on the SIFs of various groove depth, (a) $c = 6.250$, (b) $c = 9.375$ and (c) $c = 12.500$

Table 1. Effect of *r/c* on the stress intensity factors for different crack length

	a/L	*r/c*					*a/L*	*r/c*			
		4.80	2.40	1.60	1.20			8.00	4.00	2.67	2.00
r/W = 0.6	0.1	1.643	2.048	2.463	2.980	r/W = 1.0	0.1	1.584	1.914	2.264	2.723
	0.2	1.451	1.797	2.191	2.703		0.2	1.437	1.743	2.093	2.555
	0.3	1.369	1.670	2.042	2.543		0.3	1.367	1.650	1.992	2.456
	0.4	1.347	1.619	1.975	2.470		0.4	1.347	1.612	1.951	2.418
	0.5	1.367	1.625	1.974	2.472		0.5	1.367	1.624	1.964	2.442
	0.6	1.430	1.687	2.042	2.556		0.6	1.430	1.689	2.039	2.540
	0.7	1.556	1.826	2.202	2.755		0.7	1.556	1.828	2.205	2.749
	0.8	1.803	2.109	2.537	3.172		0.8	1.803	2.111	2.544	3.173
	a/L	*r/c*					*a/L*	*r/c*			
		11.2	5.60	3.73	2.80			14.4	7.20	4.80	3.60
r/W = 1.4	0.1	1.584	1.914	2.264	2.723	r/W = 1.8	0.1	1.537	1.829	2.149	2.590
	0.2	1.437	1.743	2.093	2.555		0.2	1.421	1.702	2.025	2.469
	0.3	1.367	1.650	1.992	2.456		0.3	1.362	1.629	1.952	2.400
	0.4	1.347	1.612	1.951	2.418		0.4	1.346	1.602	1.927	2.383
	0.5	1.367	1.624	1.964	2.442		0.5	1.367	1.619	1.949	2.420
	0.6	1.430	1.689	2.039	2.540		0.6	1.431	1.686	2.031	2.527
	0.7	1.556	1.828	2.205	2.749		0.7	1.557	1.827	2.199	2.742
	0.8	1.803	2.111	2.544	3.173		0.8	1.804	2.110	2.540	3.171

Table 2. Regression statistics of SIFs

Multiple *R*	0.9969
***R* Square**	0.9937
Adjusted *R* Square	0.9929
Standard Error	0.0396
Observations	128

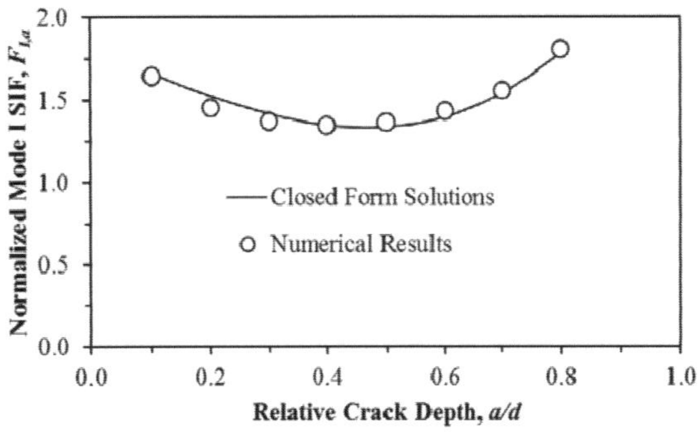

Figure 5. Comparisons between numerical and closed form solutions

5. Conclusion

The use of double edge crack in the large grooved plates are successfully applied as a numerical model using ANSYS finite element software. Different groove and crack geometries are studied on the SIFs. Based on the numerical analysis, several conclusions can be drawn. First, the SIFs obtained from the circular notches are higher than the SIFs obtained using large groove due to the different in the effect of stress concentrations. Secondly, for relatively shallow crack but depending on the groove depths, the SIFs are strongly related with *c/r*. However, when *c/r* ratio exceed a certain limit, the SIFs seem to have a single value. Last but not least, the developed closed form solution is capable to predict the SIFs well for wide ranges of groove geometries with high reliability.

Acknowledgment

Authors acknowledge Universiti Tun Hussein Onn Malaysia for financial support on this work.

References

[1] Ismail A E 2014 *Applied Mechanics and Materials* **629** 57-61
[2] Zhu X K and Joyce J A 2012 *Engineering Fracture Mechanics* **85** 1-46
[3] Radaj D 2014 *Fatigue and Fracture of Engineering Materials and Structures* **37** 1-28
[4] Kujawski D 1991 *Fatigue and Fracture of Engineering Materials and Structures* **14** 953-65
[5] Jones R and Peng D 2002 *Engineering Failure Analysis* **9** 683-702
[6] Palani G S, Iyer N R and Dattaguru B 2006 *Computers and Structures* **84** 2050-64
[7] Kuang J H and Chen L S 1993 *Engineering Fracture Mechanics* **46** 21-33
[8] Ismail A E, Tobi A L M and Nor N H M 2015 *AIP Conference Proceedings* **1660** 70027
[9] Ismail A E 2015 *AIP Conference Proceedings* **1660** 70029
[10] Ismail A E, Ariffin A K, Abdullah S and Ghazali M J 2012 *Int. J. of Automotive Technology* **12** 225-32
[11] Shih C F, Moran B and Nakamura T 1986 *Journal of Fracture* **30** 79-102
[12] Walters M C, Pauline G H and Dodds Jr. R H 2005 *Engineering Fracture Mechanics* **72** 1635-63
[13] Matsumto T, Tanaka M and Obara R 2000 *Engineering Fracture Mechanics* **65**, 683-702

Investigations of the mechanical properties of bi-layer and tri-layer fiber reinforced composites

K Jayakrishna[1*], K Balasubramani[2], M T H Sultan[3] and S Karthikeyan[4]

[1]School of Mechanical Sciences, Vellore Institute of Technology University, India
[2]Department of Mechanical Engineering, Kalasalingam University, India
[3]Aerospace Manufacturing Research Centre (AMRC), Universiti Putra Malaysia, Malaysia
[4]Department of Automobile Engineering, Kalasalingam University, India

*mail2jaikrish@gmail.com

Abstract. Natural fibers are renewable raw materials with an environmental-friendly properties and they are recyclable. The mechanical properties of bi-layer and tri-layer thermoset polymer composites have been analyzed. The bi-layer composite consists of basalt and jute mats, while the tri-layer composite consists of basalt fiber, jute fiber and glass fiber mats. In both cases, the epoxy resin was used as the matrix and PTFE as a filler in the composites. The developed tri-layer natural fiber composite can be used in various industrial applications such as automobile parts, construction and manufacturing. Furthermore, it also can be adopted in aircraft interior decoration and designed body parts. Flexural, impact, tensile, compression, shear and hardness tests, together with density measurement, were conducted to study the mechanical properties of both bi-layer and tri-layer composites. From the comparison, the tri-layer composite was found to perform in a better way in all tests.

1. Introduction

In the past few years, contemporary researchers who are working on finding sustainable solutions have focused their attention to the natural fiber-reinforced polymer composites. This situation is due to the attractive features of the natural fiber-reinforced polymer composites including light weight, low cost, moderate strength, high specific modulus, lack of health hazards, renewable and also environmentally friendly. Composite materials are defined as the structural materials that are created synthetically or artificially by combining two or more materials having dissimilar characteristics at macroscopic level [1]. Natural fiber-based composites have diverse and varied applications in aircraft, spacecraft, ships, submarines, trucks, rail vehicles, automobiles, civil engineering structures, packaging and also storage devices among others [2]. Though many studies have been conducted in which several widely used conventional materials were substituted in engineering applications with natural fiber components, there is still only a few studies done on mechanical properties of tri-layer epoxy filled composites [3-5]. Therefore, this present work has been undertaken to develop reinforced matrix composites using basalt, jute and glass fibers with epoxy resin as the matrix and to study their mechanical properties such as the tensile strength, flexural and impact strength in comparison to the bi-layer fiber reinforced composites.

2. Methodology

The methodology that was adopted in investigating the mechanical properties of bi-layer and tri-layer fiber reinforced composites is depicted in Figure 1. Fabrication and testing of the composite materials for mechanical properties in manufacturing had been selected as the main goal of this research work. A brief literature survey was conducted to investigate the research studies carried out with the focus on the mechanical properties of the composite materials. The lack of information in the previous research works in this domain were identified and used to define the research problem. The thrust activities in development of composite materials were identified for the study and subsequently, experiments were conducted. The results obtained were then analyzed to draw the conclusions.

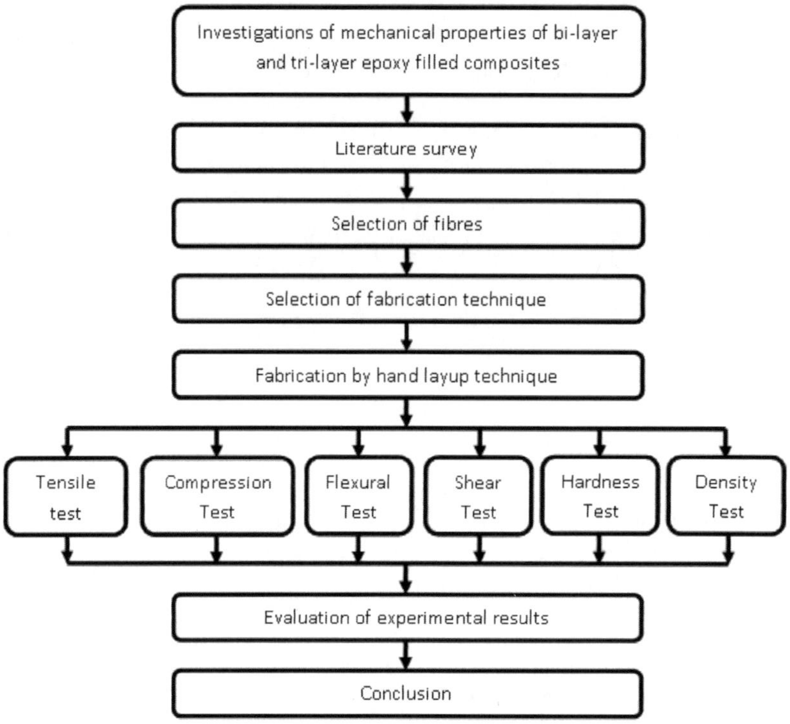

Figure 1. Methodology adopted

3. Raw materials used and their characteristics

Basalt and jute fiber mats were used as bi-layer materials. Basalt, jute and glass fiber mats, which were easily available and eco-friendly, were used as composite tri-layer materials. The epoxy resin (LY556) and industrial application hardener (HY951) were applied to fabricate the laminated sheets and hybrid combinations. LY556 epoxy resin has exceptional adhesion to different materials, high resistance to chemical and atmospheric conditions, high dimensional stability, free from internal stresses, excellent mechanical and electrical properties, and completely non-toxic with negligible shrinkage, plus it has been used as the matrix material in many composites [6]. Epoxy hardeners are not catalysts and they react with the epoxy resins, which greatly contributes to the properties of the cured epoxy resin system [7]. The performance of the epoxy hardeners in the epoxy resins system is dependent on the chemical characteristics of the latter [8].

4. Fabrication

The bi-layer and tri-layer composite plates were prepared with dimensions of 250 × 250 × 3 mm. The composite plates were prepared by a simple manual hand layup procedure. Hand lay-up is the simplest and oldest open molding method for composite fabrication process. It is a low volume, labor intensive method that is especially suited for large components [9]. The fibers were manually placed into a one-sided gel coated mould. A matrix of thermosetting resin was rolled onto the fiber using a hand roller, which also removed the entrapped air to complete the laminates structure. More layers were added and, after drying, the composite panel could be removed from the mould. For a high quality part surface, a releasing gel coat was first applied to the mold surface. After curing the composite, it was separated from the mould and the specimens were cut according to the ASTM and ISO standards using abrasive water jet cutting machine (see Figure 2). The sample specimens are shown in Figure 3.

Figure 2. Water jet cutting process of the specimen

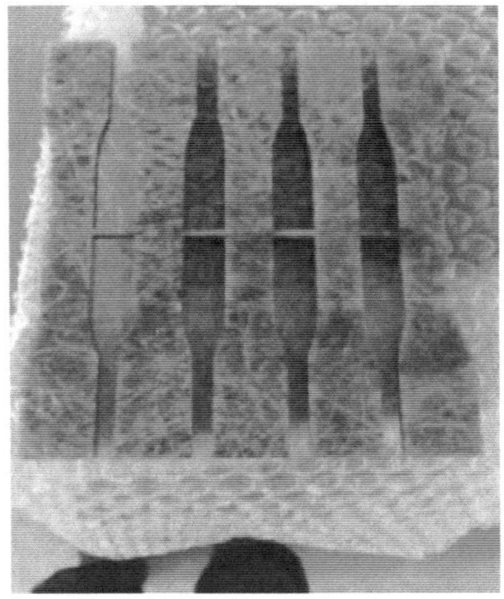

Figure 3. Composite specimens

5. Material characterization

Tensile test was performed on a computer controlled universal testing machine according to guidelines of the ASTM D638 standard. Meanwhile, the flexural tests were performed using the 3-point bending method according to the ASTM D790 standard. The standard specimens were also prepared for both compression and shear tests, which were performed using a universal testing machine. Five sample specimens were tested in each test to determine mechanical properties with average values calculated.

5.1. Tensile test

For the tensile test, the specimens used were of 20 cm length, 3 cm width and also 0.3 cm thickness in dimension [10]. Figure 4 shows a comparison of the tensile strength obtained for bi-layer and tri-layer fiber reinforced composites tested. Less fiber pull out was observed in the fractured specimen. Based on the results in Figure 4, it becomes clear that the tri-layer epoxy filled composites exhibited a higher tensile strength (74.72 MPa) than the bi-layer fiber reinforced composites (73.09 MPa).

5.2. Flexural test

The flexural tests were performed using 3-point bending method according to ASTM D790 standard with specimen size of 24 cm × 3 cm × 0.3 cm. Figure 5 shows a comparison of the flexural load for bi-layer and tri-layer epoxy filled composites tested. The experimental results revealed that the tri-layer fiber reinforced composites were a better choice than the bi-layer fiber reinforced composites because they showed 13% increase in flexural property.

5.3. Impact test

Impact tests were performed using the Charpy izod impact method. Figure 6 shows comparison of the impact energy observed by the bi-layer and tri-layer fiber reinforced composites. Once again, the tri-layer fiber reinforced composites showed more superior impact properties. The average of 6% increase was shown in the impact strength of the tri-layer composites.

IOP Conf. Series: Materials Science and Engineering **152** (2016) 012050 doi:10.1088/1757-899X/152/1/012050

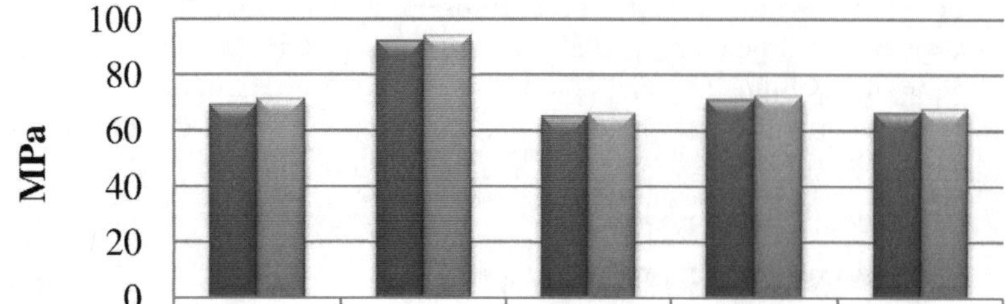

	1	2	3	4	5
Tensile Strength of Bi-Layer Composite (MPa)	69.56	92.47	65.39	71.41	66.62
Tensile Strength of Tri-Layer Composite (MPa)	71.70	94.41	66.52	72.87	68.09

Figure 4. Comparison of tensile strength for bi-layer and tri-layer fibre reinforced composites

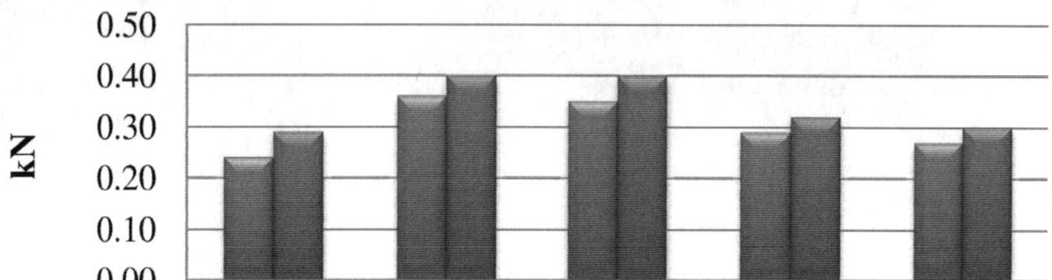

	1	2	3	4	5
Flexural Load in Bi-Layer Composite (kN)	0.24	0.36	0.35	0.29	0.27
Flexural Load in Tri-Layer Composite (kN)	0.29	0.40	0.40	0.32	0.30

Figure 5. Comparison of flexural load for bi-layer and tri-layer fibre reinforced composites

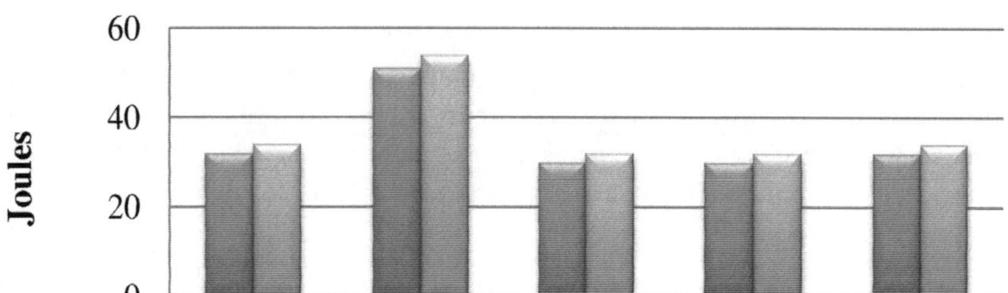

	1	2	3	4	5
Impact in Joules Bi-layer	32.00	51.00	30.00	30.00	32.00
Impact in Joules Tri-layer	34.00	54.00	32.00	32.00	34.00

Figure 6. Comparison of impact load for bi-layer and tri-layer fibre reinforced composites

5.4. Density test

Density, ρ, or more precisely the volumetric mass density of a substance, is its mass per unit volume. Mathematically, density is defined as mass divided by volume. The dimension of the test specimen for density measurements was 10 × 10 × 3 mm. Table 1 shows the density of bi-layer and tri-layer epoxy composite composites.

5.4. Hardness test

Hardness test is used to determine the hardness of a material to deformation, which can be performed either on a macroscopic or microscopic scale. In this study, D shore hardness test was used for testing. The ASTM standard for the D shore hardness test is D2240/2240. Table 2 shows the hardness values for bi-layer and tri-layer epoxy composite composites.

5.5. Shear test

Shear test was carried out using universal testing machine for specimens with the dimension of 20 cm length, 3 cm width and 0.3 cm thickness. A comparison of the composites tested under shear test is represented in Figure 7. From Figure 7, the tri-layer epoxy filled composites exhibited a higher tensile strength of the bi-layer fiber reinforced composites tested.

Table 1. Comparison of density values for the bi-layer and tri-layer fibre reinforced composites

Density (g/cm3)	
Bi-layer	**Tri-layer**
1.318	1.738

Table 2. Comparison of hardness values for the bi-layer and tri-layer fibre reinforced composites

Hardness HRL	
Bi-layer	**Tri-layer**
84.8	87.00

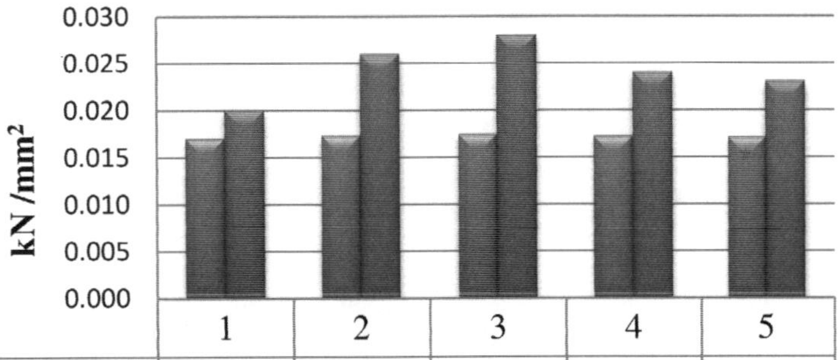

	1	2	3	4	5
Shear stress Bi-layer	0.017	0.017	0.017	0.017	0.017
Shear stress Tri-layer	0.020	0.026	0.028	0.024	0.023

Figure 7. Comparison of shear stress for bi-layer and tri-layer fibre reinforced composites

6. Conclusion

The developed tri-layer fiber reinforced composites were environmentally-friendly materials and they were also recyclable. In this study, the tri-layer composite materials were reinforced with natural fibers such as jute fibers, together with basalt and glass fibers, in the epoxy resin (LY 556) matrix and with PTFE as filler. It showed better structural properties. The developed tri-layer epoxy composites could be used in various applications in the fields of aircraft, automobile construction and manufacturing. Flexural, impact, tensile, compression, shear, hardness and density tests were conducted to study the mechanical properties of bi-layer and tri-layer composites. The tri-layer fiber reinforced composites had shown higher values in all mechanical properties in comparison with the bi-layer fiber reinforced composites. As an extension of this study, it is proposed to conduct tribological studies on the tri-layer fiber reinforced composites.

Acknowledgement
The authors would like to acknowledge Universiti Putra Malaysia for supporting this research under GP-IPB 9441503.

References

[1] Sreenivasan V S, Ravindran D, Manikandan V and Narayanasamy R 2011 *Materials and Design* **32** 2444-55
[2] Lai C Y, Sapuan S M, Ahmad M, Yahya N and Dahlan K Z H M 2005 *Polymer-Plastics Tech. Eng.* **44** 619-32
[3] Sapuan S M and Maleque M A 2005 *Materials and Design* **26** 65-71
[4] Arib M N, Sapuan S M, Hamdan M, Paridah M T and Dahlan K Z H M 2006 *Materials and Design* **27** 391-6
[5] Nimityongskul G 1987 *Journal of Ferrocement* **17** 321-8
[6] Yousif B F 2012 *Materials and Design* **40** 378-85
[7] Kumar M A 2011 *Journal of Metallurgy and Materials Science* **53** 85-95
[8] Bagghi A K and Sharma B G 1981 *Indian J. Technol.* **19** 368-72
[9] Hashim J, Looney L and Hashim M S J 1999 *Journal of Materials Processing Technology* **92-93** 1-7
[10] ASTM International 2008 *Annual book of ASTM standards*
[11] Amuthakkannan P, Manikandan V, Jappes J T W and Uthayakumar M 2013 *Science and Engineering of Composite Materials* **20** 343–50
[12] Chairman C A and Kumaresh Babu S P 2013 *Journal of Applied Polymer Science* **130** 120–30
[13] Subramanian K and Nagarajan R 2015 *Mechanical Engineering Letters* **12** 7–18

IOP Conf. Series: Materials Science and Engineering **152** (2016) 012051 doi:10.1088/1757-899X/152/1/012051

Machining analysis of natural fibre reinforced composites using fuzzy logic

K Balasubramanian[1*], M T H Sultan[2,3], F Cardona[2] and N Rajeswari[4]

[1]Department of Mechanical Engineering, Kalasalingam University, India
[2]Aerospace Manufacturing Research Center, Universiti Putra Malaysia, Malaysia
[3]Department of Aerospace Engineering, Faculty of Engineering, Universiti Putra Malaysia, Malaysia
[4]Department of Mechanical Engineering, Vellammal Engineering College, India

*bala_manu2002@yahoo.co.in

Abstract. In this work, a new composite plate with natural jute fibre as the reinforcement fibres and isophthalic polyester as the resin was manufactured and subjected to a series of end milling operation by changing three input factors namely speed, feed rate and depth of cut. During each operation, the output responses namely thrust force and torque were measured. The responses were analyzed using Taguchi method to examine the relation between the input factors and output responses, and also to know the most influencing factors on the responses. The data was also analyzed using fuzzy rule model for prediction of responses for a range of input factors. The results showed that all three factors chosen have significant effect on the responses. The fuzzy model data in comparison with the experimental values shows only a marginal error and hence the prediction was highly satisfactory.

1. Introduction

Composite materials are man-made materials, which are manufactured with the aim of replacing the conventional materials by overcoming their disadvantages. A composite material has two constituents in its structure, namely the matrix and reinforcements. Composite materials are broadly classified into three types according to their matrices: polymer matrix composites (PMC), metal matrix composites (MMC) and ceramic matrix composites (CMC). Among these composites, fibre reinforced plastics are widely used as structural elements in aircraft, marine, civil and commercial sectors. Fibre reinforced plastics replaces conventional materials in all these applications due to their tailor-made properties. Fibre reinforced plastics are made of synthetic fibres like glass, carbon, Kevlar, etc. Though they have several advantages including high strength, stiffness, fatigue life and wear resistance, they also have disadvantages like high density, high cost, and poor recycling and bio-degradable properties.

In order to overcome these disadvantages, the natural fibres taken from the plants and animals are being used as reinforcements in recent few years as alternatives to the synthetic fibres. Bio-fibres like jute, sisal, vetiver, hemp and bamboo are abundantly available at reasonable cost. These fibres, when used as reinforcements in composites, provide very good mechanical properties and they are free from environmental hazards. When the natural composites are used as structural components, they involve several machining operations like drilling, slotting, surface finishing, etc. Hence machining associated damage and failure analysis in natural composites will help the composite industries to appropriately select optimum conditions for machining and also optimum proportion of composite constituents for improved machinability. Fibre reinforced plastics have found wide applications in the field of aircraft, automotive and shipping industries owing to their better mechanical and also electrical properties. The machining of such composites attracts the researchers because of non-homogeneous internal structure. Failure of fibre reinforced composites during machining depends on improper selection of fibre, resin, fibre orientation and number of layers.

2. Literature survey

Davim *et al.* [1] reported that feed rate majorly affects the surface roughness during milling of glass fibre reinforced composites. They also demonstrated that delamination factor and surface roughness depends on feed and cutting speed during the milling process of carbon fibre reinforced composites [2]. Farahnakian *et al.* [3] made investigations on polyamide added with Nano clay and concluded that feed and speed are the major factors influencing the surface roughness. Davim *et al.* [4] studied the machining aspects of milling on medium density fibre boards and reported that feed and speed have more influence on surface roughness. Litak *et al.* [5] performed a multi scale entropy analysis on carbon fibre reinforced composites.

Rentsch *et al.* [6] performed finite element modeling on cutting of carbon fibre reinforced plastics with a disc milling cutter and reported that the calculated and experimental results of the material removal mechanism were almost the same but the cutting forces in comparison with experimental data were not satisfactory. Schule *et al.* [7], in their experiments involving spiral and wobble milling on short glass fibre reinforced composites with polyester resin, illustrated that the direction of the processing forces is directed to the center of the work piece in both methods and wobble milling is not sensitive to tool wear with respect to the process force direction. Prakash *et al.* [8] conducted a series of drilling experiments on medium density fibre boards and reported that delamination factor is influenced by both speed and feed, and the speed-feed combination is also influential. Isik *et al.* [9] reported, after a series of drilling experiments on glass fibre reinforced composites plate, that the delamination factor is higher at high feed rate and it is also increasing with the number of lutes and at high point angle. Singh *et al.* [10] performed finite element analysis on drilling of composites reinforced with glass fibre and concluded that thrust is predominantly affected by the point angle and feed. Similarly, the thrust force and damage increases with increasing feed rate while performing drilling experiments on glass fibre reinforced composites [11, 12]. Similar works have been carried out by different authors [13-16] on carbon and glass fibre reinforced composites and they reported the variation of responses like thrust, torque, delamination factor with machining parameters. During turning of a composite rod, the tool wear is found to be less at low cutting speed and cubic boron nitride tool shows the better overall performance than ceramic and tungsten carbide tools. Also, the surface quality decreases as the depth of cut increases [17, 18].

Applications of composite materials are increasing and the introduction of such new materials will enhance performance and life of the components. Hence there is always a need of machining operations for industries on composite materials. Analysis of various machining parameters reduces unnecessary experimentation and helps the end users to select optimum machining conditions. In the present work, a new composite plate made of isophthalic polyester as the matrix and natural jute as the fibre is manufactured by hand layup technique. Then, a series of 27 milling experiments were performed during which the thrust force and torque were varied. Experimental results were analyzed using the Taguchi method from which the significance of each factor on the measured responses was studied. After that, the data were analyzed using fuzzy logic for prediction of responses. Finally, the fuzzy predicted values were compared with the experimental values and the percentage of error was calculated.

3. Experimental procedure

3.1. Method and materials

Composite materials used in this study were manufactured by hand layup technique with isophthalic polyester as the resin and woven natural jute as the fibre in the ratio 1:1 by weight. The experiments were performed on a 15 mm thick and 20 mm x 20 mm square composite plate consisting of 10 fibre layers. The tool utilized was the 7 mm, 4-fluted end mill made of high speed steel. A vertical milling machine with 3HP main motor, 2HP feed motor and spindle speed range of 80 rpm to 2720 rpm was used for conducting the experiments. The output readings of thrust and torque were taken from the dynamometer.

3.2. Plan of experiments

The experiments were designed and analyzed using Taguchi's L_{27} (3^3) orthogonal array. The three levels of cutting speed, feed and depth of cut are shown in Table 1 and the combinations of input factors during all experimental trials and the measured responses are listed in Table 2.

Table 1. Assignment of levels to the machining factors

Level	Speed, N (rpm)	Feed f, (mm/rev)	Depth of cut, d (mm)
1	210	0.04	1.0
2	660	0.08	1.5
3	1750	0.15	2.0

Table 2. Experimental design and measured responses

Exp. No	Speed, N (rpm)	Feed, f (mm/rev)	Depth of Cut, d (mm)	Thrust Force	Torque
1	210	0.04	1.0	31.88	2.55
2	210	0.04	1.5	39.53	3.43
3	210	0.04	2.0	51.21	4.12
4	210	0.08	1.0	45.62	4.32
5	210	0.08	1.5	52.48	3.53
6	210	0.08	2.0	60.04	6.18
7	210	0.15	1.0	57.78	4.41
8	210	0.15	1.5	61.41	5.49
9	210	0.15	2.0	70.83	6.67
10	660	0.04	1.0	36.20	2.35
11	660	0.04	1.5	42.58	3.53
12	660	0.04	2.0	54.54	4.61
13	660	0.08	1.0	30.71	3.83
14	660	0.08	1.5	53.46	4.41
15	660	0.08	2.0	45.81	5.40
16	660	0.15	1.0	32.77	3.63
17	660	0.15	1.5	46.70	4.41
18	660	0.15	2.0	48.66	5.69
19	1750	0.04	1.0	15.40	1.28
20	1750	0.04	1.5	24.92	2.35
21	1750	0.04	2.0	26.19	3.14
22	1750	0.08	1.0	27.27	2.06
23	1750	0.08	1.5	16.19	3.34
24	1750	0.08	2.0	32.96	4.41
25	1750	0.15	1.0	28.15	3.43
26	1750	0.15	1.5	14.22	4.51
27	1750	0.15	2.0	29.04	5.10

4. Results and discussion

From the ANOVA table for the thrust force (see Table 3), the most of the influencing factors were identified as speed, depth of cut and speed- feed combinations, which all had p-value less than 0.05. In the meantime, from the ANOVA table for the torque (see Table 4), the most influencing factors were identified as speed, feed rate and depth of cut, which all had p-value less than 0.05.

Table 3. ANOVA table for thrust force

Source	Sum of squares	Degrees of freedom	Mean square	F value	P value
Model	5691.22	18	316.18	10.25	0.0011 Significant
A-Speed	3830.02	2	1915.01	62.08	< 0.0001
B-Feed	255.53	2	127.77	4.14	0.0583
C-Depth of Cut	724.72	2	362.36	11.75	0.0042
AB	523.20	4	130.80	4.24	0.0392
AC	307.53	4	76.88	2.49	0.1265
BC	50.22	4	12.56	0.41	0.7991
Residual	246.79	8	30.85		
Total	5938.00	26			

Table 4. ANOVA table for torque

Source	Sum of squares	Degrees of freedom	Mean square	F value	P value
Model	41.52	18	2.31	15.54	0.0002 Significant
A-Speed	7.37	2	3.68	24.81	0.0004
B-Feed	14.53	2	7.27	48.95	<0.0001
C-Depth of Cut	17.12	2	8.56	57.67	<0.0001
AB	1.39	4	0.35	2.34	0.1424
AC	0.53	4	0.13	0.89	0.5127
BC	0.58	4	0.14	0.97	0.4737
Residual	1.19	8	0.15		
Total	42.70	26			

Figure 1 depicts the plots for the main effects for thrust force. It is evident that the thrust force decreases as speed increases but increases with the increase in feed rate and depth of cut. Referring to Figure 2, the thrust force increases for low speed-feed combination but remains almost the same for the medium and high speed-feed combinations. Also, it was observed that the thrust force shows an increasing trend for all combinations of speed-depth of cut and at all combinations of feed-depth of cut. On the other hand, plots for the main effects for torque are illustrated in Figure 3. The torque decreases when speed increases but increases when feed rate and depth of cut are increased. The plots for the main effects for torque are rather similar to those for the thrust force. The interaction plot in Figure 4 reveals that the torque increases as the speed-feed combination increases. Similar trend was observed for two combinations factors like speed-depth of cut and for feed rate-depth of cut.

5. Fuzzy rule modeling

The fuzzy logic is used to evaluate the behavior of any system that has no analytical or numerical functions. It has high capability of understanding the more complex systems. The first stage of fuzzy is to develop the membership functions for algorithm development. In this work, the fuzzy rule was employed using the MATLAB software. The membership functions were developed by taking the appropriate range of the selected value of the factors. A triangular membership function was adopted in which three values: low, medium and high, were taken for the input and output parameters. The membership functions of the output responses are presented in Figure 5 and Figure 6.

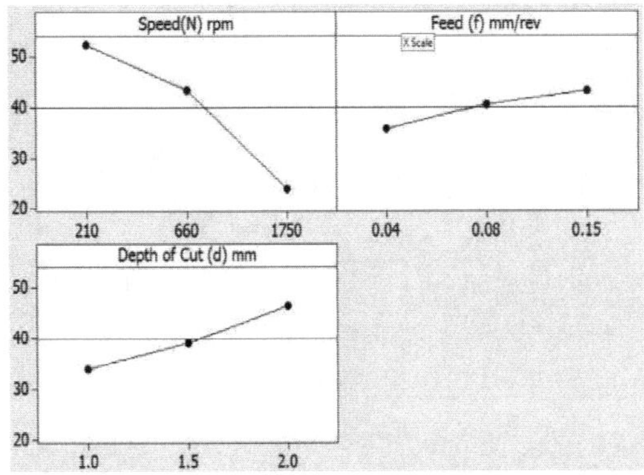

Figure 1. Plots of the main effects for thrust force

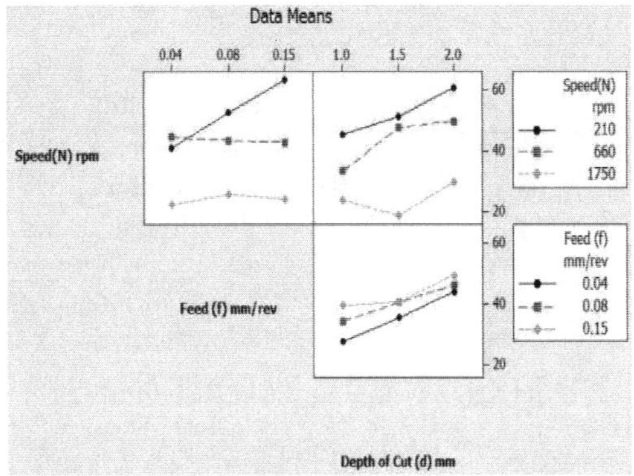

Figure 2. Plots of the interactions for thrust force

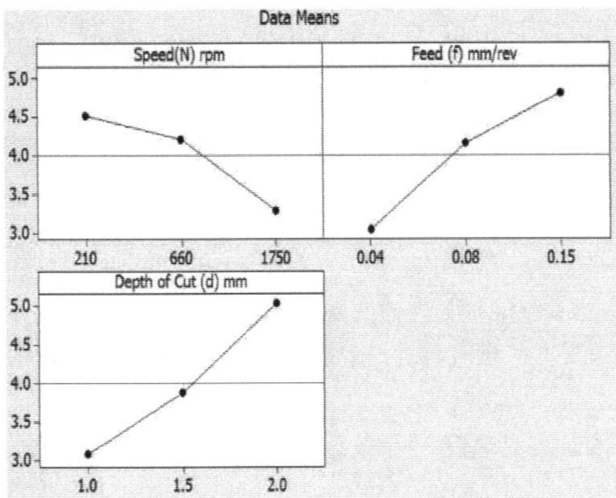

Figure 3. Plots of the main effects for torque

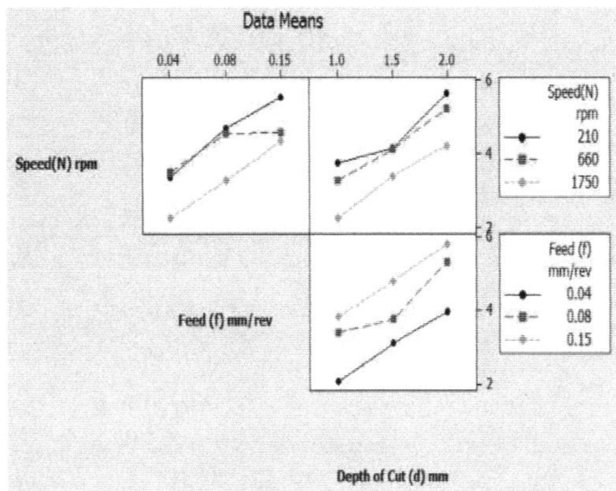

Figure 4. Plots of the interactions for torque

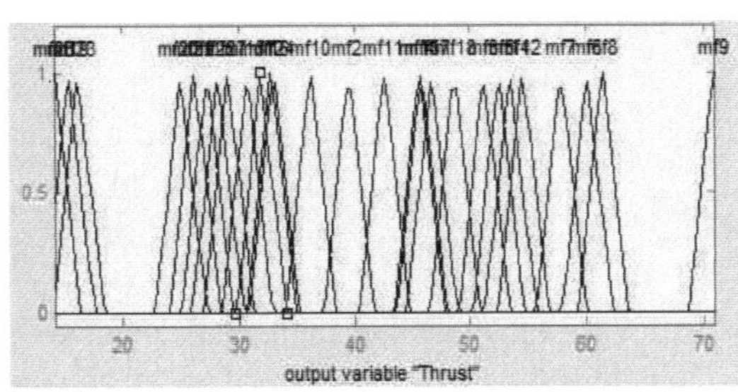

Figure 5. Membership function for thrust force

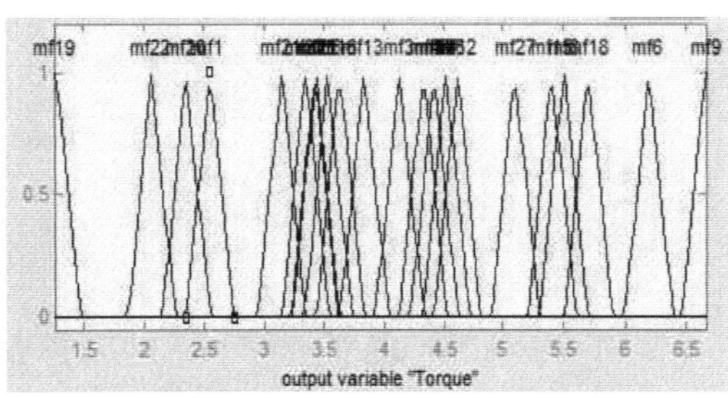

Figure 6. Membership function for torque

The next stage of fuzzy is the de-fuzzification process. This reduces the membership vector into a scalar quantity, presumably to the most representative value. Fuzzy logic works by considering a series of IF-THEN statements containing inputs and outputs. In this study, the inputs x_1, x_2 and x_3 were assigned to cutting speed, feed rate and depth of cut, respectively, while the outputs y_1 and y_2 were assigned to thrust force and torque, respectively. The experimental values and the corresponding fuzzy predicted ones are shown in Table 5. Note that the parameters' setting for each experimental run can be referred back to Table 2. From these values, the percentage error between the experimental ones and fuzzy predictions were calculated. The average percentage error for thrust and torque were found to be 0.77% and 1.22%, respectively.

Table 5. Consolidated table of experimental and fuzzy predicted values

Exp. No	Experimental Thrust	Predicted Thrust	Thrust % Error	Experimental Torque	Predicted Torque	Torque % Error
1	31.88	32.14	-0.82	2.55	2.32	9.02
2	39.53	38.55	2.48	3.43	3.39	1.17
3	51.21	50.68	1.03	4.12	4.20	-1.94
4	45.62	45.03	1.29	4.32	4.26	1.39
5	52.48	52.59	-0.21	3.53	3.52	0.28
6	60.04	60.71	-1.12	6.18	6.25	-1.13
7	57.78	57.97	-0.33	4.41	4.51	-2.26
8	61.41	60.22	1.94	5.49	5.53	-0.73
9	70.83	70.30	0.75	6.67	6.62	0.75
10	36.20	34.90	3.59	2.35	2.40	-2.13
11	42.58	41.70	2.07	3.53	3.51	0.57
12	54.54	53.60	1.72	4.61	4.47	3.04
13	30.71	30.12	1.92	3.83	3.55	7.31
14	53.46	53.20	0.49	4.41	4.15	5.90
15	45.81	47.12	-2.86	5.40	5.21	3.52
16	32.77	33.11	-1.04	3.63	3.52	3.03
17	46.70	45.89	1.73	4.41	4.38	0.68
18	48.66	48.45	0.43	5.69	5.87	-3.16
19	15.40	15.50	-0.65	1.28	1.21	5.47
20	24.92	25.00	-0.32	2.35	2.15	8.51
21	26.19	25.73	1.76	3.14	3.11	0.96
22	27.27	26.92	1.28	2.06	2.10	-1.94
23	16.19	15.98	1.30	3.34	3.29	1.50
24	32.96	33.12	-0.49	4.41	4.52	-2.49
25	28.15	27.56	2.10	3.43	3.51	-2.33
26	14.22	13.83	2.74	4.51	4.59	-1.77
27	29.04	29.00	0.14	5.10	5.11	-0.20

6. Conclusion

From the above discussions on the effects and characteristics of thrust force and torque on the natural fibre reinforced composites, the following conclusions can be drawn. Firstly, speed and depth of cut were found to be the most influencing factors for thrust force whereas speed, feed and depth of cut are the predominant factors for torque. In addition, both thrust force and torque closely behave similar as the main and predominant effects. As speed increases, both responses decreases. However, the thrust force and torque are increased when feed and depth of cut are increased. Furthermore, high speed, high feed and medium depth of cut are the optimum machining conditions while considering the thrust force factor. Meanwhile, high speed, low feed and low depth of cut are the optimum condition for torque. Moreover, thrust force and torque were modeled using fuzzy rule and the models were closely correlated with the experimental data with an average error of 0.77% and 1.22%, respectively. Hence, the fuzzy modeling of thrust and torque can be taken to be highly satisfactory. Also, fuzzy modeling helps to accurately predict the responses for any values of cutting speed, feed rate and depth of cut combinations within a specific range of values.

Acknowledgment

The authors would like to acknowledge Universiti Putra Malaysia for supporting this research under GP-IPB 9441503.

References

[1] Paulo Davim J, Reis P and Conceicao Antonio C 2004 *Composite Structures* **64** 493-500
[2] Paulo Davim J and Reis P 2005 *J. Mater. Process. Technol.* **160** 160-7
[3] Farahnakian M, Razfar M R, Moghri M and Asadnia M 2011 *Int J Adv Manuf Tech* **57** 49-60
[4] Paulo Davim J, Clemente V C and Silva S 2009 *Int J Adv Manuf Tech.* **49** 49-55
[5] Litak G, Syta A and Rusinek R 2011 *Int J Adv Manuf Tech.* **56** 445-53
[6] Rentsch R, Pecat O and Brinkmeier E 2011 *Procedia Eng.* **10** 1823-28
[7] Schulze V and Becke C 2011 *Procedia Eng.* **19** 312-7
[8] Prakash S, Palanikumar K and Manoharan N 2009 *Int J Adv Manuf Tech.* **45** 370-81
[9] Isik B and Ekici E 2010 *Int J Adv Manuf Tech.* **49** 861-9
[10] Singh I, Bhatnagar N and Viswanath P 2008 *Mater. Des.* **29** 546-53
[11] Abrao A M, Campos Rubio J C, Faria P E and Davim J P 2008 *Mater. Des.* **29** 608-13
[12] Campos Rubio J, Abrao A M, Faria P E, Esteves Correia A and Paulo Davim J 2008 *Int J Adv Manuf Tech.* **48** 715-20
[13] Gaitonde V N, Karthik S R, Campos Rubio J, Esteves Carreia A and Abrao A M 2008 *J. Mater. Process. Technol.* **203** 431-8
[14] Karthik S R, Gaitonde V N, Campos Rubio J, Esteves Carreia A, Abrao A M and Paulo Davim J 2008 *Mater. Des.* **29** 1768-76
[15] Wang X, Wang L J and Tao J P 2004 *J. Mater. Process. Technol.* **148** 239-44
[16] Abrao A M, Faria P E, Campos Rubio J C, Reis P and Paulo Davim J 2007 *J. Mater. Process. Technol.* **186** 1-7
[17] Rahman M, Ramakrishna S, Prakash J R S and Tan D C G 1999 *J. Mater. Process. Technol.* **89** 292-7
[18] Palanikumar K and Paulo Davim J 2007, *Mater. Des.* **28** 2008-14
[19] Rajasekaran T, Palanikumar K and Vinayagam B K 2011 *Journal of Prod. Eng. Res.* **5** 191-9

A study on the crushing behavior of basalt fiber reinforced composite structures

A Pandian[1*], A P Veerasimman[1], M Vairavan[1], C Francisco[2] and M T H Sultan[2]

[1]Department of Mechanical Engineering, Kalasalingam University, India
[2]Aerospace Manufacturing Research Centre (AMRC), Universiti Putra Malaysia, Malaysia

*pa_kanna@yahoo.co.in

Abstract. The crushing behavior and energy absorption capacity of basalt fiber reinforced hollow square structure composites are studied under axial compression. Using the hand layup technique, basalt fiber reinforced composites were fabricated using general purpose (GP) polyester resin with the help of wooden square shaped mould of varying height (100 mm, 150 mm and 200 mm). For comparison, similar specimens of glass fiber reinforced polymer composites were also fabricated and tested. Axial compression load is applied over the top end of the specimen with cross head speed as 2 mm/min using Universal Testing Machine (UTM). From the experimental results, the load–deformation characteristics of both glass fiber and basalt fiber composites were investigated. Crashworthiness and mode of collapse for the composites were determined from load–deformation curve, and they were then compared to each other in terms of their crushing behaviors.

1. Introduction

In structural automotive parts, fiber reinforced polymer composites can be a predominant material because of their significant weight reduction. The study of the crush behavior of the composites will be very useful for future lightweight composite vehicles. Sundarraja *et al.* [1] studied the carbon fiber reinforced polymer (CFRP) fabrics in axial strengthening of hollow square sections (HSS). It was found that thickness of the composites play major role. An increase in CFRP strips' thickness will result in the delay of the buckling and also leads to inward buckling. Mamalis *et al.* [2] investigated the crashworthy of square composite material tubes subjected to static axial compression. The mode of collapse of the composites was analyzed with both experimental and numerical methods. Shanmugavalli and Sundarraja [3] studied the behavior of the circular hollow steel under compression. It was found that CFRP composites greatly improved the load carrying capacity of the circular steel tube and also maximum deformation was controlled by CFRP layers. Kathiresan *et al.* [4] investigated the axial compression of thin-walled E-glass fiber/epoxy resin reinforced (GFRP) composite conical frusta. To study the crashworthiness of conical shells, E-glass fibre reinforced polymer (GFRP) conical specimens were fabricated with a semi-apical angle ranging from 15° to 27°. The mode of collapse was investigated through experiments as well as the FEA ABAQUS package.

Anne *et al.* [5] studied the tension and compression, and the energy of deformation of braided glass-fiber/epoxy circular tubes with polymer foam cores. It revealed that the foam filled circular tubes exhibits good energy absorbing characteristics. Ataollahi *et al.* [6] focused on natural silk/epoxy composite square tubes energy absorption and failure response of the composites. It was observed that the energy absorbed by the longest and thickest square tube took a long time to reach the compaction zone. Melo *et al.* [7] investigated the energy absorption capacity of glass/polyester composite tubes. Two types of layup were used ([0/90] and [± 45]) and it was observed that significant influence of the cross sectional geometry on energy absorbing capacity and [0/90] showed the highest specific energy absorption. Christopher *et al.* [8] studied the energy absorption and mechanical behavior of GRP composites using the compressive tests. Gauss-Legendre rule was used to evaluate the area under the stress-strain curve as a measure of amount of energy absorbed per unit volume of material. From this study, the mechanical parameters needed for the automobile components design were determined. Muralikannan *et al.* [9] studied the energy absorption of circular tubes composites under axial compression in quasi static and impact loading. The glass fiber was used with hand layup techniques. It was observed that mean load and specific energy absorption increases with the increase in tube thickness. Paolo *et al.* [10] studied the crush energy absorption of composite channel section specimens. Hand layup techniques were used to fabricate the composites with different geometries. It was found that the small corner element was the most efficient in absorbing energy per unit mass compared to those with longer flanges, particularly the square tube.

Wei *et al.* [11] numerically investigated the crushing behavior of composite structures using ABAQUS /Explicit. The numerical results agreed well with the experimental results in terms of the intra-laminar and inter-laminar damage for a range of stacking sequences. Sivakumar Palanivelu *et al.* [12] experimentally investigated the progressive deformation behavior of uni-directional pultruded composite tubes subjected to an axial impact load with three impact velocities (i.e. 9.3, 12.4 and 14 m/s). Louis *et al.* [13] studied the crushing behavior of a carbon–epoxy composite experimentally. It was found that energy absorption is independent of strain rate as the total energy absorption appeared largely associated with fiber-dominated fracture.

Based on the literatures, this current study was proposed with basalt fiber as reinforcement. The hand layup techniques were used to fabricate the square tube section composites that would be subjected to axial compression load.

2. Experimental details

In this section, the discussions are centered on the material selection and fabrication processes.

2.1. Materials used

The basalt fiber was purchased from ASA Tec. Austria while the GP polyester resin was purchased from GVR Traders, Madurai, India.

2.2. Fabrication of hollow square tube

The hollow square tube composites were fabricated using the hand layup techniques. For the fabrication of the composites, a specific mould was prepared with the required dimension of the composites. The mould was prepared with the dimension of 100 mm^2 and different heights of 100 mm, 150 mm and 200 mm as presented in Figure 1. For the fabrication, equal weight % of woven basalt fiber and polyester resin were taken (50/50 by weight), and for the curing process, the methyl ethal ketone per oxide was used and the cobalt naphthalene was taken as accelerator. Resin mixture was mixed thoroughly using mechanical stirrer and homogenous mixer resin was used. Similar procedure was followed for all the composites fabricated. The fabricated specimens are presented in Figure 2 and their details are tabulated in Table 1.

Figure 1. Mandrel for fabrication of composites

Figure 2. Fabricated composites

Table 1. Details of hollow square tube composites

Specimen	Heights (mm)	Width (mm)	Thickness (mm)	Maximum Load (kN)	Maximum Elongation (mm)	Load at peak (kN)	Compression strength (N/mm^2)
Basalt -10	100	100	3	450	45	23.76	20.042
Basalt -15	150	100	3	600	80	28.710	23.806
Basalt -20	200	100	3	300	100	16.44	14.123
Glass -10	100	100	3	300	35	5.502	5.867
Glass -15	150	100	3	300	45	7.806	7.544
Glass -20	200	100	3	300	70	2.970	2.552

2.3. Axial compression test

The axial compression test was conducted on samples with cross head speed of 2 mm/min using Universal Testing Machine (UTM) (Model – TUE CN -600) of 40 ton capacity. The UTM setup was integrated with a computer. The specimen was placed between top and bottom platen as presented in Figure 3. The mode of collapse of square tubes was observed and load–deformation curve was obtained from the software. Three specimens were tested for each of the composites. The load deformation curve was obtained from the area under the load–deformation characteristic curves.

3. Results and discussion

The axial compression loading on the specimens was carried out experimentally and the compression test values are presented in the above Table 1. The load–deformation characteristic curve obtained from axial compression test is presented in Figure 4. From the plot, it can be observed that the initial rise of load was the pre-buckling condition of the specimen, which showed a high load resistance and the deformation was minimum. However, once the composites reached the high peak load, they experienced non–linear mode of collapse [14]. This type of collapse is called as buckling. At this stage, due to axial compression load, the composites started to collapse from top towards the bottom of the specimen. After the buckling, rapid deformations began at buckled region that were non–linear mode collapse. During this rapid deformation, significant load fluctuations was observed in all specimens [4].

Figure 3. Axial compression testing of composite

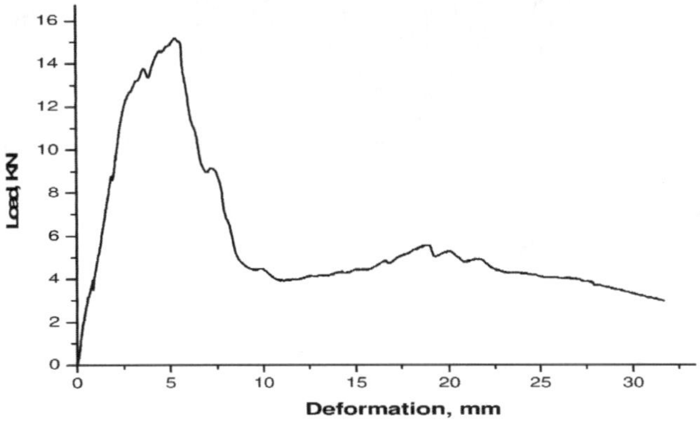

Figure 4. Load – displacement curve

From the load deformation characteristic curve, it was observed that increasing the height made larger elongation of the specimens for both basalt and glass fiber composites. For load carrying capacity, the 150 mm height composites exhibited better property than the other combinations. Compression strength of the composites also showed the same trend as that of load carrying capacity [12]. Specimen having low height attained lower energy absorption than the higher height specimens.

3.1. Modes of collapse

The modes of collapse for all basalt square tube specimens and glass fiber square tubes were analyzed and compared. They are shown in Figure 5, Figure 6, Figure 7 and Figure 8. Formation of local buckle was observed when increasing the height of square tube specimens [4]. During the local buckle, inter-laminar crack was observed at the buckled area. Similarly, by decreasing the specimen height, the outward mode of collapse was observed to be an extensive bending of lamina bundles [1].

Figure 5. Collapse of basalt fiber composites – 100 mm width and 200 mm height specimen

Figure 6. Collapse of basalt fiber composite - 100 mm width and 150 mm height specimen

Figure 7. Collapse of basalt fiber composites - basalt fiber composites – 100 mm width and 100 mm height specimen

Figure 8. Collapse of glass fiber composites – 100 mm width and 200 mm height specimen

Different modes of failures were identified on the square tube during the axial compression. The main failure of the composites was the formation of local buckling (see Figure 8) after the pre-buckling stage. The laminate bending was noticed at particular distance of the specimen from the top. During this period, symmetrical mode of collapse was evident. When it reached the buckling stage, load resistance of the specimen dropped due to further axial compression [10]. After the buckling stage, the specimens collapsed drastically in the post-buckling. At this point, it developed non-symmetrical mode of collapse and the composite tube accomplished different mode shape.

4. Conclusion

In this study, crushing behaviors of basalt fiber composite specimens and glass fiber specimens from the axial compression test were analyzed experimentally. Based on the experimental values, the steps of failure and load–deformation graphs were derived. From the load deformation curve, performance of the specimen was calculated. From the observation of the mode of collapse of the composites, it was noticed that all tested square specimens initially formed local buckling. By overall observation, it the basalt fiber specimen showed higher initial collapse load, higher average load and higher ultimate peak load than the glass fiber specimens. They also absorbed more energy before they collapsed. The basalt fiber based composites had three to five times greater energy absorption capability than glass fiber based specimens.

Acknowledgement

The authors would like to acknowledge Universiti Putra Malaysia for supporting this research under GP-IPB 9441503.

References

[1] Sundarraja M C, Sriram P and Prabhu G G 2014 *Advances in Materials Science and Engineering* **2014**
[2] Mamalis A G, Manolakos D E, Ioannidis M B, Chronopoulos D G and Kostazos P K 2009 *Composite Structures* **89** 416–23
[3] Shanmugavalli B and Sundarraja M C 2014 *IOSR Journal of Mechanical and Civil Engineering*
[4] Kathiresan M, Manisekar K and Manikandan V 2014 *Composite Structures* **108** 584–99
[5] Harte A-M, Fleck N A and Ashby M F 2000 *Eur. J. Mech. A/Solids* **19** 31–50
[6] Ataollahi S, Taher S T, Eshkoor R A, Ariffin A K and Azhari C H 2011 *Composites Part B: Engineering* **43** 542-8
[7] Melo J D D and Silva A L S 2005 *18th International Congress of Mechanical Engineering*
[8] Ihueze C C *Journal of Mechanical, Aerospace and Industrial Engineering* **3** 1-14

[9] Muralikannan R, Velmurugan R and Eswaraprasad G L 2010 *Proceedings of the IMPLAST 2010 Conference*
[10] Feraboli P, Wade B, Deleo F and Rassaian M 2009 *Composites Part A: Applied Science and Manufacturing* **40** 1248-56
[11] Tan W, Falzon B G and Price M 2015 *International Journal of Crashworthiness* **20** 60-77
[12] Palanivelu S, Paepegem W V, Degrieck J, Ackeren J V, Kakogiannis D, Wastiels J and Vantomme J 2010 *Polymer Testing* **29** 224–34
[13] Chiu L N S, Falzon B G, Ruan D, Xud S, Thomson R S, Chen B and Yan W 2015 *Composite Structures* **131** 90–8
[14] Priyadarsini R S and Kalyanaraman V *Numerical and Experimental Study of Buckling Of Advanced Fibre Composite Cylinders under Axial Compression*

IOP Conf. Series: Materials Science and Engineering **152** (2016) 012053 doi:10.1088/1757-899X/152/1/012053

Processing and characterization of redmud reinforced polypropylene composites

R Shanmugavel[1*], M Jayamani[1], R Nagarajan[1], S Irullappasamy[1], F Cardona[2] and M T H Sultan[2]

[1] Department of Mechanical Engineering, Kalasalingam University, Tamilnadu, India.
[2] Aerospace Manufacturing Research Centre (AMRC), Faculty of Engineering, Universiti Putra Malaysia, Malaysia.

*s.rajesh@klu.ac.in

Abstract. In this work, the redmud reinforced polypropylene composites were fabricated by compression molding setup. The effects of the redmud content on the mechanical, melting and crystalline behavior of the composites was investigated. The melting and crystalline behavior of the composites were investigated using Digital Scanning Calorimeter. The test results show that hardness of the composites increases with increasing redmud content while incorporation of redmud content decreases tensile and impact strength of the composites. It is determined that the addition of redmud on the polypropylene does not affect the crystalline behavior of the composites.

1. Introduction

The application of polymer based composites is unlimited and finds wide application in automotive structural parts, and they are also used as packaging materials for electronic industries among other applications [1]. To meet the demand of industrial requirements, the composites are mainly fabricated using natural and synthetic fiber. Nowadays, natural and synthetic fiber composites are finding huge application because of their good strength-to-weight ratio. Most natural composites are fabricated using thermoset and it is found that they possess good mechanical, thermal and tribological properties [2]. In case of thermoplastics, processing of composites is simple and a high production is achievable with lower production cost. Polypropylene is widely used as a matrix material because of its excellent cost to weight ratio and addition of reinforcement significantly improves the mechanical, thermal and tribological properties [3].

Recently, there is increased demand for green and renewable substitutes for organic and inorganic filler [4]. The availability of organic and inorganic filler is abundant in nature. The proper selection of filler will decide the mechanical, thermal and tribological properties of the composites. On the other hand, there are also materials that are pollutant to the environment and are not effectively utilized. Among the various industrial waste materials, the most hazardous materials are redmud. Redmud is the waste product obtained during the production of industrial alumina from bauxites. Bayer's process is the most efficient and suitable process for producing alumina. It is noted that for the production of one ton of alumina, it requires 1.5 to 2 ton of bauxite. About 0.7 ton of redmud is produced during the process, which is stored into an empty dry land [5]. The storing of the redmud causes environmental pollution and affects the land quality. To overcome these issues, there are significant attempts made by researchers to effectively use this particular waste material into useful products. Among others, some of the attempts include minerals recovery, production of low cost bricks and special cements [6]. The redmud contains huge amount of oxides of alumina, ferric and silica. These elements could be better reinforcement materials for the improvement of the mechanical properties of the composites.

The fabrication of composites is done by many ways. The most popular methods are melt mixing, injection and compression molding. In case of injection, the handling of short fiber is easier and it is possible to produce good composites. However, the handling of long fiber is difficult and the resultant properties may not be as expected. Meanwhile, compression molding is simple and the handling of long, short and particulate is not difficult. The composite obtained through this method is superior in mechanical and tribological properties. Dike *et al.* (2013) have developed the B_4C filler reinforced polypropylene composites by twin screw and injection molding method. The fabricated composites are tested for its mechanical and tribological properties, and it is found that addition of the reinforcement increases the hardness and yield strength of the composites. On contrary, the incorporation of B_4C decreases the wear resistance of the composites during the dry sliding wear test [7]. Reza *et al.* (2013) studied the effect of clay nano composites on polypropylene matrix materials. The composites are fabricated by the process of melt mixing. It was noticed that the addition of nano particles increases the Young's modulus and the yield strength of the composites [8]. Gummadi *et al.* (2012) fabricated composites using fly ash as reinforcement in polypropylene matrix. The composite fabricated by vertical injection molding process by twin heating method. It was concluded that the addition of the reinforcement reduces the flexural strength and modulus of the composites [9]. Ramadevi *et al.* (2015) fabricated treated abaca fibers and benzenediazonium chloride treated abaca stands polypropylene composites using injection molding method. The prepared composites were tested for tensile, impact and flexural properties and it was found that untreated fiber composites possess good mechanical properties than the untreated composites [10]. It was noted from the experimental result of Yang *et al.* (2007) that the tensile strength of rice-husk flour reinforced polypropylene composites decreases with addition of the reinforcement while at the same time, an increase in tensile strength of the composite was found when proper compatibilizing agent was used [11].

In this work, an attempt is made to fabricate redmud reinforced polypropylene composites using compression molding setup and the fabricated composites are tested for their mechanical, melting and crystallinity properties.

2. Experimental details

2.1. Materials

Polypropylene in the form of pellets with a density of 0.905 gm/cm^3 was purchased from Oxford Chemical Industries, Mumbai, India and it was used as the matrix materials. Redmud was used as the filler material and it was obtained from NALCO, Odisha, India.

2.2. Fabrication of composites

The received redmud was crushed in ball mill for two hours and it was sieved to obtain uniform size particulates. The size of the particulates is in the range of 4 μm. The revised polypropylene matrix materials were crushed in the pulverizer to get uniform size particulates, which were almost equal to the size of the reinforcement. The average size of the redmud reinforcement and polypropylene matrix was 4 μm. Figure 1 and Figure 2 depict the received polypropylene and redmud, respectively. Redmud reinforced polypropylene composites were then fabricated using the compression molding setup. The required quantity of matrix material and reinforcement was taken and blended for one hour to obtain better mixing. The temperature of the mold was maintained at $200+100^0C$, and it was compressed to the pressure of $200kg/cm^2$. After one hour, the composite within the die was allowed to cool to the room temperature and the final composite material was taken after 24 hours. The size of the fabricated composite was 100 x 100 x 3 mm.

Figure 1. Polypropylene

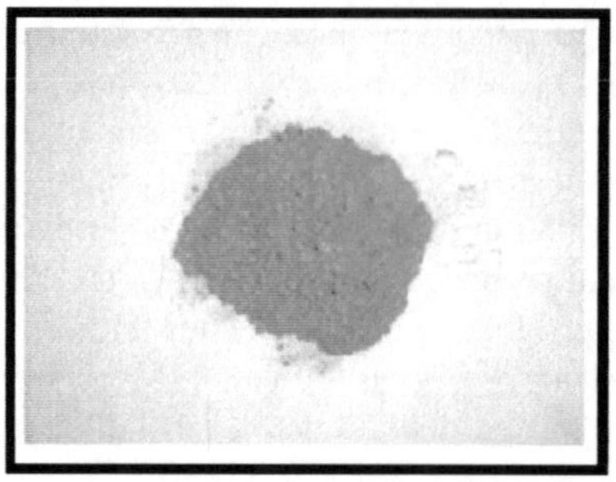

Figure 2. Redmud

2.3. Tensile testing

The tensile behavior of the composite was evaluated in accordance with the ASTM standard D3039 using DTRA 30kN twin screw, Deepak Polyplast machine. Size of the specimen used for the tensile test was 220 x 13 x 13 mm and 5 mm/min was maintained as the feed rate.

2.4. Impact testing

ASTM D256 was used to conduct the impact test for the fabricated composite material using Deepak Polyplast digital impact testing machine. Size of the composites for impact test was 65 x 15 x 3 mm. The impact strength of the composites was measured in terms of Joules.

2.5. Hardness testing

Hardness of the composites was measured as per ASTM D2240 using Shore D hardness tester, HT 6510D. The average value of three tests was reported for all the tests.

2.6. Melting temperature and degree of crystallinity

Differential Scanning Calorimeter (DSC) of METZSCH STA–1100M type was used to identify the crystallization state of the composites. Digital Scanning Calorimeter was used to study the physical transitions of the composites in terms of heat flow of the samples as function of temperature. The samples were tested at ambient condition and by gradually increasing the temperature of the sample up to 800^0C. The heating and cooling rate was maintained at 10 ^{0}C/min. Crystallinity of the composites was measured from the DSC curve, theoretical heat fusion of the unreinforced polypropylene and the mass fraction of the composites.

3. Result and Discussion

3.1. Tensile strength

Figure 3 shows the effect of redmud addition on the tensile property of the composites. It is observed from the experimental results that the addition of redmud in the polypropylene decreases the tensile strength of the composites, which was reduced from 28.55 MPa to 26.53 MPa. The reduction in tensile strength of the composites could be attributed to the lack of stress transfer ability of the reinforcement materials. Similar result was also observed by Oksman *et al.* (2009). It is established that the reduction of tensile strength is due to the poor adhesion between the matrix and reinforcement. Other reason for decreasing tensile strength of the composites is due to the agglomeration of the reinforcement in the matrix materials since size of the redmud particulate is in the range of 4 μm.

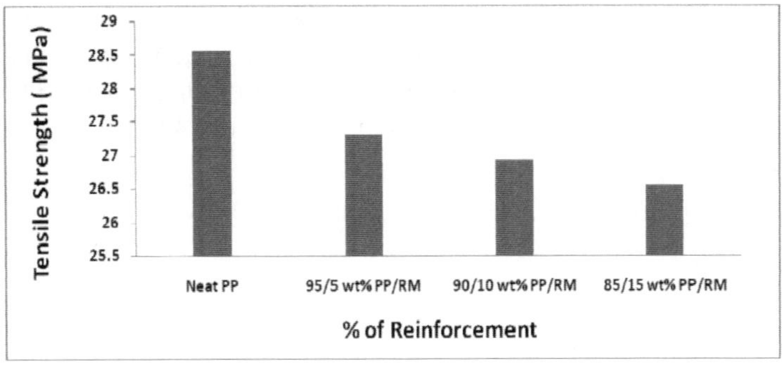

Figure 3. Effect of redmud on tensile strength of the composites

3.2. Hardness strength

Figure 4 shows the effect of redmud on the hardness of the composites. Interestingly, the addition of redmud from 0 to 15% increases the hardness of the composites in marginal level only. For instance, the hardness of the composites is increased by only 3% compared to the neat PP. It is also evident from the result of tensile modulus that the increase in tensile modulus is due to the effect of increase in hardness. As also can be clearly observed from the results of the tensile test is that the tensile strength of the composites is reduced while increasing the redmud content. The reduction in tensile strength is due to the incorporation of redmud, which influences the elongation and plasticity of the composites, and the increase in hardness is due to the effect of redmud addition into the matrix. Figure 5 shows the distribution of redmud over the neat PP.

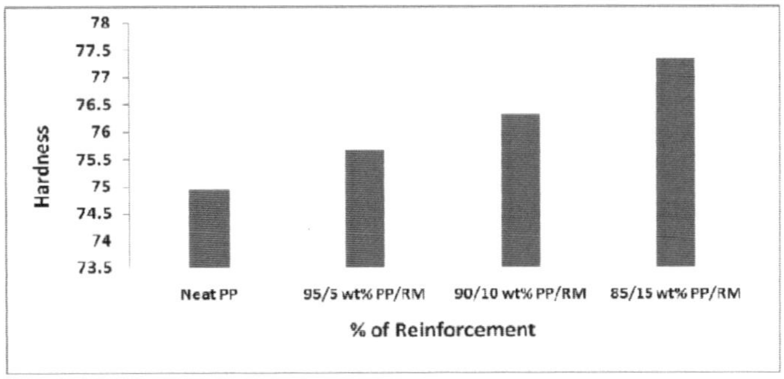

Figure 4. Effect of redmud on hardness of the composites

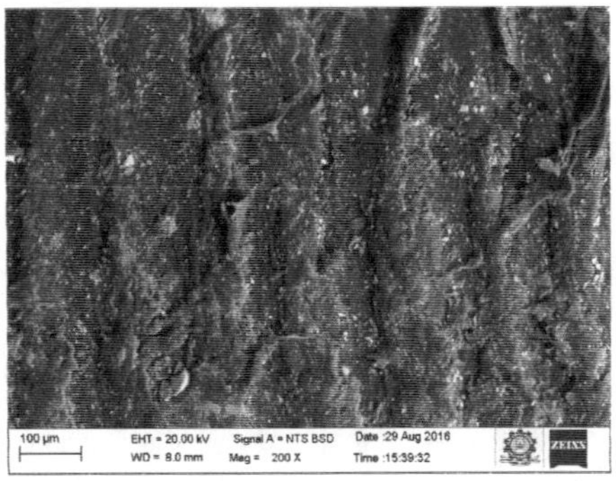

Figure 5. SEM image of redmud reinforced polypropylene composites

3.3. Impact strength

From Figure 6, it is evident that the impact strength of the composites is decreased when increasing the amount of reinforcement. The reason for this decrease in impact strength is attributed to the low affinity between the matrix and the reinforcement, which is due to low Vander Waals force between the matrix and reinforcement phase. The reduction in Vander Waals force reduces the load transfer ability of the composites, thus impact strength of the composites is reduced gradually. An increase in the redmud content also lowers the impact strength because of its braking and splitting property during loading. Due to this reason, the filler material fails to resist sudden applied force and in turn reduces the impact strength of the composites.

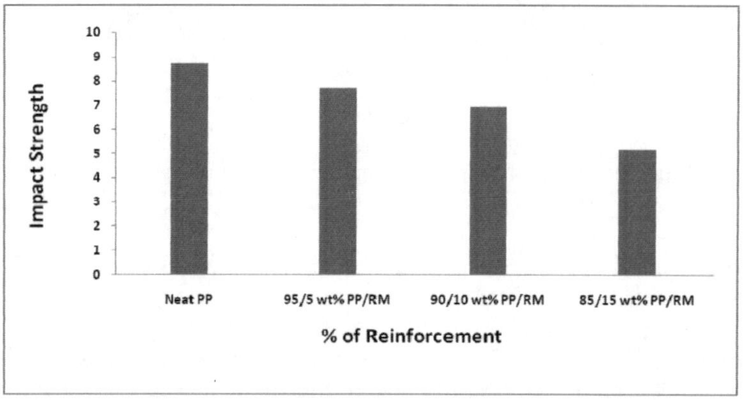

Figure 6. Effect of redmud on impact strength of the composites

3.4. Melting temperature and degree of crystallinity

Crystalline phase of the composite material plays a vital role in deciding the mechanical properties of the fabricated composite materials. The polypropylene is in the form of semi-crystalline in nature. The portion of the crystalline phase influences the mechanical properties of the composite materials. Thus it is important to study the effect of redmud on the changes in the crystalline structure. Figure 7 shows the DSC melting curves of polypropylene and its redmud reinforced polypropylene composites.

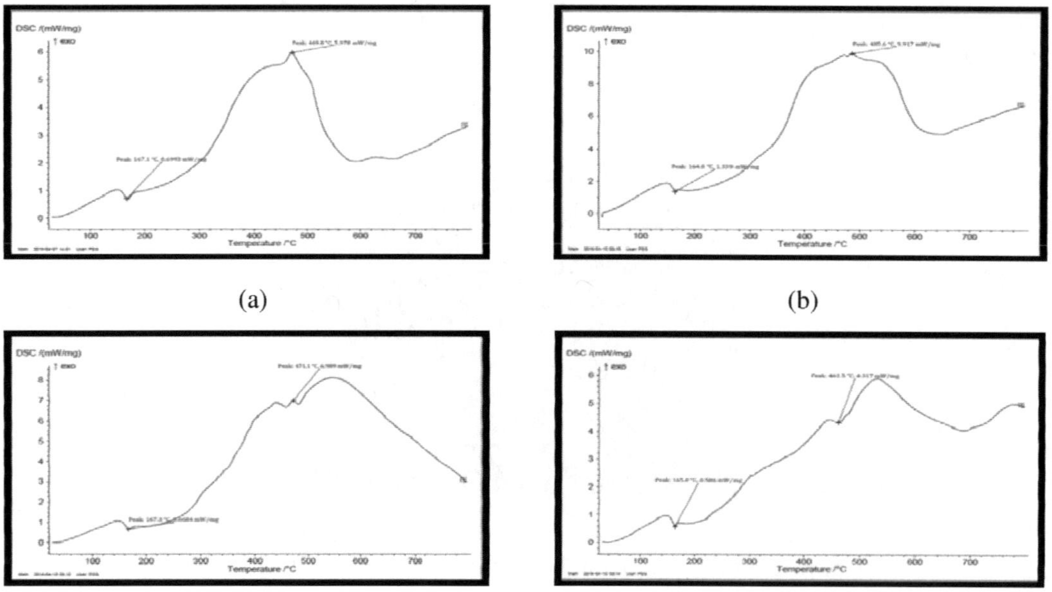

(a) (b) (c) (d)

Figure 7. (a) Neat Polypropylene, (b) 5% redmud, (c) 10% redmud, (d) 15% redmud content in the composites

The addition of redmud to polypropylene does not influence the melting properties of composite materials. It is also observed from the DSC curve that the melting temperature of redmud reinforced composite slightly shifts to lower temperature when increasing the redmud content. The reason for the slight reduction in melting temperature is the effect from the reduction of lamellar thickness of crystallites. It is realized that the addition of redmud reduces the crystallinity of the composite materials as compared to neat polypropylene. This is attributed to the more crystalline nature and less amorphous part of the polypropylene, which provides opportunity for redmud to accommodate in the interlamellar spaces.

4. Conclusion

In this work, redmud reinforced polypropylene composites were successfully fabricated using the compression molding setup. The mechanical, melting and crystalline behaviors of the composites were studied. It has been found that the addition of redmud to the polypropylene composites increases the hardness of the composites significantly, but there is a drop in tensile strength and impact strength of the composites. Furthermore, the addition of redmud reinforcement slightly reduces the melting and crystalline properties of the composites.

Acknowledgment

This research was supported by UPM GP-IPB 9441503.

References

[1] Dogan M and Bayramli E 2010 *Polymers Advanced Technologies* **22** pp. 1628 - 1632
[2] Murali Mohan Rao K and Mohana Rao K 2007 *Composite Structures* **77** pp. 288-295
[3] Bijwe J N, Indumathi J S and Fahim M 2002 *Journal of Reinforced Plastics,* **21** pp. 1221-1240
[4] ZhangY Z, Anzhen Z, Zhichao L, Fengzhu C, Paul K and Ji J 2011 *Journal of Composite Materials* **45** pp. 2811- 2816
[5] Yang J K and Xiao B 2008 *Construction and Building Materials* **22** pp. 2299 – 2307
[6] Dikea A S, Mindivanb F and Mindivanc H 2013 Effect of B_4C Content on the Mechanical and Tribological Performances of Polypropylene. *Proceedings of the 3rd International Congress.* APMAS2013 - 10.12693/APhysPolA. 125.396
[7] Reza E F, Mohammad Reza Khalili S, Ziba H and Neda S 2013 *Journal of Material Design* **53** pp. 540 -549
[8] Gummadi J K, Vijay R and Gunti G 2012 *International Journal of Modern Engineering Research* **2** pp. 2584-2590
[9] Ramadevi P, Dhanalakshmi S, Raghu P, Ranganagowda G, Basavaraju B and Chikkol V S 2015 *Journal of King Saud University - Engineering Sciences*
[10] Yang H S, Kim H J, Park H J L, Bum J H and Taek S 2007 *Composite Structures* **77** pp. 45 - 55
[11] Oksman K, Mathew A P, Långström R, Nyström B and Joseph K 2009 *Composite Science Technology* **69** pp. 1847-1853

IOP Conf. Series: Materials Science and Engineering **152** (2016) 012054 doi:10.1088/1757-899X/152/1/012054

Mechanical characterization of coir/palmyra waste fiber hybrid composites

V Arumugaprabu[1]*, M Uthayakumar[1], F Cardona[2] and M T H Sultan[2]

[1]Department of Mechanical Engineering, Kalasalingam University, India
[2]Aerospace Manufacturing Research Centre (AMRC), Universiti Putra Malaysia, Serdang, Malaysia

*v.arumugaprabu@klu.ac.in

Abstract. In the present days, the utilization of palmyra fiber in automotive and aerospace applications has increased drastically due to its high strength and low weight. This research focuses on the development of composite materials using palmyra waste and coir fiber with polyester as a matrix. The mechanical properties such as tensile, flexural and impact strength of composites were investigated. Palmyra waste fiber and coir fiber with relative varying weight percentage in the ratio of 50:50, 40:60, 30:70 and 20:80 had been considered for the study. The composites were prepared by the compression moulding method. In addition, the prepared composites were subjected to moisture studies for 24 hours, 48 hours and 72 hours to know the composite resistance to water absorption. The results showed an increase in all the mechanical properties from the addition of palmyra waste. After analysing the results obtained from the study, a suitable application in the automobile and aerospace industries is suggested for the new developed composite.

1. Introduction

The utilization of composite materials with two natural fibers as the reinforcement is finding extensive applications in automotive and aerospace industries due to their high strength and better mechanical performance when subjected to various environmental conditions. From the study on the palmyra fruit fiber reinforced with epoxy matrix that was carried out by Narayanan *et al.* [1], it is inferred that the tensile strength and tensile modulus will increase proportionally with the increase in fiber content. The use of coir fiber composites along with polymer composites is due to its high strength, modulus and rigidity as had been analysed and reviewed in detail by Verma *et al.* [2]. It is also reported that the coir fiber reinforced composites have found their applications in packaging, furnitures, etc. The effect of hybridizing oil palm fruit fiber with glass fiber has show a tremendous increase in the mechanical and physical properties as reported by Myrtha Katrina *et al.* [3]. Based on the results, a large increase of about 350% in flexural strength was obtained by the hybrid composites, apart from showing a better resistance in terms of moisture absorption.

Studies on oil palm empty fruit fiber with polypropylene as the matrix, and different manufacturing techniques that provided specific importance to the fiber loading, reported that the tensile and flexural strengths increased for 20% wt [4]. By using hand layup technique, study on mechanical performance of oil palm reinforced epoxy composites was carried out by Mohd Suri *et al.* [5,6], which inferred that the tensile and flexural strengths decreased with the increase in fiber loading. More studies continued on the same combination, with special reference to the hybrid of woven type jute fiber arrangement along with oil palm that provided higher strength [7, 8]. Mohaiman *et al.* [9] performed the hybrid composite studies on the woven kenaf/glass fiber reinforced polyester composites and noted that the reinforcement of kenaf over glass fiber increased the tensile strength and compressive strength. Sanjay *et al.* [10] carried out mechanical properties studies on jute and glass fiber reinforced epoxy matrix and reported that the results were very good in comparison to jute fiber composites alone. Also, the same hybrid composite studies of banana and sisal fiber had been done by Hemant Patel *et al.* [11].

Nonetheless, from the thorough literature study, it can be inferred that no study on the mechanical properties of palmyra waste and coir fiber hybrid composites with weighted ratio method has been carried out before.

2. Experiment details

2.1. Materials

Natural fiber coir and palmyra waste fibers were used as reinforcement. General purpose unsaturated polyester resin, which was most appropriate for indoor and outdoor application, was used as the matrix material in the composite.

2.2. Fabrication of composites

The fiber reinforced composites were been prepared by compression moulding technique. The coir and palmyra waste fibers were cut into two different lengths of 3 cm and 4 cm, respectively, which was the most optimized length obtained from previous studies. The prepared mould was coated initially with a releasing agent (paraffin wax). The general purpose unsaturated polyester resin was taken with 1:1 weight percentage to the fiber used. 1% weight percentage of accelerator (cobalt napthenate) and 1% weight percentage of catalyst (methyl ethyl ketone peroxide) were added to the unsaturated polyester resin. The mixture was then stirred until it became homogenic mixture. After that, the resin was poured over the fibre that was placed on the mould cavity. The fibers were arranged in a random manner with the layering pattern of alternate palmyra waste and coir fiber. A final coat was done after the removal of the excess resin and the mould was then closed with the match die. The mould was later kept in the compression moulding machine at room temperature for four to six hours for curing and finally, the specimen was taken out from the mould.

2.3. Mechanical testing

Tensile test was conducted on the cured composite samples using the universal testing machine with a cross head speed of 2 mm/min according to ASTM D3039. The standard size of specimen for tensile strength test is 250 x 25 x 3 mm. Flexural tests were also conducted on the cured samples using the universal testing machine with a cross head speed of 2 mm/min according to ASTM D790-10. The standard specimen for flexural strength was 127 x 13 x 3 mm. Furthermore, impact tests were done as per ASTM D256-10 using an impact tester. The standard specimen for impact strength was 65 x 13 x 3 mm. For all mechanical tests, five samples were taken and the results were averaged.

3. Results and discussions

3.1. Tensile studies on palmyra waste/coir hybrid composites

Figure 1 shows the tensile strength value of the palmyra waste/coir hybrid composites. The increase in weight percentage increased the tensile strength value slightly in the case of 3 cm fiber length up to the ratio 40:60, but then it started to decrease. Similarly, for the 4 cm length fiber, the maximum strength was observed for the same 40:60 ratio. In other words, the reinforcement of 40%wt palmyra waste and 60%wt coir fiber possesses the highest strength. The reason may be the accommodation of more short fibers created no gap and they also strongly held by the matrix. The same trend was observed from the microscopic studies image shown in Figure 2, in which the fractured surface showed few bunch fibers that were held by the matrix. This gives evidence of a high strength. On the other hand, from Figure 3, the presence of large gap demonstrated the complete failure of the composites, which resulted in a low strength.

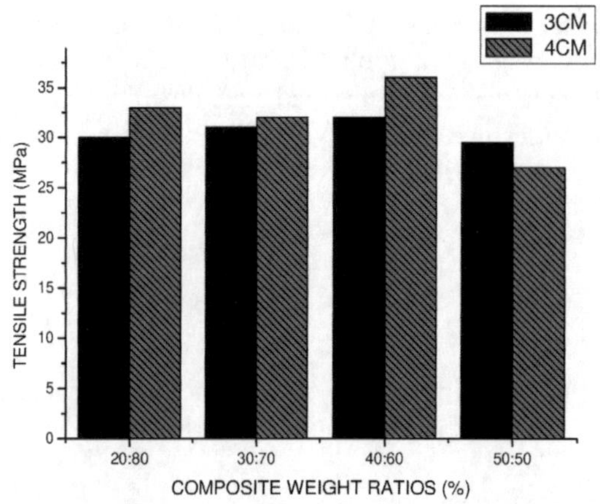

Figure 1. Tensile strength of palmyra waste/coir hybrid composites

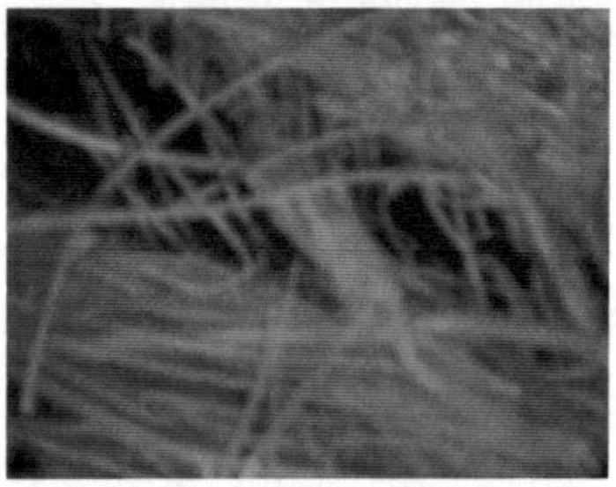

Figure 2. Microscopic image (magnification 200MP) of the fractured surface after the tensile test of palmyra waste fiber (40%wt) and coir fiber (60%wt) of fiber length 3 cm hybrid composites

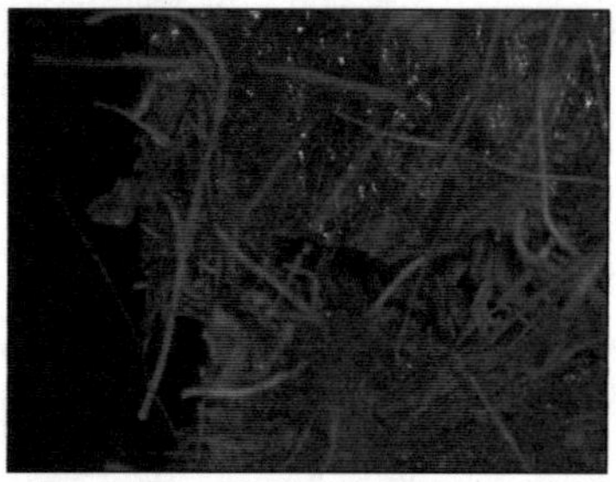

Figure 3. Microscopic image (magnification 200MP) of the fractured surface after the tensile test of palmyra waste fiber (50%wt) and coir fiber (50%wt) of fiber length 4 cm hybrid composites

3.2. Flexural studies on palmyra waste/coir hybrid composites

Figure 4 depicts the flexural strength value of the palmyra waste/coir hybrid composites. A consistent increase in flexural strength can be observed with the increase in fiber percentages for the 4 cm fiber length composites. Overall, the highest flexural strength of 55 MPa was obtained for this 4 cm fiber length hybrid composite. On the other hand, for 3 cm fiber length, there was gradual decrease in the flexural strength with increase in weight percentage ratio up to 60% and 40%. After that, it started to increase slightly. The reason for this irregular trend is due to the short fibers being randomly arranged without proper orientation, which resulted in irregular distribution of fibers. This caused the composite not being able to withstand the bending strength.

From the microscopic images shown in Figure 5, it can be observed that the fibers were arranged one above the other. The shining surface indicated that the matrix flowed over both the fibers and strongly bonded them together, which gave a high strength. The next image in Figure 6 showed the non uniform arrangement of coir and palmyra waste fiber, which led to high fiber-to-fiber interactions and dispersion problems with the increase in weight percentage, resulting in the decreased flexural strength.

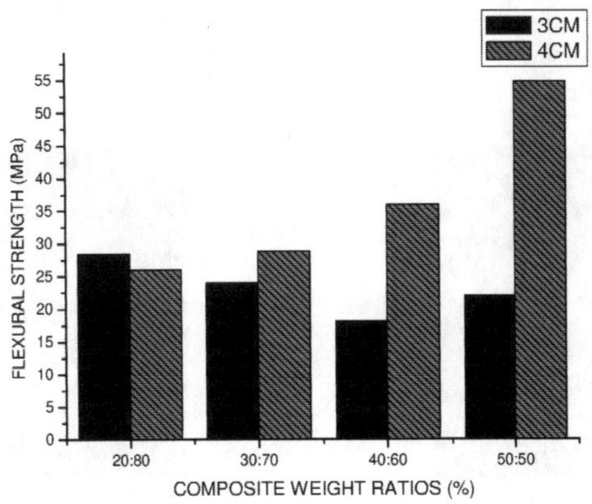

Figure 4. Flexural strength of palmyra waste/coir hybrid composites

Figure 5. Microscopic image (magnification 200MP) of the fractured surface after the flexural test of palmyra waste fiber (50%wt) and coir fiber (50%wt) of fiber length 4 cm hybrid composites

Figure 6. Microscopic image (magnification 200MP) of the fractured surface after the flexural test of palmyra waste fiber (40%wt) and coir fiber (60%wt) of fiber length 3 cm hybrid composites

3.3. Impact studies on palmyra waste/coir hybrid composites

Figure 7 shows the impact strength value of the palmyra waste/coir hybrid composites. Impact strength is the ability of the composites to absorb energy and withstand the high impact loading. Similar to the flexural strength, the composite combination of 4 cm length and 40:60 ratio possessed the highest impact strength value of 8.5 kJ/cm^2. It is expected that, if more natural fibers are arranged between each other, the composite will have an even higher strength. Since both palmyra waste and coir natural fibers were reinforced, the impact strength was very high for the high weight percentage. However, in the case of 3 cm fiber length, there was a sudden decrease in the 40:60 ratio that might be due to the arrangement pattern of the fibers.

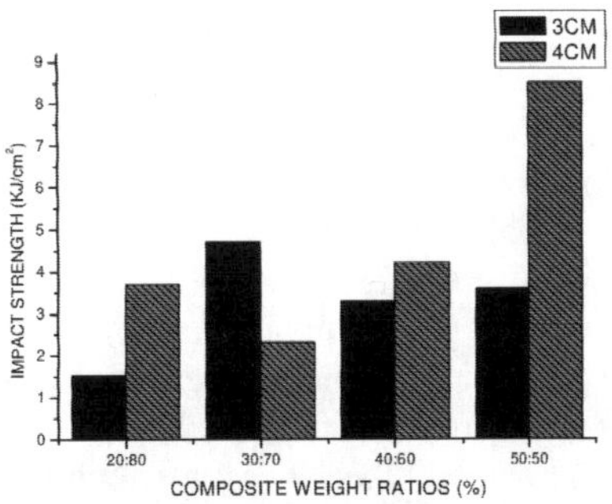

Figure 7. Impact strength of palmyra waste/coir hybrid composites

From the microscopic images shown in Figure 8 and Figure 9, it can be observed that the fibers were strongly bonded together when the matrix was spread over the layers of palmyra as well as coir fibers. These bundles of short fibers had the capability of absorbing more energy, which in terms had also a high load withstanding capability.

Figure 8. Microscopic image (magnification 200MP) of the fractured surface after the impact test of palmyra waste fiber (50%wt) and coir fiber (50%wt) of fiber length 4 cm hybrid composites

Figure 9. Microscopic image (magnification 200MP) of the fractured surface after the flexural test of palmyra waste fiber (30%wt) and coir fiber (70%wt) of fiber length 4 cm hybrid composites

3.4. Moisture studies on palmyra/coir hybrid composites

The moisture absorption studies have been carried out by measuring the weight loss of the composites before and after immersion in both distilled water and salt water. The percentage of water absorbed by the palmyra waste/coir fiber composites is shown in Table 1.

Table 1. Moisture absorption of palmyra waste/coir fiber composites

Sample No.	Fiber Length	Fibers Weight Ratios	Percentage of Water Absorbed					
			Distilled Water			Salt Water		
			24 hrs	48 hrs	72 hrs	24 hrs	48 hrs	72 hrs
1	3 cm	20:80	9	2.0	1.0	6.5	2.5	0.5
2		30:70	11	3.0	0.3	8.5	2.8	0.8
3		40:60	17	3.0	1.0	13.0	3.0	1.0
4		50:50	14	3.0	1.0	13.0	2.7	0.6
5	4 cm	20:80	7.9	2.1	0.4	5.6	2.5	0.5
6		30:70	13	2.3	0.8	9.0	2.6	0.7
7		40:60	15.3	2.6	0.6	11.0	3.0	1.0
8		50:50	12.4	2.1	0.7	10.0	2.8	0.8

Interestingly, it is noted that in all cases, as the day progresses from day 1 to day 3, the percentage of water absorption decreased. This is due to the random orientation of coir and palmyra waste fibers, resulting in the water particulates to occupy the space locations. In addition, based on this observation, the potential of the composites to be used in moisture related applications is well highlighted.

4. Conclusions

In this mechanical performance study on palmyra waste/coir fiber hybrid composites, the composites were successfully fabricated by compression moulding method. Mechanical properties such as tensile, flexural and impact strengths of the palmyra waste fiber filled hybrid composites showed a potential increase in strength by the inclusion of coir fiber. Tensile strength increased by 37% and the impact strength increased by 60% by incorporation of the coir fiber. Among various composite combinations, palmyra waste with 40% weight and coir fiber with 60% weight had the highest tensile, flexural and impact strength properties. Moisture absorption studies were also performed, in which the specimens were placed under water for 24 hours, 48 hours and 72 hours. The same ratio Palmyra waste with 40% weight and coir fiber with 60% weight composite had the highest percentage of water absorption. This type of hybrid composites were suitable for making indoors of aeroplane, aircraft dome, dies, helmets, sports good and also in moisture related applications.

Acknowledgement
The authors wish to thank the Centre for Composite Materials, Kalasalingam University (Tamilnadu, India) for providing all the facilities to carry out this research work. This research is also supported by UPM under GP-IPB 9441503.

References

[1] Nayak N C and Mishra A 2013 *Journal of Production Engineering* **16** 69-72
[2] Verma D, Gope P C, Shandilya A, Gupta A and Maheshwari M K 2013 *J. Mater. Environ. Sci.* **4** 263-76
[3] Katrina M, Onggo H, Dawam Abdullah A H and Syampurwadi A 2008 *Journal of Biological Sciences* **8** 101-6
[4] Ahmed A S, Alam M A, Piee A, Rahman M R and Hamdan S 2010 *Journal of Energy & Environment* **2** 16-21
[5] Mohamed Yusoff M Z, Salit M S, Ismail N and Wirawan R 2010 *Sains Malaysiana* **39** 87–92
[6] Mohamed Yusoff M Z, Salit M S and Ismail N 2009 *Sains Malaysiana* **38** 525–9
[7] Abdul Khalil H P S, Jawaid M and Abu Bakar A 2011 *Bio Resources* **6** 1043-52
[8] Jawaid M and Abdul Khalil H P S 2011 *Bio Resources* **6** 2309-22
[9] Mohaiman J S, Leman Z, Sultan M T H, Ishak M R and Azmah Hanim M A 2016 *International Journal of Polymer Science* **2016**
[10] Sanjay M R and Yogesha G 2016 *Journal of Minerals and Materials Characterization and Engineering* **4** 15-25
[11] Patel H, Parkhe A and Shrama P K 2016 *International Journal of Research – Granthaalayah* **4** 206-16

IOP Conf. Series: Materials Science and Engineering **152** (2016) 012055 doi:10.1088/1757-899X/152/1/012055

Effect of inter-laminar fibre orientation on the tensile properties of sisal fibre reinforced polyester composites

K Senthilkumar[1*], I Siva[1], J T Winowlin Jappes[1], S C Amico[2], F Cardona[3] and M T H Sultan[3]

[1]Centre for Composite Materials, Department of Mechanical Engineering, Kalasalingam University, Tamilnadu, India
[2]DEMAT, Federal University of Rio Grande do Sul (UFRGS), Brazil
[3]Aerospace Manufacturing Research Centre (AMRC), Universiti Putra Malaysia, Serdang, Malaysia

*kmsenthilkumar@klu.ac.in

Abstract. In this present work, effects of interlamina fibre orientation on the tensile properties of composites were studied and the results were discussed. The varying types of fibre oriented composites were prepared using the compression moulding technique at a pressure of 17 MPa. The different types of oriented composites investigated were 90°/0 ° /90 °, 0 ° /90 ° /0 °, 90 ° /0 ° /0 ° /90 °, 0 ° /45 ° /0 °, 0 ° /90 ° /45 ° /45 ° /90 ° /0 °, 0 ° /45 ° /90 ° /90 ° /45 ° /0 ° and these composites were subjected to tensile testing according to ASTM: D3039-08. The sisal fibres were arranged in various angles with the help of specially designed mould. It was found that the tensile strength of sisal fibre composites improved when 0 ° oriented fibres were positioned at the extreme layers of the composites compared to 90 ° oriented fibres. The highest tensile strength among the types of composites was observed for 0 ° /90 ° /0 °. The scanning electron microscopy (SEM) analysis was performed to understand the interphase adhesion mechanism.

1. Introduction

Increasing fear over the greenhouse effects and environmental awareness have stimulated the use of natural fibres to replace the synthetic polymeric fibres in many fields such as automotive, construction, packing and civil industries. Since natural fibres are easily available in fibrous form and cheaper to buy than the synthetic fibres, they are a good replacement of conventional synthetic fibres.

Silva *et al.* [1] studied the behaviour of high performance sisal fibres. They investigated the cross sectional areas of sisal fibres using the scanning electron microscopy technique. Moreover, Belaadi *et al.* [2] studied the tensile properties of the sisal fibres. They reported that the sisal fibres offer a good reinforcement compared to other natural fibres, owing to the less extraction cost for retting process from the plants. They found that the diameter of the fibres, strength and strain to failure and Young's modulus of about 15 different fibres. The same results have been compared with literatures [3-5]. Ku *et al.* [6] investigated the mechanical properties of natural fibre composites. Owing to their favourable characteristics like low cost, high specific strength, good mechanical properties, eco-friendly, non-abrasive and bio-degradability, the natural fibres are proposed as the alternative to synthetic/artificial fibres such as carbon, aramid, glass, etc. They also reported that the tensile properties of natural fibre composites are dependent on the interfacial bond between the fibres and the matrix. Sen and Reddy [7] described the many applications of natural fibres such as coir, bamboo, sisal and jute. They reported that amongst the many natural fibres, bamboo, sisal, coir and jute have higher impact strength and moderate flexural and tensile properties. Among the applications of sisal fibres include ropes, mats, cement reinforcements, carpets etc. Li and Shen [8] studied a single fibre pull out test for sisal fibre and analyzed its multi scale structures and dimensional parameters. In addition, the researchers also developed many theoretical models and compared the results with the experimental results.

Alavudeen *et al.* [9] studied the effect of fibre orientations in the natural fibre composites. They found that the properties of composites are influenced by varying the fibre orientations. Bennet *et al.* [10] studied the effects of lamina fibre orientation using the natural fibres. The orientations of fibre composites that were investigated include 0 °, 30 °, 45 °, 60 ° and 90 °. Among the composites, 0 ° oriented fibres composite provided the highest strength in tensile properties. Furthermore, they also studied the vibration properties of these oriented fibre composites. Almeida *et al.* [11] studied the fibre orientation on shear behaviour of synthetic fibre composites. They prepared $[0]_5$, $[90]_5$, [0/90/0/90/0] and randomly-oriented (mat) types of composite. Mortazavian and Fatemi [12] investigated the effects of fibre orientation and anisotropy properties of natural fibre composites. They found that the tensile and elastic modulus of the composites decreased with the specimen angle. Effects of fibre orientation and fibre aspect ratio on the tensile strength of carbon fibres reinforced magnesium matrix composites (Csf /Mg) were analyzed by Qi *et al.* [13]. They found that the angle between the fibres and applied load on the composites played an important role on the strength of the composites. Therefore, in this present study, natural fibre composites with varying fibre orientations were fabricated by using the compression moulding machine and subjected to tensile test.

2. Experimental details

2.1. Materials
Sisal fibre used in this study was supplied by Shiva exports, Tirunelveli, Tamilnadu, India. On the other hand, the unsaturated isophthalic polyester resin, initiator and accelerator were supplied by Vasivibala resins (P) Ltd, Chennai, Tamilnadu, India.

2.2. Fabrication of composites
Initially, the sisal fibres were combed and arranged carefully as long fibres in the mould according to the type of selected orientation of the composite. To achieve the fibre weight percentage of 50±2, the desired amount of fibres was taken into the specially designed mould. To ensure the desired angle of fibre orientation, the mould consisted of markings for every angle at the periphery. Then, 100 ml of resin, 1.5 % of catalyst and 1.5 % of accelerator were added to it. The mixture was poured over the fibres and kept in the compression moulding machine at a pressure of 17 MPa. The composite was left about 24 hours in the moulding machine for curing and were later cut according to ASTM standard.

2.3. Tensile testing
Tensile tests were carried out using Instron (Series-3382) testing machine according to ASTM: D3039-08. The specimens were cut according to the dimension of 200 × 20 × 3 mm. A 100 kN load cell was used and a speed of 2 mm/min was applied on the samples. An average of five specimens were used in each test.

3. Result and discussion
To ease the discussion, the notations tabulated in Table 1 will be used in this paper after this point to indicate the different fibre orientations for the sisal fibre oriented composites. The effect of interlamina fibre orientation on the tensile strength of sisal fibre polyester composites is shown in Figure 1. The tensile strength was determined for the different fibre orientation and it was found that the orientation 0 ° /90° /0° had the greatest tensile strength of 46.716 MPa when compared to other types of oriented composites. It was noticed that the extreme layers of fibres oriented in 0 ° at both ends of composites

(0 ° /90 ° /0 °, 0 ° /45 ° /0 °, 0 ° /90 ° /45 ° /45 ° /90 ° /0 ° and 0 ° /45 ° /90 ° /90 ° /45 ° /0 °) showed higher tensile strength than the other types of composites. This is due to the fact that 0 ° fibres were arranged longitudinally (i.e along the direction to the applied loading condition). Hence the fibres oriented at 0 ° along the applying load conditions would withstand higher load while testing. On the other hand, the lower tensile strengths were measured in 90°/0°/90 ° and 90°/0°/0°/90 ° oriented fibre composites. It was noticed that among the different types of oriented fibre composites, 90° at the extreme layered composites gave the lowest tensile strength. While testing these composites, the fibres were oriented at 90 ° that were perpendicular to the loading conditions. Therefore, the fibres did not withstand any load while testing the composites.

Table 1. Notations and descriptions

Notation	Description
1	90°/0 ° /90 °
2	0 ° /90 ° /0 °
3	90 ° /0 ° /0 ° /90 °
4	0 ° /45 ° /0 °
5	0 ° /90 ° /45 ° /45 ° /90 ° /0 °
6	0 ° /45 ° /90 ° /90 ° /45 ° /0 °

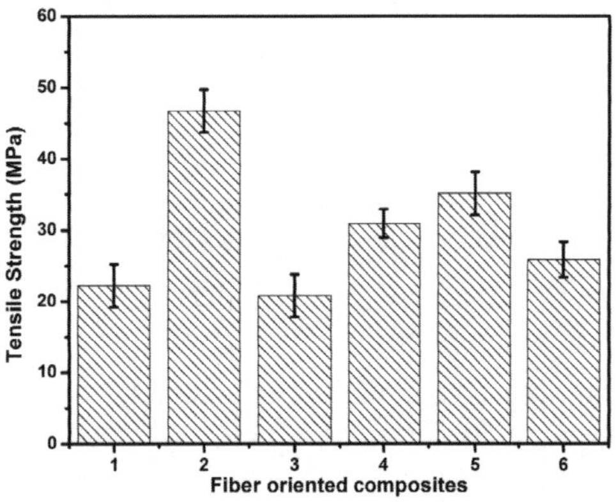

Figure 1. Effect of inter lamina fibre orientation on tensile strength of sisal fibre polyester composites

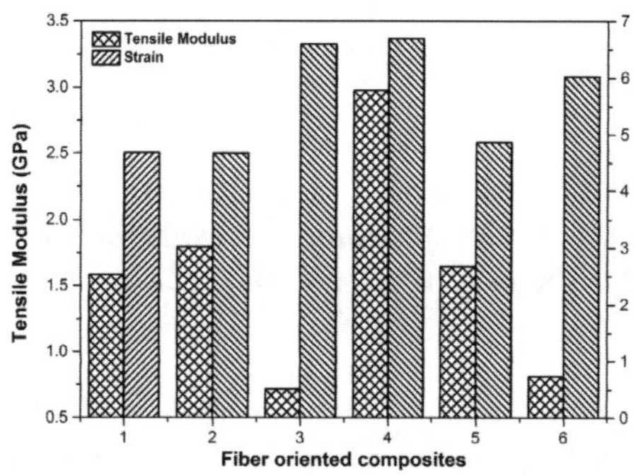

Figure 2. Tensile modulus vs strain of sisal fibre polyester composites

The tensile strengths of the panels measured at 90°/0 °/90 ° and 90 ° /0 ° /0 ° /90 ° were 22.24 MPa and 20.80 MPa, respectively. Comparing the 0 ° /90 ° /45 ° /45 ° /90 ° /0 ° types of composites with the 0 ° /45 ° /0 ° and 0 ° /45 ° /90 ° /90 ° /45 ° /0 °, they were able to withstand more loads in testing. It was confirmed that next to 0 ° oriented fibres, 90 ° oriented fibre composites had a higher strength than the 45 ° oriented fibre composites. The effects of tensile modulus and strain of fibre oriented composites are shown in Figure 2. The increase in the stiffness of the composites was observed for the 0° fibres that were positioned as the skin of composites. From Figure 2, it is shown that the tensile modulus of 90°/0°/0°/90° and 0 ° /45 ° /90 ° /90 ° /45 ° /0 ° composites was not appreciably increased. The tensile modulus of 0 ° /45 ° /0 ° oriented fibre composite was 76 % higher than that of the 90 ° /0 ° /0 ° /90 ° fibre composites. Figure 2 shows the strain of the sisal fibre reinforced polyester composites. Higher strain was observed in 0 ° /45 ° /0 ° type of fibre composites than the other types of combinations.

The SEM micrograph of the 0 ° /90 ° /0 ° sisal fibre oriented composite is shown in Figure 3(a). It is inferred from this figure that the fibre is tightly covered by the strong matrix and this strong bonded matrix could be used to hold the fibres in their position. Consequently, the fibre pull out is avoided owing to its strong interfacial bonding. Furthermore, the resin rich area shown in Figure 3(a) indicates the lack of cracking in the matrix, which can be attributed to a good packaging of materials. SEM images of the 90 ° /0 ° /0 ° /90 ° oriented sisal fibre composites taken after fracture under the tensile loading of fibre reinforced composites are shown in Figure 3(b) in which the separation of sisal fibres and hollow portion are clearly seen. This shows that the fibre pull-out have occurred in the composites.

Figure 3a. SEM studies on the effect of 0 ° /90 ° /0 ° fibre orientation on the tensile strength of sisal fibre polyester composites

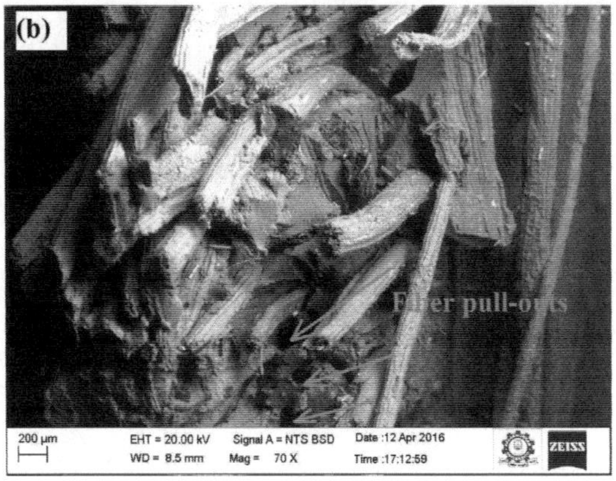

Figure 3b. SEM studies on the effect of 90 ° /0 ° /0 ° /90 ° fibre orientation on the tensile strength of sisal fibre polyester composites

4. Conclusion

The sisal fibre reinforced polyester composites were fabricated with different fibre orientations using a specially-designed mould in compressing moulding machine at a pressure of 17 MPa. Experimental results showed that 0 ° /90 ° /0 ° type of composites gave the highest strength than the other types of combinations. It was also found that the tensile strength of the composite was lower when the extreme fibres were oriented in 90° compared to the 0° oriented fibre composites. The highest tensile modulus and strain were measured for 0 ° /45 ° /0 ° oriented fibre composite.

Acknowledgments

The authors wish to thank Centre for Composite Materials, Department of Mechanical Engineering, Kalasalingam University for their kind permission to carry out the preparation and also testing of the composites, and the International Research Centre, Kalasalingam University for given kind permission to carry out the Scanning Electron Microscope analysis. This research is also supported by UPM under GP-IPB 9441503.

References

[1] Silva F A, Nikhilesh C and Romildo D 2008 *Compos. Sci. Technolo.* **68** 3438-43

[2] Belaadi A, Bezazia A, Bourchakb M and Scarpac F 2013 *Mater. Des.* **46** 76-83

[3] Munawar S S, Umemura K and Kawai S 2007 *Journal of Wood Science* **53** 108–13

[4] Filho T R D 1997 *Natural fiber reinforced mortar composites: Experimental characterization* PhD thesis PUC-Rio (Brazil), Department of Civil Engineering

[5] Bessell T J and Mutuli S M 1982 *J. Mater. Sci. Lett.* **1** 244–6

[6] Ku H, Wang H, Pattarachaiyakoop N and Trada M 2011 *Composites Part B* **42** 856-73

[7] Tara S and Jagannatha R 2011 *International Journal of Innovation, Management and Technology* **2** 186-91

[8] Li Y and Shen Y O 2015 *Biofiber reinforcement in composite materials* (Netherland, Wood Head Publishing) 165-210

[9] Alavudeen A, Rajini N, Karthikeyan S, Thiruchitrambalam M and Venkateshwaren N 2015 *Mater. Des.* **66** 246-57
[10] Bennet C, Rajini N, Winowlin Jappes J T, Venkatesh A, Harinarayanan S and Vinothkumar G 2014 *Advanced Material Research* **984-985** 172-77
[11] Almeida J H, Angrizani C C, Botelho E C and Amico S C 2015 *Mater. Des.* **65** 789-95
[12] Mortazavian S and Ali F 2014 *Composites Part B* **72** 116-29
[13] Qi L H, Ma Y Q, Zhou J M, Hou X H and Li H J 2015 *Mater. Sci. Eng. A* **652** 343-49

IOP Conf. Series: Materials Science and Engineering **152** (2016) 012056 doi:10.1088/1757-899X/152/1/012056

Pulsed Nd: YAG laser drilling of aerospace materials (Ti-6Al-4V)

N D Bahar[1*], S Marimuthu[2] and W J Yahya[1]

[1]Department of Mechanical Precision Engineering (MPE), Malaysia – Japan International Institute Technology, Universiti Teknologi Malaysia, Malaysia
[2]Department of Mechanical, Electrical and Manufacturing Engineering, Loughborough University, United Kingdom

*nadiadayana@ymail.com

Abstract. This paper studies the influence of Nd:YAG (neodymium-doped yttrium aluminium garnet) laser process parameters on laser drilled hole quality. Ti-6Al-4V of 1 mm and 3 mm thickness were used as the workpiece substrate. The principal findings are mainly based on minimising the taper angle in laser drilled holes, reducing the heat affected zone and reducing the production of spatter. Identification of key process variables associated with laser drilling process is accomplished by trial experimentation. Using the identified key process variables, further experiments were then performed with the assistance of statistical design of experiment (DOE) to find the interaction and individual effects of various laser process parameters on laser drilled hole quality. The lowest taper angle of 1.8 degrees was achieved with use of nitrogen as the assist gas. Furthermore, from the laser process observations, it was found that laser power significantly affects the quality of the laser drilled hole. Increase in laser power would increase the hole size and result in more spatter on the entry hole surfaces. The nozzle focus position substantially influenced the laser drilled hole size. The amount of spatter deposits increased with decrease in the nozzle offset. Increase in laser frequency significantly increased the exit diameter, which resulted in smaller taper angle. Number of pulse required to drill through a workpiece depends on the material properties and physical properties of the material. For 1mm Ti-6Al-4V, a minimum of two pulses was required to successfully removed the material during drilling and a minimum of 4 pulses was required to drill through the same material with 3mm thickness.

1. Introduction

The use of titanium alloy Ti-6Al-4V has been prominent in aerospace industry due to its outstanding material properties compared to other materials. The main aero components with Ti-6Al-4V are in the gas turbine engines where a large number of small holes (diameter < 1mm) [1] is required to provide cooling to the engine components. In order to produce the holes, Nd: YAG (neodymium-doped yttrium aluminium garnet) laser drilling system has been employed due to its ability to drill a large number of holes within a shorter time (approximately 340,000 holes/min for diameter less than 0.025mm). Nd: YAG laser drilling has been widely established and adopted in the aerospace industry mainly in the production line [2] to produce the cooling holes [3] due to its ability to drilled a large number of fine holes on aerospace components such as turbine engine, nozzle guide vanes, airfoils and combustion chamber [4].

The pulsed Nd:YAG laser system uses an average power of 500W with the capability to deliver enhanced peak powers and superb beam quality, resulting in an excellent quality of material removal during drilling application [2]. Millisecond pulse laser drilling of metals and alloys is used extensively in high value manufacturing to produce holes of size ranging from 0.25 mm to 1 mm and it is a well-established technology for manufacturing various components in aero-engines due to its capability to drill difficult materials at acute angles. The key trend in turbine design is to use increased numbers of complex shaped holes to maximise cooling and subsequently fuel efficiency. In the afterburner of a gas turbine, there are around 40,000 holes with a diameter of 0.5 mm. The increase in the numbers of cooling holes highlights the need for high speed-high quality laser drilling process. There were several types of laser drilling such as single pulse [5], helical [6], percussion [7] and trepanning [8]. Most industrial laser drilling is performed using a percussion or trepan laser drilling process [9], in which a millisecond laser pulse with power density of 106–109 W/cm^2 is being used to produce holes by combination of melting and vaporisation techniques. Percussion drilling process is most often used in high value manufacturing industries due to its ability to produce holes of better quality at high speed. In general, millisecond laser drilling process is also associated with a number of undesirable defects including hole taper, recast layer, heat affected zone, oxide layer and spatter. Taper and heat affected zone formation are considered very critical for rotating components like the turbine blades as it can lead to initiation of fatigue cracks and ultimately, the failure of the laser drilled components.

Prior work on percussion laser drilling has been done on Ti-6Al-4V with thickness of 4 mm and 8 mm by Bandyopadhyay *et al.* [10]. The study revealed that laser drilling defects including geometrical characteristics (hole size, taper angle and aspect ratio) and metallurgical characteristics (heat affected zone, recast layer and micro-cracking) can be minimised by optimising the laser parameters such as average power, pulse energy, pulse duration, pulse frequency and focal positions. Low *et al.* [11] also reported that the amount of spatter depositions can be reduced by having the right combination of laser parameters, i.e. shorter pulse width, low peak power and high pulse frequency. Yilbas [12] reported that the mean diameter depends significantly on the material thickness and focus position. This was agreed by Bandyopadhyay *et al.* [10] and they reported that the hole size decreased with the increase of substrate thickness. Further parameter studies done by Yilbas [12] discovered that the amount of re-solidified material was mainly contributed by inappropriate pulse length and the interaction with the focal position setting. Bandoyadhyay *et al.* [10] used nitrogen as assist gas for drilling of Ti-6Al-4V. The use of assist gas during the drilling process is to prevent the laser optic from being contaminated by the debris ejection. It also helps to remove the material and speed up the breakthrough [13]. Thus choosing the proper assist gas may also result in better drilling and improvement of the hole quality. There are several types of assist gas used in laser drilling process. The typical types of gas jet used are oxygen (O_2), nitrogen (N_2), argon (Ar) and compressed air. The oxygen gas is mainly used for reactive fusion while nitrogen is for high speed and thin sheets, and also to prevent oxidation scales on steel. Argon is mainly used in aerospace application. Compressed air is a low cost gas jet compared to other gas jet used in drilling. The disadvantage of the compressed air is that it increases oxidation and dross. Attentive reminder has also been given on the utilisation of oxygen on metals like titanium as it gets burnt easily [13]. Other laser parameters such as beam focal position contributes to the effect of the final shape and the finishing of the hole [14].

Chien *et al.* [15] investigated the recast layer formation during Nd:YAG laser drilling of Inconel 718 and showed that gas pressure, peak power and focal position had significant influence on the recast layer thickness. They also showed that increasing the pulse energy and reducing the speed can result in reduced recast layer. Wang *et al.* [16] used electrochemical polarisation tests to assess the corrosion performance of the recast layer produced during laser drilling of stainless steel. The result showed that the use of nitrogen as the assist gas provides better corrosion resistance performance, followed by argon and oxygen. Noddin *et al.* [17] proposed a new drilling method to achieve good quality with two different lasers and trepanned motions, in which the second trepanned motion will smooth the edges of first trepanned motion. Lozier *et al.* [18] suggested a new laser drilling technique to achieve high quality, in which second trepanning orbit will be performed in opposite direction to the first one. Yilbas *et al.* [12] reported that the mean diameter can be controlled by material thickness and the position of focus point. Jacobs *et al.* [19] studied the effect of number of trepanning orbit and reported that the increase in the number of orbits can result in increased hole quality.

This paper investigates the effect of laser drilling process parameters on laser drilled hole quality. Experiments were performed on the aerospace material Ti-6Al-4V with thickness of 1 mm and 3 mm. The study took into consideration the main laser parameters including average power, pulse energy, pulse frequency, pulse width, focal position, gas pressure, number of pulse and type of assist gas used. Based on the considered range of parameters input, the angle of taper was calculated. Initial selection was performed based on the lowest tapered angle.

2. Experimental

2.1. Laser drilling experiments

The laser drilling experiments were performed using Nd: YAG solid state laser (class 4) with 550 Watts maximum output power and 1.064μm wavelength. The pulse length was within 0.5 – 20 ms and the maximum repetition rate was up to 500 Hz. The raw beam size was 10 mm focused beam, which resulted in an approximate beam size of 0.5 mm. Figure 1 shows the general construction of an Nd: YAG solid-state laser and the experimental set-up used for this investigation.

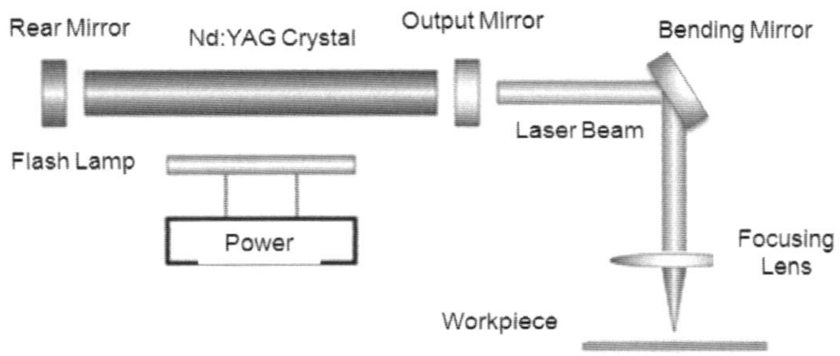

Figure 1. General construction of Nd: YAG solid-state laser used for laser drilling

The drilling process has been done in percussion drilling mode at three different focal positions: focused on the surface, and defocused in and out of the work piece with distance from the nozzle tips to the work piece of 2.829 mm, 2.329 mm and 3.329 mm, respectively. This drilling process was a photo thermal process that can result in thermal defects such as heat affected zone (HAZ), recast layer and taper angle. Ti-6Al-4V was used for the experiments due to its wide use in aerospace applications. Initial experiments were carried out prior to full experiments, which involved seven laser parameters (pulse width, pulse frequency, pulse energy/power, type of assist gas, gas pressure, focal position and number of pulse). This was done to ensure that only appropriate range of parameters was used and simultaneously reduced the time and energy consumption due to arbitrary parameter setting. The initial experiments mainly involved pulse frequency, pulse width and power. The parameters were initially set from low to high to identify the lowest range for the Ti-6Al-4V to be drilled through. The process has been iterated with the pulse width and power. Appropriate laser power had to be identified in order to achieve the desired material removal rate [20]. After the initial laser parameters ranges had been identified, combination of all laser parameters was then structured using statistical tools. Combination of seven parameters resulting 150 runs of experiments. In the 150 runs of experiments, four types of assist gas were evaluated including nitrogen (N_2), oxygen (O_2), argon (Ar) and compressed air. The laser drilling was set at peak power within 3 kW to 7.33 kW. The effect to the hole size on the entry and exit was evaluated and simultaneously, the taper angle of the holes were calculated. In order to have a deep understanding on the effect of number of pulse in the laser drilling process, two different thicknesses of same material (Ti-6Al-4V) had been examined during the experiments to obtain number of pulses required to enable the substrate to be drilled through. Different range of gas pressure from 2 bars to 6 bars was used during the experiment to study the effect to the material removal. Pulse width and pulse frequency were also varied during the experiments. Optimum laser parameter for Ti-6Al-4V of 1 mm thickness was mainly selected based on the hole taper angle. The combination of best laser parameters was obtained through the Design of Experiment (DOE).

Further analysis on hole characterization including heat affected zone and recast layer was executed by studying the significant difference in terms of hole size, spatter pattern and taper angle. The sample preparation was carried out by sectioning the sample and cutting through the middle of the hole. The sample was then mounted for its surface to be grinded and it was finished by finely polished 200μm diamond paste. After that, the sample was etched with Kellar etching agent to reveal the heat affected zone and recast layer microstructure when viewed under the microscope. The observation of the heat affected zone and the recast layer was performed using the microscope up to 100x Nikon lens and also Scanning Electron Microscope (SEM). It is known that laser drilling process will contribute to the range of severity effects such as holes tapering, spatter, recast layer, micro cracking and dross with different used of laser parameter. Taper angle is calculated by using Equation 1 [21]. This is vital as the optimum laser parameters are mainly selected by the corresponding lowest taper angle obtained. The range of taper mainly depends on the entry and exit diameters. Small difference between the entry and exit hole size leads to a lower taper angle, and vice versa.

$$Taper\ (\theta) = Tan^{-1}\left[\frac{d_{entry}-d_{exit}}{2t}\right] \tag{1}$$

where (θ) is the taper angle, d is the hole diameter and t is the material thickness. The other vital parameter in hole quality is the circularity of the hole, which can be computed by using Equation 2. This is performed by taking the ratio of the maximum and the minimum diameters of the same hole either the entry or the exit.

$$C_1 = \frac{D_{min}}{D_{max}} \tag{2}$$

where C_1 is the hole circularity and D is the hole diameter.

2.2. Laser parameters and ranges

The experiment was carried out with respect to the laser parameters within the stated range shown in Table 1. The range of laser parameters given was selected based on the initial experiments. The initial experiments were performed to ensure that only appropriate range of laser parameters was used for the laser drilling studies. Identified laser parameters were then simplified by DoE through combination of all laser parameters, which resulted with 150 numbers of runs. In the 150 runs, it consisted of different combinations of laser parameters within their identified range.

Table 1. Range of laser parameters used for the experiments

Laser parameters	Range
Pulse energy (J)	6.0 – 11.0
Number of pulse	2.0 – 4.0
Pulse width (ms)	1.5 – 2.0
Pulse frequency (Hz)	4.0 – 6.0
Gas pressure (bar)	2.0 – 5.0
Focal position	- 0.5 – 0.5
Type of assist gas*	2.0 – 4.0

*1.0 – 5.0 represents no gas, nitrogen, argon, oxygen and compressed gas, respectively.

3. Results and discussions

3.1. Effect of number of pulse (NOP)

The number of pulse used for drilling imposed significant effect on the drilling quality. Figure 2 shows the effect of the number of pulse to the drilling depth. 1 mm and 3 mm thickness of Ti-6Al-4V were drilled with similar other laser parameters except for number of pulse, resulting in a minimum of two pulses required to drill through 1 mm thickness of Ti-6Al-4V and a minimum of four pulses to drill through 3 mm thickness of the same material.

3.2. Effect of pulse width and pulse frequency

Range of pulse width from 1.5 ms to 2.0 ms was used during the experiments. It was observed that there were significant effects on the entry and exit size of the hole. Initially, when pulse width utilised was too low, the laser drilling process unsuccessfully removed the material. The diameters for both entry and exit started to stabilise approximately at pulse length of 1.5 ms. Han [20] reported that the presence of the laser supported absorption (LSA) wave could occur from the exceeding long pulse width. This phenomenon caused the hole to be unsuccessfully drilled through. Pulse width is always kept as low as possible in order to produce high peak power. Increase in pulse frequency significantly increased the exit diameters of the laser drilling holes, which simultaneously improved the taper angle as shown in Figure 3. Han [20] reported that higher frequency will result in straighter sidewall profile because of less melted material re-solidified and more molten material is pushed in the laser beam direction. However, the effect of pulse frequency may not be significant to all workpiece thickness, Ghoreishi *et al.* [22] reported hole entrance diameter of stainless steel was found significantly affected by the pulse energy while not in drilling the mild steel.

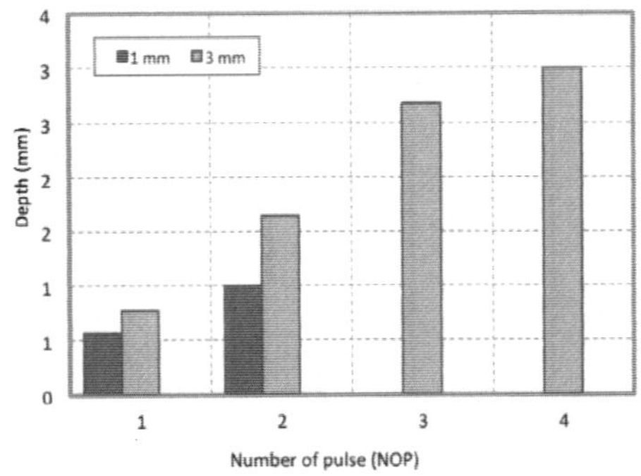

Figure 2. Number of pulses required to drill 1mm and 3 mm of Ti-6Al-4V, f = 4 Hz, fpp =-0.5, peak power = 4 kW, gas pressure = 3 bars, assist gas = nitrogen

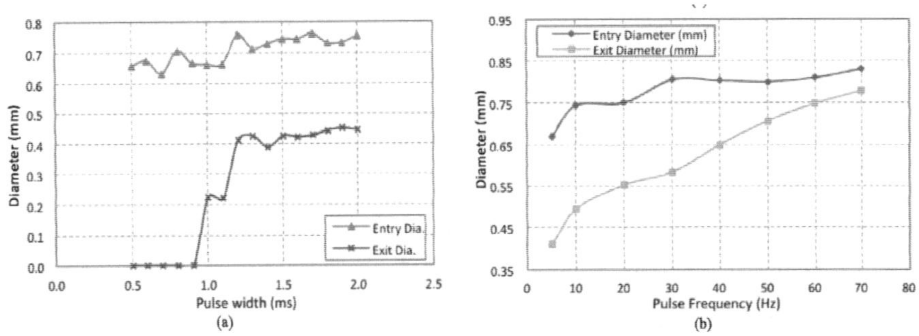

Figure 3. Entry and exit diameter affected by (a) various pulse width: f = 5 Hz, fpp = 0, gas pressure = 6 bars, assist gas = argon and (b) various pulse frequencies: pulse width = 1ms, fpp = 0, gas pressure = 6 bars, assist gas = argon

3.3. Effect of assist gas

Four types of assist gas including nitrogen (N_2), argon (Ar), oxygen (O_2) and compressed air were used in the experiments with remaining similar laser parameters. The taper angle was determined by using Equation 1. Figure 4 illustrates that compressed air produced the lowest taper angle with 8.11 degrees, followed by the nitrogen with 9.79 degrees. Argon and oxygen produced an even higher taper angle, within 10 to 11 degrees, respectively. The entry hole sizes were obtained within 700µm to 800µm for all assist gases used except for the oxygen, which reached up to 2.5 mm. The material removed was observed to be not in circular shape. Figure 5 illustrates the shape of the Ti-6Al-4V, which was drilled using oxygen. The substrate was severely destroyed due to the over burnt during the drilling process. Dahotre *et al.* [13] study showed that use of oxygen as assist gas in laser machining produced the best quality holes in iron, nickel and their alloys as it assists burning the metal away. However, attentive reminder has been given when dealing with metals like titanium because it can burn easily. Nitrogen is the second assist gas, which produced the second lowest taper angle after compressed air. Further investigation and study was carried by changing the laser parameters in order to minimise the taper angle currently produced.

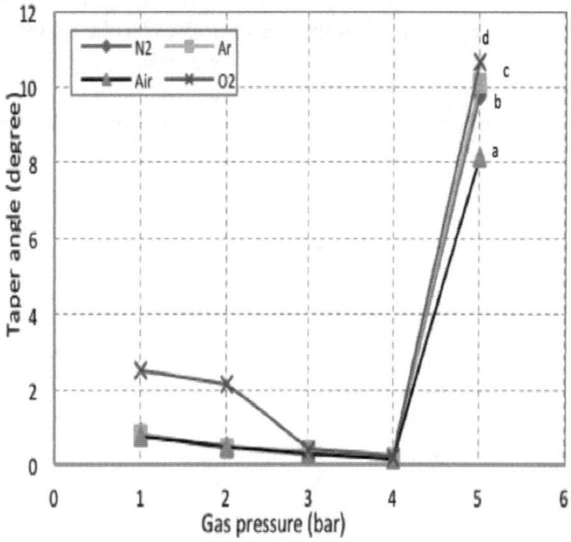

Figure 4. Taper angle at different gas pressures: (a) compressed air, (b) nitrogen, (c) argon and (d) oxygen, with f = 4 Hz, fpp = -0.5, peak power = 4kW, gas pressure = 3 bars

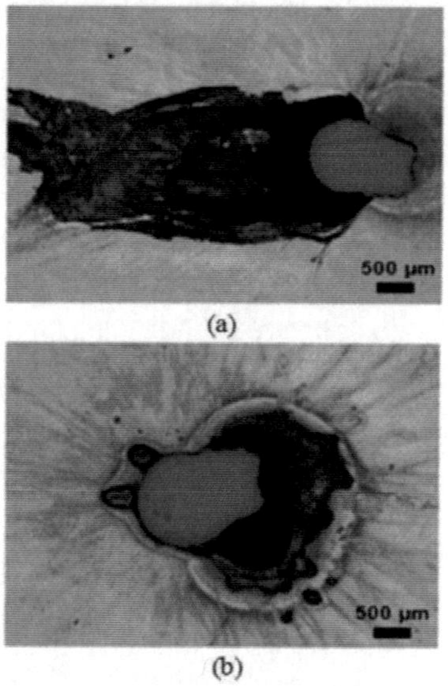

Figure 5. The hole drilled by the oxygen assist gas: (a) entry diameter and (b) exit diameter

3.4. Effect of gas pressure

The effect of gas pressure was evaluated from a minimum of 2 bars to the maximum of 6 bars. The investigation showed that setting appropriate gas pressure helped to achieve a better hole quality. As noticed from Figure 6, insufficient gas pressure led to the failure in removing the material. However, excessive gas pressure can also result in increased hole diameters, mainly at the entrance and hence increasing the taper. Higher assist gas also reduced the recast layer but can also result in increased material distortion [22].

3.5. Effect of the focal positions

Experiments were performed at three different focal positions (at the surface, above the surface and also below the surface). This was performed by adjusting the vertical distance between the nozzle and the work piece. The defocused positions were mainly focusing in the middle of the substrate and far above the substrate. The distance from the tips of the nozzle to the work piece was as stipulated below:

- The zero focal position was the condition where the laser was focussed on the surface of the workpiece and the distance of the nozzle tip to workpiece was 2.829 mm with the uncertainty of ± 0.002 mm
- 0.5 focal position was the condition where the laser beam was zoomed out from the focussed surface position and the distance of the nozzle tip to workpiece was increased to 3.329 mm with the uncertainty of ± 0.003 mm
- -0.5 focal position was the condition where the laser beam was focused at the middle of the workpiece and the distance of the nozzle tip to workpiece was 2.329 mm with the uncertainty of ± 0.013 mm

As shown in Figure 7, focal position maintained on the surface produced the smallest entrance and exit hole size, and also produced lower taper angle compared to the other beam positions. Increase in the nozzle offset resulted in high taper angle. However, the amount of spatter deposition was observed to be in opposite trend. The amount of spatter deposition on the workpiece was found to increase with the offset distance.

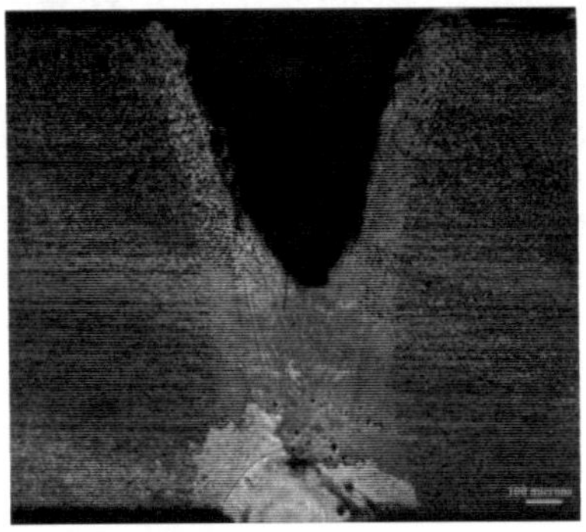

Figure 6. Ti-6Al-4V 1 mm thickness drilled with 2 bars gas pressure, f = 4 Hz, fpp = -0.5, gas pressure = 3 bars, assist gas = nitrogen, nop = 2, peak power = 4kW

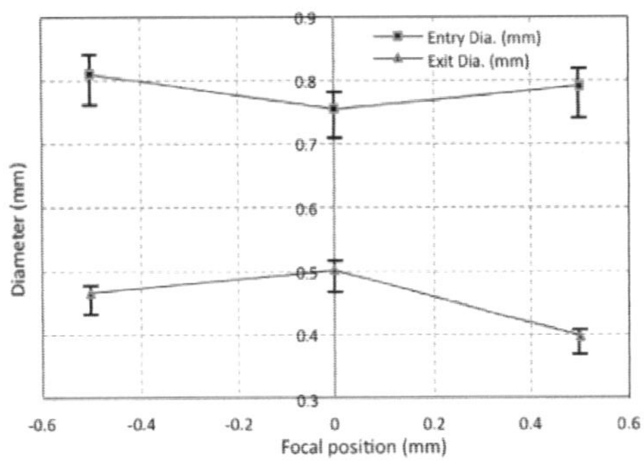

Figure 7. Entry and exit diameters at different focal positions, f = 4 Hz, gas pressure = 3 bars, assist gas = nitrogen, nop = 2, peak power = 4kW

3.6. Effect of average power

Three different average power settings, ranging from 24 W, 44 W and 66 W, were investigated. Both entry and exit hole size showed significant effect when the power was increased from 24 W to 66 W. The entry hole diameter was about 790μm at 24 W power and increased to approximately 910μm at 66 W. Similarly, the exit hole was increased from 400μm at 24 W to roughly 750μm at 66 W. However, the difference between the entry and exit hole diameters was smaller at the highest power level and hence resulting in lower taper angle. The other defects including spatter deposition on the surface were increased with laser power. The patterns of deposition were observed to be distinctly different from each other. Figure 8 shows the hole drilled at 24 W, 44 W and 66 W. Deposition of spatter has been analysed and observed to be increasing with laser power. At higher power, the spatter was observed to be accumulated around the entry hole. In severe cases, depending on the substrate properties, removal of overlapped spatter from drilling with inert gas would have to compromise with undesired substrate surface and hole geometry modification [11]. Ng *et al.* [23] found less spatter when holes were drilled with low laser energy and high power. Similar phenomenal was also observed in Figure 8, where the hole drilled with 24 W of power shows less spatter deposition and vice versa.

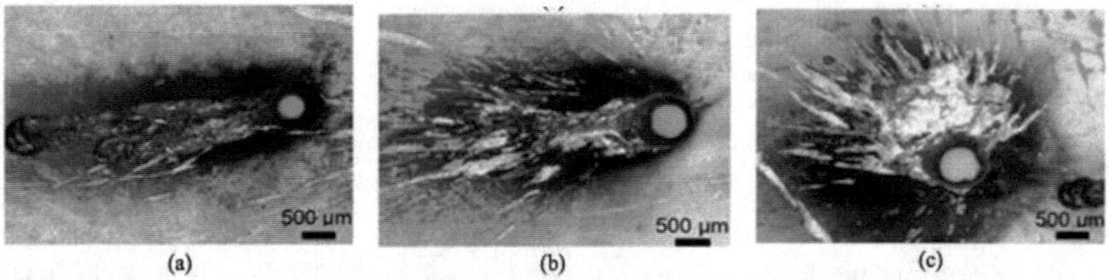

Figure 8. The deposition of spatter at different ranges of power: (a) 24 Watts, (b) 44 Watts and (c) 66 Watts, f = 4 Hz for 24 W and 44 W and f = 6 Hz for 66 W, fpp = 0.5, gas pressure = 3 bars, assist gas = nitrogen, nop = 2 for 24 W and 66 W and nop = 4 for 66 W

In all experiments, the diameter of hole was larger at the entrance than at the exit. As can be seen from Figure 9(i) for 66 Watts laser power, the top and bottom diameters were 850µm and 460µm, respectively. The entrance diameter for 24 Watts and 44 Watts indicated by (a) and (b), respectively, in Figure 9(i) were obtained at approximately 763µm. HAZ along the hole depth showed an increasing trend with increased power. The HAZ observed with 24 Watts, 44 Watts and 66 Watts is illustrated in Figure 9(ii). HAZ were found to be smaller in 24 Watts power and increased with laser power. The diameter ranges only started to vary when it passed 200µm in depth with approximately 100µm apart. Distinctive shape of hole also was observed for 66 Watts power where the hole diameter diverged back from approximately at 600µm depth to approximately 37µm. This is because, in Figure 9(i)(c), higher frequency was used compared to (a) and (b), thus increased the exit diameter. This improved the taper angle due to the effect of pulse frequency. Han *et al.* [20] reported the increased laser peak power had less effect on the entrance hole size than the exit diameter.

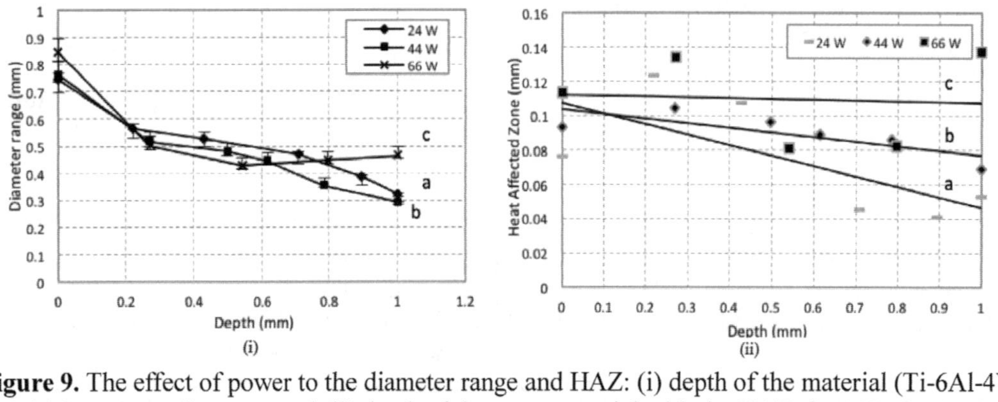

Figure 9. The effect of power to the diameter range and HAZ: (i) depth of the material (Ti-6Al-4V, 1mm thickness) vs. diameters and (ii) depth of the same material with the HAZ, f = 4 Hz for 24 W and 44 W and f = 6 Hz for 66 W, fpp = 0.5, gas pressure = 3 bars, assist gas = nitrogen, nop = 2 for 24 W and 66 W and nop = 4 for 66 W

3.7. Optimum laser drilling parameters

Figure 10 illustrates the hole drilled with the optimum laser parameters. The entry and exit hole size of 752µm and 689µm resulted in the lowest taper angle of 1.8 degrees. The hole was drilled with nitrogen gas with pressure set to 3 bars. Laser power was set to 66 W with respectively 11 Joule of pulse energy and 6 Hz repetition rate. Additional two pulses from the actual requirement in number of pulse for the substrate to be drilled through, making the total of four pulses utilised during the laser drilling process.

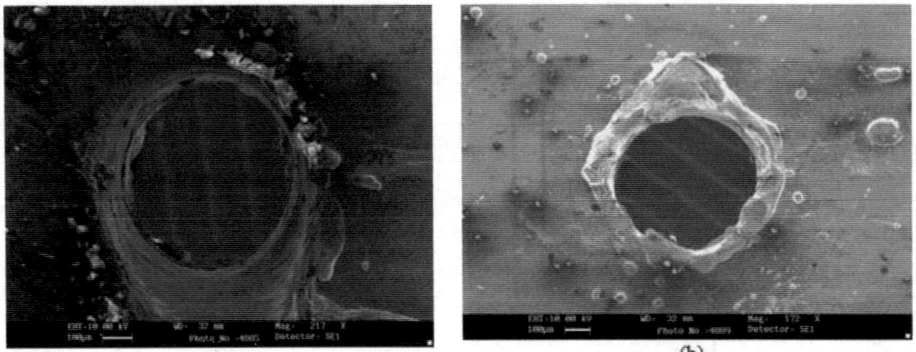

(a) (b)

Figure 10. Ti-6Al-4V, 1mm thickness holes obtained by the optimum laser parameters taken by SEM: (a) entry hole and (b) exit hole, with f = 4 Hz, fpp = -0.5, gas pressure = 3 bars, peak power = 5.5kW, nop = 4, assist gas = nitrogen

4. Conclusions

At the optimal parameters setting, laser drilling of 1mm thickness Ti-6Al-4V produced a taper angle of 1.8 degrees. Higher power led to larger hole size but the difference between entry and exit diameters were reduced, thus decreased the taper angle compared to that obtained from lower power. The amount of spatter deposition increased with increased of power. Heat affected zone (HAZ) was also found to increase with increased power. Percussion drilling with low pulse width resulted in better drill quality but high pulse frequency resulting straighter hole geometry and subsequently lower taper angle. The use of nitrogen and compressed air as assist gas produced low taper angle whereas oxygen was not the best assist gas to be used for drilling Titanium Ti-6Al-4V 1mm thickness. Furthermore, it was found that the amount of spatter deposition increased with increase in nozzle offset. Sufficient gas pressure was vital in the laser drilling process to ensure that the vaporised material was successfully removed. However, excessive gas pressure might lead to larger hole size.

Acknowledgement

This research was fully supported by the Mechanical, Electrical and Manufacturing Engineering Department, Loughborough University. We thank our colleagues Dave Britton and Ray Owens from Department of Optical Engineering and Laser and Material Department Laboratory, whom had provided insight and expertise that has greatly assisted from the beginning to the end of this research. Also not forgotten the laboratory supports given by the staff on the equipment handling and operation.

References

[1] Naeem M, Lane C and Valley S 2008 *Proceedings of PICALEO*
[2] Bandyopadhyay S, Gokhale H, Sarin Sundar J K, Sundararajan G and Joshi S V 2005 *Optics and Lasers in Engineering* **43** 163-82
[3] Yeo C, Tam S, Jana S and Lau M W 1994 *Journal of Materials Processing Technology* **42** 15-49
[4] Bandyopadhyay S, Sarin Sundar J K, Sundararajan G and Joshi S V 2002 *Journal of Materials Processing Technology* 127 83-95
[5] Voisey K T and Clyne T W 2004 *Surface and Coatings Technology* **176** 296-306
[6] Zabel A and Heilmann M 2012 *CIRP Annals - Manufacturing Technology* **61** 111-4
[7] Okasha M M, Mativenga P T, Driver N and Li L 2010 *CIRP Annals - Manufacturing Technology* **59** 199-202
[8] Sezer H K, Li L, Schmidt M, Pinkerton A J, Anderson B and Williams P 2006 *International Journal of Machine Tools and Manufacture* **46** 1972-8

[9] Emer G, Edlund D, Vanasse S and Dizzine M L 1998 *Laser drilling holes in components by combined percussion and trepan drilling* Google Patents
[10] Bandyopadhyay S, Sundar J S, Sundararajan G and Joshi S 2002 *Journal of Materials Processing Technology* **127** 83-95
[11] Low D, Li L and Corfe A 2000 *Applied Surface Science* **154** 689-95
[12] Yilbas B 1997 *Journal of Materials Processing Technology* **70** 264-73
[13] Dahotre N B and Harimkar S 2008 *Laser Fabrication and Machining of Materials* (New York: Springer US)
[14] Dhar S, Saini N and Purohit R 2006 *Proceedings of the International Conference of Advances in Mechanical Engineering*
[15] Chien W-T and Hou S-C 2007 *The International Journal of Advanced Manufacturing Technology* **33** 308-16
[16] Wang X, Ng G, Liu Z, Li L, Skeldon P and Bradley L 2003 *Journal of Corrosion Science & Engineering* **6**
[17] Noddin D B 1999 *Method of sequential laser processing to efficiently manufacture modules requiring large volumetric density material removal for micro-via formation* Google Patents
[18] Lozier T B and Kelley JG 1993 *Method and apparatus for laser trepanning* Google Patents
[19] Jacobs P 2008 *Precision Trepanning with a Fiber Laser* (Smithfield: LFI Inc.)
[20] Han W 2004 *Computational and experimental investigations of laser drilling and welding for microelectronic packaging* (Cambridge: Draper Laboratory)
[21] Jiao L, Ng E, Wee L and Zheng H 2011 *Applied Physics A* **104** 1081-4
[22] Ghoreishi M, Low D and Li L 2002 *International Journal of Machine Tools and Manufacture* **42** 985-95
[23] Ng G and Li L 2001 *Optics & Laser Technology* **33** 393-402

Three dimensional numerical investigations on the heat transfer enhancement in a triangular facing step channels using nanofluid

K A Mohammed[1*], A R Abu Talib[1,3], N Abdul Aziz[4] and K A Ahmed[1]

[1]Department of Aerospace Engineering, Universiti Putra Malaysia, Malaysia
[2]Department of Mechanical Engineering, College of Engineering, Universiti of Anbar, Anbar, Iraq
[3]Aerospace Manufacturing Research Centre, Universiti Putra Malaysia, Malaysia
[4]Department of Mechanical and Manufacturing Engineering, Universiti Putra Malaysia, Malaysia

*kafelazeez@yahoo.com

Abstract. In this paper, laminar flow for the distilled and SiO_2–water nanofluid flow and heat transfer were numerically investigated in three-dimensional triangular facing-step channel. The nanoparticle volume fraction and Reynolds number considered are in the range of 0–1% and 100–1500, respectively. Numerical solutions are obtained by using finite difference method to solve the governing equations. The effects of the volume fraction of nanoparticle, triangular facing-step channel amplitude height, wavelength and Reynolds number on local skin-friction coefficient, average Nusselt number and enhancement of heat transfer are presented and discussed. The results show that the Nusselt number and friction coefficient increases as the amplitude height of triangle channel increases. As the nanoparticle volume fraction increases, the Nusselt number is also found to be significantly increased, accompanied by only a slight increase in the friction coefficient. In addition, it is found that the heat transfer enhancement mainly depends on the amplitude height of the triangle wall, nanoparticle volume fraction and Reynolds number rather than the wavelength.

1. Introduction

Heat transfer improvement in the heat exchanger is essential in developing a compact one that has high thermal efficiency, light weight, small size and low manufacturing cost. It is known that deployment of corrugations walls can potentially lead to a better performance in heat transfer enhancement due to the mixing of the fluid by the corrugations surfaces. Investigation on the method for improvement of heat transfer in these devices has become necessary. In recent years, it is revealed that thermal conductivity of conventional heat transfer fluids such as oil, water, and ethylene glycol can be improved by adding nanoparticles. Aung [1] performed experimental work of two-dimensional (2D) laminar flow through a downstream facing-step. It was observed that a small fluctuation occurred at low values of Reynolds number (*Re*). The observed fluctuation was the onset of transition to the separated boundary layers. In addition, there was also a tendency towards a secondary separation at the upper wall due to velocity inflection profiles with high values of *Re*. Kherbeet *et al.* [2] reported their experimental heat transfer results of laminar flow of air passing through a facing-step channel with backward under constant wall temperature. It was concluded that the heat transfer under laminar flow enhances in the direction of stream wise as compared to smooth channel and the optimum heat transfer is at the re-attachment point downstream. The velocity of separated airflow under laminar three-dimensional (3D) with a facing-step has been measured by Armaly *et al.* [3]. It was found that it has no recirculation flow region near the wall for low *Re*. On the other hand, a region of small recirculation flow appears at the upper corner of the sidewall for high *Re*. Nie and Armaly [4] measured the velocity near walls for 3D backward facing step flow. A comparison of the 2D at the downstream wall has been conducted for 3D at the channel center. Results of 3D were significantly lower for transition flow regime but slightly lower at regimes of turbulent and laminar flow. Hattori and Nagano [5] have studied boundary layer turbulent flow through a forward facing step. The separation regions have occurred on the step and in front of the forward facing step flow.

Many experimental studies of convective heat transfer of conventional fluid in corrugated channels at laminar flow condition have been carried out in the past [6, 7, 8, 9]. Rushet *et al.*[10] experimentally performed the study on the heat transfer and flow characteristics under sinusoidal wavy laminar flows through a channel. Water tunnel was used to study the flow field using visualization methods while wind tunnel has been used to conduct heat transfer experiments in the range of *Re* from 100 to 1000. It was observed that the geometry of the channel and *Re* have directly affected the location of the mixing onset. Heat transfer has significantly increased with onset of macroscopic mixing. The performance of the corrugated channel of 5 mm height spacing is better than that of the 10 mm height spacing. The studies on convective heat transfer through corrugated channels under turbulent flow conditions have been carried out by a number of researchers [11, 12, 13]. Oyakawa *et al.* [14] experimentally studied the influence of the channel height on the heat transfer improvement and pressure drop in a sinusoidal-wavy channel over the range of *Re* of 4000-200000. The wavelength and amplitude of wavy channel were 80 mm and 20 mm, respectively. Their results showed that the optimal performance of the wavy channel could be achieved when the ratio of wavelength to channel height was two. Additionally, the separation point of the flow was observed to be highly dependent on the channel height, while the reattachment point of the flow was independent of it. The flow characteristics and heat transfer of air flow in the corrugated channels under the conditions of constant surface temperature and *Re* ranging from 3220 to 9420 were studied experimentally by Elshafei *et al.* [15, 16]. The results indicated the influence of channel spacing variations on the heat transfer and the friction factor was dependent than the phase shift variation especially at high *Re*.

Nanofluid is metallic or non-metallic nanoparticles with a size of less than 100 nm dispersing in the liquids. A novel heat transfer fluid, prepared by suspending nanopowders in conventional heat transfer fluid to obtain higher thermal conductivity, is called as nanofluid [17]. Many attempts were conducted to suspend metal particles in conventional fluid as reported by Maxwel [18] and others but the main challenge was the large size of particles and weight. Nanofluids, on the other hand, have small sized nanoparticles with less sedimentation of particles suspension.

It has been acknowledged that efficiency of heat exchanger is still subject to further investigation. Therefore, this study considers new configurations that have not been investigated before with the aim of getting a better enhancement in heat exchanger. Furthermore, the study involves establishing the effects of variation in geometrical parameters such as amplitude height, wave length, Reynolds number and nanofluid on the Nusselt number, friction factor and pressure drop. Overall, the objectives are to enhance heat transfer using corrugated surface in back ward facing step model, investigate the effect of pressure drop and friction factor of flow region in corrugated facing step channel, and assess the effect of nanofluids with corrugated facing step channel.

2. Mathematical model

2.1. Physical domain definition

The geometry of the shape under investigation in the present problem is shown in Figure 1. It consists of a lower triangular corrugated plate with wavelength, L_w and amplitude height, *a*. The triangular wall is isothermally heated to 24000 W/m^2. The triangular wall's total length is 20 cm, i.e. there are 2, 4, 5, and 10 cm waves along the triangular wall. For creating appropriate boundary conditions for both inlet and outlet of the triangular wall, two smooth adiabatic sections are considered: one is before the triangular wall section and the other is after it. The length of each smooth section is 70 cm and 20 cm, respectively, and the height of step is 1 cm. For this study, the following geometric dimensionless parameters are applied: triangular wavelength, L_w of 0, 2, 4, 5 and 10 cm, and triangular amplitude height, a of 0, 0.1, 0.2,0.3and 0.4 cm.

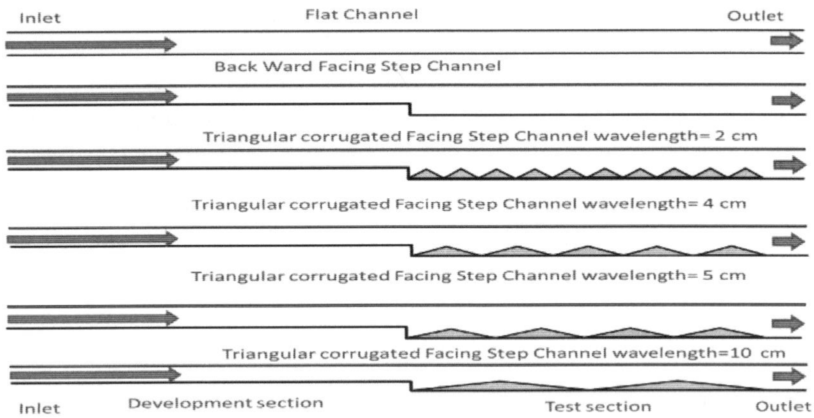

Figure 1. Schematic of the computational domain of flat channel, backward facing-step, triangular with wavelengths of 2, 4, 5 and 10 cm

2.2. Governing equations and problem assumptions in present study
The nanofluid is taken as Newtonian fluid and the flow is considered to be laminar, incompressible, steady and three-dimensional. Moreover, it is presumed that the mixture of solid particles and water enters the channel at similar temperature and velocity. Based on the assumptions, the non dimensional governing stream function, energy and vorticity transport equations for the mixture of nanoparticles and liquid are modelled and presented as (ANSYS Fluent, 14.5).

Continuity equation:
$$\nabla \cdot (\rho \vec{V}) = 0 \tag{1}$$

Momentum equation:
$$\nabla \cdot \left(\vec{V}\vec{V}\right) = -\frac{1}{\rho}\nabla p + v\nabla^2 \vec{V} \tag{2}$$

Energy equation:
$$\nabla \cdot \left(\vec{V}T\right) = \frac{1}{\rho_{\mathrm{nf}} c_{p_{\mathrm{nf}}}} k_{\mathrm{nf}} \nabla^2 T \tag{3}$$

where ρ is the density (kg/m^3), V is the velocity (m/s), p is the pressure (Pa), v is dynamic viscosity (kg/m.s), c_p is specific heat capacity (J/kg.K), k is thermal conductivity (W/m.K), T is temperature (K) and nf is nanofluid.

2.3. Thermophysical properties of nanofluids
These properties are calculated using the following equations. For effective thermal conductivity [19]:

$$k_{eff=}k_{static} + k_{Brownian} \tag{4}$$

$$k_{static} = k_{bf}\left[\frac{\left(k_{np}+2k_{bf}\right)-2\phi\left(k_{bf}+2k_{np}\right)}{\left(k_{np}+2k_{bf}\right)+\phi\left(k_{bf}+2k_{np}\right)}\right] \tag{5}$$

where k_{bf} and k_{np} are the base fluid and the thermal conductivity of the nanoparticle, respectively. The Brownian motion thermal conductivity is presented by Vajjha and Das [20] as:

$$k_{\text{Brownian}} = 5\times 10^{4}\emptyset\beta\rho_{\text{bf}}C_{\text{p,bf}}\sqrt{\frac{KT}{2\rho_{\text{np}}R_{\text{np}}}}f(T,\emptyset) \tag{6}$$

where K is the Boltzmann constant, T is the fluid temperature and T_0 is the reference temperature. The term of $f(T,\emptyset)$ is a function of the particle volume fraction and temperature. The correlation of β is a function of the liquid volume that departures with a particle material, expressed in Table 1 as given by Vajjha and Das [21].The effective dynamic viscosity is given by Corcione [22]:

$$\frac{\mu_{eff}}{\mu_{bf}} = \frac{1}{1-34.8\left(\frac{d_{np}}{d_{bf}}\right)^{-0.3}\phi^{1.03}} \tag{7}$$

where $d_{bf} = \left[6M/N\pi\rho_{bf}\right]^{1/3}$ μ_{eff} and μ_{bf} are the effective dynamic viscosity of nanofluid and base fluid respectively, d_{np} is the nanoparticle diameter, d_{bf} and ϕ are the base fluid equivalent diameter and the nanoparticle volume fraction, respectively. M is the base fluid molecular weight.

Table 1. The value of β particles with its boundary conditions [21]

Type of particles	β	Concentration	Temperature
SiO_2	$1.9526(100\phi)^{-1.4594}$	$1\% \le \phi \le 10\%$	$298\text{ K} \le T \le 363\text{ K}$

The effective density is given as [22]:

$$\rho_{eff} = \left(1-\phi\right)\rho_{bf} + \phi\rho_s \tag{8}$$

where ρ_{bf} and ρ_{eff} are the base fluid and nanofluid densities, respectively, and ρ_s is the density of the nanoparticle. The effective nanofluid's specific heat at constant pressure, $(C_p)_{eff}$ is calculated using Equation 9 [22].

$$\left(Cp\right)_{eff} = \frac{(1-\phi)\left(\rho Cp\right)_{bf} + \phi\left(\rho Cp\right)_s}{\left(1-\phi\right)\rho_{bf} + \phi\rho_s} \tag{9}$$

where Cp_s is the heat capacity of the solid particles and Cp_{bf} is the heat capacity of the base fluid. Following [23], *Nu*, *Re* and friction factor are dimensionless parameters that can be calculated as follows.

$$Nu = \frac{hD_h}{k} \tag{10}$$

where h and k are the average heat transfer coefficient of fluid and thermal conductivity, respectively. The Reynolds Number *Re* is expressed as:

$$Re = \frac{\rho u_m D_h}{\mu} \tag{11}$$

where ρ, u_m and μ are density, mean fluid velocity over the cross-section of fluid and dynamic-viscosity, respectively. Meanwhile, the hydraulic diameter (D_h) is defined as in Equation 12.

$$D_h = \frac{4A}{P} \tag{12}$$

where P is the wetted perimeter and A is the area of the cross-section. The friction factor for fully developed flow is expressed as in Equation 13.

$$f = \frac{2\,\Delta p\, D_h}{L\,\rho\, u_m^2} \tag{13}$$

For example, Table 2 lists out the thermophysical properties of distilled water based fluids and SiO_2 nanoparticles [22].

Table 2. Thermophysical properties of distilled water based fluids and SiO_2 nanoparticles at T = 300 K

Thermophysical Properties	SiO_2	Water
ρ (kg/m^3)	2220	996.5
c_p (J/kg·K)	745	4181
k (W/m·K)	1.4	0.613
μ (Ns/m^2)	-	1 x 10^{-3}
β (1/K)	5.5 x 10^{-6}	2.75 x 10^{-4}

3. Numerical analysis

Cartesian coordinate system (x,y) of the non-dimensional form of stream function, vorticity transport and energy equations for the liquid–solid mixture are transformed into non-orthogonal curvilinear coordinate system (f,g). The governing equations with appropriate boundary conditions are discretized using the finite volume method. For the current study, the Poisson equation is solved to develop the computational mesh. The energy equations and vorticity transport are solved using the simple explicit method based on the time marching technique whereas the stream function equation is solved using relaxation method [24, 25]. As the iteration started, the stream function is calculated, and the energy and vorticity transport equations are solved to determine the temperature and vorticity for all nodes in computational domain. The iteration is continued until the sum of absolute residual for all parameters in computational domain is less than 10^{-5}. The constant time step used for all calculations is 10^{-3}.

4. Grid independence test and validation of numerical methods

In order to assess the grid independent result for water at Re=100, expansion ratio of 2, and heat flux of 24000 W/m^2 for triangular corrugated facing step. Five different sizes of grid are tested to check the effect of the grid density on the computational results as shown in Table 3. The results show that the grid size of 3000000 ensures grid-independent test and it is consequently employed during the course of this study.

Table 3. Grid independence test

Grid number	Average Nusselt Number	Relative Error*	% Relative Error
46875	8. 1311889	----------------	------------------------
375000	8. 1460963	1.83*10-3	0.18 %
1736110	8. 1489158	3.46*10-4	0.034 %
3000000	8. 1496036	8.44*10-5	0.008 %
5859375	8. 1501611	6.84*10-5	0.0068 %

*Relative error = $(Nu_{new} - Nu_{previous})/Nu_{new}$

The validation studies are carried out for laminar distilled water and 5% copper-water nanofluid flow in the straight channel. The numerical results are then compared with that of Santra *et al.* [26], which include the inlet temperature of 293K, upper and lower wall preserve at uniform wall temperature condition, height of channel (0.01 mm), channel length and the diameter of nanoparticles (100 nm). The result shows a good agreement as shown in Figure 3.

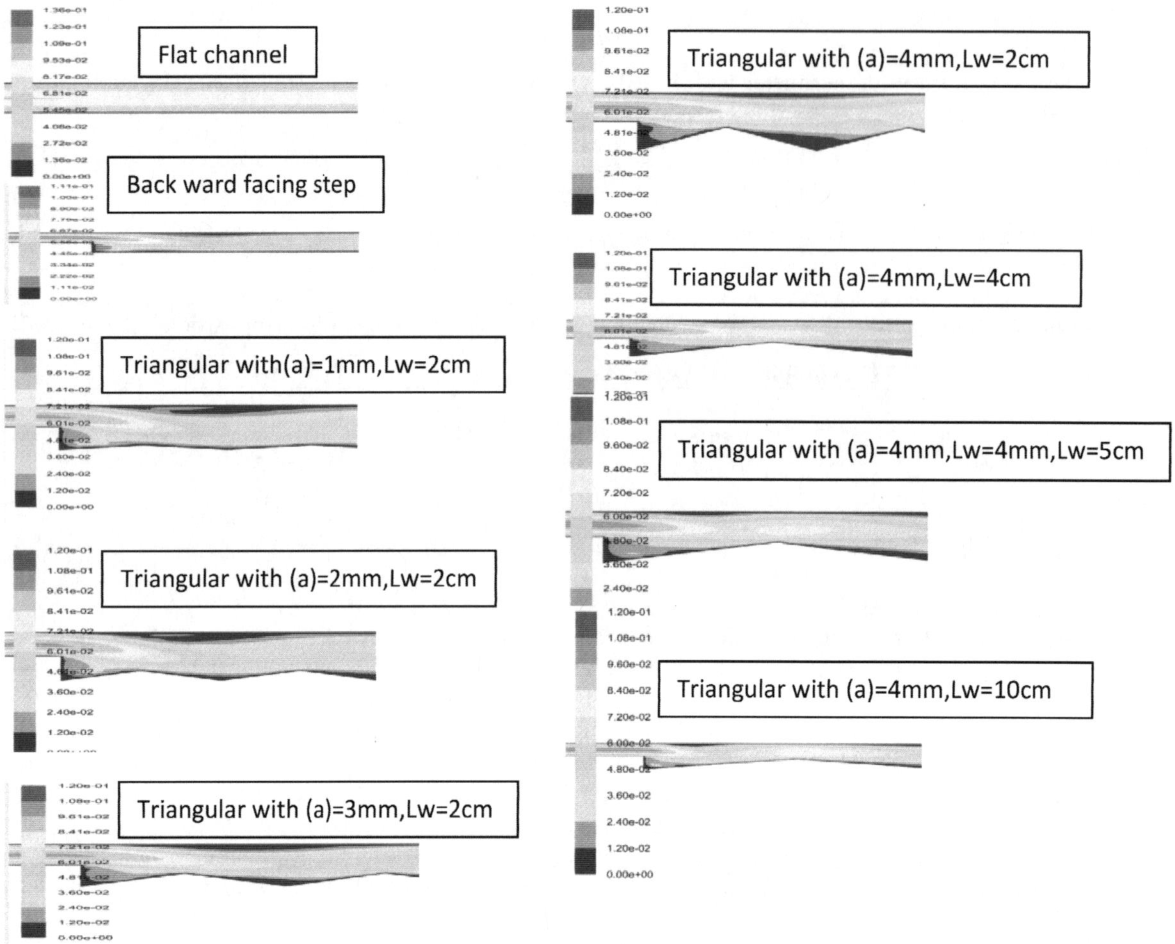

Figure 2. Streamlines (top) and isotherms (bottom) for nine different geometries at *Re* = 750, without amplitude height (*a*), and wave length (L_w)

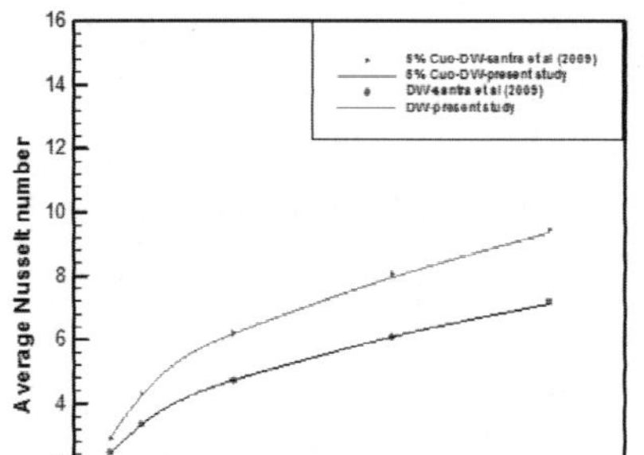

Figure 3: Comparison of the average *Nu* for present study with previous numerical study in [26]

Figure 4 shows the comparison between numerical results of this study and experimental data of Mohammed *et al.* [27]. The average Nusselt number for 1% SiO_2-water nanofluid at L_w = 20 mm and d_p = 30 nm triangular corrugated channel with amplitude length of 1.5 mm is shown to be in a good agreement between the two data with only 3% of error percentage.

5. Results and discussion

Figure 5 shows the effect of channel shape, triangular amplitude height and *Re* on *Nu* in the laminar flow region. It is found that the highest *Nu* is obtained by the triangular with amplitude height of 4mm, followed by triangular with amplitude heights of 3, 2, 1, 0 mm and the lowest *Nu* optioned by flat channel. It can be observed the enhancement of thermal conductivity due to the area increase as the amplitude height for triangular shape rises. Moreover, the effect of different *Re* in the range of 100 to 1500 shows that the Nusselt number increases as Reynolds number increases. Figure 6 shows the effect of the geometry parameters and *Re* on the pressure drop in the laminar flow region. It is clearly shown that by increasing *Re* from 100 to 1500, the pressure dropped significantly from the inlet of the duct to the outlet. It is also found that the lowest pressure drop corresponds to the flat channel while the highest pressure drop is obtained by the triangular having amplitude height of 4 mm, followed by triangular having amplitude heights of 3, 2, 1 and 0 mm. Figure 7 shows the effect of amplitude height and Reynolds number on the friction factor along the triangle corrugated facing step channel. It can be seen that the friction decreases with the increase in Reynolds number. On the other hand, the friction factor increases with increasing amplitude height.

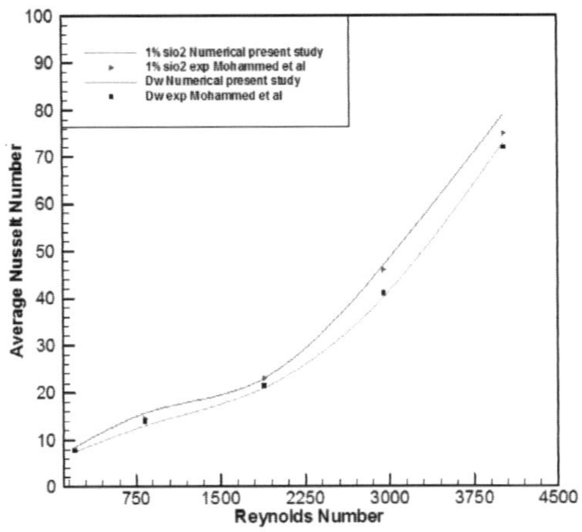

Figure 4. Comparison between numerical present study results and measured data in [27]

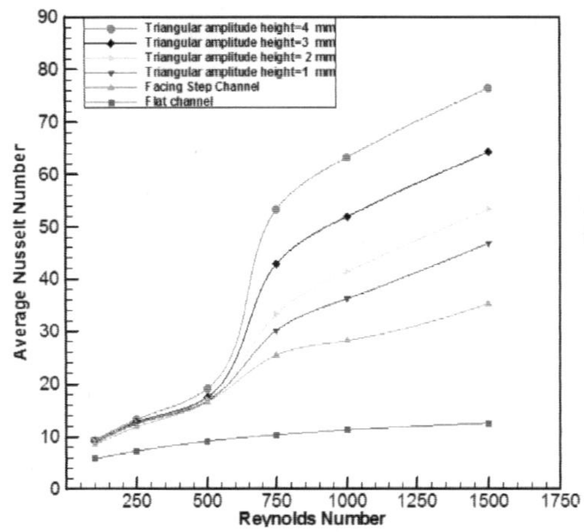

Figure 5. Effect of channel shape and triangular amplitude height to *Nu* in laminar flow region

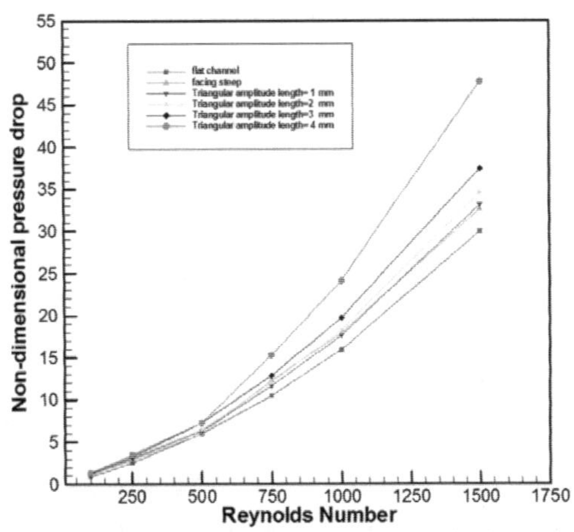

Figure 6. Effect of channel shape and triangular amplitude length to pressure drop in the laminar flow region

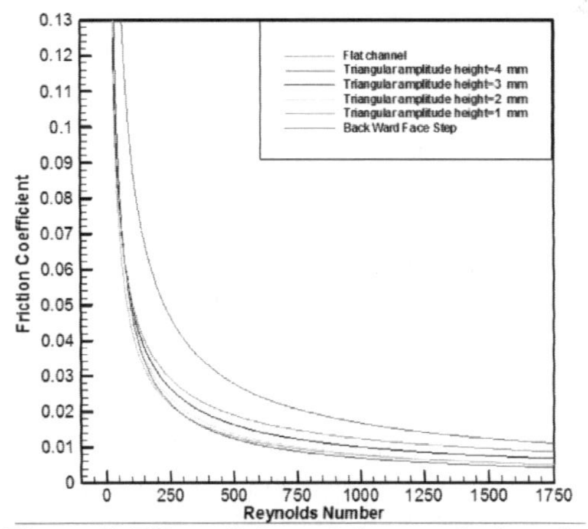

Figure 7. Effect of channel shape and triangular amplitude height to friction factor

Figure 8 shows the effects of wavelength to the Nusselt number for four different wavelengths, L_w from 0, 2,4,5 and 10 cm. The result shows that the wavelength of 2 cm gives a better enhancement and *Nu* decreases with increasing wavelength. Meanwhile, Figure 9 shows the effect of adding 1% SiO_2 nanofluid to the base fluid for three types of channels: flat, backward and triangular facing-step. It can be seen that *Nu* increases with addition of nanoparticle to the base fluid and the triangular channel with amplitude height of 4 mm gives better enhancement.

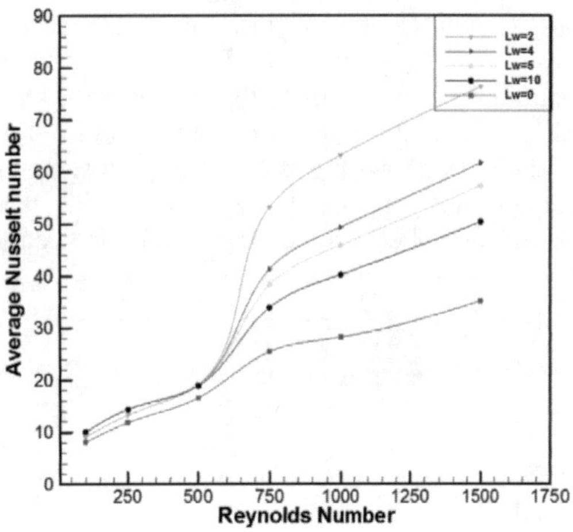

Figure 8. The effect of different wavelength, L_w with *Re* to *Nu* in laminar flow region

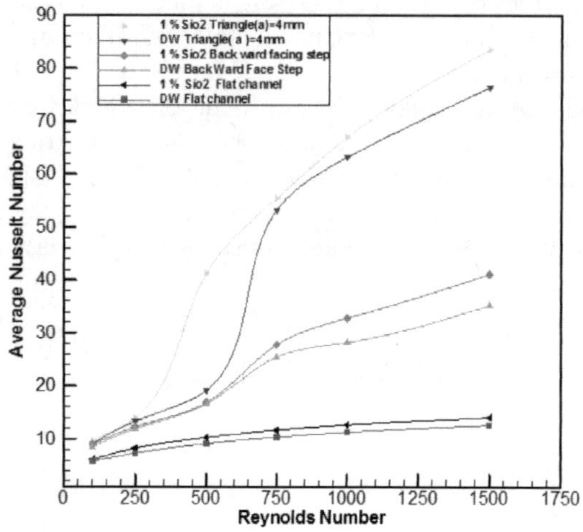

Figure 9. Effect of added 1% SiO_2 nanofluid to *Nu* for different shapes of channel, triangular amplitude length 4 mm in laminar flow region

6. Conclusion

In the current study, three dimensional in three types of channels are examined using finite difference method for a range of Reynolds number, *Re* between 100 to 1500 and nanoparticle volume fraction range of 0–1%. The effect of Reynolds number *Re,* nanoparticle volume fraction, height of amplitude and wavelength of the triangular wall on the average *Nu*, local skin-friction coefficient, heat transfer enhancement and pressure drop are presented and analyzed. The results show that the skin-friction coefficient for the base fluid increases with the increase in the amplitude height of triangular wall. In addition, it is found that the improvement in heat transfer essentially depends on amplitude height, *Re* and nanoparticle volume fraction of the triangular wall rather than the wavelength. Finally, the results of this study ensures that using nanofluid instead of distilled water in the triangular facing step channel can lead to further improvement in the thermal performance of heat exchangers.

References

[1] Aung W 1983 *Journal of Heat Transfer* **105** 823-9
[2] Kherbeet A S, Mohammed H and Salman B 2012 *International Journal of Heat and Mass Transfer* **55** 5870-81
[3] Li A and Armaly B F 2000 *Proceedings of the ASMEIMECE conference*
[4] Armaly B F, Li A and Nie J 2003 *International Journal of Heat and Mass Transfer* **46** 3573-82
[5] Hattori H and Nagano Y 2010 *International Journal of Heat and Fluid Flow* **31** 284-94
[6] Sparrow E and Comb J 1983 *International Journal of Heat and Mass Transfer* **26** 993-1005
[7] Sawyers D R, Sen M and Chang H-C 1998 *International Journal of Heat and Mass Transfer* **41** 3559-73
[8] Pehlivan H, Taymaz I and İslamoğlu Y 2013 *International Communications in Heat and Mass Transfer* **46** 106-11
[9] Patel V, Chon J T and Yoon J 1991 *Journal of Fluids Engineering* **113** 579-86

[10] Rush T, Newell T and Jacobi A 1999 *International Journal of Heat and Mass Transfer* **42** 1541-53
[11] Islamoglu Y and Parmaksizoglu C 2003 *Applied Thermal Engineering* **23** 979-87
[12] Parsazadeh M, Mohammed H and Fathinia F 2013 *International Communications in Heat and Mass Transfer* **46** 97-105
[13] Namburu P K, Das D K, Tanguturi K M and Vajjha R S 2009 *International Journal of Thermal Sciences* **48** 290-302
[14] Oyakawa K, Teruya I, Senaha I, Yaga M and Mabuchi I 1996 *Trans Jpn Soc Mech Eng Ser B* **62** 1104-10
[15] Elshafei E, Awad M, El-Negiry E and Ali A 2010 *Energy* **35** 101-10
[16] Elshafei E, Awad M, El-Negiry E and Ali A 2008 *Second International Conference on Thermal Issues in Emerging Technologies*
[17] Nguyen C, Desgranges F, Roy G, Galanis N, Maré T, Boucher S and Mintsa H A 2007 *International Journal of Heat and Fluid Flow* **28** 1492-1506
[18] Yu W and Choi S U S 2003 *Journal of Nanoparticle Research* **5** 167-71
[19] Ghasemi B and Aminossadati S 2009 *Numerical Heat Transfer, Part A: Applications* **55** 807-23
[20] Vajjha R S and Das D K 2009 *International Journal of Heat and Mass Transfer* **52** 4675-82
[21] Vajjha R S, Das D K and Kulkarni D P 2010 *International Journal of Heat and Mass Transfer* **53** 4607-18
[22] Corcione M 2010 *International Journal of Thermal Sciences* **49** 1536-46
[23] Eiamsa-Ard S and Promvonge P 2009 *International Communications in Heat and Mass Transfer* **36** 705-11
[24] Anderson J D 1995 *Computational Fluid Dynamic: The Basics with Applications* (New York: McGraw-Hill)
[25] Anderson J D and Wendt J 1995 *Computational Fluid Dynamics* (Berlin: Springer)
[26] Santra A K, Sen S and Chakraborty N 2009 *International Journal of Thermal Sciences* **48** 391-400
[27] Mohammed A, Yusoff M, Ng K and Shuaib N 2015 *Case Studies in Thermal Engineering* **6** 77-92

IOP Conf. Series: Materials Science and Engineering **152** (2016) 012058 doi:10.1088/1757-899X/152/1/012058

Computational investigations and grid refinement study of 3D transient flow in a cylindrical tank using OpenFOAM

F Mohd Sakri*, M S Mat Ali and S A Z Sheikh Salim

Wind Engineering, Malaysia-Japan International Institute of Technology, Universiti Teknologi Malaysia, Kuala Lumpur,Malaysia

*eypasakri@gmail.com

Abstract. The study of physic fluid for a liquid draining inside a tank is easily accessible using numerical simulation. However, numerical simulation is expensive when the liquid draining involves the multi-phase problem. Since an accurate numerical simulation can be obtained if a proper method for error estimation is accomplished, this paper provides systematic assessment of error estimation due to grid convergence error using OpenFOAM. OpenFOAM is an open source CFD-toolbox and it is well-known among the researchers and institutions because of its free applications and ready to use. In this study, three types of grid resolution are used: coarse, medium and fine grids. Grid Convergence Index (GCI) is applied to estimate the error due to the grid sensitivity. A monotonic convergence condition is obtained in this study that shows the grid convergence error has been progressively reduced. The fine grid has the GCI value below 1%. The extrapolated value from Richardson Extrapolation is in the range of the GCI obtained.

1. Introduction

Recently, the use of liquid storage tanks has been necessary for various industries. They can be found in space vehicles and rockets for their propellant storage in the liquid fuel system, water tank for daily use and also in oil and gas industries for long-short term storages, among others. All these applications come with a similar flow phenomenon problem, which is liquid draining that often induces an air-core vortex formation. The air-core is generated when liquid drains through the outlet and dip is developed on the surface as the surface level reaches a certain critical height, Hc. Then, the dip further grows into a vortex with an air core, which lengthens to the bottom of the tank. This air-core vortex will reduce the cross-sectional area and affect the discharge rates.

At this moment, there are various types of passive flow control and air-core vortex suppression mechanisms. The most popular one is the vortex breaker in liquid propellant of launch vehicles [1]. It decreases the critical height, Hc (as shown in Figure 1) by using its circular flat plate with blades or porous wall. In other studies, inclined base [2] and eccentric drain outlet [3] have been used in order to change the pressure at the center of the outlet. Consequently, this will affect and reduce the formation of air-core vortex. Additionally, Ramamurthi and Tharakan [4] stated in their studies that by placing net and turf in the tank also can reduce the development of surface dip and air-core vortex. However, these mechanisms cannot completely suppress the air-core vortex. This problem has been and still continues to be investigated, including in this study. In order to get a comprehensive data, this paper applies numerical simulation in the study of liquid draining inside the tank. Since mesh shows a very important result of numerical simulation [1], it is important to estimate the grid convergence error due to the sensitivity of the resolution.

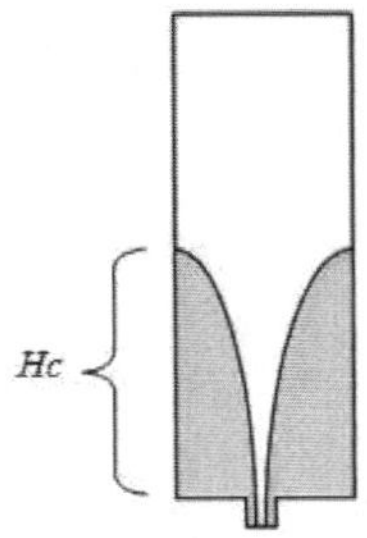

Figure 1. Liquid storage tank with the generation of an air-core vortex

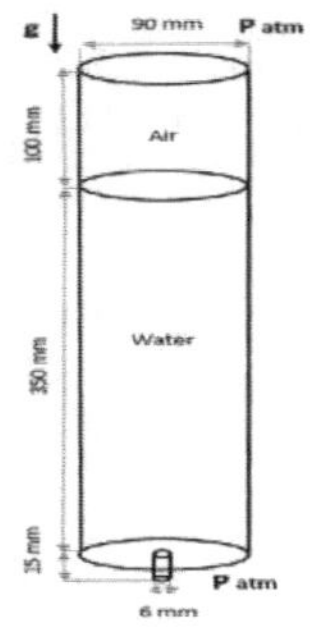

Figure 2. Schematic diagram

The purpose of this study is to determine the Grid Convergence Index (GCI) of three types of grid resolution for the liquid draining. It is the most familiar and reliable technique for quantification of the numerical uncertainty [5, 6, 7]. The result of this investigation, especially for the finer grid, is very important for further analysis in order to correctly replicate the air-core vortex formation inside the tank. Figure 2 shows the schematic diagram for the case under investigation, which is a cylindrical tank partially filled with water. The diameter of the tank is 90 mm, with 450 mm in length, and the initial height of the water is 350 mm. The diameter of the outlet nozzle is 6 mm, with 15 mm in length. The liquid is drained by gravity, g with open to the atmospheric condition on the top and bottom of the tank [2].

2. Solution methodology

2.1. Navier - Stoke equations

The following continuity and momentum equations are obtained from the Navier-Stoke equations after using the incompressible condition [3, 4]:

$$\frac{\partial \rho u_j}{\partial x_j} = 0 \tag{1}$$

$$\frac{\partial \rho u_i}{\partial t} + \frac{\partial \rho u_j u_i}{\partial x_j} = -\frac{\partial p}{\partial x_i} + \frac{\partial}{\partial x_j}\left(\mu \frac{\partial u_i}{\partial x_j}\right) \tag{2}$$

2.2. Turbulence model

In this study, the Reynolds Averaged Navier-Stokes equations (RANS) and two-equation $k - \varepsilon$ model are employed to simulate the transient flows.

2.3. Discretization of the transport equation

Transformation from PDE to linearized algebraic equation needs to be done to solve the equations that describe the flow transport. Equation 3 shows the standard form of the transport equation for the field variable, ϕ [4]:

$$\frac{\partial \rho \phi}{\partial t} + \nabla . \rho \mathrm{U} \phi - \nabla . (\rho \Gamma_\phi \nabla_\phi) = S_\phi(\phi) \tag{3}$$

All terms in Equation 3 are integrated over time step ranging from $t \to t + \Delta t$ and control volume, V_p as shown in Equation 4 [4].

$$\int_t^{t+\Delta t}\left[\frac{\partial}{\partial t}\int_{V_p} \rho dV + \int_{V_p} \nabla . (\rho U \phi) dV - \int_{V_p} \nabla . (\rho \Gamma_\phi \nabla \phi)\right] dt = \int_t^{t+\Delta t}\left[\int_{V_p} S(\phi) dV\right] dt \tag{4}$$

Here, U is the velocity field, ρ is the density, Γ is the diffusion coefficient, S is all source terms and index p describes the midpoint of the control volume. Table 1 describes numerical schemes for non-swirl case.

Table 1. Order of convergence and numerical scheme of differential operator in non-swirl case [5]

Term	Scheme	Order of Convergence	Interpolation Scheme
$\frac{\partial}{\partial t}, \frac{\partial^2}{\partial^2 t}$	Euler	$O(h)$	
∇	Gauss	$O(h^2)$	Linear
$\nabla.(\rho\phi, U)$	Gauss	$O(h^2)$	Linear
$\nabla.(\phi, \alpha)$	Gauss	$O(h^2)$	Van Leer
$\nabla.(\phi, k)$	Gauss	$O(h)$	upwind
$\nabla.(\phi, \varepsilon)$	Gauss	$O(h)$	upwind
∇^2	Gauss	$O(h^2)$	linear corrected

2.4. Discretization of governing equations

2.4.1. Convection scheme. Using Gauss' theorem, the generalized linearization form of the convective term in Equation 4 is given in Equation 5 [4]:

$$\int_{V_p} \nabla.(\rho U\phi)dV = \sum_f S.(\rho U\phi)_f = \sum_f S.(\rho U)_f\phi_f = \sum_f F\phi_f \tag{5}$$

where F is the mass flux through the cell's faces, as given in Equation 6.

$$F = S.(\rho U)_f = S.\mathrm{u}_f \tag{6}$$

Here, S is the surface area element pointing outward, u is the local velocity in three spatial component (x, y and z) and index f describes the value at the surface control volume. It can be assumed that the mass flux from the Equation 6 can be calculated from the interpolated values of U and ρ. In this study, Second Order Upwind (SOU) differencing scheme is applied. This scheme is conditionally bounded and is more advanced compared to Central Differencing Scheme (CDS) [6, 7, 8]. For SOU, the critical cell Reynolds number of the result of purely convective transport of a scalar field is free from ∞ and it is infinite [6, 9] when the cell Reynolds number is over two. The face value is determined according to the polynomial in Equation 7 [10]:

$$u_e = \begin{cases} u_P + (u_P - u_W)\Delta x_{Pn} \text{for } F \geq 0; \\ u_N + (u_N - u_{NN})\Delta x_{nN} \text{ for } F < 0, \end{cases} \tag{7}$$

$$\Delta x_{Pn} = \frac{x_n - x_P}{x_P - x_W} \tag{8}$$

$$\Delta x_{nN} = \frac{x_N - x_n}{x_{NN} - x_N} \tag{9}$$

2.4.2. Diffusion scheme. Using Gauss' theorem, a generalized linearization form of the diffusion term in Equation 4 is given as Equation 10 [4]:

$$\int_{V_p} \nabla.(\rho\Gamma_\phi\nabla\phi)\, dV = \sum_f S.(\rho\Gamma_\phi\nabla\phi)_f = \sum_f(\rho\Gamma_\phi)_f S.(\nabla\phi)_f \tag{10}$$

The following Equation 11 is possible to be used for the orthogonal mesh, such that the vectors S and d are parallel [4]:

$$S.(\nabla\phi)_f = |\mathrm{S}|\frac{\phi_N - \phi_P}{|\mathrm{d}|} \tag{11}$$

According to [4], the face gradient of ϕ can be collected from two values around the face, which are:

- Cell-centred gradient for the cells sharing the face: $(\nabla\phi)_p = \frac{1}{V_p}\sum_f S\phi_f$ (12)
- Cell-centred gradient for interpolate it to the face: $(\nabla\phi)_f = f_x(\nabla\phi)_P + (1 - f_x)(\nabla\phi)_N$ (13)

In this study, the diffusion terms are discretized using the 2nd order deferred correction scheme. As Equation 11 is the 1st order accurate uncorrected, the diffusion term is corrected by using Equation 14 when the vectors N and P are not coordinated [11]. Thus the term is separated into non-orthogonal and orthogonal parts as follows [12]:

$$S.(\nabla\phi_f) = \delta\left(\frac{\phi_E-\phi_P}{|d|}\right)_f + k.\left(\frac{\phi_E-\phi_P}{|d|}\right)_f \tag{14}$$

The first part of Equation 14 is for the orthogonal whereas the second part is for the non-orthogonal. Vector δ is parallel to the distance vector, d and vector k is aligned to the face of the gradient, $\nabla\phi_f$. In addition, Equation 14 must satisfy the following requirement [4]:

$$S = \delta + k \tag{15}$$

The limiting Equation 16 is recommended for stability as the non-orthogonal part is much bigger than the orthogonal part [12]:

$$K_L\left[k.\left(\frac{\phi_E-\phi_P}{|d|}\right)\right]_f = S.(\nabla\phi_f) - \delta\left(\frac{\phi_E-\phi_P}{|d|}\right)_f \tag{16}$$

where $K_L$is user-specified parameter with $0 < K_L < 1$.

2.4.3. Gradient term. The gradient term is discretized using bounded central differencing scheme that is a 2nd order Gaussian method. As in Equation 17, the gradient is calculated using integrals over faces [4]:

$$\int_{V_P} \nabla\phi dV = \oint_{\partial V_P} dS\,\phi = \sum_f S_f\phi_f \tag{17}$$

The face value can be evaluated from the cell center values as follows:

$$\phi_f = f_x\phi_P + (1-f_x)\phi_N \tag{18}$$

$$f_x = \frac{\overline{fN}}{\overline{fP}} \tag{19}$$

Figure 3 shows the interpolation factor of f_x is the ratio between the distances $\overline{fP}$ and $\overline{fN}$.

2.4.4. Temporal/time scheme. For transient flow, the physical properties of the flow are changing over time. Thus volume integrals from Equation 4 can be converted into surface integrals by applying the Gauss Theorem. They can then be written as the sum over the regarded control volume [4, 13] and the expression is called as "semi-discretized" form of the transport equation as shown in Equation 20 [14]:

$$\int_t^{t+\Delta t}\left[\left(\frac{\partial\rho\phi}{\partial t}\right)_P + \sum F\phi_f - \sum(\rho\Gamma_\phi)_f S.(\nabla\phi)_f\right]dt = \int_t^{t+\Delta t}\left[S_u v_p + S_p V\phi_p\right]dt \tag{20}$$

This time discretization can be split into two parts: explicit and implicit methods. The explicit method needs a low computational effort as it only uses the previous time level in the discretization. However, when the Courant-Friedrichs-Lewy number is bigger than ∞, there is a problem with the stability [8]. On the other hand, the implicit method needs more computational effort as it uses the new time level in discretization. Robust solutions are normally retrieved when the coupling between flow properties is bigger than in the explicit method [8].

The Euler implicit scheme is 1st order accurate linear approximation and it guarantees boundedness. The coupling in the system is more stable than the explicit method, even if the Courant number limit is violated [14]. As shown in Equation 21 and Equation 22, this method considers the face values in the new time level cell [4].

$$\phi_f = f_x\phi_P^n + (1-f_x)\phi_N^n \tag{21}$$

$$S.(\nabla\phi_f) = |\delta|\frac{\phi_N^n-\phi_P^n}{|d|} + k.(\nabla\phi_f) \tag{22}$$

2.5. Boundary conditions

Boundary conditions set down the series of faces in the computational mesh, which correspond to the boundaries of physical domain. They are separated into numerical and physical boundary conditions. Numerical boundary conditions are divided into the Von Neumann and Dirichlet boundary conditions. These boundary conditions stipulate the gradient of the variable normal to the boundary and the value of the variable on the boundary (or fixed/or constant value), respectively [4]. Meanwhile, the physical boundary conditions for the incompressible flow are explained as follows:

- Inlet: the velocity value is determined and the pressure condition is specified as zero gradient
- Outlet: the outlet boundary is prescribed similar to total mass balance. This can be perfomed in two ways:
  a) when the velocity dissemination is projected from the inside of the domain, the boundary condition on pressure is set as *zeroGradient*
  b) when the pressure dissemination is stated, boundary condition on pressure and velocity is set as *fixedValue* and *zeroGradient*, respectively.
- Impermeable no-slip wall: the flow velocity on the wall is the same as the wall. Thus the flux through the wall is zero, so that the pressure gradient is set as *zeroGradient*
- Slip wall: in scalar case, the variable is set as *zeroGradient*, but in vector magnitude case, the variable that normal and tangential to the wall is set as *zero* and *zeroGradient*, respectively

In this study, three different patches have been constructed in the computational domain. The top surface is represented as an inlet and the bottom surface is represented as an outlet. Moreover, the side surface is represented as a wall. A *fixedValue* boundary condition is set up at the outlet patch while that at the wall patch is *zeroGradient*. In this case, the liquid is drained normally by the gravity. Other details of boundary conditions are listed in Table 2 and the description of each boundary is explained in Table 3.

Table 2. Boundary conditions for the non-swirl case

Flow Properties	Type of Patches		
	Outlet	**Inlet**	**Walls**
α	zeroGradient	inletOutlet	zeroGradient
U	zeroGradient	pressureInletOutletVelocity	fixedValue
ρ rgh	fixedValue	totalPressure	bouyantPressure
k	zeroGradient	inletOutlet	kqRWallFunction
nuT	calculated	calculated	nutkWallfunction
nuTilda	zeroGradient	inletOutlet	zeroGradient
ε	zeroGradient	inletOutlet	epsilonWallFunction

Table 3. Explanation of boundary conditions

Type	Description of boundary conditions
zeroGradient	Normal gradient of ϕ is zero
fixedValue	Value of ϕ is specified
pressureInletOutletVelocity	Combination of *pressureInletVelocity* and *inletOutlet*(pressureInletVelocity : When $\mathcal{P}$ is known at inlet, U is evaluated from the flux, normal to the patch)
totalPressure	Total pressure $\mathcal{P}_0 = \mathcal{P} + \frac{1}{2}\rho\|U\|^2$is fixed; when U changes, $\mathcal{P}$ is adjusted accordingly
inletOutlet	Switches U and between *fixedValue* and *zeroGradient* depending on direction of U.
bouyantPressure	Sets fixedGradient pressure based on the atmospheric pressure gradient
calculated	Boundary field ϕ derived from other fields
nutkWallfunction	On corresponding patches in the turbulent fields k and nut
kqRWallFunction	On corresponding patches in the turbulent fields k, q and R
epsilonWallFunction	On corresponding patches in the epsilon field

2.6. Multiphase solvers
Multiphase solvers are available in OpenFOAM. They are listed as follows [13]:

- interFoam, LTSinterFoam and InterDyMFoam: these are based on Volume of Fluid (VOF) method
- twoPhaseEulerFoam, multiphaseEulerFoam and multiphaseInterFoam: these are based on the Eulerian-Eulerian method [17]

2.6.1. InterFoam. In this study, the interFoam solver is applied because it is capable to simulate flow of two fluids (liquid and air). In this solver, only one momentum and one mass conservation equation is determined for both fluids. Therefore, the viscosity and density of both fluids are averaged based on the volume fractions in the cell. In this condition, momentum and mass transfer are inconsiderate [13].

2.6.2. Volume of fluid (VOF). Volume of Fluid (VOF) method, a Eulerian-Eulerian model, is applied in averaging procedure for the liquid and air (multiphase) flow system. The two phases of fluids are modelled by adopting RANS equations. These governing equations are calculated numerically based on the OpenFOAM numerical simulation system. The parameter of α in the Volume of Fluid (VOF) method is a function that indicates the relative fraction between liquid and gas in each cell of physical domain. α equals to one when the fluid is in the liquid phase and α equals to 0 when the fluid is in the gas phase. The change of their properties in each cell is calculated in the Equation 23 [18].

$$\rho = \alpha \rho_l + (1-\alpha)\rho_g, \ \mu = \alpha \mu_l + (1-\alpha)\mu_g, \ \bar{U} = \alpha \bar{U_l} + (1-\alpha)\bar{U}_g \qquad (23)$$

Here, the subscripts l and g indicate the liquid and gas, while ρ is the density and μ is the dynamic viscosity. $\bar{U}$ is the mean velocity of fluid and it is transported with the function of α.

The VOF method can also be recognised as an isothermal fluid system as shown in Equation 24. Only a single set of governing equations is needed in this method and it is cheaper than the Eulerian-Lagrangian model, which has a number of governing equations and relying on the number of phases.

$$\frac{\partial \rho}{\partial t} + \nabla.(\rho\,\bar{U}) = 0 \ ; \ \text{Incomp.} \Rightarrow \frac{\partial \bar{U}_i}{\partial x_i} = 0 \ ; \ i = x, y, z$$

$$\begin{gathered}
\frac{\partial \bar{U}_i}{\partial t} + \bar{U}_j \frac{\partial \bar{U}_i}{\partial x_j} = -\frac{1}{\rho}\frac{\partial \bar{p}}{\partial x_i} + \frac{\partial}{\partial x_j}\left(v\frac{\partial \bar{U}_i}{\partial x_j} - \overline{U'_i U'_j}\right) + f_i + \sigma\xi\nabla\alpha \\
\frac{\partial k}{\partial t} + \bar{U}_j\frac{\partial k}{\partial x_j} = -\overline{U'_i U'_j}\frac{\partial \bar{U}_i}{\partial x_j} - \varepsilon + \frac{\partial}{\partial x_j}\left[\left(v + v_T\right)\overline{U'_i U'_j}\right)\frac{\partial k}{\partial x_j}\right] \\
\frac{\partial \varepsilon}{\partial t} + \bar{U}_j\frac{\partial \varepsilon}{\partial x_j} = -C_{1\varepsilon}\frac{\varepsilon}{k}\overline{U'_i U'_j}\frac{\partial \bar{U}_i}{\partial x_j} - C_{2\varepsilon}\frac{\varepsilon^2}{k} + \frac{\partial}{\partial x_j}\left[\left(v + \frac{v_T}{\gamma_0}\right)\frac{\partial \varepsilon}{\partial x_j}\right] \\
\frac{\partial \alpha}{\partial t} + \nabla.(\alpha\,\bar{U}) + \nabla.\left(\alpha(1-\alpha)(\bar{U}_l - \bar{U}_g)\right) = 0 \\
\xi = -\nabla.\left(\frac{\nabla\alpha}{|\nabla\alpha|}\right)
\end{gathered} \qquad (24)$$

where k is the turbulent kinematic energy, v_T is the turbulent kinematic viscosity, v is the kinematic viscosity, σ is the surface tension, ξ is the curvature of free surface, ε is the dissipation rate of the turbulent kinematic energy and $C_{1\varepsilon}$, $C_{2\varepsilon}$ and γ_0 are characteristic constants for the $k-\varepsilon$ model. In this study, following Ref. [18], the constants $C_{1\varepsilon}$, $C_{2\varepsilon}$ and γ_0 are set as 1.44, 1.92 and 1.30, respectively.

2.7. Algebraic equations for the Navier-Stokes equations
The discretization of the transport equations creates a system of algebraic equations that is as indicated by Equation 25.

$$[A][\phi] = [R] \qquad (25)$$

Here, $[A]$ is a sparse matrix, $[\phi]$ is the vector of ϕ–s for all control volumes and $[R]$ is the source term vector. Matrix $[A]$ consists of all coefficients of the variables that are saved in the column vector $[\phi]$ and $[R]$ controls all right hand side terms [4]. The final form of the discretization process is shown by the equations below [4].

$$a_P U_P = H(U) - \sum_f S(p)_f \tag{26}$$

$$\sum_f S.\left[\left(\frac{1}{a_P}\right)_f (\nabla p)_f\right] = \sum_f S.\left(\frac{H(U)}{a_P}\right)_f \tag{27}$$

where the flux is given by Equation 28 as follows:

$$F = S.U_f = S.\left[\left(\frac{H(U)}{a_P}\right)_f - \left(\frac{1}{a_P}\right)_f (\nabla p)_f\right] \tag{28}$$

The discretization of Equation 26 and Equation 27 needs a pressure-velocity coupling treatment to solve the equations [4]. In this study, Pressure Implicit Split Operator (PISO) is applied in order to solve the governing equations. PISO is a pressure-velocity treatment for the transient flow and it was introduced by Isaa [19] to improve Semi-Implicit Method for Pressure-Linked Equations (SIMPLE) sheme [20]. In particular, when standard PISO algorithm is set, the complete procedure of interFOAM solver contains of the following steps [13, 21] (see Figure 3 for annotations):

- VOF equations are solved on basis of the old velocity field from the previous time step. Thus, this produces new values for the density and the volumetric of each cell.
- The PISO algorithm starts with the initial values. On the momentum predictor stage, previous time step is used to produce a new velocity field by the Equation 26.
- Then to satisfy the continuity equation, the value of the pressure field is corrected.
- From momentum equation, continuity equation is reconstructed to combine the new velocity field U^*:

$$U_f = S.\left[\left(\frac{H(U^*)}{a_P}\right)_f - \left(\frac{1}{a_P}\right)_f (\nabla p^*)_f\right] \tag{29}$$

$$\sum_f S{U^*}_f = 0 \tag{30}$$

$$\sum_f S.\left[\left(\frac{H(U^*)}{a_P}\right)_f - \left(\frac{1}{a_P}\right)_f (\nabla p^*)_f\right] = 0 \tag{31}$$

In the explicit velocity correction stage, the new corrected pressure p^* is used to produce the new corrected velocity field, U^{**}. This step continues to obtain two corrected processes (n = 2) by applying the new operator $H(U^{**})$. These two corrected processes are adequate to attain the robust solution [8, 19].

3. Result and discussion

3.1. Grid refinement

The grid refinement study is applied to the three grid resolutions. Case A has the finest grid, Case B has the medium resolution and Case C has the coarsest resolution. These cases are shown in Table 4 for its specification and Figure 4 for its visualization. Richardson Extrapolation is acknowledged as "the deferred approach to the limit ($h \to 0$)". It determines a higher-order estimate of flow fields from a series of lower-order discrete values $(f_1, f_2, \ldots, f_n)$. The method of Richardson Extrapolation is shown in Equation 32 [22].

$$f = f[exact] + g_1 h + g_2 h^2 + g_3 h^3 + \cdots \tag{32}$$

where f is discrete solutions, h is grid spacing, and g_1 and g_2 are functions that are illustrated in the continuum and do not lean on any discretization. The quantity f is considered "second-order" when $g_1 = 0$. The $f_{h=0}$ is the continuum value at zero grid spacing.

IOP Conf. Series: Materials Science and Engineering **152** (2016) 012058 doi:10.1088/1757-899X/152/1/012058

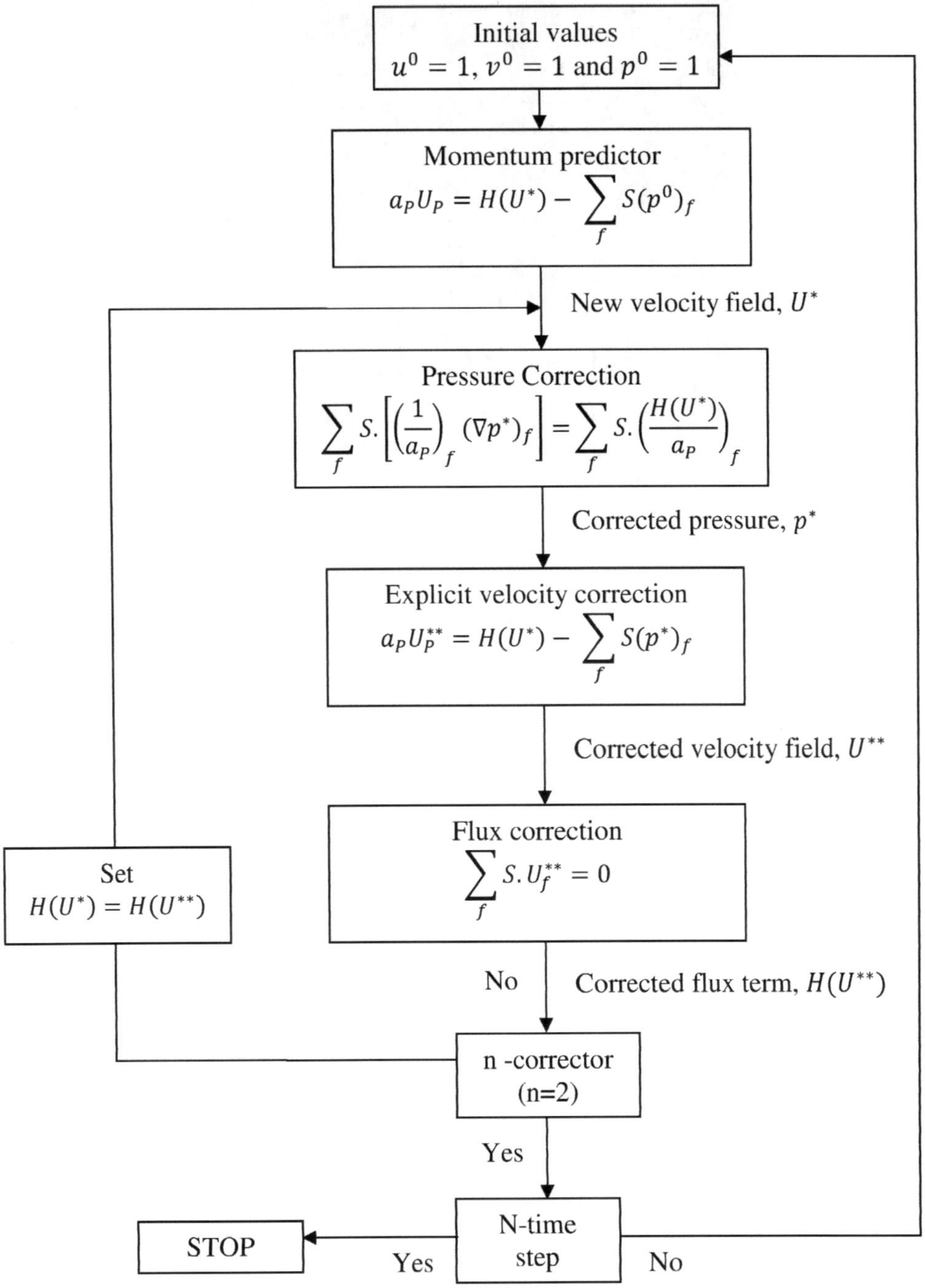

Figure 3. PISO algorithm [8]

Figure 4. Draining tank mesh and geometry structure for different cases: (a)fine grid, (b)medium grid and (c) coarser grid

Table 4. Grid parameter for cases A, B and C

CASE	A (Fine)	B (Medium)	C (Coarse)
Number of Cells	38 620	25 853	17 032

Without considering the non-appearance of odd power in Equation 32, Richardson Extrapolation can be generalized to pth order methods and r-value of grid ratio as follows:

$$f_{exact} \approx f_1 + \frac{(f_1 - f_2)}{r^p - 1} \tag{33}$$

In this study, the grid refinement ratio, r for the uniformed meshed is described as:

$$r = \frac{Total\ no.of\ grid\ (Fine)}{Total\ no.of\ grid\ (Medium)} = \frac{Total\ no.of\ grid\ (Medium)}{Total\ no.of\ grid\ (Coarse)} = 1.54 \tag{34}$$

The refinement ratio is higher than the minimum value of 1.3. Equation 32 can be measured for order-of-accuracy by applying Equation 35.

$$p = \frac{\ln(\frac{\varepsilon_{32}}{\varepsilon_{21}})}{\ln(r)} \tag{35}$$

$$\varepsilon_{i+1,i} = f_{i+1} - f_i \tag{36}$$

Convergence conditions of this system must be clarified first in order to assess the extrapolated value from the equations above. The convergence conditions are listed as follow:

- Monotonic convergence: $0 < R < 1$
- Oscillatory convergence: $R < 1$
- Divergence: $R > 1$

where R is the convergence ratio: $R = \frac{\varepsilon_{21}}{\varepsilon_{32}}$ (37)

For monotonic convergence, generalized Richardson Extrapolation is applied to estimate the errors and uncertainties, which are explained in Equation 37. For oscillatory convergence, the results show to exhibit some oscillations. Lastly, for divergence, the results diverge while errors and uncertainties are impossible to be determined [23]. Table 5 shows the results for order of accuracy for draining time, t_{drain} and flow rate, Q. The results are calculated from three different types of mesh as defined in the previous Table 3.

Table 5. Grid Convergence Index (GCI) and order of accuracy for parameter: draining time, t_{drain} and flow rate, Q

	ε_{32}	ε_{21}	R	p	GCI_{32} [%]	GCI_{21} [%]
t_{drain}	8.000	6.500	0.8125	0.477	0.833	0.602
Q	0.033	0.208	0.7230	0.748	0.522	0.427

** Index 1, 2 and 3 represent case A, B and C, respectively*

The convergence conditions for draining time, t_{drain} and flow rate, Q are monotonic as the value of convergence ratio, R is more than zero and less than one for both cases. Hence these two parameters are applicable in the Grid Convergence Index (GCI) study. The draining time, t_{drain} can be expressed theoretically with the following Equation 38.

$$t_{drain} = \frac{\sqrt{h_0} - \sqrt{h}}{\sqrt{\frac{g}{2}}} \left(\frac{d_t}{d_n}\right)^2 \tag{38}$$

Here, d_t is the tank diameter, d_n is the nozzle diameter, t is the draining time, h_o is the initial water level, h is the water level at a t time and g is gravitational acceleration. This equation is derived with the assumption that the flow is irrotational and inviscid. On another hand, the equation for the flow rate, Q can be estimated following in Equation 39, where v is flow velocity and A_n is cross-sectional area inside the outlet nozzle.

$$Q = \iint_{A_n} v \, . \, dA_n \tag{39}$$

Roache [22] has done a valuable contribution to systematic and most common methodology for the grid refinement studies, which is called Grid Convergence Index (GCI) method. GCI is based on grid refinement error estimator derived from the generalized Richardson Extrapolation. The percentage of differences between the computed and asymptotic values is calculated by using GCI. It demonstrates how far is the error of the computed value with the asymptotic value. How much the solution of the computed value would change with further refinement also can be illustrated with GCI. A small value of GCI percentage shows the computed value is approaching asymptotic range. The GCI for fine grid can be interpreted as shown in Equation 40.

$$GCI_{i+1,i} = F_s \frac{|\varepsilon_{i+1,i}|}{f_i(r^p - 1)} \tag{40}$$

where F_s is safety factor. As three different grids are used in this study, the safety factor should be $F_s = 1.25$ [24]. Otherwise, for the refinement with two grids, the safety factor is $F_s = 3.0$. Table 6 shows that GCI for parameter draining time, t_{drain} and flow rate, Q from three different meshes is good with the decrement value from GCI_{32} to GCI_{21} ($GCI_{21} < GCI_{32}$). The results demonstrate that the dependency of numerical method on the mesh size has been decreased when the GCI for the finer grid (GCI_{21}) is lower than the coarser grid(GCI_{32}). Therefore, the results of simulation will not be of much difference as further refinement of the grid is changed.

Table 6. GCI and order of accuracy for parameter: draining time, t_{drain} and flow rate, Q

	ε_{32}	ε_{21}	R	p	GCI_{32} [%]	GCI_{21} [%]
t_{drain}	8	6.5	0.8125	0.477	0.833	0.602
Q	0.033	0.208	0.723	0.748	0.522	0.427

*Index 1, 2 and 3 represent case A, B and C, respectively

Based on the results in Table 7, Figure 5 is plotted and it displays the monotonic convergence. It shows that the values obtained for three different meshes (fine, medium and coarse grids) for draining time, t_{drain} are just slightly lower than the Richardson extrapolated value. On the other hand, based on the results in Table 8, Figure 6 is plotted and it shows a different line pattern from Figure 5. This demonstrates that the Richardson extrapolated value is slightly lower than the values obtained for three different meshes for flow rate, Q. However, both of these solutions are still in the range of finer GCI as shown in Figure 5 and Figure 6.

Table 7. Calculations of global results normalized by the Richardson extrapolated value for parameter draining time, t_{drain}

Case	A(Fine), f_1	B(Medium),f_2	C(Coarse), f_3	f_{RE}
t_{drain}	58.5	52	44	70.45
	$\frac{f_1}{f_{RE}}$	$\frac{f_2}{f_{RE}}$	$\frac{f_3}{f_{RE}}$	$\frac{f_{RE}}{f_{RE}}$
	0.8303	0.738	0.525	1

Table 8. Calculations of global results normalized by the Richardson extrapolated value for parameter flow rate, Q

Case	A(Fine), f_1	B(Medium), f_2	C(Coarse), f_3	f_{RE}
Q	0.1843	0.2085	0.242	0.141
	$\frac{f_1}{f_{RE}}$	$\frac{f_2}{f_{RE}}$	$\frac{f_3}{f_{RE}}$	$\frac{f_{RE}}{f_{RE}}$
	1.3	1.489	1.728	1

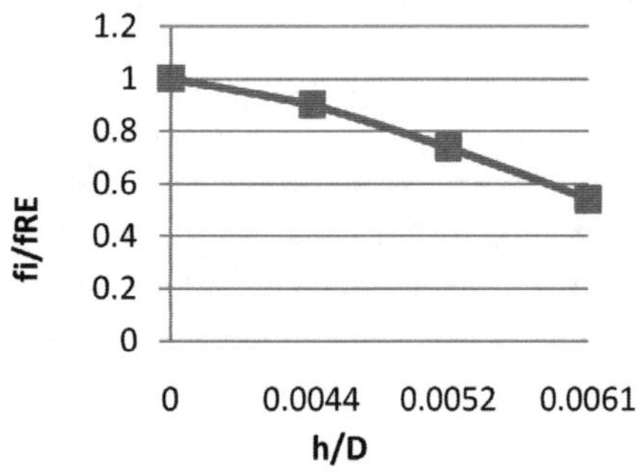

Figure 5. Comparison of parameter draining time, t_{drain} between three different meshes and Richardson Extrapolation estimation

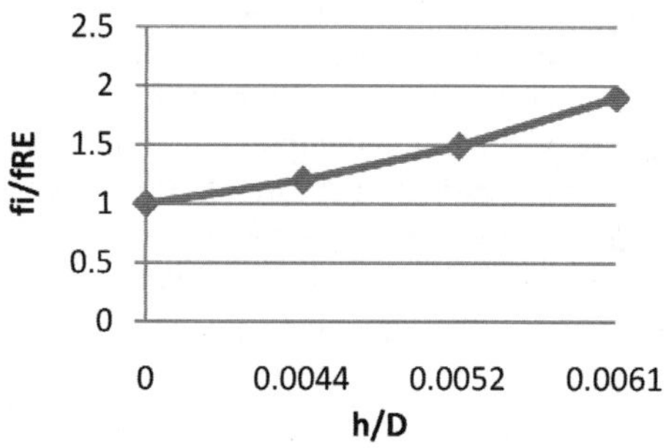

Figure 6. Comparison of parameter flow rate, Q between three different meshes and Richardson Extrapolation estimation

Besides, error of these two parameters (draining time, t_{drain} and flow rate, Q) can be defined from this Richardson extrapolated value and discrepancy between simulation value. The error is determined by Equation 41.

$$E_i = \left|\frac{f_i - f_{RE}}{f_{RE}}\right| \tag{41}$$

Based on the calculation of RMS error, Figure 7 shows relative error between Richardson extrapolated value with the finer grid is only 0.169 % for the parameter draining time, t_{drain} and 0.216% for the parameter of flow rate, Q. These illustrate that the finer grid has approached asymptotic value, where the error due to the spatial discretization has been reduced significantly.

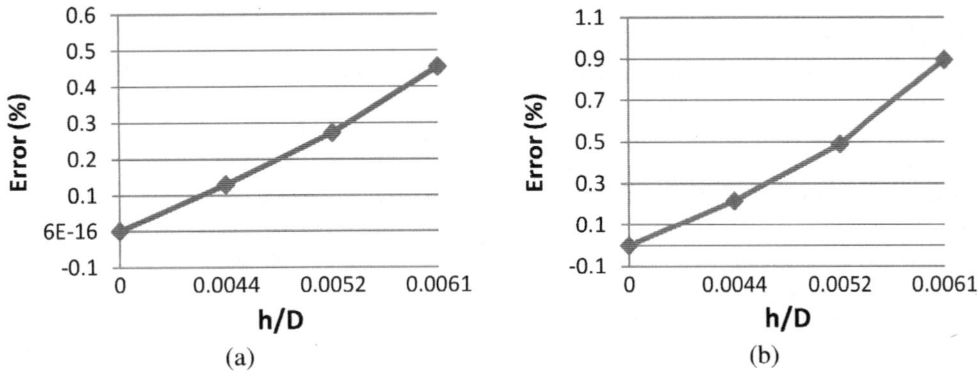

Figure 7. Percent of RMS error for parameter:(a) draining time, t_{drain} and (b) flow rate, Q with three different meshes and Richardson Extrapolation estimation

3.2. Comparison of numerical data from Case A (fine grid) with previous data

Table 9 shows the comparison of draining time result for liquid level obtained in this study with that from studies by Park and Sohn [2]. It highlights that an excellent agreement is achieved between the results, especially with the experimental result from Park and Sohn.

Table 9. Comparison of current study with previous studies by Park and Sohn [2]

Case	Draining Time, t_{drain} [s]
Park & Sohn (Experimental)	58.16
Park & Sohn (Theoretical)	69.08
Park & Sohn (Numerical)	63.18
Current study (Case A)	58.50

4. Conclusion

This grid refinement study was successfully achieved by using OpenFOAM framework. OpenFOAM displays a good result and simultaneously saves cost since no license is required. The flow behavior was simulated via PISO algorithm with first order accuracy of discretization scheme. The multiphase solver – interFoam and Volume of Fluid (VOF) were adopted to track the multiphase problem. The value of Grid Convergence Index (GCI) from coarser to finer grid was found to be decreasing as the grid was refined. The finer grid (Case A) was relevant to be applied in the future studies as the value of GCI is less than 1% .

Acknowledgements

This research was financially supported by Malaysia Ministry of Higher Education (MOHE) under Research University Grant (RUG) project of Universiti Teknologi Malaysia FRGS PY/2015/05383 and TWAS 13-272, and the use of High-Performance Computer (HPC) Universiti Teknologi Malaysia facilities was also appreciated.

References

[1] Zubair M, Abdullah M Z and Ahmad K A 2013 *Computational and Mathematical Methods in Medicine* **2013**

[2] Park I S and Sohn C H 2011 *International Communications in Heat and Mass Transfer* **38** 1044-9

[3] Versteeg H K and Malalasekera W 2008 *An Introduction to Computational Dynamics - The Finite Volume Method* (Essex: Longman Scientific & Technical)

[4] Jasak H 1996 *Error Analysis and Estimation for the Finite Volume Method with Applications to Fluid Flows* PhD Thesis University of Zagreb

[5] Rechiman L M, Centero M I and Dari E A 2014 *de Mecanica Computacional* **XXXIII** 2927-38

[6] Yu B, Tao W Q, Zhang D S and Wang Q W 2001 *Numerical Heat Transfer, Part B: Fundamentals* **40** 343-65
[7] Al-Rabeh A 1993 *Computer Methods in Applied Mechanics and Engineering* **109** 131-41
[8] Ali M S M 2011 *Flow and Noise Associated with The Interaction of A square Cylinder with A Downstream Flat Plate* PhD Thesis University of Adelaide
[9] Shyy W 1985 *Journal of Computational Physics* **57** 415-38
[10] Papadakis G and Bergeles G 1995 *International Journal of Numerical Methods for Heat & Fluid Flow* **5** 49-62
[11] Ferziger J H and Peric M 1996 *Computational Methods for Fluid Dynamics* (Berlin: Springer)
[12] Robertson E, Choudhury V and Walters D 2015 *Computer and Fluids* **123** 122-45
[13] Shulze L and Thorenz C 2014 *11th International Conference on Hydroscience & Engineering*
[14] Hirsch C 1991 *Numerical Computation of Internal and External Flow* (Oxford: Butterworth-Heinemann)
[15] Courant R, Friedrichs K and Lewy H 1967 *IBM J. Res. Dev.* **11** 215-34
[16] Jasak H, Weller H and Gosman A 1999 *International Journal for Numerical Methods in Fluids* **31** 431-49
[17] Rusche H 2002 *Computational Fluid Dynamics of Dispersed Two-phase Flows at High Phase Fractions* PhD Thesis Imperial College London
[18] Rechiman L M, Centero M I and Dari E A 2014 *de Mecanica Computacional* **XXXIII** 2927-38
[19] Isaa R I 1986 *Journal of Computational Physics* **62** 40-65
[20] Patankar S 1980 *Numerical Heat Transfer and Fluid Flow* (Boca Raton: CRC Press)
[21] Damian S M 2013 *An Extended Mixture Model for the Simultaneous Treatment of Short and Long Scale Interfaces* PhD Thesis Universidad Nacional Del Litoral
[22] Roache P J 1994 *Journal of Fluids Engineering* **116** 405-41
[23] Stern F, Wilson R V, Coleman H W and Paterson E G 2001 *Journal of Fluids Engineering* **123** 793-802
[24] Wilcox D C 2006 *Turbulence Modelling for CFD* (California: D C W Industries)

Effects of gaseous nitriding AISI4140 alloy steel on corrosion and hardness properties

L Tamil Moli[1*], N Wahab[2], M Gopinathan[3], K Karmegam[4] and M Maniyarasi[1]

[1]Department of Mechanical Engineering, Polytechnic of Banting Selangor, Malaysia
[2]Faculty of Mechanical Engineering, Universiti Teknologi Mara Shah Alam, Malaysia
[3]Department of Mechanical Engineering, Universiti Tenaga Nasional, Malaysia
[4]Faculty of Medicine and Health Sciences, Universiti Putra Malaysia, Malaysia

*tamilmoli@yahoo.com

Abstract. Corrosion is one of the major problems in the industry especially on machinery since it weakens the structure of the machinery part and causes the mechanical failure. This will stop the production and increase the maintenance cost. In this study, the corrosion behaviour of gas nitriding on a screw press machine shaft made from AISI 4140 steel was investigated. Pitting corrosion was identified as a major cause of the shaft failure and this study was conducted to improve the corrosion resistance on the AISI 4140 alloy steel shaft by gas nitriding as a surface hardening treatment. Gas nitriding was performed with composition of 15% ammonia and 85% nitrogen at temperatures of 525 °C, 550 °C and 575 °C and with the soaking time of 30, 45 and 60 minutes, respectively. The samples were prepared as rectangular sized of 30mm x 12mm x 3mm for immersion testing. The results showed that corrosion rate of untreated samples was 77% higher compared to the nitrided samples. It was also found that hardness of the nitrided samples was higher than untreated sample. All in all, it can be concluded that gaseous nitriding can significantly improve the surface hardness and the corrosion resistance of the shaft made of AISI 4140 alloy steel, hence reduces the pitting that is the root cause of failure.

1. Introduction

Gas nitriding is a thermochemical case hardening process used to increase wear resistance [1], surface hardness [2] and fatigue life by dissolution of nitrogen and hard nitride precipitation [3]. This process involves the atomic nitrogen from dissociating ammonia gas diffusing into the steel surface in the temperatures range 450-590 °C [4]. Gas nitriding represents one of the factors in enhancing the service life of the shaft in a screw press machine. The present study has shown the existence of γ-Fe_3N phases in all compound zones of the nitrided specimen surface.

This research is mainly carried out to investigate the frequent occurrences of shaft failure in a screw press machines in a Malaysia palm oil mill. It had been reported that the shaft was operating in a corrosive environment due the facts that the extracted palm oil contains high concentration of palmitic and oleic acids [6]. The chemical reaction between those acids and metal surface under high operating temperature leads to deterioration of the shaft, which eventually causing the metal shaft to experience corrosive wear and pits formation. Pitting is one of the mechanisms in triggering widespread fatigue crack initiation and reducing life of the shaft. The pitting corrosion was identified as the main root causes of the premature shaft failures [6]. The capability of surface heat treatment such as nitriding to overcome the pitting corrosion has been reported and confirmed. The nitrided components have an increased resistance against atmospheric corrosion and corrosion in neutral aqueous media [7, 8]. The corrosion resistance of nitride layers increases in the order γ'-nitride, ε-nitride, ε-carbonitride, and can be improved by post-oxidation [7].

2. Experimental procedure

2.1. Gas nitriding

The screw press machine shaft made from AISI 4140 alloy steel was machined as rectangular sized 30 mm x 12 mm x 3 mm in length, width and thickness, respectively, by using a horizontal band saw and lathe machine. The gas nitriding treatment was applied in a tube furnace (as shown in Figure 1) at the temperatures of 525 °C, 550 °C and 575 °C, with soaking time of 30, 45 and 60 minutes, respectively, as in Figure 2. A gas mixture of 85% nitrogen and 15% ammonia flow was maintained throughout this process, which was controlled by a gas control panel. The nitrogen flow started from the beginning of this process while ammonia started to flow when the temperature of furnace reached about 300°C. The ramp time to reach the peak temperature, the soaking time and the cooling time were all set up before carrying out the process. In this study, the ramp time was set up to 2 hours with increments of 4°C per minute while the soaking time followed the parameters set in the machine. Normal cooling condition was carried out in the furnace until it reached the ambient temperature.

Figure 1. Tube furnace and specimen holder

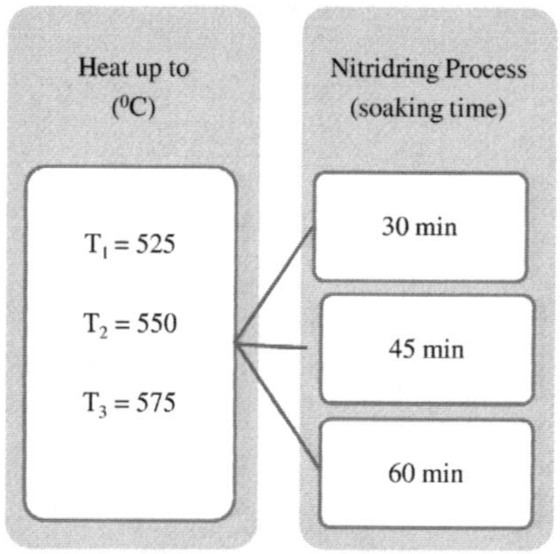

Figure 2. Selected temperature and soaking time for gas nitriding

2.2. Corrosion testing

Corrosion tests were carried out under static or stagnant position on untreated and nitrided samples in aqueous solution containing dirty crude palm oil, which was adapted from ASTM G-31. All specimens were immersed in the dirty crude palm oil at 80 °C for 720 hours or 30 days and they are shown in Figure 3. The temperature of 80°C was selected because the shaft may be operating at that particular temperature in the screw press machine. A hole was made on the sample and a string was attached to it. Before testing was carried out, the initial weight of the samples was measured using microbalance. After the immersion process, the corroded samples were removed and cleaned using ultrasonic cleaner. Following the cleaning stage, samples were immersed in nitric acid (HNO_3) for 2-3 minutes to remove the corroded part. The specimens were rinsed by distilled water and weighed to obtain final weight as shown in Figure 4. The weight loss is converted to corrosion rate as calculated using Equation 1 [9].

$$CR = \frac{weight\ loss\ (g) \times K}{Density\left(\frac{g}{cm^3}\right) \times Exposed\ Area\ (A) \times Exposure\ Time\ (hr)} \tag{1}$$

where CR is corrosion rate (mm/year), K is constant factor, 8.76 x 10^4 and A is exposed area (cm^2).

Figure 3. Immersed samples

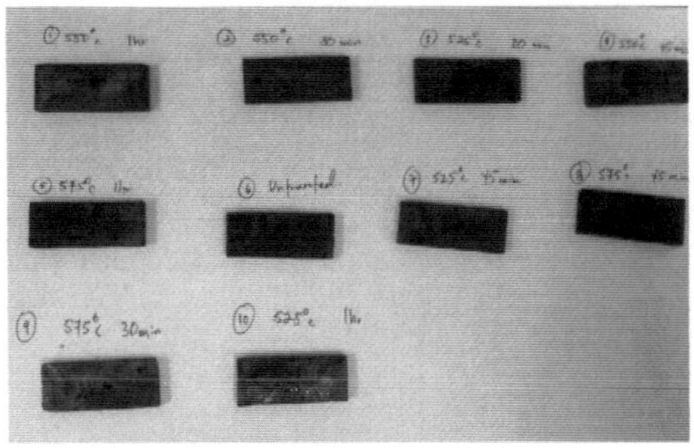

Figure 4. After cleaning the samples

2.3. Microhardness testing
After undergoing the nitriding process, the specimens from each different parameters of heat treatment were sectioned at the middle to a cross-section with thickness of 3 mm and mounted with the phenolic powder. Then, the surfaces were ground with grit paper size 240. The specimens were later polished with alumina powder size of 1 μm to make sure the surfaces were smooth and flat prior to performing the hardness test. The hardness testing was carried out using Mitutoyo Vickers microhardness tester machine Model AVK-1C with indenter load of 30 kgf value. The indentations were uniformly spaced through from the nitrided layer towards to the core with an interval of 0.8 mm as shown in Figure 5. Three indentations were produced at each point and the average was recorded.

Figure 5. Location of the points for microhardness measurement

3. Results and discussion
Corrosion rate is the speed at which any metal in a specific environment deteriorates [10]. Table 1 shows the corrosion weight loss of the untreated and the gaseous nitrided AISI 4140 alloy steel after immersion in dirty crude palm oil solution for 720 hours. An accurate assessment and/or prediction of corrosion is required if the damage prediction models are to be reliably used to predict both the rate and severity of damage [11]. Corrosion rate, in unit of millimeter per year (mm/year), was calculated using Equation 1 as tabulated in Table 1.

IOP Conf. Series: Materials Science and Engineering **152** (2016) 012059 doi:10.1088/1757-899X/152/1/012059

Table 1. Result for immersion testing using dirty crude palm oil as the solution

Sample No.	Description	Area (mm^2)	Initial Weight (gram)	After Immersion Weight (gram)	Weight loss(gram)	Corrosion Rate (mm/yr)	% of Reduction
1	Untreated	821.11	8.53	8.48	0.05	8.87	0
2	525°C 30min	834.72	8.55	8.51	0.03	5.95	33
3	525°C 45min	830.17	8.69	8.66	0.03	5.60	37
4	525°C 60min	827.90	8.97	8.95	0.02	4.12	54
5	550°C 30min	889.34	9.33	9.30	0.03	5.75	35
6	550°C 45min	851.32	9.51	9.49	0.02	4.53	49
7	550°C 60min	876.81	9.07	9.06	0.02	3.18	64
8	575°C 30min	826.78	8.83	8.80	0.03	5.25	41
9	575°C 45min	821.08	9.26	9.23	0.02	4.37	51
10	575°C 60min	847.69	9.21	9.20	0.01	2.01	77

Nitriding the sample at temperature of 575°C for 60 minutes gave a huge impact on the reduction of the corrosion rate by a value of 77%. As depicted in Figure 6, higher nitriding temperature resulted in lower corrosion weight loss, which in turns led to lower corrosion rate when compared to the untreated sample. During the test, gas bulbs were continuously emitted from the surface of untreated AISI 4140 alloy steel sample. Upon completion of the test, the sample surface possessed a rough morphology and was covered with appreciable dark-coloured corrosion products. All these evidences suggested that the sample AISI4140 alloy steel had deteriorated corrosion resistance in the diluted dirty crude palm oil at temperature of 80^0C (working temperature of shaft) but with a variation in the rate of reactivity. The increase in corrosion potential with increasing nitriding time could possibly be attributed to increasing amount of nitrides, which were formed at high nitriding temperature. Nitriding improves the corrosion resistance due to the formation of the compact nitride compound layer[12]. The compound layer at the surface consists of iron nitrides, whilst the diffusion layer beneath the compound layer consists of an interstitial solution of nitrogen in the α-Fe lattice. A thicker and denser nitrided layer would enhance the corrosion resistance [13, 14]. Results of the immersion tests that were applied to the liquid nitrided samples also pointed out that the presence of white layer was another important factor affecting the characteristics of pit formation [15]. This could be due to the fact that the white layer containing the maximum concentration of nitrogen aids to prevent formation of pits at the surface of the immersed sample. In other words, pit formation is very rare at the surface of samples that contain white layer after the application of simple immersion tests, which can again be tied up to the maximum nitrogen concentration reached in white layer.

The degree of corrosion resistance in the immersion test is related to the amount of iron dissolved in a defined time by the test solution. The corrosion resistance of the layer increases with formation of γ-nitride. This is valid for resistance against pitting corrosion, as well as for uniform corrosion, that is characterized by the free corrosion potential and the immersion test [7]. This is also supported by the findings of Ebrahimi *et al.* [12], who mentioned that pitting was formed in both untreated and nitrided samples, however the pits in untreated samples were finer and higher in number [12]. In addition, the corrosion potential of the nitrided samples was shifted to the noble direction with the increment of the nitriding time[16]. Increase in nitriding temperature and soaking time had resulted in an increase of nitrogen diffusivity, leading to the formation of a thicker compound layer. Thick oxide layer can seal the pores in the compound layer more effectively, and positively affect the corrosion resistance of the AISI 4140 alloy steel. As a result, corrosion resistance of the shaft was significantly improved by gas nitriding process. It is important to emphasize the fact that the compound layer is very beneficial in the case of the shaft with regards to wear and corrosion resistance if the porosity in the nitride layer is under control.

The hardness profile of nitrided samples with nitriding temperature and soaking time ranging from 525 ^{0}C to 575^{0}C and 30 minutes to 60 minutes, respectively, is presented in Figure 7. Surface hardness of the samples had increased with increment of nitriding temperature and soaking time, while the core possessed hardness in the range of 250–300 VHN. The highest thickness of nitride layer with a value of 438.6μm was recorded when the sample was nitrided at 550 °C temperature for 60 minutes of the soaking time. The sample possessed surface hardness of 340.8 HV whilst the untreated sample without the presence of the nitride layer recorded a surface hardness of 263.5 HV. Previous work has also reported that hardness value of the compound and diffusion layer is greatly influenced by the nitride layer formed at different temperatures of the nitriding process[15]. Higher surface hardness has been found to lead to higher corrosion resistance. Apart from an increasing surface hardness, the compound layer has been reported to be responsible for excellent wear and corrosion resistance of the nitrided parts [16].

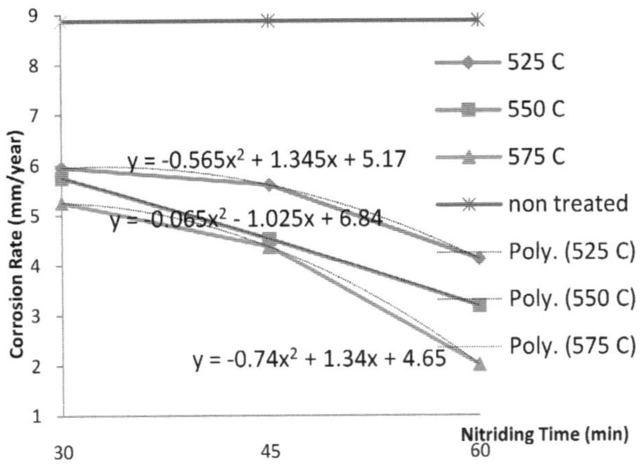

Figure 6. The graph of corrosion rate versus nitriding time

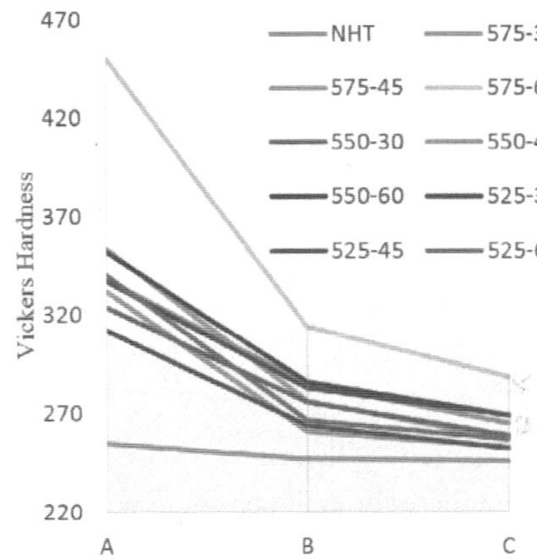

Figure 7. Hardness value of nitride and untreated sample

4. Conclusion

The immersion test showed that gas nitriding had significantly improved the corrosion resistance of AISI 4140 alloy steel by 77% as opposed to the untreated sample. This is due to the formation of the nitride layer that protects the surface from harsh environment when the shaft is working in a corrosive environment and exposed to the dirty crude palm oil, which have contamination of free fatty acid. Diffusion of nitrogen into steel formed a nitrided layer that enhanced the surface hardness. Optimum corrosion resistance and hardness were achieved at a nitriding temperature of 575°C at 60 minutes.

Acknowledgement

The authors would like to express their gratitude to the Ministry of Higher Education Malaysia for the financial support provided under the FRGS grant.

References

[1] Farrahi G H and Ghadbeigi H *Journal of Materials Processing Technology* **174** 318-24
[2] Straffelini G 2015 *Friction and Wear* (Springer International Publishing) 201-23
[3] Jupp J 2012 *28th International Congress of the Aeronautical Sciences*
[4] Akhtar S S Arif F M and Yilbas B 2010 *The International Journal of Advanced Manufacturing Technology* **47** 687-98
[5] Maleque M A, Masjuki H H and Haseeb A S M A 2000 *Wear* **239** 117-25
[6] Loganathan T M, Purbolaksono J, Inayat-Hussain J I, Muthaiyah G and Wahab N 2010 *Engineering Failure Analysis* **17** 1086-93

[7] Spies H J 2015 *Thermochemical Surface Engineering of* Steels (Oxford Woodhead Publishing) 267-309
[8] Durst O, Ellermeier J and Berger C 2008 *Surface and Coatings Technology* **203** 848-54
[9] Khan P F, Shanthi V, Babu R K, Muralidharan S and Barik R C 2015 *Journal of Environmental Chemical Engineering* **3** 10-9
[10] Bell T *Corrosion Rate Calculator*
[11] Otieno M, Beushausen H and Alexander M 2011 *Prediction of Corrosion Rate in RC Structures - A Critical Review in Modelling of Corroding Concrete Structures* (Springer Netherlands) 15-37
[12] Ebrahimi M, Sohi M H, Raouf H and Mahboubi F 2010 *Surface and Coatings Technology* **205** S261-S266
[13] Wen D C 2009 *Applied Surface Science* **256** 797-804
[14] Ensinger W, Schröer A and Wolf G K 1992 *Surface and Coatings Technology* **51** 217-21
[15] Genel K, Demirkol M and Gülmez T 2000 *Materials Science and Engineering* **288** 91-100
[16] Liu W 2008 *The Effects of Contaminants on the Gas Nitriding of Nitralloy-135* Master of Science Thesis Worcester Polytechnic Institute

IOP Conf. Series: Materials Science and Engineering **152** (2016) 012060 doi:10.1088/1757-899X/152/1/012060

Effects of different processing techniques on multi-walled carbon nanotubes/silicone rubber nanocomposite on tensile strength properties

N Mazlan[1*], M Jaafar[2], A Aziz[2], H Ismail[2] and J J C Busfield[3]

[1]Department of Aerospace Engineering, Faculty of Engineering, Universiti Putra Malaysia, Malaysia
[2]School of Materials and Mineral Resources Engineering, Universiti Sains Malaysia, Malaysia
[3]Department of Materials, Queen Mary, University of London, United Kingdom

*norkhairunnisa@upm.edu.my

Abstract. In this work, two different processing techniques were approached to identify the properties of the multi-walled carbon nanotubes (MWCNT) reinforced polydimethylsiloxane (PDMS). The MWCNT was dispersed in the polymer by using the ultrasonic and twin screw extruder mixer. The final composite showed different manner of dispersed tubes in the silicone rubber matrix. High shear twin screw extruder tends to fragment the tubes during processing compound, which can be observed by scanning electron microscope (SEM). Tensile strength of the extrusion MWCNT/PDMS nanocomposites was found to be higher compared to ultrasonic MWCNT/PDMS nanocomposites.

1. Introduction

One dimensional (1D) tubular structure of carbon nanotubes (CNT) has stimulated intensive research due to their unique structure and also improved mechanical, thermal and electronic properties when combined with polymers. The remarkable properties of the CNT are well known and their exploitation for a wide range of applications can replace existing materials. Silicone rubber is a versatile material that makes it flexible above its glass transition temperature (approximately -127^{0}C). In aerospace applications, polydimethylsiloxane (PDMS) has been used for adhesion for silica thermal protection tile, heat sinks on printed circuit board and also strain isolation pad for space shuttle. In addition, the elasticity of the silicone rubber can be changed by varying its molecular structure, network density, molecular size and functional contents. This further creates various forms of silicone rubber for wider applications.

Recently, great interests have been focused on the development of method for the preparation of CNT composites in exploiting their remarkable properties. The effective utilization of CNT in polymer depends on the homogeneity of the tubes in the matrix resin and on the interfacial adhesion between the tubes wall and the polymer matrix [1, 2]. Unfortunately, it is difficult to disperse CNT in polymeric materials due to nano size of the tubes, close together with high surface energy that induces intrinsic van der Waals interactions between individual tubes that makes the tubes entangled together [2]. Many attempts have been done on dispersing the CNT uniformly in the matrix such as modifying the CNT surfaces by oxidizing it with mixture of acids or attachment with free radicals on the tube surfaces, physically mix with ultrasound or use high shear mixing instrument and mixture with surfactant or solvents, etc.

Most of the researchers compound the CNT in rubber with aid of solvent by means of ultrasonic to disperse the CNT in the suspension [3, 4]. Besides ultrasonic, others have managed to compound the CNT in rubber by using high shear mixer [5]. It is important to ensure that the processing technique should be less damaging to the integrity of the tube surfaces. If the CNT was badly digested during dispersions, the conductivities of the CNT composites could be poor [6]. Inadequate dispersion not only reduces the filler-matrix interaction but also shows different concentrations of matrix and CNT boundaries in the composites, which reduce the desirable properties of the CNT composite. A common method for preparing CNT polymer composite is simply by mixing the CNT with solvent or aqueous solutions of the polymer and evaporating the solvents in a controlled way. Ultrasonic can be used to disperse the CNT in various solvent or liquid polymer [7]. The low viscosity of the solution allows the tubes to move freely through the matrix during the sonicated process. Hernandez *et al.* [8] used both types of ultrasonic, which are ultrasonic bath and ultrasonic tips, to prepare stock dispersion of CNT in polyvinyl alcohol (PVA)-water solution. The drawback of the sonication process is not only limited to polymers, which are freely dissolved in common solvents, but also take a long time to break down the aggregate materials as to uniformly disperse through the liquid medium [9, 10]. Lack of processing time reduces the efficiency to break down the aggregate CNT into individual tubes and less uniformity of the tubes dispersion [11]. However, long sonication process was known to open the tube end and fragment the tube length [4, 6, 12]. Thus an optimum time for CNT sonication is desirable to minimize the structural damage [13].

High shear mixing is one of the most preferable techniques in dispersing the CNT in matrix phase. Hammel *et al.* [14] extruded nanofibers without addition of modifiers in Polypropylene (PP) resin and compound for 5-30 minutes. The aggregate of fibers after the compounding process was found to be individually dispersed. As a result, they achieved improvement in Young Modulus of about 90% after reinforcing the fiber in PP. Li *et al.* [15] compounded 0.5 to 10 wt% of CNT in polyethylene (PET) and observed that the CNTs were homogeneously dispersed in PET matrix. As CNT concentration increased, the viscosity of the melt composite increased and led to stronger shearing thinning effect. However, by prolonging the extrusion process, the fiber tend to break down due to the shear process that destroyed the filler agglomeration and conductive network [16].

In this paper, the MWCNT/PDMS nanocomposite was compounded by using ultrasonic and twin screw extruder to understand the effects of the different processing techniques. The CNT sonication process was performed in liquid medium to break down the CNT aggregates and disperse them well in the suspension. Use of high shear twin screw mixer also tends to fragment the entangled CNT into short tube length. The tensile property was characterized by using Instron and the resulting micro-structure was observed under Scanning Electron Microscope (SEM).

2. Experimental

2.1. Materials

Polydimethylsiloxane (PDMS) type PP2-OE^{TM}41, which consists of two parts, was obtained from Gelest, Morrisville, PA and used as the matrix. Raw multi-walled CNT (Nanocyl 7000) with diameter of 9.5nm, length of 1.5μm and surface area of 250-300m^2/g was supplied by Nanocyl S. A. (Belgium). They are produced by CVD process and their purity is 90%. Toluene, which was supplied by Sigma Aldrich, was used as dispersing medium for the tubes when compound using ultrasonic method.

2.2. Preparation procedures

Ultrasonic and high shear twin screw extrusion were used to prepare the MWCNT/PDMS composites. Firstly, ultrasonic is one of the preferable methods. It delivers high levels of vibrational energy to the suspension. In this work, the as received MWCNT (>90% purity) was added into toluene using the required volume fraction loading. The suspension was then sonicated using ultrasonic tip until the MWCNT was completely dispersed in the toluene solvent. PDMS part A was then added inside the MWCNT/toluene suspension and the sonication process was continued for another 40 minutes. Next, the MWCNT/PDMS/toluene mixture was stirred vigorously using mechanical stirrer at 70^{0}C so as to remove the toluene. In next step, part B that consists of the curing agent was added with continuous stirring. The uncured composite was then cast on the mould and placed under vacuum to remove the trapped air. After de-gassing, the solution was then pressed and cured to desired thickness by using compression molding at 55^{0}C for 4 hours.

Another preferable method is to use high shear mixer such as twin screw extruder. Solvent was not required when using the twin screw extruder. PDMS part A was injected inside the DSM Micro 15 mini extruder. The screw speed was set to 200 rpm and maintained for 2 minutes. Then the desired amount of carbon nanotubes were mixed together with PDMS part A. The mixture was mixed together for 10 minutes. After that, part B (curing agent) was injected into the extruder machine and the process was continued for another 5 minutes. After that, the uncured composite was extruded on the mould, followed by vacuum as to release the trapped air. The uncured composite was then pressed and cured with compression molding at 55^{0}C for 4 hours.

2.3. Characterization of sample
Tensile tests were conducted on Instron 5566, equipped with a video extensometer. The gauge length and cross-head speed used for the cured composite were 40mm and 50mm/min, respectively, at room temperature. The morphologies of fractured surface of the composite were observed with a scanning electron microscope (SEM) (JEOLJSM-6300F).

3. Results and discussion
Typical stress strain diagram are given in Figure 1 and Figure 2. For both processing systems, presence of MWCNT was shown to act as reinforcing filler that tend to increase the strength of the composites. A reinforcing effect was achieved by an efficient filler-matrix stress transfer, which can be obtained by a good adhesion between filler and matrix [17, 18]. The stress and strain level were found to increase at the same time by the addition of MWCNT. At a low strain modulus ≤5%, the result indicated the filler-filler networks in the composite. High filler loading with high surface area provided higher filler networking [19]. From this result, it is deduced that the reinforcing effect of MWCNT in the PDMS is obvious. As for ultrasonic processing composite, the strain of the composite was found to increase up to 1.5 vol% addition of filler loading. However, the strain slightly reduced when reaching 2.0 volume % of filler loading. A possible reason is at high filler concentration, the composite becomes stiffer due to large entanglement of CNT in the silicone resin. High shear mixer tends to break the CNT bundle into smaller size of aggregates or individual tubes in the silicone rubber. As can be seen from Figure 2, the strength of the MWCNT/PDMS increased proportionally with the filler content. High shear twin screw extruder is capable on breaking down high volume of CNT aggregates and disperses them well in the silicone rubber. Therefore, high energy is needed to separate a CNT bundle and reduce the tube entanglement [20].

Table 1 tabulates the mechanical properties for ultrasonic and mini extruder compounding process of the MWCNT/PDMS nanocomposites. The elongation at break at modulus 100 indicates that during stretching takes place, there is an interfacial slippage occurred between the tubes and silicone rubber matrix [21]. Strong bonding between filler and matrix needs higher strain to pull apart the filler from the matrix. A good filler dispersion exposed the tubes surface area to adhere with the silicone rubber and hence, prevent aggregation of the tubes. The property of elastomer is controlled by the degree of the cross-links in the network. The MWCNT in this work was used as received and therefore, there is no possible chemical reactions can occur between the tubes and the silicone rubber. Interactions may occur between the MWCNT and rubber molecules through entanglements and classical non-bonded interactions [3].

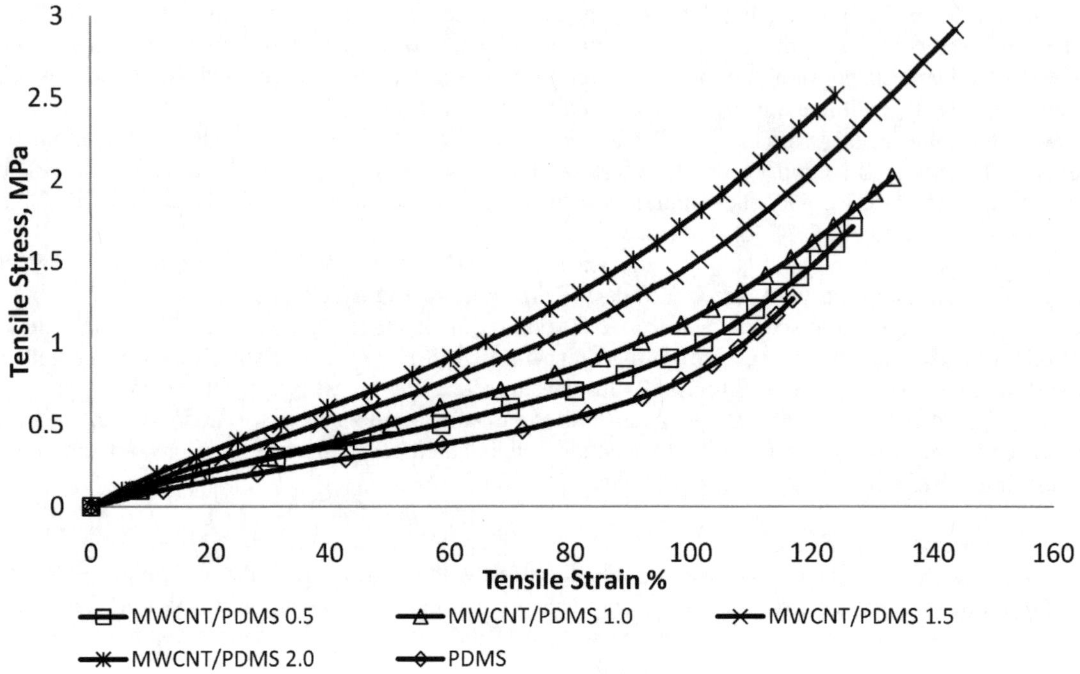

Figure 1. Typical stress-strain curves for MWCNT/PDMS nanocomposite by sonication process ranging from 0.5 to 2.0 vol% of filler content

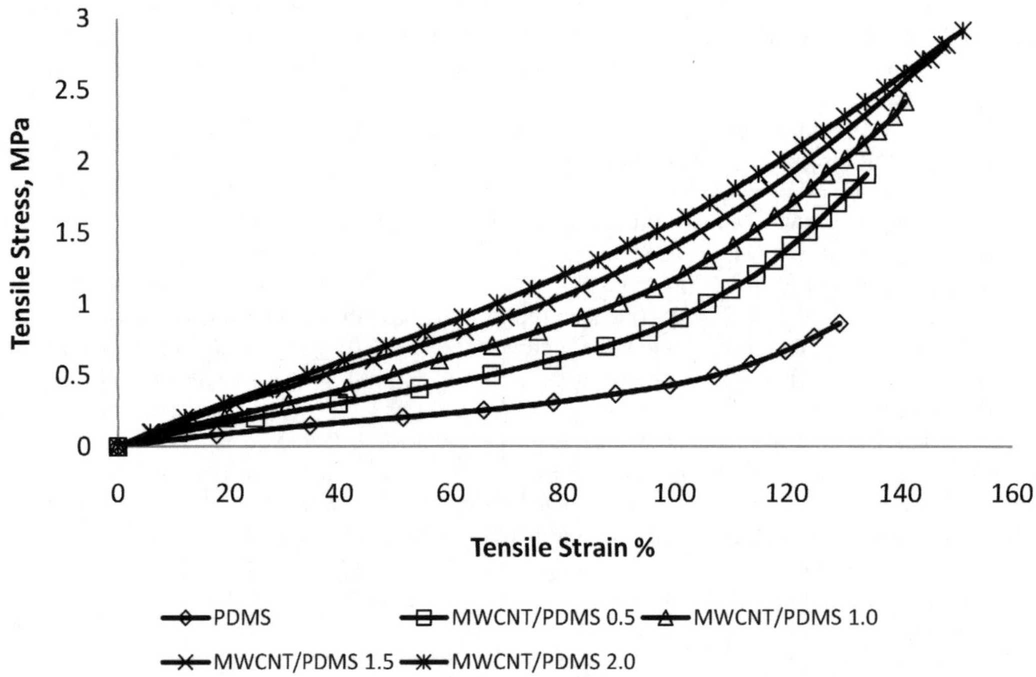

Figure 2. Typical stress-strain curves for MWCNT/PDMS nanocomposite by twin screw extruder process ranging from 0.5 to 2.0 vol% of filler content

Table 1. The mechanical properties of the MWCNT/PDMS composites

Method	Nanocomposites	Modulus at 100% elongation (MPa)	Elongation at break (%)
	PDMS	0.6735 ± 0.03	129.67 ± 5.83
Ultrasonic	MWCNT/PDMS 0.5vol%	0.9557 ± 0.05	132.69 ± 2.51
	MWCNT/PDMS 1.0vol%	1.0048 ± 0.03	139.71 ± 4.61
	MWCNT/PDMS 1.5vol%	1.4785 ± 0.02	140.48 ± 3.81
	MWCNT/PDMS 2.0vol%	1.4061 ± 0.02	101.90 ± 12.38
Mini extruder	MWCNT/PDMS 0.5vol%	0.9461 ± 0.02	132.09 ± 5.29
	MWCNT/PDMS 1.0vol%	1.1805 ± 0.02	132.43 ± 3.25
	MWCNT/PDMS 1.5vol%	1.5265 ± 0.04	141.10 ± 2.72
	MWCNT/PDMS 2.0vol%	1.6081 ± 0.03	149.34 ± 3.85

The tensile fractured surfaces of the composites are presented in Figure 3. Figure 3a and Figure 3b indicate the fracture surface of MWCNT/PDMS compound by means of ultrasonic. As can be seen, the aggregate MWCNT is broken down at low filler content. The presence of a high amount of MWCNT seemed to make the dispersion in silicone rubber much difficult, which is as apparent in Figure 3d. As a consequence, the tensile strength slightly reduced due to the presence of cluster MWCNT. Compared with the SEM images in Figure 3c and Figure 3d, the MWCNT was found dispersed throughout the silicone rubber matrix. The obvious difference that can be observed from ultrasonic and twin screw extruder method is the tube length. High shear process induced shortening of tube length, which tends to reduce the surface area of the tubes. As a matter of fact, it increases the possibility to disperse the short tubes throughout the resin during compounding process.

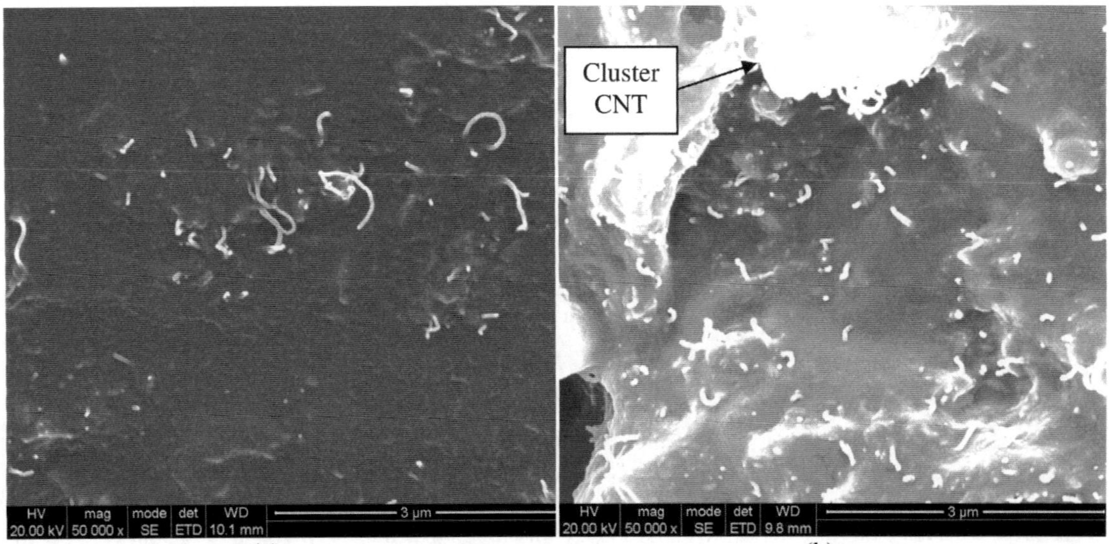

(a) (b)

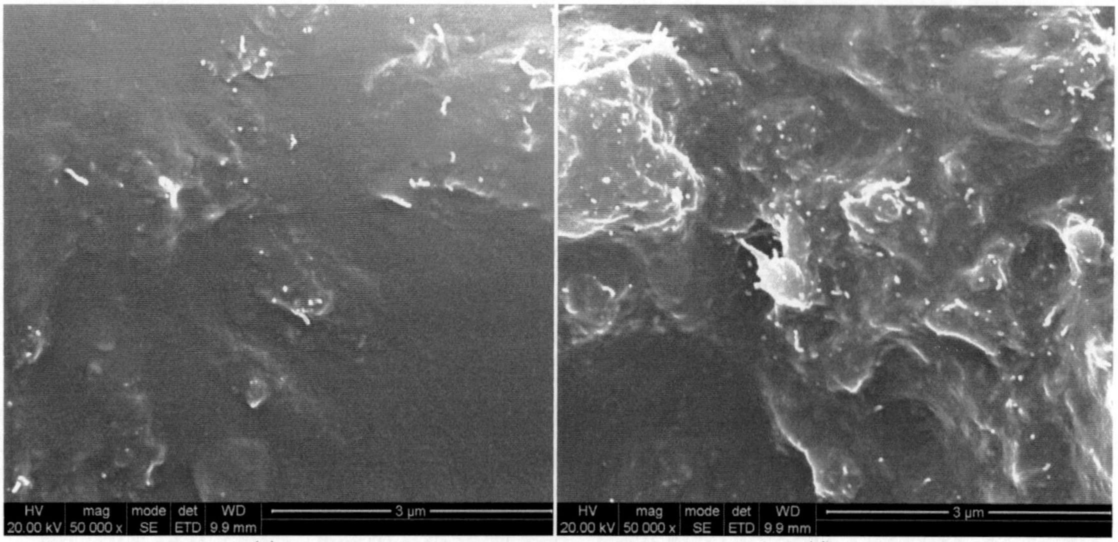

(c) (d)

Figure 3. SEM micrograph of the fracture surface of: (a) 0.5 vol% MWCNT/PDMS nanocomposite compound by ultrasonic, (b) 2.0 vol% MWCNT/PDMS nanocomposite compound by ultrasonic, (c) 0.5 vol% MWCNT/PDMS nanocomposite compound by twin screw extruder, (d) 2.0 vol% MWCNT/PDMS nanocomposite compound by twin screw extruder

Figure 4 illustrates the schematic drawing on the behavior of the nanotubes in silicone rubber. The nanocomposites prepared by sonication method showed that the tubes intercalated within the silicone matrix (Figure 4a). Addition of toluene reduces the viscosity of the silicone resin and allows the silicone molecules to move freely through the galleries of nanotubes. The agglomerate tubes can expand and break down into small aggregates or single tubes as the silicone molecules pass through the entangle tubes. Therefore, this method tends to give better conductivity result as the tubes performed conductive network, which can be explained in the conductivity session. Compared to the extrusion method, the CNT degrades to a large extent. Twin screw extruder mixer is capable of cutting the tube length into short segments. Therefore, the tube was found to be dispersed very well in the silicone matrix. However, shortening the tube length can reduce the aspect ratio of the tubes due to the gradual breakage of CNT during high shear process. As a consequence, less interconnect tubes and less formation of conductive path (discussed in conductivity session).

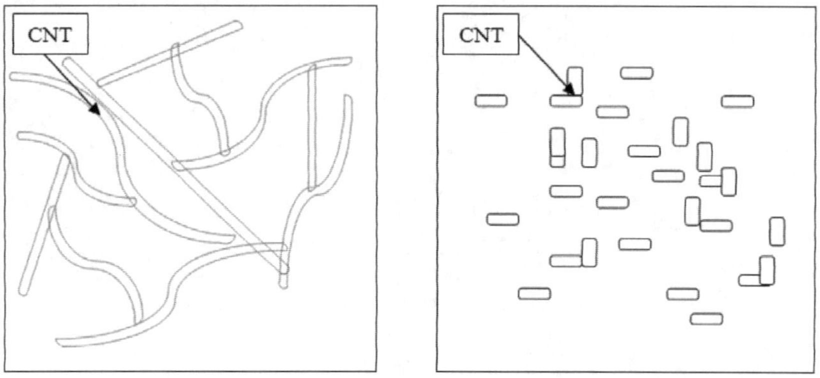

(a) (b)

Figure 4. Schematic drawing of: (a) ultrasonic compound MWCNT/PDMS nanocomposite, (b) twin screw extruder compound MWCNT/PDMS nanocomposite

4. Conclusion

MWCNT/PDMS nanocomposites have been successfully prepared using ultrasonic and twin screw extruder methods. The dispersion of the tubes was examined by SEM, which indicates dispersion and random orientation of the tubes. Moreover, high shear extrusion process shortened the length of the tubes compared to ultrasonic process. The strength of the extrusion MWCNT/PDMS nanocomposites was found higher due to the good dispersibility of the tubes in the silicone matrix. However, excessive cutting of the tubes tend to reduce the thermal conductivity.

Acknowledgement

The authors acknowledge the British Council for the financial support, Queen Mary, University of London and Universiti Sains Malaysia for providing necessary facilities.

References

[1] Loos M R, Pezzin S H, Amico S C, Bergmann C P and Coelho L A F 2008 *J. Mater. Sci.* **243** 6064-69
[2] Shen J, Huang W, Wu L, Hu Y and Ye M 2007 *Compos. Sci. Technol.* **67** 3041-50
[3] Zhao Q, Tannenbaum R and Jacob KI 2006 *Carbon* **44** 1740-45
[4] Fakhru'l-Razi A, Atieh M A, Girun N, Chuah T G, El-Sadig M and Biak DRA 2006 *Compos. Struct.* **75** 496-500
[5] Kim Y A, Hayashi T, Endo M, Gotoh Y, Wada N and Seiyama J 2006 *Scripta Mater.* **54** 31-5
[6] Wang Y, Wu. J and Wei F 2003 *Carbon* **41** 2939-48
[7] Jiang M–J, Dang Z –M, Yao S–H and Bai J 2008 *Chem. Phys. Lett.* **457** 352-6
[8] Hernandez Y R, Gryson A, Blighe F M, Cadek M, Nicolosi V, Blau W J, Gun′ko Y K and Coleman J N 2008 *Scripta Mater.* **58** 69-72
[9] Biercuk M J, Llaguno M C, Radosavljevic M, Hyun J K and Johnson A T 2002 *Appl. Phys. Lett.* **80** 2767
[10] Lioudakis E, Kanari C, Othonos A and Ioannis A 2008 *Diam. Relat. Mater.* **17** 1600-3
[11] Sandler J, Shaffer M S P, Prasse T, Bauhofer W, Schulte K and Windle A H 1999 *Polymer* **40** 5967-71
[12] Cipiriano B H, Kashiwagi T, Raghavan S R, Yang Y, Grulke E A, Yamamoto K, Shields J R and Douglas J F 2007 *Polymer* **48** 6086-96
[13] Yang C, Hu X and Zhang Y 2006 *IEEE Conf. on Nano/Micro Engineered and Molecular Systems* 83-6
[14] Hammel E, Tang X, Trampert M, Schmitt T, Mauthner K, Eder A and Pötschke P 2004 *Carbon* **42** 1153-8
[15] Li Z, Luo G, Wei F and Huang Y 2006 *Compos. Sci. Technol.* **66** 7-8
[16] Alig I, Lellinger D, Engel M, Skipa T and Pötschke P 2008 *Polymer* **49** 1902-9
[17] Silva V P, Gonçalves M C and Yoshida I V P 2006 *J. Appl. Polym. Sci.* **101** 290-9
[18] Sui G, Zhong W H, Yang X P and Yu Y H 2008 *Mater. Sci. Eng. A-Struct.* **485** 524-31
[19] Fröhlich J, Niedermeier W and Lugisland H–D 2005 *Compos. Part A-Appl S* **36** 449-60
[20] Huang Y Y and Terentjev E M 2008 *Int. J. Mater. Form* **1** 63-74
[21] Yue D, Liu Y and Shen Z 2006 *J. Mater. Sci.* **41** 2541-4

IOP Conf. Series: Materials Science and Engineering **152** (2016) 012061 doi:10.1088/1757-899X/152/1/012061

A study on flexural and water absorption of surface modified rice husk flour/E-glass/polypropylene hybrid composite

K Rassiah*, T W Sin and M Z Ismail

Department of Mechanical Engineering, Politeknik Port Dickson (PPD), Negeri Sembilan, Malaysia.

*kannan@polipd.edu.my

Abstract. This work is to study the effects of rice husk (RH)/E-Glass (EG)/polypropylene (PP) hybrid composites in terms of flexural and water absorption properties. The tests conducted are the flexural test and also the water absorption test using two types of water: distilled and sea water. The hybrid composites are prepared with various ratios of fibre weight fractions and the rice husk is treated using 2% Sodium Hydroxide (NaOH) to improve interaction and adhesion between the non-polar matrix and the polar lignocellulosic fibres. It was found that the content of rice husk/E-Glass fillers affected the structural integrity and flexural properties of hybrid composites. In addition, a higher ratio of rice husk contributes to higher water absorption in the hybrid composites.

1. Introduction

Common polymer matrices used in composites can be classified into thermoplastics, thermosets or elastomers. Polypropylene (PP) belongs to the group of thermoplastics, which is the synthetic polymer with many applications. Due to its high-volume and low-cost, PP is produced in large quantities and widely used in fabrication of automotive parts since 1959. As of now, most parts are produced using petroleum-based plastics but alternatives are needed to supplant the plastics materials while giving the suitable mechanical execution and ideally reducing the cost for the final item. With the increase in the environmental cognizance, ordinary fibres have growing interest to be used as reinforcing resources for fibre-strengthened composites due to their low cost, low density, renewability, biodegradability, and high particular stiffness and strength. Synthetic or natural fibres such as coconut coir, bamboo, banana, sugar cane, kenaf and palm oil are being used by many researchers [1-6]. Fabricating particle reinforced composites is much easier and with less cost than fabricating fibre reinforced thermoplastic composites. With polymeric matrices, the particles are easy to be added into the polymer melt in an internal mixture apparatus during polymer processing. Presence of particles can decrease the ductility and permeability of the matrix, and increase the modulus of the matrix. Filler is generally associated with a low cost material that is added into the formulation to decrease the cost of matrix materials [7].

Rice husk is an alternative natural fibre for production of composite and hybrid composites. In general, the fibres usage is also not thoroughly free from issues. A natural fibre, which is hydrophilic material, does not have good interaction with wide range polymers, corresponding to sago starch and thermoplastic [8]. Rice husk is a product from rice mills and it creates disposal and pollution issues. A few papers were published with discussions on the rice husk [7, 9-13]. Efficiency and performance of the composite materials rely on the properties of each material component and also their interface similarity. To make sure suitable interface interactions between them, their surface properties must be modified thoroughly. The essential ingredients of rice husk are ash, lignin and cellulose [10]. Figure 1 shows the microstructure picture of untreated rice husk.

Researches in rice husk fibre-strengthened composites have targeted the development of material properties through chemical response. Studies thus far have investigated several treatment agents that can modify rice husk surfaces to reinforce its efficiency. Chemical treatment of cellulosic materials usually changes both chemical and physical structures of the fibre through the reaction of OH groups on the cellulose fibres. There have been reports confirming that chemically-handled natural fibres are extra stable and can face up to a greater extent of heat compared to untreated fibres [1-6] during the course of thermal cycling. If the treatment improves the distribution and orientation of the fibres, then the fibre surface becomes more hydrophobic.

The objective of this work is to explore the effects of treated flour rice husk loading on flexural strength and water absorption behaviour of the RH/EG/PP hybrid composite. The hot press surface of the hybrid composite is also examined to study fibre-matrix interaction and distribution of fibre in the matrix.

2. Material

The materials utilized were "Polypropylene" (polymer) and rice husk (natural fibre) as a reinforcing of glass fibres (synthetic fibre).

2.1. Polypropylene

Polypropylene has been utilized as the matrix materials in the production of polymeric materials. The Polypropylene used in this study was of a "pellet" type, which was Polypropylene TITANPRO 6331 with melt stream rate at 230°C and thickness of 0.9g/cm^3.

2.2. Rice husk

Rice husk was obtained from the rice field in Kedah, Malaysia. The average granule size was 800 μm and the moisture content was 13 wt%.

2.3. E-Glass

The manufactured fibre in this study was E-Glass fibre (E2-4533 6000), which was supplied with the guide of DK composite. The primary ingredients of E-Glass fibre were 4.5 wt.% magnesium oxide, 8 wt.% boron oxide, 17 wt.% calcium oxide, 15 wt.% aluminum oxide and 54 wt.% silicon oxide. The E-Glass fibre has an elastic modulus of 72.5 GPa and a density of 2.59 gm/cc. The selected fibre glass size was cut in to 6 mm.

2.4. Formulation

The components of rice husk, E- Glass and polypropylene were separated into six compositions as shown in Table 1.

Figure 1. Untreated rice husk

Table 1: Composition ratio of rice husk (RH), E-Glass (EG) an Polypropylene (PP)

Sample	RH (wt. %)	EG (wt. %)	PP (wt. %)
1	0	0	100
2	0	60	40
3	60	0	40
4	30	30	40
5	15	45	40
6	45	15	40

2.5. Treatment of rice husk

The washed rice husk was boiled in water with temperature of 100°C for roughly 60 minutes. Figure 2 shows the treatment procedure of the boiled rice husks. The boiled rice husk was soaked for 120 minutes with 2% concentration of sodium hydroxide, and it was then dried under direct sunlight for 4320 minutes. Figure 3 depicts the microstructure of the boiled and the NaOH treated rice husks. The NaOH treated rice husk showed rougher surfaces than that of the boiled rice husk.

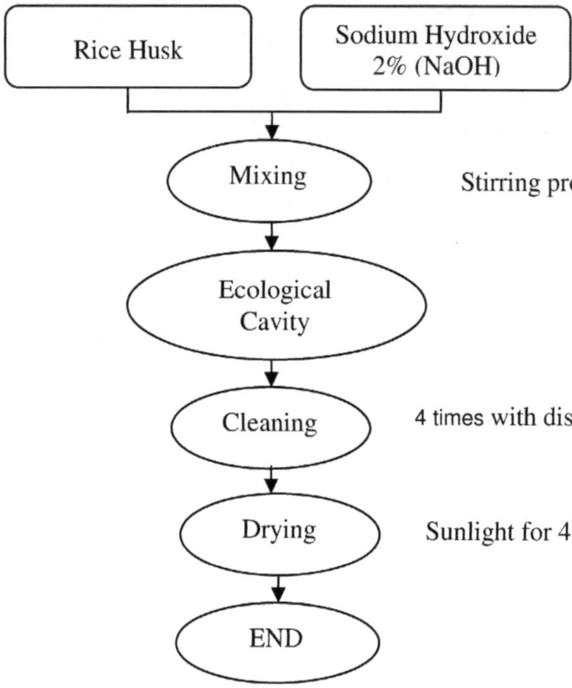

Figure 2. The treatment process of rice husk

Figure 3. Microstructure of: (a) treated boiling rice husk, (b) NaOH treated rice husk

3. Experiment

A thermal mixing method was implemented using HAAKE™ Rheomix Lab Mixers for the HAAKE™ PolyLab™ 600 OS System as the internal mixer at a temperature of 180°C and 480 seconds with a rotor speed of 50 rpm. By means of this approach, the mixture of substances was predicted to be better and more uniform before entering the crusher machine to produce an average particle measurement of 0.2 cm length x 0.2 cm wide. The last phase of the composite development in this study was the hot pressing (Figure 4). Hot pressing was completed at a temperature of 180°C for 600 seconds and the blend was then cooled below the room temperature for 900 seconds. The resultants were plates with measurements of 20 cm length x 20 cm wide x 0.2 cm thick and they were arranged into the cutting pieces as required by ASTM D790 and D570 standard for testing.

3.1. Flexural testing

Three point bending flexural test was performed using the Shimadzu Universal Testing Machine in accordance to ASTM D790. Dimension of the samples was 6 cm length x 1.27 cm wide x 0.2 cm thick and they were tested at a 10 mm/ min of crosshead speed (for seven samples from each composition). The tests were completed in controlled environments with 50% of humidity at 25 ± 3 °C. Figure 5 shows the process while Equation 1 and Equation 2 were used in the solution method.

$$\sigma_f = \frac{3PL}{2bd^2} \qquad (1)$$

where σ = stress at midpoint in outer fibres (MPa), P = load at a given factor on the burden deflection curve (N), L = support span (mm), b = width of experiment beam (mm) and d = depth of demonstrated beam (mm).

$$E_B = \frac{L^3 m}{4bd^3} \qquad (2)$$

where E_B = modulus of elasticity in bending (MPa), L = support span (mm), b = width of experiment beam (mm), d = depth of demonstrated beam (mm) and m = slope of the tangent to the initial straight-line portion of the load deflection curve (N/mm).

Figure 4. Hot press sample

Figure 5. The position of specimen during flexural test

3.2. Water absorption testing

The water absorption test was done by ASTM standard D 570. It involved a whole immersion of the seven specimens in distilled and sea water at room temperature. Specimens with measurements 3 cm length x 1 cm wide x 0.2 cm thick were removed from the panel using a hand saw. All the specimens were dried in an oven at 50 °C for 720 minutes and then kept in desiccators. The water absorption test was done by weighing the specimens at regular intervals using a weighing scale. The percent moisture content/weight gain, Mi is calculated by Equation 3.

$$M_i = \frac{W_i - W_b}{W_b} \times 100 \qquad (3)$$

where M_i = weight gain in percentage, W_i = weight of the specimen and W_b = oven dry specimen weight or baseline weight.

3.3. Observation of Scanning Electron Microscope (SEM)

The microstructure of materials was obtained using the FEI/Philips XL-30 Field Emission ESEM that activated from 20 kV till 30 kV. The fractured specimens from flexural test were initially dried and coated with the gold using (E-5100) Sputter Coater Polaron classification. Scattering condition of the RH/EG fibre and break type of the hybrid composites under the hot press processing were determined.

4. Results and discussion

4.1. Flexural test

As shown in Figure 6 and Figure 7, flexural strength and modulus were reduced by the accumulation of natural fillers. Addition of 60 wt. % rice husk fibre to 40 wt. % PP resulted in decrease of flexural strength and modulus. However, addition of fibre EG increases the flexural strength and modulus of the hybrid composite. PP 40 wt. %, EG 45 wt. % and RH 15 wt. % mixture was observed and the flexural strength and modulus with composition of EG were higher compared to other compositions. It is proven by the result where the 45 wt. % of rice husk shows a lower value in hybrid composition. According to Ismail *et al.* [12] and Yang *et al.* [13], this is because the incapability of the fibre to support the stresses transferred from the matrix polymer to reinforcement, and negative interfacial bonding caused partly areas between polymer and reinforcement, and thus generated a weak structure.

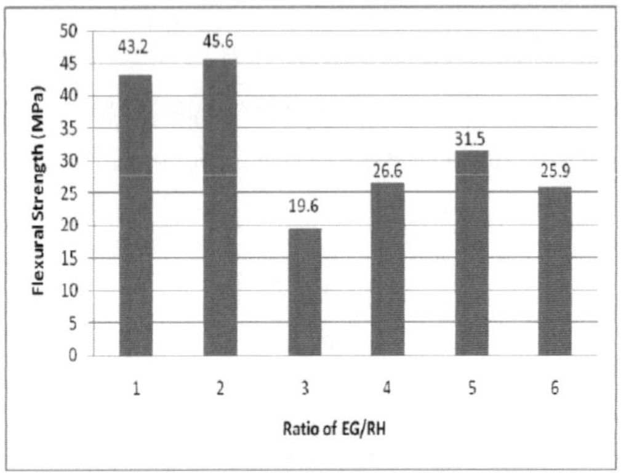

Figure 6. Flexural strength against ratio composition of EG/RH

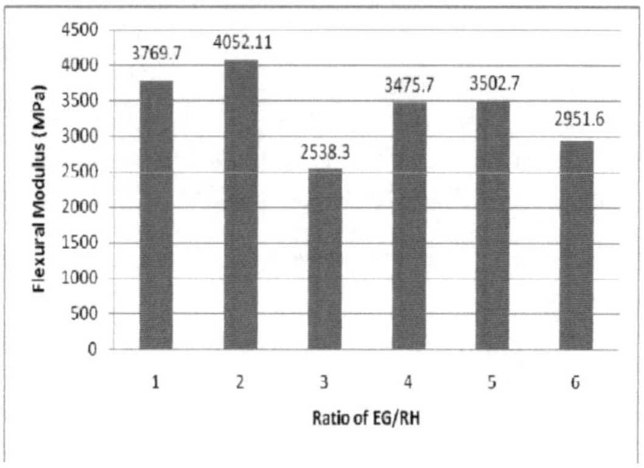

Figure 7: Flexural modulus against ratio composition of EG/RH

4.2. Water absorption analysis

Figure 8 and Figure 9 exhibit the percent of two types of water absorption. They show the trend of two distinct forms of distilled and sea water in composite material. In the beginning, all specimens linearly increased in moisture absorption and achieved their immersion state at the highest moisture. From Figure 8 and Figure 9, composites with greater rice husk filling had the greatest absorption percentage. It was reported that fillers such as chopped rice husk, kenaf and wood correspond to much greater absorption when the fibre loading is increased [11, 14-15]. The addition of the EG fibre decreased the percentage of distilled and sea water absorption. The less amount of rice husk cellulose interaction with NaOH will give a low OH content [16].

The hydrophilic nature of rice husk causes water absorption in hybrid composites. Generally, the natural fibre strengthened composites exhibit large water absorption due to the presence of micro-voids at the reinforcement/matrix interface [17]. The constant polypropylene content of 40 wt% in all formulations and the different water absorptions among all hybrid composites can be used to illustrate the role of RH/EG ratio. The hybrid composite shows a moderate water absorption, which is 10.8%, 9.3%, 8.2% for RH 45 wt% / EG15 wt%, RH 30 wt% / EG 30 wt% and RH 15 wt% / EG 45 wt%, respectively for distilled water and 11.9%, 10.8%, 9.3% for RH 45 wt%/ EG 15 wt%, RH 30 wt% / EG 30 wt% and RH 15 wt% / EG 45 wt%, respectively, for sea water. This shows that the hybrid compound of fibre with treated rice husk fibre will limit the absorption of moisture into the distilled and sea water.

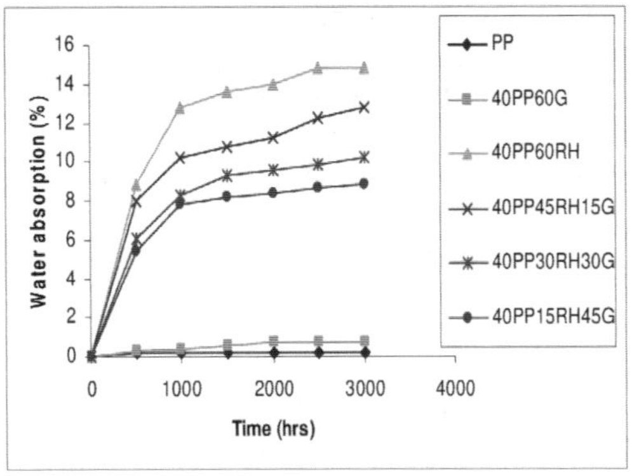

Figure 8. Graph of water absorption for distilled water

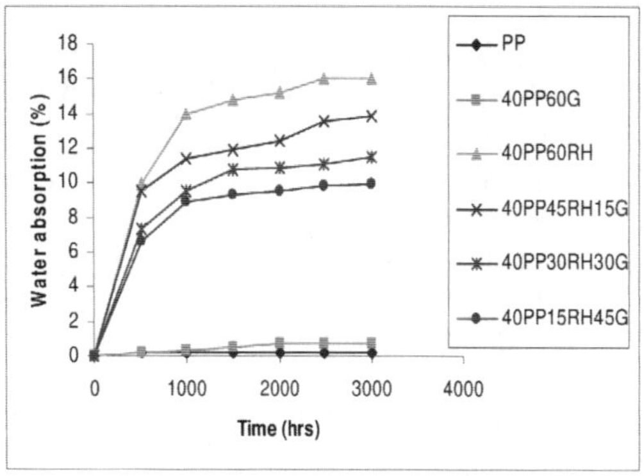

Figure 9: Graph of water absorption for sea water

4.3. Morphology of cracked specimens

The scanning electron microscope (SEM) was used to analyse the surface morphology of hot press specimens. The bonds surrounded by the RH/EG/PP hybrid composite structures were recognized by using the surface morphology. It is further used to observe the variations in the composite after mixing accordingly in hybrid form. Figure 10, Figure 11 and Figure 12 show results of surface morphology. The scanning electron microscope photos in Figure 10 and Figure 11 show the treated rice husk 30 wt. % and 15 wt. %. The RH/EG/PP images prove that integration mixed easily and have useful bonding surrounded by them.

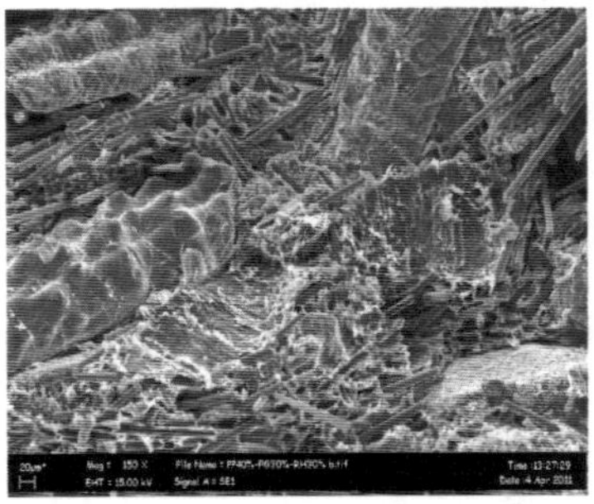

Figure 10. Mixture of 30 wt.% RH, 30 wt.% EG, 40 wt.% PP

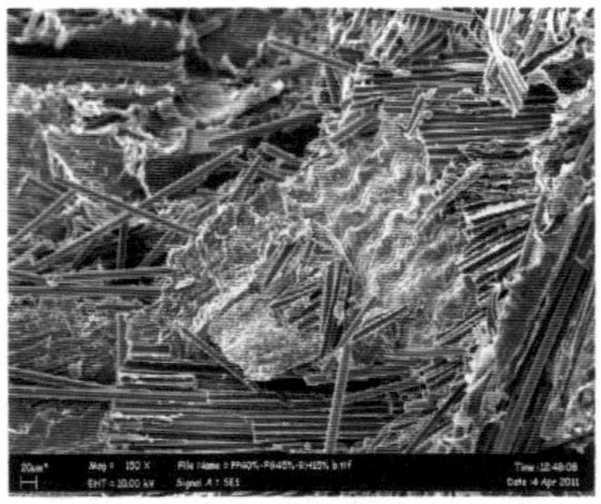

Figure 11. Mixture of 15 wt.% RH, 45 wt.% EG, 40 wt.% PP

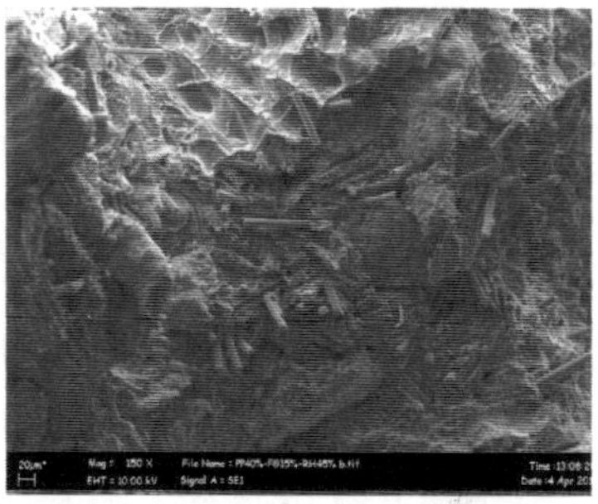

Figure 12. Mixture of 45 wt.% RH, 15 wt.% EG, 40 wt.% PP

5. Conclusion

In this study, it was found that the content of RH/EG filler affected the structural integrity, flexural strength and water absorption behaviours of composites. Based on the observation, increasing the amount of the rice husk in PP/EG resulted in a decrease in flexural values. The water absorption value of the RH/EG filled PP was measured and it was observed that 15 wt. % rice husk had less absorption for distilled and sea water. It is thought that the hot press effects of the RH/EG led to an increased rate in orientation of the polymer. Rice husk in minimum content was taken to be an alternative additive to the PP/EG. If a suitable hybrid composition could be added to rice husk/E-Glass and PP mixtures, the flexural strength and water absorption properties can possibly be increased.

Acknowledgments

The authors acknowledge Jabatan Pendidikan Politeknik for association with this study effort, and Department of Mechanical Engineering, Polytechnic Port Dickson and Faculty of Manufacturing Engineering, Universiti Teknikal Malaysia Melaka for allowing consent to use the apparatus.

References

[1] Cao X V, Ismail H, Azura A R and Takeichi T V 2011 *BioResources* **6** 3260-71
[2] Kittinaovarat S and Suthamnoi W 2009 *Journal of Metals, Materials and Minerals* **19** 9-15
[3] Wirawan R, Zainudin E S and Sapuan S M 2009 *Sains Malaysiana* **38** 531-5
[4] Majhi S K, Nayak S K, Mohanty S and Unnikrishnan L 2010 *International Journal of Plastics Technology* **14** 57-75
[5] Kannan R, Mohd As'ri C, Mohd Yuhazri Y and Sihombing H 2011 *International Journal of Integrated Engineering* **3** 63-7
[6] Ahmadzadeh A and Zakaria S 2009 *Iranian Polymer Journal* **18** 247-55
[7] Premalal H G B, Ismail H and Baharin A 2002 *Polymer Testing* **7** 833-9
[8] Rohani A M, Hanafi I and Razaina M T 2010 *Iranian Polymer Journal* **7** 501-10
[9] Kim H S, Yang H S, Kim H J and Park H J 2004 *Journal of Thermal Analysis and Calorimetry* **76** 395-404
[10] Panthapulakkal S, Sain M and Law S 2005 *Polymer International* **54** 137-42
[11] Naori R M, Dogouri J F, Oromiehie A and Langroudi E A 2006 *Iranian Polymer Journal* **15** 757-66
[12] Ismail H, Nizam J M and Khalil H P S A 2001 *Polymer Testing* **20** 125-33
[13] Yang H S, Kim H J, Son J, Park H J and Lee B J H 2004 *Composite Structures* **63** 305-12
[14] Nosbi N, Akil H M, Ishak Z A M and Abubaker A 2011 *BioResourses* **6** 950-60
[15] Najafi K S and Kordkheili Y H 2011 *European Journal of Wood and Wood Products* **69** 553-6
[16] Matuana L M, Balatinecz J J, Sodhi R N and Park C B 2001 *Wood Science and Technology* **35** 191-201
[17] Kannan R, Sasitharan N and Jidin M R 2012 *Canadian Journal on Mechanical Sciences & Engineering* **3** 19-24

IOP Conf. Series: Materials Science and Engineering **152** (2016) 012062 doi:10.1088/1757-899X/152/1/012062

CFD analysis of flow through Venturi tube and its discharge coefficient

A Tukimin[1*], M Zuber[2] and K A Ahmad[2]

[1]Department of Mechanical and Manufacturing Engineering, Faculty of Engineering, Universiti Putra Malaysia, Malaysia
[2]Department of Aerospace Engineering, Faculty of Engineering, Universiti Putra Malaysia, Malaysia

*agustinnie@gmail.com

Abstract. Venturi tube plays a very important role in different fields of engineering. It has a number of industrial applications in which its design is an essential factor. Venturi tube used in gas measurement applications provides an accurate critical gas flow measurement. There is a need to design Venturi tube with an effective analytical tool or software. In this work, two parameters: pressure drop and velocity discharge nozzle were analyzed using Computational Fluid Dynamics (CFD). The results obtained were then analyzed for accurate determination of the Venturi tube's discharge coefficient, Cd. It was found that there is less than 1% difference between the average values of the discharge coefficient obtained from the numerical analysis and experimental results.

1. Introduction

Due to the gaining popularity of Venturi tubes in the gas and wet gas measurement applications, there are increasing studies being developed to understand its behaviours in the application in order to fully benefit from its use. Earlier works have shown that the performance of Venturi tubes in gas is very different from that in water [1, 2]. As the understanding of Venturi tubes' performance in dry gas is the priority before its performance in wet gas can be analysed, the National Engineering Laboratory (NEL) in the United Kingdom, supported by Elf Exploration, Jordan Kent Metering Systems/ Seiko, Mobil North Sea, and Shell International, had conducted the experimental and numerical work to investigate the Venturi tubes' discharge coefficient behaviour in dry gas [3]. The results of the work indicated that the discharge coefficient numerical models are very different between gas and water applications. In addition to the numerical models, further analysis also indicated the Venturi tube surface roughness could impact the discharge coefficient error by more than 5%.

In general, most of the Venturi tubes have to be calibrated either using dry gas or water for the discharge coefficient determination prior to being used in the actual application in order to improve their measurement accuracy. This paper presents the CFD analysis that was carried out to verify the flow through Venturi meter and the discharge coefficient determined by the experiment using dry gas at the high-pressure wet-gas test facility at National Engineering Laboratory (NEL) in Glasgow, United Kingdom. The objectives are to investigate the most effective models for numerical analysis for discharge coefficient determination and to develop faster and cheaper methods for discharge coefficient determination for practical application, especially for cases where big numbers of Venturi meters are involved. The Venturi tube used in this study is manufactured for the well testing loop on a well head platform for PETRONAS Carigali Sdn. Bhd. (PCSB) SK316 Field Development project.

The CFD analysis is the first step for subsequent analysis of the Venturi tube design for the wet gas application.

2. Experimental setup

2.1. Overview of test loop

The high-pressure wet-gas test facility at NEL is based around an 200 mm nominal bore flow loop. A schematic diagram of the nominal facility arrangement for wet-gas tests is provided in Figure 1. Although nominally 200 mm diameter, the test section can accommodate line sizes ranging from 100 mm to 300 mm. The gas used for testing is oxygen-free nitrogen, supplied in 230 bar gauge cylinder banks. The facility typically operates at a nominal temperature of 20°C over a nominal pressure range of 10 to 63 bar gauge, which corresponds to a gas density range of 12.76 to 74.54 kg/m^3.

In Figure 1, the gas is driven around the flow loop by a 200 kW fully-encapsulated gas blower. In wet gas operation, the gas is drawn from the gas-liquid separator outlet by the blower and is then cooled using a chilled water supply flowing through shell and tube heat exchanger. The gas then passes through the reference gas flow meter and into the test line. The gas flow rate is controlled by varying the speed of the blower while the liquid flow rate is controlled by using two variable flow control valves located downstream of the liquid pumps. The maximum achievable dry gas volumetric flow rate is dependent on the size and type of reference or test flow meter installed.

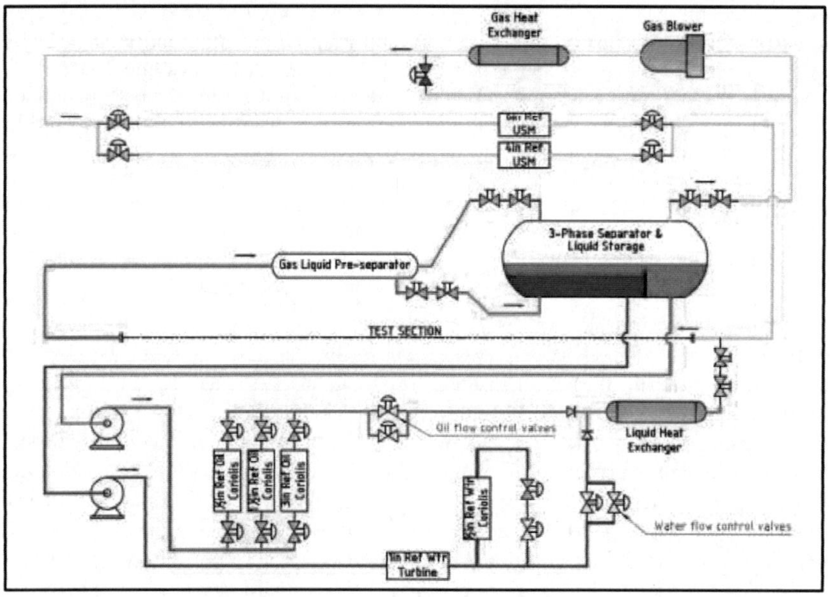

Figure 1. Flow test loop schematic

In Figure 2, the Venturi tube is installed horizontally with the differential pressure, pressure and temperature transmitters mounted closely to the Venturi tube. All the transmitters are connected to NEL data acquisition and control system and the test is automatically run by the system.

2.2. Venturi details

Dimensions of the Venturi used in the experiment as indicated in Figure 3 below. The Venturi is sized according to ISO 5167:2003 [4]. The length of downstream pipe spool, which is located at the end of divergent cone, is 6D in order to get full pressure recovery profile at the downstream end of the Venturi tube. All the units in Figure 3 below are in mm.

Figure 2. Test line setup of Venturi tube

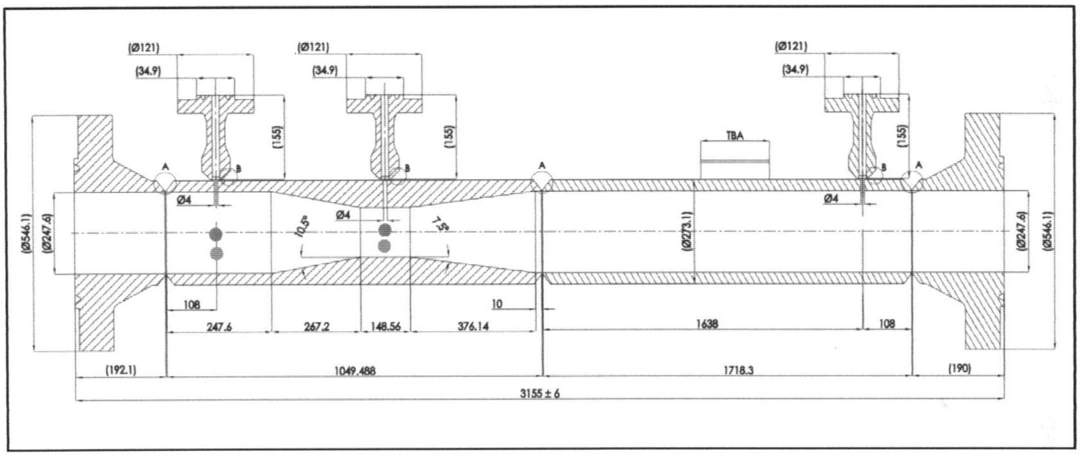

Figure 3. Venturi tube dimensions

The beta ratio, which is the ratio of internal throat diameter to the internal upstream pipe diameter, is 0.6. The nominal diameter of the Venturi is 250 mm and wall thickness of 12.75 mm, with pressure rating of ASME 900 lbs. The convergent angle of the inlet cone is 10.5° and divergent angle of the outlet cone is 7.5°. The two red dots on Figure 3 mark the positions where the pressure measurements are taken to obtain the differential pressure value. The yellow dots mark the positions where pressure measurements are taken to obtain the total permanent pressure loss across the Venturi tube. The green dots mark the positions where velocity measurement is taken for the volumetric flow rate calculation.

2.3. Test matrix

Table 1 gives the selected pressure and temperature test points, which are within the available range provided by the test facility. The selected test points are based on typical flow rates, pressure and temperature ranges for Venturi tubes used in the well test applications at offshore installation in Malaysia [5]. Although the result in this analysis will be used in natural gas application, nitrogen was selected for the test as it is considerably easier to handle and cost effective. In the earlier work by NEL, CFD did not pick up any appreciable effect from changing the nitrogen to other gases and the experiment did reveal the effect was small enough as to be largely ignored [6].

Table 1. Test matrix for the experiment

Test Point	Pressure	Temp	Gas Flowrate	Nitrogen Density	Gas superficial velocity
(-)	(bar gauge)	(°C)	(m^3/h)	(kg/m^3)	(m/s)
1	60.37	20.04	1213	70.921	7.00
2	60.36	20.06	875	70.904	5.05
3	60.30	19.97	520	70.860	3.00

2.4. Discharge coefficient
Discharge coefficient is calculated using Equation (1):

$$Cd = \frac{4m\sqrt{1-\beta^4}}{\pi\varepsilon d^2\sqrt{200000Dp_1\rho_1}} \quad (1)$$

where m is the reference gas mass flow rate (kg/s), β is the diameter ratio (-), ε is the gas expansibility (-), d is the throat diameter (m), ρ_1 is the upstream gas density (kg/m^3) and Dp_1 is the measure of upstream-to-throat differential pressure (bar) [7].

3. CFD setup

3.1. Venturi model
The Venturi model is drawn using ANSYS Design Modeller. The model consists of the internal section of the Venturi tube. For analysis, 2D symmetry has been used for the numerical simulation. Pressure based solver and SIMPLE pressure-velocity coupling have been used as these are applicable for wide range of flow regimes and applicable for this analysis. It also requires less memory (storage) and allows flexibility in the solution procedure.

3.2. Turbulence model
For the analysis, SKE turbulence model has been selected for the numerical simulation. The SKE model solves for two variables: turbulent kinetic energy, k and rate of dissipation of kinetic energy, epsilon. The wall functions are used in this model. The SKE model is selected because of its good convergence rate and relatively low memory requirements [8].

3.3. Near wall meshing
As the differential pressure is measured from the pressure taps at the wall of the Venturi tube, near wall numerical simulation is important. Inflation has been used for the near wall meshing. Based on y+ equals to 1, the first layer thickness for the inflation setup is calculated for each of the test cases. The selected first layer thickness is 0.0011 mm as this is the smallest calculated first layer thickness from the test cases.

3.4. Convergence analysis
The numbers of elements ranges from 100,000 to 1,350,000 are used in the numerical analysis. The convergence for both cases has shown a decrease in residuals by two orders of magnitude for epsilon and continuity which indicates good qualitative convergence. The number of elements of 650,000 has been selected for the economics of computation as shown in Figure 4.

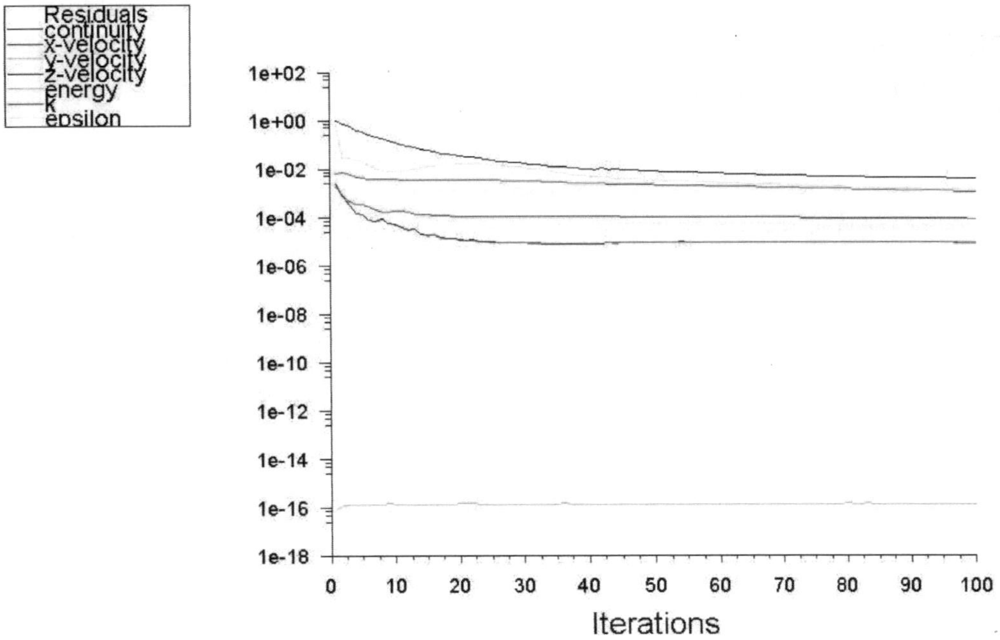

Figure 4. Convergence for 650,000 elements

4. Results and discussion

From Table 2, the error between the experimental and numerical results is only 0.937%. This shows that the model selected in the numerical analysis has been verified to be suitable in the subsequent simulation in designing the Venturi tube for the gas applications.

Table 2. Calculated discharge coefficient from CFD-post data

Venturi Under Test	Average Discharge Coefficient	Average Discharge Coefficient
	From experiment	From CFD-post
Venturi 1	0.99366	0.984347

4.1. Pressure contour charts

Pressure contours in Figure 5 are to plot pressure distribution across the Venturi tube. Flow direction for all the plots in Figure 5 is from left to right. The results clearly show the effect of mass flow rate of gas on the pressure gradient as obtained for various test cases. The pressure values at the upstream of the Venturi and at the centre point of the throat cylinder are recorded to determine the differential pressure across the Venturi tube, which are then used in the flow calculation.

4.2. Velocity contour charts

It is clear from Figure 6 that the velocity is at maximum at the throat of the Venturi. The velocity also decreases according to the decrease in flow rates. Flow direction for all plots in Figure 6 is from left to right. Velocity values recorded from the contours below are used to calculate the volumetric flow rates.

IOP Conf. Series: Materials Science and Engineering **152** (2016) 012062 doi:10.1088/1757-899X/152/1/012062

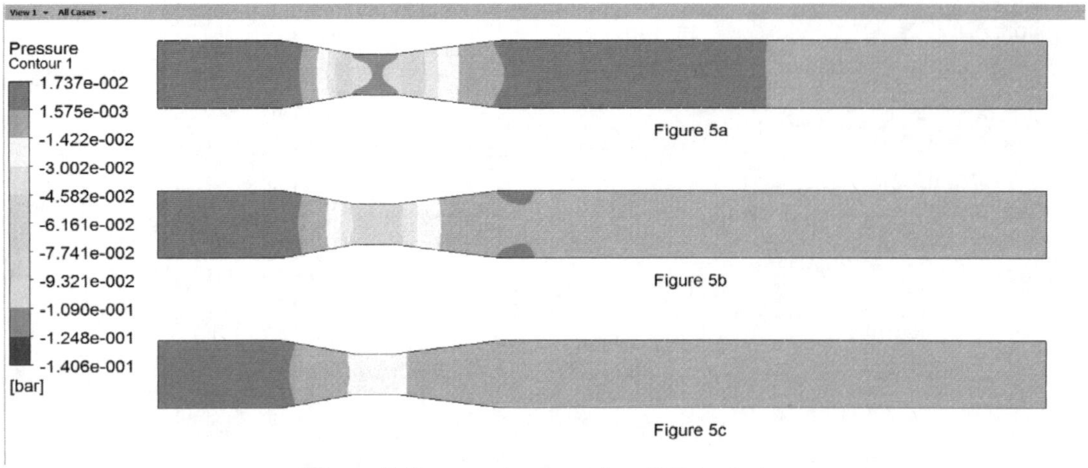

Figure 5. Pressure contours for all three tests

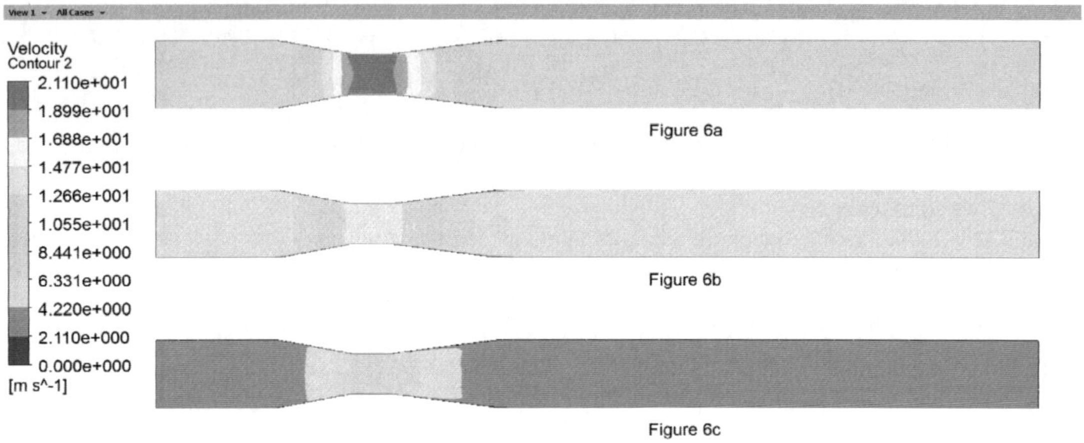

Figure 6. Velocity contours for all three tests

4.3. Discharge coefficient and other observations

The recorded differential pressure and velocity values are tabulated for the discharge coefficient calculation as shown in Table 3.

Table 3. Calculated discharge coefficient from CFD-post data

Test	Dp_1	Press.	Temp.	Density	Expansibility	Velocity	Mass Flow Rate	Discharge Coefficient, Cd
	(From CFD-post)					(From CFD-post)		
	(bar)	(bar gauge)	(°C)	(kg/m^3)		(m/s)	(kg/hr)	
1	0.12286	60.37	20.04	70.92	0.9987	7.00	86,030	0.97551
2	0.06220	60.36	20.06	70.90	0.9993	5.05	62,057	0.98850
3	0.02191	60.30	19.97	70.85	0.9997	3.00	36,855	0.98902

In Table 3, while the pressure, temperature, density and expansibility are almost constant, it is interesting to note that decrease in flow rates resulting in the increase of calculated Cd values. It may suggest that, with additional test points at much lower flow rates may bring the Cd value closer to the result obtained by experiment.

In Table 4, although mass flow rate is directly proportional to Cd, the increase in mass flow rate does not register any increase in Cd. This indicates that the Dp_l has a significant effect on the Cd calculation.

Table 4. Mass flowrate versus discharge coefficient (or Dp_l versus discharge coefficient)

Test	Dp_l	Dp_l	Mass Flow Rate	Mass Flow Rate	Discharge Coefficient	Discharge Coefficient, Cd
	(From CFD-post)					
	(bar)	(% change)	(kg/hr)	(% change)		(% change)
1	0.12286	98%	86,030	39%	0.97551	-1.33%
2	0.06220	184%	62,057	68%	0.98850	-0.05%
3	0.02191		36,855		0.98902	

In Table 5, analysis on pressure gradient increment against the mass flow rate increment indicates that for every 1% increase in mass flow rate, it will register 2.5 times percentage increase in pressure gradient. The analysis also indicates that for every 1% increase in mass flow rate, it will register 1% increase in velocity gradient. This pattern is consistent between test 1 and test 2, and between test 2 and test 3. This indicates that pressure has more effect on the computation compared to velocity.

Table 5. Pressure and velocity gradients

Test	Pressure	Mass Flowrate	Pressure Gradient	Velocity Gradient
	(bar gauge)	(kg/hr)	(bar/m)	(s^{-1})
1	60.37	86,030	-0.26	56.51
2	60.36	62,057	-0.13	40.77
3	60.30	36,855	-0.05	24.23

From Table 5, it shows that increase in pressure will exponentially increase the pressure gradient. This means for every bar increase in pressure, pressure gradient will exponentially become steeper and this in return will increase flow velocity at the Venturi throat. By understanding the correlation between pressure gradient and pressure, the convergent cone of the Venturi can be designed to operate within the velocity limit at Venturi throat.

In Table 6, the recorded PPL follows similar trend as compared to the Dp_l. The PPL obtained is approximately 11% of Dp_l for each of the tests for 7.5° divergent angle. This follows the same observation recorded in the previous analysed data [9]. The knowledge on this correlation between Dp_l and PPL can be used as validation check on the CFD model.

Table 6. Permanent Pressure Loss (PPL)

Test	Dp_l	Mass Flow Rate	PPL
	(From CFD-post)		(From CFD-post)
	(bar)	(kg/hr)	(bar)
1	0.12286	86,030	0.01323
2	0.06220	62,057	0.00699
3	0.02191	36,855	0.00266

5. Conclusion

In the investigation, the flow through Venturi tube was analysed with SKE turbulence model. The results from the numerical simulation had been used to obtain the Cd values for each of the tests. The

Cd values obtained from simulation for test 1, test 2 and test 3 were 0.97551, 0.98850 and 0.98902, respectively, which averaged to 0.984347. Test 1 resulted in the lowest Cd value whereas Test 3 provided the maximum Cd value. The error between the experimental and numerical results of 0.937% indicates the SKE turbulence model agreed with the experimental data. The analysis also indicated that pressure has more computational impact compared to velocity, and by understanding the correlation between pressure gradient and pressure, the convergent cone of the Venturi can be designed to operate within the velocity limit at Venturi throat. From the analysis and results obtained, it is shown that the model selected in the numerical had been verified to be suitable in the subsequent simulation in designing the Venturi tube for the gas applications.

Acknowledgments
Authors acknowledge the assistance from FRGS grant 5524612, as well as support from KROHNE (M) Sdn. Bhd. and NEL.

References
[1] Jamieson A W, Johnson P A, Spearman E P and Sattary J A 1996 *Proc. 14th North Sea Flow Measurement, Peebles, Scotland*
[2] Van Weers T, Van Der Beek M P and Landheer I J 1998 *Proc. 9th International Conference on Flow Measurement, FLOMEKO, Lund, Sweden* 203-7
[3] Reader-Harris M J, Barton N, Brunton W C, Gibson J J, Hodges D, Nicholson I G and Johnson P 2000 *Proc. 18th North Sea Flow Measurement, Perthshire, Scotland*
[4] International Organization for Standardization 2003 *Measurement of fluid flow by means of pressure differential devices inserted in circular cross-section conduits running full, Part 4: Venturi tubes* BS EN ISO 5167-4:2003
[5] KROHNE (M) Sdn. Bhd. 2015 *Venturi tubes installation reference list in Malaysia*
[6] Reader-Harris M J, Gibson J J and Hodges D 2007 *Proc. 6th South East Asia Hydrocarbon Flow Measurement, Kuala Lumpur, Malaysia*
[7] Fang L and Zhang T 2008 *Chinese Journal of Chemical Engineering* **16** 320-4
[8] Sowjanya V and Jie C 2010 *Applied Mathematical Modelling* **34** 1458-66
[9] Reader-Harris M J 2015 *Orifice plates and venturi tubes* (Cham: Springer International Publishing)

Detection of missing rods of 4-directional carbon preform from images

Y H Kim and J R Lee*

Department of Aerospace Engineering, Korea Advanced Institute of Science and Technology, Republic of Korea

*leejrr@kaist.ac.kr

Abstract. Poor quality of the carbon preform including missing rods can affect the desired properties of the tailored composite. The carbon preform is currently manufactured manually and therefore, an effective solution for guaranteeing the quality of carbon preform is needed. This paper proposes new inspection method that can detect the missing rods of carbon preform by just taking a picture of it. The new method found all missing rods and marked missing place with red circle. By wide use of smart devices, it becomes very easy to take a picture nowadays and the picture quality has improved dramatically. Moreover, the inspection method suggested in this paper can be easily converted into a form of android application and it is expected to be applied effectively to the fields of manufacture of the carbon preform.

1. Introduction

Carbon/carbon (C/C) composite consists of carbon preform and carbon matrix. Therefore, its specific strength and stiffness hardly change with environmental temperature, even when the temperature goes up extremely high compared to the normal temperature. Additionally, C/C composite has several other outstanding properties, including great ablation resistance, low thermal expansion coefficient and high thermal shock resistance. Because of these superior engineering properties, C/C composite has become one of the essential materials in the field of aerospace engineering. However, C/C composite also has some drawbacks. The most critical drawback is its high cost. In addition, another drawback is the fact that C/C composite is anisotropic. 1-D (unidirectional) and 2-D (two-directional) C/C composites are really anisotropic, and show weakness in bearing some forces acting in certain directions. This is why multi-directional C/C composite is often used in the structure of aerospace engineering. For example, the multi-directional C/C composite can be used for a rocket nozzle, leading edge and nose cap of a space shuttle. Two most frequently used structures of multi-directional C/C composite are 3-D (three-directional) and 4-D (four-directional). The 4-D structure is commonly known to be more isotropic and efficient structure than the 3-D structure. This is because 4-D structure has one more radial axis of carbon rods. As a result, 4-D structure consists of one axial direction and three radial directions. This is well displayed in Figure 1, which is rendering result from a three-dimensional sketch program.

The type of structure for the C/C composite is determined at the manufacturing stage of the carbon preform. The carbon preform is an essential framework of tailored composite. However, the carbon preform is manually manufactured now, hence many mistakes can occur in its manufacturing stage. In particular, a common mistake is the omission of carbon rod. A poor quality of carbon preform affects some of the desired properties like superior strength and good ablation performance for the tailored composite. Thus an effective solution for guaranteeing the quality of the carbon preform is needed.

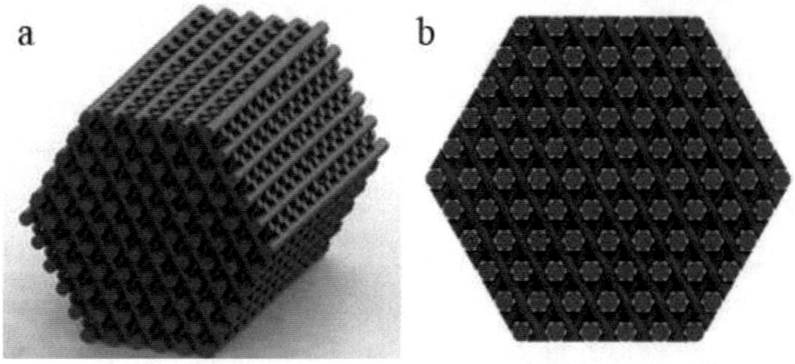

Figure 1. The four-directional structure: (a) overall view and (b) top view

The goal of this paper is to propose and show that by just taking a picture of the carbon preform, it can be a powerful method of quality inspection. For convenience, in this paper, the mistakes stated above are restricted to the omission or missing of carbon rods only. The carbon preform pictures used in this paper are supplied by DACC. There are so many carbon rods in the picture and their diameter is 1.7mm, so they are almost fibre. In order to detect missing carbon rods that are very thin, some image processing techniques are utilized. The typically used technique is binarization, which is a mid-process that makes the picture much simpler.

2. Image processing technique

First of all, the carbon preform picture should be converted into a gray scale image to reduce the size of data. In the RGB image, there are three values at each pixel but in the gray scale image, there is only intensity value at each pixel. This makes the gray scale image is much simpler to handle than the original image. In addition, if the picture is high-definition, the run time will be very long. Therefore, it is necessary to convert the RGB image into the gray scale image in the case of high-definition.

2.1. Binarization by adaptive thresholding

The main goal of image processing is to separate the object from the background. To do that, it needs a clear criterion. In the binarization, the criterion is threshold value of the intensity. A bright pixel with high intensity value gets 1 whereas a dark pixel with low intensity value gets 0. Generally, the object is brighter than the background such as it is possible to classify the pixels into the object group and the background group theoretically.

There are two types of threshold. The first type is global threshold, which is calculated once and applied to all pixels at a time. If the picture is quite simple, this type of threshold is very effective and efficient to classify the pixels. Another type is local threshold that is calculated at every pixel using the information of surrounding pixels. Because this type of threshold is calculated adaptively, it is possible to cope with a change of brightness and illuminance. The carbon preform pictures can be taken under various environments, hence the second type of threshold is adopted for binarization in this study. This can be expressed in Equation 1 and it is called adaptive thresholding method.

$$b(i,j) = \begin{cases} 1, & f(i,j) \geq t(i,j) \\ 0, & other\ wise \end{cases} \tag{1}$$

The $b(i,j)$ means i-th row and j-th column element of the binary image. In similar manner, the $f(i,j)$ means i-th row and j-th column element of the original image. The value $t(i,j)$ is locally computed threshold of i-th row and j-th column pixel.

2.2. Integral image

Unfortunately, the adaptive thresholding method has a big problem. It takes too long time because the threshold has to be computed at every pixel using the information of surrounding pixels. Therefore, it is necessary to reduce the run time using an integral image method. The integral image is defined as Equation 2.

$$I(x,y) = \sum_{i=1}^{x}\sum_{j=1}^{y} f(i,j) \tag{2}$$

The $I(x,y)$ is x-th row and y-th column element of integral image and it means a simple summation of all left and upper values of (x, y) pixel of the original image. Using Equation (2), these three following Equation 3, Equation 4 and Equation 5 can be easily derived and they help the integral image to be computed fast.

$$I(1,y) = f(1,y) + I(1,y-1) \tag{3}$$

$$I(x,1) = f(x,1) + I(x-1,1) \tag{4}$$

$$I(x,y) = f(x,y) + I(x,y-1) + I(x-1,y) - I(x-1,y-1) \tag{5}$$

2.3. Local sum

After getting the integral image, it is possible to calculate the local sum, $s(x,y)$ using Equation (6). The local sum refers to a simple summation of certain area of the original image. In image processing, the area is called window. Equation 6 implies that the local sum value can be computed using only four elements of the integral image. This is depicted as Figure 3.

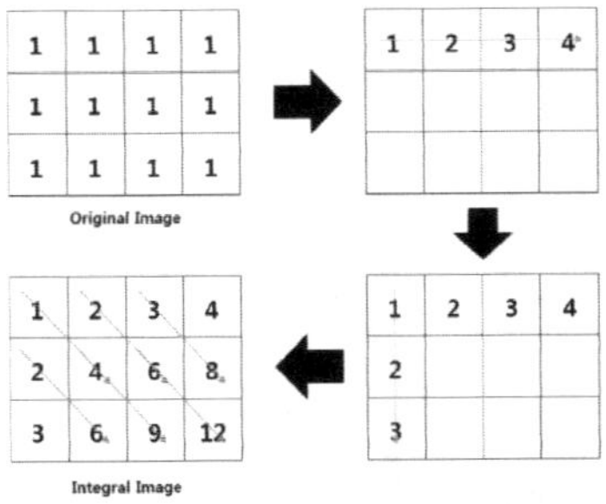

Figure 2. Calculation of integral image

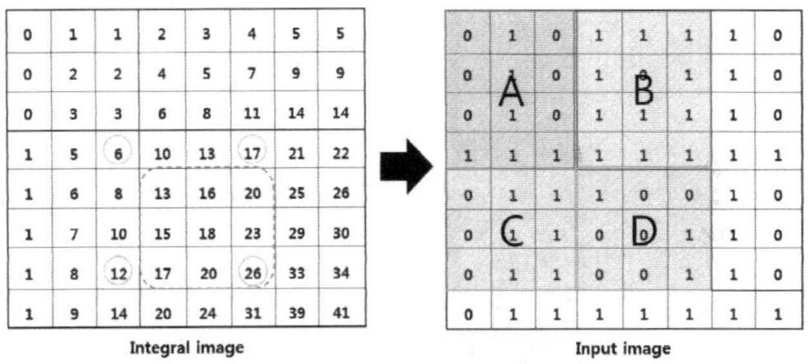

Figure 3. Calculation of local sum

$$\begin{aligned} s(x,y) &= \sum_{x=x_1}^{x_2}\sum_{y=y_1}^{y_2} f(x,y) \\ &= I(x_2,y_2) - I(x_2,y_1-1) - I(x_1-1,y_2) + I(x_1-1,y_1-1) \end{aligned} \tag{6}$$

Finally, using the local sum, $s(x,y)$, the threshold, $t(i,j)$ can be computed by Equation 7. The first fraction of Equation 7 is average value of intensity of surrounding pixels while the second fraction of is a kind of margin. If p is positive, the threshold becomes lower than the average of the surrounding pixels. On the other hand, the threshold becomes higher than the average when p is negative. So it is possible to control the local threshold level by changing the value p.

$$\mathrm{t(i,j)} = \frac{s(i,j)}{w^2}\frac{100-p}{100} \tag{7}$$

It is very simple to calculate the local threshold because the local sum already has the information of surrounding pixels in the form of a simple summation. If the integral image and local sum are not utilized for getting the local threshold, it will take extremely long time because the process of getting the local threshold is a kind of 2-dimensional convolution or image convolution.

3. Results

In order to create a circumstance of missing rods to evaluate the proposed method, several rods are randomly chosen and covered by neighboring background color. After erasing some rods, the size of picture is adjusted. The size of the original carbon preform picture is 4032 by 3024. It means that the original picture has about 12 million pixels. However, most smart devices have a much lower display resolution and thus the size of the picture is adjusted to 2048 by 1536, which is about 3 million pixels.

Figure 4 is the resultant picture of the detection and it is composed of only zeros and ones because it is binary image. There are many red circles but only two of them are located inside of the carbon preform. The other red circles are located at the outside and also at the boundary of the carbon preform. Therefore, the two red circles are real detection result and the other red circles are errors. Figure 5 is an enlarged image of Figure 4. In Figure 5, there are many white circles and there is also one red circle. The white circles are real carbon rods while the red circle signifies missing carbon rod. The red circle is located at the exact position of the missing carbon rod.

4. Conclusion

In this paper, a new detection method for missing rods of carbon preform is proposed. In the result of the detection, red circles mark all missing rods without omission. In addition, all errors are located at the outside and boundary of the carbon preform, so they do not interfere in finding the missing carbon rods. Furthermore, this new method to detect the missing rods can be converted into a form of android application that can be used to find the missing carbon rod during manufacturing phase. All in all, this method is good at detecting missing rods and it is really expected to be applied to the manufacturing site of the carbon preform such that it can be produced much faster and more accurate than before.

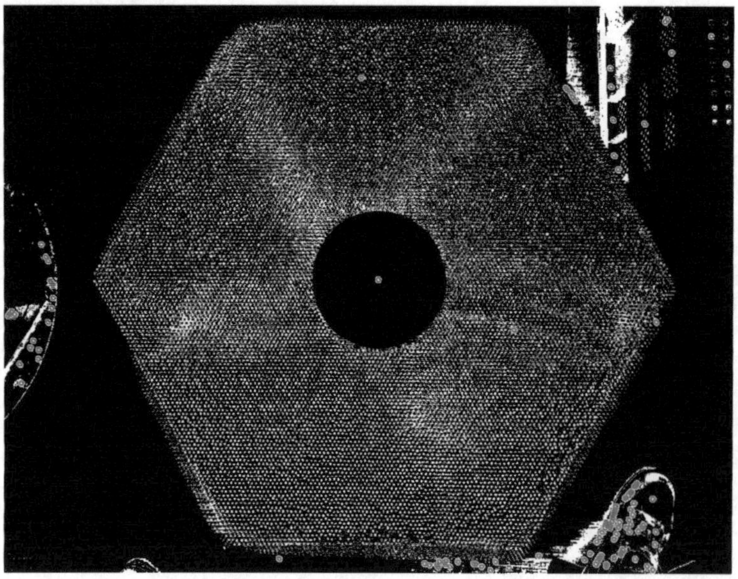

Figure 4. Detection result of the carbon preform

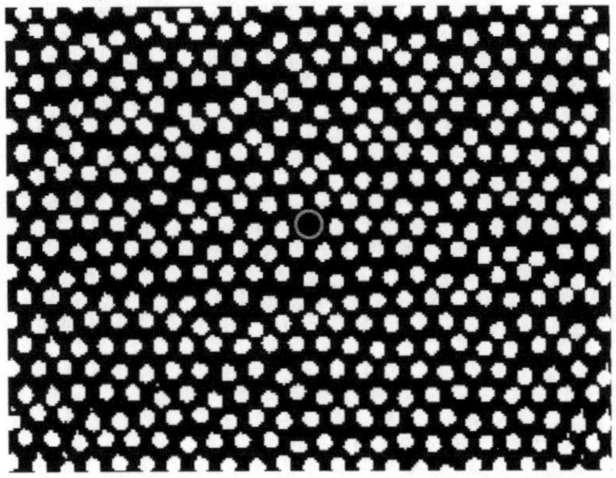

Figure 5. Enlarged result of detection

Acknowledgments
This research has been supported by the Leading Foreign Research Institute Recruitment Program (2011-0030065) through the National Research Foundation of Korea and it is funded by the Ministry of Science, ICT and Future Planning.

References

[1] Fitzer E 1997 *Carbon* **5** 163-190
[2] Fitzer E and Manocha L M 1998 *Carbon Reinforcements and Carbon/Carbon Composites* (Berlin: Spriger-Verlag)
[3] Venkatesh S, Mohan V M and Karthikeyan T V 2014 *International Journal of Advancements in Mechanical and Aeronautical Engineering* **1** 1-5

IOP Conf. Series: Materials Science and Engineering **152** (2016) 012064 doi:10.1088/1757-899X/152/1/012064

Obstacle-avoiding robot with IR and PIR motion sensors

R Ismail[1*], Z Omar[1] and S Suaibun[2]

[1]Electrical Engineering Department, Polytechnic Kota Bharu, Malaysia
[2]Mechanical Engineering Department, Polytechnic Kota Bharu, Malaysia

*roslindaismail.poli@1govuc.gov.my

Abstract. Obstacle avoiding robot was designed, constructed and programmed which may be potentially used for educational and research purposes. The developed robot will move in a particular direction once the infrared (IR) and the PIR passive infrared (PIR) sensors sense a signal while avoiding the obstacles in its path. The robot can also perform desired tasks in unstructured environments without continuous human guidance. The hardware was integrated in one application board as embedded system design. The software was developed using C++ and compiled by Arduino IDE 1.6.5. The main objective of this project is to provide simple guidelines to the polytechnic students and beginners who are interested in this type of research. It is hoped that this robot could benefit students who wish to carry out research on IR and PIR sensors.

1. Introduction

Many robots for automation and navigation have been developed in recent years like wall-following, edge-following, human following and obstacle avoiding robots. The obstacle avoiding robot will evade obstacles it encounters in its path towards its operational goal. Due to the reliability, accessibility and cost effectiveness of using mobile robot in industry and technical applications, the obstacle avoiding robots are very important in factory floor. On the other hand, Unmanned Aerial Vehicles (UAVs) are playing a vital role in defence as well as civilian applications [14]. The military applications include reconnaissance, surveillance, battle damage assessment and communications. Meanwhile, civilian applications include disaster management, remote sensing, traffic monitoring, etc. Many of the UAVs applications need the capability to navigate in urban environment or unknown terrains that have many obstacles of different types and sizes. Basic requirement of autonomous UAVs is to detect obstacles in its path and avoid them. This paper proposed an example of the obstacle avoiding robot algorithm and design of the robot base using IR and PIR sensors. The developed robot can be used as a platform for several applications in educational, research or industrial.

The behaviour of a mobile robot is dictated by the interaction between the program running on the robot (the 'task'), physical hardware of the robot (the way its sensors and motor work) and the terrain (environment) [4]. These three fundamental components are as shown in Figure 1. The development in this paper is aimed to highlight the hardware and software implementation to design an autonomous platform. It serves to achieve research purposes like developing software modules to realize obstacle avoidance and investigating the interaction between the robot task and terrain. This platform serves for achieving educational purposes to reveal the latest technology for different sensors and actuators that is used in mobile robot design [19]. Besides that, it also provides the opportunity to study on how to program the microcontroller chip to acquire sensors data and react autonomously with environment in order to perform tasks.

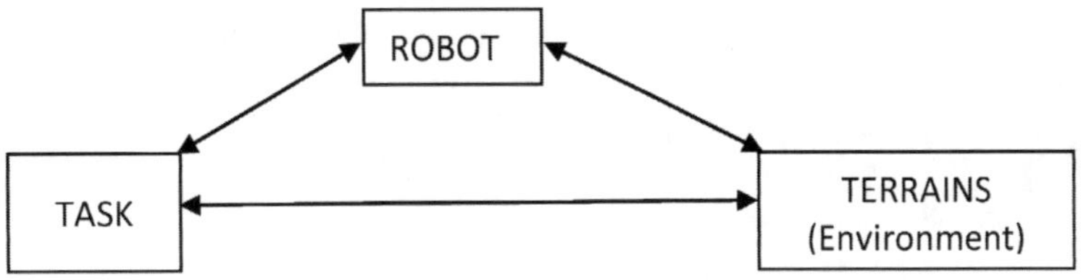

Figure 1. Obstacle avoidance fundamental component

2. Literature review

Robots need miscellaneous of sensors to obtain information about the world around them. Sensors will help detect position, velocity, acceleration and range for the object in the robot's workspace. There is a variety of sensors used to detect the range of an object. One of the most common range finders is the ultrasonic transducer. Vision systems are also used to greatly improve the robot's versatility, speed and accuracy for its complex and difficult task. Varun *et al.* [13] developed obstacle avoidance equipped with pan-tilt mounted vision system. In this case, the robot uses histograms of images obtained from a monocular and monochrome camera to detect and avoid obstacles in a dynamic environment. Kim *et al.* [1] presented the detection of moving obstacles (particularly walking humans) using single camera attached to a mobile robot. Detection of object that moves near the robot is searched by block-based motion estimation. Shoval *et al.* [8] described the use of mobile robot obstacle avoidance system as a guidance device for blind and visually impaired people. Electronic signals are sent to a mobile robot's motor controllers and auditory signals can guide the blind traveller around the obstacles. Kumar [3] proposed an alternative design for cost effective and simplified version of the obstacle avoidance robot using three ultrasonic sensors. Borenstein *et al.* [9] built a mobile robot system, capable of performing various tasks for the physically disabled people. The developed robot uses ultrasonic range finder for detection and mapping to avoid collision with the unexpected obstacles. Sathiyanaranayan *et al.* [10] developed self-controlled robots for military purposes. It uses GPS and magnetic compass, and adjusts strategies based on the surroundings using path planning and obstacle detection algorithms. Bhanu *et al.* [12] built a system for obstacle detection during rotorcraft low altitude flight. The requirements of an obstacle detection system for rotorcraft in low altitude flight based on various rotorcraft motion constraints is analysed in details. Huballi *et al.* [11] discussed on Smart Distance Measurement using IR sensor. From the research, they found that a major drawback of IR based sensors is their capability of detecting only in short distance.

3. Sensors selection

The objective of having obstacle avoiding robot is to enable autonomous functioning without human supervision. A lot of sensors are available for obstacles detection such as ultrasonic sensor, infrared sensor, camera and LIDAR (laser based sensor system), which has been considered as one of the most accurate schemes for generating spatial information about the shape and surface characteristics of any object [18]. In this paper, infrared sensors are used to accomplish evading obstacle. IR and PIR sensors have been placed on prototype for obstacle detection as shown in Figure 2.

Accuracy with low cost is the constraint of this research, hence the IR sensor was selected for the design. Stereo camera vision does not perform well under some environment conditions such as plain wall, glass surfaces or poor lighting conditions whereas IR sensor can be used to improve the overall vision system of mobile robot [17]. IR sensors are widely used for measuring distances, so they can be used in robotics for obstacles avoidance. IR sensors are also faster in response time than ultrasonic sensors [17]. In addition, the power consumption of IR sensor is lower than ultrasonic sensors [11]. Active Infrared (IR) sensors can be an emitter and detector, which operate at the same wavelength. It is also known as photoelectric sensor working with reflective surfaces [15]. IR sensor can be categorized as retro-reflective sensors and diffuse reflection sensors. Retro-reflective sensors are proper for harsh environment conditions and have much larger detection range than the diffuse reflective sensor [15]. IR sensors use a specific light sensor that can detect a selective light wavelength in the IR spectrum. When an object is close to the sensor, the light from the LED bounces off the object and into the light sensor as shown in Figure 3.

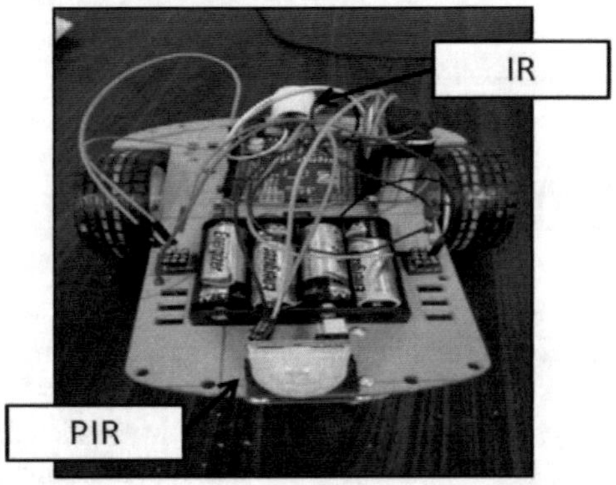

Figure 2. Position of IR and PIR sensor

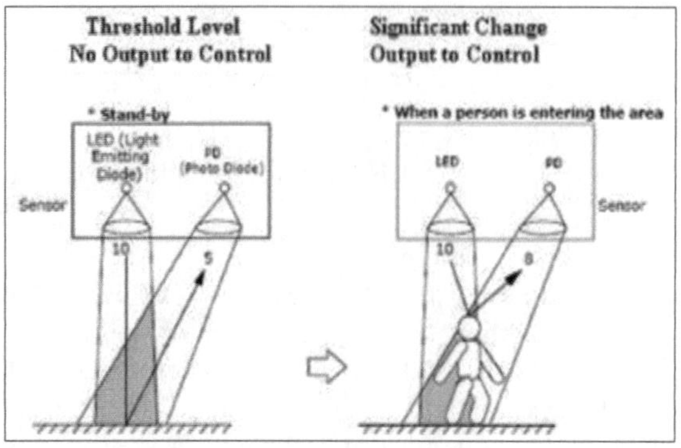

Figure 3. IR sensor object detection

3.1. IR and PIR sensors operation

Infrared sensors detect the object's distance with infrared radiation. When the beam detects an object, the light beam returns to the receiver with an angle after reflection. The method of triangulation is as shown in Figure 4. PIR sensors are also known as Pyroelectric Infrared sensor, Passive Infrared sensor or IR motion sensor, which detect the difference in temperature, thermal radiation, human body or an animal [15]. PIR sensor operates with the radiation of body heat as shown in Figure 5. The hotter the detected object, there will be more emission occurs in PIR sensor.

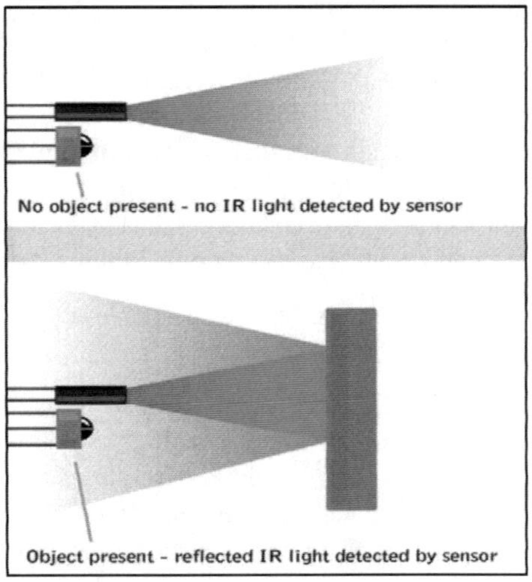

Figure 4. IR sensor working principle [15]

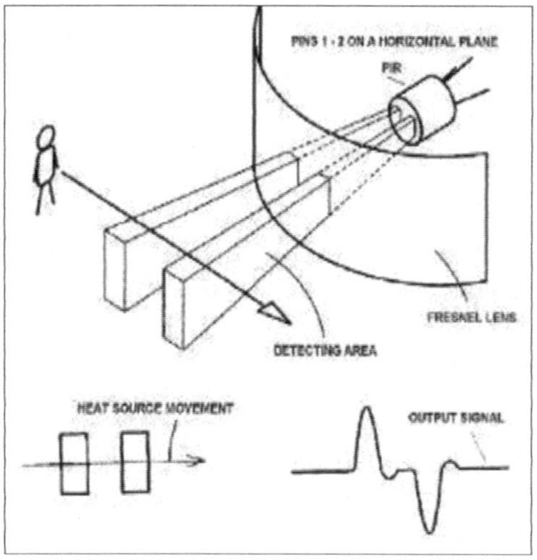

Figure 5. PIR sensor object detection

Object temperature calculation is based on Stefan-Boltzmann Law.

$$T_e = \sqrt{T_s^4 + \frac{\Phi}{A\sigma\varepsilon\varepsilon_s}},$$ [T_s = sensor's surface temperature T_e = object's temperature in Kelvin]

[Φ = magnitude of net thermal radiation flux Φ, ε = emissivity of the object]

PIR sensor relatively has a lower power consumption compared to IR sensor. PIR sensor also senses accurate detection in narrow areas and is compatible to work in microcontrollers [15]. The IR and PIR sensors can act as a transducer since they use infra-red signal as the input and convert it to analogue electrical output signal [15].

3.2. Limitation of IR and PIR sensor

Performance of IR sensors has been limited by their poor tolerance to light reflections such as ambient light or bright object colours. No object recognition at the dead zone area, for example Sharp GP2D12 IR distance sensor dead zone between 0 to 4 cm. IR sensors also give inaccurate detection result with transparent or bright colour materials. Detection results also depend on the weather conditions and the sensing reliability of IR sensors decreases with moisture and humidity. Furthermore, IR sensors can sense IR radiation from the sunlight, which can cause correctable or non-correctable errors at output. Besides that, if analogue IR sensor is used, signal losses will occur at the amplifier circuit. Meanwhile, PIR motion sensor needs a long calibration time and is sensitive to thermal radiation. Besides that, PIR sensor is insensitive to very slow motions or to objects in standing mode.

4. Experiment set up

Circuit diagram of the obstacle avoiding robot is given in Figure 6. The hardware developed consists of Microcontroller Atmel ATMEGA 32, IR sensor, two DC motors as differential driving system, PIR (Passive Infra-Red) sensor and motor driver L293D. The microcontroller ATMEGA 32 is the central brain of the mobile robot. Figure 7 shows the chassis of the robot. The IR detector (rear) is connected to PD6 of ATMEGA32. If any object is located at the rear part of the robot frame, the IR sensor output will alert the microcontroller that an obstacle is detected. The IR sensor used as shown in Figure 8. PIR sensor (front) is connected to PD7 I/O pins. If any object is moving in front of the robot, the PIR sensor output will alert the microcontroller that an obstacle is detected. Motion sensor has 90 degrees field of view and it will be triggered when a warm object moves across the area it is facing. It is very sensitive and will trigger with just a hand movement. PIR motion sensor is shown in Figure 9.

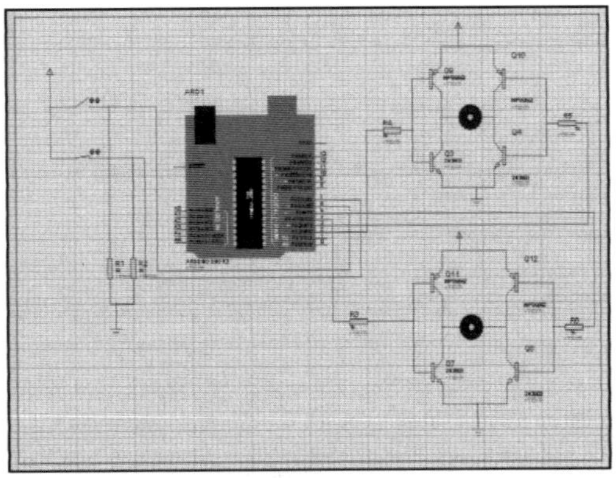

Figure 6. Circuit diagram of obstacle avoiding robot

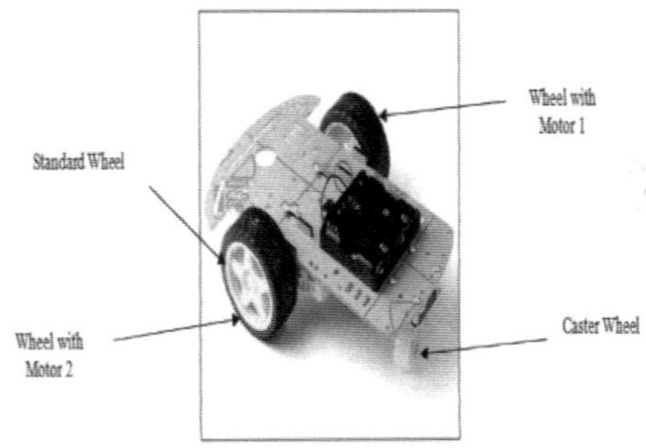

Figure 7. Chassis of obstacle avoiding robot

Figure 8. IR sensor

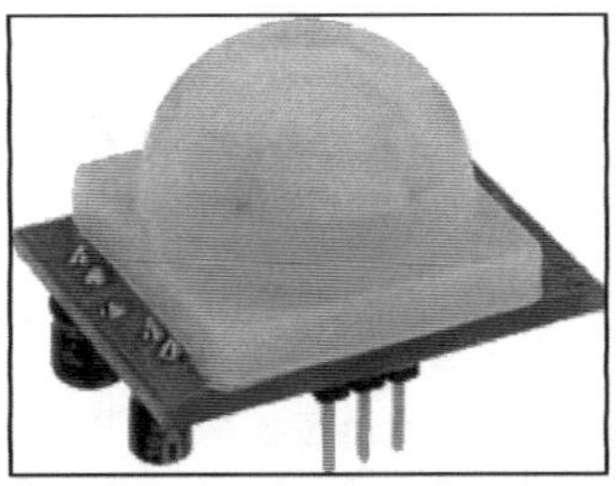

Figure 9. PIR sensor

Motor driver L293D (H-bridge) decides which motor will be moved or stopped in accordance to the incoming signal from the microcontroller ATmega32. The selected dimension of the mobile base is suitable to install sensors to interact with the environment. The wheel used in Figure 7 will provide sufficient traction and stability for the robot to cover all of the desired terrain. Standard wheel and caster wheel was used in this design. Standard wheel with rotation around motorized axle and rotation around the contact point. Meanwhile, caster wheel rotation around the offset steering joint and rotation around the contact point. The software is written using Arduino Integrated Development Environment or Arduino Software (IDE). It contains a text editor for writing code, a message area, a text console, a toolbar with buttons for common functions and a series of menus. It connects to the Arduino hardware to upload programs and communicate with them. Once the compile file (Hex code) is obtained, it can be downloaded into microcontroller.

5. Result

A brief overview of the software functions and the system architecture is shown in Figure 10. From the flowchart, the calling sequence and the relationship between the functions are visualized.

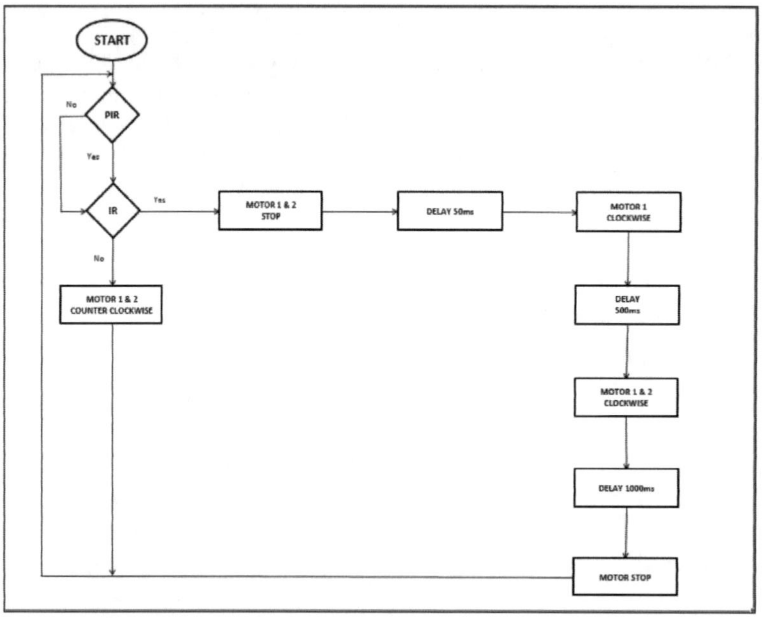

Figure 10. Flow chart of the obstacle avoidance robot movement

The working principle of the robot is transmitting sensed signal to the microcontroller to control the DC motors for obstacle avoidance. The H-bridge L293D controls the direction of the motors to move either clockwise or anti-clockwise directions as provided by the microcontroller. Figure 10 shows that, if PIR sensor detect a moving object while IR sensor does not detect any object, the robot will move backward (motor 1 and motor 2 counter clockwise). On the other hand, if PIR detect object and IR sensor also detect object, the robot will stop (motor 1 and motor 2 OFF). After 50 ms, motor 1 will move clockwise and the robot will turn left. After 500 ms, the robot will move forward (motor 1 and motor 2 clockwise) and after 1000ms, both motors will stop. From the flow chart in Figure 10, it shows that the IR and PIR sensors are very effective in sensing signals in their path for the obstacle avoiding robots to evade obstacles in its path.

7. Discussion and conclusion

The developed robot platform was not designed for specific task but as a general wheeled autonomous platform. It can therefore be used for educational, research or industrial implementation. Students can use it to learn the microcontroller programming using C++, Arduino Uno 1.6.5 compiler, IR and PIR sensors characteristics, motor driving circuit and signal condition circuit design. Research on obstacle avoidance robot at the polytechnic level can help students to develop communication, technical skills and teamwork. The design of such robot is very flexible and various methods can be adapted for another implementation. From this paper, it shows that PIR sensors are more sensitive compared to IR sensors while detecting human being.

8. Limitation and further improvement

The robot has been successful in avoiding obstacle but it also has certain limitations as well. Further improvement can be achieved by adding sensors on the left and right side of the robot. Besides that, computer vision with camera features can be implemented for monitoring applications. For further improvement, to implement an obstacle avoidance in aerospace, well-suited sensors should be used to gather the accurate information about the environment and obstacles. The laser based (LIDAR) sensor system is robust especially in off-road outdoor environments [18]. LIDAR based mapping gives most accurate scheme for generating information about the shape and surface characteristics of any object. The recent advancement in LIDAR technologies allows researchers and practitioners to examine the environment with higher precision, accuracy and flexibility than before. LIDAR sensor is considered as an effective solution to the problem of obstacle detection and recognition. However, the obstacle avoidance poses challenges to the image processing using LIDAR sensor.

References

[1] Jeongdae K and Yongtae D 2012 *Procedia Engineering* **41** 911-6
[2] Kumar R C, Saddam Khan M, Kumar D, Birua R, Mondal S and Parai M K 2013 *International Journal of Advanced Research in Electrical, Electronics and Instrumentation Engineering* **2** 1430-4
[3] Dey M K 2016 *International Journal of Soft Computing and Engineering* **6** 66-9
[4] Ali W G 2011 *Journal of King Saud University-Engineering Sciences* **23** 131-8
[5] Medina-Santiago A, Camas-Anzueto J L, Vazquez-Feijoo J A, Hernandez-de Leon H R and Mota-Grajales R 2014 *Journal of Applied Research and Technology* **12** 104-10
[6] Syed Ali A, Zanziger Z, Debose D and Stephens B 2016 *Building and Environment* **100** 114-26
[7] Candelas F A, Garcia G J, Puente S, Pomares J, Jara C A, Perez J, Mira D and Torres F 2015 *IFAC Workshop on Internet Based Control Education*
[8] Shoval S, Borenstein J and Koren Y 1994 *Proceeding of 1994 IEEE Robotics and Automation Conference*
[9] Borenstein J and Koren Y 1988 *Journal of Robotics and Automation* **4** 213-8
[10] Sathiyanarayanan M, Azharuddin S, Santhosh K and Khan G 2014 *International Journal For Technological Research in Engineering* **1** 1075-7
[11] Hubballi P, Bannikoppa S, Bhairi V and Jahagirdar R G *Indian J. Sci. Res.* **12** 155-8
[12] Bhanu B, Das S, Roberts B and Duncan D 1996 *IEEE Transaction on Aerospace and Electronic Systems* **32** 875-97
[13] Raj K V, Bidargaddi S V, Krisnanand K N and Ghose D 2007 *13th National Conference on Mechanisms and Machines*
[14] Gade M M, Hangal S, Krishnan D and Arya H 2016 *International Federation of Automatic Control*
[15] Saracoglu B 2009 *A Guide to IR/PIR Sensor Set-up and Testing* Michigan State University
[16] Gwee J *Electronic Review* **20** 16-8
[17] Mohamed T 2009 *International Journal of Mechanical, Aerospace, Industrial, Mechatronic and Manufacturing Engineering* **3** 267-72
[18] Yaghi M, Onder Efe M and Dincer K 2015 *Ankara International Aerospace Conference*
[19] Nehmzow U 2001 *Proc. Trends in Robotics*

Electro pneumatic trainer embedded with programmable integrated circuit (PIC) microcontroller and graphical user interface platform for aviation industries training purposes

I Burhan[1*], A A Azman[2] and R Othman[1]

[1]Aircraft Maintenance Department, Politeknik Banting Selangor, Selangor, Malaysia
[2]Electrical Engineering Department, Politeknik Sultan Salahuddin Abdul Aziz Shah, Selangor, Malaysia

[*]bibrahim1971@yahoo.com

Abstract. An electro pneumatic trainer embedded with programmable integrated circuit (PIC) microcontroller and Visual Basic (VB) platform is fabricated as a supporting tool to existing teaching and learning process, and to achieve the objectives and learning outcomes towards enhancing the student's knowledge and hands-on skill, especially in electro pneumatic devices. The existing learning process for electro pneumatic courses conducted in the classroom does not emphasize on simulation and complex practical aspects. VB is used as the platform for graphical user interface (GUI) while PIC as the interface circuit between the GUI and hardware of electro pneumatic apparatus. Fabrication of electro pneumatic trainer interfacing between PIC and VB has been designed and improved by involving multiple types of electro pneumatic apparatus such as linear drive, air motor, semi rotary motor, double acting cylinder and single acting cylinder. Newly fabricated electro pneumatic trainer microcontroller interface can be programmed and re-programmed for numerous combination of tasks. Based on the survey to 175 student participants, 97% of the respondents agreed that the newly fabricated trainer is user friendly, safe and attractive, and 96.8% of the respondents strongly agreed that there is improvement in knowledge development and also hands-on skill in their learning process. Furthermore, the *Lab Practical Evaluation* record has indicated that the respondents have improved their academic performance (hands-on skills) by an average of 23.5%.

1. Introduction

Graphical User Interface (GUI) has become one of the potential tools in teaching and learning process, especially in engineering discipline for better understanding of complexity in engineering processes. The students have the opportunity to engage in active learning environment by conducting simulated experiments, gathering results, collecting data, drawing conclusions and troubleshooting. Furthermore, GUI or computer simulation can enhance the student's active involvement in teaching and learning process [1]. Computer simulations such as Electronics Workbench, LegoCAD and Car Builder help trainees to understand events, processes and activities that can be replicated into real world problems. Besides that, it engages the students with activities that are equal to the real-life hands-on laboratory experiences effectively [2]. This contributes towards greater understanding and skill enhancement as compared to typical teaching and learning process (chalk and talk).

PIC that belongs to the microcontrollers family is manufactured by Microchip Technology Inc., the leading microcontroller and analogue semiconductor manufacturer. Microcontroller is essentially a microprocessor that is equipped with Input/Output (I/O) circuitry and built-in peripheral, allowing for direct interface with real-world devices such as switch, pneumatic solenoid, motors, sensors and other devices [3]. PIC simplifies the design of logic and control system, permitting the design of complex behaviour to be embedded into a piece of electronics instruments. Meanwhile, an understanding on embedded system is an area of study that requires more in-depth exposure so as to meet the global industry's needs [4, 5].

The role of control engineering discipline in the polytechnic education and aviation industries is very significant. Competent aviation's technician, semi-skill trained technician or technical manpower is trained in programs offered at Electrical Engineering Department, Polytechnic Sultan Salahuddin Abdul Aziz Shah (PSA) and Aircraft Maintenance Engineering Department, Polytechnic of Banting Selangor (PBS). Students should receive hands-on experience with the support of simulation and real devices. It is anticipated that there is better demand and wider job opportunities for technical graduates who are exposed to, and understand electro pneumatic control and embedded principles [6]. Hence, teaching and learning need to be supported with more advanced hands-on experiences and graphical simulations in order to make the learning more interesting and helpful compared to typical classroom method.

The aviation industries involve huge number of electro pneumatic devices that are used in aircraft system such as pneumatic actuator for moving the flight control surfaces, spoiler control, pneumatic actuator for landing gear of small aircraft, de-icing boot for small and older aircraft, etc. The need to produce competent aerospace/aircraft technician is vital for the future safety of aircraft manufacturer, line maintenance and airline operators. Better understanding of knowledge and higher opportunities for hands-on training for the aircraft technicians will help to minimize aircraft catastrophes caused by human errors [7]. The booming aviation industries stress on production of skilled aviation technicians or engineers with hands-on experience to meet the demands from aviation industries. Unfortunately, not many education providers or institutions are well equipped to produce these competent aviation technicians or engineers. As a matter of fact, there is also a lack of aviation institutions that could produce quality, competent and experienced trainees to meet the industries' needs [8]. As such, it is hoped that these 'issues' could be resolved with the introduction and usage of enhanced teaching aids such as the electro pneumatic trainer in the teaching and learning process of the students.

The development of this electro pneumatic trainer is focused on students studying *Instrumentation* (DEJ 3143), *Pneumatic System* (DEA 4043) for Diploma in Electronics (Control) at PSA and also *Pneumatic and Hydraulic* (DWM 4302) courses for Diploma in Aircraft Maintenance at PBS. It is also useful for other related programs offered in the Malaysian polytechnics system [9]. The development also helps to meet the demands from other aviation training institutions that produce competent and skilled human resources needed by the aviation maintenance organisations as forecasted by relevant bodies [10].

2. Execution of existing electro pneumatic course

Learning is best conducted when teaching and learning is focused on simulation, hands-on experiences with real instrument, machines or in automation system [11]. During the establishment of Sultan Salahuddin Abdul Aziz Shah Polytechnic in 1997 and Banting Polytechnics in 2012, there was no suitable device prepared for teaching and learning. The students had to relate to the theory they learned in the classroom with experimental activities by using the individual electro pneumatic apparatus only. The apparatus need to be wired individually in order to trigger the operation of the devices as shown in Figure 1. For instance, students need to connect the air supply to the air motor in order to operate the motor and disconnect the air supply to stop the air motor operation. The existing method involves high risk because students have to deal with the 240 VAC motor switches frequently to manually operate the air motor. No wiring connection is available between two or more apparatus such as semi rotary motor and air motor, and this needs to be tested individually and manually. Furthermore, there is no graphical simulation available before the students can conduct the hands-on lesson during the teaching and learning process. Besides that, aviation industries also require more skilled personnel with hands-on experience dealing with real devices [12]. Nowadays, the aviation industries require more skilled workers that can operate real devices. In order to achieve this aviation industry demand, educational sector needs to be aware and come out with more practical courses that can equip the students with more hands-on skills.

Figure 1. Manually execution of individual electro pneumatic apparatus

This electro pneumatic trainer is embedded with PIC and VB, and it is fabricated based on the problems encountered by lecturers and students at the Electrical Engineering Department, Sultan Salahuddin Abdul Aziz Shah Polytechnic and Banting Polytechnic, and also the aviation industrial needs. Studying electro pneumatic courses is best when applied with simulation and hands-on work. Facilities at Electrical Engineering Department are unable to meet the practical requirement outlined in the educational syllabus. With regard to this, lecturers often face problems in complying with syllabus requirement and meeting industrial needs. The existing approach also does not enhance the students' apparatus wiring and programming skills. The use of VB and PIC in teaching is considered vital as currently, there is limited equipment that can be used by students for gaining knowledge and hands-on skills [13]. Most trainers in the market can hardly meet the requirement of the syllabus. Moreover, the existing technique only applies a minimum hands-on wiring skill and students are assessed at low level of competency. There is a great need to reform the teaching methods and assessment tools, laboratory equipment and experiments, in enhancing students' hands-on skills and practical ability, laying a good foundation for their future job, and improving their social competitiveness and the quality of personnel training [14]. In light of this deficiency, this innovation is designed and fabricated to meet the overall objective in enhancing knowledge and hands-on skill for electro pneumatic courses and introducing various types of output apparatus to a trainer.

3. Overall system process and design requirement for electro pneumatic trainer

At preliminary stage, it is required to come up with the overall process and design requirement for the electro pneumatic trainer. Figure 2(a) shows the flow chart of the overall system process meanwhile Figure 2(b) shows the designed requirementand the sequence step for the start-up and troubleshooting process for the electro pneumatic trainer. There are some criteria that need to be considered prior to the overall design requirement – attractive design (to attract the students' attention during the teaching and learning session), effective (more integration of output components compared to the availability of the trainer in market) and user friendly (easy for installation and use by instructor and trainees during the teaching and learning and hands-on session). This consideration will be looked into in driving the designing stage prior to the actual execution and fabrication of the trainer.

First stage is the fabrication of PICs interface circuit, which provides analogue and digital series of input and output that can be used to control the electro pneumatic trainer. The block diagram of PIC circuits as interface to control the electro pneumatic apparatus is shown in Figure 3. It consists of PIC module, input section, output section and power supply. The input section is controlled through GUI programming (VB platform) while output section consists of one unit of air motor, one unit of single acting cylinder, one unit of double acting cylinder, one unit of semi rotary air motor and one unit of linear drive. The microprocessor used in this circuit fabrication is 40-pin PIC16F877A.

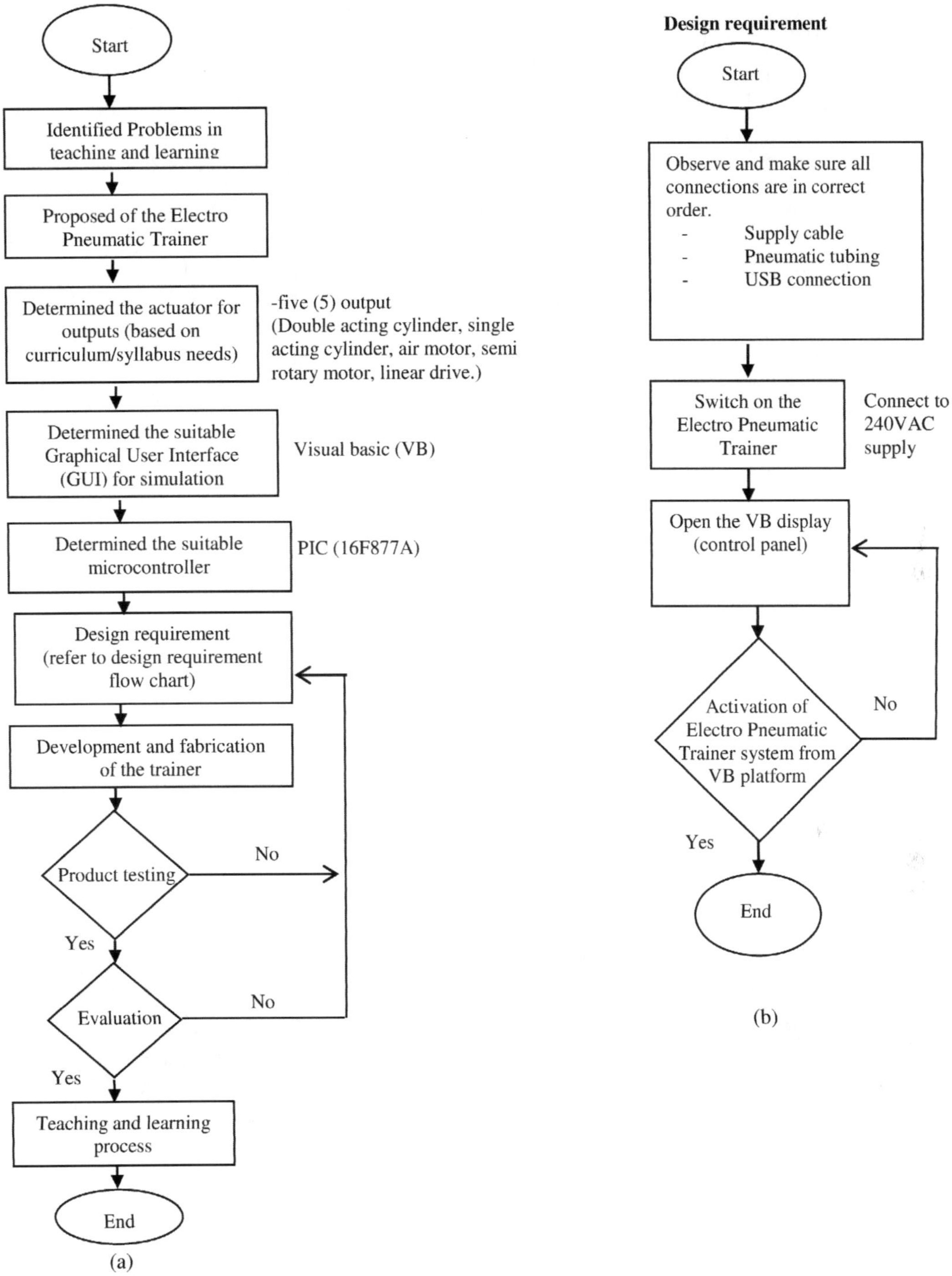

Figure 2. (a) Flow chart for the overall system process, (b) design requirement and sequence of trainer usage

IOP Conf. Series: Materials Science and Engineering **152** (2016) 012065 doi:10.1088/1757-899X/152/1/012065

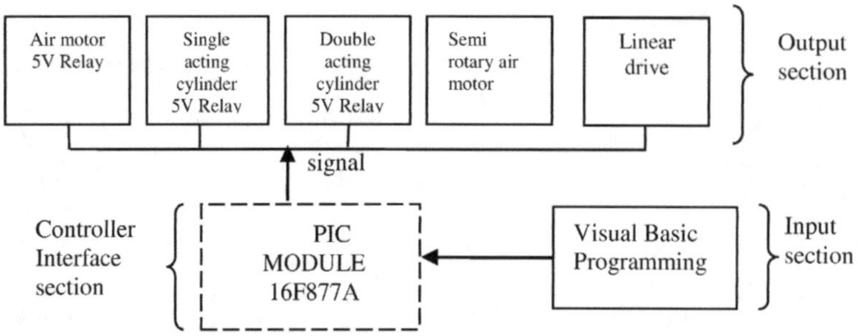

Figure 3. Block diagram for PIC control of electro pneumatic apparatus

4. Building the electro pneumatic trainer

Power supply distribution for electro pneumatic trainer is as shown in Figure 4. From the 240VAC power supply to power-up the air compressor unit, the wiring is then reduced to 12 VDC for electro pneumatic apparatus and lastly 5 VDC for PIC interface board and relay unit. The fabrication process of this electro pneumatic trainer is divided into three main stages: arrangement of electro pneumatic apparatus, PIC module (interface) and GUI (VB programming) as shown in Figure 5**Figure**.

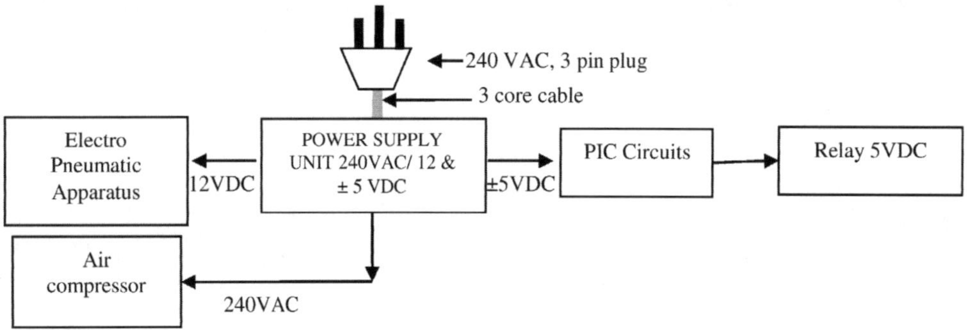

Figure 4. Power supply distribution for electro pneumatic trainer

OUTPUT INTERFACE INPUT

1st. stage: Electro pneumatic apparatus (real world segment)

2nd. stage: PIC module (control interface segment)

3rd. stage: VB (GUIsegment)

Figure 5. Stages of electro pneumatic trainer fabrications

C code was used in programming the PIC through MPLAB since it is the assembly (or assembler) language (ASM). Figure 6(a) shows the compiled code for programming the electro pneumatic trainer while Figure 6(b) shows schematic diagram for PIC interface circuit. USB cable was used to download the HEX code generated by compiler from computer to PIC microchip when the programming process completed. Visual Basic 6.0 was used as the graphical user interface (GUI) input to trigger the output (electro pneumatic apparatus). If the user clicks 'ON' (HOLD) switch for a single acting cylinder, the actual single acting cylinder apparatus will *expand* and when releasing 'ON' (HOLD) switch the actual single acting cylinder will return to its original position. Same operation goes on when each of the 'buttons' acting as switches on VB display is clicked as shown in Figure 7.

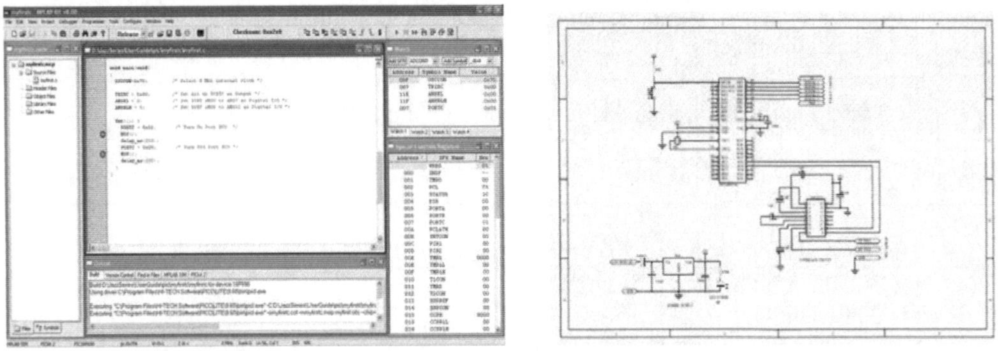

Figure 6. (a) Programming code of PIC 16F877 using MPLAB platform, (b) Schematic diagram of PIC interface board

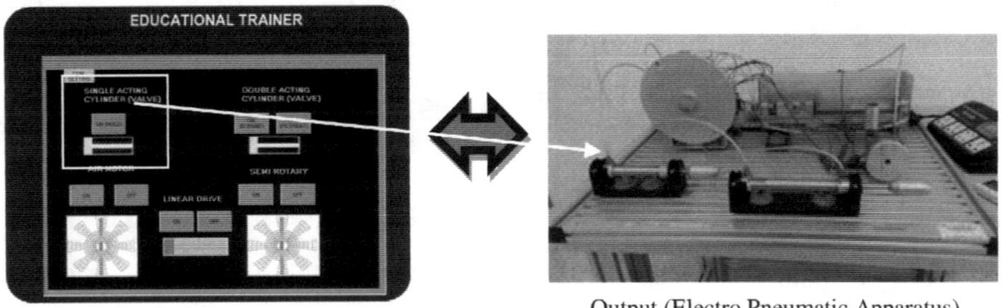

Input (VB Display)

Output (Electro Pneumatic Apparatus)

Figure 7. VB (input) display controlling the electro pneumatic (output) apparatus

5. Implementation and measurement stages

In order to measure the enhancement of knowledge, hands-on skills, and layout of the newly fabricated electro pneumatic trainer, a questionnaire will be distributed to selected respondents. The criteria for selection of respondents are based on the requirement to undergo for related electro pneumatic courses, they have electrical engineering background and they have had direct hands-on experience with the newly fabricated Electro Pneumatic Trainer. Four classes from the Electrical Engineering Department, DJK 3A, 3B, 3C and 3D that have completed the *Instrumentation* (DEJ 3143) lab course were selected as respondents. Before the questionnaires were distributed to the respondents, they need to accomplish the required hands-on practical. They need to complete existing hands-on practical (manual connection of individually apparatus) and continued with using the newly fabricated electro pneumatic trainer. **Table** 1 1 shows a total of 175 respondents were involved in the questionnaire for July 2015 session cohort of students.

Table 1. Total number of respondents

Course Name	Programme Name	Class	No. of Students
Instrumentation (DEJ3143)	Diploma in Electrical Engineering (Control)	DJK 3A	43
		DJK 3B	45
		DJK 3C	45
		DJK 3D	42
	TOTAL		**175**

Table 2 displays the summary of the distributed questionnaire. It was separated into two different sections: Section A is to measure the hardware and simulation section of the electro pneumatic trainer whereas Section B is to gauge the knowledge and also hands-on skill of the respondents. Section A includes three main questions regarding the attractiveness, safety and user friendliness while Section B includes eight questions, mainly on understanding, hands-on skill improvement, critical thinking and interest.

Table 2. Summary of finding for execution of electro pneumatic trainer

No.	Items	Frequency Yes	Frequency No	Yes (%)	No (%)
	Section A: Hardware and VB development for Electro Pneumatic Trainer				
1	It is easy to use the trainer during hands-on because it is user friendly.	170	5	97.1	2.9
2	Are safety features emphasized in the newly developed trainer?	171	4	97.7	2.3
3	Various apparatus of electro pneumatic on this trainer attracts me to study about *Control System*.	169	6	96.6	3.4
		Average (%)		**97.1**	**2.9**
	Section B: Knowledge and Hands-on skill				
1	My understandingson control of electro pneumatic apparatus (input and output devices) isenhanced.	169	6	96.6	3.4
2	It is easy to understand the connection from input to output.	172	3	98.3	1.7
3	Using this trainer can enhance the understanding on electro pneumatic courses.	168	7	96.0	4.0
4	My wiring skill in connecting the electro pneumatic apparatus is improved.	170	5	97.1	2.9
5	Hands-on experience by using this trainer can provide some ideas for my *Final Year Project* (FYP)	165	10	94.3	5.7
6	Using this trainer can enhance my hands-on skills	170	5	97.1	2.9
7	My interest in learning about electro pneumatic is enhanced when using this trainer.	169	6	96.6	3.4
8	It is easy to understand the theoretical part when using this trainer.	172	3	98.3	1.7
		Average (%)		**96.8**	**3.2**

From the questionnaires, it shows that majority of the respondents are satisfied with the hardware fabrication and simulation development for electro pneumatic trainer. About 97% of the respondents agreed that it is user friendly, attractive and safe during the execution of the hands-on practical. The electro pneumatic trainer is considered attractive as more students give attention during the teaching and learning process, and also during the hands-on session (Section A: Item 3: 96.6%). It is also more effective because additional integration of various output components is embedded in this trainer as to uphold the curriculum requisite (i.e. five different output components). Furthermore, it is also user friendly because it is easily installed and operated by the instructor and students during the teaching and learning, and hands-on session as indicated in the survey.

Meanwhile, 96.8% of the total respondents agreed that the newly fabricated trainer enhanced their knowledge and hands-on skill in learning *electro pneumatic* courses. This indicates a good preparation

IOP Conf. Series: Materials Science and Engineering **152** (2016) 012065 doi:10.1088/1757-899X/152/1/012065

for the students before entering real-work environment. 94% of the respondents also agreed that it would enhance their critical thinking in early preparation for their final year project (FYP).

The overall cost for the development of this Electro Pneumatic Trainer was MYR 350 per trainer and this includes fabrication of the interface module and input module (VB platform) only. There are some recycled components that were used in this development such as the output module (i.e. double acting cylinder, single acting cylinder, air motor, semi rotary motor and linear drive) and the air compressor machine. This is cost effective in enhancing the teaching and learning process since the institution only needs to invest a very minimum amount for the production of this trainer. If the training institution wants to produce the same computer simulation and hands-on training apparatus, they need to purchase the software and hardware with an average cost between MYR 10,000 to MYR 25,000 as market price. Luckily, with the implementation of this newly developed electro pneumatic trainer, the cost is already reduced by using available expertise and materials. Table 3 and Table 4 compare the respondents' outcome based (students' performance) on identical practical's tasks that has been assigned using the manually connected electro pneumatic apparatus method as well as using the newly developed Electro Pneumatic Trainer.

Table 3. Respondents' marks using manually connected electro pneumatic devices

Lab Practical Assessment Item For DEJ3143: Instrumentations						
	Using Manually Connected Electro Pneumatic Apparatus					
CLASS	**Tubing connection (25%)**	**Solenoid valve connection skill (25%)**	**Discussion (20%)**	**Conclusion (20%)**	**Reports (10%)**	**Total (100 %)**
DJK3A	15	15	13	15	9	67
DJK 3B	14	15	12	15	9	65
DJK 3C	15	16	13	16	9	69
DJK3D	15	16	14	16	8	69
Average	14.75	15.5	13	15.5	8.75	67.5

Table 4. Respondents' marks using newly developed electro pneumatic trainer

Lab Practical Assessment Item For DEJ3143: Instrumentations							
	Using Electro Pneumatic Trainer (Simulation And Hardware)						
Class	**Tubing connection (25%)**	**Solenoid valve connection skill (25%)**	**Discussion (20%)**	**Conclusion (20%)**	**Reports (10%)**	**Total (100%)**	**Improvement (%)**
DJK 3A	23	22	18	18	9	90	23
DJK 3B	23	23	18	17	9	90	25
DJK 3C	24	23	19	18	9	93	24
DJK 3D	23	22	18	19	9	91	22
Average	23.25	22.5	18.25	18	9	91	23.5

In order to measure the academic performance (practical's skills)in teaching and learning process, the four selected classes were asked to conduct the existing electro pneumatic practical by observing manually operating of electro pneumatic apparatus for *Instrumentation* (DEJ 3143) laboratory course. In measuring the students' practical skills, the students were first assessed based on typical tubing and solenoid valve connection between individual electro pneumatic devices and relays, and manually tested the operation of the apparatus one by one. Once finished, they were required to submit a report based on manually tested practical. Next, the students were given the newly electro pneumatic trainer and were asked to carry out the connection of power supply, which was the connection of 240VAC from bench supply to electro pneumatic trainer, powered for air compressor, connection of solenoids valve toward all apparatus by using the *banana plug cable* on electro pneumatic trainers, and opened VB graphic on computer. Students had to observe and relate the VB graphical command with actual acting of the devices (i.e. operation of air motor, expand and retreat of solenoid cylinder, turning of 180°semi rotary motor and movement of linear drive) and once again they were required to submit a written report. On average, respondents scored 67.5% when using the manually connected electro pneumatic apparatus as compared to 91% when using the newly developed electro pneumatic trainer. This shows that there is an improvement in the overall students' performance at an average of 23.5% on their marks when using the electro pneumatic trainer. Table 3 also shows that the score on their practical skills, which was connecting the air tubing and solenoid valve, also increased at an average of 8.5% and 7.0%, respectively. Since items for reports, conclusion and discussion (refer Table 3 and Table 4) are not relevant for assessing their practicals skills, the findings will not be discussed in this paper. The practical skills were assessed based on standard rubric for *Laboratory Practical Evaluation* that has been standardized throughout the Electrical Engineering department.

6. Conclusion

The newly developed electro pneumatic trainer helps the students to achieve the objective and learning outcomes by enhancing the teaching and learning (cognitive), and psychomotor skill aspect through circuit installation, testing and troubleshooting. The development of electro pneumatic trainer also has been designed to expose the students by showing more apparatus related to electro pneumatic control in which five different output apparatus can be controlled using the VB platform. This is supported by questionnaires distributed to 175 respondents in which 97% of them agreed that the newly fabricated trainer is user friendly, safe and attractive. In addition, 96.8% respondents strongly agreed that there is improvement in knowledge development and hands-on skill in their learning process. Furthermore, through the *Lab Practical Evaluation* record, it is evident that the students improved their academic performance by an average of 23.5%. It also showed a high satisfaction among the respondents in perceptions and an increase of academic performance with regard to the usage of the newly fabricated electro pneumatic trainer.

Acknowledgements

The authors wish to express our thanks to the management of Electrical Engineering Department, Sultan Salahuddin Abdul Aziz Shah Polytechnic and also Aircraft Maintenance Department, Banting Selangor Polytechnic for allowing us to use their materials and tools for development, fabrication and implementation of this research.

References

[1] Ngabonziza Y and Delcham H 2014 *Conference of the American Society for Engineering Education - (ASEE Zone 1)*

[2] Burhan I, Azman A A, Talib S and Ab. Aziz A A 2015 *Third International Conference on Artificial Intelligence, Modelling and Simulation (AIMS2015)*

[3] Meiklejohn D 2013 *Introduction to PIC Programming Baseline to Enhanced Mid-Range Architecture* Gooligum Electronics

[4] Neilsen M L, Lenhert D H, Mizuno M, Singh G, Zhang N and Gross A B 2002 *Proceedings of the American Society for Engineering Education Annual Conference and Exposition*

[5] Caspy P, Sangiovanni-Vincentelli A, Almeida L, Benveniste A and Bouyssounouse B 2005 *ACM Transactions Embedded Computing Systems* **4** 587–611

[6] Kheiralla A F, Siddig O, Mokhtar A A E, Esameldeen M and Addalla O 2007 *International Conference on Engineering Education – ICEE 2007*

[7] Hobbs A 2008 *An overview of human factors in aviation maintenance* Australian Transport

Safety Bureau
[8] Sidhu B 2013 *MAE-SR Technics tie-up?* The Star Malaysia
[9] Division C D and E 2011 Curriculum Document
[10] Williams M J and Rhoades D L 2006 *World Review of Science, Technology and Sustainable Development* **3** 62–70
[11] Burhan I, Talib S and Azman A A 2012 *IEEE 8th International Colloquium on Signal Processing and its Applications*
[12] Vaananen M, Horelli J and Katajisto J 2010 *2nd International Conference on Education Technology and Computer (ICETC)*
[13] Dieter A and James R 2006 *Journal of Information Systems Education* **17** 163–9
[14] Zhen R, Meng H, Wu X L, Li J and Liu S M 2011 *Advanced Materials Research* **271-273** 1231–4

Towards an aerogel-based coating for aerospace applications: reconstituting aerogel particles via spray drying

N Bheekhun[1*], A R Abu Talib[1], S Mustapha[2], R Ibrahim[3] and M R Hassan[4]

[1]Propulsion and Thermo-Fluids Group, Department of Aerospace Engineering, Faculty of Engineering, Universiti Putra Malaysia, Malaysia
[2]Department of Chemical and Environmental Engineering, Faculty of Engineering, Universiti Putra Malaysia, Malaysia
[3]Advanced Materials Research Centre, 09000 Hi-Tech Park, Kulim, Kedah, Malaysia
[4]Department of Mechanical and Manufacturing Engineering, Faculty of Engineering, Universiti Putra Malaysia, Malaysia

*minadiir@gmail.com

Abstract. Silica aerogel is an ultralight and highly porous nano-structured ceramic with its thermal conductivity being the lowest than any solids. Although aerogels possess fascinating physical properties, innovative solutions to tackle today's problems were limited due to their relative high manufacturing cost in comparison to conventional materials. Recently, some producers have brought forward quality aerogels at competitive costs, and thereby opening a panoply of applied research in this field. In this paper, the feasibility of spray-drying silica aerogel to tailor its granulometric property is studied for thermal spraying, a novel application of aerogels that is never tried before in the academic arena. Aerogel-based slurries with yttria stabilised zirconia as a secondary ceramic were prepared and spray-dried according to modified Taguchi experimental design in order to appreciate the effect of both the slurry formulation and drying conditions such as the solid content, the ratio of yttria stabilised zirconia:aerogel added, the amount of dispersant and binder, inlet temperature, atomisation pressure and feeding rate on the median particle size of the resulting spray-dried powder. The latter was found to be affected by all the aforementioned independent variables at different degree of significance and inclination. Based on the derived relationships, an optimised condition to achieve maximum median particle size was then predicted.

1. Introduction

Aerogels [1], invented by Samuel Stephens Kistler 85 years ago, are highly porous nanostructured materials with an open-cell network. Aerogel is generally characterized as a gel in which the liquid phase has been replaced by a gas in such a way that the solid framework is being retained with only a slight or no shrinkage in the gel. It can be argued that there is no austere definition for aerogels. For example, according to the International Union of Pure and Applied Chemistry (IUPAC), an aerogel is a gel which is comprised of a microporous (< 2nm) solid in which the dispersed phase is a gas, but yet mesoporous (2-50nm) aerogels are acknowledged within the community. As time evolves, different types of aerogels have been synthesized from the very first silica aerogel produced at the stage of invention in 1931. Within the classes of aerogels, silica aerogel is still considered to possess the most fascinating physical properties such as low thermal conductivity, ultra low bulk density, optical transparency in the visible spectrum, high specific surface area, low dielectric constant, low refractive index and low sound velocity. Up to now, the latter is the solid with the lowest thermal conductivity ever existed but no more the lightest material as this property has been surpassed after the invent of graphene aerogel in 2013. In terms of commercialisation, silica aerogel is lagging behind significantly because of its relatively high manufacturing cost due to overpriced precursor and high energy required in supercritical drying. As a result, driven by the fact that the potential applications of silica aerogels are almost illimitable, extensive studies ranging from modelling to experiments, in both academia and industries, have been conducted during the past two decades in the quest for affordable silica aerogels.

After tremendous efforts, numerous producers have brought forward different forms of aerogels that are derived from different sources of silica and modified drying conditions. Silica aerogel is now experiencing a renaissance and is tracing its way into various applications like thermal and acoustical insulation, kinetic energy absorption, electronics, optics and biomedicine, amongst others, with the former as the most preferred and versatile one. Consequently, industries are perpetually developing innovative solutions to existing problems by the use of silica aerogel as a super insulative material. High-performance insulation or super insulation is attributed to aerogel based on its attractive R-value, which is usually the factor considered when taking into account the thermal insulation performance of a system. By definition, R-value is a measure of resistance to heat flow through a given thickness of material. The higher the R-value, the more thermal resistance the material has and therefore, the better its insulating value. This value is inversely proportional to thermal conductivity and hence, being the solid with the lowest thermal conductivity, implies that silica aerogel possesses the highest R-value than any material for a fixed length. Furthermore, it should be appreciated that a lower thermal conductivity means the same insulation performance is achieved with a thinner layer or that the same thickness of super insulation can be used to ensure far better insulation performance. Up to now, two thermal insulative solutions are being used widely: in centimetre-thick flexible blankets wherein silica aerogel particles are dispersed into a reinforcing fibre matrix such as glasswool and in millimetre-thick paints as additives wherein aerogel granules are mixed with the alcohol-based solvent. It is worthwhile to note that aerogel-based blankets are already in use in astronautical applications whereas aerogel-based paints are mostly opted for architectural uses. Due to the space and weight constraints, thick insulation is not favoured in aeronautics and thus it is imagined that aerogels could be integrated as a form of micrometre-thick protective coating that can resist substantial mechanical loads and thermal cycling. The issues, challenges and applications of aerogels in aerospace have been detailed in [2].

As per authors' knowledge, no investigation has been carried out until now to study the feasibility of developing an aerogel-based thermally sprayed coating. This paper addresses this technological gap partly since the complete research encompasses different phases before the thermal spraying. The main purpose of carrying out the spray-drying intermediate step prior to deposition is to reconstitute the particles of aerogels to obtain an adhering coating. In consequence, this study employed the Taguchi design to investigate the effect of the different independent variables in terms of the slurry formulation and drying conditions on the median particle size in order to predict the optimum condition to produce aerogel-based powders with maximum particle size with a spherical morphology.

Unlike the spheroidisation, fusing-crushing-cladding or temperature sintering methods [3], spray-drying is a procedure that can reconstitute the granulometric or the morphological properties of a nanostructured ceramic without compromising its microstructures. It is by which a water or organic-based suspension is transformed into dry powder by atomising the slurry into a hot drying medium [4].

2. Experimental specifics

2.1. Materials

Ecological GEAT® 0.125 aerogel was supplied from Green Earth Aerogel Technologies whilst the commercialized yttria stabilized zirconia (YSZ), ZrO_2 $8Y_2O_3$ (Metco 240NS-G) was purchased from Oerlikon Metco (previously Sulzer Metco). Polyvinyl alcohol (Mowiol® 4-88) and polyethylene glycol, $(C_2H_4O)nH_2O$, molar weight 400 were bought from Sigma-Aldrich. On the other hand, Darvan 821-A was provided by Vanderbilt Minerals, LLC. Distilled water was available at AMREC. All the materials were used as-obtained with no further treatment.

2.2. Slurry formulation and spray drying
For a fixed volume of slurry, Darvan 821-A was firstly added to distilled water at 95°C and stirred for 10 minutes. PVA as a hydro-soluble binder was then dropped and mixed for another 20 minutes at the same temperature until it completely dissolved. Next, a specific amount of plasticizer PEG of 0.5 wt.% was dipped and left for mixing until 10 minutes had elapsed. The solution was then left to be cooled at room temperature and was weighted. The loss of water during the heating was compensated. Yttria stabilized zirconia was then mixed gradually and the suspension was blended for another 15 minutes at room temperature. Finally, aerogel was added and commixed for additional 15 minutes. The resulting mixture was ball-milled for longer than 24 hours to attain a well-stabilized slurry. The weight ratio of slurry-to-ball was kept 1:2 at all times.

A LabPlant SD-05 spray-dryer with a two-fluid nozzle was used in this investigation. The heated air was used as the drying medium in the drying chamber. Drying conditions such as inlet temperature, atomization pressure and feeding rate were controlled during the spray drying. Finally, the feedstock was collected in the collecting chamber from the cyclone. In order to study the relationship between slurry properties (powder content, YSZ: aerogel, dispersant and binder) and aforementioned drying conditions upon the particle size of feedstock, an experimental method was designed accordingly.

2.3. Design of experiment
Taguchi's Design of Experiment was used (as in Table 1) as the statistical tool for screening design to ensure a systematic and time-consuming procedure in the pursuit for better morphological properties of aerogels. It is a powerful tool being used within the scientific community for design of high quality systems through screening and optimization by using a strategically designed experiment that is being derived mathematically and empirically. A modified orthogonal array L18 was used to examine seven factor system (A-G) with a combined number of levels wherein one of the factors consists of six levels and the remaining with three levels.

Table 1. Modified L18 Taguchi design

Symbol	Factors	Levels					
		1	2	3	4	5	6
A	Powder Content (vol.%)	15	20	25	30	35	40
B	YSZ: Aerogel (wt)	1:1	1:3	1:5			
C	Dispersant (wt.%)	0.3	0.6	0.9			
D	Binder (wt.%)	4	6	8			
E	Inlet Temperature (°C)	150	175	200			
F	Atomization Pressure (bar)	0.9	1.0	1.1			
G	Feeding Rate (ml/min)	18	21	24			

Note: Each experiment was carried out in triplicate.

2.4. Characterisation
A laser particle size analyser (Mastersizer 2000 E by Malvern Instruments) was utilized to obtain the median particle size. The technique, laser diffraction uses Mie theory of light scattering to calculate the particle size distribution, assumes a volume equivalent of sphere model.

2.5. Facilities
Formulation of slurry and spray-drying were conducted at AMREC, SIRIM while characterization of spray-dried powders was done at the Materials Characterisation Laboratory, UPM.

3. Results and discussion

Figure 1 presents the effect of each control factor, viz. the solid content, the weight ratio YSZ:Aerogel, dispersant, binder, inlet temperature, atomization pressure and feeding rate on the median particle size, D_{50} of prepared spray-dried aerogel-YSZ powders. It was found that the latter ranged from 16.17±0.59 μm to 27.27±0.67 μm with a mean coefficient of variation of 3.90%, hence attesting high consistency and precision among the 18 samples by being under the acceptable value of 15%.

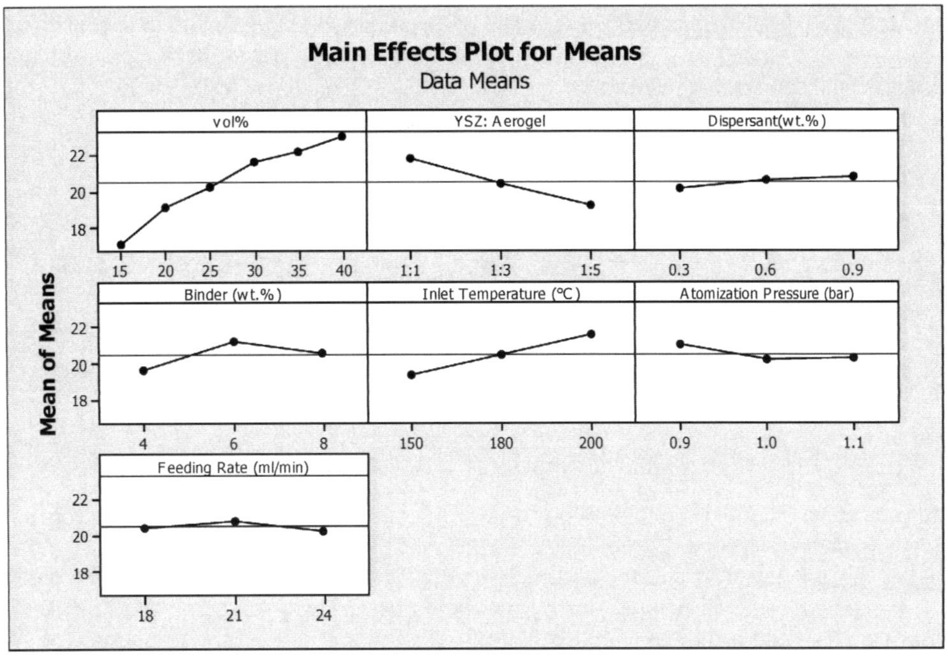

Figure 1. Effect of slurry formulation and spray-drying operating variables on D_{50}

In general, granulation process can be divided into three steps: (1) atomization, where the droplets were formed after the slurry had passed by the nozzle that was positioned at the top of drying chamber; (2) drying gas and droplet contact, where evaporation took place and the liquid feed was turned into droplets in the drying chamber; (3) powder recovery, where the dried feedstock was separated from the drying gas stream in the collecting chamber. From Figure 1, it can be seen that each control factor had an effect on the spray-dried granules at different degree of significance and inclination.

Atomisation pressure and feeding rate contributed in the first step of spray drying. It was observed that the particle size was affected in such a way that an increase in pressure led to a decrease in D_{50}. Figure 2 depicts the schematic diagram of a two-fluid nozzle with an external mix and the sequence involved during the formation of droplets. First, the slurry was forced into the two-fluid nozzle by the aid of the compressed air, which caused the formation of an air cone covered by liquid membrane, dictated by the cone angle or spray angle, θ. For a constant feeding rate, an increase in the atomisation pressure would enlarge the spray angle, causing the thickness of the liquid membrane to decrease. Consequently, the diameter of droplet would decrease, which in turn would reduce the diameter of the spray-dried particle, in other words, D_{50}, as observed in Figure 1. Increasing the feeding rate of the slurry from 18 ml/min to 21 ml/min, on the other hand, would increase the thickness of the membrane leading to the formation of larger particle size, as pointed out in the corresponding diagram. However, when the feeding rate was augmented to 24 ml/min, D_{50} shrank. This phenomenon could be probably because more atomizing air and a higher atomizing air pressure drop was required due to the fact that the mixing and atomization of liquid was taking place outside the nozzle.

The second step in spray drying process was related to the slurry formulation and inlet temperature. It could be observed that increasing the volume fraction of the solid content in the slurry had a strong influence on D_{50}. This could be explained by the fact that a rise in the powder content results in an increase of the suspension viscosity, as proposed by Cao *et al.* [5]. In addition, the particle size is intrinsically the summation of the hollow centre part and the wall-thickness. It was reported in the quoted study that the lower the solid content in the slurry, the larger is the ballooning but the thinner the wall thickness. Adding more solid content could be translated as increasing the density of the solid content as well. YSZ had an apparent density of 2.3±0.2 g/cm^3 while the aerogel was considerably less dense with a value of 0.2±0.01 g/cm^3. Therefore, the decrease in D_{50} with an increase in the ratio of YSZ:Aerogel could be possibly justified by the fact that the suspension is less viscous.

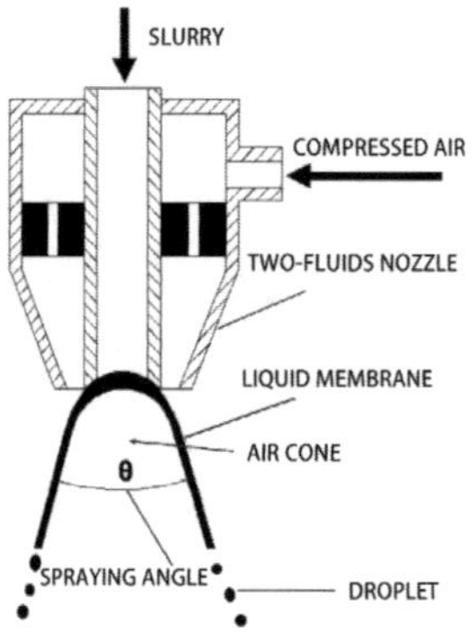

Figure 2. Schematic diagram of two-nozzle with external mix and formation of droplets

Changing the amount of dispersant from 0.3 wt.% to 0.9 wt.% increased the median particle size negligibly compared to the other three slurry formulation parameters: solid content, YSZ:Aerogel and binder. Besides, it should be pointed out that ball-milling were carried out for a minimum of 24 hours, which could have enhanced the dispersion of the particles and making it a homogeneous suspension. Mahdjoub *et al.* [6] studied the effect of the slurry formulation upon the morphology of spray-dried YSZ particles by measuring the relative sediment height (RSH) that represents the state of dispersion of the suspension and is a key factor controlling the droplet drying. It was found that hollow granules were formed in the case of dispersed slurries (low Ratio Sediment Height – RSH < 53%) whereas full granules were obtained with flocculated slurries (high RSH > 62%). Adding binder in the slurry had two effects; D_{50} was increased when Polyvinyl alcohol (PVA) was added with quantity of 0.4 wt.% to 0.6 wt.% whilst a decrease was observed when the amount was brought to 0.8 wt.%. The binder is practically required in order to produce solid compact granules. It was shown in [7] that the RSH value can be drastically modified by the addition of binder and subsequently affecting the state of dispersion of the suspension being stable or unstable and hence the overall granule morphology upon spray-drying. When particles are fully dispersed in a slurry with no interaction between each other, which is the case of a basic suspension (without binder or dispersant), they are very mobile and can pack together. In consequence, particles and binder move to form a dense shell leaving behind an internal void. Because of the difference of pressure between the internal void and the ambient atmosphere, one part of the granule collapses generating hollow powder. In contrast, when the slurry is flocculated, the particles can be considered as immobile; the shell does not appear and granules are formed. Hence, the dual effect due to an addition of PVA in the aerogel-based slurry can be appreciated.

Increasing the inlet temperature resulted in a larger particle size. When the aqueous aerogel-based suspension was forced into the drying chamber, the compressed air converted it into small droplets which contracted as fast evaporation of the water in the slurry took place due to the heated air. During this process of moisture moving from the inside to the outside of the droplet, the solid content, binder and dispersant were also carried along. Inside the droplet, a hollow centre part would be produced when the evaporation rate of moisture inside the droplet was higher than the diffusion rate through the droplet surface. Overall, for spray drying, increasing the drying temperature resulted in greater loss of water of resultant powder due to the higher rate of heat transfer into particles that is causing faster water removal. However, if inlet temperature was set above 200°C in this study, it could be anticipated that the evaporation rate of moisture inside the droplet would become higher than the diffusion rate through the droplet surface, which would cause a decrease in the resulting spray-dried powder.

4. Conclusion

An attempt was made to tailor the granulometric property of silica aerogel via spray-drying for thermal spray applications. Taguchi experimental design was employed to systematically analyse the effect of the slurry formulation and spray-drying process on the median particle size and to predict the optimal condition to achieve maximum median particle size, which had been found to be 40 vol.%, 1:1, 0.9 wt.% 6 wt.%, 200°C, 0.9 bar and 21 ml/min for solid content, YSZ:Aerogel, dispersant, binder, inlet temperature, atomisation pressure and feeding rate, respectively.

Acknowledgments

This work was supported by the Fundamental Research Grant Scheme (FRGS), Grant No. 5524611. The first author wishes to express his appreciation to the Ministry of Education of Malaysia for their financial support. Technical assistance provided by Mr. Ahmad Nizam at the Advanced Materials Research Centre is gratefully acknowledged by the team.

References

[1] Kistler S S 1931 *Nature* **127** 741
[2] Bheekhun N, Abu Talib A R and Hassan M R 2013 *Advances in Materials Science and Engineering* 18
[3] Johari N, Ibrahim R, Ahmad M A, Suleiman Ahmad M J and Abu Talib A R 2014 *Advanced Materials Research* **2014** 139-43
[4] Masters K 1979 *Spray drying handbook*
[5] Cao X Q, Vassen R, Schwartz S, Jungen W, Tietz F and Stöever D 2000 *Journal of the European Ceramic Society* **20** 2433-9
[6] Mahdjoub H, Roy P, Filiatre C, Bertrand G and Coddet C 2003 *Journal of the European Ceramic Society* **23** 1637-48
[7] Bertrand G, Filiatre C, Mahdjoub H, Foissy A and Coddet C 2003 *Journal of the European Ceramic Society* **23** 263-71

Analysing generator matrices G of similar state but varying minimum determinants

H Harun[1,2*], M F Razali[1,2] and N A Abdul Rahman[2]

[1]Avionics Section, UniKL MIAT, Malaysia
[2]Aerospace Engineering Department, Universiti Putra Malaysia, Malaysia

*harlisya@unikl.edu.my

Abstract. Since Tarokh discovered Space-Time Trellis Code (STTC) in 1998, a considerable effort has been done to improve the performance of the original STTC. One way of achieving enhancement is by focusing on the generator matrix **G**, which represents the encoder structure for STTC. Until now, researchers have only concentrated on STTCs of different states in analyzing the performance of generator matrix **G**. No effort has been made on different generator matrices **G** of similar state. The reason being, it is difficult to produce a wide variety of generator matrices **G** with diverse minimum determinants. In this paper a number of generator matrices **G** with minimum determinant of four (4), eight (8) and sixteen (16) of the same state (i.e., 4-PSK) have been successfully produced. The performance of different generator matrices **G** in term of their bit error rate and signal-to-noise ratio for a Rayleigh fading environment are compared and evaluated. It is found from the MATLAB simulation that at low SNR (<8), the BER of generator matrices **G** with smaller minimum determinant is comparatively lower than those of higher minimum determinant. However, at high SNR (>14) there is no significant difference between the BER of these generator matrices **G**.

1. Introduction

The rapid growth of data usage, as well as information and communications technology services, has forged in increasing the information capacity. Hitherto, a reliable high-speed wireless communication channels is imperative for reliable transmission without increasing the transmitted power or sacrificing the bandwidth [1, 2]. A great deal of research has been carried out to improve wireless communication transmission [3-7] since the transmitted signals in wireless environment are received through multiple paths where they are attenuated by fading phenomenon as well as interference from other applications that are sharing the same transmission medium. These situations result in performance degradations. The increase in demand for data intensive applications has also created a need for high transmission data rate, spectral efficiency and also transmission reliability. Data rate describes how fast data can be transmitted, whilst spectral efficiency is indicated by the data rate over a certain frequency band. The transmission reliability, on the other hand, is measured by the probability that the transmitted data can be correctly detected. Diversity techniques are broadly applied to increase the information capacity as well as to decrease the effects of multiple path fading.

The diversity techniques can be divided into time, frequency and space diversity. These techniques require replicas of transmitted signal, which uncorrelated or independent to each other at the receiving terminal. The main foundation of the diversity technique is that, if two or more independent signals are received, they will fade in an uncorrelated way such that some signals are severely attenuated whilst the others will be less attenuated. The proper combination of the transmitted signals' replicas at the receiver will greatly scale down the fading by increasing overall received signal-to-noise ratio (SNR), thus improving the reliability of the transmissions [1, 2, 8, 9]. In space diversity or so-called antenna diversity technique, antenna arrays are arranged in space in proper distance to obtain the uncorrelated signals to mitigate loss in bandwidth efficiency. Space diversity is generally classified into "Transmit Diversity" and "Receive Diversity". Transmit diversity is more practical than receive diversity for enhancing the downlink (i.e. bottleneck in broadband asymmetric applications like internet browsing and downloading) to preserve the small size and low power consumption features of the user terminal. It is more practical to consider transmit diversity for the downlink (i.e. from base stations to mobile) communication because it is easy to install multiple transmit antennas at the base station, as well as providing them the extra power for multiple transmissions.

The space-time coding was first introduced by Tarokh *et al.* [8, 9] in 1998 as a novel approach for providing transmit diversity for multiple-antenna fading channel. Space-Time Trellis Codes (STTCs) is one of the coding structures in the space-time coding technique that can be utilised to improve the performance of the wireless communication systems over fading channels. STTC has been proven to efficiently utilise transmit diversity and effectively exploit the effects of multipath fading to increase the information capacity of the multiple antenna systems [8-10]. STTCs rely on transmitting multiple, redundant copies of a data stream to the receiver in the hope that at least some of them will survive the physical path between transmission and reception in a good state to allow reliable decoding. STTCs combat the effects of fading, hence achieving high spectral efficiencies and performance (capacity) gains [8-10]. STTCs are based on the trellis structure in which its encoding architecture relies on the generator matrix **G**. Thus, all possible combinations of the coefficients in the generator matrix **G** is imperative in the STTC design so that resultant STTCs are capable of optimising both diversity and coding gains. This implies that, the quality of the transmission of STTC communication system depends on the generator matrix **G**. From its early introduction in 1998 [9] until recently, techniques in evaluating generator matrix **G** by performing a full code search on the coefficients on the generator matrix **G** have been developed and improved [3-5, 7, 11-15, 18]. Mostly, the interest is on the different states, to evaluate the overall STTC performance. Realising the importance in analysing the behaviour of different generator matrices within the same state in improving the performance of STTC, a few generator matrices **G** with different minimum determinant has been produced in this paper. Thorough analyses on the impact of bit error rate (BER) and SNR towards different generator matrix **G** with different minimum determinants within the same state have also been explored in this paper.

2. STTC design criteria

The design of optimal STTC is based on rank and determinant criterion (RDC). Tarokh's contribution in designing optimal STTCs in quasi-static flat Rayleigh fading channels is determined by matrices constructed from pairs of distinct code sequences [9]. The minimum rank (rank criterion) among these matrices quantifies the diversity gain while the minimum determinant (determinant criterion) of these matrices quantifies the coding gain. RDC is developed by Tarokh to minimise the pairwise error probability (PEP). The pairwise error probability (PEP), denoted as $p(X \rightarrow E)$, is the probability for mistaking a transmit **x** for another matrix **e**. It depends on the distance between the two matrices after transmission through the fading and noisy channel. The upper bound of PEP is used instead as the index of code performance.

PEP is minimised when STTC achieved full diversity (i.e. full rank) and high coding gain. Full diversity is achieved when the distance or difference matrix has full rank for all the codewords. STTC has a full rank if the distance or difference matrix between any codewords has full rank equals to the number of transmit antenna, n_T. STTC design that guarantees full diversity is based on these two heuristics design rules [9, 12, 14, 18], which are:

- Diverging branches from the same node differ in the second symbol while remaining constant in the first symbol.
- Merging branches differ in the first symbol but stays constant in the second symbol.

Since a two transmit antenna model has been used in this paper and the maximum rank is equal to two, the STTC in this paper can be considered to achieve full diversity. The rank order of this code must be two, since any pair of codewords for the two rows of the difference matrix are linearly independent. This agrees with the first rule such that $(e^1_{t=1} - x^1_{t=1}) = 0$ and $(e^2_{t=1} - x^2_{t=1}) \neq 0$; and the second rule which is $(e^1_{t=l} - x^1_{t=l}) \neq 0$ and $(e^2_{t=l} - x^2_{t=l}) = 0$. Therefore, the difference matrix, **B(x,e)** will have the form as in Equation 1 [9, 10, 19]:

$$B(x,e) = \begin{bmatrix} 0 & *\ldots* & e^1_l - x^1_l \\ e^2_1 - x^2_1 & *\ldots* & 0 \end{bmatrix} \tag{1}$$

To achieve the minimum error probability, minimum rank and minimum product of all eigenvalues of distance matrix **A(x, e) = B^H(x, e) B(x, e)** should be maximised. The design criterion focuses on maximising the minimum coding gain, which is established on the combinations of all **x** and **e**. The coding gain distance (CGD) between the coded sequences x and e is defined as **d²(x,e) = det(A(x,e))** where det(.) is the determinant operator. Generator matrix **G** is used to facilitate the code search procedure (i.e. through RDC) to obtain an optimal STTC (i.e full diversity and high coding gain). High coding gain is obtained by maximising the minimum determinant. Consequently, techniques in evaluating generator matrix **G** is crucial in calculating the minimum determinant effectively to obtain high coding gain [3-5, 7, 11-15, 18].

3. Generator matrix G

Generator matrix G will define the STTC encoder structure [10], where the branch coefficients are arranged alternatively in the generator matrix as in Equation 2.

$$G = \begin{bmatrix} g^1_{0,1} & g^2_{0,1} & g^1_{1,1} & g^2_{1,1} \\ g^1_{0,2} & g^2_{0,2} & g^1_{1,2} & g^2_{1,2} \end{bmatrix}^T \tag{2}$$

In this study, the QPSK STTC encoder with two transmit antennas model is used. Figure 1 illustrates the QPSK STTC encoder structure. It is shown in Figure 1 that generator matrix G is the multiplying factor to the current input bit and the stored bit in the encoder structure. This is further described in Equation 3 [8-10].

$$x^i_t = \sum_{k=1}^{2} \sum_{j=0}^{v_k} g^k_{j,i} c^k_{t-j} \quad \text{mod } 4, \quad i = 1,2 \tag{3}$$

Upon multiplication with the generator matrix G (as in Equation 3), the binary digits (input and stored bit) will be converted into non-binary output (the space-time modulated symbol), which are then used to obtain the trellis diagram. This is shown in Figure 2.

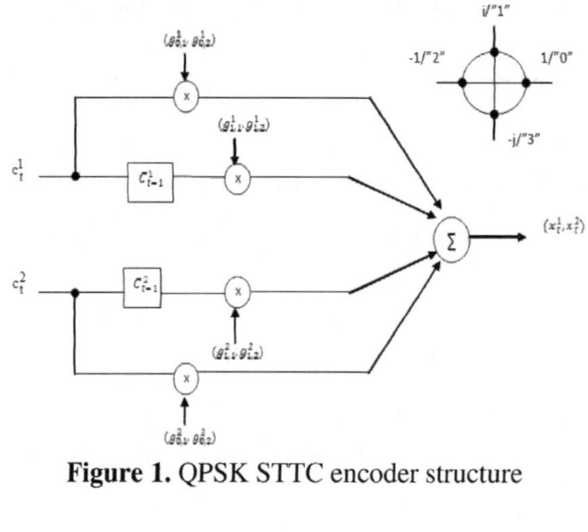

Figure 1. QPSK STTC encoder structure

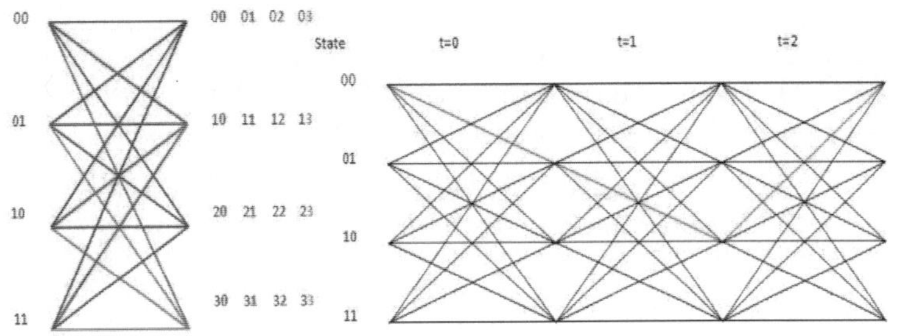

Input bits	01	10	00
Space time symbol transmitted from first transmit antenna	0	1	2
Space time symbol transmitted from second transmit antenna	1	2	0

Figure 2. Continuous path through the trellis that defines the transmitted symbols [9]

As illustrated in Figure 2, path movement of the input bits can be traced with the help of the trellis diagram. Defining the path movement is significant in order to minimize the error probability (hence minimum determinant) [3, 5, 8, 10]. The optimal STTC is the one that minimizes the error probability. Hence generator matrix G is used to facilitate the code search procedure to obtain an optimal STTC.

4. Different generator matrices G of similar state

Realizing the importance of the generator matrix G in producing optimal STTC, a few techniques have been developed in producing the generator matrix G [3-9, 12-15]. The focus thus far is on the STTC performance analysis for different states, such as 4, 8, 16 and 32 states. For example, it has been found that as the number of states increases, the STTC performance improves [6]. So far, there is no effort yet in producing a few generator matrices G with different minimum determinants within the same state. In this study, a few generator matrices G with different minimum determinants, namely four (4), eight (8) and also sixteen (16) of the 4-state QPSK, have been produced to analyse the BER and SNR performance of STTC. The generation of the generator matrices G for minimum determinant eight (8) and sixteen (16) was done using Lisya tree structure [5]. Table 1 shows the summary of the generator matrix G that was used in this paper.

Table 1. Summary of generator matrices G with their minimum determinant

Code	Generator Matrix G	Determinant	Code	Generator Matrix G	Determinant
Tarokh *et al.* [9]	$G=\begin{pmatrix}0&0&2&1\\2&1&0&0\end{pmatrix}$	4.0	Baro *et al.* [12]	$G=\begin{pmatrix}2&0&1&3\\2&2&0&1\end{pmatrix}$	8.0
Tao *et al.* [15]	$G=\begin{pmatrix}0&2&1&2\\2&1&2&0\end{pmatrix}$	4.0	Lisya16-31 [5]	$G=\begin{pmatrix}0&0&1&2\\0&0&2&0\end{pmatrix}$	16.0
Bahador [16]	$G=\begin{pmatrix}0&2&1&2\\2&3&2&0\end{pmatrix}$	4.0	Lisya16-63 [5]	$G=\begin{pmatrix}0&0&2&1\\0&0&0&2\end{pmatrix}$	16.0
Lisya8-2 [5]	$G=\begin{pmatrix}3&3&3&0\\1&3&1&1\end{pmatrix}$	8.0	Lisya16-100 [5]	$G=\begin{pmatrix}0&0&3&2\\0&0&2&0\end{pmatrix}$	16.0
Yuan *et al.* [17]	$G=\begin{pmatrix}0&2&1&0\\2&2&0&1\end{pmatrix}$	8.0			

A two-transmit antenna model has been used, with number of iterations for this simulation is 1000 for each generator matrices G. The same random data is applied throughout the simulation process for each generator matrices G. The flow chart in analysing the generator matrices G of similar state but with varying minimum determinants is shown in Figure 3.

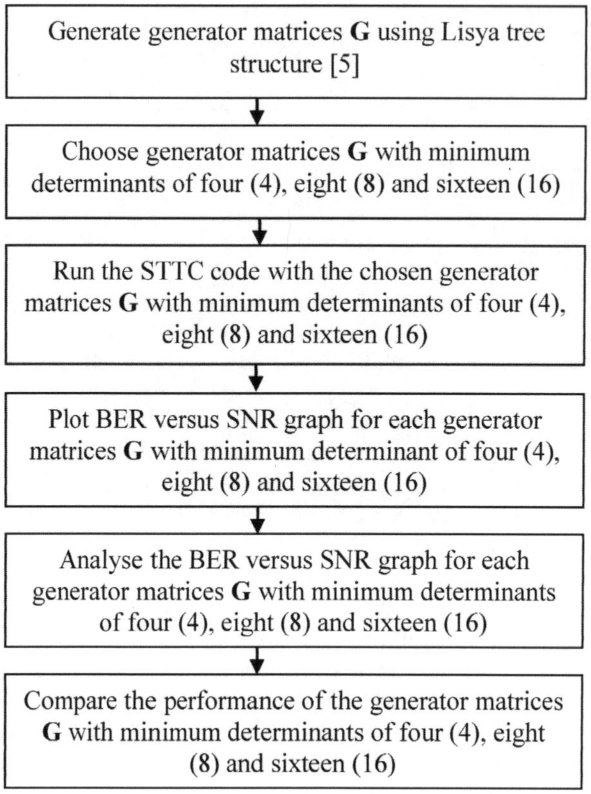

Figure 3. Flow chart in analysing generator matrices G of similar state but varying minimum determinants

5. Results and discussion

The performance of Tarokh [9], Baro [12] and Lisya16-31 [5] for minimum determinant of four (4), eight (8) and sixteen (16) is illustrated in Figure 4. The details of the generator matrix G can be found in Table 1. It is shown in Figure 4 that the BER for Lisya16-31 with the minimum determinant of 16 is comparatively lower than Baro with the minimum determinant of 8. The result is significant since with the same kind of errors, Lisya16-31 can send more data (i.e. four times more as compared to Baro). Hence Lisya16-31 should be preferred compared to Baro. Usually, the BER of generator matrices G with smaller minimum determinant is comparatively lower than those of higher minimum determinant.

BER versus SNR performance for Bahador [11], Lisya8-2 [5] and Lisya16-100 [5] for minimum determinant of four (4), eight (8) and sixteen (16) is found in Figure 5. Again, it is clearly presented in Figure 5 that the BER for Lisya16-100 with the minimum determinant of 16 is comparatively lower than Lisya8-2 with the minimum determinant of 8. This implies that Lisya 16-100 can send more data as compared to Lisya 8-2.

From Figure 6, it is also shown that the generator matrix output for Yuan [17] with the minimum determinant of 8 has a higher BER value compared to Lisya16-63 [5] with the minimum determinant of 16. The result is consistent with Figure 4 and 5, which justifies that the generator matrices G with the minimum determinant 16 from Lisya tree structure perform better, particularly at high SNR value (i.e. at SNR > 12).

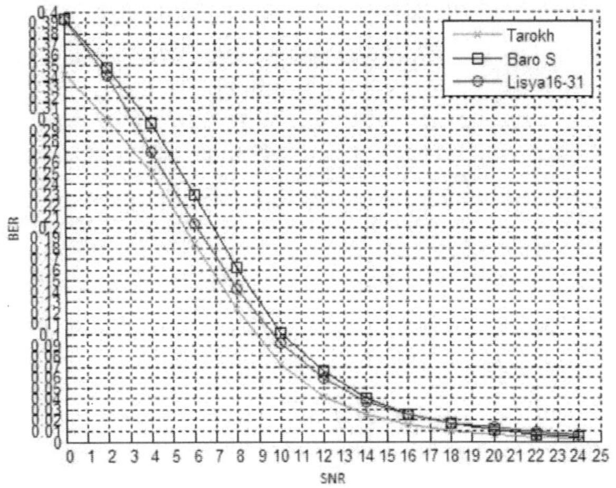

Figure 4. A graph of BER vs SNR performance for Tarokh [9], Baro [12] and Lisya16-31 [5], with minimum determinant of four (4), eight (8) and sixteen (16)

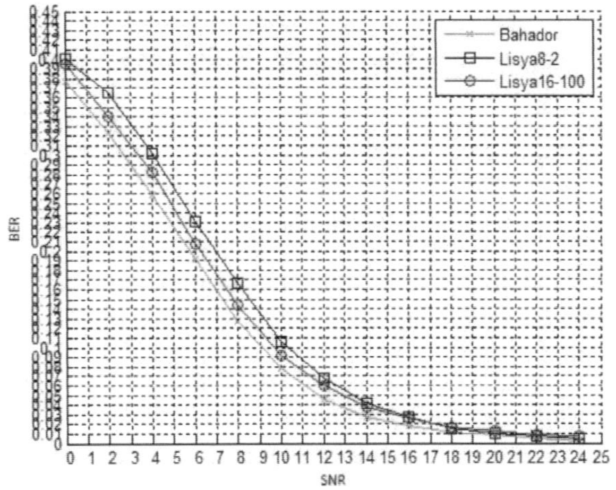

Figure 5. A graph of BER vs SNR performance for Bahador [16], Lisya8-2 [5] and Lisya16-100 [5], with minimum determinant of four (4), eight (8) and sixteen (16)

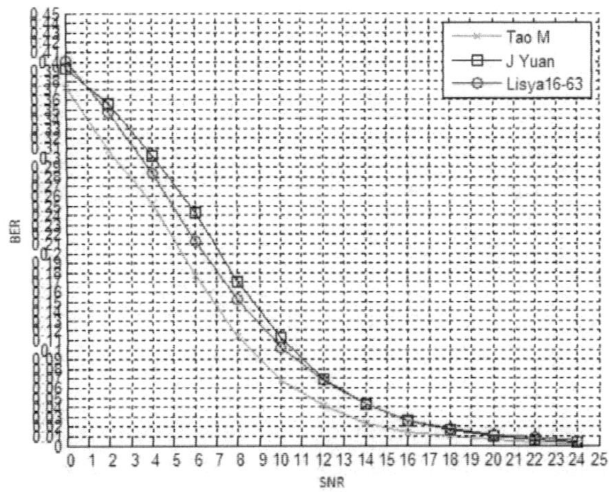

Figure 6. A graph of BER vs SNR performance for Tao [14], Yuan [17] and Lisya16-63 [5], with minimum determinant of four (4), eight (8) and sixteen (16)

6. Conclusions

This paper compares the performance of different generator matrices G at the same state. It provides an analysis of how the generator matrices of varying minimum determinants would behave when the transmission is established in a similar state of trellis. With different generator matrices G, how would it impact BER and SNR. This research is imperative since no effort has been done yet to evaluate the performance of generator matrix G within the same state. So far, the evaluation is on different states. It is found that the generator matrix G with higher minimum determinant produced by the Lisya tree structure performs better compared to the other generator matrices with lower minimum determinant, especially at higher SNR (i.e., SNR > 12). This research can be used as a platform for future studies that employ multiple receiving antennas.

References

[1] Simon M K and Alouini M S 2000 *Digital communication over fading channels: a unified approach to performance analysis* (United Kingdom: John Wiley and Sons)

[2] Rappaport T S 2002 *Wireless communications: principles and practice* (USA: Prentice-Hall)

[3] Harun H, Dimyati K and Ungku Chulan U A 2013 *Aerospace Science and Technology* **24** 136-40

[4] Harun H, Ungku Chulan U A and Khazani K 2011 *15th International Conference on Information Visualization*

[5] Harun H 2010 *An improved algorithm for fast evaluation of space-time trellis code (STTC) generator matrix* PhD Thesis University of Malaya, Malaysia

[6] Yuen N 2003 *Performance analysis of space-time trellis codes* MEng Thesis University of British Columbia, Canada

[7] Banerjee S and Agrawal M 2013 *Wireless Personal Communications* **75** 1141-54

[8] Tarokh V, Seshadri N and Calderbank A R 1997 *IEEE International Conference on Communications*

[9] Tarokh V, Seshadri N and Calderbank A R 1998 *IEEE Transactions on Information Theory* **44** 744-65

[10] Vucetic B and Yuan J 2003 *Space-time coding* (United Kingdom: John Wiley and Sons).

[11] Harun H, Ungku Chulan U A I and Ungku Chulan U A 2013 *IEEE Computing, Communications and IT Applications Conference*

[12] Baro S, Bauch G and Hansmann A *Communications Letters* **4** 20-2

[13] Abdool-Rassool B, Nakhai M R, Heliot F, Revelly L and Aghvami H 2004 *IEE Proceeding on Communications* **151** 25-31

[14] Yuan J, Vucetic B, Chen Z and Firmanto W 2003 *IEEE Transactions on Communications* **51** 1991-6

[15] Tao M and Cheng R S 2001 *IEEE Communications Letters* **5** 313-5

[16] Amiri B 2006 *Space-time trellis codes*

[17] Yuan J, Vucetic B, Xu B and Chen Z 2001 *IEEE Vehicular Technology Conference*

[18] Fukuda T, Otsu S, Tokunaga Y and Zhao H 2008 *23rd International Technical Conference on Circuits/Systems, Computers and Communications*

[19] Jankiraman M 2004 *Space-time Codes and MIMO Systems* (USA: Artech House)

Flammability of self-extinguishing kenaf/ABS nanoclays composite for aircraft secondary structure

S Karunakaran[1], D L Majid[1, 2*] and M L Mohd Tawil[2]

[1]Department of Aerospace Engineering, Faculty of Engineering, Universiti Putra Malaysia, Malaysia
[2] Laboratory of Biocomposite Technology, Institute of Tropical Forestry and Forest Products, Universiti Putra Malaysia, Malaysia

*dlaila@upm.edu.my

Abstract. This study investigates the flammability properties of kenaf fiber reinforced acrylonitrile butadiene styrene (ABS) with nanoclays composites. Natural fiber is one of the potential materials to be used with thermoplastic as a composite due to its attractive properties such as lightweight and strong. In this paper, flammability properties of this material are evaluated through Underwriters Laboratory 94 Horizontal Burning (UL94 HB), which has been conducted for both controlled and uncontrolled conditions, smoke density and limiting oxygen index tests (LOI). These flammability tests are in compliance with the Federal Aviation Regulation (FAR) requirement. The results from UL94 HB and smoke density tests show that the presence of nanoclays with effective composition of kenaf fiber reinforced ABS has enhanced the burning characteristics of the material by hindering propagation of flame spread over the surface of the material through char formation. Consequently, this decreases the burning rate and produces low amount of smoke during burning. On contrary, through LOI test, this material requires less oxygen to burn when exposed to fire, which hinders the enhancement of burning characteristics. This is due to burning mechanism exhibited by nanoclays that catalyzes barrier formation and flame propagation rate over the surface of the biocomposite material. Overall, these experimental results suggest that this biocomposite material is capable of self-extinguishing and possesses effective fire extinction. The observed novel synergism from the result obtained is promising to be implemented in secondary structures of aircraft with significant benefits such as cost-effective, lightweight and biodegradable self-extinguishing biocomposite.

1. Introduction

In recent years, the novel approach of using natural fibre reinforced thermoplastic material is receiving increased attention in many applications such as housing, automotive, aerospace and other low cost, high volume applications [1]. Hence natural fibre composites are emerging as realistic alternatives to synthetic fibre composites. This transformation is due to attractive properties exhibited by the natural composites fibre such as lightweight, environmental friendly, vast source of production, and reasonable strength and stiffness [2, 3]. However, knowledge on flammability properties of natural fibre reinforced thermoplastic material has not been broadly researched. Flammability is defined as the reaction of particles in order to

IOP Conf. Series: Materials Science and Engineering **152** (2016) 012068 doi:10.1088/1757-899X/152/1/012068

produce fire processes of a material, and this reaction can be classified into several different phases such as ignitability, flame spread, heat release, and resulting products from fire process [4]. Since fire issue is becoming the major concern in aerospace industry, a new biocomposite material that has the capability to act as a self-extinguisher is being popularly explored by researchers. Attention in reducing the fire risk and hazard to both human and structures is leading to the development of newer composite materials that can reduce fire accidents [5]. Flammability is a serious issue in aerospace industry applications, which needs a meticulous attention particularly in the area of structures since small or confined space is exposed to fire hazard [6]. Fire risk in aviation sector has been identified ever since the Wright's brother built the aircraft. Looking back to past occurrences, the first fatal accident in aviation occurred due to an uncontrolled in-flight fire [7, 8]. Even though continuous improvement on fire safety of aircraft such as airworthiness requirement, manufacturing and maintenance standards, design of protective and emergency equipment, and improved procedures and training for flight crew and cabin crew has been taken, in-flight fire and smoke still occur in transport aircraft [7]. Aviation history proves that, apart from the loss of control, controlled flight into terrain and specific component failure, aircraft fire accidents is one of the top four leading fatalities [8]. Statistic shows that from 1990 to 2010, 18 major in-flight fire accidents occurred that lead to 423 fatalities [9].

There are different types of flammability testing techniques for investigating flammability properties of a material. However, aerospace sectors are subjected to strict rules and regulation when it comes to safety matter. The requirement for interior application must satisfies the FAR Part 25 Airworthiness Standards in Transport Category Airplane Subpart D regulation for interior application, which are FAR 25.853 (fire protection for compartment interiors) and FAR 25.856 (fire protection for thermal and acoustic insulation materials) [10]. Besides that, according to the requirements for biomaterials and biodegradable materials TA11-01-0, the material must conform to FAR 25.853 and FAR 25.856 for interior application [11]. Therefore, flammability testing techniques that comply with the standard requirement of FAR are the Underwriters Laboratory 94 horizontal burning (UL94 HB), smoke density and limiting oxygen index tests (LOI). These fire testing techniques evaluate various burning characteristics of biocomposites by considering the targeted applications. The burning characteristics that are being evaluated in these tests include burning rates, flame propagation, flame drips, smoke production and oxygen concentration [12].

A study has stated that kenaf fiber is one of the commemorate constituents of natural fibre reinforced plastic composites in Malaysia [13]. The physical property of kenaf, which is in long fibre form, enhances the mechanical properties and this becomes an advantage as it can be used in many industrial applications such as insulators seals [14]. However, since kenaf fibre constitutes of cellulose, hemicellulose and lignin, it is highly flammable and has a poor fire resistance due to the poor compatibility of hydrophobic polymer matrices [3-14]. Recently, kenaf has been used in conjunction with various matrices like polypropylene (PP), polyethylene (PE), high density polyethylene (HDPE), polylactic acid (PLA), polyvinyl chloride (PVC), phenol formaldehyde (PF), polyester and epoxy(EP) [15]. Due to high demands on plastic industry for petroleum-based material production, a vast investigation and research study is carried out on plastic composite including ABS [3-16]. ABS is a type of engineered polymers with an advantage of higher mechanical and thermal properties. When pure ABS is exposed to fire, it is easily burned with a luminous yellow flame, producing black soothe and smoke, and continues to burn even after removal of the ignition fire source [17]. Table 1 shows flammability data of polymeric materials based on LOI and UL 94 rating.

The major responsibility of flame retardant agent is to prevent, cease or minimize the combustion of a material. There are many types of flame retardant including phosphorus, halogen, silicon, nanometric particles and mineral additives. In this present work, nanoclay is being used as flame retardant and it can be classified as nanometric particles. Bonati [18] stated that as flame retardant agent, nanoclays possessed the capability to inhibit or stop the polymer combustion as it can either act physically such as by cooling through formation of a protective layer or chemically such as reaction in the condensed or gas phase.

Chigwada [19] researched the effects of fire behavior on ABS, PE and epoxy with incorporation of the nanoclays through cone calorimeter. The result showed that with nanoclay, the maximum heat release rate decreased while the ignition time increased during fire. On the other hand, Saba [20] concluded that there are two main flame retardant mechanisms in ABS polymer and kenaf with nanoclays composites. First mechanism is the formation of a barrier against heat and volatiles through the migration of the nanoclays towards the surface of the material. The second mechanism is the char formation between the surface of the material and the flame [21, 22]. Furthermore, Gacitua [22] stated that a small amount of nanoclays incorporation with the biocomposites can improve the flammability properties of biocomposite. Therefore, it can be considered as high-potential filler material for improving burning characteristic of biocomposite. It is thus concluded that addition of small amount of nanoclay into the polymer matrix can improve the flammability properties of biocomposites.

Table 1. Flammability data of polymeric material based on LOI and UL 94 rating

Polymer	LOI (%)	UL 94 Rating
ABS	18	HB
PE	17	HB
PVC	45	V
PP	17	HB
EP	19	HB

HB = Horizontal Burning, V = Vertical

The aim of this paper is to investigate the flammability properties of kenaf fiber reinforced acrylonitrile butadiene styrene (ABS) with nanoclays composites. The flammability properties of this material were evaluated through standard flammability testing such as Underwriters Laboratory 94 horizontal burning (UL94 HB), which was conducted for controlled and uncontrolled conditions, smoke density and limiting oxygen index tests (LOI). The performance of the material during UL 94 HB test in controlled condition was compared to that in uncontrolled condition. Results from three types of fire tests are presented.

2. Methodology

2.1. Materials and preparation

Kenaf fiber used in this study was composed of kenaf core and bast fibre obtained from fresh whole stem supplied by National Kenaf and Tobacco Board, Malaysia [24]. Meanwhile, ABS was supplied by Toray Plastics (Malaysia) with specific melt flow rate of 25 g/10 min at 220^0C/10 kg. Both ABS and kenaf were pulverized with apparent density of 0.51 g/cm^3 and 0.14 g/cm^3, respectively, in accordance to the ISO 60:1977. Prior to extrusion, 1 phr of ULTRA-PLAST TP01 was added during the compounding process using a co-rotating twin screw extruder with twin screw speed of 5 Hz [23]. Table 2 shows the fabricated material composition.

C-100 was set as the control formulation while C-50/50 was the optimized formulation for the material in terms of mechanical properties such as tensile strength (TS), flexural strength (FS), Young's modulus (YM) and flexural modulus (FM) as reported in [23]. This optimized formulation was added with 1% and 3% of nanoclays, respectively, in order to investigate the effect on the burning property of the material. On the other hand, C-90/10 was selected as part of the investigation compared to other composition due to its lowest value of TS, YM and FM. Although it showed the highest FS value, there was no remarkable differences compared to the C-50/50. Table 3 shows the material compositions that are being used in this flammability test.

Table 2. Material composition and code

Composition		Code
ABS (%)	Kenaf (%)	
100	0	C- 100
90	10	C- 90/10
80	20	C- 80/20
70	30	C- 70/30
60	40	C- 60/40
50	50	C- 50/50

Table 3. Material composition

Composition		Nano Clay Filler (%)
ABS (%)	Kenaf (%)	
100	0	-
90	10	-
50	50	-
Optimized Material Composition		
50	50	1
50	50	3

2.2. UL94 horizontal burning
This flammability test is to serve as preliminary indication of their acceptability with respect to flammability for a particular application. This method determines the burning rate, dripping time, and flame propagation.

2.2.1 Controlled condition. The burning rates of all samples were investigated by UL94 HB test according to the UL94 standards. The experimental setup with the dimensions for the test is as shown in Figure 1. Flammability test was conducted for all five samples. The specimen was clamped on the retort stand and flame was applied with transverse axis inclined at 45 ±2 degrees to the free end of specimen for 30 ±1 seconds without changing its position. The burner was then removed after 30 ±1 seconds, or as soon as the combustion at the front of the specimen reaches the 25 mm mark [24]. The timing device started when the combustion front reached the 25 mm mark and time taken for the flame to propagate to the benchmark was measured. The sample was classified as HB rating if the burning rate did not exceed 40 mm per minutes or it self-extinguished before 100 mm benchmark [24].

2.2.2. Uncontrolled condition. The main purpose of this test is to verify the burning characteristics of the material in real life application. In real life fire incident, the conditions are random, uncontrolled and the source of fire can even be anything. Therefore, it is important to analyze the differences between the material's performance in uncontrolled and standard controlled flammability tests. Generally, similar procedure from the standard control test was applied to this uncontrolled test except for the conditions and source of fire. This test was conducted in open environment with random conditions in which the relative humidity, air temperature and wind speed were subjected to the environmental condition during the experimental work. Apart from that, the source of fire used in this test was the Piezo-igniter cassette gas and a handmade fume cupboard was used as the test chamber during the flammability testing. The test chamber is shown Figure 2.

2.3. Smoke Density
The purpose of the smoke density test is to observe the visibility level through smoke generation during combustion, especially in confined spaces. The specimens were tested within a 300 x 300 x 790 mm test chamber [25]. The test chamber was instrumented with a light source, a photoelectric cell and also a meter to horizontally measure the light absorption across the 300 mm light beam path that will reflect on the amount of smoke produced [25]. This is shown in Figure 3. If the material dripped excessively, a second burner was introduced into the test chamber. The second burner was positioned in the test chamber such that its flame was directed towards the center of the steel collection tray.

2.4. Limiting oxygen index

The purpose of limiting oxygen index (LOI) test is to measure the minimum concentration of oxygen, in which is expressed in volume percentage that is required in order to support combustion process or flame propagation [26]. The schematic representation of the LOI test is shown in Figure 4. The specimen was placed in the holder at the center of the base of the test column. The flow valves were adjusted to obtain the desired initial oxygen concentration and the total flow rate. The oxygen and nitrogen flowed into the dispersion chamber and through the glass bead bed, hence the gases were mixed and dispersed evenly over the cross section of the test column [26]. The specimen was ignited such that the entire top tip of the specimen was burning like a candle. A gas flame at the end of a tube with small orifice was used to ignite the specimen. The samples were tested with a small pilot flame to find the minimum oxygen concentration required to just sustain the combustion of the sample. The result is expressed as shown in the Equation 1 [26]:

$$n = \frac{100 \times O_2}{O_2 + N_2}\% \qquad (1)$$

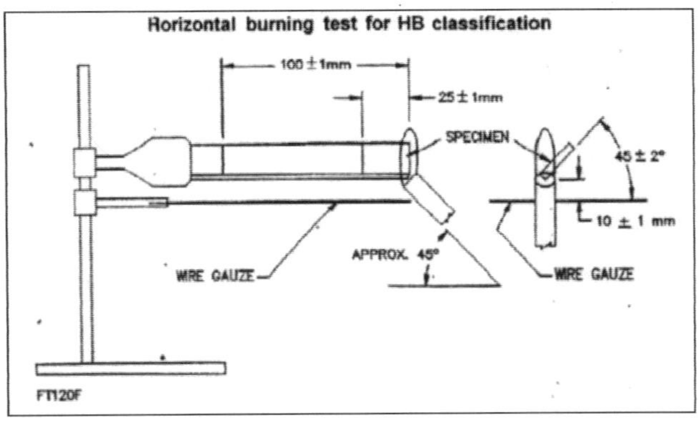

Figure 1. UL94 horizontal burning test experimental set-up [24]

Figure 2. Handmade fume chamber

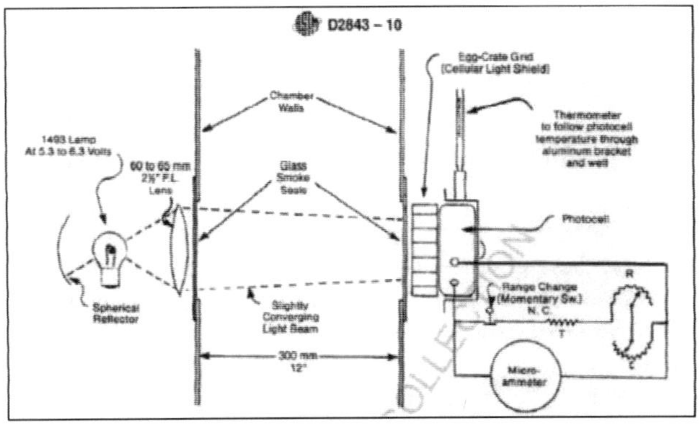

Figure 3. Layout for smoke density test chamber photometer [25]

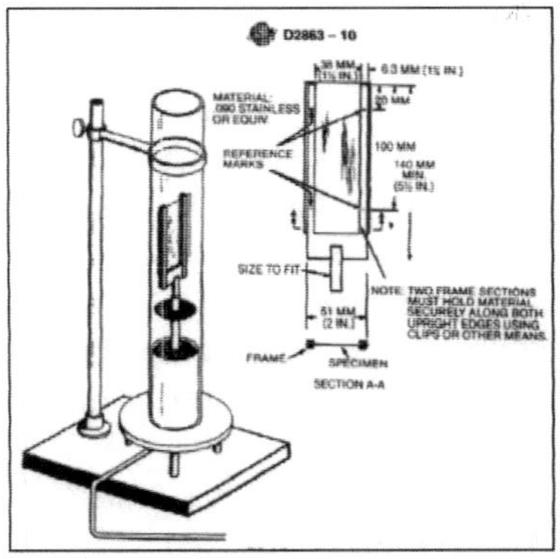

Figure 4. Schematic representation of a limiting oxygen index (LOI) test setup [26]

IOP Conf. Series: Materials Science and Engineering **152** (2016) 012068 doi:10.1088/1757-899X/152/1/012068

3. Result and discussion

3.1. UL94 horizontal burning

From Figure 5 and Figure 6, it can be concluded that burning rate was decreasing with increasing kenaf fibre loading. This can be seen through samples ABS/KF 90/10 and ABS/KF 50/50. This result was due to the rapid formation of char layer by kenaf fibre that prevented heat and volatiles from penetrating the flammable zones. Pure ABS and ABS/KF 50/50 with 3% obtained the value of 19.67 and 19, respectively, for controlled condition and the value of 25.28 and 20.33 for uncontrolled condition. Overall, all samples could be classified as HB since all test specimens did not have burning rate exceeding 40 mm per minute.

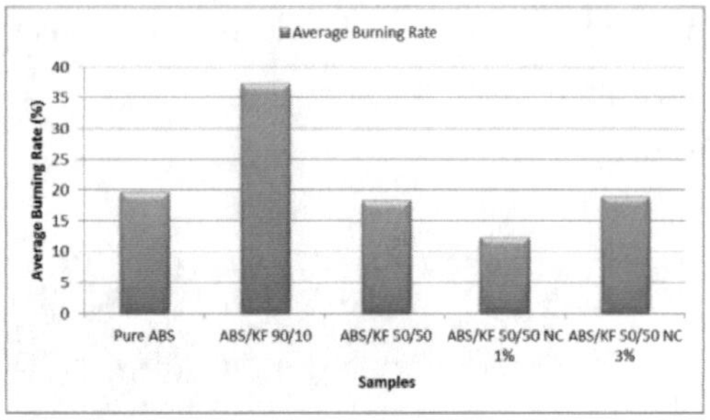

Figure 5. Graph of UL 94 horizontal burning rates (controlled condition)

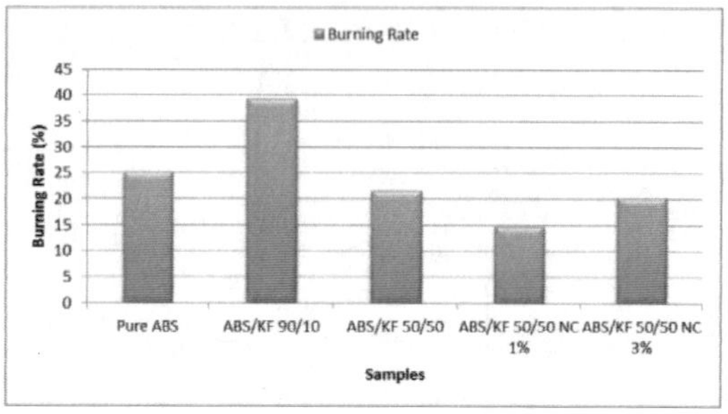

Figure 6. Graph of UL 94 horizontal burning rates (uncontrolled condition)

From the analysis of graphs above, relatively large percentage difference can be seen between ABS/KF 90/10 and ABS/KF 50/50 1%. This is due to the presence of nanoclays in ABS/KF 50/50. Nanoclays play a role as flame retardant in which it changes the decomposition path of ABS matrix. It promotes reduction in viscosity or diluting the flame volatile when exposed to flame. This process increases the formation of a protective char barrier towards the surface of the material. This barrier is created through a roundabout path of migration of nanoclays. It will resist the oxygen diffusion or volatile flames from source of fire from entering into the material. The data suggests that the mechanism of nanoclays as flame retardant is exhibited in two different ways, which are through heat release reduction and also prevention of volatile flame dripping. Besides that, both graphs highlight that the burning rate for sample ABS/KF 90/10 showed the highest burning rate and this proves that addition of kenaf fibre, which serves as source of fuel, will increase the flammability properties of the material. Kenaf fibre is highly flammable because it is made up of cellulose, hemicellulose and lignin that affect the decomposition of the material constituent at different temperatures. Furthermore, it is hydrophilic that results in a poor compatibility with hydrophobic polymer matrices and has a poor fire resistance [3].

As stated previously, UL 94 horizontal burning test were conducted in two different conditions, which were controlled and uncontrolled conditions. The aim of this approach was to observe the difference in the flammability properties of the materials that were subjected to controlled environment and uncontrolled environment. Here, the uncontrolled conditions represented the real fire scenario. Based on the percentage differences that are tabulated in Table 4, it shows that the burning rates for the uncontrolled conditions are slightly higher than the controlled condition. This might be due to possible effects from the environmental conditions that may promote increased burning rate. However, the trend of the data for both controlled and uncontrolled conditions is similar. The propagation of flame spread was captured during the flammability testing and the visual observation is explained in Table 5.

Table 4. Percentage difference between controlled and uncontrolled conditions

Samples	UL 94 Horizontal Burning Rates (Controlled Condition) %	UL 94 Horizontal Burning Rates (Uncontrolled Condition) %	Percentage Difference %
Pure ABS	19.67	25.28	28.52
ABS/KF 90/10	37.33	39.47	5.73
ABS/KF 50/50	18.33	21.71	18.44
ABS/KF 50/50 1%	12.33	14.85	20.44
ABS/KF 50/50 3%	19.00	20.33	7.00

Table 5. Physical observation on the flame propagation

Pure ABS

This sample ignited at a fast rate and a rapid spread of fire could be seen through the captured images. During the burning process, dripping and melting of the ABS polymer could be observed at a shorter time interval. It was noted that the dripping mass of volatile flame burned vigorously and produced black soothe.

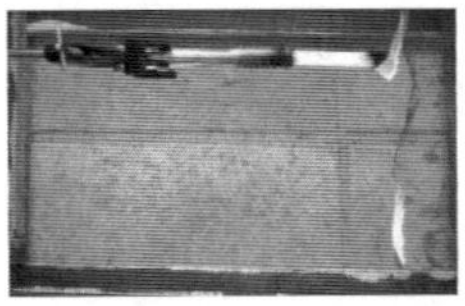

ABS/Kenaf 90/10

This sample ignited at a faster rate than the pure ABS and a rapid spread of fire could be seen through the captured images. During the burning process, dripping and melting of the ABS/KF could be observed at shorter time interval. The dripping mass of volatile flame burned more vigorously than the pure ABS until the material's end. Smoke production was also higher during the burning and there was less black soothe production.

ABS/Kenaf 50/50

This sample ignited at a slow rate but a rapid spread of fire could be seen through the captured images. During the burning process, dripping and melting of the ABS/KF could be observed but at a delayed time interval. It was noted that the dripping mass of volatile flame extinguished once it dropped and a little ash could be seen.

Table 5. Physical observation on the flame propagation (continued)

ABS/Kenaf/Nanoclay 50/50 (1%)
This sample ignited at the slowest rate with the slowest spread of fire, which could be seen from the captured images. During the burning process, at the initial stage, the material did not drip and melt. Instead it turned out as ash before it fell down. This mechanism of the ABS/KF could be observed in the left image. Less amount of smoke production could be seen and no black soothe was produced.

ABS/Kenaf/Nanoclay 50/50 (3%)
This sample ignited at a slower rate with a slower spread of fire, which could be seen through the captured images. During the burning, from the initial stage until the end, the material did not drip and melt. Instead, it turned out as ash before it finally fell down. This mechanism of the ABS/KF could be observed in both images below. Less amount of smoke production could be seen and no black soothe was produced.
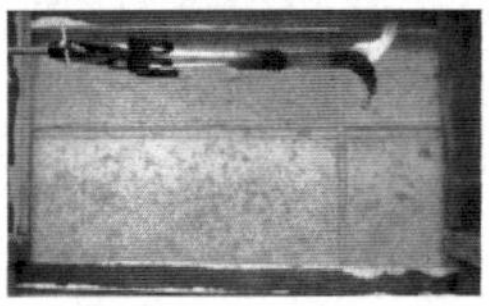 

3.2. Smoke density

Figure 7 shows that the presence of kenaf fibre increases the smoke density rating. Pure ABS recorded a rating of 82. On the other hand, ABS/KF 90/10 recorded the highest rating of 96, which is a 17.07% rise in value compared to the pure ABS. This is due to the characteristics of kenaf fibre that promote smoke production through char formation when exposed to flame. However, looking into the trend in Figure 7 for samples 3, 4 and 5, it shows that the smoke density rating for sample 3 slightly decreased with increasing kenaf fibre loading. The rating value recorded by ABS/KF 50/50 was 93. Apparently, with the presence of nanoclays, ABS/KF 50/50 1% and 3% recorded the same rating value of 91, which was slightly lower than the optimized composition ABS/KF 50/50 (or a 2.15% value drop to be exact). Hence the addition of a small amount of nanoclays resulted in decrement of smoke density rating.

The rate of production of flammable volatiles mass and also the degree of conversion of the volatiles to smoke for ABS/KF 90/10 were higher. Therefore, it had a higher density of smoke. On the other hand, for the pure ABS, the mass of flammable volatiles released during combustion and the degree of smoke generation were lower. Thus the density of smoke of pure ABS was the lowest. Typically, thermoplastics such as ABS have a low smoke emission when exposed to fire as they yield more char and black soothe formation. ABS/KF 50/50 1% and 3% recorded an optimum smoke density along with the optimized composition ABS/KF 50/50. This is because higher fiber contents and the presence of nanoclays in the material reduce the smoke generation. This result suggests that the presence of organic material in the material composition promotes smoke generation.

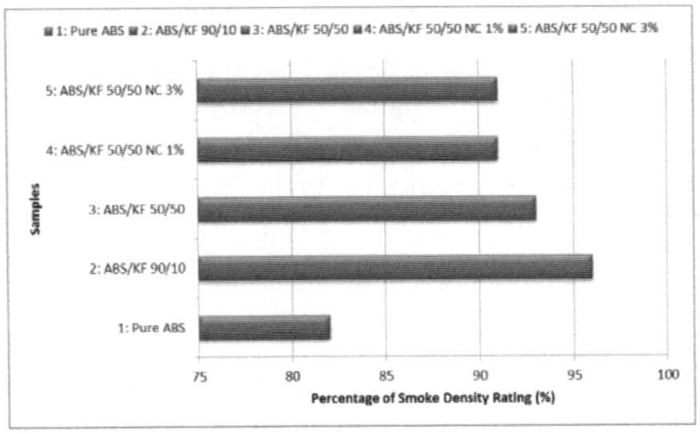

Figure 7. Graph of smoke density

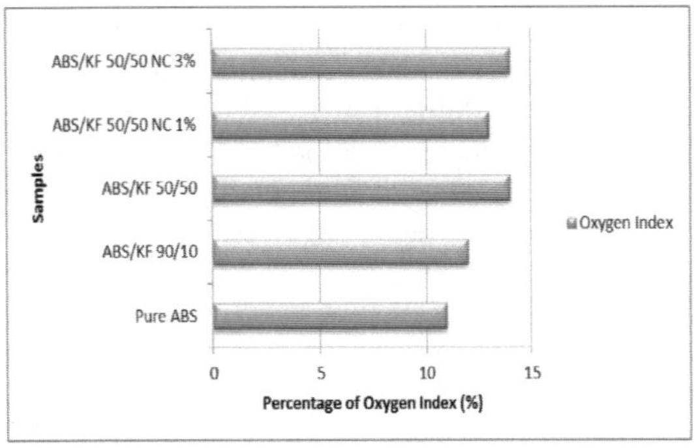

Figure 8. Graph of limiting oxygen index

3.3. Limiting oxygen index

Figure 8 shows that the volume of oxygen is increased with increasing kenaf fibre loading. The highest LOI value was obtained by ABS/KF 50/50, which was 14. Meanwhile, the lowest value was obtained by pure ABS, which was 11. LOI value obtained by ABS/KF 90/10 was 12. Apparently, with the presence of nanoclays, ABS/KF 50/50 1% obtained LOI value of 13 with 7.1% drop in value compared to optimized composition ABS/KF 50/50. However, for ABS/KF 50/50 3%, there was no change in volume. It could be seen that LOI value for ABS/KF 50/50 3% was higher compared to ABS/KF 50/50 1%. This might due to the melting and dripping of material during the test that caused the specimen to extinguish, thus mislead the LOI values [3]. Overall, an addition of small amount of nanoclays resulted in decreased LOI value.

Looking through the standard requirement for LOI, materials with LOI value less than 21 percent are defined as combustible material. In contrast, materials that have LOI value of more than 21 percent are defined as self-extinguishing. This is because, based on standard air composition, the oxygen level in air is 21 percent and thus materials with more than this value are not able to support the combustion at ambient temperature without an external source of fire. However, the flammability data obtained in this experiment does not correlate with the 'real fire scenario' [27]. This is because the flammability measurement was obtained at room temperature. LOI values decrease with increasing temperature, says under intense fire condition, in which materials with high LOI values may burn without extinguishing itself. Besides, if the materials self-extinguish through melting or dripping, this can mislead the decision for LOI values [2-4]. Therefore, this method cannot be considered as s fast rule in determining self-extinguishing property.

By making a comparison for the result obtained from the experimental test and standard evaluation for LOI, it can be concluded that all five samples are classified as highly combustible. However, this can be justified through the limitation applied to LOI for nanocomposites. The presence of nanoclays in ABS/KF 50/50 decreases the LOI values due to two factors related to the burning mechanism. The first factor is the increased flame propagation rate over the surface of the specimen when it is exposed to flame. Moreover, the second factor is the barrier formation by the nanoclays from the inside of the material that suppresses the flame spread.

4. Conclusion

Flammability properties of kenaf fibre reinforced ABS with nanoclays are evaluated in this study. It can be concluded that the presence of nanoclays in kenaf fibre reinforced ABS helps to enhance the burning characteristics of the material when exposed to source of fire. Flammability studies are carried out through three tests: UL 94 horizontal burning, smoke density and limiting oxygen index. The results suggest that this material possesses low burning rate, slow flame propagation, low amount of smoke generation and requires less oxygen concentration to support the burning process (subject to limitation). Hence, an ideal formulation of ABS/KF 50/50 with 1% nanoclays is applicable for industrial usage, especially in confined spaces in aircraft as it can protect the underlying structure from risk of fire due to its capability as self-extinguishing material. For the better understanding of the burning mechanism, further investigation is required to develop correlation between these three tests in order to reach an optimized conclusion. Future work will be emphasized on developing a systematic approach for predicting fire behavior in real fire scenario by varying parameters such as relative humidity, temperature and wind speed.

Acknowledgment
The authors like to acknowledge the research funding from the Ministry of Education Malaysia under the Fundamental Research Grant Scheme (FRGS) – 02-02-13-1382FR. The authors gratefully acknowledge all staff in Laboratory of Biocomposite Technology, Institute of Tropical Forestry and Forest Products (INTROP), Universiti Putra Malaysia, Malaysian Nuclear Agency, SIRIM Berhad and also Faculty of Engineering, Universiti Putra Malaysia for all technical guidance, financial and facilities provided.

References
[1] Greene E 1999 *Marine Composites* (Annapolis: Eric Greene Associates, Inc.)
[2] Lyon R N 2008 *Flammability of Polymer Composites* (Springfield: FAA)
[3] Mfiso E M, Maya M J, Jacobs V and Luyt A S 2014 *Carbohydrate Polymers* **111** 149-82
[4] Mahadev B, Alagirusamy R and Das A 2015 *Fibers and Polymers* **16** 705-17
[5] Edwards T 2008 *Wealth Creation* **36** 24-8
[6] Horner A 2000 *Aircraft Materials Fire Test Handbook* (Washington: FAA)
[7] Quintiere J G, Walters R N and Crowley S 2007 *Flammability Properties of Aircraft Carbon- Fiber Structural Composite* (Springfield: FAA)
[8] Boeing 2005 *Statistical Summary of Commercial Jet Airplane Accidents: Worldwide Operations 1959–2004*
[9] Flight Safety Foundation *Aviation Safety Network*
[10] FAA1993 *Advisory Circular* US Department of Transportation.
[11] Vialas N, Amould L, Parlevliet P P, De Vries H P J and Vigoureus A 2009 *Requirement for Biomaterials and Biodegradable Materials TA11-01-01*
[12] Chai M W, Bickerton S, Bhattacharyya D and Das R 2012 *Composite Part B: Engineering* **43** 2867-74
[13] Lee C H, Salit M S and Hassan M R 2014 *Advances in Materials Science and Engineering* **2014**
[14] Begum K and Islam M A 2013 *Research Journal of Engineering Sciences* **2** 46-53
[15] Salit M S, Mustapha F, Majid D L, Leman Z, Ariff A H M, Ariffin M K A, Zuhri M Y M, Ishak M R and Sahari J 2011 *Key Engineering Materials* **471-472** 563-567
[16] Saba N, Md Tahir P and Jawaid M 2014 *Polymers* **6** 2247-73
[17] Yen C S 2006 *Mechanical, Chemical and Flammability* Universiti Teknologi Malaysia
[18] Bonati A, Merusi F, Bochicchio G, Tessadri B, Polacco G, Filippi S and Giuliani F 2013 *Construction and Building Materials* **47** 990-1000.
[19] Chigwada G, Panchatapa J, Jiang D and Wilkie C A 2005 *Polymer Degradation and Stability* **89** 85–100
[20] Faruk O, Bledzki A K, Fink H and Sain M 2012 *Progress in Polymers Science* **37** 1552-96
[21] Cuppoletti J 2011 *Nanocomposites and Polymers with Analytical Methods* (Croatia: InTech)
[22] Gacitua W, Ballerini A and Zhang J 2005 *Maderas. Ciencia y tecnología* **7** 159-78
[23] Lufti M T M, Majid D L, Faizal A R M and Norkhairunnisa M 2015 *Advanced Materials Research* **1119** 263-7

IOP Conf. Series: Materials Science and Engineering **152** (2016) 012068 doi:10.1088/1757-899X/152/1/012068

[24] Underwriters Laboratories Inc. 1996 *Test for Flammability of Plastic Material for Parts in Devices and Appliances*
[25] Intertek Testing Services Inc. 2007 *Standard Test Method for Density of Smoke from the Burning or Decomposition of Plastics*
[26] Department of Defense *Standard Test Method for Measuring the Minimum Oxygen Concentration to Support Candle like Burning of Plastics*
[27] Carty P and White S 1994 *Polymer* **35** 5595-96

IOP Conf. Series: Materials Science and Engineering **152** (2016) 012069 doi:10.1088/1757-899X/152/1/012069

A generalized self-consistent method for estimating effective shear properties of unidirectional composites comprising cylindrical orthotropic constituents

M Minhat*, N Z Mahmud Zuhudi, M H Shamsudin, M D Isa and M K Harun

Universiti Kuala Lumpur - Malaysian Institute of Aviation Technology, Malaysia

*mulia@unikl.edu.my

Abstract. A micromechanics model based on generalized self-consistent method is proposed to estimate shear elastic properties of unidirectional fibre composites comprising cylindrical orthotropic layers. The features of the generalized self-consistent method are briefly presented and the homogenization schemes in determining the effective shear moduli of transversely isotropic medium are demonstrated. Numerical examples on the prediction of axial and transverse shear moduli of polymer composite reinforced with nanostructure hybrid fibres are illustrated. The close agreement between the prediction results and the results obtained from the available experimental data and finite element study validates the solutions produced by the proposed model.

1. Introduction

Recent advances in nano-material sciences have led to the development of different types of advanced multifunctional composite materials, which are designed for simultaneous enhancement in various structural functions including stiffness, strength, vibrational damping, fracture, electrical and thermal conductivities [1, 2]. One of the examples of such advanced composites currently developed for such intended multi-functionality purposes is nanostructures hybrid fibres reinforced polymeric material. Basically, these fibres have radially aligned nanostructures that directly grown on the fibre surfaces. In most cases, the nanostructures are in form of nanowires and carbon nanotubes (CNTs). The presence of nanowires coated fibre have been shown to improve shear, damping and electrical characteristics of the polymeric composites when compared to the conventional fibre composite materials [3, 4, 5]. In case of carbon nanotubes, the fibre system is also known as 'fuzzy' fibre and similar enhancements including thermal and electrical behaviours were observed when such fibre system is embedded into polymeric materials [6-8,27]. These materials, from the practical perspective, could be very beneficial for the next generation multi-functional advanced fibre composite materials.

It is interesting to note that due to the presence of radially aligned nanostructures surrounding the main fibre, a fictitious reinforced interphase layer comprising the nanostructures and matrix material is formed. This layer is responsible for the enhancement in the shear characteristics of fibre composite material, which are well known to exhibit the poor performance for the conventional fibre reinforced composites. Therefore, it is of interest to estimate the effective shear properties of such propitious advanced composite materials.

To model the effective shear moduli of the unidirectional nanostructures hybrid fibre composite material, three distinct constituents are considered, which include base carbon fibre, polymer matrix material and reinforced interphase layer. As such, an analysis on a multiphase composite is necessary. From the material point of view, each constituent has different characteristic. Typically, the polymer matrix is modelled as an isotropic material and carbon fibre as a transversely isotropic material with axis of symmetry parallel to the principle axis of fibre. Interestingly, the reinforced interphase layer can be treated as nano-composite medium and it assumes the cylindrical orthotropic property [9, 10] considering that they are radially grown on the surface of fibre. From general modelling perspective, all constituents in such fibre composite system can be generalized to have orthotropic property where isotropic, as well as transversely isotropic property, is considered as a special case of the orthotropic property having one or more plane of material symmetry.

In the research and engineering practices, estimating the effective elastic properties of multiphase composites with known properties of the constituents is essential for initial and associated analysis. In general, such prediction requires micromechanics modelling and in the first approximation analysis, the most popular classical micromechanics models used include composite cylindrical assemblage (CCA) method [11], self-consistent method [12], generalized self-consistent (GSC) method or three-phase model [13], differential scheme [14] and also Mori-Tanaka's method [15]. The GSC method has been critically assessed and validated based on the most elusive property of transverse shear modulus, showing that estimations were most in agreement to numerical simulations and experimental data up to the typical high volume fraction of inclusion [13, 16, 17, 18, 19]. In fact, many modern analytical or semi-analytical micromechanics approaches used the results of GSC method for base comparison in justifying the accuracies of their proposed models [18, 19]. It should be noted that majority of these classical models were originally developed to homogenize the effective elastic properties of a two-phase composite having constituent to possess either isotropic or transversely isotropic property with plane of symmetry perpendicular to the principle axis of fibre. In later period, some of these models were upgraded for the multiphase composite applications, i.e. to study the effect of interphase layer, coating layer and voids [20, 21, 22].

Therefore, in this paper, the GSC method will be applied to predict the axial and transverse shear moduli of multiphase unidirectional composite system comprising cylindrical orthotropic constituents. Some mathematical preliminaries and the features of GSC method are presented. For illustrations, few numerical examples on nanowire and CNTs hybrid fibre composites are studied here and for further validation, some results are compared with the existing experimental findings and numerical studies.

2. The formulation of generalized self-consistent method

In this section, the mathematical preliminaries related to linear classical elasticity and homogenization techniques are briefly reviewed. The features and governing equations related to the GSC method are presented.

2.1. Mathematical preliminaries

Constitutive equation of Hooke's law for orthotropic linear elastic material having nine independent constants written compactly in cylindrical coordinate system can be written as Equation 1.

$$\begin{pmatrix} \sigma_{rr}^{(i)} \\ \sigma_{\theta\theta}^{(i)} \\ \sigma_{zz}^{(i)} \\ \sigma_{\theta z}^{(i)} \\ \sigma_{rz}^{(i)} \\ \sigma_{r\theta}^{(i)} \end{pmatrix} = \begin{bmatrix} C_{rr}^{(i)} & C_{r\theta}^{(i)} & C_{rz}^{(i)} & 0 & 0 & 0 \\ C_{r\theta}^{(i)} & C_{\theta\theta}^{(i)} & C_{\theta z}^{(i)} & 0 & 0 & 0 \\ C_{rz}^{(i)} & C_{\theta z}^{(i)} & C_{zz}^{(i)} & 0 & 0 & 0 \\ 0 & 0 & 0 & G_{\theta z}^{(i)} & 0 & 0 \\ 0 & 0 & 0 & 0 & G_{rz}^{(i)} & 0 \\ 0 & 0 & 0 & 0 & 0 & G_{r\theta}^{(i)} \end{bmatrix} \begin{pmatrix} \varepsilon_{rr}^{(i)} \\ \varepsilon_{\theta\theta}^{(i)} \\ \varepsilon_{zz}^{(i)} \\ 2\varepsilon_{\theta z}^{(i)} \\ 2\varepsilon_{rz}^{(i)} \\ 2\varepsilon_{r\theta}^{(i)} \end{pmatrix} \quad (1)$$

where σ_{pq} and ε_{pq} are the second order linear stress and strain tensors, respectively, and C_{pq} is the fourth-order stiffness tensors written in compact form. The superscript index is given as $i = 1, 2, ..., N$ and N is the total number of constituents in a composite. For transversely isotropic material with its axis of symmetry parallel to $z-$ axis, the stress-strain equation assumes the form as in Equation 2.

$$\begin{pmatrix} \sigma_{rr}^{(i)} \\ \sigma_{\theta\theta}^{(i)} \\ \sigma_{zz}^{(i)} \\ \sigma_{\theta z}^{(i)} \\ \sigma_{rz}^{(i)} \\ \sigma_{r\theta}^{(i)} \end{pmatrix} = \begin{bmatrix} C_{rr}^{(i)} & C_{r\theta}^{(i)} & C_{rz}^{(i)} & 0 & 0 & 0 \\ C_{r\theta}^{(i)} & C_{rr}^{(i)} & C_{rz}^{(i)} & 0 & 0 & 0 \\ C_{rz}^{(i)} & C_{rz}^{(i)} & C_{zz}^{(i)} & 0 & 0 & 0 \\ 0 & 0 & 0 & G_{rz}^{(i)} & 0 & 0 \\ 0 & 0 & 0 & 0 & G_{rz}^{(i)} & 0 \\ 0 & 0 & 0 & 0 & 0 & \frac{C_{rr}^{(i)}-C_{r\theta}^{(i)}}{2} \end{bmatrix} \begin{pmatrix} \varepsilon_{rr}^{(i)} \\ \varepsilon_{\theta\theta}^{(i)} \\ \varepsilon_{zz}^{(i)} \\ 2\varepsilon_{\theta z}^{(i)} \\ 2\varepsilon_{rz}^{(i)} \\ 2\varepsilon_{r\theta}^{(i)} \end{pmatrix} \tag{2}$$

where the stiffness tensor consists of five independent elastic constants only. As will be seen later, it is useful to define stiffness tensor for transversely isotropic medium with its axis of symmetry parallel to $r-$ axis, which can be written as Equation 3.

$$\begin{pmatrix} \sigma_{rr}^{(i)} \\ \sigma_{\theta\theta}^{(i)} \\ \sigma_{zz}^{(i)} \\ \sigma_{\theta z}^{(i)} \\ \sigma_{rz}^{(i)} \\ \sigma_{r\theta}^{(i)} \end{pmatrix} = \begin{bmatrix} C_{rr}^{(i)} & C_{r\theta}^{(i)} & C_{r\theta}^{(i)} & 0 & 0 & 0 \\ C_{r\theta}^{(i)} & C_{\theta\theta}^{(i)} & C_{\theta z}^{(i)} & 0 & 0 & 0 \\ C_{r\theta}^{(i)} & C_{\theta z}^{(i)} & C_{\theta\theta}^{(i)} & 0 & 0 & 0 \\ 0 & 0 & 0 & \frac{C_{\theta\theta}^{(i)}-C_{\theta z}^{(i)}}{2} & 0 & 0 \\ 0 & 0 & 0 & 0 & G_{r\theta}^{(i)} & 0 \\ 0 & 0 & 0 & 0 & 0 & G_{r\theta}^{(i)} \end{bmatrix} \begin{pmatrix} \varepsilon_{rr}^{(i)} \\ \varepsilon_{\theta\theta}^{(i)} \\ \varepsilon_{zz}^{(i)} \\ 2\varepsilon_{\theta z}^{(i)} \\ 2\varepsilon_{rz}^{(i)} \\ 2\varepsilon_{r\theta}^{(i)} \end{pmatrix} \tag{3}$$

Finally, the stress-strain relationship for isotropic material is described by Equation 4.

$$\begin{pmatrix} \sigma_{rr}^{(i)} \\ \sigma_{\theta\theta}^{(i)} \\ \sigma_{zz}^{(i)} \\ \sigma_{\theta z}^{(i)} \\ \sigma_{rz}^{(i)} \\ \sigma_{r\theta}^{(i)} \end{pmatrix} = \begin{bmatrix} C_{rr}^{(i)} & C_{r\theta}^{(i)} & C_{r\theta}^{(i)} & 0 & 0 & 0 \\ C_{r\theta}^{(i)} & C_{rr}^{(i)} & C_{r\theta}^{(i)} & 0 & 0 & 0 \\ C_{r\theta}^{(i)} & C_{r\theta}^{(i)} & C_{rr}^{(i)} & 0 & 0 & 0 \\ 0 & 0 & 0 & \frac{C_{rr}^{(i)}-C_{r\theta}^{(i)}}{2} & 0 & 0 \\ 0 & 0 & 0 & 0 & \frac{C_{rr}^{(i)}-C_{r\theta}^{(i)}}{2} & 0 \\ 0 & 0 & 0 & 0 & 0 & \frac{C_{rr}^{(i)}-C_{r\theta}^{(i)}}{2} \end{bmatrix} \begin{pmatrix} \varepsilon_{rr}^{(i)} \\ \varepsilon_{\theta\theta}^{(i)} \\ \varepsilon_{zz}^{(i)} \\ 2\varepsilon_{\theta z}^{(i)} \\ 2\varepsilon_{rz}^{(i)} \\ 2\varepsilon_{r\theta}^{(i)} \end{pmatrix} \tag{4}$$

where only two independent elastic constants are required to fully describe such material behaviour. The Cauchy small strain-displacement written in the cylindrical coordinate system has the relations as follow.

$$\begin{gathered} \varepsilon_{rr} = \frac{\partial u_r}{\partial r}, \quad \varepsilon_{\theta\theta} = \frac{1}{r}\left(\frac{\partial u_\theta}{\partial \theta} + u_r\right), \quad \varepsilon_{zz} = \frac{\partial u_z}{\partial z}, \\ \varepsilon_{rz} = \frac{1}{2}\left(\frac{\partial u_z}{\partial r} + \frac{\partial u_\theta}{\partial z}\right), \quad \varepsilon_{\theta z} = \frac{1}{2}\left(\frac{\partial u_\theta}{\partial z} + \frac{1}{r}\frac{\partial u_z}{\partial \theta}\right), \quad \varepsilon_{r\theta} = \frac{1}{2}\left(\frac{1}{r}\frac{\partial u_r}{\partial \theta} + \frac{\partial u_\theta}{\partial r} - \frac{u_\theta}{r}\right) \end{gathered} \tag{5}$$

where u_p is displacements in respective coordinate system. Corresponding equations of equilibrium in cylindrical coordinate system are given as follows:

$$\begin{aligned}
&\frac{\partial \sigma_{rr}}{\partial r}+\frac{1}{r}\frac{\partial \sigma_{r\theta}}{\partial \theta}+\frac{\sigma_{rr}-\sigma_{\theta\theta}}{r}+\frac{\partial \sigma_{rz}}{\partial z}=0,\\
&\frac{\partial \sigma_{r\theta}}{\partial r}+\frac{1}{r}\frac{\partial \sigma_{\theta\theta}}{\partial \theta}+\frac{2\sigma_{r\theta}}{r}+\frac{\partial \sigma_{\theta z}}{\partial z}=0,\\
&\frac{\partial \sigma_{rz}}{\partial r}+\frac{1}{r}\frac{\partial \sigma_{\theta z}}{\partial \theta}+\frac{\sigma_{rz}}{r}+\frac{\partial \sigma_{zz}}{\partial z}=0.
\end{aligned} \tag{6}$$

Most micromechanics methods employ energy principle in homogenizing the heterogeneous properties of composite materials. Thus, it is necessary to describe the relationship between energy, stress and strain in linear elastic material, which can be represented as in Equation 7.

$$W(\varepsilon_{pq})=\tfrac{1}{2}\sigma_{pq}\varepsilon_{pq}=\tfrac{1}{2}C_{pqrs}\varepsilon_{rs}\varepsilon_{pq} \tag{7}$$

where W designates the strain energy density in a material.

The concept of representative volume element (RVE) is commonly used in many micromechanics models. RVE can be thought as the smallest possible unit volume element over which homogenization takes place and the average properties are obtained to represent the effective properties of a whole material. It should be noted that RVE must be carefully chosen in such a way that its size is notably large enough to account for sufficient number of various nanostructures but extremely small compared to the size of a whole composite material so that accurate estimation of the effective properties can be achieved. Furthermore, such RVE size requirement is also required for the effective properties to be independent of the surface values of displacement or traction vector as long as these values remain macroscopically homogeneous [12]. In addition, any quantity measured inside the RVE is considered as an average value and this value is normally represented by $\langle\bullet\rangle=\frac{1}{V}\int_V(\bullet)dV$ where V is the volume of RVE. For example, the respective average strain energy density, strain and stress in RVE can be written as:

$$\langle W\rangle=\tfrac{1}{2V}\int_V\sigma_{pq}\varepsilon_{pq}\,dV,\qquad \langle\varepsilon_{pq}\rangle=\tfrac{1}{V}\int_V\varepsilon_{pq}\,dV,\qquad \langle\sigma_{pq}\rangle=\tfrac{1}{V}\int_V\sigma_{pq}\,dV. \tag{8}$$

Furthermore, with the presence of ideal contact surface between the constituents and macroscopically uniform boundary conditions, the respective average strain and stress in the RVE is identical to the respective homogenous strain and stress applied on its boundary, for example:

$$\langle\varepsilon_{pq}\rangle=\varepsilon^0_{pq},\qquad \langle\sigma_{pq}\rangle=\sigma^0_{pq} \tag{9}$$

where ε^0_{pq} and σ^0_{pq} are the homogenous strain and stress applied on the boundary surface of RVE respectively, which are also equal to the macroscopically homogenous boundary conditions imposed on the macroscopic composite material. Equalities shown in Equation 9 are known as the average strain and average stress theorems respectively.

Finally, in averaging the final properties of unidirectional fibre composite material, the fibres are typically assumed as infinitely long cylinders with circular cross-section and they are also randomly distributed within the composite ply. In this approach, the principle axis of fibre is oriented to z-axis as such transversely isotropic composite with its plane of the symmetry perpendicular to z-axis is obtained. In this case, the behaviour of such composite system can be defined by five effective engineering moduli, which include plane strain bulk modulus κ^{eff}_{12}, axial Young's modulus E^{eff}_{33}, major Poisson's ratio ν_{31}^{eff}, axial shear modulus μ_{23}^{eff} and transverse shear modulus μ_{12}^{eff}. Nevertheless, in this paper, only the shear properties are of concern.

2.2. The generalized self-consistent method

The GSC method employs RVE based on the physical model adopted from the composite cylindrical assemblage (CCA) method [11] where it contains single inclusion of fibre only embedded in a matrix system in the form of concentric cylinders. However, RVE in the GSC formulation treats fibre, matrix and, in this case, reinforced interphase layer as inclusions. These inclusions in concentric cylindrical assemblage form is surrounded by an equivalent homogenized medium having unknown properties that we wish to determine. Such RVE can be considered as a multi-layered or $N-$ layered cylindrical orthotropic unidirectional composite system (see Figure 1a). In this study, $N=3$ where the fibre is designated by $i=1$, $i=2$ for reinforced interphase layer and for matrix, $i=3$. Naturally, the fibre volume fraction can be defined as the ratio of circumferential areas of the fibre over matrix, i.e. $V_f = r_1^2 / r_N^2$. As for the equivalent homogenized medium, it is designated by $N+1$ and has an infinite outer radius where homogenous deformation is imposed.

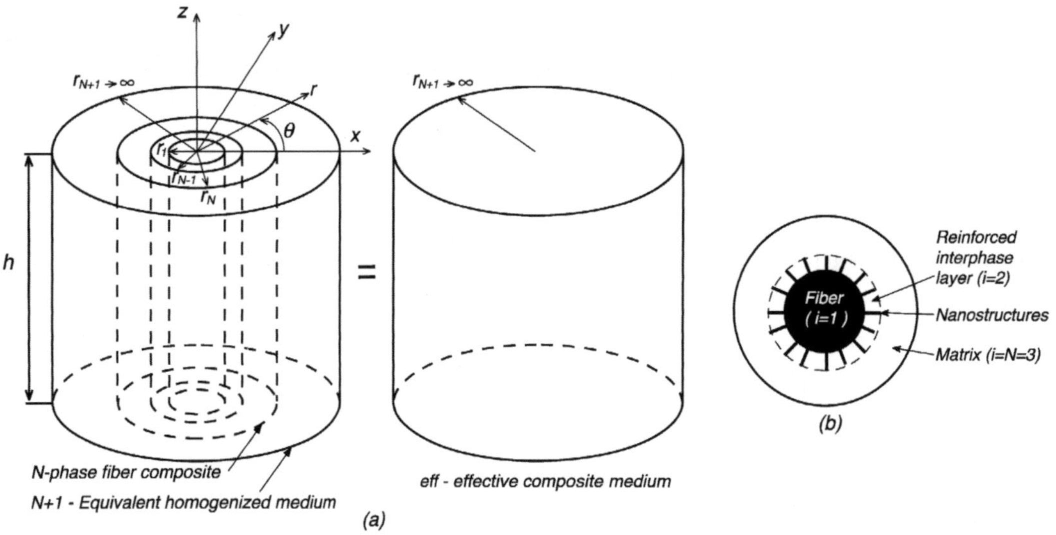

Figure 1. Schematic RVEs of (a) cylindrical composite assemblage in the GSC method and (b) nanostructures hybrid fibre composite

The GSC method originally proposed by Christensen and Lo [13] was developed on the basis of energy principle. It is important for us to understand first that, when the RVE containing the inclusions is homogenized, the homogenized RVE can be represented as an effective composite medium, which is designated as $eff-$medium and this medium also has the same unknown effective properties of the equivalent homogenized medium, i.e. $\mu_{23}^{N+1} = \mu_{23}^{eff}$ and $\mu_{12}^{N+1} = \mu_{12}^{eff}$. The equivalent RVE for effective composite medium is as shown in Figure 1a. Based on the energy principle, it is proposed that the difference in average strain energy densities of RVE containing inclusions and RVE of an effective composite medium equals

$$\langle W^{RVE} \rangle = \langle W^{eff} \rangle + \langle W^{INT} \rangle = \langle W^{eff} \rangle + \tfrac{1}{2V} \int_S \left(\sigma_{ij}^{N+1} u_j^{eff} - \sigma_{ij}^{eff} u_j^{N+1} \right) dS \tag{10}$$

where $\langle W^{RVE} \rangle$ is the average strain energy density of a medium or RVE with inclusions, $\langle W^{eff} \rangle$ is the average strain energy density of a medium without the inclusion or RVE of an effective composite medium, $\langle W^{INT} \rangle$ is the average surface interaction strain energy density at contact surface S between the Nth – layer and $(N+1)^{th}$ – layer, σ_{pq}^{N+1}, u_q^{N+1} are the components of stress tensor and displacement vector on the contact surface of $(N+1)^{th}$ – medium respectively, σ_{pq}^{eff}, u_q^{eff} assume the respective stress tensor and displacement vector on the contact surface of an effective composite medium. Christensen and Lo [13] further postulated that the average strain energy densities of RVE containing the inclusions and RVE of an effective composite medium are equal (see Equation 11).

$$\langle W^{RVE} \rangle = \langle W^{eff} \rangle \tag{11}$$

Using Equation 11, Equation 10 is reduced to:

$$\tfrac{1}{2V}\int_S \left(\sigma_{pq}^{N+1} u_q^{eff} - \sigma_{pq}^{eff} u_q^{N+1}\right) dS = 0. \tag{12}$$

It can be concluded that Equation 12 appears as a result of average strain energy density equivalency between medium with inclusions and effective medium. The importance of this equation will be illustrated in the next section when solving for the unknown effective shear moduli.

3. Homogenization schemes in the GSC method

In this section, the homogenization procedures based on GSC method in determining the effective axial and transverse shear moduli are demonstrated.

3.1. Effective axial shear modulus

The allowable displacement field that satisfies equilibrium equations under specific axial shear loading deformation applied at outer boundaries of RVE for every orthotropic phase can be written in the form of Equation 13.

$$u_z^{(i)}(r,\theta) = \left(D_1^{(i)} r^{\lambda_{(i)}} + D_2^{(i)} r^{-\lambda_{(i)}}\right)\cos\theta \tag{13}$$

where index i identifies the phase in RVE, $D_1^{(i)}$ and $D_2^{(i)}$ are the unknown constants associated with each phase or layer in a composite(see Appendix A.1) and $\lambda_{(i)} = \sqrt{G_{\theta z}^{(i)} / G_{rz}^{(i)}}$ is a material constant related to the material property of that particular phase, i.e. $\lambda_{(i)}$ equals 1 for isotropic and transversely isotropic material plane of isotropic perpendicular to the principle axis of main fibre.

Substituting Equation 13 into Equation 5 and subsequently, into Equation 6, the allowable stress fields for every orthotropic phase can be obtained, which are written as:

$$\begin{aligned} \sigma_{\theta z}^{(i)}(r,\theta) &= -G_{\theta z}^{(i)}\left(D_1^{(i)} r^{\lambda_{(i)}-1} + D_2^{(i)} r^{-\lambda_{(i)}-1}\right)\sin\theta, \\ \sigma_{rz}^{(i)}(r,\theta) &= G_{rz}^{(i)}\left(\lambda_{(i)} D_1^{(i)} r^{\lambda_{(i)}-1} - \lambda_{(i)} D_2^{(i)} r^{-\lambda_{(i)}-1}\right)\cos\theta. \end{aligned} \tag{14}$$

For $(N+1)$ – medium, which possesses transversely isotropic property with its plane of isotropy perpendicular to the principle axis of main fibre, Equation 13 and Equation 14 transform into:

$$u_z^{N+1}(r,\theta) = \left(D_1^{N+1} r + D_2^{N+1} r^{-1}\right)\cos\theta, \tag{15}$$

$$\begin{aligned} \sigma_{\theta z}^{N+1}(r,\theta) &= -\mu_{23}^{N+1}\left(D_1^{N+1} + D_2^{N+1} r^{-2}\right)\sin\theta, \\ \sigma_{rz}^{N+1}(r,\theta) &= \mu_{23}^{N+1}\left(D_1^{N+1} - D_2^{N+1} r^{-2}\right)\cos\theta. \end{aligned} \tag{16}$$

Notice that Equation 16 contains the unknown effective axial shear modulus μ_{23}^{N+1} that is intended to be determined. Since $N=3$, a total of eight unknown constants with one unknown effective shear modulus must be determined - $D_1^{(1)}, D_2^{(1)}, D_1^{(2)}, D_2^{(2)}, D_1^{(3)}, D_2^{(3)}, D_1^{N+1}, D_2^{N+1}$ and μ_{23}^{N+1}.

Initially, it is imperative to understand that the solution to this boundary problem ceases to exist at the origin $r_0=0$. Thus, for the solution to exist the non-singularity condition must be obeyed such that the second unknown constant of the first layer must vanish, i.e. $D_2^{(1)}=0$.Subsequently, the remaining seven unknown constants and one unknown effective modulus are determined from the continuity conditions at ideal contact surfaces between the phases:

$$\begin{aligned} u_z^{(i)}(r_{(i)},\theta) &= u_z^{(i+1)}(r_{(i)},\theta), \qquad (i=1,2,...N) \\ \sigma_{rz}^{(i)}(r_{(i)},\theta) &= \sigma_{rz}^{(i+1)}(r_{(i)},\theta), \end{aligned} \tag{17}$$

and from the external boundary condition imposed at outer radius of RVE ($r_{N+1} \to \infty$) :

$$u_z^{N+1}(r_{N+1}) = 2\varepsilon_0 r_{N+1} \cos\theta \tag{18}$$

and finally, from the surface integral equation given by Equation 8:

$$\int_S \left(\sigma_{rz}^{N+1} u_z^{eff} - \sigma_{rz}^{eff} u_z^{N+1} \right)_{r=r_N} dS = 0 \tag{19}$$

To integrate Equation 19, the displacement and stress fields for effective composite medium need to be defined first. Prior to that, it must be emphasized that the admissible displacement and stress fields for $(N+1)-$medium are also applicable to effective composite medium where the superscript $N+1$ is simply replaced by *eff*. Furthermore, the non-singularity solution requirement as well as the imposed external boundary condition of Equation 18 are also applicable to that composite medium; as such based on Equation 15 and Equation 16, the required displacement and stress fields for effective composite medium can be written as:

$$\begin{aligned} u_z^{eff}(r,\theta) &= D_1^{eff} \cos\theta, \\ \sigma_{\theta z}^{eff}(r,\theta) &= -\mu_{23}^{eff} D_1^{eff} \sin\theta, \\ \sigma_{rz}^{eff}(r,\theta) &= \mu_{23}^{eff} D_1^{eff} \cos\theta. \end{aligned} \tag{20}$$

It easy to show that with Equation 16, Equation 18 and Equation 20, the integration of Equation 19 leads to $D_2^{(N+1)}=0$ and subsequently, by Equation 15 and Equation 18, $D_1^{(N+1)}=2\varepsilon_0$. Finally, the six remaining unknowns including the effective axial shear modulus $\mu_{23}^{eff}=\mu_{23}^{N+1}$ can now be solved from the six simultaneous equations generated from Equation 17, which are listed as follows:

$$\begin{aligned}
&1)\quad D_1^{(1)} r_1^{\lambda_{(1)}} - D_1^{(2)} r_1^{\lambda_{(2)}} - D_2^{(2)} r_1^{-\lambda_{(2)}} = 0, \\
&2)\quad D_1^{(2)} r_2^{\lambda_{(2)}} + D_2^{(2)} r_2^{-\lambda_{(2)}} - D_1^{(3)} r_2^{\lambda_{(3)}} - D_2^{(3)} r_2^{-\lambda_{(3)}} = 0, \\
&3)\quad D_1^{(3)} r_3^{\lambda_{(3)}} + D_2^{(3)} r_3^{-\lambda_{(3)}} - 2\varepsilon_0 r_3 = 0, \\
&4)\quad C_{55}^{(1)} \lambda_{(1)} D_1^{(1)} r_1^{\lambda_{(1)}-1} - C_{55}^{(2)} \lambda_{(2)} D_1^{(2)} r_1^{\lambda_{(2)}-1} + C_{55}^{(2)} \lambda_{(2)} D_2^{(2)} r_1^{-\lambda_{(2)}-1} = 0, \\
&5)\quad C_{55}^{(2)} \lambda_{(2)} D_1^{(2)} r_2^{\lambda_{(2)}-1} - C_{55}^{(2)} \lambda_{(2)} D_2^{(2)} r_2^{-\lambda_{(2)}-1} - C_{55}^{(3)} \lambda_{(3)} D_1^{(3)} r_2^{\lambda_{(3)}-1} + C_{55}^{(3)} \lambda_{(3)} D_2^{(3)} r_2^{-\lambda_{(3)}-1} = 0, \\
&6)\quad C_{55}^{(3)} \lambda_{(3)} D_1^{(3)} r_3^{\lambda_{(3)}-1} - C_{55}^{(3)} \lambda_{(3)} D_2^{(3)} r_3^{-\lambda_{(3)}-1} - 2\varepsilon_0 \mu_{23}^{N+1} = 0.
\end{aligned} \tag{21}$$

It is interesting to point out that the last expression shown in Equation 21 gives explicit expression for final effective axial shear modulus and this can be rewritten as Equation 22.

$$\mu_{23}^{eff} = \frac{1}{2\varepsilon_0} C_{55}^{(N)} \left[\lambda_{(N)} D_1^{(N)} r^{\lambda_{(N)}-1} + (-\lambda_{(N)}) D_2^{(N)} r^{-\lambda_{(N)}-1} \right], \tag{22}$$

where $\lambda_{(N)} = \sqrt{G_{\theta z}^{(N)} / G_{rz}^{(N)}}$. It is possible to get different expression of effective modulus from the one obtained in Equation 22, and this is obtained through Equation 10 or Equation 11. With the help of Equation 7 and Equation 8, the term on the left hand-side of Equation 11 leads to:

$$\langle W^{eff} \rangle = \frac{1}{2V} \int_V \sigma_{pq}^{eff} \varepsilon_{pq}^{eff} dV = 2\mu_{23}^{eff} \varepsilon_0^2 \tag{23}$$

while the term on the right hand-side gives

$$\langle W^{RVE} \rangle = \frac{1}{2r_{N+1}^2} \left[\sum_{i=1}^{N} C_{44}^{(i)} \left(\frac{\left(D_1^{(i)}\right)^2}{\lambda_{(i)}} \left(r_{(i)}^{2\lambda_{(i)}} - r_{(i-1)}^{2\lambda_{(i)}} \right) + \frac{\left(D_2^{(i)}\right)^2}{-\lambda_{(i)}} \left(r_{(i)}^{-2\lambda_{(i)}} - r_{(i-1)}^{-2\lambda_{(i)}} \right) \right) + 4\mu_{23}^{N+1} \varepsilon_0^2 r_{N+1}^2 - 4\mu_{23}^{N+1} \varepsilon_0^2 r_N^2 \right]. \tag{24}$$

Thus, equating Equation 23 and Equation 24 yields

$$\mu_{23}^{eff} = \frac{1}{4\left(\varepsilon_0 r_N\right)^2} \sum_{i=1}^{N} C_{44}^{(i)} \left(\frac{\left(D_1^{(i)}\right)^2}{\lambda_{(i)}} \left(r_{(i)}^{2\lambda_{(i)}} - r_{(i-1)}^{2\lambda_{(i)}} \right) + \frac{\left(D_2^{(i)}\right)^2}{-\lambda_{(i)}} \left(r_{(i)}^{-2\lambda_{(i)}} - r_{(i-1)}^{-2\lambda_{(i)}} \right) \right). \tag{25}$$

Although the expressions found in Equation 22 and Equation 25 are different, they produce identical results when numerical calculations are performed.

3.2. Effective transverse shear modulus

The admissible displacement fields for each orthotropic layer under specific transverse shear loading condition imposed at the outer radius of RVE can be written as

$$\begin{aligned} u_r^{(i)}(r) &= \left(\phi_1^{(i)} D_1^{(i)} r^{\lambda_1^{(i)}} + \phi_2^{(i)} D_2^{(i)} r^{\lambda_2^{(i)}} + \phi_3^{(i)} D_3^{(i)} r^{\lambda_3^{(i)}} + \phi_4^{(i)} D_4^{(i)} r^{\lambda_4^{(i)}} \right) \sin 2\theta, \\ u_\theta^{(i)}(r) &= \left(D_1^{(i)} r^{\lambda_1^{(i)}} + D_2^{(i)} r^{\lambda_2^{(i)}} + D_3^{(i)} r^{\lambda_3^{(i)}} + D_4^{(i)} r^{\lambda_4^{(i)}} \right) \cos 2\theta. \end{aligned} \tag{26}$$

where

$$\phi_j^{(i)} = 2\left(C_{\theta\theta}^{(i)} + G_{r\theta}^{(i)} - \lambda_j^{(i)} (C_{r\theta}^{(i)} + G_{r\theta}^{(i)}) \right) \left(C_{\theta\theta}^{(i)} + 4G_{r\theta}^{(i)} - (\lambda_j^{(i)})^2 C_{rr}^{(i)} \right)^{-1} \tag{27}$$

and the material constants $\lambda_j^{(i)}$ are obtained from the following characteristic equation, which is given as

$$T\left(\lambda_j^{(i)}\right)^4 - U\left(\lambda_j^{(i)}\right)^2 + V = 0 \tag{28}$$

where

$$\begin{aligned} T &= C_{rr}^{(i)} G_{r\theta}^{(i)}, \\ U &= \left(C_{rr}^{(i)} + C_{\theta\theta}^{(i)} - 8C_{r\theta}^{(i)} \right) G_{r\theta}^{(i)} + 4\left(C_{rr}^{(i)} C_{\theta\theta}^{(i)} - \left(C_{r\theta}^{(i)}\right)^2 \right), \\ V &= 9C_{\theta\theta}^{(i)} G_{r\theta}^{(i)}. \end{aligned}$$

Here, it is noted that real positive values are obtained for $\lambda_1^{(i)}$ and $\lambda_3^{(i)}$ and real negative values are for $\lambda_2^{(i)}$ and $\lambda_4^{(i)}$.

Similarly as before, with Equation 5, Equation 6, Equation 26, Equation 27 and Equation 28, and the corresponding allowable stress fields for every orthotropic layer can be obtained as:

$$\sigma_{\theta\theta}^{(i)}(r)=\left(\sum_{j=1}^{4} D_j^{(i)} r^{\lambda_j^{(i)}-1}\left(C_{r\theta}^{(i)}\phi_j^{(i)}\lambda_j^{(i)}+C_{\theta\theta}^{(i)}(\phi_j^{(i)}-2)\right)\right)\sin 2\theta,$$
$$\sigma_{r\theta}^{(i)}(r)=G_{r\theta}\left(\sum_{j=1}^{4} D_j^{(i)} r^{\lambda_j^{(i)}-1}\left(2\phi_j^{(i)}+\lambda_j^{(i)}-1\right)\right)\cos 2\theta. \tag{29}$$

For $N+1$ medium, the admissible displacement and stress fields are found as:

$$u_r^{N+1}(r,\theta)=\tfrac{1}{4\mu_{12}^{N+1}}\left(2r+\left(\eta^{N+1}+1\right)D_2^{N+1}r^{-1}+D_4^{N+1}r^{-3}\right)\sin 2\theta,$$
$$u_\theta^{N+1}(r,\theta)=\tfrac{1}{4\mu_{12}^{N+1}}\left(2r+\left(\eta^{N+1}-1\right)D_2^{N+1}r^{-1}-D_4^{N+1}r^{-3}\right)\cos 2\theta, \tag{30}$$

$$\sigma_{rr}^{N+1}(r,\theta)=\left(1-\Gamma^{N+1}D_2^{N+1}r^{-2}-\tfrac{3}{2}D_4^{N+1}r^{-4}\right)\sin 2\theta,$$
$$\sigma_{r\theta}^{N+1}(r,\theta)=\left(1+D_2^{N+1}r^{-2}+\tfrac{3}{2}D_4^{N+1}r^{-4}\right)\cos 2\theta \tag{31}$$

where $\eta^{N+1}=1+2K_{12}^{N+1}/\mu_{12}^{N+1}$, $\Gamma^{N+1}=1+\left(K_{12}^{N+1}/\mu_{12}^{N+1}\right)^2$.

In deriving the expressions shown in Equation 30 and Equation 31, the following equalities for unknown constants of $D_1^{N+1}=1/2\mu_{12}^{N+1}$ and $D_3^{N+1}=0$ were obtained. Furthermore, the unknown constants of $D_2^{(1)}$ and $D_4^{(1)}$ must also vanish due to the requirement of non-singularity solution at origin, i.e. $D_2^{(1)}=D_4^{(1)}=0$. As before, necessary simultaneous equations need to be established to solve for twelve unknown constants — $D_1^{(1)}, D_3^{(1)}, D_1^{(2)}, D_2^{(2)}, D_3^{(2)}, D_4^{(2)}, D_1^{(3)}, D_2^{(3)}, D_3^{(3)}, D_4^{(3)}, D_2^{N+1}, D_4^{N+1}$ and one unknown desired effective modulus of $\mu_{12}^{N+1}=\mu_{12}^{eff}$. It should be noted that although the unknown effective plane strain bulk modulus ($K_{12}^{N+1}=K_{12}^{eff}$) appears in Equation 30 and Equation 31, the knowledge of such modulus is not required. To eliminate the unwanted effective modulus from system of equations, the surface integral equation is employed, which is given as Equation 32.

$$\int_S\left(\sigma_{rr}^{N+1}u_r^{eff}+\sigma_{r\theta}^{N+1}u_\theta^{eff}-\sigma_{rr}^{eff}u_r^{N+1}-\sigma_{r\theta}^{eff}u_\theta^{N+1}\right)_{r=r_N}dS=0. \tag{32}$$

Note again that Equation 30 and Equation 31 as well as the requirement of non-singularity condition are also valid for effective composite medium; as such the allowable displacement and stress fields for such medium can be expressed as:

$$u_r^{eff}(r,\theta)=\tfrac{r}{2\mu_{12}^{eff}}\sin 2\theta, \qquad u_\theta^{eff}(r,\theta)=\tfrac{r}{2\mu_{12}^{eff}}\cos 2\theta,$$
$$\sigma_{rr}^{eff}(r,\theta)=\sin 2\theta, \qquad \sigma_{r\theta}^{eff}(r,\theta)=\cos 2\theta. \tag{33}$$

Using Equation 30, Equation 31 and Equation 33, the surface integral of Equation 32 gives $D_2^{N+1}=0$, which will then eliminate the unnecessary effective modulus from the equations. Hence, the remaining twelve unknowns including the effective transverse shear modulus can now be found with the help of continuity conditions at every perfect contact surface between phases:

$$u_r^{(i)}(r_{(i)},\theta)=u_r^{(i+1)}(r_{(i)},\theta), \qquad u_\theta^{(i)}(r_{(i)},\theta)=u_\theta^{(i+1)}(r_{(i)},\theta),$$
$$\sigma_{rr}^{(i)}(r_{(i)},\theta)=\sigma_{rr}^{(i+1)}(r_{(i)},\theta), \qquad \sigma_{r\theta}^{(i)}(r_{(i)},\theta)=\sigma_{r\theta}^{(i+1)}(r_{(i)},\theta), \qquad (i=1,2,...N); \tag{34}$$

It can be seen that Equation 34 provides twelve necessary equations to solve for effective transverse shear modulus. Nevertheless, no explicit expression can be found since the established formulations in this loading case contain nonlinear equations. As a result, an iterative method is required to solve for the unknown constants and effective modulus.

4. Numerical examples and discussions

In previous section, the homogenization schemes employed by GSC method in determining the effective axial and transverse shear properties of unidirectional fibre composite comprising cylindrical orthotropic constituents were shown. To demonstrate the capability of such method to predict the required effective properties, two problems involving different types of multifunctional advanced nanostructures hybrid fibre composites will be given. As mentioned earlier, to validate the solutions of the proposed applied model, the obtained results will be compared to the available experimental data and the results of finite element studies found in the existing literatures.

The first problem involves the estimation of effective shear properties of zinc oxide (ZnO) nanowires hybrid IM7 carbon fibre reinforced with epoxy polymer and in the second problem, the effective shear moduli of CNTs hybrid T65 carbon fibre reinforced epoxy composite are analysed. In either case, the RVE for modelling purpose comprises of four layers of fibre, reinforced interphase layer, matrix material and equivalent homogenized medium. As has been mentioned earlier, carbon fibre is normally assumed as a transversely isotropic material with its plane of isotropy perpendicular to the principle axis of fibre and physically, the properties are defined by five engineering constants, which are experimentally and/or numerically determined. From those physical properties, five elastic constants can be derived and used in the stiffness tensor shown in Equation 2. The transformation equations from physical engineering constant to elastic constants are given in the Appendix B. As for reinforced interphase layer, it exhibits cylindrically orthotropic property; however, transversely isotropic property with plane of isotropy perpendicular to principle axis of nanostructure is assumed in this analysis for simplicity. As such, the stiffness tensor shown in Equation 3 is applicable to this material. Nevertheless, this interphase layer is a composite layer, which its final properties must be obtained first through homogenization based on the properties of nanostructures and matrix material. The method and formula in determining the required effective moduli will be described next prior to numerical examples on each problem. Once the effective engineering properties are found, then the necessary elastic stiffness constants can be determined. In the case of epoxy polymer, the stiffness tensor shown in Equation 4 is applicable.

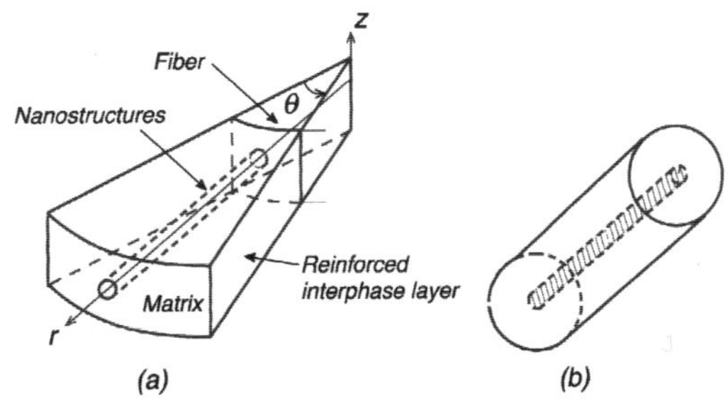

Figure 2. The RVE of (a) reinforced interphase layer and (b) equivalent geometrical model

4.1. Effective properties of reinforced interphase layer

Actual RVE of reinforced interphase layer is shown in Figure 2a and it indicates the layer has gradient property along the length of nanostructures. Nevertheless, for ease of analysis, the actual RVE can be approximated into an equivalent geometrical model as illustrated in Figure 2b. Such approach is acceptable considering the relatively small thickness of reinforced interphase layer. Here, the reinforced interphase layer is considered as a two-phase composite material where both constituents behave as isotropic material. Thus, the following commonly used formula based on the CCA and GSC methods can be used to estimate the effective transversely isotropic property of reinforced interphase layer where the following effective properties are defined: axial Young's modulus $E_{11}^{eff-Int}$, Poisson's ratio $v_{12}^{eff-Int}$, plane strain bulk modulus $\kappa_{12}^{eff-Int}$, axial shear modulus $\mu_{12}^{eff-Int}$ and transverse shear modulus $\mu_{23}^{eff-Int}$:

$$
\begin{aligned}
&E_{11}^{eff-Int} = c_b E_b + (1-c_b) E_m + \frac{4c_b(1-c_b)(v_b - v_m)^2 \mu_m}{(1-c_b)\mu_m/(k_b+\mu_b/3) + c_b\mu_m/(k_m+\mu_m/3)+1},\\
&v_{12}^{eff-Int} = c_b v_b + (1-c_b) v_m + \frac{c_b(1-c_b)(v_b-v_m)(\mu_m/(k_m+\mu_m/3) - \mu_m/(k_b+\mu_b/3))}{(1-c_b)(\mu_m/(k_b+\mu_b/3)) + c_b(\mu_m/(k_m+\mu_m/3))+1},\\
&K_{23}^{eff-Int} = k_m + \frac{\mu_m}{3} + \frac{c_b}{1/(k_b - k_m + \frac{1}{3}(\mu_b-\mu_m)) + (1-c_b)/(k_m+4\mu_m/3)},\\
&\frac{\mu_{12}^{eff-Int}}{\mu_m} = \frac{\mu_b(1+c_b)+\mu_m(1-c_b)}{\mu_b(1-c_b)+\mu_m(1+c_b)},\\
&A\left(\frac{\mu_{23}^{eff-Int}}{\mu_m}\right)^2 + 2B\left(\frac{\mu_{23}^{eff-Int}}{\mu_m}\right) + C = 0
\end{aligned}
\tag{35}
$$

where E = Young's modulus, μ = shear modulus, v = Poisson's ratio, k = bulk modulus, c = volume fraction, $\eta = 3-4v$, and the subscripts b and m represents nanostructures and matrix, respectively, with:

$$
\begin{aligned}
A =\ & 3c_b(1-c_b)^2(\mu_b/\mu_m - 1)(\mu_b/\mu_m - \eta_b) + \\
& + \left[\mu_b/\mu_m \eta_m + \eta_b\eta_m - (\eta_m\mu_b/\mu_m - \eta_b)c_b^3\right]\left[c_b\eta_m(\mu_b/\mu_m - 1) - (\eta_m\mu_b/\mu_m + 1)\right],\\
B =\ & -3c_b(1-c_b)^2(\mu_b/\mu_m - 1)(\mu_b/\mu_m + \eta_b) + \\
& + \frac{1}{2}\left[\eta_m\mu_b/\mu_m + (\mu_b/\mu_m - 1)c_b + 1\right]\left[(\eta_m - 1)(\mu_b/\mu_m + \eta_b) - \right.\\
& \left. - 2(\eta_m\mu_b/\mu_m - \eta_b)c_b^3\right] + \frac{c_b}{2}(\eta_m+1)(\mu_b/\mu_m - 1)\left[\mu_b/\mu_m + \eta_b + (\eta_m\mu_b/\mu_m - \eta_b)c_b^3\right],\\
C =\ & 3c_b(1-c_b)^2(\mu_b/\mu_m - 1)(\mu_b/\mu_m + \eta_b) + \\
& + \left[\eta_m\mu_b/\mu_m + (\mu_b/\mu_m - 1)c_b + 1\right]\left[\mu_b/\mu_m + \eta_b + (\eta_m\mu_b/\mu_m - \eta_b)c_b^3\right].
\end{aligned}
$$

The volume fraction of nanostructures in this case is defined as:

$$c_b = M_b^2 d_b^2 / 4\pi(l_b + D)D \tag{36}$$

where M_b is the allowable number of nanostructures grown radially on the surface of fibre, d_b is the diameter of nanostructures, l_b is the length of nanostructures and D is the diameter of fibre. It is interesting to note such approximate model with Equation 35 and Equation 36 can be used to analysed the effects of length, diameter, density of nanostructures on the effective properties of such hybrid fibre composite material [22, 23].

4.2. Effective shear moduli of ZnO nanowires hybrid fibre composite

The configurations of the constituents used in this test problem are as follows: IM7 carbon fiber with diameter of 5.2 μm, high-density fully coated ZnO nanowires (NWs) with diameter of 50 nm and length of 500 nm, and epoxy matrix material. The mechanical properties of the constituents are derived from the work of Lurie et al [22]. The effective axial and transverse shear moduli found from the proposed method are depicted in Figure 3. As can be seen, the presence of ZnO NWs significantly enhanced the shear characteristics of such composite when compared to the properties of composite without the nanostructures. At 50% fiber volume fraction, the axial and transverse shear modulus are improved by 53% and 56% respectively. Specifically on the effective axial shear modulus, the proposed developed model gives accurate estimations on the effective property when compared to experimental data produced by Ehlert [25] for both composite having 50% fibre concentration with and without the ZnO NWs. Unfortunately, no experimental data is found on transverse shear modulus for proper comparison.

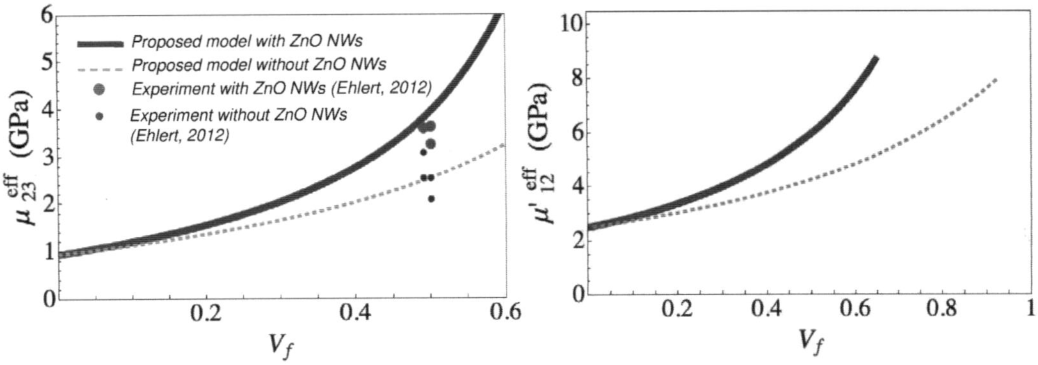

Figure 3. Estimation of (a) effective axial shear modulus, μ_{23}^{eff} and (b) effective transverse shear modulus, μ_{12}^{eff} of ZnO nanowires hybrid fibre and conventional fibre reinforced composites

4.3. Effective shear moduli of fuzzy fibre composite

In this test problem, the fuzzy fibre composite comprises T650 carbon fibre, Epikote 862 epoxy polymer and single-walled CNTs. The configurations and mechanical properties of the constituents used in this test problem are obtained from the work of Chatzigeorgiouet. al. [26]. In their work, a final element model based on asymptotic homogenization method was developed to study the effective properties of such composite system. The RVE employed in their work is based on hexagonal unit cell, which closely simulate random distribution of fibre within the composite [11]. The comparison in axial and transverse shear values is depicted in Table 1. It can be seen that the analytical results are in excellent agreement with the results produced by finite element modelling.

Table 1. Comparison on the effective properties of fuzzy fiber composites

Model	Axial shear modulus (GPa)	Transverse shear modulus (GPa)
Analytical-GSCM	2.43	7.05
FE-Hexagonal [26]	2.55	6.92

5. Conclusions

This paper presents the estimation of the effective axial and transverse shear moduli of cylindrical orthotropic composite based on the generalized self-consistent method. The concepts, features and procedures of GSCM in solving such problem are also presented. The applied model has the capability to solve the cylindrical multi-layered composites problem with isotropic, transversely isotropic and orthotropic constituents.

In estimating the effective properties, the admissible displacement fields for the two loading cases of axial and transverse shear deformation are described, from which the stress fields can be obtained. Each expression of these fields contains a set of unknown constants possessing unique values for each phase and one of the stress fields contains the respective effective shear properties. These constants, together with effective modulus, are determined with the help of non-singularity provision, continuity requirements, boundary conditions and surface integral equation derived from the GSC method. Explicit expression for the effective axial shear modulus was obtained but not for the transverse shear modulus.

Several test problems are performed and the results are evaluated and validated against available the existing data, both experimentally and numerically. It is found that the proposed model is capable to capture the behaviour of nanostructures hybrid fibre composites and predict the effective elastic properties of ZnO nanowires hybrid fibre and fuzzy fibre composites reliably and accurately.

6. Appendices

A. Derivation of admissible displacement vectors

In this appendix section, the derivations of admissible displacement fields for every constituent in the RVE are shown for axial and transverse shear loading condition cases.

A.1. Axial shear deformation

Figure A.1 illustrates the axial shear loading condition experienced by RVE where axial shear deformation is imposed at its outer boundary. It is assumed that the only solution of displacement function can exist in such loading condition is $u_z(r,\theta) \neq 0$ while $u_r = u_\theta = 0$.

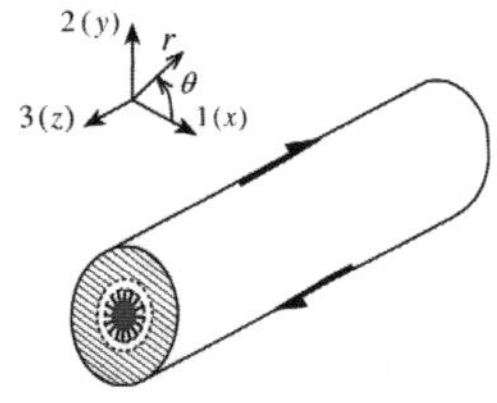

Figure A.1. Axial shear loading condition

Next, the general solution of displacement function is substituted into Equation 5, from which the following strain-displacement functions are obtained

$$\varepsilon_{zz} = \partial u_z / \partial z, \quad 2\varepsilon_{\theta z} = \tfrac{1}{r}\partial u_z / \partial \theta, \quad 2\varepsilon_{rz} = \partial u_z / \partial r. \tag{A.1}$$

Substituting Equation A.1 into Equation 1, the stress-strain functions are found to be

$$\sigma_{\theta z} = G_{\theta z}\left(\tfrac{1}{r}\partial u_z / \partial \theta\right), \qquad \sigma_{rz} = G_{rz}\left(\partial u_z / \partial r\right). \tag{A.2}$$

Finally, rearranging Equation A.2 into Equation 6, the following partial differential equation is obtained as

$$r^2\left(\partial^2 u_z / \partial r^2\right) + r\left(\partial u_z / \partial r\right) + \lambda^2\left(\partial^2 u_z / \partial \theta^2\right) = 0 \tag{A.3}$$

where $\lambda^2 = G_{\theta z} / G_{rz}$. Solving the partial differential given in Equation A.3 gives us

$$u_z(r,\theta) = \left(D_1 r^{\lambda} + D_2 r^{-\lambda}\right)\cos\theta. \tag{A.4}$$

where D_1 and D_2 are the unknown constants. The displacement field shown in Equation A.4 is admissible since it satisfies the equilibrium equations.

A.2. Transverse shear deformation

The specific transverse shear loading condition is shown in Figure A.2 below, in which the general solutions of displacement functions are assumed to have the following form

$$u_r = u(r)\sin 2\theta, \qquad u_\theta = w(r)\cos 2\theta, \qquad u_z = 0. \tag{A.5}$$

where $u(r)$ and $w(r)$ are some unknown functions that depends on the radius of RVE only.

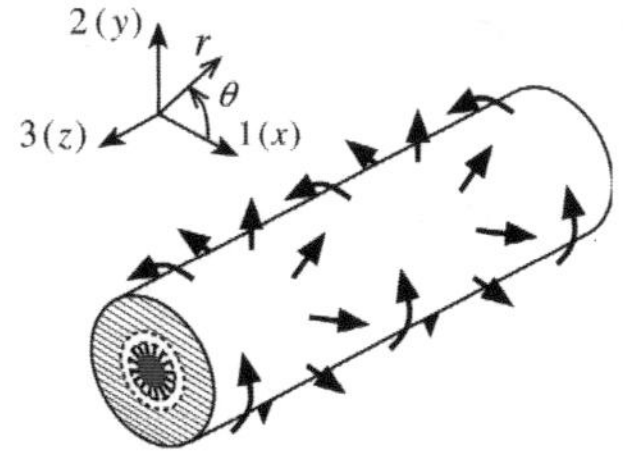

Figure A.2. Transverse shear loading condition

Substituting Equation A.5 into Equation 5, the relations of strain-displacement are obtained as

$$\begin{aligned} \varepsilon_{rr} &= u'(r)\sin 2\theta, \\ \varepsilon_{\theta\theta} &= \left((u(r) - 2w(r))/r\right)\sin 2\theta, \\ 2\varepsilon_{r\theta} &= \left((2u(r) - w(r))/r + w'(r)\right)\cos 2\theta. \end{aligned} \tag{A.6}$$

where superscript prime notation represents derivative with respect to variable r. The subsequent substitution of Equation A.6 into Equation 1 provides us the stress-strain relationships, which can be written as

$$\begin{aligned} \sigma_{rr} &= \left[C_{rr}u'(r) + C_{r\theta}(u(r) - 2w(r))r^{-1}\right]\sin 2\theta, \\ \sigma_{\theta\theta} &= \left[C_{r\theta}u'(r) + C_{\theta\theta}(u(r) - 2w(r))r^{-1}\right]\sin 2\theta, \\ \sigma_{r\theta} &= G_{r\theta}\left((2u(r) - w(r))r^{-1} + w'(r)\right)\cos 2\theta. \end{aligned} \tag{A.7}$$

Finally, substituting Equation A.7 into Equation 6, the two partial differential equations are obtained and expressed as

$$C_{rr}\left((r^2u''(r) + ru'(r)\right) - \left(C_{\theta\theta} + 4G_{r\theta}\right)u(r) - 2\left(C_{r\theta} + G_{r\theta}\right)rw'(r) + 2\left(C_{\theta\theta} + G_{r\theta}\right)w(r) = 0 \tag{A.8}$$

$$2\left(C_{r\theta} + G_{r\theta}\right)ru'(r) + 2\left(C_{\theta\theta} + G_{r\theta}\right)u(r) + G_{r\theta}\left(r^2w''(r) + rw'(r)\right) - \left(4C_{\theta\theta} + G_{r\theta}\right)w(r) = 0 \tag{A.9}$$

Next, it is proposed that the general solutions to the second order partial differential equations given in Equation A.8 and Equation A.9 have the following solutions

$$u(r) = Fr^\lambda, \qquad w(r) = Dr^\lambda. \tag{A.10}$$

where F and D are some constants that are dependent with each other. We are now required to determine the characteristic equation that will define the value of material constant λ so that nontrivial solutions exist for the second order partial differential equations when taking into account the relations given in Equation A.10. Substitute Equation A.10 together with their first and second derivatives functions into Equation A.8 and Equation A.9, the following expressions are found.

$$\begin{aligned} &\left[C_{rr}\lambda^2 - (C_{\theta\theta} + 4G_{\theta\theta})\right]F - \left[2\left(C_{r\theta} + G_{r\theta}\right)\lambda - 2\left(C_{\theta\theta} + G_{r\theta}\right)\right]D = 0, \\ &\left[2\left(C_{r\theta} + G_{r\theta}\right)\lambda + 2\left(C_{\theta\theta} + G_{r\theta}\right)\right]F + \left[G_{r\theta}\lambda^2 - \left(4C_{\theta\theta} + G_{r\theta}\right)\right]D = 0. \end{aligned} \tag{A.11}$$

It is possible to combine the two expressions given in Equation A.11 and represent them as one single equation as in Equation A.12.

$$C_{rr}G_{r\theta}\lambda^4 - \left[\left(C_{rr} + C_{\theta\theta} - 8C_{r\theta}\right)G_{r\theta} + 4\left(C_{rr}C_{\theta\theta} - C_{r\theta}^{\ 2}\right)\right]\lambda^2 + 9C_{\theta\theta}G_{r\theta} = 0 \tag{A.12}$$

Notice that Equation A.12 has the same form as the characteristic equation given in Equation 27. Again, the solution to this equation will provide the values for the required material constants - $\lambda_1, \lambda_2, \lambda_3$ and λ_4. The general forms of solutions to the second order partial differential equations can be written as

$$u(r) = F_1 r^{\lambda_1} + F_2 r^{\lambda_2} + F_3 r^{\lambda_3} + F_4 r^{\lambda_4},$$
$$w(r) = D_1 r^{\lambda_1} + D_2 r^{\lambda_2} + D_3 r^{\lambda_3} + D_4 r^{\lambda_4}, \tag{A.13}$$

where, as mentioned earlier, $\lambda_1^{(i)}$ and $\lambda_3^{(i)}$ are found to have real positive values and for $\lambda_2^{(i)}$ and $\lambda_4^{(i)}$, are real negative values. Since the unknown constants of F_j and D_j are dependent with each other, their relationships can be represented in the form of

$$F_j = \phi_j D_j \tag{A.14}$$

where $\phi_j = 2\left(C_{\theta\theta} + G_{r\theta} - \lambda_j (C_{r\theta} + G_{r\theta})\right)\left(C_{\theta\theta} + 4G_{r\theta} - (\lambda_j)^2 C_{rr}\right)^{-1}$. Based on Equation A.14, Equation A.13 and Equation A.5, the final form of admissible displacement fields under transverse shear loading conditions can be expressed as

$$u_r(r) = \left(\sum_{j=1}^{4} \phi_j D_j r^{\lambda_j}\right) \sin 2\theta, \qquad u_\theta(r) = \left(\sum_{j=1}^{4} D_j r^{\lambda_j}\right) \cos 2\theta. \tag{A.15}$$

Note that the expressions shown in Equation A.15 are identical to the forms given in Equation 26.

B. Transformation equations

For main fibre material, the transformation equations from the engineering stiffness constants obtained from the experiments to elastic stiffness constants for transversely isotropic medium with its plane of symmetry perpendicular to the z-axis are provided as follows:

$$\begin{aligned}
&C_{11} = C_{22} = C_{rr} = C_{\theta\theta} = E_T\left(1 - v_{TL} v_{LT}\right)\Upsilon, \\
&C_{33} = C_{zz} = E_L\left(1 - v_T^2\right)\Upsilon, \qquad C_{12} = C_{r\theta} = E_T\left(v_T + v_{TL} v_{LT}\right)\Upsilon, \\
&C_{13} = C_{23} = C_{rz} = C_{\theta z} = E_T\left(v_{LT} + v_T v_{LT}\right)\Upsilon = E_L\left(v_{TL} + v_T v_{TL}\right)\Upsilon, \\
&C_{44} = C_{55} = G_{rz} = G_{\theta z} = \mu_L, \qquad C_{66} = G_{r\theta} = \mu_T.
\end{aligned} \tag{B.1}$$

where $\Upsilon = (1 - v_T^2 - 2v_{TL}v_{LT} - 2v_T v_{TL} v_{LT})^{-1}$, $E_T = E_{11} = E_{22} = E_{rr} = E_{\theta\theta}$ - transverse Young's modulus, $E_L = E_{33} = E_{zz}$ - axial Young's modulus, $v_T = v_{12} = v_{21} = v_{r\theta} = v_{\theta r}$ - transverse Poisson's ratio, $v_{LT} = v_{31} = v_{32} = v_{zr} = v_{z\theta}$ - major Poisson's ratio, $v_{LT} = v_{13} = v_{23} = v_{rz} = v_{\theta z}$ - minor Poisson's ratio, $\mu_L = \mu_{13} = \mu_{31} = \mu_{23} = \mu_{32} = \mu_{rz} = \mu_{zr} = \mu_{\theta z} = \mu_{z\theta}$ - axial shear modulus and $\mu_T = \mu_{12} = \mu_{21} = \mu_{r\theta} = \mu_{\theta r}$ $= E_T / 2(1 + v_T)$ - transverse shear modulus. Further useful relations are given as follows: $v_{31}/E_{33} = v_{13}/E_{11}$, $v_{32}/E_{33} = v_{23}/E_{22}$.

For reinforced interphase layer, the transformation equations for transversely isotropic medium with its axis of symmetry parallel to the r-axis are given as follows:

$$\begin{aligned}
&C_{11} = C_{rr} = E_{11}^{eff-Int} + 4(v_{12}^{eff-Int}) K_{23}^{eff-Int}, \\
&C_{12} = C_{13} = C_{r\theta} = C_{rz} = 2K_{23}^{eff-Int} v_{12}^{eff-Int}, \\
&C_{22} = C_{33} = C_{\theta\theta} = C_{zz} = \mu_{23}^{eff-Int} + K_{23}^{eff-Int}, \\
&C_{23} = C_{\theta z} = -\mu_{23}^{eff-Int} + K_{23}^{eff-Int}, \\
&C_{66} = C_{55} = G_{r\theta} = C_{rz} = \mu_{12}^{eff-Int} = \mu_{13}^{eff-Int}.
\end{aligned} \tag{B.2}$$

As for isotropic matrix or other constituent material, the following relations are defined as

$$C_{11} = C_{22} = C_{33} = C_{rr} = C_{\theta\theta} = C_{zz} = E(1-\nu)/(1+\nu)(1-2\nu),$$
$$C_{12} = C_{13} = C_{23} = C_{r\theta} = C_{rz} = C_{\theta z} = E\nu/(1+\nu)(1-2\nu) \quad \text{(B.3)}$$

and the shear modulus can be defined as $\mu = E/2(1+\nu)$.

Acknowledgement
The principle author is highly indebted to Prof Dr Sergey Lurie and Dr Natasha Tuchkova, Institute of Applied Mechanics, Russian Academy of Sciences, Moscow for the fruitful discussion in this research area and the supports they have extended throughout these years.

References

[1] Baur J and Silverman E 2007 *MRS Bulletin* **32** 328-34
[2] Gibson R E 2010 *Compos.Struct.* **92** 2793-810
[3] Lin Y, Ehlert G J and Sodano H A 2009 *Adv. Funct. Mater.* **19** 2654-60
[4] Agnihotri P, Basu S and Kar K K 2011 *Carbon* **49** 3098-106
[5] Liao Q, Zhang Z, Zhang X, Mohr M, Zhang Y, Fecht H-J 2014 *Nano Research* **7** 917-28
[6] Garcia E J, Wardle B L, Hart A J and Yamamonj N 2008 *Compos. Sci. Technol.* **68** 2034-41
[7] Steiner S A, Li R and Wardle B L 2013 *ACS Appl. Mater. Interf.* **5** 4892–903
[8] Yamamoto N, Villoria R G and Wardle B L 2012 *Compos. Sci. Technol.* **72** 2009-15
[9] Lurie S and Minhat M 2014 *Compos. B* **61** 26-40
[10] Guz I A, Rodger A A, Guz A N and Rushchitsky J J 2008 *Philos Trans R Soc* A **366** 1827-33
[11] Hashin Z and Rosen B W 1964 *J. Appl. Mech.* **31** 223-32
[12] Hill R 1965 *J. Mech. Phys. Solids* **13** 213–22
[13] Christensen R M and Lo K H 1979 *J. Mech. Phys. Solids* **27** 315–30
[14] Norris A N 1985 *Mech. Mater.* **4** 1–16
[15] Benveniste Y, Dvorak G J and Chen T 1987 *Mech. Mater.* **6** 147-57
[16] Christensen RM 1990 *J. Mech. Phys. Solids* **38** 379–404
[17] Lurie S, Minhat M, Tuchkova N and Soliaev J 2014 *Appl. Compos. Mater.* **21** 179-96
[18] Nemat-Nasser S and Hori M 1999 *Micromechanics: Overall properties of heterogeneous materials* (Amsterdam: Elsevier)
[19] Aboudi J, Arnold S M and Bednarcyk B A 2013 *Micromechanics of composite materials: A generalized multiscale analysis approach* (Oxford, Massachusetts: Butterworth-Heinemann)
[20] Herve E and Zaoui 1995 *Int. J. Eng. Sci* **33** 1419-33
[21] Norris A N 1989 *J. Appl. Mech.* **56** 83-8
[22] Hashin Z 2002 *J. Mech. Phys. Solids* **50** 2509-37
[23] Lurie S and Minhat M 2014 *Compos. B* **61** 26-40
[24] Lurie S, Minhat M. and Tuchkova N 2015 *J. Eng. Math.* **95** 7-29
[25] Ehlert G J 2012 *Development of zinc oxide nanowire interphase for enhanced structural composites* PhD Thesis University of Florida
[26] Chatzigeorgiou G, Efendiev Y and Lagoudas D C 2011 *Int. J. Solids Struct.* **48** 2668-80
[27] Pozegic T R, Hamerton I, Anguita J V, Tang W, Ballocchi P, Jenkins P and Silva S R P 2014 *Carbon* **74** 319-28